The Neurobiology, Physiology, and Psychology of Pain

The Neurobiology, Physiology, and Psychology of Pain

The Neuroscience of Pain, Anesthetics, and Analgesics

Edited by

Rajkumar Rajendram
Department of Medicine, King Abdulaziz Medical City, King Abdulaziz International Medical Research Center, Ministry of National Guard - Health Affairs, Riyadh, Saudi Arabia

College of Medicine, King Saud bin Abdulaziz University of Health Sciences, Riyadh, Saudi Arabia

Vinood B. Patel
Centre for Nutraceuticals, School of Life Sciences, University of Westminster, London, United Kingdom

Victor R. Preedy
King's College London, London, United Kingdom

Colin R. Martin
Institute for Health and Wellbeing, University of Suffolk, United Kingdom

ELSEVIER

ACADEMIC PRESS
An imprint of Elsevier

Academic Press is an imprint of Elsevier
125 London Wall, London EC2Y 5AS, United Kingdom
525 B Street, Suite 1650, San Diego, CA 92101, United States
50 Hampshire Street, 5th Floor, Cambridge, MA 02139, United States
The Boulevard, Langford Lane, Kidlington, Oxford OX5 1GB, United Kingdom

Notices
Knowledge and best practice in this field are constantly changing. As new research and experience broaden our understanding, changes in research methods, professional practices, or medical treatment may become necessary.

Practitioners and researchers must always rely on their own experience and knowledge in evaluating and using any information, methods, compounds, or experiments described herein. In using such information or methods they should be mindful of their own safety and the safety of others, including parties for whom they have a professional responsibility.

To the fullest extent of the law, neither the Publisher nor the authors, contributors, or editors, assume any liability for any injury and/or damage to persons or property as a matter of products liability, negligence or otherwise, or from any use or operation of any methods, products, instructions, or ideas contained in the material herein.

Library of Congress Cataloging-in-Publication Data
A catalog record for this book is available from the Library of Congress

British Library Cataloguing-in-Publication Data
A catalogue record for this book is available from the British Library

ISBN 978-0-12-820589-1

SET ISBN 978-0-12-821066-6

For information on all Academic Press publications
visit our website at https://www.elsevier.com/books-and-journals

Publisher: Nikki Levy
Acquisitions Editor: Natalie Farra
Editorial Project Manager: Timothy Bennett
Production Project Manager: Paul Prasad Chandramohan
Cover Designer: Miles Hitchen

Typeset by STRAIVE, India

Printed in the United States of America
Last digit is the print number: 10 9 8 7 6 5 4

Contents

4. Genetics of chronic widespread musculoskeletal pain

María Jesús Álvarez-Cubero, Sergio Cuenca-López, Verónica Arenas-Rodríguez, Fernando Estévez-López, and Luis Javier Martínez-González

5. Fentanyl: Polymorphisms, and adverse events

Mongi Benjeddou and Ana M. Peiró

6. Propofol anesthesia and molecular changes in the brain

Desanka Milanović, Željko Pavković, and Vesna Pešić

7. Protein kinase G is a molecular switch for pain

Ying-Ju Sung and Richard T. Ambron

15. Nociception and sweet solutions: Applications to inflammatory pain

Khawla Q. Nuseir, Manal Kassab, and Ahmad Altarifi

16. Interlinking interleukin-33 (IL-33), neuroinflammation and neuropathic pain

Camila Rodrigues Ferraz, Fernanda Soares Rasquel-Oliveira, Sergio Marques Borghi, Anelise Franciosi, Thacyana Teixeira Carvalho, Telma Saraiva-Santos, Nayara Anitelli Artero, Rubia Casagrande, and Waldiceu A. Verri, Jr.

17. Neurons of the parabrachial nucleus, nociceptive input, and pain pathways

Yosuke Arima, Yoshinori Otani, and Masashi Fujitani

18. Anterior cingulate cortex, pain perception, and pathological neuronal plasticity during chronic pain

Fernando Kasanetz, Mario A. Acuña, and Thomas Nevian

Part II
Physiology, imaging and physical recordings

28. Neurocognition and placebo analgesia: Linking in functional magnetic resonance imaging

Sara Palermo

29. Linking the cortex, functional spectroscopy, and pain: Features and applications

Wolnei Caumo, Janete S. Bandeira, and Jairo Alberto Dussan-Sarria

30. Muscle origins of myofascial pain syndrome

Eva María Martínez-Jiménez, Daniel López-López, Carlos Romero-Morales, Victoria Mazoteras-Pardo, Marta Elena Losa-Iglesias, David Rodríguez-Sanz, Marta San-Antolín-Gil, Ricardo Becerro-de-Bengoa-Vallejo, and César Calvo-Lobo

Part III
Psychology and behavior

31. Behavioral markers of pain: Understanding the cognitive, motor, and societal interactions in the pain experience

S.A. Holmes, A. Quinlan, and M.E. Pierce

Contributors

Numbers in paraentheses indicate the pages on which the authors' contributions begin.

Mario A. Acuña (193), Department of Physiology, University of Bern, Bern, Switzerland

Rohit Aiyer (287), Interventional Pain Management and Pain Psychiatry, Henry Ford Health System, Detroit, MI, United States

Sergiu Albu (375), Guttmann Brain Health Institute, Barcelona, Spain

Ahmad Altarifi (161), Department of Pharmacology, Faculty of Medicine, Jordan University of Science and Technology, Irbid, Jordan

Duygun Altıntaş Aykan (217), Department of Pharmacology, Kahramanmaras Sutcu Imam University, Faculty of Medicine, Kahramanmaras, Turkey

María Jesús Álvarez-Cubero (33), Department of Biochemistry and Molecular Biology III, Faculty of Medicine, PTS, University of Granada; GENYO (Pfizer-University of Granada-Andalusian Government Centre for Genomics and Oncological Research), Granada, Spain

Richard T. Ambron (67), Department of Pathology, Anatomy, and Cell Biology, Vagelos College of Physicians and Surgeons, Columbia University, New York, NY, United States

Verónica Arenas-Rodríguez (33), GENYO (Pfizer-University of Granada-Andalusian Government Centre for Genomics and Oncological Research), Granada, Spain

Yosuke Arima (183), Department of Anatomy and Neuroscience, Faculty of Medicine, Shimane University, Izumo, Shimane, Japan

Nayara Anitelli Artero (171), Department of Pathology, Centre of Biological Sciences, Londrina State University, Londrina, Paraná, Brazil

Janete S. Bandeira (319), Pain & Neuromodulation Laboratory, Hospital de Clinicas de Porto Alegre, Porto Alegre, RS, Brazil

Ricardo Becerro-de-Bengoa-Vallejo (337, 405), Department of Nursing, School of Nursing, Physiotherapy and Podiatry, University Complutense of Madrid, Madrid, Spain

Eric J. Bellefroid (23), ULB Neuroscience Institute (UNI), Free University of Brussels, Gosselies, Belgium

Mongi Benjeddou (45), Department of Biotechnology, University of the Western Cape, Cape Town, South Africa

Sergio Marques Borghi (171), Center for Research in Health Sciences, University of Northern Paraná, Londrina, Paraná, Brazil

César Calvo-Lobo (337, 405), Department of Nursing, School of Nursing, Physiotherapy and Podiatry, University Complutense of Madrid, Madrid, Spain

Taís de Campos Lima (139), Department of Physiology, Biomedical Sciences Institute, Federal University of Uberlândia, Uberlândia, Minas Gerais, Brazil

Yolanda Campos-Jurado (91), Department of Pharmacy and Pharmaceutical Technology and Parasitology, School of Pharmacy, University of Valencia, Burjassot, Valencia, Spain

Thacyana Teixeira Carvalho (171), Department of Pathology, Centre of Biological Sciences, Londrina State University, Londrina, Paraná, Brazil

Rubia Casagrande (171), Department of Pharmaceutical Sciences, Centre of Health Sciences, Londrina State University, Londrina, Paraná, Brazil

Wolnei Caumo (319), Pain & Neuromodulation Laboratory, Hospital de Clinicas de Porto Alegre, Porto Alegre, RS, Brazil

Rok Cerne (117), Institute of Pathophysiology, Faculty of Medicine, University of Ljubljana, Ljubljana, Slovenia; Laboratory of Antiepileptic Drug Discovery, Peyton Manning Hospital for Children, Ascension St. Vincent, Indianapolis, IN, United States

Krishnan Chakravarthy (395), Division of Pain Medicine, Department of Anesthesiology, University of

California San Diego, La Jolla; VA San Diego Health Care, San Diego, CA, United States

James M. Cook (117), Department of Chemistry and Biochemistry, University of Wisconsin-Milwaukee, Milwaukee, WI, United States

Fabiano V. Costa (387), Department of Biochemistry and Molecular Biology School of Life Sciences, Federal University of Santa Maria, Santa Maria, Rio Grande do Sul, Brazil

Sergio Cuenca-López (33), GENYO (Pfizer-University of Granada-Andalusian Government Centre for Genomics and Oncological Research), Granada, Spain

Javier Cuitavi (91), Department of Pharmacy and Pharmaceutical Technology and Parasitology, School of Pharmacy, University of Valencia, Burjassot, Valencia, Spain

Maya D'Eon (395), UCSD Department of Psychiatry, La Jolla, CA, United States

Rodolfo Delgado-Lezama (129), Departamento de Fisiología, Biofísica y Neurociencias, Cinvestav, Zacatenco, Mexico City, Mexico

Simon Desiderio (23), Institute for Neurosciences of Montpellier, University of Montpellier, INSERM U1051, Montpellier, France

Jairo Alberto Dussan-Sarria (319), Universidade FEEVALE; Clínica Hamburguesa de Anestesiologia, Novo Hamburgo, RS, Brazil

Fernando Estévez-López (33), Department of Pediatrics, Wilhelmina Children's Hospital, University Medical Center Utrecht, Utrecht University, Utrecht, The Netherlands

Camila Rodrigues Ferraz (171), Department of Pharmaceutical Sciences, Centre of Health Sciences, Londrina State University, Londrina, Paraná, Brazil

Janet L. Fisher (117), Department of Pharmacology, Physiology & Neuroscience, School of Medicine, University of South Carolina, Columbia, SC, United States

Anelise Franciosi (171), Department of Pathology, Centre of Biological Sciences, Londrina State University, Londrina, Paraná, Brazil

Úrzula Franco-Enzástiga (129), Neurobiology of Pain Laboratory, Departamento de Farmacobiología, Cinvestav, South Campus, Mexico City, Mexico

Masashi Fujitani (183), Department of Anatomy and Neuroscience, Faculty of Medicine, Shimane University, Izumo, Shimane, Japan

Soudeh Ghafouri-Fard (13), Department of Medical Genetics, Shahid Beheshti University of Medical Sciences, Tehran, Iran

Lalit K. Golani (117), Department of Chemistry and Biochemistry, University of Wisconsin-Milwaukee, Milwaukee, WI, United States

Shaila Gowda (255), Department of Neurology, Baylor College of Medicine, Houston, TX, United States

Vinicio Granados-Soto (129), Neurobiology of Pain Laboratory, Departamento de Farmacobiología, Cinvestav, South Campus, Mexico City, Mexico

Semih Gungor (287), Department of Anesthesiology, Critical Care and Pain Management, Hospital for Special Surgery, Weill Medical School of Cornell University, New York, NY, United States

Mogge Hajiesmaeil (13), Department of Biology and Biotechnology "Charles Darwin", Sapienza University of Rome, Rome, Italy

Hee Chul Han (227), Department of Physiology, College of Medicine and Neuroscience Research Institute, Korea University, Seoul, South Korea

Natalie Hellman (359), The University of Tulsa, Department of Psychology, Tulsa, OK, United States

Luis Eduardo Hernández-Reyes (129), Departamento de Fisiología, Biofísica y Neurociencias, Cinvestav, Zacatenco, Mexico City, Mexico

Lucia Hipólito (91), Department of Pharmacy and Pharmaceutical Technology and Parasitology, School of Pharmacy, University of Valencia, Burjassot, Valencia, Spain

S.A. Holmes (349), Department of Anesthesiology, Critical Care and Pain Medicine, Boston Children's Hospital; Instructor in Anesthesia, Harvard Medical School, Boston, MA, United States

Allan V. Kalueff (387), School of Pharmacy, Southwest University, Chongqing, China; Ural Federal University, Yekaterinburg, Russia

Fernando Kasanetz (193), Universidad de Buenos Aires, CONICET, Instituto de Fisiología y Biofísica (IFIBIO) Bernardo Houssay, Grupo de Neurociencia de Sistemas, Buenos Aires, Argentina

Manal Kassab (161), Department of Maternal and Child Health, Faculty of Nursing, Jordan University of Science and Technology, Irbid, Jordan; University of Technology, Sydney (UTS), Sydney, NSW, Australia

Seonghoon Kim (203), Department of Neurology, Uijeongbu St. Mary's hospital, College of Medicine, The Catholic University of Korea, Seoul, Republic of Korea

Johann Klein (299), Department of Neurosurgery, University Hospital Carl Gustav Carus, of the Technical University of Dresden, Dresden, Germany

Daniel E. Knutson (117), Department of Chemistry and Biochemistry, University of Wisconsin-Milwaukee, Milwaukee, WI, United States

Hannu Kokki (239), School of Medicine, University of Eastern Finland, Kuopio, Finland

Merja Kokki (239), Department of Anaesthesia and Operative Services, Kuopio University Hospital, Kuopio, Finland

Yarim E. De la Luz-Cuellar (129), Neurobiology of Pain Laboratory, Departamento de Farmacobiología, Cinvestav, South Campus, Mexico City, Mexico

Natalia Landsberg (91), Department of Pharmacy and Pharmaceutical Technology and Parasitology, School of Pharmacy, University of Valencia, Burjassot, Valencia, Spain

Hans Linsenbardt (375), Battlefield Pain Management, U. S. Army Institute of Surgical Research, San Antonio, TX; Oak Ridge Institute for Science and Education, Oak Ridge, TN, United States

Daniel López-López (337, 405), Research, Health and Podiatry Group, Department of Health Sciences, Faculty of Nursing and Podiatry, University of A Coruña, Ferrol, Spain

Jesús D. Lorente (91), Department of Pharmacy and Pharmaceutical Technology and Parasitology, School of Pharmacy, University of Valencia, Burjassot, Valencia, Spain

Marta Elena Losa-Iglesias (337, 405), Department of Nursing, School of Nursing, Physiotherapy and Podiatry, University Complutense of Madrid, Madrid; Department of Nursing, Faculty of Health Sciences, Universidad Rey Juan Carlos, Alcorcón, Spain

Celina Monteiro da Cruz Lotufo (139), Department of Physiology, Biomedical Sciences Institute, Federal University of Uberlândia, Uberlândia, Minas Gerais, Brazil

Luis Javier Martínez-González (33), GENYO (Pfizer-University of Granada-Andalusian Government Centre for Genomics and Oncological Research), Granada, Spain

Eva María Martínez-Jiménez (337, 405), Department of Nursing, School of Nursing, Physiotherapy and Podiatry, University Complutense of Madrid, Madrid, Spain

Victoria Mazoteras-Pardo (337, 405), Research Group "ENDOCU", Department of Nursing, Physiotherapy and Occupational Therapy, Faculty of Physiotherapy and Nursing of Toledo, University of Castilla-La Mancha, Toledo, Spain

Mary W. Meagher (375), Department of Psychology, Texas A&M University, College Station, TX, United States

Desanka Milanović (57), Department of Neurobiology, Institute for Biological Research "Siniša Stanković"—National Institute of Republic of Serbia, University of Belgrade, Belgrade, Serbia

Keisuke Miyamoto (151), Department of Neuropharmacology, Graduate School of Pharmaceutical Sciences, Nagoya City University, Nagoya, Japan

Janet Murbartián (129), Departamento de Farmacobiología, Cinvestav, South Campus, Mexico City, Mexico

Thomas Nevian (193), Department of Physiology, University of Bern, Bern, Switzerland

Khawla Q. Nuseir (161), Clinical Pharmacy Department, Faculty of Pharmacy, Jordan University of Science and Technology, Irbid, Jordan

Masahiro Ohsawa (151), Department of Neuropharmacology, Graduate School of Pharmaceutical Sciences, Nagoya City University, Nagoya, Japan

Yoshinori Otani (183), Department of Anatomy and Neuroscience, Faculty of Medicine, Shimane University, Izumo, Shimane, Japan

Antonella Paladini (275), Department MESVA, University of L'Aquila, L'Aquila, Italy

Sara Palermo (307), Neuroradiology Unit, Neurological Institute C. Besta IRCCS Foundation, Milan, Italy

Eui Ho Park (227), Department of Physiology, College of Medicine and Neuroscience Research Institute, Korea University, Seoul, South Korea

Jeong-wook Park (203), Department of Neurology, Uijeongbu St. Mary's hospital, College of Medicine, The Catholic University of Korea, Seoul, Republic of Korea

Vinood B. Patel (419), School of Life Sciences, University of Westminster, London, United Kingdom

Željko Pavković (57), Department of Neurobiology, Institute for Biological Research "Siniša Stanković"—National Institute of Republic of Serbia, University of Belgrade, Belgrade, Serbia

Ana M. Peiró (45), Neuropharmacology on Pain and Functional Diversity (NED), Alicante Institute for Health and Biomedical Research (ISABIAL Foundation); Department of Clinical Pharmacology, General Hospital of Alicante, Alicante, Spain

Vesna Pešić (57), Department of Neurobiology, Institute for Biological Research "Siniša Stanković"—National Institute of Republic of Serbia, University of Belgrade, Belgrade, Serbia

M.E. Pierce (349), Translational Research Center for TBI & Stress Disorders, VA Boston Health Care System, Boston University School of Medicine, Boston, MA, United States

Witold H. Polanski (299), Department of Neurosurgery, University Hospital Carl Gustav Carus, of the Technical University of Dresden, Dresden, Germany

Victor R. Preedy (419), Department of Nutrition and Dietetics, Life Course Sciences, Faculty of Life Science and Medicine, King's College London, London, United Kingdom

A. Quinlan (349), Department of Anesthesiology, Critical Care and Pain Medicine, Boston Children's Hospital, Boston, MA, United States

Rajkumar Rajendram (419), College of Medicine, King Saud bin Abdulaziz University for Health Sciences; Department of Medicine, King Abdulaziz Medical City, King Abdullah International Medical Research Center, Ministry of National Guard Health Affairs, Riyadh, Saudi Arabia

Fernanda Soares Rasquel-Oliveira (171), Department of Pathology, Centre of Biological Sciences, Londrina State University, Londrina, Paraná, Brazil

Guadalupe Raya-Tafolla (129), Departamento de Fisiología, Biofísica y Neurociencias, Cinvestav, Zacatenco, Mexico City, Mexico

Martina Rekatsina (275), Whipps Cross University Hospital, Barts Health NHS Trust, London, United Kingdom

Jamie L. Rhudy (359), The University of Tulsa, Department of Psychology, Tulsa, OK, United States

David Rodríguez-Sanz (337, 405), Department of Nursing, School of Nursing, Physiotherapy and Podiatry, University Complutense of Madrid, Madrid, Spain

Carlos Romero-Morales (337, 405), Faculty of Sport Sciences, Universidad Europea de Madrid, Madrid, Spain

Robert Root-Bernstein (79), Professor of Physiology, Michigan State University, East Lansing, MI, United States

Luiz V. Rosa (387), Department of Biochemistry and Molecular Biology School of Life Sciences, Federal University of Santa Maria, Santa Maria, Rio Grande do Sul, Brazil

Denis B. Rosemberg (387), Department of Biochemistry and Molecular Biology School of Life Sciences, Federal University of Santa Maria, Santa Maria, Rio Grande do Sul, Brazil

Thomas Rutledge (395), VA San Diego Healthcare System/UC San Diego Department of Psychiatry, Psychology Service, San Diego, CA, United States

Giorgia Saltelli (275), Department of Anesthesiology and Intensive Care, Sant'Andrea Hospital, "La Sapienza University", Rome, Italy

Marta San-Antolín-Gil (337, 405), Department of Psychology, Universidad Europea de Madrid, Madrid, Spain

Débora de Oliveira Santos (139), Bucomaxillofacial Pathology Laboratory, Odontology Faculty, Federal University of Uberlândia, Uberlândia, Minas Gerais, Brazil

Telma Saraiva-Santos (171), Department of Pathology, Centre of Biological Sciences, Londrina State University, Londrina, Paraná, Brazil

Jamie Sleigh (265), Department of Anesthesia, Waikato Clinical Campus, University of Auckland, Hamilton, New Zealand

Jodi L. Smith (117), Laboratory of Antiepileptic Drug Discovery, Peyton Manning Hospital for Children, Ascension St. Vincent, Indianapolis, IN, United States

Maryam Sobhani (13), Blood Transfusion Research Center, High Institute for Research and Education in Transfusion Medicine, Tehran, Iran

Ying-Ju Sung (67), Department of Pathology, Anatomy, and Cell Biology, Vagelos College of Physicians and Surgeons, Columbia University, New York, NY, United States

Charles Akos Szabó (255), Department of Neurology, UT Health, San Antonio, TX, United States

Jorge E. Torres-López (129), Mechanisms of Pain Laboratory, División Académica de Ciencias de la Salud, Universidad Juárez Autónoma de Tabasco; Hospital Regional de Alta Especialidad "Dr. Juan Graham Casasús", Villahermosa, Tabasco, Mexico

Giustino Varrassi (275), Paolo Procacci Foundation, Rome, Italy

Vishal Varshney (395), St. Paul's Hospital, Providence Healthcare, Vancouver, BC, Canada

Simon Vermeiren (23), ULB Neuroscience Institute (UNI), Free University of Brussels, Gosselies, Belgium

Waldiceu A. Verri, Jr. (171), Department of Pathology, Centre of Biological Sciences, Londrina State University, Londrina, Paraná, Brazil

Logan Voss (265), Department of Anesthesia, Waikato District Health Board, Hamilton, New Zealand

Jeffrey M. Witkin (117), Laboratory of Antiepileptic Drug Discovery, Peyton Manning Hospital for Children, Ascension St. Vincent, Indianapolis, IN; Department of Chemistry and Biochemistry, University of Wisconsin-Milwaukee, Milwaukee, WI; Departments of Neuroscience and Trauma Research, Ascension St. Vincent Hospital, Indianapolis, IN, United States

Zizhen Wu (3), Department of Neuroscience Cell Biology and Anatomy, University of Texas Medical Branch at Galveston, Galveston, TX, United States

Qing Yang (3), Department of Neuroscience Cell Biology and Anatomy, University of Texas Medical Branch at Galveston, Galveston, TX, United States

Fatemeh Yazarlou (13), Department of Medical Genetics, School of Medicine, Tehran University of Medical Sciences, Tehran, Iran

Zainul Amiruddin Zakaria (103), Department of Biomedical Sciences, Faculty of Medicine and Health Sciences, Universiti Putra Malaysia, Serdang, Selangor, Malaysia

Preface

The etiology of pain is complex and multifactorial. This complexity is magnified by the use of analgesics and local or general anesthetics. While analgesics can reduce pain, general anesthetics reduce consciousness and local anesthetics reduce localized pain. Analgesics may be used after surgery performed under anesthesia. However, anesthesia is not without risk. Depending on patient-related factors such as comorbid disease, anesthesia may be associated with significant morbidity or even mortality. Adverse events during or after anesthesia may necessitate the use of other pharmacological agents such as vasopressors, inotropes, sedatives, antiarrhythmics, or antiemetics.

The perception of pain itself results from a multifaceted interaction between illness beliefs, age, gender, time of onset, stress, socioeconomic status, and other factors. To a certain extent, one could argue that pain itself is helpful in treating disease. It can be considered the "sixth vital sign" as it indicates the need for assessment by a healthcare professional and the need for clinical investigations, diagnosis, and appropriate medication. The pain associated with myocardial infarction is a good example of this. One needs to consider though that some acute and chronic pain can lead to psychological distress and reduced quality of life. In the long term, chronic persistent pain can impact significantly on the family unit and many diseases present with pain. There are a plethora of pharmacological agents currently available for pain management. Furthermore, studies showing the beneficial effects of plant or natural extracts provide the foundation for further rigorous studies in clinical trials.

The neuroscience of pain in one condition may be relevant to understanding the pain observed in other conditions. The onset of pain, the cause of the pain, and the administration of analgesia or anesthesia is a continuing scientific spectrum. At each point, there is a firm scientific basis with established literature and also an ongoing drive to discover new facts and data. Hitherto, such material on the neuroscience of pain, anesthesia, and analgesia has been sporadic and/or written for the experts who specialize in narrow and focused areas. For example, the expert in the use of general anesthetics in the surgical setting may not necessarily be an expert in the molecular biology of neurons activated in the pain process. The cellular biologist may be aware of neither the science underpinning the provision of anesthesia nor the adverse outcomes that the anesthesiologist may encounter in the clinical setting. To address the aforementioned issues, the editors have compiled *The Neuroscience of Pain, Anesthetics, and Analgesics.*

The Neuroscience of Pain, Anesthetics, and Analgesics is divided into three books:

Book 1: *Features and Assessments of Pain, Anesthesia, and Analgesia*
Book 2: *Treatments, Mechanisms, and Adverse Reactions of Anesthetics and Analgesics*
Book 3: *The Neurobiology, Physiology, and Psychology of Pain*

This book, *The Neurobiology, Physiology, and Psychology of Pain*, is divided into three parts. *Part I, Molecular and Cellular Aspects*, covers topics on Fentanyl, polymorphisms, adverse events, genetics of fibromyalgia and pain, protein kinase G and pain, ATP-sensitive potassium channels in pain, nociception and sweet solutions, antinociceptive glucagon-like peptides, propofol anesthesia, and molecular changes in the brain. *Part II, Physiology, Imaging, and Physical Recordings*, covers topics on anesthetic and proconvulsant effects of ketamine on EEG, linking heart function and analgesia, computed tomography–guided procedures for epidural injections, brain stimulation, and sensory functional MRI. *Part III, Psychology and Behavior*, covers topics on behavioral markers of pain, adverse life events and chronic pain risk, cognitive-affective modulation of pain, implantable pain devices, and psychosocial aspects.

Each chapter has:

- *An Abstract (published online)*
- *Key Facts*
- *Mini-Dictionary of Terms*
- *Applications to Other Areas*
- *Other Agents of Interest*
- *Summary Points*

The section *Key Facts* focuses on areas of knowledge written for the novice. The *Mini-Dictionary of Terms* explains terms that are frequently used in the chapter. The *Applications to Other Areas* describes other fields of neuroscience, pain science, analgesia, and anesthesia that the chapter may be relevant to. The *Summary Points* encapsulates each chapter in a succinct way.

The Neuroscience of Pain, Anesthetics, and Analgesics is designed for research and teaching purposes. It is suitable for neurologists and anesthesiologists as well as those interested in pain and its interlink with pain relief. It is valuable as a personal reference book and also for academic libraries that cover the neuroscience of pain, anesthetics, and analgesics. Contributions are from leading national and international experts including those from world-renowned institutions.

Rajkumar Rajendram, Victor R. Preedy, Vinood B. Patel, and Colin R. Martin (Editors)

Part I

Molecular and cellular aspects

Chapter 1

KCNQ/Kv7 channels as therapeutic target to treat neuropathic pain

Zizhen Wu and Qing Yang

Department of Neuroscience Cell Biology and Anatomy, University of Texas Medical Branch at Galveston, Galveston, TX, United States

Abbreviations

CaM calmodulin
DRG dorsal root ganglion
LQTS fatal long-QT syndrome
PIP2 phosphatidylinositol-4,5-bisphosphate
PIPN paclitaxel-induced peripheral neuropathic
SCI spinal cord injury

Introduction

Excitable cells of all animals are consistently sensing the changes in the environment and communicate with each other by electrical signal. Such communications are largely dependent on the orchestrated movement of ions across the plasma membrane, which is mediated by ion channels that are highly permeable to selective ions with sophisticated gating mechanisms.

Potassium channels are highly selective to K^+ over smaller ions such as Na^+ and Li^+ (Thompson et al., 2009). Although these channels share a highly conserved K^+ selective filter in the pore, they are diverse in kinetics, genetics, topology, and subunits combination, thus are classified into several different subfamilies. Potassium channels are widely expressed in all types of cells of multiple-cell live organisms. Different subsets of potassium channels are present in different type of cells, and same types of channels appear at different locations of the same cell (McKeown, Swanton, Robinson, & Jones, 2008), thus underpinning their unique role in cell functional attributes. The physiological functions of ion channels are sometimes revealed by genetic diseases in which their expression and kinetics are altered. This genetic approach leads to the discovery of *KCNQ/Kv7* gene family where their mutations cause diverse human diseases including cardiac arrhythmia (Wang et al., 1996), deafness (Neyroud et al., 1997), and neonatal epilepsy (Charlier et al., 1998).

KCNQ/Kv7 channel family members

KCNQ proteins and function

In 1996, KCNQ1 (previously called KvLQT1) was discovered at chromosomal loci 11p15 when searching for genes responsible for the inherited fatal long-QT syndrome (LQTS) (Wang et al., 1996). Subsequently, four homologs (KCNQ2, 3, 4, and 5) have been identified at chromosomal loci 20q13, 8q24, 1p34, and 6q13, which share between 30% and 65% identity in their sequence (Soldovieri, Miceli, & Taglialatela, 2011). These genes encode Kv7 potassium channel subunits with the topology of six transmembrane domains and a single pore loop (Soldovieri et al., 2011). The proteins are ranged from 676 (KCNQ1) to 932 (KCNQ5) in the order of KCNQ1 < KCNQ4 < KCNQ3 < KCNQ2 < KCNQ5 (Robbins, 2001). All subtypes are highly homologous in their transmembrane regions (Robbins, 2001). The fourth transmembrane domain contains a peculiar sequence of 4 or 6 positively charged amino acids, which probably works as the voltage sensor. The last two transmembrane domains form the poor loop with the presence of potassium channel signature sequence TxxTxGYG (Robbins, 2001). Both N-terminus and C-terminus locate in the cytoplasm. While the N-terminus has only about 100 amino acids in length, the C-terminus of KCNQ proteins is very long and comprised of four distinct helices

The Neurobiology, Physiology, and Psychology of Pain. https://doi.org/10.1016/B978-0-12-820589-1.00001-4

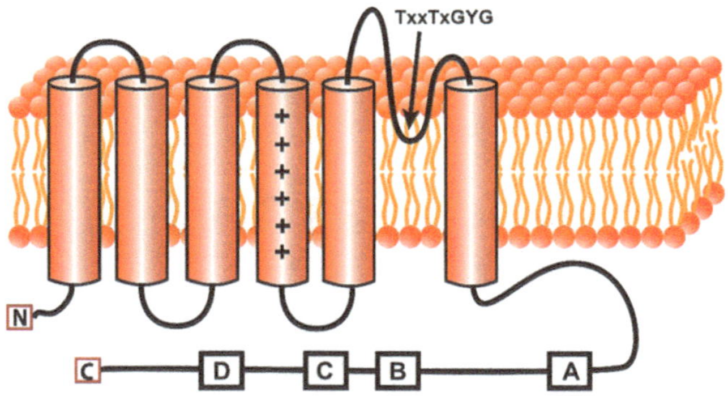

FIG. 1 Topology of KCNQ proteins. KCNQ proteins have six transmembrane domains and a single pore loop. The fourth transmembrane domain contains four or six positively charged amino acids. The last two transmembrane domains form the poor loop with the presence of potassium channel signature sequence TxxTxGYG. The long intracellular C-terminus contains four distinct helices.

(A–D) that determine subunit composition and channel modulation (Wiener et al., 2008). While A and B proximal helices are responsible for calmodulin (CaM) binding, the distal helixes C and D form crucial sites for phosphatidylinositol-4,5-bisphosphate (PIP2) binding and subunit assembly (Wiener et al., 2008) (Fig. 1). It is thus not surprising to find that many disease-causing mutations are identified in the C-terminus of KCNQ proteins (Haitin & Attali, 2008).

KCNQ1 is found to be expressed in the heart, cochlea, intestine, and kidney (Neyroud et al., 1997; Schroeder, Waldegger, et al., 2000; Wang et al., 1996; Zheng et al., 2007). In addition to causing inherited fatal LQTS, KCNQ1 mutations can also result in a form of inherited deafness by impairing potassium cycling in the inner ear (Neyroud et al., 1997; Ning et al., 2003). The expression of KCNQ4 is restricted to the inner ear and certain brainstem nuclei (Kharkovets et al., 2000; Kubisch et al., 1999). While KCNQ1 extremely expresses in the cochlear marginal cells to contribute markedly to the potassium-rich endolymph of the cochlear stria vascularis, KCNQ4 locates on the basolateral membrane of auditory and vestibular hair cells to unload accumulated potassium from these cells. Consistent with this location, mutations of KCNQ4 cause autosomal dominant nonsyndromic hearing impairment (DFNA2) by impairing the homeostasis of potassium in auditory hair cells (Coucke et al., 1999). KCNQ2, KCNQ3, and KCNQ5 are mainly present in neuronal tissues. Mutation in either KCNQ2 or KCNQ3 can both result in the same phenotype of benign familial neonatal seizures (Charlier et al., 1998), probably because they combine together to form a single functional ion channel in neuronal cells. While KCNQ5 was mapped to human chromosome 6q14 that is associated to several retinal diseases (Schroeder, Hechenberger, Weinreich, Kubisch, & Jentsch, 2000), no human disease is paired with this gene. In addition to their expression in the brain, immunohistochemistry analysis shows that KCNQ2, 3, and 5 are expressed in primary sensory neurons (Passmore et al., 2003; Rose et al., 2011), indicating they might participate in pain perception.

KCNQ channel compositions

Functional KCNQ channels are tetramer, either homomeric or heteromeric. While all subtypes of KCNQ subunits can form homomeric channels in vitro, they generally form heteromers in certain combinations in vivo.

KCNQ1 generally interacts with auxiliary subunits, KCNEs, to form functional ion channels. In the heart and inner ear, KCNQ1 is associated with KCNE1 to repolarize cardiac action potential and maintain potassium homeostasis in the endolymph of inner ear (Sanguinetti et al., 1996; Vetter et al., 1996). However, KCNQ1 interacts with KCNE3 in intestinal crypt cells to participate in the potassium recycling at the basolateral membrane (Schroeder, Waldegger, et al., 2000). In vitro studies indicate that KCNQ1 can also interact with KCNE2, 4, and 5 (Angelo et al., 2002; Grunnet et al., 2002; Tinel, Diochot, Borsotto, Lazdunski, & Barhanin, 2000), although the physiological role of such combination in vivo is not clear. KCNQ1 channel gating can be profoundly altered by its interaction with KCNE subunits. The interaction with KCNE1 slows activation kinetics and increases current amplitude (Sanguinetti et al., 1996). KCNE2 or KCNE3 locks KCNQ1 channel in an open configuration, leading to instantaneous activation (Schroeder, Waldegger, et al., 2000; Tinel et al., 2000). In contrast, coexpression of KCNE4 or KCNE5 with KCNQ1 results in decreased voltage sensitivity of the channel (Angelo et al., 2002; Grunnet et al., 2002).

KCNQ3 is generally coexpressed with KCNQ2 and KCNQ5 in neurons and KCNQ4 in hair cells (Robbins, 2001), where it forms heteromultimers with other subtypes. The coexpression of KCNQ2 and KCNQ3 increases the surface expression of the heteromultimers as compared with each homomer (Schwake, Pusch, Kharkovets, & Jentsch, 2000). However, the outcomes resulted from the interaction between KCNQ5 and KCNQ3 are complicate depending on the expression level of KCNQ3. Low level of KCNQ3 increases the amplitude of KCNQ3/5 channel–mediated currents, whereas a large amount of KCNQ3 results in decreased current amplitude (Schroeder, Hechenberger, et al., 2000). Considering that KCNQ5 competes with KCNQ2 to catch KCNQ3, which decreases the number of more efficient KCNQ2/3 heteromultimers, it is thus very difficult to predict the overall outcomes when the expression of one single subtype of KCNQ protein in neurons is altered under pathological condition. Additionally, it also has been reported that KCNQ2 and KCNQ3 can probably interact with KCNE1 or KCNE2 (Tinel et al., 2000; Yang et al., 1998). KCNQ4 currents are enhanced by KCNE1, KCNE2, and KCNE4 and attenuated by KCNE3 when coexpressed in *Xenopus* oocytes (Robbins, 2001). However, the physiologically relevant to the interaction between KCNQ2–4 and KCNEs has not been detected.

KCNQ channels in primary sensory neurons and their contribution to neuropathic pain

Primary sensory neurons and neuropathic pain

Living organisms are very often exposed to noxious stimuli and respond with a series of motivational and affective activities designed to avoid injury. Dorsal root ganglion (DRG) neurons that innervate the skin play an essential role in initiating multiple sensory modalities, including touch and nociception (pain sensation). DRG neurons are heterogeneous populations of cells with distinct physiological and morphological properties in terms of molecular compositions, cell size, myelination, and terminal projections. Subgroup of DRG neurons, primarily neurons with unmyelinated C-fiber and thinly myelinated Aδ afferents, are generally considered to be nociceptors since they are responsible for initiating nociceptive reflexes and pain sensation upon noxious stimuli (Dirajlal, Pauers, & Stucky, 2003).

Acute nociception is usually transient and is necessary for our escaping from dangers. However, under pathological conditions, such as peripheral nerve injury and inflammation, pain can be chronic and annoying in the absence of original causes and purpose. Neuropathic pain that can be produced by a lesion or disease of the somatosensory system is one of the major causes of disability in our society. Multiple sites along pain pathways are altered after lesion or disease of the somatosensory system, including immune system in the spinal cord and brain, central nervous system, and peripheral nervous system (Campbell & Meyer, 2006). Accumulated evidence indicates that the plasticity intrinsic to primary sensory neurons (especially pain-sensing nociceptors) is required for the development of neuropathic pain (Bedi et al., 2010; Vaso et al., 2014; Yang et al., 2014) (Fig. 2). For example, nearly all amputees experience chronic phantom limb pain, but it can be rapidly and reversibly abolished by local delivery of anesthetic lidocaine to the DRG surface (Vaso et al., 2014). The majority of spinal cord injury (SCI) patients also experience slowing developing neuropathic pain. Contusive SCI at T10 results in increased spontaneous firing in small diameter nociceptors (Bedi et al., 2010). The increased spontaneous firing of DRG neurons is significantly correlated with reflexive behavioral hypersensitivity (Bedi et al., 2010). Eliminating spontaneous firing in most nociceptors in the lumbar region by knockdown Nav1.8, a voltage-gated Na^+ channel whose expression is restricted to DRG neurons and is necessary for spontaneous firing (Roza, Laird, Souslova, Wood, & Cervero, 2003; Shields et al., 2012), can potently reduce SCI-induced hypersensitivity of hindlimb withdrawal responses (Yang

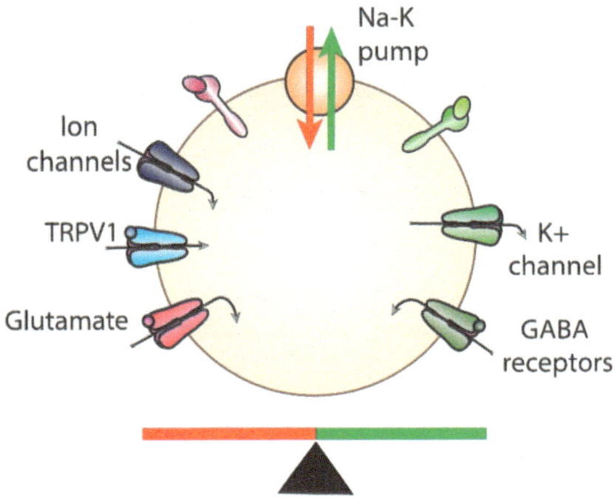

FIG. 2 A model of neuronal plasticity. The neuronal excitability is controlled by inhibitory and excitatory factors. Breaking the balance will cause a variety of pathological consequences.

et al., 2014). Central sensitization and spinal cord glial activation also occur in various neuropathic pain models, which probably are induced and maintained by ongoing activity in nociceptors (Xie, Strong, & Zhang, 2009; Yang et al., 2014). Thus, altered neuronal activity occurring in primary sensory neurons, especially nociceptors, is assumed to be at the core for both the development and maintenance of neuropathic pain.

KCNQ channels in primary sensory neurons

Many rat DRG neurons, including nociceptors, express KCNQ2, KCNQ3, and KCNQ5 (Passmore et al., 2003; Rose et al., 2011). In this regard, a specific KCNQ channel activator, retigabine, hyperpolarizes DRG neurons (Wu et al., 2017) (Fig. 3), mitigates the inputs of Aδ and C-fibers to dorsal horn neurons (Passmore et al., 2003), and attenuates heat stimulation–induced Aδ- and C-fiber discharges (Lang, Fleckenstein, Passmore, Brown, & Grafe, 2008; Passmore et al., 2012). While immunohistochemistry presents controversial outcomes for the distribution of KCNQ2 and KCNQ5 in nociceptors (King & Scherer, 2012; Passmore et al., 2003, 2012), electrophysiology reveals clearly that the excitability of myelinated fibers and Aδ fibers, but not C-fibers, is increased by blocking KCNQ channels with XE-991 (Passmore et al., 2012), a KCNQ channel blocker targeting efficiently KCNQ2/3 over KCNQ5 (Schroeder, Hechenberger, et al., 2000). It has assumed that KCNQ2 contributes to fast action potential propagation since it presents in nodes of Ranvier of peripheral myelinated fibers (Devaux, Kleopa, Cooper, & Scherer, 2004), but recent studies indicate that such process is probably mainly mediated by TREK-1 and TRAAK leak potassium channels (Kanda et al., 2019).

Plasticity of KCNQ channels and their contribution to neuropathic pain

Stimuli from the outworld are encoded into the electrical signals by activating ion channels expressed in sensory neurons. Under pathological conditions, the expression and/or activity of some ion channels enriched in nociceptors is altered, resulting in sensitization of DRG neurons that are described physiologically as hypersensitivity (the decreased threshold for activation and increased response to stimuli) and hyperexcitability (including spontaneous firing), thereby driving allodynia, hyperalgesia, and spontaneous pain. KCNQ channels show no inactivation with a threshold for activation below −60 mV, which allows them to maintain strong control of overexcitability in a variety of central and peripheral neurons (Passmore et al., 2003). Thus, if the expression or activity of the KCNQ channel is altered, the excitability/sensitivity of DRG neurons will be significantly changed. In this regard, suppressed expression of KCNQ2 (and KCNQ3) in DRG neurons is observed in both bone cancer and nerve injury–induced neuropathic pain animals (Laumet et al., 2015; Rose et al., 2011; Zheng et al., 2013), accompanying enhanced excitability of DRG neurons and amplified inputs to spinal cord (Cai et al., 2015; Zheng et al., 2013). Several signaling pathways underlying the nerve injury–induced downregulation of KCNQ expression in the DRG neurons are observed. It has suggested that repressor element 1-silencing transcription factor that is profoundly overexpressed after nerve injury strongly suppresses KCNQ expression in the DRG neurons (Rose et al., 2011). Epigenetic modification of KCNQ transcripts by euchromatic histone-lysine *N*-methyltransferase-2 is also reported to be critical to the suppressed KCNQ expression during the acute-to-chronic pain transition (Laumet et al., 2015). However, neither the expression of KCNQ proteins nor KCNQ channel activity is modified in DRG neurons after spinal cord injury (SCI) (Wu et al., 2017), although the excitability of primary sensory neurons is significantly increased (Bedi et al., 2010). Similar results are also observed in paclitaxel-induced peripheral neuropathic (PIPN) animals (Li et al., 2019). Thus, the contribution of KCNQ channels to neuropathic pain needs to be evaluated case by case.

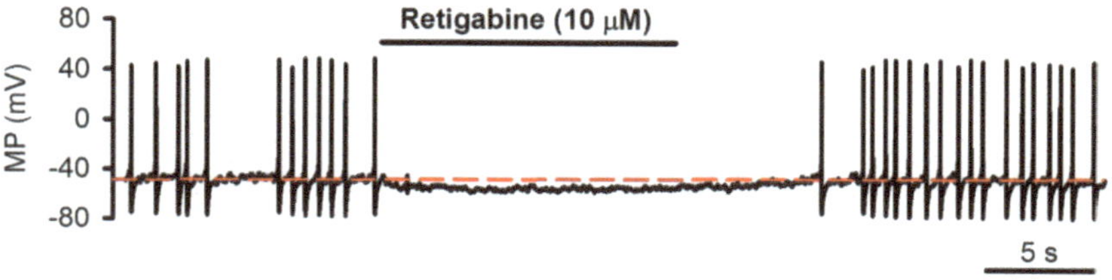

FIG. 3 Opening KCNQ channels can effectively decrease neuronal excitability. Representative recording shows retigabine (10 μM) reversibly hyperpolarizes membrane potential and attenuates the spontaneous firing of a DRG neuron from rats. *(Reprinted with permission from Wu, Z., Li, L., Xie, F., Du, J., Zuo, Y., Frost, J. A., et al. (2017). Activation of KCNQ channels suppresses spontaneous activity in dorsal root ganglion neurons and reduces chronic pain after spinal cord injury.* Journal of Neurotrauma, 34, *1260–1270.)*

KCNQ channels as therapeutic target to treat neuropathic pain

KCNQ channels as therapeutic target to treat established pain

KCNQ channels are widely expressed in DRG neurons including nociceptors (Passmore et al., 2003; Rose et al., 2011). They show no inactivation with a threshold close to resting membrane potential. Since the hyperexcitation of primary sensory neurons is critical for both the development and maintenance of neuropathic pain, KCNQ channels thus can be a valuable target to treat neuropathic pain. KCNQ channels in DRG neurons could only be an effective target for treating neuropathic pain if sufficient channels remain available. Although KCNQ channel expression and function are unchanged in SCI pain and PIPN pain animal models (Li et al., 2019; Wu et al., 2017), KCNQ expression in DRG neurons is decreased upon nerve injury or bone cancer (Laumet et al., 2015; Rose et al., 2011; Zheng et al., 2013). Despite the suppressed expression in sensory neurons in some neuropathic pain models, KCNQ channel openers can still effectively attenuate neuropathic pain behaviors (Laumet et al., 2015; Rose et al., 2011; Wu et al., 2017; Zheng et al., 2013) (Fig. 4). This indicates that enough KCNQ channels are still available as therapeutic targets to treat neuropathic pain. In addition to effectively reverse reflex hypersensitivity evoked by mechanical or heat stimulation (Laumet et al., 2015; Rose et al., 2011; Wu et al., 2017; Zheng et al., 2013), KCNQ channel activators also effectively suppress DRG spontaneous firing and spontaneous pain after SCI and PIPN (Li et al., 2019; Wu et al., 2017) (Fig. 5). Interesting, KCNQ channel openers display little or no effect on sensory fiber discharge and reflex sensitivity of naïve animals (Li et al., 2019; Roza & Lopez-Garcia, 2008; Wu et al., 2017, 2020).

KCNQ channels also present in the spinal cord and brain (Cai et al., 2015; Jentsch, 2000). Intrathecal injection of KCNQ channel opener can also attenuate pain behaviors (Cai et al., 2015), but it is difficult to identify the major site of action since the whole pain pathway from dorsal root to brain is exposed to the drug in this case. It is unknown whether brain KCNQ channels are involved in the analgesic effect of their openers for neuropathic pain, but the effect of brain KCNQ on the relief of inflammatory pain has been reported by intracerebroventricular injection of retigabine (Xu et al., 2010). However, intracerebroventricular administration of XE991 failed to reverse systemic applied retigabine-mediated reduction of paw withdrawal responses to heat stimulation (Hayashi, Iwata, Tsuchimori, & Matsumoto, 2014). The primary sensory neurons play a critical role in the development of neuropathic pain (Bedi et al., 2010; Vaso et al., 2014; Yang et al., 2014). As compared with the spinal cord and brain, DRGs lack effective vascular permeability barrier (Jimenez-Andrade et al., 2008).

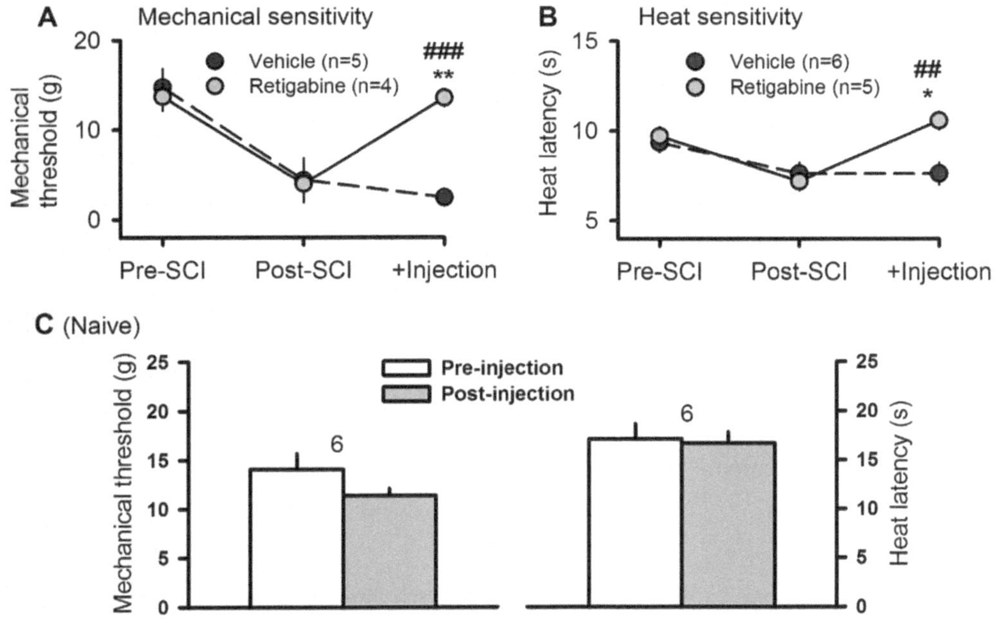

FIG. 4 Retigabine attenuates established neuropathic pain of SCI animals. (A) and (B) Reflexive hypersensitivity to both mechanical and thermal stimulation is developed after spinal cord injury. Retigabine (10 mg/kg, i.p.), but not its vehicle, reversed the evoked responses in these animals. (C) No detectable effect was observed when naïve rats received retigabine. *(Reprinted with permission from Wu, Z., Li, L., Xie, F., Du, J., Zuo, Y., Frost, J. A., et al. (2017). Activation of KCNQ channels suppresses spontaneous activity in dorsal root ganglion neurons and reduces chronic pain after spinal cord injury.* Journal of Neurotrauma, 34, *1260–1270.)*

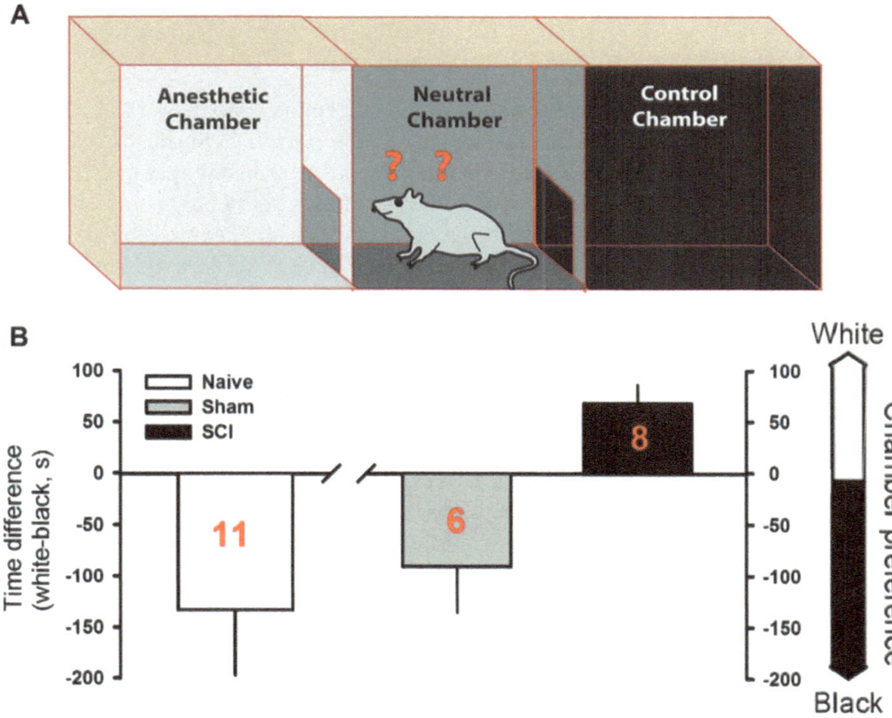

FIG. 5 The conditioned place preference (CPP) tests for SCI-induced spontaneous pain. (A) Three-chamber device was used for CPP tests, which include *black*, *gray*, and *black* chambers, with the *gray* one in the middle. Animals accepted 3 days of conditional treatments with two sessions for each day. In morning session, animals received vehicle injection and restricted to the *black* chamber. In another daily session in the afternoon, same animals received retigabine and were kept in *white* chambers. (B) On testing day, animals were placed in *gray* chamber and allowed to run freely through all the chambers without any injection. SCI rats, but not naïve rats or sham rats (laminectomy only), show preference to the *white* chamber that is linked to retigabine (analgesic). *(Reprinted with permission from Wu, Z., Li, L., Xie, F., Du, J., Zuo, Y., Frost, J. A., et al. (2017). Activation of KCNQ channels suppresses spontaneous activity in dorsal root ganglion neurons and reduces chronic pain after spinal cord injury.* Journal of Neurotrauma, 34, *1260–1270.)*

Systemically applied retigabine should have better access to sensory neurons in the DRGs than to neurons within the CNS. It is thus possible that KCNQ openers, especially at a relative lower concentration, mainly act on the channels in primary sensory neurons for the analgesic effect.

KCNQ channels as therapeutic target to prevent the development of neuropathic pain

Neuropathic pain is refractory to current available therapy once it transits to the chronic stage, and preventing the development of neuropathic pain is thus a more attractive strategy. It has been demonstrated that pretreatment of retigabine decreases cisplatin-induced electrophysiological impairment of primary sensory fibers (Nodera, Spieker, Sung, & Rutkove, 2011). Consistent with this observation, both evoked reflexive hypersensitivity and spontaneous pain failed to develop when retigabine and paclitaxel are coapplied (Fig. 6). The spinal cord glia activation and peripheral nerve degeneration are also attenuated (Li et al., 2019). Also, the application of retigabine at an earlier stage of acute SCI attenuates the development of pain behaviors (Wu et al., 2020). These studies suggest that acute excitation of sensory neurons probably plays a critical role in the acute-to-chronic transition of neuropathic pain.

Challenges

KCNQ channel openers show encourage effect on established neuropathic pain treatment in preclinical studies (Laumet et al., 2015; Rose et al., 2011; Wu et al., 2017; Zheng et al., 2013). However, retigabine has failed in a clinical trial against established neuropathic pain in patients who suffered postherpetic neuralgia for more than 6 months after healing of herpes zoster skin rash. The preclinical data thus need to be evaluated very carefully. Pain is a motivational and affective response. It is noted that reflexive signs are analyzed to evaluate pain of animals in the majority of preclinical studies. However, pain levels in patients with neuropathic pain are not always correlated with reflexive responses (Kalliomaki et al., 2011). Thus,

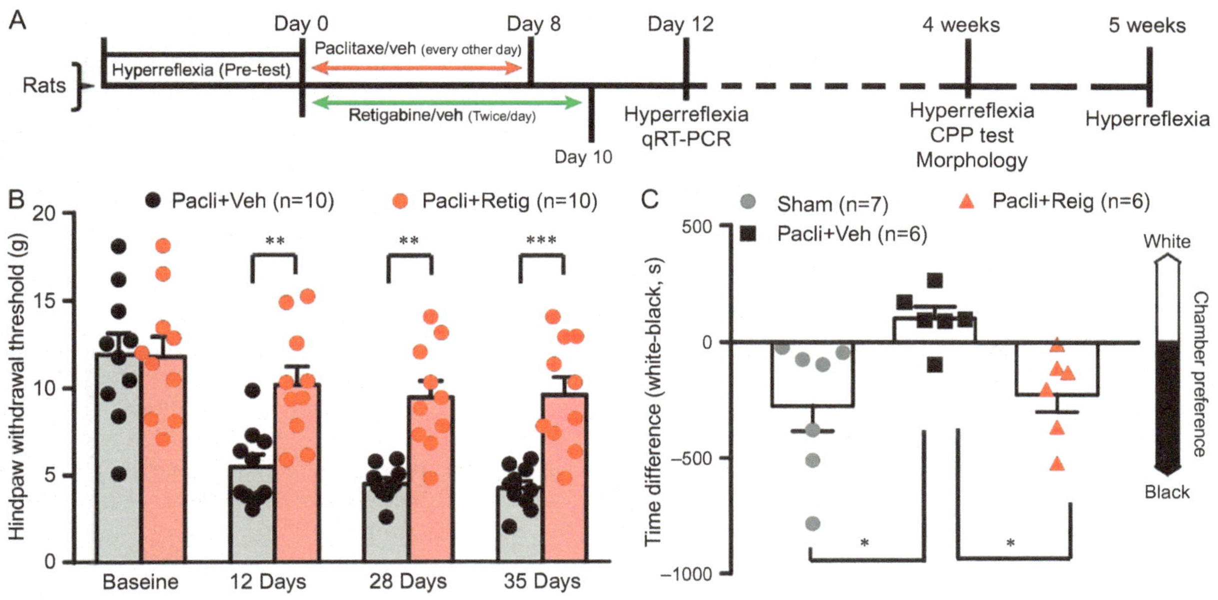

FIG. 6 Application of retigabine at early stage of paclitaxel infusion prevents the development of neuropathic pain. (A) Illustration of the experimental timeline. (B) As compared with the vehicle control group, the group received coapplication of retigabine with paclitaxel failed to develop mechanical hypersensitivity even long after the cease of retigabine treatment. (C) CPP tests indicates spontaneous pain was developed in paclitaxel + vehicle animals, but not paclitaxel + retigabine animals and vehicle only animals. *(Reprinted with permission from Li, L., Li, J., Zuo, Y., Dang, D., Frost, J. A., Yang, Q. (2019). Activation of KCNQ channels prevents paclitaxel-induced peripheral neuropathy and associated neuropathic pain.* Journal of Pain, 20, *528–539.)*

more clinic-relevant pain assessment approaches need to be developed. KCNQ channel openers flupirtine and retigabine were approved for clinical use in the United States and/or Europe. However, the twin molecules show different side-effect profiles. While flupirtine causes liver injury, long-term application of retigabine is linked to reversible discoloration of the skin and eye (Mathias & Abou-Khalil, 2017). The disparity suggests that it is the molecules themselves, but not KCNQ channels, that cause these side effects. Several ligands targeting KCNQ channels are under development and are offering promising features (Kumar et al., 2016). It is thus possible that in the near future, more specific molecules with fewer side effects can be available to treat neuropathic pain.

Applications to other areas

KCNQ channels are widely expressed in neurons. Similar to what happens in the DRG neurons after nerve injury (Rose et al., 2011), KCNQ2/3 expression in the dorsal cochlear nucleus is also decreased after noise exposure, which contributes to the generation of tinnitus (Li, Kalappa, & Tzounopoulos, 2015). Thus, targeting KCNQ channels with specific channel agonists is used to prevent the development of this disease (Li et al., 2015). Neuronal hyperexcitation is probably also involved in neurodegeneration. In addition to mitigating the development of paclitaxel-induced peripheral fiber degeneration and chronic pain (Li et al., 2019), recent evidences show that decreasing neuronal excitability by enhancing the activity of KCNQ channels also improves amyotrophic lateral sclerosis–induced spinal cord motor neuron loss (Wainger et al., 2014), traumatic brain injury–induced dysfunction (Vigil et al., 2020), and SCI-induced functional and pathological alterations (Wu et al., 2020). Thus, although neuronal injury can be induced by different causes in different locations, neuronal hyperpolarization after neuronal injury is likely a shared mechanism to induce neurodegeneration and plasticity. In addition to prevent the development of some neurodegenerative diseases, KCNQ channels are also targeted to correct dysfunction of some disease. For example, enhancing the activity of KCNQ channels by retigabine was used to treat epilepsy since some neonatal epilepsies result from KCNQ2/3 mutations (Charlier et al., 1998). Parkinson's disease is characterized by progressive loss of dopaminergic neurons in the brain. KCNQ channel blocker XE-991 was used to boost the release of dopamine, thus improving the symptoms (Chen et al., 2018). However, considering the possible role of hyperexcitability of neurons to the degeneration as revealed by many studies, such boosting strategy needs to be evaluated very carefully.

Other agents of interest

KCNQ channel agonists flupirtine and retigabine have been approved for clinical use in the United States and Europe. However, flupirtine can cause liver damage and is thus limited to short-term use. Retigabine is linked to tissue discoloration and was discontinued. Fortunately, several more specific agonists for KCNQ2 are developed. By modifying retigabine, some of its analogs are synthesized and evaluated by Tzounopoulos and Wipf's labs, including SF0034 and RL648_81 (Kumar et al., 2016). SF0034 displays increased potency in activating KCNQ2/3 channel currents that are five times stronger than retigabine, and RL648_81 is three times more potent than SF0034. Different from retigabine, these modified compounds do not activate KCNQ4 or KCNQ5 homomeric channels that are overexpressed in CHO cells (Kumar et al., 2016). As compared with retigabine, SF0034 induces more potent effects in anticonvulsant and tinnitus prevention with less toxic outcomes (Kalappa et al., 2015).

As we discussed, primary sensory neurons play a critical role in the initiation and maintenance of neuropathic pain (Bedi et al., 2010; Vaso et al., 2014; Yang et al., 2014). Focal delivery of analgesics to DRGs can effectively reverse the phantom limb pain of amputees (Vaso et al., 2014). Thus, activating brain KCNQ2/3 is not necessary for the treatment of neuropathic pain except to cause side effects. Current available compounds can all penetrate the blood-brain barrier, and the central nervous system side effect is thus unavoidable when utilizing these compounds to treat neuropathic pain. So, developing compounds that can specifically target KCNQ2/3 but cannot cross the blood-brain barrier would be very attractive, which would lead to better efficacy on neuropathic pain treatment with a less toxic profile.

Mini-dictionary of terms

Homomultimer: Functional protein consisting of identical components.

Heteromultimer: Functional protein consisting of two or more different components.

Retigabine-induced discoloration: A reversible side effect caused by long-term retigabine application in which discoloration appears in the hard palate, lips, nails, and the conjunctiva.

DRG neurons: Pseudo-unipolar neurons that relay sensory signals from the periphery to the spinal cord.

Nociceptors: Subgroup of DRG neurons and their afferent fibers, primarily unmyelinated C-fiber and thinly myelinated Aδ afferents that are capable of initiating nociceptive reflexes and pain sensation upon noxious stimuli.

Key facts of KCNQ/Kv7 channels and neuropathic pain

- Mutations in KCNQ channels cause several human diseases.
- KCNQ2, KCNQ3, and KCNQ5 are widely expressed in neuronal system.
- KCNQ channels play critical role in modulating neuronal excitability.
- The expression level of KCNQ2 in primary sensory neurons is decreased in some neuropathic pain models.
- Targeting neuronal KCNQ channels can effectively prevent or reverse neuropathic pain.

Summary points

- KCNQ proteins have five family members, and their mutations can cause human diseases.
- Functional KCNQ channels are tetramers with restricted combination and distribution.
- KCNQ2, KCNQ3, and KCNQ5 are expressed in primary sensory neurons, including nociceptors.
- Plasticity in primary sensory neurons plays a critical role in the development and maintenance of neuropathic pain.
- Targeting neuronal KCNQ channels can effectively prevent or reverse neuropathic pain.

References

Angelo, K., Jespersen, T., Grunnet, M., Nielsen, M. S., Klaerke, D. A., & Olesen, S. P. (2002). KCNE5 induces time- and voltage-dependent modulation of the KCNQ1 current. *Biophysical Journal*, *83*, 1997–2006.

Bedi, S. S., Yang, Q., Crook, R. J., Du, J., Wu, Z., Fishman, H. M., et al. (2010). Chronic spontaneous activity generated in the somata of primary nociceptors is associated with pain-related behavior after spinal cord injury. *Journal of Neuroscience*, *30*, 14870–14882.

Cai, J., Fang, D., Liu, X. D., Li, S., Ren, J., & Xing, G. G. (2015). Suppression of KCNQ/M (Kv7) potassium channels in the spinal cord contributes to the sensitization of dorsal horn WDR neurons and pain hypersensitivity in a rat model of bone cancer pain. *Oncology Reports*, *33*, 1540–1550.

Campbell, J. N., & Meyer, R. A. (2006). Mechanisms of neuropathic pain. *Neuron*, *52*, 77–92.

Charlier, C., Singh, N. A., Ryan, S. G., Lewis, T. B., Reus, B. E., Leach, R. J., et al. (1998). A pore mutation in a novel KQT-like potassium channel gene in an idiopathic epilepsy family. *Nature Genetics*, *18*, 53–55.

Chen, X., Xue, B., Wang, J., Liu, H., Shi, L., & Xie, J. (2018). Potassium channels: A potential therapeutic target for Parkinson's disease. *Neuroscience Bulletin*, *34*, 341–348.

Coucke, P. J., Van Hauwe, P., Kelley, P. M., Kunst, H., Schatteman, I., Van Velzen, D., et al. (1999). Mutations in the KCNQ4 gene are responsible for autosomal dominant deafness in four DFNA2 families. *Human Molecular Genetics*, *8*, 1321–1328.

Devaux, J. J., Kleopa, K. A., Cooper, E. C., & Scherer, S. S. (2004). KCNQ2 is a nodal K+ channel. *Journal of Neuroscience*, *24*, 1236–1244.

Dirajlal, S., Pauers, L. E., & Stucky, C. L. (2003). Differential response properties of IB(4)-positive and -negative unmyelinated sensory neurons to protons and capsaicin. *Journal of Neurophysiology*, *89*, 513–524.

Grunnet, M., Jespersen, T., Rasmussen, H. B., Ljungstrom, T., Jorgensen, N. K., Olesen, S. P., et al. (2002). KCNE4 is an inhibitory subunit to the KCNQ1 channel. *Journal of Physiology*, *542*, 119–130.

Haitin, Y., & Attali, B. (2008). The C-terminus of Kv7 channels: A multifunctional module. *Journal of Physiology*, *586*, 1803–1810.

Hayashi, H., Iwata, M., Tsuchimori, N., & Matsumoto, T. (2014). Activation of peripheral KCNQ channels attenuates inflammatory pain. *Molecular Pain*, *10*, 15.

Jentsch, T. J. (2000). Neuronal KCNQ potassium channels: Physiology and role in disease. *Nature Reviews. Neuroscience*, *1*, 21–30.

Jimenez-Andrade, J. M., Herrera, M. B., Ghilardi, J. R., Vardanyan, M., Melemedjian, O. K., & Mantyh, P. W. (2008). Vascularization of the dorsal root ganglia and peripheral nerve of the mouse: Implications for chemical-induced peripheral sensory neuropathies. *Molecular Pain*, *4*, 10.

Kalappa, B. I., Soh, H., Duignan, K. M., Furuya, T., Edwards, S., Tzingounis, A. V., et al. (2015). Potent KCNQ2/3-specific channel activator suppresses in vivo epileptic activity and prevents the development of tinnitus. *Journal of Neuroscience*, *35*, 8829–8842.

Kalliomaki, M., Kieseritzky, J. V., Schmidt, R., Hagglof, B., Karlsten, R., Sjogren, N., et al. (2011). Structural and functional differences between neuropathy with and without pain? *Experimental Neurology*, *231*, 199–206.

Kanda, H., Ling, J., Tonomura, S., Noguchi, K., Matalon, S., & Gu, J. G. (2019). TREK-1 and TRAAK are principal K(+) channels at the nodes of Ranvier for rapid action potential conduction on mammalian myelinated afferent nerves. *Neuron*, *104*, 960–971. e967.

Kharkovets, T., Hardelin, J. P., Safieddine, S., Schweizer, M., El-Amraoui, A., Petit, C., et al. (2000). KCNQ4, a K+ channel mutated in a form of dominant deafness, is expressed in the inner ear and the central auditory pathway. *Proceedings of the National Academy of Sciences of the United States of America*, *97*, 4333–4338.

King, C. H., & Scherer, S. S. (2012). Kv7.5 is the primary Kv7 subunit expressed in C-fibers. *Journal of Comparative Neurology*, *520*, 1940–1950.

Kubisch, C., Schroeder, B. C., Friedrich, T., Lutjohann, B., El-Amraoui, A., Marlin, S., et al. (1999). KCNQ4, a novel potassium channel expressed in sensory outer hair cells, is mutated in dominant deafness. *Cell*, *96*, 437–446.

Kumar, M., Reed, N., Liu, R., Aizenman, E., Wipf, P., & Tzounopoulos, T. (2016). Synthesis and evaluation of potent KCNQ2/3-specific channel activators. *Molecular Pharmacology*, *89*, 667–677.

Lang, P. M., Fleckenstein, J., Passmore, G. M., Brown, D. A., & Grafe, P. (2008). Retigabine reduces the excitability of unmyelinated peripheral human axons. *Neuropharmacology*, *54*, 1271–1278.

Laumet, G., Garriga, J., Chen, S. R., Zhang, Y., Li, D. P., Smith, T. M., et al. (2015). G9a is essential for epigenetic silencing of K(+) channel genes in acute-to-chronic pain transition. *Nature Neuroscience*, *18*, 1746–1755.

Li, S., Kalappa, B. I., & Tzounopoulos, T. (2015). Noise-induced plasticity of KCNQ2/3 and HCN channels underlies vulnerability and resilience to tinnitus. *Elife*, *4*.

Li, L., Li, J., Zuo, Y., Dang, D., Frost, J. A., & Yang, Q. (2019). Activation of KCNQ channels prevents paclitaxel-induced peripheral neuropathy and associated neuropathic pain. *Journal of Pain*, *20*, 528–539.

Mathias, S. V., & Abou-Khalil, B. W. (2017). Ezogabine skin discoloration is reversible after discontinuation. *Epilepsy & Behavior Case Reports*, *7*, 61–63.

McKeown, L., Swanton, L., Robinson, P., & Jones, O. T. (2008). Surface expression and distribution of voltage-gated potassium channels in neurons (review). *Molecular Membrane Biology*, *25*, 332–343.

Neyroud, N., Tesson, F., Denjoy, I., Leibovici, M., Donger, C., Barhanin, J., et al. (1997). A novel mutation in the potassium channel gene KVLQT1 causes the Jervell and Lange-Nielsen cardioauditory syndrome. *Nature Genetics*, *15*, 186–189.

Ning, L., Moss, A. J., Zareba, W., Robinson, J., Rosero, S., Ryan, D., et al. (2003). Novel compound heterozygous mutations in the KCNQ1 gene associated with autosomal recessive long QT syndrome (Jervell and Lange-Nielsen syndrome). *Annals of Noninvasive Electrocardiology*, *8*, 246–250.

Nodera, H., Spieker, A., Sung, M., & Rutkove, S. (2011). Neuroprotective effects of Kv7 channel agonist, retigabine, for cisplatin-induced peripheral neuropathy. *Neuroscience Letters*, *505*, 223–227.

Passmore, G. M., Reilly, J. M., Thakur, M., Keasberry, V. N., Marsh, S. J., Dickenson, A. H., et al. (2012). Functional significance of M-type potassium channels in nociceptive cutaneous sensory endings. *Frontiers in Molecular Neuroscience*, *5*, 63.

Passmore, G. M., Selyanko, A. A., Mistry, M., Al-Qatari, M., Marsh, S. J., Matthews, E. A., et al. (2003). KCNQ/M currents in sensory neurons: Significance for pain therapy. *Journal of Neuroscience*, *23*, 7227–7236.

Robbins, J. (2001). KCNQ potassium channels: Physiology, pathophysiology, and pharmacology. *Pharmacology & Therapeutics*, *90*, 1–19.

Rose, K., Ooi, L., Dalle, C., Robertson, B., Wood, I. C., & Gamper, N. (2011). Transcriptional repression of the M channel subunit Kv7.2 in chronic nerve injury. *Pain*, *152*, 742–754.

Roza, C., Laird, J. M., Souslova, V., Wood, J. N., & Cervero, F. (2003). The tetrodotoxin-resistant Na+ channel Nav1.8 is essential for the expression of spontaneous activity in damaged sensory axons of mice. *Journal of Physiology*, *550*, 921–926.

Roza, C., & Lopez-Garcia, J. A. (2008). Retigabine, the specific KCNQ channel opener, blocks ectopic discharges in axotomized sensory fibres. *Pain, 138*, 537–545.

Sanguinetti, M. C., Curran, M. E., Zou, A., Shen, J., Spector, P. S., Atkinson, D. L., et al. (1996). Coassembly of K(V)LQT1 and minK (IsK) proteins to form cardiac I(Ks) potassium channel. *Nature, 384*, 80–83.

Schroeder, B. C., Hechenberger, M., Weinreich, F., Kubisch, C., & Jentsch, T. J. (2000). KCNQ5, a novel potassium channel broadly expressed in brain, mediates M-type currents. *Journal of Biological Chemistry, 275*, 24089–24095.

Schroeder, B. C., Waldegger, S., Fehr, S., Bleich, M., Warth, R., Greger, R., et al. (2000). A constitutively open potassium channel formed by KCNQ1 and KCNE3. *Nature, 403*, 196–199.

Schwake, M., Pusch, M., Kharkovets, T., & Jentsch, T. J. (2000). Surface expression and single channel properties of KCNQ2/KCNQ3, M-type K+ channels involved in epilepsy. *Journal of Biological Chemistry, 275*, 13343–13348.

Shields, S. D., Ahn, H. S., Yang, Y., Han, C., Seal, R. P., Wood, J. N., et al. (2012). Nav1.8 expression is not restricted to nociceptors in mouse peripheral nervous system. *Pain, 153*, 2017–2030.

Soldovieri, M. V., Miceli, F., & Taglialatela, M. (2011). Driving with no brakes: Molecular pathophysiology of Kv7 potassium channels. *Physiology (Bethesda), 26*, 365–376.

Thompson, A. N., Kim, I., Panosian, T. D., Iverson, T. M., Allen, T. W., & Nimigean, C. M. (2009). Mechanism of potassium-channel selectivity revealed by Na(+) and Li(+) binding sites within the KcsA pore. *Nature Structural & Molecular Biology, 16*, 1317–1324.

Tinel, N., Diochot, S., Borsotto, M., Lazdunski, M., & Barhanin, J. (2000). KCNE2 confers background current characteristics to the cardiac KCNQ1 potassium channel. *EMBO Journal, 19*, 6326–6330.

Vaso, A., Adahan, H. M., Gjika, A., Zahaj, S., Zhurda, T., Vyshka, G., et al. (2014). Peripheral nervous system origin of phantom limb pain. *Pain, 155*, 1384–1391.

Vetter, D. E., Mann, J. R., Wangemann, P., Liu, J., McLaughlin, K. J., Lesage, F., et al. (1996). Inner ear defects induced by null mutation of the isk gene. *Neuron, 17*, 1251–1264.

Vigil, F. A., Bozdemir, E., Bugay, V., Chun, S. H., Hobbs, M., Sanchez, I., et al. (2020). Prevention of brain damage after traumatic brain injury by pharmacological enhancement of KCNQ (Kv7, "M-type") K(+) currents in neurons. *Journal of Cerebral Blood Flow and Metabolism, 40*, 1256–1273.

Wainger, B. J., Kiskinis, E., Mellin, C., Wiskow, O., Han, S. S., Sandoe, J., et al. (2014). Intrinsic membrane hyperexcitability of amyotrophic lateral sclerosis patient-derived motor neurons. *Cell Reports, 7*, 1–11.

Wang, Q., Curran, M. E., Splawski, I., Burn, T. C., Millholland, J. M., VanRaay, T. J., et al. (1996). Positional cloning of a novel potassium channel gene: KVLQT1 mutations cause cardiac arrhythmias. *Nature Genetics, 12*, 17–23.

Wiener, R., Haitin, Y., Shamgar, L., Fernandez-Alonso, M. C., Martos, A., Chomsky-Hecht, O., et al. (2008). The KCNQ1 (Kv7.1) COOH terminus, a multitiered scaffold for subunit assembly and protein interaction. *Journal of Biological Chemistry, 283*, 5815–5830.

Wu, Z., Li, L., Xie, F., Du, J., Zuo, Y., Frost, J. A., et al. (2017). Activation of KCNQ channels suppresses spontaneous activity in dorsal root ganglion neurons and reduces chronic pain after spinal cord injury. *Journal of Neurotrauma, 34*, 1260–1270.

Wu, Z., Li, L., Xie, F., Xu, G., Dang, D., & Yang, Q. (2020). Enhancing KCNQ channel activity improves neurobehavioral recovery after spinal cord injury. *Journal of Pharmacology and Experimental Therapeutics, 373*, 72–80.

Xie, W., Strong, J. A., & Zhang, J. M. (2009). Early blockade of injured primary sensory afferents reduces glial cell activation in two rat neuropathic pain models. *Neuroscience, 160*, 847–857.

Xu, W., Wu, Y., Bi, Y., Tan, L., Gan, Y., & Wang, K. (2010). Activation of voltage-gated KCNQ/Kv7 channels by anticonvulsant retigabine attenuates mechanical allodynia of inflammatory temporomandibular joint in rats. *Molecular Pain, 6*, 49.

Yang, W. P., Levesque, P. C., Little, W. A., Conder, M. L., Ramakrishnan, P., Neubauer, M. G., et al. (1998). Functional expression of two KvLQT1-related potassium channels responsible for an inherited idiopathic epilepsy. *Journal of Biological Chemistry, 273*, 19419–19423.

Yang, Q., Wu, Z., Hadden, J. K., Odem, M. A., Zuo, Y., Crook, R. J., et al. (2014). Persistent pain after spinal cord injury is maintained by primary afferent activity. *Journal of Neuroscience, 34*, 10765–10769.

Zheng, Q., Fang, D., Liu, M., Cai, J., Wan, Y., Han, J. S., et al. (2013). Suppression of KCNQ/M (Kv7) potassium channels in dorsal root ganglion neurons contributes to the development of bone cancer pain in a rat model. *Pain, 154*, 434–448.

Zheng, W., Verlander, J. W., Lynch, I. J., Cash, M., Shao, J., Stow, L. R., et al. (2007). Cellular distribution of the potassium channel KCNQ1 in normal mouse kidney. *American Journal of Physiology. Renal Physiology, 292*, F456–F466.

Chapter 2

A new mutation in *NTRK1* gene is associated with congenital insensitivity to pain without anhidrosis

Mogge Hajiesmaeil[a], Fatemeh Yazarlou[b], Maryam Sobhani[c], and Soudeh Ghafouri-Fard[d]
[a]*Department of Biology and Biotechnology "Charles Darwin", Sapienza University of Rome, Rome, Italy,* [b]*Department of Medical Genetics, School of Medicine, Tehran University of Medical Sciences, Tehran, Iran,* [c]*Blood Transfusion Research Center, High Institute for Research and Education in Transfusion Medicine, Tehran, Iran,* [d]*Department of Medical Genetics, Shahid Beheshti University of Medical Sciences, Tehran, Iran*

Abbreviations

ACMG	American College of Medical Genetics
CIPA	congenital insensitivity to pain with anhidrosis
CONDEL	consensus deleteriousness score of missense mutations
FTT	failure to thrive
HGMD	Human Gene Mutation Database
HSAN4	hereditary sensory and autonomic neuropathy type 4
LM-PCR	ligation-mediated PCR
NGF	nerve growth factor
NTRK1	neurotrophic tyrosine receptor kinase-1
NVC	nerve conduction velocity
PVFD	population variation frequency database

Introduction

Congenital insensitivity to pain with anhidrosis (CIPA) is a very rare and infrequent autosomal recessive disease. The disease has been described for more than 50 years ago (Swanson, 1963), but the genetic basis was unknown approximately until 2000 (Indo et al., 1996; Miura et al., 2000). The disease is considered as a hereditary sensory and autonomic neuropathy type 4 (HSAN4) (Haga, Kubota, & Miwa, 2013). Greatest of the previously reported CIPA cases are from Japan where the total occurrence is estimated to be one in 600,000–950,000 individuals (Haga et al., 2013). There is an inability to feel pain and temperature, and decreased or absent sweating (anhidrosis). The signs and symptoms of CIPA usually appear at birth or during infancy. The inability to feel pain and temperature often leads to repeated, severe injuries, and unintentional self-injury is common. People with CIPA may also heal slowly from skin and bone injuries, which can lead to chronic bone infections (osteomyelitis) or a condition called Charcot joints. Absent sweating can cause recurrent, high fevers (hyperpyrexia) and seizures brought on by high temperature (febrile seizures). Other features may include dental caries, difficulty controlling urine and defecating (urine and fecal incontinence), behavioral or emotional problems, and intellectual disability (Lv et al., 2017). The genetic basis for CIPA is unknown. Nerve growth factor (NGF) induces neurite outgrowth and promotes survival of embryonic sensory and sympathetic neurons. CIPA has been associated with germline mutations in the neurotrophic tyrosine receptor kinase-1 (*NTRK1*) gene (Miura et al., 2000). NTRK1 codes for a tropomyosin-related kinase A (TrkA) that binds to the nerve growth factor (NGF). Based on the role of NGF in enhancement of neurite outgrowth and reservation of the embryonic sensory and sympathetic neurons, mutations in TrkA and its human homolog are associated with CIPA (Indo et al., 1996). In human, the *NTRK1* gene resides on chromosome

The Neurobiology, Physiology, and Psychology of Pain. https://doi.org/10.1016/B978-0-12-820589-1.00002-6

1q21-q22 and has 17 exons. As NTRK1 protein is the receptor for NGF, mutations in NTRK1 destruct the NGF signal transduction, resulting in apoptosis of sympathetic ganglion neurons and neurons responsible for pain sensation (Indo, 2001). Three founder mutations (mutations with high frequency in an isolated population, in which one or more of the ancestors carried the mutation) in NTRK1 (p.Phe284TrpfsTer36 (IVS7-33T > A), p.Arg554GlyfsTer104 (exon 14), and p.Asp674Tyr (exon 15)) represent more than two-third of pathogenic variants in the Japanese population (Indo, 2001). However, studies in other populations are not adequate to find probable founder mutations. In the current case report, we describe the first case of CIPA in Iranian population with a phenotypic variation.

The patient was a 10-month-old girl born to a healthy consanguineous Iranian parent (Fig. 1). She was delivered by normal vaginal delivery at 40 gestational weeks after an unremarkable pregnancy. The birth weight, height, and head circumstance were within the normal range. Afterward, parents reported mild developmental delay. Parents reported that the girl did not feel pain from any harmful stimuli such as inflammation and heat, from her birth. They also reported repeated injuries and delayed wound healing. She had history of several hospital admissions for recurrent infections and failure to thrive (FTT). She had normal hidrosis. The most notable sign was insensitivity to pain which was evaluated by applying painful stimuli and lack of body moving or discomfort expression by the patient. Furthermore, several deep ulcerative lesions were detected in her mouth and tongue. Nerve conduction velocity (NCV) exam showed generalized small fiber neuropathy (Fig. 2). Complete blood count analysis showed hypochromic microcytic anemia. Cytogenetic analysis revealed normal female karyotype in all assessed cells. Two-phase bone scintigraphy showed normal radiotracer activity (Fig. 3). Delayed images also showed homogenous and physiologic uptake throughout the body. Skin biopsy showed unremarkable epidermis, preserved eccrine glands in dermis, and hypodermal fat. Based on the consanguinity of parents and presence of similar affected members in the family, genetic analyses were performed. For this reason, blood samples were collected from the patients and parents in EDTA tubes. Informed consents were obtained from parents based on the protocol approved by the local ethics committee. DNA was extracted from all samples using the standard salting out method. Ligation-mediated PCR (LM-PCR) method was used for amplification of DNA. Whole-exome sequencing was performed using a custom-designed Nimblegen chip capturing method (target region capture with BGI59M whole-exome capture chip) and next-generation sequencing (Illumina HiSeq 2000 system, BGI Clinical Laboratories, Shenzhen, China). The Illumina basecalling Software 1.7 was used for image processing. The nucleotide sequences were produced by 90 bp

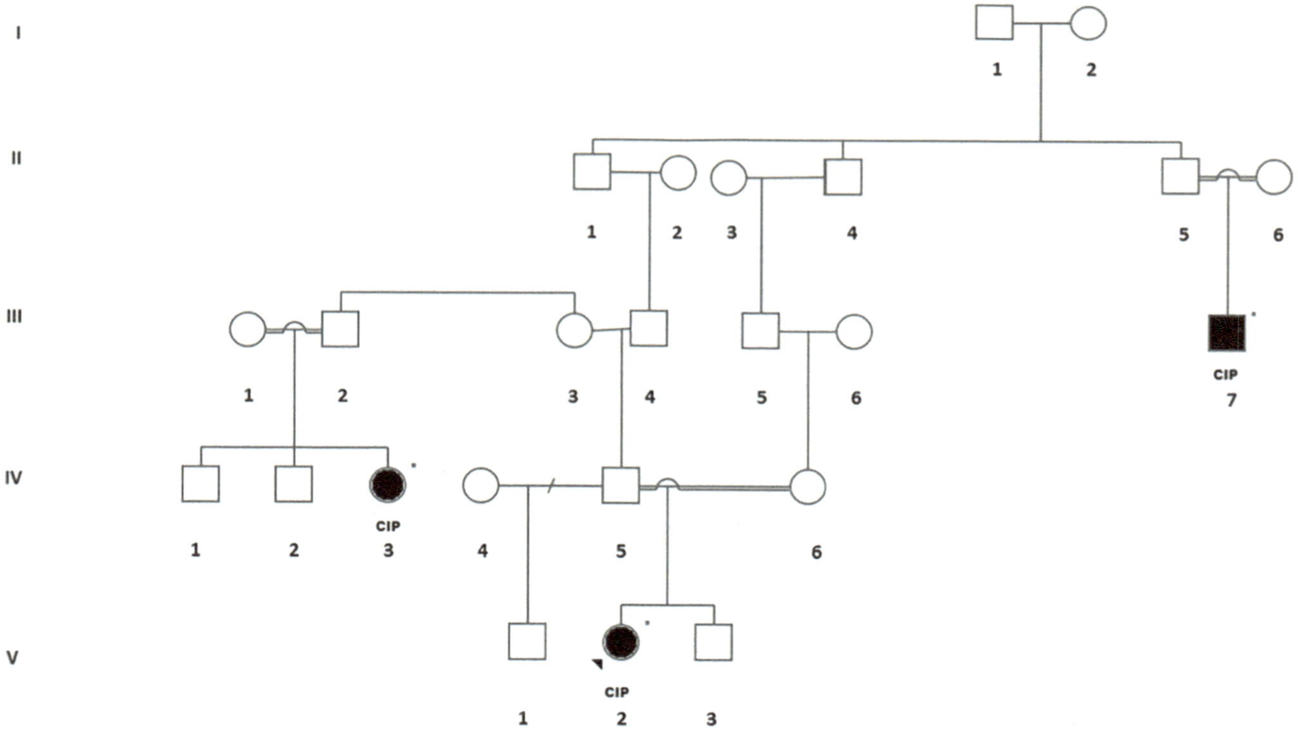

FIG. 1 Pedigree of family indicative of three cases with similar phenotypes including insensitivity to pain (CIP, congenital insensitivity to pain). Individuals V-2, IV-5, and IV-6 are the affected girl, her father, and her mother, respectively. Two other affected members exist in the pedigree (III-7 and IV-3) (Hajiesmaeila, Yazarlou, Sobhani, & Ghafouri-Fard, 2019).

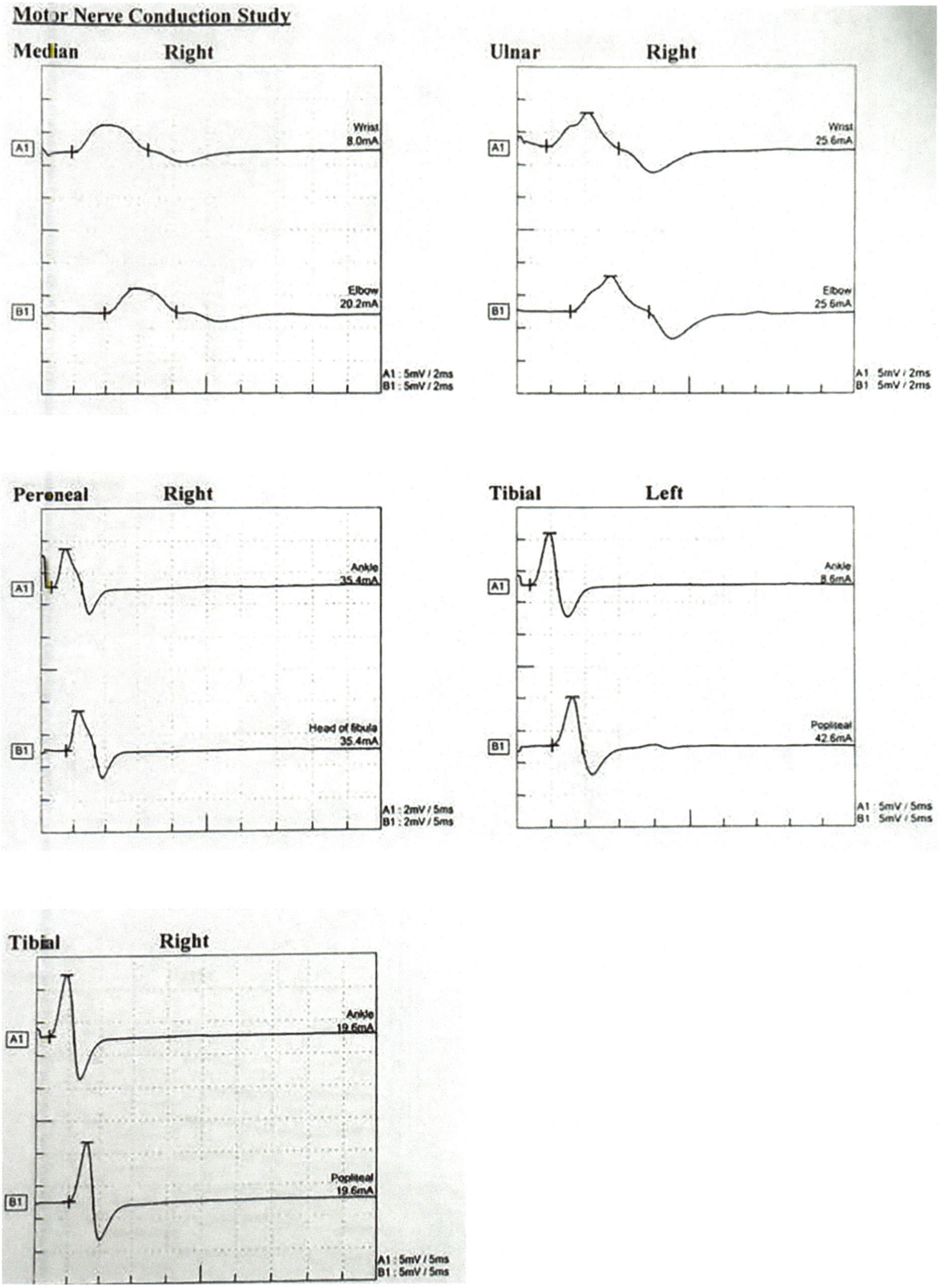

FIG. 2 Nerve conduction velocity (NCV) exam compatible with generalized small fiber neuropathy (Hajiesmaeila et al., 2019).

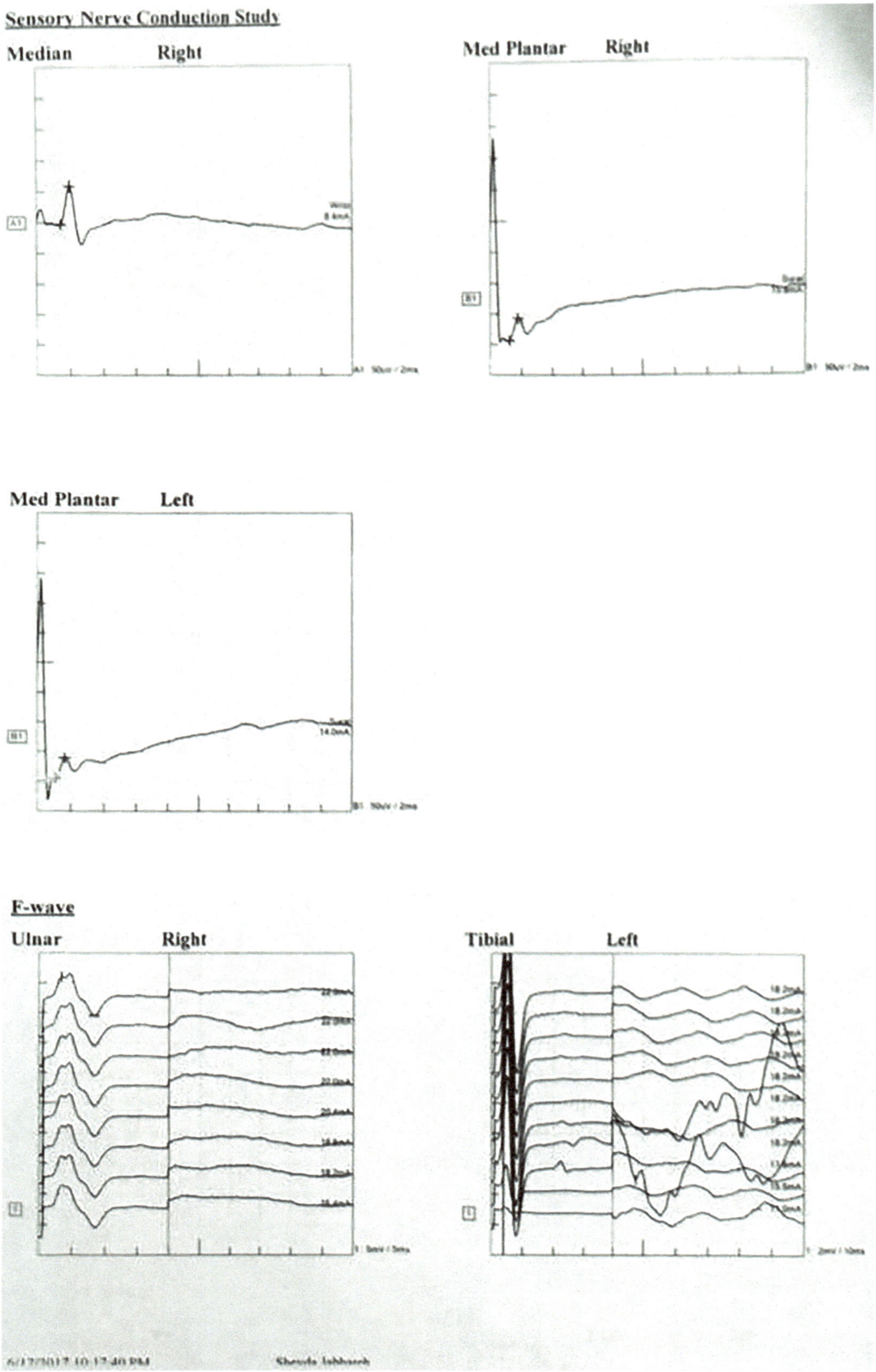

FIG. 3 Two-phase bone scintigraphy of patient.

(Continued)

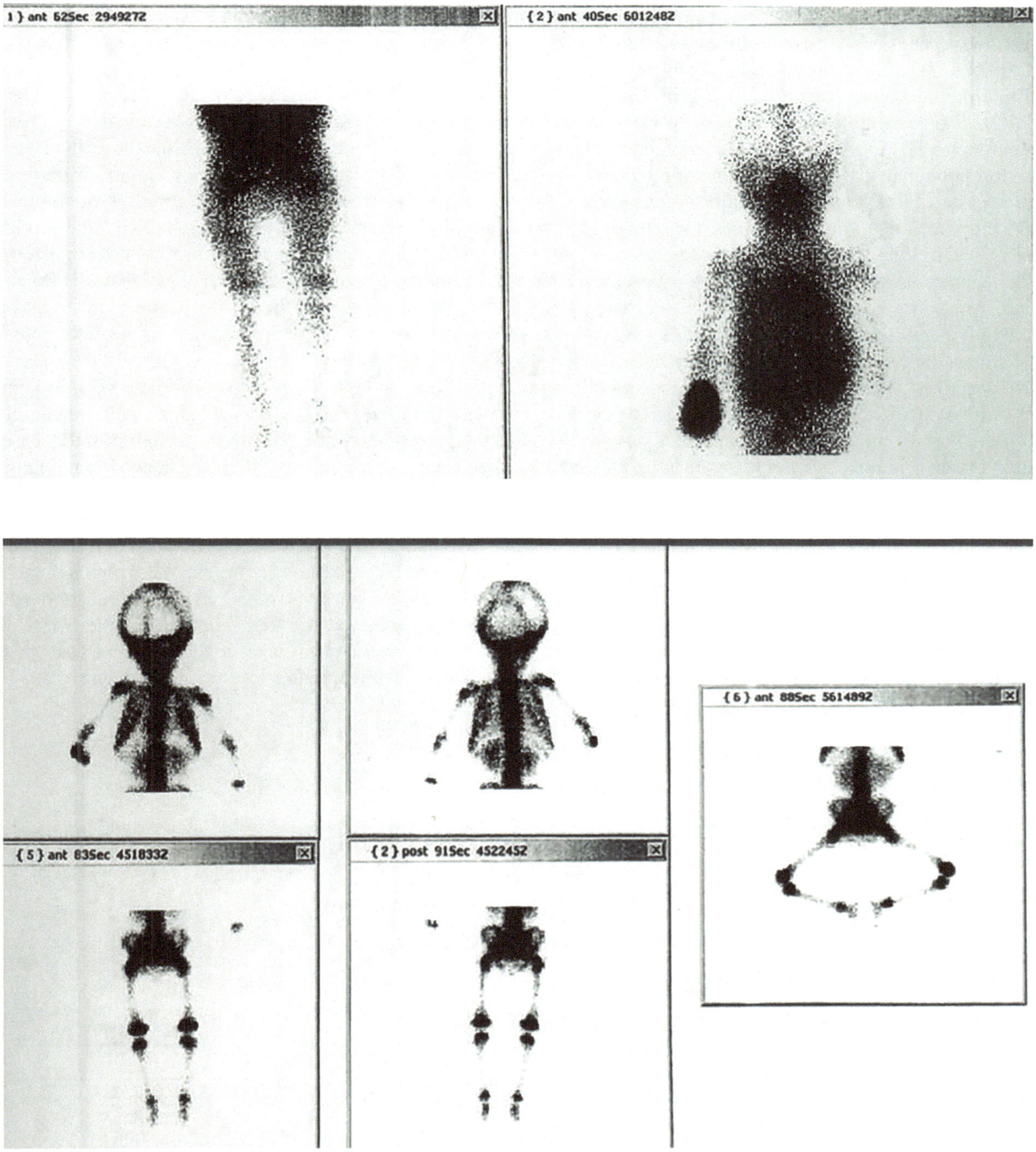

FIG. 3 CONT'D

pair-end reads. SREads were mapped on the reference genome using OAPaligner/SOAP2 tool. The test coverage was 99.78% with the depth of 164.11 ×. Subsequently, the sequencing results were analyzed using international mutation and polymorphism databases (dbSNP, the 1000 genome database and Exome Sequencing Project 6500) in addition to the BGI local database (population variation frequency database (PVFD)). Common nonpathogenic variants and recurrent artifact variant calls were filtered. All identified mutations were compared with gene mutation database (The Human Gene Mutation Database (HGMD), http://www.hgmd.cf.ac.uk). Nonsense mutations, frameshift mutations, classic splice site mutations, or in-frame changes of $\geq$5 amino acids were regarded as pathogenic. Consensus deleteriousness score of missense mutations (Condel) was measured for assessment of missense mutation pathogenicity (Gonzalez-Perez & Lopez-Bigas, 2011). The results revealed the presence of a new likely pathogenic mutation in the exon 13 of *NTRK1* gene (NM_002529.3) in the proband in homozygote state (c.1524_1531dupGGACATCG, p.Val511Glyfs*39). Sequence variants interpretation was performed using standards and guidelines provided by American College of Medical Genetics (ACMG). The frameshift mutation leads to early termination of the coding sequence, which is anticipated to affect the protein function. This mutation has not been reported for its pathogenicity. Based on the BGI local database, the frequency of this mutation in heterozygote state in normal population was very low. The list of other identified variants is presented in Supplementary File. A total of 25 homozygous variants were detected (12 missense variants, 6 coding synonymous variants, 5 intronic variants, and 2 frameshift variants). The other frameshift variant was located in HLA-DRB1 and was known as the rs768015363. Missense variants were assessed for their pathogenicity using prediction tools and recognized to be nonpathogenic. Sanger sequencing using the previously described primers for exon 13 (Mardy et al., 1999) (ABI Prism3130 Genetic Analyzer, Applied Biosystems, Foster City, CA, United States) confirmed the results in the proband and other affected members of the family. In addition, the parents were shown to carry the same mutation in heterozygote state. Fig. 4 shows the results of Sanger sequencing.

We described the first case of *NTRK1* gene mutation in Iranian population. The patient was born to a consanguineous family with other affected members. Based on the rarity of mutations in this gene, it is expected that a significant number of cases were born to consanguineous parents. Consistent with this expectation, Axelrod showed that half of CIPA cases are children of consanguineous parents (Axelrod, 2002). *NTRK1* gene spans approximately 23 kb and codes for a protein with 760 amino acids (NP_001007793). The detected mutation in this study is located in a region that codes for tyrosine kinase domain (Indo, 2001), so it is expected to destruct tyrosine kinase activity of the protein. The identified mutation in the current study was a novel frameshift mutation leading to early termination of translation. So, it is pathological. Previous reports have identified two frameshift and one missense mutations as founder mutations in the Japanese population (Miura et al., 2000). A single study in Chinese population has shown compound heterozygous mutations in two unrelated patients, with two of them being intronic variants (Lv et al., 2017). Of note, only one of the four reported mutations in Chinese population was novel (Lv et al., 2017). A more recent study in Chinese population has been conducted in 36 CIPA patients from 34 discrete families. The study reported 27 mutations in *NTRK1* gene comprising 15 novel ones (Geng et al., 2018).

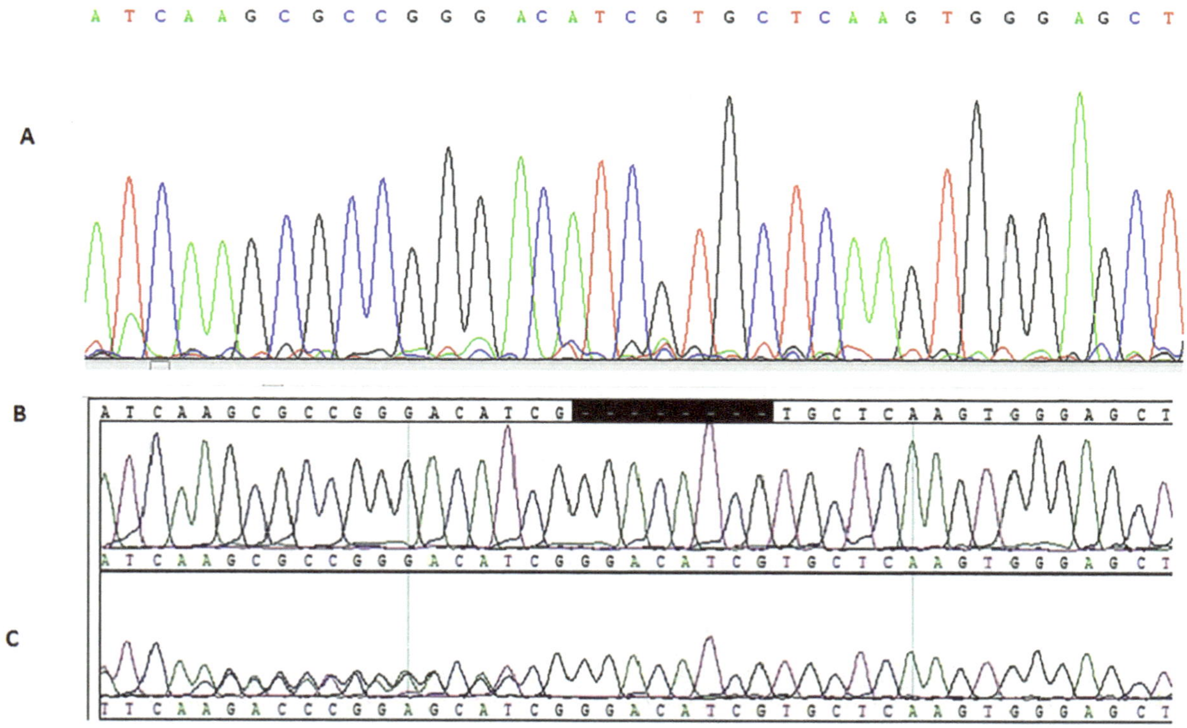

FIG. 4 The results of Sanger sequencing: normal NTRK1 sequence in IV-1 (A), homozygote c.1524_1531dupGGACATCG in V-2 (B), and heterozygote state in IV-5 (C). As demonstrated in B section, eight nucleotides have been duplicated. Heterozygote duplication has been recognized by double peak chromatogram (C) (Hajiesmaeila et al., 2019).

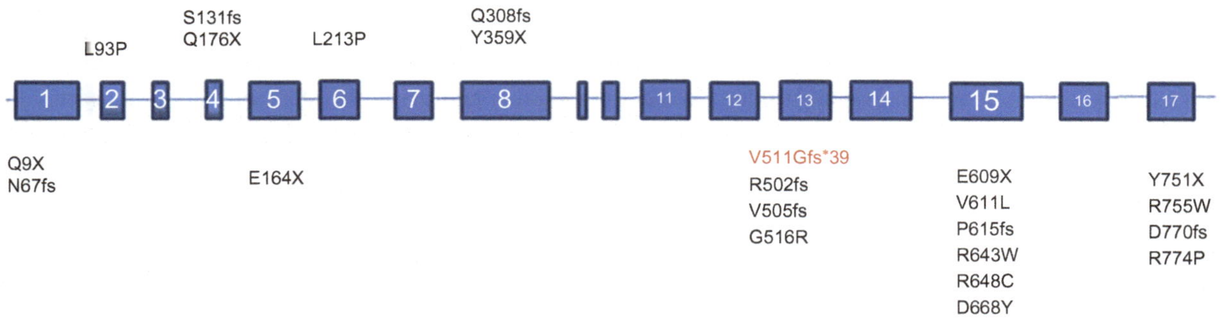

FIG. 5 The position of the detected mutation in the current study and a number of other identified mutations (Hajiesmaeila et al., 2019; Indo, 2001).

Moreover, another recent study in three Chinese CIPA patients led to identification of four different NTRK1 mutations (Wang et al., 2018). Houlden et al. (2001) have identified Y359C mutation in a boy, from a consanguineous Pakistani family, with distal loss of pain and temperature perception and general anhidrosis. Taken together, more than 100 NTRK1 mutations have been identified in CIPA patients from different populations. Fig. 5 shows the position of the detected mutation in the current study and a number of other identified mutations. The currently reported patient has a distinctive phenotype of normal hidrosis as confirmed by skin biopsy. Moreover, despite the previously reported CIPA patients, she did not suffer from bone fractures and arthropathy which might be due to the fact that diagnosis was performed in the age before ambulation. Self-mutilation behavior, high bone resorption, and low bone mass density as reported in Chinese patients (Lv et al., 2017) were also absent in the currently reported patient (Hajiesmaeila et al., 2019). Such phenotypic variations can be attributed to different identified mutations or presence of gene-gene interactions that modifies the effects of certain mutations.

In brief, the current study shows a different phenotypic variant of CIPA in Iranian population and adds to the repository NTRK1 mutations. Identification of such disease-causing mutations would facilitate carrier testing and prenatal diagnosis.

The applications to other areas

Mutations in the *NTRK1* gene are frequently found in people with a common type of thyroid cancer called papillary thyroid carcinoma.

Mini-dictionary of terms

Congenital insensitivity to pain with anhidrosis (CIPA) is a rare autosomal recessive disorder of the nervous system that prevents the feeling of **pain** or temperature, and prevents a person from sweating. Cognitive disorders are commonly coincident.

Hereditary sensory and autonomic neuropathy type IV (**HSAN4** or **HSAN IV**) is a rare **genetic** disorder that usually begins in infancy and is characterized by an inability to feel pain and an inability to sweat (anhidrosis). Affected individuals also cannot feel temperature and cannot distinguish between hot and cold.

Anhidrosis: Anhidrosis is the inability to sweat normally. When you don't sweat (perspire), your body can't cool itself, which can lead to overheating and sometimes to heatstroke—a potentially fatal condition. Anhidrosis—sometimes called hypohidrosis—can be difficult to diagnose. Mild anhidrosis often goes unrecognized.

Osteomyelitis (OM) is an infection of bone. Symptoms may include pain in a specific bone with overlying redness, fever, and weakness. The long bones of the arms and legs are most commonly involved in children, while the feet, spine, and hips are most commonly involved in adults.

A **febrile seizure** is a **convulsion** in a child caused by a spike in body temperature, often from an infection. They occur in young children with normal development without a history of neurologic symptoms. It can be frightening when your child has a **febrile seizure**, and the few minutes it lasts can seem like an eternity.

A **germline mutation**, or germinal **mutation**, is any detectable variation within germ cells (cells that, when fully developed, become sperm and ovum). **Mutations** in these cells are the only **mutations** that can be passed on to offspring, when either a **mutated** sperm or oocyte comes together to form a zygote.

Urine and fecal incontinence: Urinary and fecal incontinences are pelvic floor disorders. They result in involuntary loss of control of a bodily function whether it's the normal voiding reflex for urine or the control of a normal bowel movement.

Founder mutations: In **genetics**, a founder mutation is a mutation that appears in the DNA of one or more individuals which are founders of a distinct population. Founder mutations initiate with changes that occur in the DNA and can be passed down to other generations.

A ***frameshift mutation*** is a type of mutation involving the insertion or deletion of a nucleotide in which the number of deleted base pairs is not divisible by three.

Consanguineous marriage is traditional and respected in most communities of North Africa, Middle East, and West Asia, where intrafamilial unions collectively account for 20% to 50+% of all marriages.

Key facts

- Congenital insensitivity to pain is suggested in an infant or young child presenting with a history of multiple traumas (e.g., burns, abrasions) that are unnoticed by the patient and, at least initially, also by the parents or guardians.
- Repeated traumas to joints and surrounding structures, resulting from a lack of protective proprioceptive feedback, lead to an appearance of advanced joint disease.
- Developing physes are particularly vulnerable to stress fractures, resulting in significant fragmentation and other architectural distortion.
- A loose periosteum in infants and children and subperiosteal hemorrhage may lead to remarkable periosteal reaction.
- Severe burns may be associated with joint contractures.

Summary points

- Congenital insensitivity to pain is a rare autosomal recessive disease characterized by varying degrees of autonomic dysfunction and sensory loss, including nociceptive hyposensitivity.
- The molecular genetic defect involves loss of function mutations in the *NTRK1* gene, which encodes the tyrosine kinase receptor A gene. This autonomic dysfunction makes both general and neuraxial anesthesia challenging.
- Testing the block during neuraxial anesthesia may be difficult, and the patient may be prone to hypotension and bradycardia.

References

Axelrod, F. B. (2002). Hereditary sensory and autonomic neuropathies. Familial dysautonomia and other HSANs. *Clinical Autonomic Research*, *12*(Suppl. 1), 12–14. 12102459. Epub 2002/07/10. eng.

Geng, X., Liu, Y., Ren, X., Guan, Y., Wang, Y., Mao, B., et al. (2018). Novel NTRK1 mutations in Chinese patients with congenital insensitivity to pain with anhidrosis. *Molecular Pain*, *14*. 1744806918781140 29770739. 6009080.

Gonzalez-Perez, A., & Lopez-Bigas, N. (2011). Improving the assessment of the outcome of nonsynonymous SNVs with a consensus deleteriousness score, Condel. *American Journal of Human Genetics*, *88*(4), 440–449. 21457909. 3071923.

Haga, N., Kubota, M., & Miwa, Z. (2013). Epidemiology of hereditary sensory and autonomic neuropathy type IV and V in Japan. *American Journal of Medical Genetics. Part A*, *161A*(4), 871–874. WOS:000316631300036. English.

Hajiesmaeila, M., Yazarlou, F., Sobhani, M., & Ghafouri-Fard, S. (2019). A new mutation in NTRK1 gene is associated with congenital insensitivity to pain without anhidrosis. *Meta Gene*, *20*, 1–7.

Houlden, H., King, R. H., Hashemi-Nejad, A., Wood, N. W., Mathias, C. J., Reilly, M., et al. (2001). A novel TRK A (NTRK1) mutation associated with hereditary sensory and autonomic neuropathy type V. *Annals of Neurology*, *49*(4), 521–525. 11310631. Epub 2001/04/20. eng.

Indo, Y. (2001). Molecular basis of congenital insensitivity to pain with anhidrosis (CIPA): Mutations and polymorphisms in TRKA (NTRK1) gene encoding the receptor tyrosine kinase for nerve growth factor. *Human Mutation*, *18*(6), 462–471. 11748840. Epub 2001/12/19. eng.

Indo, Y., Tsuruta, M., Hayashida, Y., Karim, M. A., Ohta, K., Kawano, T., et al. (1996). Mutations in the TRKA/NGF receptor gene in patients with congenital insensitivity to pain with anhidrosis. *Nature Genetics*, *13*(4), 485–488. 8696348. Epub 1996/08/01. eng.

Lv, F., Xu, X. J., Song, Y. W., Li, L. J., Wang, O., Jiang, Y., et al. (2017). Recurrent and novel mutations in the NTRK1 gene lead to rare congenital insensitivity to pain with anhidrosis in two Chinese patients. *Clinica Chimica Acta*, *468*, 39–45. WOS:000403121800008. English.

Mardy, S., Miura, Y., Endo, F., Matsuda, I., Sztriha, L., Frossard, P., et al. (1999). Congenital insensitivity to pain with anhidrosis: Novel mutations in the TRKA (NTRK1) gene encoding a high-affinity receptor for nerve growth factor. *American Journal of Human Genetics*, *64*(6), 1570–1579. 10330344. PMC1377900. Epub 1999/05/20. eng.

Miura, Y., Mardy, S., Awaya, Y., Nihei, K., Endo, F., Matsuda, I., et al. (2000). Mutation and polymorphism analysis of the TRKA (NTRK1) gene encoding a high affinity receptor for nerve growth factor in congenital insensitivity to pain with anhidrosis (CIPA) families. *Human Genetics*, *106*(1), 116–124. 10982191. Epub 2000/09/12. eng.

Swanson, A. G. (1963). Congenital insensitivity to pain with anhydrosis. A unique syndrome in two male siblings. *Archives of Neurology*, *8*, 299–306. 13979626. Epub 1963/03/01. eng.

Wang, W. B., Cao, Y. J., Lyu, S. S., Zuo, R. T., Zhang, Z. L., & Kang, Q. L. (2018). Identification of a novel mutation of the NTRK1 gene in patients with congenital insensitivity to pain with anhidrosis (CIPA). *Gene*, *679*, 253–259.

Chapter 3

Prdm12, a key transcriptional regulator of the nociceptive lineage

Simon Vermeiren[a], Simon Desiderio[b], and Eric J. Bellefroid[a]
[a]*ULB Neuroscience Institute (UNI), Free University of Brussels, Gosselies, Belgium,* [b]*Institute for Neurosciences of Montpellier, University of Montpellier, INSERM U1051, Montpellier, France*

Abbreviations

BHLH	basic helix-loop-helix
CIP	congenital insensitivity to pain
CNS	central nervous system
DRG	dorsal root ganglia
HMT	histone methyltransferase
HSAN	hereditary sensory and autonomic neuropathies
MiTES	midface toddler excoriation syndrome
NGF	nerve growth factor
PNS	peripheral nervous system
PR	PRDI-BF1-RIZ1 homologous domain
Prdm	PR domain containing methyltransferase
SJG	superior-jugular ganglia
TG	trigeminal ganglia
Trk	tyrosine receptor kinase
TRP	transient receptor potential channel
VAG	vestibuloacoustic ganglia

Introduction

Pain perception is an evolutionarily conserved warning mechanism that alerts us from environmental dangers and potential damaging stimuli, and is therefore essential for our survival. Pain is also causing untold misery and suffering when the pain-sensing system becomes over- or erroneously activated. Chronic inflammatory and neuropathic pain is a major public health burden, estimated to affect one-fifth of the population worldwide. Despite its prevalence, currently available pain therapeutics often display limited efficacity and/or have serious side effects. There is thus an important need for more effective analgesics. The development of such innovative pain therapeutics requires a better understanding of the mechanisms leading to the development of pain as a pathological condition (Price & Gold, 2018; Van Hecke, Torrance, & Smith, 2013).

Pain is a complex unpleasant experience that results from the initial activation of a subset of sensory neurons called nociceptors. The cell bodies of nociceptors as well as those of other somatic sensory neurons (proprioceptors, mechanoreceptors) are located in specialized ganglia of the peripheral nervous system (PNS): the trigeminal (TG), superior-jugular (SJG), and accessory ganglia in the head region, and the dorsal root ganglia (DRG) found in a metameric pattern on each side of the spinal cord in the body region. Nociceptors are slow conductive thinly myelinated Aδ- or unmyelinated C-fibers. They are pseudo-unipolar neurons with a unique process attached to the cell body that bifurcates into two branches. One branch, referred to as the distal process, projects to cutaneous or deep peripheral tissues. The other one, referred to as the proximal process, stereotypically projects onto interneurons in specific laminae in the dorsal horn of the spinal cord or in sensory nuclei of the brainstem to process and transmit the nociceptive signal to higher brain centers. Nociceptors can be subdivided into several subtypes according to several features such as the stimuli they detect, their underlying molecular

The Neurobiology, Physiology, and Psychology of Pain. https://doi.org/10.1016/B978-0-12-820589-1.00003-8

characteristics, and innervation pattern. Two main groups can be distinguished: peptidergics and nonpeptidergics. Peptidergic nociceptors secrete neuropeptides such as substance P and calcitonin gene-related peptide (CGRP), and express the nerve growth factor receptor TrkA. Nonpeptidergic nociceptors express receptors such as Ret or P2X3. Anatomically, nonpeptidergic nociceptors preferentially innervate the skin, whereas peptidergics innervate most parts of the body (Yang et al., 2013). They can be further classified on the basis of their functional specialization. Their ability to detect various specific, generally intense, stimuli, of thermal, mechanical, or chemical nature, is conferred by the combinatorial expression of modality specific receptors. Among them the transient receptor potential (TRP) ion channels that respond to specific ranges of temperature and also to some chemicals (Table 1; Basbaum, Bautista, Scherrer, & Julius, 2009; Emery & Ernfors, 2018; Usoskin et al., 2015; Zeisel et al., 2018). While nociceptor sensitization after injury drives behavioral changes that enhance survival, the persistent and excessive excitation of nociceptors is also a key process in the development of neuropathic and chronic pain syndromes. However, currently available analgesics mostly target central nervous system (CNS) neurons and are often inappropriate or inefficient (Colloca et al., 2017).

Painlessness genes

Rare pain insensitivity disorders provide a unique opportunity to identify key players in pain perception and novel druggable targets. Congenital insensitivity to pain (CIP) is an extremely rare phenotype characterized by the inability to perceive pain from birth. Genetic analyses on patients with CIP have identified genes whose mutation is responsible for their inability to sense pain. Until now, CIP has been associated with either defects in nociceptor function or development (Fig. 1; reviewed in Nahorski, Chen, & Woods, 2015). Mutations in the voltage-gated sodium channel genes *Scn9a* (coding for Nav1.7) and *Scn11a* (Nav1.9) lead to nociceptors being unable to trigger action potentials. Nav1.7 is considered as a very promising target for the development of novel analgesics, but drug developers are struggling to unlock its therapeutic promise (Kingwell, 2019). Mutations in the *nerve growth factor* (*NGF*) gene or in the *Ntrk1* gene encoding its receptor TrkA lead to a specific loss of nociceptors by apoptosis, highlighting the importance of the NGF-TrkA signaling pathway for the survival and maturation of nociceptors during embryonic development. While NGF-TrkA signaling dependence for survival is lost after birth, NGF is still involved in nociception in adults and acts as a mediator of pain in inflammatory conditions (Denk, Bennett, & McMahon, 2017; Mantyh, Koltzenburg, Mendell, Tive, & Shelton, 2011). More recently, two new CIP-causing genes were identified in patients with uncharacterized CIP, *CLTCL1* and *Prdm12*. *CLTCL1* codes for Clathrin Heavy Chain 22 associated with vesicular transport whose specific role in developing and mature nociceptors remains to be addressed (discussed in Nahorski et al., 2015). *PRDM12* codes for a transcriptional regulator

TABLE 1 Major categories and some functional subtypes of DRG small size sensory neurons.

DRG categories of small sensory neurons[a]	Example of molecular markers[a,b]	Biological function[c]
Peptidergic (P)	TrKA, TrpVl, CGRP, Substance P, Nav1.8	Innervate skin, viscera and bones, response to noxious mechanical and heat stimuli, sensitization in inflammatory conditions
Nonpeptidergic (NP)	Runxl, P2X3, Nav1.8	Innervate skin essentially
NP1	MrgprD, GFRα2	Response to noxious mechanical stimuli
NP2	MrgpA3, TrkA	Itch
NP3	Somatostatin, neurotensin, TrpVl	Itch
Cold-sensing	TrpM8, Nav1.8	Innervate skin and deep tissue, response to cold sensation, allodynia cold pain
C-LTMR	TH, Piezo2, Vglut3, bhlha9[d]	Innervate skin hair follicle, responsible for pleasant touch, and involved in pain

[a]*Usoskin et al. (2015).*
[b]*Zeisel et al. (2018).*
[c]*Emery and Ernfors (2018).*
[d]*Bohic, M., Marics, I., Santos, C., Malapert, P., Ben-Arie, N., Salio, C., Reynders, A., Le Feuvre, Y., Saurin, A. J., & Moqrich, A. (2020). Loss of bhlha9 impairs thermotaxis and formalin-evoked pain in sexually dimorphic manner.* Cell Reports, *30(3), 602–610.*

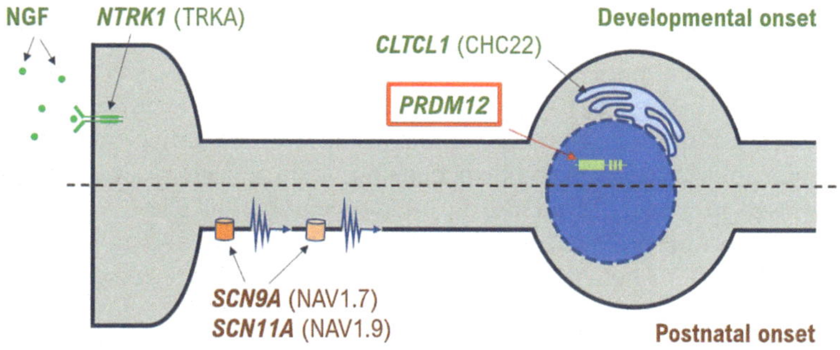

FIG. 1 Genes associated with congenital insensitivity to pain. The schematic diagram shows the location of the gene products within the nociceptors, and whether the mutation of these genes leads to defective nociceptor development or postnatal nonfunctional nociceptors. *(Adapted from Nahorski, M. S., Chen, Y., & Woods, C. G. (2015). New Mendelian disorders of painlessness.* Trends in Neurosciences, *38(11), 712–724.)*

(Chen et al., 2015). In the following paragraphs of this chapter, we will focus on PRDM12. We will discuss what is known and what remains to be discovered about its implication in pain disorders and potential as a novel pain therapeutic target.

Mutations in *PRDM12* cause CIP and midface toddler excoriation syndrome

Prdm12 is an epigenetic regulator of the *Prdm* gene family. Members of this family are characterized by the presence of a N-terminal PR (PRDI-BF1 and RIZ1 homology) domain that is related to the SET (Suppressor of variegation 3–9, Enhancer of zeste, and Trithorax) catalytic domain found in many proteins with histone methyltransferase (HMT) activity. Prdm proteins have a variable number of C_2H_2 zinc finger repeats that mediate sequence-specific DNA binding and/or the recruitment of interacting partners. They often exist as different isoforms, resulting from either pre-mRNA alternative splicing or expression of alternative promoters. Prdm factors are involved in many developmental processes such as adipogenesis, hematopoiesis, and neurogenesis by regulating cell proliferation and differentiation. Mutations in *PRDM* genes are also associated with many congenital diseases and cancers in humans (Hohenauer & Moore, 2012).

PRDM12 is a five-exon gene encoding an evolutionarily conserved PR-domain–containing protein with three zinc finger motifs and a C-terminal poly-alanine tract (only found in mammals; Thelie et al., 2015). Despite the presence of a PR domain, Prdm12 lacks intrinsic HMT activity. However, experiments in P19 cells have shown that Prdm12 recruits the histone methyltransferase G9a/EHMT2 via its second zinc finger motif to methylate histone H3 on lysine 9 (Yang & Shinkai, 2013). More than 10 different homozygous mutations have been identified in the *PRDM12* gene in patients with CIP (Chen et al., 2015; Navya, Pramod, Sujatha, & Ashok, 2019; Saini et al., 2017). From their birth, these CIP patients have been unable to feel any pain. They all present severe injuries mainly due to automutilation (self-biting of fingers, lips, …) and unnoticed trauma (ulcers, burns, Charcot joints, …). Meanwhile, other sensory modalities such as touch or proprioception appear unaffected. Analyses of sural nerve and skin biopsies from these patients have revealed a decrease or absence of innervation associated with small diameter Aδ- and C-fibers characteristic of nociceptive neurons, suggesting either axonal degeneration or defective development of these afferences. The identified mutations are widely distributed throughout the gene and alter conserved residues. Mechanistically, how these different mutations impact the function of PRDM12 is not clear. While all the mutations identified appear to reduce the HMT activity of PRDM12, only one of them, the p.His289leu mutation, found in the second zinc finger motif, significantly reduces the binding to G9a. These other mutations are thus likely to interfere with Prdm12 HMT capacity via other mechanisms, by impacting the structure of the protein, its stability, and/or its subcellular localization. Intriguingly, some of these mutants unable to methylate H3K9 still induce neurogenesis when overexpressed in neuralized *Xenopus* animal cap explants, suggesting that Prdm12 may modulate the expression of its targets via alternative mechanisms (Chen et al., 2015; Nagy et al., 2015).

Recently, another disorder, milder and more localized than CIP, designated midface toddler excoriation syndrome (MiTES), has been described that appears to be caused by some mutations in the *PRDM12* gene. Patients with MiTES bear scars and depigmentation marks at the level of the forehead, nose, and around the orbitals, due to autoinflicted chronic scratching in response to exacerbated itch specifically at those face regions. They present the first symptoms in their very young age, and the itch sensations decrease when they grow old. All MiTES patients present biallelic polyalanine expansions in PRDM12 to 17–18 residues (the average number of alanine residues in the human population being 12–13, whereas 19 causes CIP; Moss et al., 2018; Noguera-Morel, Ortiz-Cabrera, Campos, Hernández-Martín, & Torrelo, 2020; Srinivas, Gowda, Owen, Moss, & Hiremagalore, 2017). The impact of this poly-alanine expansion on PRDM12 protein stability and activity still remains to be investigated.

Prdm12 is selectively expressed in developing somatosensory ganglia in the nociceptive lineage

Prdm12 shows a highly restricted evolutionarily conserved expression pattern in the CNS and PNS of zebrafish, *Xenopus*, chicken, and mouse embryos. In the developing CNS, *Prdm12* is expressed in different brain regions, including the lateral ventricular zone and septum region of the dorsal telencephalon, postmitotic cells adjacent to the Zli (zona limitans intrathalamica), differentiating hypothalamic neurons in the diencephalon, and progenitors of the dorsal mesencephalon. In the spinal cord, *Prdm12* is expressed in a specific population of progenitors that gives rise to V1 inhibitory interneurons involved in the speed of locomotion movements, and is required for their specification (Hael, Rojo, Orquera, Low, & Rubinstein, 2020; Kinameri et al., 2008; Thelie et al., 2015; Yildiz, Downes, & Sagerström, 2019; Zannino, Downes, & Sagerström, 2014).

In the developing PNS, *Prdm12* expression is restricted to the cranial TG, SJG, vestibuloacoustic ganglia (VAG), and body DRG where it is detected from the first steps of their development (Kinameri et al., 2008; Thelie et al., 2015; Zannino et al., 2014). In *Xenopus*, *Prdm12* is expressed in profundal/trigeminal and lateral line placodes and is absent from epibranchial placodes which give rise to viscerosensory neurons. It is also absent in primary Rohon-Beard cells that constitute a transient population of mechanosensory neurons found in the spinal cord of fish and amphibian embryos. In mice, immunostainings and single-cell RNA sequencing have revealed that in TG and DRG, *Prdm12* expression is restricted to the nociceptors, and that it is the most early specific marker of this lineage, being detectable from progenitors to postmitotic neurons. DRG neurons arise in two successive and overlapping waves of neurogenesis, involving respectively the proneural factors Neurog2 and Neurog1, either directly from neural crest cells or indirectly via the neural crest–derived boundary cap cells. Prdm12 has been detected in cells derived from those three different contributing lineages, thus indicating that it is expressed during the differentiation of all DRG nociceptive neurons (Fig. 2; Bartesaghi et al., 2019; Chen et al., 2015; Desiderio et al., 2019; Soldatov et al., 2019; Ventéo et al., 2019).

Prdm12 is essential for the emergence of the entire nociceptive lineage

First evidence for a role of Prdm12 in somatosensory neurogenesis has been obtained in *Xenopus*. In this system, *Prdm12* knockdown using antisense morpholinos decreased the expression of trigeminal placode neuronal markers. Conversely, overexpression of *Prdm12* in pluripotent *Xenopus* animal cap explants resulted in the upregulation of somatosensory neuronal markers (Chen et al., 2015; Matsukawa, Miwata, Asashima, & Michiue, 2015; Nagy et al., 2015).

In mice, two recent studies reported the generation and characterization of *Prdm12* knock-out (KO) mice (Bartesaghi et al., 2019; Desiderio et al., 2019). In *Prdm12* KO embryos, no SJG was detected, and TG and DRG were hypoplastic. Both studies found that the loss or reduced size of these ganglia was the consequence of the specific absence of nociceptive neurons, while the generation of mechano-/proprioceptive neurons was unaffected. Interestingly, *Prdm12* conditional KO mice in which *Prdm12* is depleted as soon as sensory neurons become postmitotic (using *Advillin*-Cre driver mice) also have reduced somatosensory ganglia due to loss of nociceptive neurons (Desiderio et al., 2019). These results demonstrate that Prdm12 is required from the neural progenitor to the postmitotic neuron stage for the generation of the entire nociceptive lineage. They suggest that the insensitivity to pain observed in CIP patients carrying PRDM12 mutations is likely the consequence of a failure to generate nociceptors.

How does Prdm12 function in the specification and maturation of nociceptive neurons?

The mechanisms by which Prdm12 functions in somatosensory precursors to promote nociceptor development remain largely to be resolved. These mechanisms are likely to be different in TG, SJG, and DRG given the distinct timing and severity of the hypoplasia observed in these different ganglia. The complete absence of SJG observed in *Prdm12* KO embryos suggests that Prdm12 might be important for very early steps of the formation of these ganglia, such as progenitor migration/aggregation or survival. In TG and DRG, how Prdm12 operates remains controversial. While Prdm12 has been shown to be essential for nociceptor survival through its critical role in the initiation and maintenance of the expression of the NGF receptor TrkA (Desiderio et al., 2019), Prdm12 could also be necessary for the proliferation of neuronal progenitors (Bartesaghi et al., 2019).

Whether Prdm12 acts in nociceptor specification as a transcriptional activator or repressor is also unclear. In DRG of E11.5 embryos, loss of *Prdm12* upregulates the boundary cap cell marker *Krox20*/*Egr2* and the glial marker *Mpz* (*myelin protein zero*; Desiderio et al., 2019). In chicken embryos overexpressing *Prdm12*, Prdm12$^+$ cells almost systematically had a neuronal identity, but never became mechano- or proprioceptors (expressing TrkB or TrkC), in contrast to control *GFP*-electroporated cells that exhibited any somatosensory neuron subtypes and could become glial cells (Bartesaghi et al.,

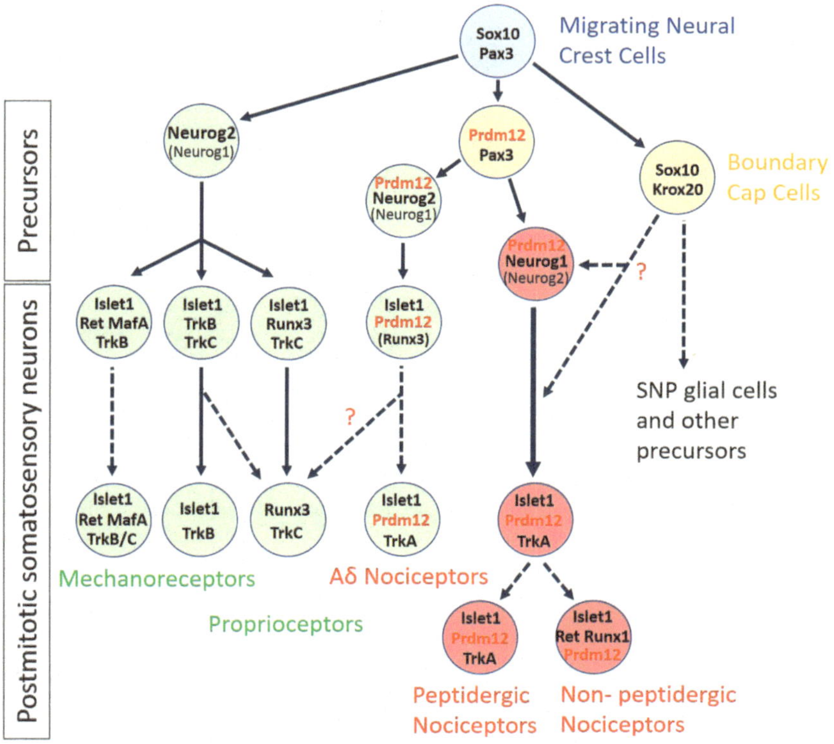

FIG. 2 Model of Prdm12 action in the genetic cascade controlling DRG sensory neurogenesis. Sensory neurons and glial cells found in DRG are generated during two successive waves from three types of precursors derived from neural crest progenitors (Sox10$^+$, Pax3$^+$). During the first wave, some neural crest progenitors start to express at high-level *Neurog2*. Neurog2$^+$/Prdm12$^-$ neural precursors generate mechanoreceptors and proprioceptors, characterized by the expression of the receptors TrkB and/or TrkC and/or Ret. Some Prdm12$^+$/TrkA$^+$ nociceptive neurons corresponding to the subpopulation of Aδ nociceptors are also generated during this first wave from Neurog2$^+$/Prdm12$^+$ neural precursors. Most of the nociceptors that constitute 80% of DRG neurons are generated during the second wave from neural crest progenitors and derived boundary cap cells that start to express *Neurog1* at high level. These Neurog1$^+$/Prdm12$^+$ neural precursors give rise to differentiating nociceptors (Prdm12$^+$, Islet1$^+$, TrkA$^+$) that later segregate into peptidergic and nonpeptidergic nociceptors. *(Adapted from Lallemend, F., & Ernfors, P. (2012). Molecular interactions underlying the specification of sensory neurons.* Trends in Neurosciences, 35*(6), 373–381.)*

2019). The percentage of TrkA$^+$-traced neurons remained the same in Prdm12$^+$ and GFP$^+$ controls, indicating that Prdm12 is not sufficient to induce TrkA$^+$ neurons. However, in *Xenopus* animal cap explants fated to the sensory lineage by overexpressing the proneural factors Neurog1 or Neurog2 as well as in human-induced pluripotent stem cells differentiated into sensory neurons, *Prdm12* overexpression was found to promote TrkA expression at the expense of TrkC only or TrkB and TrkC, respectively (Desiderio et al., 2019).

Like other members of the family, Prdm12 is thought to bind indirectly to DNA, via other DNA-binding proteins. Neurog1/2 could be Prdm12 partners in somatosensory neuron precursors (Desiderio et al., 2019). Whether Prdm12 needs G9a or any other cofactors to down- or upregulate its targets (Fig. 3) remains to be investigated. The acquisition of Prdm12 ChIP-seq data will be crucial to understand how Prdm12 operates in the specification and maturation of nociceptive neurons.

Does Prdm12 play a role in mature nociceptors?

Prdm12 remains expressed in TG, SJG, and DRG of adult mice. It is detected in the majority of mature nociceptors of both categories, peptidergic (TrkA$^+$, CGRP$^+$) and nonpeptidergic (Runx1$^+$, Ret$^+$), as well as in some C-fiber low-threshold mechanoreceptors (C-LTMRs, TH$^+$) that beyond playing a role in pleasant touch modulate pain (Desiderio et al., 2019; Goswami et al., 2014; Kambrun et al., 2018; Kupari, Häring, Agirre, Castelo-Branco, & Ernfors, 2019; Mazzone et al.,

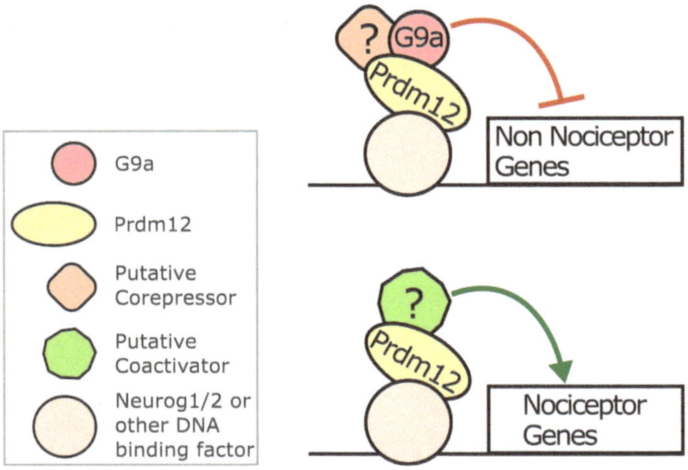

FIG. 3 Model of Prdm12 mechanism of action in DRG somatosensory precursors. Prdm12 is thought to bind indirectly to DNA, perhaps via Neurog1/2. It promotes a nociceptive fate by repressing nonnociceptor target genes, via the recruitment of G9a or other corepressors and/or by forming an activator complex to induce some nociceptor target genes.

2019; Usoskin et al., 2015; Wang et al., 2017; Zeisel et al., 2018). This suggests that Prdm12 could modulate the function of some mature nociceptors.

Different sensory neuron subtypes respond to different stimuli and can be sensitized differently depending on the type of injury (Emery & Ernfors, 2018; Gold & Gebhart, 2010). Prdm12 might thus have specific functions depending on the considered nociceptor subtype. For example, Prdm12 may still be involved in the regulation of TrkA expression in peptidergic nociceptors, or regulate specific TRP receptors or other ion channels in some subtypes of nonpeptidergic nociceptors. The specific removal of Prdm12 from adult mice needs to be done to find out what is its regulatory role in mature nociceptors. Whether Prdm12 expression is modified during inflammation or after nerve injury needs also to be investigated to approach its possible contribution to chronic and neuropathic pain.

Applications to other areas

Pharmacotherapies targeting the NGF pathway are considered as promising methods for the treatment of pain conditions. Several anti-NGF methodological approaches have been investigated. Among them, NGF-sequestering antibodies appear the most promising. Clinical studies have been performed showing that they are effective for the relief of some pain conditions such as osteoarthritis. However, joint damages have been reported in patients treated with anti-NGF antibodies, leading to joint replacement surgery. Thus, the safety profile of the current NGF antagonists still needs to be established (Dakin et al., 2019; Schnitzer et al., 2019).

If Prdm12 plays a role in the excitability of mature nociceptors and is misregulated in chronic or neuropathic pain conditions, it may constitute an interesting target for the development of novel pain therapeutics, particularly given its close link with NGF and the importance of epigenetic mechanisms in the induction and maintenance of chronic pain (Crow, Denk, & McMahon, 2013; Denk & McMahon, 2012; Laumet et al., 2015). Prdm12 is also a promising novel target based on its restricted expression in the nervous system of adult mice, almost exclusive to somatosensory neurons, suggesting that any molecules targeting its action would likely have little side effects. Prdm12 is however expressed in some nuclei of the adult brain, such as the A13 nucleus of the *zona incerta* hypothalamic area and the arcuate nucleus of the ventromedial hypothalamus (Romanov et al., 2017; Saunders et al., 2018). In this latter nucleus, Prdm12 is required for the specification of POMC$^+$ anorexigenic neurons during development and may be important for the maintenance of POMC expression in adult mice. Loss of *Prdm12* in these neurons causes POMC downregulation and leads to increased food intake, adiposity, and obesity (Chen et al., 2020; Hael et al., 2020). Thus, great care will have to be taken to avoid undesired secondary effects linked to its expression in these brain regions.

Other agents of interest

To develop Prdm12 selective drugs that inhibit its action and thus may reduce pain symptoms, it will be necessary to identify its cofactors in mature nociceptors. G9a is a candidate Prdm12 cofactor as it is broadly expressed in DRG neurons of adult mice. Targeting Prdm12 and G9a interaction may be a promising route for the development of novel analgesics as G9a is upregulated upon nerve injury and function in the repression of K^+ channels associated with neuropathic pain (Laumet et al., 2015). Targeting Prdm12 interactions with the transcription factors that likely mediate its binding to specific genomic sites in mature nociceptors could be also used to develop drugs reducing pain. Whether Prdm12 forms complexes with basic helix-loop-helix (bHLH) transcription factors in mature nociceptors in adults as it seems to do during spinal cord and somatosensory neurogenesis remains to be investigated (Desiderio et al., 2019; Thelie et al., 2015; Yildiz et al., 2019).

In conclusion, Prdm12 appears to be a promising candidate for the development of new drugs relieving pain. Therefore, it is urgent to tackle its implication in the physiology of adult nociceptors.

Mini-dictionary of terms

Boundary cap cells: A transient neural crest–derived stem cell population localized at entry and exit points of peripheral nerve roots that contributes to DRG neurons and glial cells of the secondary wave of migration of neural crest to the PNS.

Congenital insensitivity to pain (CIP, also known as **congenital analgesia)**: A rare condition in which a person is unable to feel pain from birth, due to lack or malfunction of nociceptors.

Epigenetic regulation: Control of gene expression through the modification of the local state of the chromatin.

Midface toddler excoriation syndrome (MiTES): A newly recognized autosomal recessive condition arising in the first year of life and characterized by deep, self-inflicted excoriations largely confined to the medial cheeks, nasal bridge, and central forehead.

Neural crest: A transient embryonic cell population in vertebrates that arises from the dorsal most region of the closing neural tube, migrates into the embryo, and generates a large number of differentiated cell types, including neurons and glia of the peripheral nervous system.

Nociceptor: A primary sensory neuron specialized to sense potentially tissue-damaging stimuli and relaying the information to second-order neurons in the central nervous system.

Proneural factor: bHLH transcription factors that initiate the differentiation of neural progenitors into neurons in vertebrates.

Key facts showing the importance of Prdm12 in nociceptors

- In mouse somatosensory ganglia, Prdm12 expression is specific to the nociceptive lineage.
- In human, mutations in *PRDM12* are associated with CIP and MiTES.
- In mouse, inactivation of *Prdm12* leads to the loss of the entire nociceptive lineage.
- In *Xenopus*, chick- and human-induced pluripotent stem cells, forced expression of Prdm12 represses alternate nonnociceptive somatosensory fates.

Summary points

- Prdm12 is an evolutionarily conserved transcriptional regulator selectively expressed in the nociceptive neuron lineage, from precursors to mature nociceptors.
- Prdm12 is required for the survival of nociceptive precursors and the maturation of developing nociceptors.
- Prdm12 acts in the emergence of nociceptors in conjunction with proneural factors via the initiation and maintenance of the expression of the NGF receptor TrkA.
- Prdm12 remains expressed in nociceptors of adult animals, suggesting a physiological function in these cells and that it could constitute an interesting target for the development of novel pain therapeutics.

References

Bartesaghi, L., Wang, Y., Fontanet, P., Wanderoy, S., Berger, F., Wu, H., et al. (2019). PRDM12 is required for initiation of the nociceptive neuron lineage during neurogenesis. *Cell Reports*, *26*(13), 3484–3492.

Basbaum, A. I., Bautista, D. M., Scherrer, G., & Julius, D. (2009). Cellular and molecular mechanisms of pain. *Cell*, *139*(2), 267–284.

Chen, Y. C., Auer-Grumbach, M., Matsukawa, S., Zitzelsberger, M., Themistocleous, A. C., Strom, T. M., et al. (2015). Transcriptional regulator PRDM12 is essential for human pain perception. *Nature Genetics*, *47*(7), 803–808.

Chen, X., Wyler, S. C., Li, L., Arnold, A. G., Wan, R., Xu, P., et al. (2020). Transcriptomic profiling of developing melanocortin neurons reveals a role for Prdm12 in energy balance. *Journal of Neuroscience*. https://doi.org/10.1523/JNEUROSCI.0155-20.2020. JN-RM-0155-20. [Epub ahead of print].

Colloca, L., Ludman, T., Bouhassira, D., Baron, R., Dickenson, A. H., Yarnitsky, D., et al. (2017). Neuropathic pain. *Nature Reviews Disease Primers*, *3*, 17002.

Crow, M., Denk, F., & McMahon, S. B. (2013). Genes and epigenetic processes as prospective pain targets. *Genome Medicine*, *5*(2), 12.

Dakin, P., DiMartino, S. J., Gao, H., Maloney, J., Kivitz, A. J., Schnitzer, T. J., et al. (2019). The efficacy, tolerability, and joint safety of fasinumab in osteoarthritis pain: A phase IIb/III double-blind, placebo-controlled, randomized clinical trial. *Arthritis and Rheumatology*, *71*(11), 1824–1834.

Denk, F., Bennett, D. L., & McMahon, S. B. (2017). Nerve growth factor and pain mechanisms. *Annual Review of Neuroscience*, *40*(1), 307–325.

Denk, F., & McMahon, S. B. (2012). Chronic pain: Emerging evidence for the involvement of epigenetics. *Neuron*, *73*(3), 435–444.

Desiderio, S., Vermeiren, S., Van Campenhout, C., Kricha, S., Malki, E., Richts, S., et al. (2019). Prdm12 directs nociceptive sensory neuron development by regulating the expression of the NGF receptor TrkA. *Cell Reports*, *26*(13), 3522–3536.

Emery, E. C., & Ernfors, P. (2018). Dorsal root ganglion neuron types and their functional specialization. In Wood J.N (Ed.), *The Oxford Handbook of the Neurobiology of Pain* Oxford University Press.

Gold, M. S., & Gebhart, G. F. (2010). Nociceptor sensitization in pain pathogenesis. *Nature Medicine*, *16*(11), 1248–1257.

Goswami, S. C., Mishra, S. K., Maric, D., Kaszas, K., Gonnella, G. L., Clokie, S. J., et al. (2014). Molecular signatures of mouse TRPV1-lineage neurons revealed by RNA-seq transcriptome analysis. *Journal of Pain*, *15*(12), 1338–1359.

Hael, C. E., Rojo, D., Orquera, D. P., Low, M. J., & Rubinstein, M. (2020). The transcriptional regulator PRDM12 is critical for Pomc expression in the mouse hypothalamus and controlling food intake, adiposity, and body weight. *Molecular Metabolism*, *34*, 43–53.

Hohenauer, T., & Moore, A. W. (2012). The Prdm family: Expanding roles in stem cells and development. *Development*, *139*, 2267–2282.

Kambrun, C., Roca-Lapirot, O., Salio, C., Landry, M., Moqrich, A., & Le Feuvre, Y. (2018). TAFA4 reverses mechanical allodynia through activation of GABAergic transmission and microglial process retraction. *Cell Reports*, *22*(11), 2886–2897.

Kinameri, E., Inoue, T., Aruga, J., Imayoshi, I., Kageyama, R., Shimogori, T., et al. (2008). Prdm proto-oncogene transcription factor family expression and interaction with the Notch-Hes pathway in mouse neurogenesis. *PLoS One*, *3*(12), 1–10.

Kingwell, K. (2019). Nav1.7 withholds its pain potential. *Nature Reviews Drug Discovery*. https://doi.org/10.1038/d41573-019-00065-0.

Kupari, J., Häring, M., Agirre, E., Castelo-Branco, G., & Ernfors, P. (2019). An atlas of vagal sensory neurons and their molecular specialization. *Cell Reports*, *27*(8), 2508–2523.

Laumet, G., Garriga, J., Chen, S., Zhang, Y., Li, D., Smith, T. M., et al. (2015). G9a is essential for epigenetic silencing of K(+) channel genes in acute-to-chronic pain transition. *Nature Neuroscience*, *18*(12), 1746–1755.

Mantyh, P. W., Koltzenburg, M., Mendell, L. M., Tive, L., & Shelton, D. L. (2011). Antagonism of nerve growth factor-TrkA signaling and the relief of pain. *Anesthesiology*, *115*(11), 189–204.

Matsukawa, S., Miwata, K., Asashima, M., & Michiue, T. (2015). The requirement of histone modification by PRDM12 and Kdm4a for the development of pre-placodal ectoderm and neural crest in Xenopus. *Developmental Biology*, *399*, 164–176.

Mazzone, S. B., Tian, L., Moe, A. A. K., Trewella, M. W., Ritchie, M. E., & McGovern, A. E. (2019). Transcriptional profiling of individual airway projecting vagal sensory neurons. *Molecular Neurobiology*, *57*(2), 949–963.

Moss, C., Srinivas, S. M., Sarveswaran, N., Nahorski, M., Gowda, V. K., Browne, F. M., et al. (2018). Midface toddler excoriation syndrome (MiTES) can be caused by autosomal recessive biallelic mutations in a gene for congenital insensitivity to pain, PRDM12. *British Journal of Dermatology*, *179*(5), 1135–1140.

Nagy, V., Cole, T., Van Campenhout, C., Khoung, T. M., Leung, C., Vermeiren, S., et al. (2015). The evolutionarily conserved transcription factor PRDM12 controls sensory neuron development and pain perception. *Cell Cycle*, *14*(12), 1799–1808.

Nahorski, M. S., Chen, Y., & Woods, C. G. (2015). New Mendelian disorders of painlessness. *Trends in Neurosciences*, *38*(11), 712–724.

Navya, M., Pramod, G., Sujatha, G., & Ashok, L. (2019). Congenital insensitivity to pain in a 1-year-old boy. *Journal of Indian Society of Pedodontics and Preventive Dentistry*, *37*(3), 308–310.

Noguera-Morel, L., Ortiz-Cabrera, N. V., Campos, M., Hernández-Martín, Á., & Torrelo, A. (2020). A case of mid-face toddler excoriation syndrome (MiTES). *Pediatric Dermatology*, *37*(2), 345–346.

Price, T. J., & Gold, M. S. (2018). From mechanism to cure: Renewing the goal to eliminate the disease of pain. *Pain Medicine*, *19*(8), 1525–1549.

Romanov, R. A., Zeisel, A., Bakker, J., Girach, F., Hellysaz, A., Tomer, R., et al. (2017). Molecular interrogation of hypothalamic organization reveals distinct dopamine neuronal subtypes. *Nature Neuroscience*, *20*(2), 176–188.

Saini, A. G., Padmanabh, H., Sahu, J. K., Kurth, I., Voigt, M., & Singhi, P. (2017). Hereditary sensory polyneuropathy, pain insensitivity and global developmental delay due to novel mutation in PRDM12 gene. *Indian Journal of Pediatrics*, *84*(4), 332–333.

Saunders, A., Macosko, E. Z., Wysoker, A., Goldman, M., Krienen, F. M., de Rivera, H., et al. (2018). Molecular diversity and specializations among the cells of the adult mouse brain. *Cell*, *174*(4), 1015–1030.

Schnitzer, T. J., Easton, R., Pang, S., Levinson, D. J., Pixton, G., Viktrup, L., et al. (2019). Effect of tanezumab on joint pain, physical function, and patient global assessment of osteoarthritis among patients with osteoarthritis of the hip or knee: A randomized clinical trial. *Journal of the American Medical Association*, *322*(1), 37–48.

Soldatov, R., Kaucka, M., Kastriti, M. E., Petersen, J., Chontorotzea, T., Englmaier, L., et al. (2019). Spatiotemporal structure of cell fate decisions in murine neural crest. *Science*, *364*. eaas9536.

Srinivas, S. M., Gowda, V. K., Owen, C. M., Moss, C., & Hiremagalore, R. (2017). Mid-face toddler excoriation syndrome (MiTES): A new paediatric diagnosis. *Clinical and Experimental Dermatology*, *42*(1), 68–71.

Thelie, A., Desiderio, S., Hanotel, J., Quigley, I., Van Driessche, B., Rodari, A., et al. (2015). Prdm12 specifies V1 interneurons through cross-repressive interactions with Dbx1 and Nkx6 genes in Xenopus. *Development*, *142*(19), 3416–3428.

Usoskin, D., Furlan, A., Islam, S., Abdo, H., Lönnerberg, P., Lou, D., et al. (2015). Unbiased classification of sensory neuron types by large-scale single-cell RNA sequencing. *Nature Neuroscience*, *18*(1), 145–153.

Van Hecke, O., Torrance, N., & Smith, B. H. (2013). Chronic pain epidemiology and its clinical relevance. *British Journal of Anaesthesia*, *111*(1), 13–18.

Ventéo, S., Desiderio, S., Cabochette, P., Deslys, A., Carroll, P., & Pattyn, A. (2019). Neurog2 deficiency uncovers a critical period of cell fate plasticity and vulnerability among neural-crest-derived somatosensory progenitors. *Cell Reports*, *29*(10), 2953–2960.

Wang, J., Kollarik, M., Ru, F., Sun, H., McNeil, B., Dong, X., et al. (2017). Distinct and common expression of receptors for inflammatory mediators in vagal nodose versus jugular capsaicin-sensitive/TRPV1-positive neurons detected by low input RNA sequencing. *PLoS One*, *12*(10), e0185985.

Yang, C.-M., & Shinkai, Y. (2013). Prdm12 is induced by retinoic acid and exhibits anti-proliferative properties through the cell cycle modulation of P19 embryonic carcinoma cells. *Cell Structure and Function*, *38*, 197–206.

Yang, F. C., Tan, T., Huang, T., Christianson, J., Samad, O. A., Liu, Y., et al. (2013). Genetic control of the segregation of pain-related sensory neurons innervating the cutaneous versus deep tissues. *Cell Reports*, *5*(5), 1353–1364.

Yildiz, O., Downes, G. B., & Sagerström, C. G. (2019). Zebrafish prdm12b acts independently of nkx6.1 repression to promote eng1b expression in the neural tube p1 domain. *Neural Development*, *14*(1), 5.

Zannino, D. A., Downes, G. B., & Sagerström, C. G. (2014). Prdm12b specifies the p1 progenitor domain and reveals a role for V1 interneurons in swim movements. *Developmental Biology*, *390*(2), 247–260.

Zeisel, A., Hochgerner, H., Lönnerberg, P., Johnsson, A., Memic, F., van der Zwan, J., et al. (2018). Molecular architecture of the mouse nervous system. *Cell*, *174*(4), 999–1014.

Chapter 4

Genetics of chronic widespread musculoskeletal pain

María Jesús Álvarez-Cubero[a,b], Sergio Cuenca-López[b], Verónica Arenas-Rodríguez[b], Fernando Estévez-López[c], and Luis Javier Martínez-González[b]

[a]*Department of Biochemistry and Molecular Biology III, Faculty of Medicine, PTS, University of Granada, Granada, Spain;* [b]*GENYO (Pfizer-University of Granada-Andalusian Government Centre for Genomics and Oncological Research), Granada, Spain;* [c]*Department of Pediatrics, Wilhelmina Children's Hospital, University Medical Center Utrecht, Utrecht University, Utrecht, The Netherlands*

Abbreviations

ADRA1A	adrenoceptor alpha 1A
ADRB2	adrenoceptor beta 2
ADRB3	adrenoceptor beta 2
AR	androgen receptor
ATP	adenosine triphosphate
BECN1	autophagy-related gene, beclin-1
BRCA	BReast CAncer gene
CHMP1A	charged multivesicular body protein 1A
CI	confidence interval
COMT	catechol-*O*-methyltransferase
DNA	deoxyribonucleic acid
DRD3	dopamine receptor D3
GABRB3	gamma-aminobutyric acid type A receptor subunit beta3
GBP1	guanylate binding protein 1
GWAS	genome-wide association study
IASP	International Association for the Study of Pain
MTHFR	methylene tetrahydrofolate reductase
MYT1L	myelin transcription factor 1 like
NGS	next-generation sequencing
NRXN3	neurexin 3
RNA	ribonucleic acid
SCN9A	sodium voltage-gated channel alpha subunit 9
SE	standard error
SNAP25	synaptosomal-associated protein, 25 kDa
SNPs	single nucleotide polymorphisms
T2DM	type 2 diabetes mellitus
TAAR1	trace amine-associated receptor 1
TACR1	tachykinin receptor 1
TRPV2	transient receptor potential cation channel subfamily V member

Introduction

Chronic pain is a leading cause of disability worldwide (Vos et al., 2012). According to the IASP, "pain is chronic when it persists beyond normal tissue healing time, which is assumed to be 3 months" (Vervest & Schimmel, 1988). Chronic pain has been consistently associated with women, older age, lower educational levels, and poorer socioeconomic status (Fayaz, Croft, Langford, Donaldson, & Jones, 2016; Jackson et al., 2015; Kamerman et al., 2020). Although its prevalence increases

The Neurobiology, Physiology, and Psychology of Pain. https://doi.org/10.1016/B978-0-12-820589-1.00004-X

with age, chronic pain is also common in children and adolescents (van den Heuvel, Jansen, Bindels, Bierma-Zeinstra, & van Middelkoop, 2020).

The most common type of chronic pain is musculoskeletal (Breivik, Collett, Ventafridda, Cohen, & Gallacher, 2006), which is a type of pain that comes from muscles, bones, joints, or tendons (Perrot et al., 2019). Chronic musculoskeletal pain is often characterized by a considerable impact on people's functioning, involving diverse spheres such as psychological (e.g., increased levels of anxiety and depression), social (e.g., reduced network of friends), and physical (interference in daily life activities and reduced participation in social roles) functioning (Briggs et al., 2016; Nyvang, Hedström, & Gleissman, 2016). In most of the cases, a root underlying cause of chronic musculoskeletal pain is not identified by clinical tests (e.g., X-ray) (Perrot et al., 2019). Unfortunately, this unknown origin of chronic musculoskeletal pain has resulted in the stigmatization of patients (Kool, van Middendorp, Boeije, & Geenen, 2009; Santiago, Marques, Kool, Geenen, & da Silva, 2017).

According to its location, chronic musculoskeletal pain is either local (e.g., low back pain) or diffuse (i.e., widespread musculoskeletal pain) (Perrot et al., 2019). Chronic widespread musculoskeletal pain is often defined as pain that simultaneously exists in the axial skeleton, above and below the waist, and two contralateral limbs (Docking et al., 2015; Fayaz et al., 2016). This definition of chronic widespread musculoskeletal pain was originally developed to diagnose fibromyalgia (Wolfe et al., 1990, 2010).

Because chronic widespread musculoskeletal pain has been inconsistently defined in the past literature (Butler, Landmark, Glette, Borchgrevink, & Woodhouse, 2016), in the present chapter, we focus on fibromyalgia which is a well- and robustly defined population in which chronic widespread musculoskeletal pain is the cardinal sign. Therefore, the "Heritability of chronic widespread musculoskeletal pain" section summarizes the heritability of fibromyalgia; the "Candidate gene studies in chronic widespread musculoskeletal pain" section summarizes the evidence provided by candidate gene studies on the genetics of fibromyalgia susceptibility (case-control studies) and the genotype-phenotype association in people with fibromyalgia; the "GWAS" section summarizes genome-wide studies; the "Interactions (gene-gene and gene-environment)" section summarizes the evidence on gene-gene and gene-environment interactions; and the "Future perspectives (mainly, research agenda)" section provides information on future perspectives.

Heritability of chronic widespread musculoskeletal pain

Fibromyalgia often aggregates in families, which provides a reasonable support to the hypothesis that it may have a genetic origin (Arnold et al., 2004, 2013). Although early studies were conducted in small samples, they were inspiring and stimulated further research. For instance, Buskila et al. conducted a study in 20 families and observed familial aggregation (Buskila, Neumann, Hazanov, & Carmi, 1996). The preliminary findings from small samples were later confirmed by a large cohort; namely, the fibromyalgia family study (Arnold et al., 2004, 2013).

To rule out the potential role of the environment, a study using data from the Finnish Twin Cohort was conducted and the heritability of fibromyalgia was 51% (95% CI from 45% to 56%) (Markkula et al., 2009). However, this finding must be interpreted with caution because fibromyalgia was defined by a cluster of self-reported symptoms (e.g., tender points, stiffness, and tiredness) instead of using the American College of Rheumatology criteria. Recently, in a genome-wide association study of over 25,000 participants, the heritability of fibromyalgia (using the fibromyalgia-2011 criteria) was 8.6% (SE = 3.8) (Dutta et al., 2020). Interestingly, this study also found that the heritability of fibromyalgia is considerably higher in younger people; the prevalence in people up to 50 years old was 12.4% (SE = 6.1) while in those older than 50 years old was 6.5 (SE = 5.8%). Thus, in younger people, fibromyalgia might be more likely determined by genotype while, in older people, fibromyalgia might be more likely a consequence of other nongenetic factors such as lifestyle.

Candidate gene studies in chronic widespread musculoskeletal pain

The analysis or study of "candidate genes" as the genes with the highest aggregate frequency is very common in many fields of medicine. The most common goal of candidate genes studies is to obtain a better classification of the disease in its early stages or during its management. Genomics science allows the identification of biomarkers by the use of genetic markers. Many of them, focused on SNPs variants, known as DNA sequence that occurs commonly in the population (e.g., 1%) in which a single nucleotide base differs between members of the population (e.g., people without vs people with chronic widespread musculoskeletal pain) (Esposito, Azhari, & Borelli, 2018).

Case-control studies

Case-control studies are a type of epidemiological study in which people who have a disease or outcome of interest (cases) are compared to people who do have it (controls) (Chung et al., 2019). Case-control studies are helpful for different purposes such as, for instance, answering questions about the etiology of a disease, its prognosis, or providing data on therapeutic effectiveness. This section presents a summary of the findings from case-control studies, using the candidate gene approach, in relation to fibromyalgia susceptibility, symptoms, or potential mechanisms (Dupépé, Kicielinski, Gordon, & Walters, 2019).

ADRA1A, ADRB2, and ADRB3

The defining characteristics of fibromyalgia could be explained by a pathogenesis known as "sympathetically maintained pain" (Vargas-Alarcón et al., 2009; Zhang et al., 2018). Naturally occurring sympathetic neurotransmitters are catecholamines known as norepinephrine, epinephrine, and dopamine. Catecholamines act by binding to ARs. ARs are parts of the sympathetic nervous system that are fundamental for maintenance of homeostasis. Such receptors are broadly classified as α-AR and β-AR. The α-AR are more directly related to vasoconstriction, whereas the β-AR increase cardiac output and vasodilation. ARs are subject to many regulatory factors, including desensitization, downregulation, and internalization. This plasticity may compromise homeostasis. Investigations in animals demonstrated that low COMT activity leads to increased pain sensitivity via AR activation. Furthermore, recent studies proved that sustained stimulation of ADRB2 and ADRB3 produces functional pain and neuroinflammation that persist for weeks after removal of the causal stimulus.

CHMP1A

Despite central nervous disorders, also dysfunctional mitochondria as well as morphological changes of muscle tissue could be related to the fibromyalgia. In blood mononuclear cells and skin biopsies of fibromyalgia patients, a decrease of COQ10, an electron carrier in the mitochondrial respiratory chain, and a reduction of mitochondrial membrane potential have been observed. Both could lead to reduced activity of the essential enzymes ATP and citrate synthase that are necessary for generation of energy and disposal of oxygen radicals, in turn followed by an increase of lactate production and a fast fatigue of muscles. Furthermore, Cordero et al. observed an upregulation of Becn1 and Map1Lc3b, two essential autophagic genes, in blood mononuclear cells of affected patients (Cordero, de Miguel, et al., 2010; Cordero, Moreno-Fernández, et al., 2010). Autophagy is a crucial process involved in degradation of damaged cellular components and in balancing energy sources.

Accordingly, Oezel et al. observed mitochondrial dysfunction and autophagy in blood mononuclear cells of animal models with fibromyalgia (Oezel et al., 2016). In humans, Cordero et al. suggested that massive mitophagy may promote cell-bioenergetics imbalance affecting cells functioning of people with fibromyalgia (Cordero, de Miguel, et al., 2010). Recently, Estévez-López et al. identified that a genetic marker of multivesicular body sorting of proteins to the interiors of lysosomes, the rs6860 polymorphism of the CHMP1A gene, is associated with fibromyalgia susceptibility in southern Spanish women (Estévez-López et al., 2018).

COMT

The sympathetic nervous system plays an intricate and incompletely understood role in the pathogenesis of chronic and acute pain. One aspect of this role is related to the observed decrease in serum and cerebrospinal fluid levels of catecholamine metabolites, observed in fibromyalgia patients. This finding has been interpreted as indicating a reduction in the sympathetic tone, which may be related to reduced pain inhibition. With this background, numerous studies have focused on genetic markers related to the metabolism of catecholamines, with special emphasis on the gene coding for COMT (Ablin & Buskila, 2015; Vargas-Alarcón et al., 2007). This enzyme participates in degrading catecholamines and several other neurotransmitters and, therefore, in modulating pain perception by the central nervous system. Findings of the previous literature regarding COMT gene and susceptibility to fibromyalgia are inconclusive. Therefore, research has aimed at identifying new candidate genes.

Recent studies developed in 2020 by Hatami et al. did not prove the role of 472G > A (Val158Met) polymorphism as a risk factor for developing fibromyalgia (Hatami et al., 2020).

DRD3

Lowered pain thresholds and poor pain inhibition are two main features of fibromyalgia. Potvin et al. investigated the relationship of certain polymorphisms with the pain inhibition system, known as diffuse noxious inhibitory controls and the thermal pain thresholds (Potvin et al., 2009). This study showed that the DRD3 Ser9Gly polymorphism influences the efficacy of diffuse noxious inhibitory controls and thermal pain thresholds and that the latter relationship is present only in fibromyalgia patients.

GABRB3, GBP1, and TAAR1

To make a comprehensive and in-depth evaluation of the majority of genes known to influence persistent pain conditions, Smith et al. carried out a case-control study of candidate genes, which evaluated more than 350 genes (Smith et al., 2012). Three previously unsuspected genes, GABRB3, TAAR1, and GBP1, harbored SNPs that differed in frequency between patients with FM and healthy control subjects; TAAR1 displayed evidence for association in a second cohort. The available meagre evidence suggests that receptors for trace amines might influence the development of chronic pain by modulating the activity of dopaminergic receptors. TAAR-induced alterations in dopamine bioavailability and function may increase pain sensitivity, a hallmark of fibromyalgia.

GCH1

GCH1 is the rate-limiting enzyme in the synthesis of tetrahydrobiopterin, which is an essential cofactor in the production of nitric oxide. Kim et al. carried out a case-control study to determine whether single nucleotide polymorphisms in the GCH1 gene affect susceptibility and/or pain sensitivity in fibromyalgia syndrome (Kim et al., 2013). Their study provided evidence that certain GCH1 haplotypes may be protective against susceptibility and pain sensitivity in fibromyalgia and suggested that it is not responsible for pain sensitivity in fibromyalgia pathogenesis.

Studies developed by Smith et al. include the existence of epistatic interactions between the COMT (rs4680-val(158) met) and polymorphisms in GCH1 (rs10483639) and ESR1 (rs3020377) (Smith et al., 2014). These data reveal that the GCH1 minor G allele confers a protective advantage among met carriers in COMT gene. Furthermore, these data suggest that the ability to predict the downstream effects of genetic variation on COMT activity is critically important to understand the molecular basis of chronic pain conditions.

HTR2A

Serotonin (5-hydroxytryptamine, 5-HT) is a key neurotransmitter in the central nervous system. The effect of serotonin is mediated by different 5-HT receptor subtypes. The 5-HT2A receptor is located in the cortex, caudate nucleus, and all the intestines, and it may play a role in the etiology of several neuropsychiatric diseases and pain perception. A silent polymorphism in the 5-HT2A receptor gene is defined by a T to C transition at position 102. Although the 5-HT2A 102T/C polymorphism does not result in alteration of the amino acid sequence of the protein, the 5-HT2A receptor 102T/C polymorphism has been determined to confer susceptibility to fibromyalgia (Lee, Choi, Ji, & Song, 2012).

Fujioka et al. have proven that there was an association between the polymorphism of the HTR2A gene (rs6311) and fibromyalgia (Fujioka, Stahlberg, Ochi, & Olmarker, 2016). Women with GG genotype or G allele presented greater risk of developing fibromyalgia and that these patients have lower limb muscle strength compared to the control group (de Lima et al., 2019). Moreover, gene expression profiling of NOS1 and HTR2A seems to be a useful tool for detecting new potential pharmaceutical targets for spinal pain syndromes.

MTHFR

MTHFR gene mutations affect homocysteine metabolism and provoke hyperhomocysteinemia. This increases the homocysteine levels, which in turn causes activity in cytokine, lipid peroxidation accompanied by vascular endothelial damage, prothrombotic process, aterothrombogenesis, thromboembolism, and systemic vascular occlusive diseases. The increases in the homocysteine levels have been reported as a risk factor in vascular stiffness, hypercoagulability, and for thrombotic complications. Also, MTHFR polymorphisms influence folic acid-B12-B6 metabolism and are implicated in vascular disease associated with aging. Since gene polymorphisms are of great importance in understanding the basis of multifactorial diseases, Inanir et al. assessed the effects of MTHFR (C677T) gene polymorphisms in fibromyalgia patients (Inanir et al., 2015). Thus, they determined an association between MTHFR C677T polymorphism and feeling of stiffness and dry eye which are among the clinical characteristics of fibromyalgia.

MYT1L

In order to elucidate the genetic susceptibility factors of fibromyalgia, Docampo et al. carried out a GWAS analysis and copy number variants of fibromyalgia cases with low comorbidity (Docampo et al., 2014). In this study, no SNP reached the

GWAS association threshold, but 21 of the most associated SNPs were chosen for replication. The second study provided a polymorphism of MYT1L gene associated with fibromyalgia with low comorbidities, Rs11127292.

Several potential candidate genes have been suggested by D'Agnelli et al. to be associated with fibromyalgia such as SLC64A4, TRPV2, MYT1L, and NRXN3; but with not many conclusive data (D'Agnelli et al., 2019).

SCN9

The dorsal root ganglia are nodules that lie along the spinal column and they play a key role in pain perception. Under normal circumstances, dorsal root ganglia have scant sympathetic innervation. Nevertheless, trauma and/or infection trigger sympathetic sprouting within DRG via nerve growth factor overexpression. Sodium channels play a pivotal role in this hyperexcitability. These mechanisms are the basis of the sympathetically maintained pain concept. Sodium channels located in dorsal root ganglia act as molecular gatekeepers of pain detection at peripheral nociceptors. An isoform of sodium channels (Nav1.7) encoded in gene SCN9A is predominantly expressed in the dorsal root ganglia pain-sensing neurons and sympathetic ganglia neurons.

Several sodium "channelopathies" have been associated with rare painful dysautonomic syndromes such as primary erythromelalgia and paroxysmal extreme pain disorder (formerly familial rectal pain syndrome). Moreover, a nucleotide polymorphism in the SCN9A gene (Rs6746030), which makes dorsal root ganglia neurons hyperexcitable, was related to pain perception intensity in individuals with different painful diseases such as fibromyalgia (Vargas-Alarcon et al., 2012; Vargas-Alarcón et al., 2009).

SLC6A4

This gene encodes an integral membrane protein that transports the neurotransmitter serotonin (serotonin transporter, 5-HTT) from synaptic spaces into presynaptic neurons. Human 5-HTT is encoded by a single gene (SLC6A4) mapped to chromosome 17q12. A polymorphism in the transcriptional region, composed of a 44-basepair insertion ("long allele" (L)) or deletion ("short allele" (S)), has been described. In vitro studies of the functional effects of this polymorphism showed that the long variant is associated with a threefold increase in transcriptional activity. It was also found that levels of serotonin transporter messenger RNA (mRNA) and serotonin uptake capacity were reduced in lymphoblastoid cell lines derived from individuals with one or two copies of the short allele. Therefore, by regulating the magnitude and duration of serotonergic responses, 5-HTT is central to the fine-tuning of brain serotonergic neurotransmission.

In humans, the short variant of the 5-HTT polymorphic region is more frequent among patients with bipolar and unipolar affective disorders and has also been related to anxiety-related traits, late-onset Alzheimer's disease, autism, severe alcohol dependence, migraine, and obsessive-compulsive disorders. Some of these disorders are known to be linked to fibromyalgia. However, the relationship between this gene and its polymorphisms with fibromyalgia has not yet been demonstrated (Gursoy, 2002; Offenbaecher et al., 1999).

SNAP25

SNAP-25 protein is contributory to plasma membrane and synaptic vesicle. SNAP-25 forms complexes with synaptobrevin in synaptic vesicles and with syntaxin in the plasma membrane. In simple terms, SNAP-25 protein is critical in neurotransmission for fusion of plasma membrane and synaptic vesicle. Several studies investigated the relationship between SNAP-25 gene polymorphism and personality disorders, schizophrenia, and attention deficit and hyperactivity disorder; these studies reported that the SNAP-25 gene might influence development of these disorders. Furthermore, SNAP-25 protein might be relevant to other different neurotransmitters due to its involvement in vesicle membrane transition and fusion.

Balkarli et al. showed increased SNAP-25 Ddel T/C genotype in patients with fibromyalgia compared to healthy patients, which is related to behavioral symptoms, personality, and psychological disorders in fibromyalgia patients (Balkarli, Sengül, Tepeli, Balkarli, & Cobankara, 2014).

TACR1

TACR1 gene is located on the 2p11 region of chromosome 2. TACR1 belongs to the family 1 of G protein-coupled receptors and is widely expressed at both the central and the peripheral level, and it is present in neurons, vascular endothelial cells, muscle and different types of immune cells, sharing the same structural motif. The activation of the tachykinin receptors has been implicated in a wide variety of biological actions, such as the modulation of pain perception, regulation of emotional behavior, including stress response, motivation and reward/aversion circuits, modulation of smooth muscle motility, visceral sensitivity, induction of neurogenic inflammation, activation of the immune system, regulation of hematopoiesis, and regulation of endocrine secretion.

Rodriguez-Rodriguez et al. analyzed in subjects diagnosed with fibromyalgia the association of different genetic variants with the presence of various comorbid symptoms and syndromes, and with the severity of the disease (Norheim et al., 2014). They observed that the rs3771863 SNP from the TACR1 gene was significantly associated with the presence of sicca syndrome. The sicca syndrome is characterized by ocular and oral dryness of no apparent cause, and lack of autoimmune abnormalities, such as the presence of autoantibodies, or a pathologic labial salivary gland biopsy. Last, Warner et al. demonstrated the contribution of TACR1 gene to modulating pain sensitivity in osteoarthritis (Warner et al., 2017).

Association genotype-phenotype

Studies aimed at linking the variation in the human genome to observable characteristics of an individual (phenotypes) have progressed at a tremendous rate in recent years and have shown that most human traits are modified by a large number of genetic variants throughout the genome (Young, Benonisdottir, Przeworski, & Kong, 2019).

In the recently past years, genetic analyses were performed using phenotypes in families, without genetic data. But the discovery of easily measurable genomic markers allowed the identification of disease genes through the analysis of links in Mendelian diseases such as cystic fibrosis, Huntington's disease, as well as breast cancer genes (Young et al., 2019). After that, researchers have detected SNPs associated with several traits and diseases such as Alzheimer's disease and height (Lee, Wang, & Xing, 2017). Nevertheless, most phenotypes are not determined by single alleles so it remains a challenge to find genetic variants that influence complex diseases caused by multiple genetic and environmental factors (Mitteroecker, Cheverud, & Pavlicev, 2016). However, it is due to advances in genome sequencing that it is now more affordable to generate extensive datasets needed to establish the complex associations between genotypes, transcripts, and phenotypes (Lee et al., 2017).

People with fibromyalgia frequently experience chronic pain and other nonpainful symptoms such as fatigue symptoms. In this illness, lifestyle and specifically physical activity play a key role in the development and treatment of the disease. However, it is relevant to consider the interplay of genetic and lifestyle factors from a clinical and public health perspective (Estévez López, 2018). For example, it has been studied that variations in neurotransmitter-related genes increase susceptibility to FM via hypersensitivity to peripheral painful stimulus by the central nervous system (Docampo et al., 2014). COMT is the most studied gene in fibromyalgia because of its link to pain modulation as it is related to the degradation of catecholamines and several other neurotransmitters (Estévez-López et al., 2018). On the other hand, the SCN9A gene shows very robust associations with the characteristic phenotype of FM (Estévez López, 2018).

SNPs that has been studied in only one analysis in fibromyalgia

Table 1 summarizes additional genes related to fibromyalgia.

TABLE 1 Summary of additional genes tested in previous case-control studies in fibromyalgia.

Gene	Studies	Participants	Population	Year	Other related diseases
APOE	Kling et al. (2008)	47 cases 41 controls	Brazilian	2010	Alzheimer's disease, atherosclerosis, obesity, depression
	Reeser, Payne, Kitchner, and McCarty (2011)	151 cases 300 controls	North American	2011	
	Pellesi et al. (2019)	69 MOH cases 42 controls	Italian	2019	
CNR1	Smith et al. (2012)	496 cases 348 controls	North American and Caucasian	2012	Depression, obesity, anxiety
GPX1	Akbas, Inanir, Benli, Onder, and Aydogan (2014)	127 cases 56 controls	Turkish	2014	Cancer, psoriasis, cardiomyopathy

TABLE 1 Summary of additional genes tested in previous case-control studies in fibromyalgia—cont'd

Gene	Studies	Participants	Population	Year	Other related diseases
MAOA	Su, Chen, Lai, Chen, and Tsai (2007)	62 cases 101 controls	Taiwanese	2006	Stress disorder, cancer, anxiety
	Gürsoy et al. (2008)	107 cases 90 controls	Turkish	2008	
NOS3	de Luca et al. (2015)	89 cases 196 controls	Italian	2015	Cardiovascular disease, inflammatory disease, rheumatoid arthritis
SOD2	Akbas et al. (2014)	127 cases 56 controls	Turkish	2014	Cancer, Alzheimer's disease, diabetes

GWAS

GWAS involve testing genetic variants across the genomes of many individuals to identify genotype-phenotype associations. Since the first GWAS developed in macular degeneration disease and age (2005), there are more than 50,000 GWAS developed. Although genome-wide SNP arrays were originally designed to interrogate common genetic variation, there are studies using exome-centered custom arrays identifying rare and low-frequency coding variants associated with various complex traits, including blood lipid levels, blood pressure level, height, or T2DM (Tam et al., 2019). Here, we have to include the case of fibromyalgia or musculoskeletal pain. A multisite chronic pain GWAS with data from the large-scale UK Biobank found 76 independent multisite chronic pain-associated variants at 39 loci and an estimated heritability of 10.2% (Dutta et al., 2020). Genetic correlations were observed between multisite chronic pain and a range of psychiatric, autoimmune and anthropometric traits, including major depressive disorder, asthma, and body mass index. Additionally, a polygenic risk score for multisite chronic pain was found to significantly predict chronic widespread pain (pain all over the body), indicating the existence of genetic variants contributing to both of these pain phenotypes (Docampo et al., 2014). Moreover, the genetic correlation of fibromyalgia scores with several psychiatric, personality, and autoimmune traits using publicly available GWAS summary statistics. These last studies support the notion that there are (at least) two different forms of fibromyalgia: one that occurs in younger individuals and is strongly genetically driven, and another one occurs in older individuals and can be driven by a variety of nongenetic factors and other conditions that cause pain (Dutta et al., 2020).

Interactions (gene-gene and gene-environment)

It is known that environment is full of potentially hazardous factors, but these factors only affect people with specific genetic predispositions that render them more susceptible to harmful environmental influences (Belsky & Pluess, 2009). Variations in a certain quality (e.g., height) within the population that are not accounted for by genetic factors could be due to environmental variables, a combination of both factors, known as $G \times E$ interactions. Last, a closely linked concept that is intuitive to the idea of $G \times E$ interaction is phenotypic variance. Phenotypic variance typically accounts for the combinatorial effects of genetic and environmental variance in explaining a specific presenting trait (Esposito et al., 2018).

One of the main factors that play a relevant role in gene-environmental interactions is epigenetic alteration such as DNA methylation, or noncoding RNAs. In fibromyalgia, there are hypomethylated DNA patterns in genes implicated in stress response, DNA repair, autonomic system response, and subcortical neuronal abnormalities. As well as in chronic pain, there are also proved epigenetic mechanisms that play an important role as mediators of long-term changes in central and peripheral nervous system as epigenetic mechanisms. These changes epigenetic patterns such as miRNAs expression in pain-related regions appear to occur in the presence of peripheral inflammation and nerve injury.

These environmental components observed in fibromyalgia pathogenesis highlight a possible role of the gene-environment interaction in the development of the disease (D'Agnelli et al., 2019).

Moreover, it is relevant to have the identification of environmental factor(s) of interest and having a clear conceptualization of what that environment, and the resulting $G \times E$ interaction, would represent. Environmental hazards are defined as substances or events which have the potential to threaten one's environment and, in so doing, adversely affect

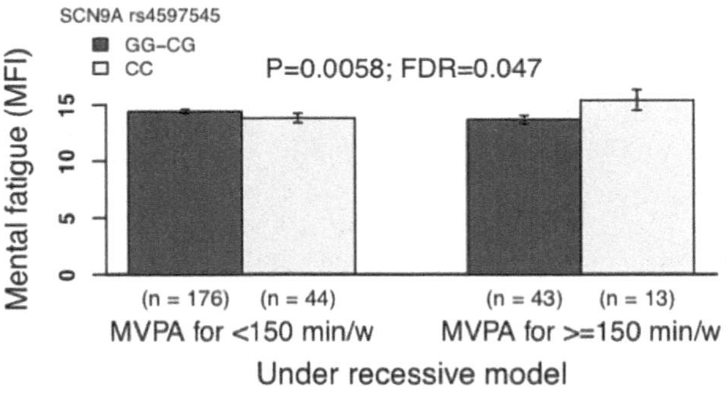

FIG. 1 Additive association gene-physical activity with mental fatigue from the al-Ándalus project (Estévez López, 2018). *MVPA*, moderate to vigorous physical activity; *MFI*, multidimensional fatigue inventory (scores range 0–20). Physical activity was objectively measured using accelerometers and was dichotomized using the minimum physical activity recommendations ($\geq$150 min/week of MVPA in bouts of, at least, 10 min of length). According to the p- and false discovery rate (FDR) values, the gene-physical activity interaction was significant.

health (Esposito et al., 2018). The most common environmental effects smoke or alcohol exposures, nutrition habits or even lifestyle, or any other chemical exposure; all of them can interfere in the genome, metabolome or DNA methylation; and so modulating the disease (Esposito et al., 2018).

Among the environmental factors that potentially may modulate the genetic predisposition to fibromyalgia, physical activity is one of the most promising (Ablin, Cohen, & Buskila, 2006; McBeth & Mulvey, 2012; Rahman, Underwood, & Carnes, 2014).

The al-Ándalus project has provided evidence that the COMT, GABRB3, MTHFR, and SCN9A genes were, with physical activity behaviors levels, additively related to pain-, fatigue-, and psychological resilience-related outcomes. For instance, Fig. 1 shows the significant interaction between the genotype of the SCN9A gene and physical activity with mental fatigue (Estévez López, 2018).

Future perspectives (mainly, research agenda)

Mainly due to a revolution in genetic technology, in genomic analysis, there is a wide range of possibilities that can be developed in molecular analysis. So it is expected that future perspective in genomic analysis could include NGS strategies such as whole exome analysis or new biomarkers in microvesicles such as exosomes (Steyaert et al., 2018).

One of the best options for a good monitoring of the disease is having a good biomarker that could help the stratification, monitoring, and managing of the disease. The success in many others pathologies such as cancer has been reached by the discovery of efficient biomarkers, such as BRCA1/2 in breast cancer screening, KRAS mutation status as a predictive biomarker in patients with advanced CRC, among others (Tejpar et al., 2010). However, there is no one when talking about fibromyalgia or pain, being one of the big challenges of the disease.

Medicine and disease management is changing from the traditional "one size fits all" strategy to a new "personalized" medicine (i.e., the right drug to the right patient at the right time, using the right dose and schedule), where biomolecular biomarkers' tests in combination with clinical reports play the most relevant role.

Molecular biology has rich one of the highest powerful strategies by the use of "omics" technologies (Mordente, Meucci, Martorana, & Silvestrini, 2015). These methodologies could be one of the main tools for rare diseases or those where many of the molecular implications are unknown, such as fibromyalgia.

For example, in recent years, there is a growing appreciation for the role of the gut microbiome in host health and disease. It is known as the diverse and dynamic community of microorganisms that inhabit the digestive tract. Gut microbiome has been highlighted as one of the most exciting future perspectives for better mechanistic understanding of fibromyalgia, for the development of diagnosis, and potentially for new therapeutic modalities (Minerbi et al., 2019). There are also data about metabolomics analysis conducted to compare differentially expressed metabolites in women with and without fibromyalgia. These studies reported that lipid and amino acid metabolites reflecting heightened oxidative stress, inflammation, and tryptophan degradation in patients with fibromyalgia (Menzies et al., 2020).

Moreover, we should not forget to include naïve ones such as weight, body temperature, and the heart rate for fibromyalgia syndrome (Antoranz, Sakellaropoulos, Saez-Rodriguez, & Alexopoulos, 2017) that, in combination with molecular ones, could greatly help in the management of the disease.

Applications to other areas

In this chapter, we have reviewed the current evidence the genetics of chronic widespread musculoskeletal pain. There is an overlap in the symptomatology of people with different diseases characterized by persistent physical symptoms (e.g., irritable bowel syndrome and chronic fatigue syndrome). Moreover, it is commonly hypothesized that central sensitization may play an important role in emergence and persistence of physical symptoms. The present chapter highlights that many genes are related to neurotransmission. Thus, future research analyzing to what extent the role that genetics play in persistent physical symptoms is similar or general for different symptoms (e.g., musculoskeletal pain vs disabling levels of fatigue).

Mini-dictionary of terms

Physical activity is any bodily movement that increases energy expenditure above resting energy expenditure.

Phenotype refers to the observable physical properties of an organism; these include the organism's appearance, development, and behavior.

Genotype is the set of genes the organism carries, as well as by environmental influences upon these genes.

Candidate gene is an approach used in genetics research to associate specific prespecified genes of interest with phenotypes.

GWAS is an approach used in genetics research to associate specific genetic variations with particular diseases. The method involves scanning the genomes from many different people and looking for genetic markers that can be used to predict the presence of a disease.

Key facts

- The most common type of chronic pain is musculoskeletal, which is a type of pain that comes from muscles, bones, joints, or tendons.
- In most of the cases, a root underlying cause of chronic musculoskeletal pain is not identified by clinical tests (e.g., X-ray).
- The unknown origin of chronic musculoskeletal pain has resulted in the stigmatization of patients.
- It is commonly hypothesized that central sensitivity may play an important role in chronic widespread musculoskeletal pain.
- Chronic widespread musculoskeletal pain has been inconsistently defined in the past literature. Thus, the present chapter focused on fibromyalgia which is a well- and robustly defined population in which chronic widespread musculoskeletal pain is the cardinal sign.

Summary points

- Fibromyalgia often aggregates in families, which provides a reasonable support to the hypothesis that it may have a genetic origin.
- Recently, in a genome-wide association study, the heritability of fibromyalgia was around 9%.
- In younger people, fibromyalgia might be more likely determined by genotype.
- In older people, fibromyalgia might be more likely a consequence of other nongenetic factors such as lifestyle.
- In fibromyalgia, it has been showed that the genotype of the COMT, GABRB3, MTHFR, and SCN9A genes were, along with physical activity, additively related to pain-, fatigue-, and psychological-related outcomes.

References

Ablin, J. N., & Buskila, D. (2015). Update on the genetics of the fibromyalgia syndrome. *Best Practice and Research: Clinical Rheumatology*, *29*(1), 20–28. Bailliere Tindall Ltd https://doi.org/10.1016/j.berh.2015.04.018.

Ablin, J. N., Cohen, H., & Buskila, D. (2006). Mechanisms of disease: Genetics of fibromyalgia. *Nature Clinical Practice Rheumatology*, *2*(12), 671–678. https://doi.org/10.1038/ncprheum0349.

Akbas, A., Inanir, A., Benli, I., Onder, Y., & Aydogan, L. (2014). Evaluation of some antioxidant enzyme activities (SOD and GPX) and their polymorphisms (MnSOD2 Ala9Val, GPX1 Pro198Leu) in fibromyalgia. *European Review for Medical and Pharmacological Sciences*, *18*(8), 1199–1203. http://europepmc.org/article/med/24817295.

Antoranz, A., Sakellaropoulos, T., Saez-Rodriguez, J., & Alexopoulos, L. G. (2017). Mechanism-based biomarker discovery. *Drug Discovery Today*, *22* (8), 1209–1215. Elsevier Ltd https://doi.org/10.1016/j.drudis.2017.04.013.

Arnold, L. M., Fan, J., Russell, I. J., Yunus, M. B., Khan, M. A., Kushner, I., & Iyengar, S. K. (2013). The fibromyalgia family study: A genome-wide linkage scan study. *Arthritis and Rheumatism*, *65*(4), 1122–1128. https://doi.org/10.1002/art.37842.

Arnold, L. M., Hudson, J. I., Hess, E.v., Ware, A. E., Fritz, D. A., Auchenbach, M. B., … Keck, P. E. (2004). Family study of fibromyalgia. *Arthritis and Rheumatism*, *50*(3), 944–952. https://doi.org/10.1002/art.20042.

Balkarli, A., Sengül, C., Tepeli, E., Balkarli, H., & Cobankara, V. (2014). Synaptosomal-associated protein 25 (Snap-25) gene polymorphism frequency in fibromyalgia syndrome and relationship with clinical symptoms. *BMC Musculoskeletal Disorders*, *15*(1), 191. https://doi.org/10.1186/1471-2474-15-191.

Belsky, J., & Pluess, M. (2009). Beyond diathesis stress: Differential susceptibility to environmental influences. *Psychological Bulletin*, *135*(6), 885–908. https://doi.org/10.1037/a0017376.

Breivik, H., Collett, B., Ventafridda, V., Cohen, R., & Gallacher, D. (2006). Survey of chronic pain in Europe: Prevalence, impact on daily life, and treatment. *European Journal of Pain*, *10*(4), 287. https://doi.org/10.1016/j.ejpain.2005.06.009.

Briggs, A. M., Cross, M. J., Hoy, D. G., Sànchez-Riera, L., Blyth, F. M., Woolf, A. D., … Briggs, A. (2016). Musculoskeletal health conditions represent a global threat to healthy aging: A report for the 2015 World Health Organization world report on ageing and health background and context. *Gerontologist*, *56*(S2), 243–255. https://doi.org/10.1093/geront/gnw002.

Buskila, D., Neumann, L., Hazanov, I., & Carmi, R. (1996). Familial aggregation in the fibromyalgia syndrome. *Seminars in Arthritis and Rheumatism*, *26* (3), 605–611. https://doi.org/10.1016/S0049-0172(96)80011-4.

Butler, S., Landmark, T., Glette, M., Borchgrevink, P., & Woodhouse, A. (2016). Chronic widespread pain—The need for a standard definition. *Pain*, *157* (3), 541–543. https://doi.org/10.1097/j.pain.0000000000000417.

Chung, J. W. Y., So, H. C. F., Yan, V. C. M., Kwok, P. S. T., Wong, B. Y. M., Yang, J. Y., & Chan, A. P. C. (2019). A survey of work-related pain prevalence among construction workers in Hong Kong: A case-control study. *International Journal of Environmental Research and Public Health*, *16*(8). https://doi.org/10.3390/ijerph16081404.

Cordero, M. D., de Miguel, M., Moreno Fernández, A. M., Carmona López, I. M., Garrido Maraver, J., Cotán, D., … Sánchez Alcázar, J. A. (2010). Mitochondrial dysfunction and mitophagy activation in blood mononuclear cells of fibromyalgia patients: Implications in the pathogenesis of the disease. *Arthritis Research and Therapy*, *12*(1). https://doi.org/10.1186/ar2918.

Cordero, M. D., Moreno-Fernández, A. M., Carmona-López, M. I., Sánchez-Alcázar, J. A., Rodríguez, A. F., Navas, P., & de Miguel, M. (2010). Mitochondrial dysfunction in skin biopsies and blood mononuclear cells from two cases of fibromyalgia patients. *Clinical Biochemistry*, *43*(13–14), 1174–1176. https://doi.org/10.1016/j.clinbiochem.2010.06.013.

D'Agnelli, S., Arendt-Nielsen, L., Gerra, M. C., Zatorri, K., Boggiani, L., Baciarello, M., & Bignami, E. (2019). Fibromyalgia: Genetics and epigenetics insights may provide the basis for the development of diagnostic biomarkers. *Molecular Pain*, *15*. https://doi.org/10.1177/1744806918819944. SAGE Publications Inc.

de Lima, L. O., Zicarelli, C. A. M., Matsumura, A. S., Moroti-Perugini, L. R., de Castro Teixeira, D., Fernandes, K. B. P., … Poli-Frederico, R. C. (2019). Lower limb muscle strength and serotonin receptor gene polymorphism as factors associated in women with fibromyalgia. *Advances in Rheumatology*, *59*(1), 59. https://doi.org/10.1186/s42358-019-0101-9.

de Luca, C., Gugliandolo, A., Calabrò, C., Currò, M., Ientile, R., Raskovic, D., … Caccamo, D. (2015). Role of polymorphisms of inducible nitric oxide synthase and endothelial nitric oxide synthase in idiopathic environmental intolerances. *Mediators of Inflammation*, *2015*, 245308. https://doi.org/10.1155/2015/245308.

Docampo, E., Escaramís, G., Gratacòs, M., Villatoro, S., Puig, A., Kogevinas, M., … Rabionet, R. (2014). Genome-wide analysis of single nucleotide polymorphisms and copy number variants in fibromyalgia suggest a role for the central nervous system. *Pain*, *155*(6), 1102–1109. https://doi.org/10.1016/j.pain.2014.02.016.

Docking, R. E., Beasley, M., Steinerowski, A., Jones, E. A., Farmer, J., Macfarlane, G. J., & Jones, G. T. (2015). The epidemiology of regional and widespread musculoskeletal pain in rural versus urban settings in those $\geq$55 years. *British Journal of Pain*, *9*(2), 86–95. https://doi.org/10.1177/2049463714527438.

Dupépé, E. B., Kicielinski, K. P., Gordon, A. S., & Walters, B. C. (2019). What is a case-control study? *Neurosurgery*, *84*, 819–826. https://doi.org/10.1093/neuros/nyy590.

Dutta, D., Brummett, C. M., Moser, S. E., Fritsche, L. G., Tsodikov, A., Lee, S., … Scott, L. J. (2020). Heritability of the fibromyalgia phenotype varies by age. *Arthritis & Rhematology*, *72*(5), 815–823. https://doi.org/10.1002/art.41171.

Esposito, G., Azhari, A., & Borelli, J. L. (2018). Gene × environment interaction in developmental disorders: Where do we stand and what's next? *Frontiers in Psychology*, *9*(OCT), 26. https://doi.org/10.3389/fpsyg.2018.02036.

Estévez López, F. (2018). *Gene-physical activity interaction with pain, fatigue, and resilience in women with fibromyalgia*. University of Granada. http://hdl.handle.net/10481/51163.

Estévez-López, F., Camiletti-Moirón, D., Aparicio, V. A., Segura-Jiménez, V., Álvarez-Gallardo, I. C., Soriano-Maldonado, A., … Álvarez-Cubero, M. J. (2018). Identification of candidate genes associated with fibromyalgia susceptibility in southern Spanish women: The al-Ándalus project. *Journal of Translational Medicine*, *16*(1). https://doi.org/10.1186/s12967-018-1416-8.

Fayaz, A., Croft, P., Langford, R. M., Donaldson, L. J., & Jones, G. T. (2016). Prevalence of chronic pain in the UK: A systematic review and meta-analysis of population studies. *BMJ Open*, *6*(6). https://doi.org/10.1136/bmjopen-2015-010364.

Fujioka, Y., Stahlberg, A., Ochi, M., & Olmarker, K. (2016). Expression of inflammation/pain-related genes in the dorsal root ganglion following disc puncture in rats. *Journal of Orthopaedic Surgery*, *24*(1), 106–112. https://doi.org/10.1177/230949901602400124.

Gursoy, S. (2002). Absence of association of the serotonin transporter gene polymorphism with the mentally healthy subset of fibromyalgia patients. *Clinical Rheumatology*, *21*(3), 194–197. https://doi.org/10.1007/s10067-002-8284-5.

Gürsoy, S., Erdal, E., Sezgin, M., Barlas, I.Ö., Aydeniz, A., Alaşehirli, B., & Şahin, G. (2008). Which genotype of MAO gene that the patients have are likely to be most susceptible to the symptoms of fibromyalgia? *Rheumatology International*, *28*(4), 307–311. https://doi.org/10.1007/s00296-007-0454-y.

Hatami, M., Sobhan, M. R., Dastgheib, S. A., Jarahzadeh, M. H., Jafari, M., Yadegari, A., … Neamatzadeh, H. (2020). Association of catechol-O-methyltranferase 472G>A (Val158Met) polymorphism with susceptibility to fibromyalgia syndrome. *Journal of Orthopaedics*, *20*, 257–260. https://doi.org/10.1016/j.jor.2020.01.013.

Inanir, A., Yigit, S., Tekcan, A., Pinarli, F. A., Inanir, S., & Karakus, N. (2015). Angiotensin converting enzyme and methylenetetrahydrofolate reductase gene variations in fibromyalgia syndrome. *Gene*, *564*(2), 188–192. https://doi.org/10.1016/j.gene.2015.03.051.

Jackson, T., Thomas, S., Stabile, V., Han, X., Shotwell, M., & McQueen, K. (2015). Prevalence of chronic pain in low-income and middle-income countries: A systematic review and meta-analysis. *The Lancet*, *385*, S10. https://doi.org/10.1016/s0140-6736(15)60805-4.

Kamerman, P. R., Bradshaw, D., Laubscher, R., Pillay-van Wyk, V., Gray, G. E., Mitchell, D., & Chetty, S. (2020). Almost 1 in 5 South African adults have chronic pain: A prevalence study conducted in a large nationally representative sample. *Pain*, *161*(7), 1629–1635. https://doi.org/10.1097/j.pain.0000000000001844.

Kim, S. K., Kim, S. H., Nah, S. S., Hyun Lee, J., Hong, S. J., Kim, H. S., … Lee, S. S. (2013). Association of guanosine triphosphate cyclohydrolase 1 gene polymorphisms with fibromyalgia syndrome in a Korean population. *Journal of Rheumatology*, *40*(3), 316–322. https://doi.org/10.3899/jrheum.120929.

Kling, A., Seddighzadeh, M., Arlestig, L., Alfredsson, L., Rantapää-Dahlqvist, S., & Padyukov, L. (2008). Genetic variations in the serotonin 5-HT2A receptor gene (HTR2A) are associated with rheumatoid arthritis. *Annals of the Rheumatic Diseases*, *67*(8), 1111–1115. https://doi.org/10.1136/ard.2007.074948.

Kool, M. B., van Middendorp, H., Boeije, H. R., & Geenen, R. (2009). Understanding the lack of understanding: Invalidation from the perspective of the patient with fibromyalgia. *Arthritis Care and Research*, *61*(12), 1650–1656. https://doi.org/10.1002/art.24922.

Lee, Y. H., Choi, S. J., Ji, J. D., & Song, G. G. (2012). Candidate gene studies of fibromyalgia: A systematic review and meta-analysis. *Rheumatology International*, *32*(2), 417–426. https://doi.org/10.1007/s00296-010-1678-9.

Lee, S., Wang, H., & Xing, E. P. (2017). Backward genotype-transcript-phenotype association mapping. *Methods*, *129*, 18–23. https://doi.org/10.1016/j.ymeth.2017.09.004.

Markkula, R., Järvinen, P., Leino-Arjas, P., Koskenvuo, M., Kalso, E., & Kaprio, J. (2009). Clustering of symptoms associated with fibromyalgia in a Finnish Twin Cohort. *European Journal of Pain*, *13*(7), 744–750. https://doi.org/10.1016/j.ejpain.2008.09.007.

McBeth, J., & Mulvey, M. R. (2012). Fibromyalgia: Mechanisms and potential impact of the ACR 2010 classification criteria. *Nature Reviews Rheumatology*, *8*(2), 108–116. https://doi.org/10.1038/nrrheum.2011.216.

Menzies, V., Starkweather, A., Yao, Y., Thacker, L. R., Garrett, T. J., Swift-Scanlan, T., … Lyon, D. E. (2020). Metabolomic differentials in women with and without fibromyalgia. *Clinical and Translational Science*, *13*(1), 67–77. https://doi.org/10.1111/cts.12679.

Minerbi, A., Gonzalez, E., Brereton, N. J. B., Anjarkouchian, A., Dewar, K., Fitzcharles, M. A., … Shir, Y. (2019). Altered microbiome composition in individuals with fibromyalgia. *Pain*, *160*(11), 2589–2602. https://doi.org/10.1097/j.pain.0000000000001640.

Mitteroecker, P., Cheverud, J. M., & Pavlicev, M. (2016). Multivariate analysis of genotype–phenotype association. *Genetics*, *202*(4), 1345–1363. https://doi.org/10.1534/genetics.115.181339.

Mordente, A., Meucci, E., Martorana, G. E., & Silvestrini, A. (2015). Cancer biomarkers discovery and validation: State of the art, problems and future perspectives. *Advances in Experimental Medicine and Biology*, *867*, 9–26. Springer New York LLC https://doi.org/10.1007/978-94-017-7215-0_2.

Norheim, K. B., le Hellard, S., Nordmark, G., Harboe, E., Gøransson, L., Brun, J. G., … Omdal, R. (2014). A possible genetic association with chronic fatigue in primary Sjögren's syndrome: A candidate gene study. *Rheumatology International*, *34*(2), 191–197. https://doi.org/10.1007/s00296-013-2850-9.

Nyvang, J., Hedström, M., & Gleissman, S. A. (2016). It's not just a knee, but a whole life: A qualitative descriptive study on patients' experiences of living with knee osteoarthritis and their expectations for knee arthroplasty. *International Journal of Qualitative Studies on Health and Well-Being*, *11*. https://doi.org/10.3402/qhw.v11.30193.

Oezel, L., Then, H., Jung, A. L., Jabari, S., Bonaterra, G. A., Wissniowski, T. T., … di Fazio, P. (2016). Fibromyalgia syndrome: Metabolic and autophagic processes in intermittent cold stress mice. *Pharmacology Research & Perspectives*, *4*(5), 248. https://doi.org/10.1002/prp2.248.

Offenbaecher, M., Bondy, B., de Jonge, S., Glatzeder, K., Krüger, M., Schoeps, P., & Ackenheil, M. (1999). Possible association of fibromyalgia with a polymorphism in the serotonin transporter gene regulatory region. *Arthritis and Rheumatism*, *42*(11). https://doi.org/10.1002/1529-0131(199911)42:11<2482::AID-ANR27>3.0.CO;2-B.

Pellesi, L., Bellei, E., Guerzoni, S., Cainazzo, M. M., Baraldi, C., Monari, E., & Pini, L. A. (2019). Exploration of candidate serum biomarkers potentially related to the chronic pain condition in Medication-overuse headache. *BMC Neurology*, *19*(1). https://doi.org/10.1186/s12883-019-1469-1.

Perrot, S., Cohen, M., Barke, A., Korwisi, B., Rief, W., & Treede, R. D. (2019). The IASP classification of chronic pain for ICD-11: Chronic secondary musculoskeletal pain. *Pain*, *160*(1), 77–82. Lippincott Williams and Wilkins https://doi.org/10.1097/j.pain.0000000000001389.

Potvin, S., Larouche, A., Normand, E., de Souza, J. B., Gaumond, I., Grignon, S., & Marchand, S. (2009). DRD3 Ser9Gly polymorphism is related to thermal pain perception and modulation in chronic widespread pain patients and healthy controls. *Journal of Pain*, *10*(9), 969–975. https://doi.org/10.1016/j.jpain.2009.03.013.

Rahman, A., Underwood, M., & Carnes, D. (2014). Fibromyalgia. *BMJ*, *348*. https://doi.org/10.1136/bmj.g1224. BMJ Publishing Group.

Reeser, J. C., Payne, E., Kitchner, T., & McCarty, C. A. (2011). Apolipoprotein E4 genotype increases the risk of being diagnosed with posttraumatic fibromyalgia. *PM and R*, *3*(3), 193–197. https://doi.org/10.1016/j.pmrj.2010.12.009.

Santiago, M. G., Marques, A., Kool, M., Geenen, R., & da Silva, J. A. P. (2017). Invalidation in patients with rheumatic diseases: Clinical and psychological framework. *Journal of Rheumatology*, *44*(4), 512–518. https://doi.org/10.3899/jrheum.160559.

Smith, S. B., Maixner, D. W., Fillingim, R. B., Slade, G., Gracely, R. H., Ambrose, K., … Diatchenko, L. (2012). Large candidate gene association study reveals genetic risk factors and therapeutic targets for fibromyalgia. *Arthritis and Rheumatism*, *64*(2), 584–593. https://doi.org/10.1002/art.33338.

Smith, S. B., Reenilä, I., Männistö, P. T., Slade, G. D., Maixner, W., Diatchenko, L., & Nackley, A. G. (2014). Epistasis between polymorphisms in COMT, ESR1, and GCH1 influences COMT enzyme activity and pain. *Pain*, *155*(11), 2390–2399. https://doi.org/10.1016/j.pain.2014.09.009.

Steyaert, W., Callens, S., Coucke, P., Dermaut, B., Hemelsoet, D., Terryn, W., & Poppe, B. (2018). Future perspectives of genome-scale sequencing. *Acta Clinica Belgica: International Journal of Clinical and Laboratory Medicine*, *73*(1), 7–10. Taylor and Francis Ltd https://doi.org/10.1080/17843286.2017.1413809.

Su, S. Y., Chen, J. J. H., Lai, C. C., Chen, C. M., & Tsai, F. J. (2007). The association between fibromyalgia and polymorphism of monoamine oxidase A and interleukin-4. *Clinical Rheumatology*, *26*(1), 12–16. https://doi.org/10.1007/s10067-006-0213-6.

Tam, V., Patel, N., Turcotte, M., Bossé, Y., Paré, G., & Meyre, D. (2019). Benefits and limitations of genome-wide association studies. *Nature Reviews Genetics*, *20*(8), 467–484. Nature Publishing Group https://doi.org/10.1038/s41576-019-0127-1.

Tejpar, S., Bertagnolli, M., Bosman, F., Lenz, H., Garraway, L., Waldman, F., … Roth, A. (2010). Prognostic and predictive biomarkers in resected colon cancer: Current status and future perspectives for integrating genomics into biomarker discovery. *The Oncologist*, *15*(4), 390–404. https://doi.org/10.1634/theoncologist.2009-0233.

van den Heuvel, M. M., Jansen, P. W., Bindels, P. J. E., Bierma-Zeinstra, S. M. A., & van Middelkoop, M. (2020). Musculoskeletal pain in 6-year-old children: The Generation R Study. *Pain*, *161*(6), 1278–1285. https://doi.org/10.1097/j.pain.0000000000001797.

Vargas-Alarcon, G., Alvarez-Leon, E., Fragoso, J. M., Vargas, A., Martinez, A., Vallejo, M., & Martinez-Lavin, M. (2012). A SCN9A gene-encoded dorsal root ganglia sodium channel polymorphism associated with severe fibromyalgia. *BMC Musculoskeletal Disorders*, *13*, 23. https://doi.org/10.1186/1471-2474-13-23.

Vargas-Alarcón, G., Fragoso, J. M., Cruz-Robles, D., Vargas, A., Martinez, A., Lao-Villadóniga, J. I., … Martínez-Lavín, M. (2009). Association of adrenergic receptor gene polymorphisms with different fibromyalgia syndrome domains. *Arthritis and Rheumatism*, *60*(7), 2169–2173. https://doi.org/10.1002/art.24655.

Vargas-Alarcón, G., Fragoso, J. M., Cruz-Robles, D., Vargas, A., Vargas, A., Lao-Villadóniga, J. I., … Martínez-Lavín, M. (2007). Catechol-O-methyltransferase gene haplotypes in Mexican and Spanish patients with fibromyalgia. *Arthritis Research and Therapy*, *9*(5). https://doi.org/10.1186/ar2316.

Vervest, A. C. M., & Schimmel, G. H. (1988). Taxonomy of pain of the IASP. *Pain*, *34*(3), 318–321. Pain https://doi.org/10.1016/0304-3959(88)90129-7.

Vos, T., Flaxman, A. D., Naghavi, M., Lozano, R., Michaud, C., Ezzati, M., … Murray, C. J. L. (2012). Years lived with disability (YLDs) for 1160 sequelae of 289 diseases and injuries 1990-2010: A systematic analysis for the Global Burden of Disease Study 2010. *The Lancet*, *380*(9859), 2163–2196. https://doi.org/10.1016/S0140-6736(12)61729-2.

Warner, S. C., Walsh, D. A., Laslett, L. L., Maciewicz, R. A., Soni, A., Hart, D. J., … Valdes, A. M. (2017). Pain in knee osteoarthritis is associated with variation in the neurokinin 1/substance P receptor (TACR1) gene. *European Journal of Pain (United Kingdom)*, *21*(7), 1277–1284. https://doi.org/10.1002/ejp.1027.

Wolfe, F., Clauw, D. J., Fitzcharles, M. A., Goldenberg, D. L., Katz, R. S., Mease, P., … Yunus, M. B. (2010). The American College of Rheumatology preliminary diagnostic criteria for fibromyalgia and measurement of symptom severity. *Arthritis Care and Research*, *62*(5), 600–610. https://doi.org/10.1002/acr.20140.

Wolfe, F., Smythe, H. A., Yunus, M. B., Bennett, R. M., Bombardier, C., Goldenberg, D. L., … Sheon, R. P. (1990). The American College of Rheumatology 1990 criteria for the classification of fibromyalgia. *Arthritis and Rheumatism*, *33*(2), 160–172. https://doi.org/10.1002/art.1780330203.

Young, A. I., Benonisdottir, S., Przeworski, M., & Kong, A. (2019). Deconstructing the sources of genotype-phenotype associations in humans. *Science*, *365*(6460), 1396–1400. American Association for the Advancement of Science https://doi.org/10.1126/science.aax3710.

Zhang, X., Hartung, J. E., Bortsov, A.v., Kim, S., O'Buckley, S. C., Kozlowski, J., & Nackley, A. G. (2018). Sustained stimulation of β 2- and β 3-adrenergic receptors leads to persistent functional pain and neuroinflammation. *Brain, Behavior, and Immunity*, *73*, 520–532. https://doi.org/10.1016/j.bbi.2018.06.017.

Chapter 5

Fentanyl: Polymorphisms, and adverse events

Mongi Benjeddou[a] and Ana M. Peiró[b,c]

[a]*Department of Biotechnology, University of the Western Cape, Cape Town, South Africa* [b]*Neuropharmacology on Pain and Functional Diversity (NED), Alicante Institute for Health and Biomedical Research (ISABIAL Foundation), Alicante, Spain* [c]*Department of Clinical Pharmacology, General Hospital of Alicante, Alicante, Spain*

Abbreviations

ABCB1	P-glycoprotein coding gene, ATP-binding cassette subfamily B membrane
ADRs	adverse drug reactions
AE	adverse event
Asn	asparagine
Asp	aspartic acid
CNS	central nervous system
COMT	catechol-*O*-methyltransferase
CP	chronic pain
CYP	cytochrome
CYP2D6	isoform 2D6 of cytochrome P450
DDD	defined daily doses
DNA	deoxyribonucleic acid
EM	extensive metabolizer
KCNJ6	potassium ion channels named Kir3.2
OPRD1	δ opioid receptor coding gene (in English, δ opioid receptor)
OPRM1	gene coding for the opioid receptor μ (in English, opioid receptor μ)
PGx	pharmacogenetics
SL	slow metabolizer
SLCO	solute carrier organic anion transporter family member
SNP	single nucleotide polymorphism
UM	ultra-fast metabolizer
US	United States

Opioid fentanyl use in pain management

The goal of pain treatment is to achieve adequate control in the shortest possible time, which must be both effective and safe. However, the response to analgesic treatments, as for fentanyl opioid prescription, is characterized by having a wide interindividual variability. Thus, despite having better treatments, more powerful diagnostic tools and a more detailed knowledge of the pathophysiology of pain, patients with chronic pain (CP) continue to experience pain. This makes difficult to predict and establish a balance between the effectiveness and the occurrence of adverse events (AEs), which remain the first cause of therapy abandonment (Woodcock, 2009).

Potential reasons of this variability can be linked to different functional capacity and quantity of phase I enzymes (such as different isoforms of CYP P450), receptors (such as OPRM1 for opioids as fentanyl) and transporters (such as ABCB1) involved in the metabolic destiny of each drug, all of these are dictated by DNA and RNA variations that will be discussed in next section (Fig. 1).

The Neurobiology, Physiology, and Psychology of Pain. https://doi.org/10.1016/B978-0-12-820589-1.00005-1

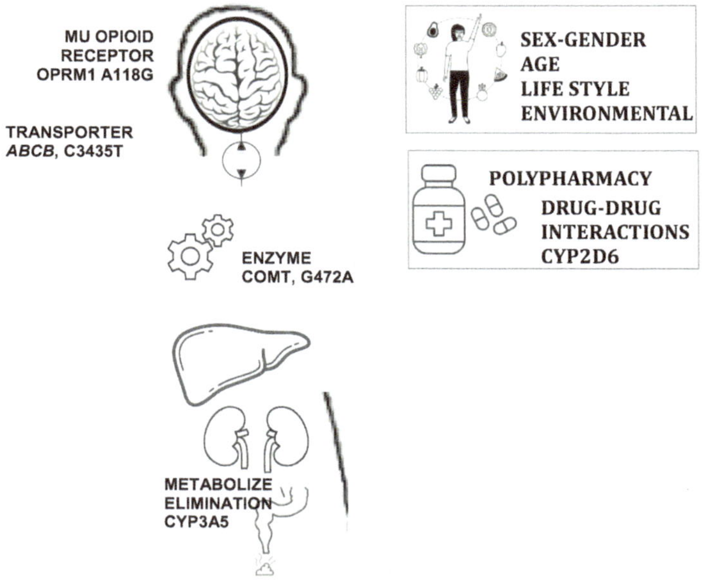

FIG. 1 Pharmacogenetics applied to fentanyl opioid treatment. Pharmacogenetics and other factors than can influence fentanyl therapy results on pain management.

Fentanyl pharmacology

Fentanyl is one of the most frequently used opioids for the treatment of moderate-severe chronic and acute cancer-related pain. It is a synthetic derivative of phenylpiperidine that binds to the mu receptor opioid (OPRM), more potent than morphine. Due to their high affinity for μ-opioid receptors, larger doses of naloxone are required to reverse the effects than are commonly used. Furthermore, fentanyl can increase dopamine levels in the central nervous system. This increase in dopamine produces a state of relaxation, relieves pain, decreases the perception of suffering, and promotes a feeling of well-being (euphoria). It is used to manage pain during surgery and to treat moderate-to-severe of several CP syndromes being the first choice in renal impairment. It is very fat-soluble, unlike morphine, which determines its enormous potency and shorter duration of action. It can work within minutes to relieve pain and produce sedation with a short duration of effects around 30–90 min (Nelson & Schwaner, 2009).

Fentanyl is extensively metabolized by *N*-dealkylation to norfentanyl that is mediated by CYP3A4 and -3A5 (CYP3A). Approximately 70% of fentanyl is excreted in the urine, mainly as metabolites. However, little is known, however, about the identity and formation of human fentanyl metabolites (Labroo, Paine, Thummel, & Kharasch, 1997).

Medically prescribed fentanyl is available in a variety of formulations, including lozenges, lollipops, oral and nasal sprays, and injections. Routes of administration: transdermal, endovenous, and transmucosal. When administered as a lozenge for oral transmucosal absorption, swallowed fentanyl is subject to first pass metabolism in the liver and possibly small intestine. For continuous delivery, fentanyl can be administered through a transdermal patch that adheres to the skin. The patch works by slowly releasing fentanyl through the skin into the bloodstream over 48–72 h. Transdermal fentanyl has proven to be as effective as subcutaneous morphine while maintaining a similar rate of AE (perhaps less constipation and daytime sleepiness, although more night-time sleep disorders), showing good compliance by patients (Ahn et al., 2017). In general, fentanyl is well tolerated in patients receiving opioids chronically, although some of patients have nausea (frequent but transient), topical skin reactions, dry mouth, asthenia, sweating, and drowsiness can appear (Manchikanti et al., 2017).

Fentanyl's effectiveness is dependent on the individual's size, weight, overall state of health, the amount that is taken, whether fentanyl is taken in combination with other drugs, and whether the person is used to taking opioids. However, this effectiveness and tolerability profiles' variability are underrepresented in medical literature. Also, patterns of prescription per opioid vary considerably per physician specialty and per country. An increase in prescriptions was seen for codeine, oxycodone, and tramadol, while other opioids such as fentanyl had decreased in the last few years, mainly due to the opioid US epidemic crisis. The most common overdose symptoms are coma, lethargy, respiratory depression, and arrest. Although naloxone, an opioid receptor antagonist, is the standard drug for fentanyl overdose rescue, attempts to revive patients with naloxone could be unsuccessful, due to the rapid onset of fentanyl's action (Kuczynska, Grzonkowski, Kacprzak, & Zawilska, 2018).

Although fentanyl is mostly dosed by titration, nevertheless in specific situations, such as a rotation to fentanyl from another opioid or a change in comedication, under- or overdosing may occur. Therefore, knowledge of factors that influence the variability of fentanyl pharmacokinetics is important.

Fentanyl safety profile

CP treatment includes the use of different category of drugs as monotherapy of in combination (polytherapy). The latter is more frequent, especially to treat comorbidities, which may interfere with the pharmacological action of analgesic drugs. The use of CYP3A4 inhibitors and inducers, impaired liver function, and heating of the patch potentially influences fentanyl pharmacokinetics in a clinically relevant way. In elderly patients, current data suggest that we should carefully dose fentanyl due to alterations in absorption and metabolism (Geist et al., 2019). The influence of body mass index or gender on fentanyl pharmacokinetics can be relevant but continue questionable, most probably due to a large heterogeneity in the published studies.

Adverse events

The main AEs associated with opioids are related to their pharmacological effects and are therefore dose dependent. An AE is any undesirable event experienced by a patient, regardless of whether or not the medication administered is suspected. Even more, suspected adverse drug reaction (ADRs) is a special type of AE in which a causative relationship can be shown. It is estimated that approximately 0.5% of hospital emergency and primary care visits are due to a suspicion ADR; however, only 1% of them are notified to pharmacovigilance centers (Narumol, Arunrot, & Krska, 2015). This is essential for promoting a safe use of analgesic drugs in "real world" because presence of any AE can define the safety profile of a drug and can either aggravate an underlying clinical situation, or produce clinical manifestations that can cause misdiagnosis and cause more medications to be added to a patient who already tends to be polymedicated. As an example, elderly people are more likely to have constipation (up to 30%), nausea, and dizziness than younger individuals. Therefore, detecting AEs is essential to prevent these situations. Above all, because there is usually an underreporting of RAM due to care overload, low motivation or even ignorance of notification programs (Hooman Khademi, Farin Kamangar, Paul Brennan, & Reza Malekzadeh, 2016).

Among the most common analgesics AEs are: gastrointestinal events (nausea 21%, vomiting 10%, constipation 15%) (Camilleri, Lembo, & Katzka, 2017), nervous system (dizziness 14%, confusion 14%), or dermatological (pruritus 13%). Other AEs include sedation, tolerance, dependence, respiratory depression, and addiction (Xu & Johnson, 2013). Next are more prevalent by system affected:

- *Gastrointestinal:* Nausea and vomiting are the most frequent (30%–50%), but generally a rapid tolerance is created. These discomforts are caused by the stimulation of the chemoreceptor trigger zone of the last area. Constipation (15%) is also frequent, due to the direct action on the opioid receptors of the intestinal wall, which cause an increase in gastrointestinal muscle tone, a decrease in intestinal peristalsis and gastric emptying.
- *Nervous system:* There is rapid tolerance to sedation and drowsiness, but special care should be taken in the elderly and patients taking other drugs that potentiate these effects. Other AEs related to the central nervous system are drowsiness, dizziness, headache, cognitive disorders, hallucinations, delusions, myoclonus, and hyperalgesia. It should be clarified that opioids do not generate persistent cognitive impairment. Exciting side effects, such as allodynia, hyperalgesia, myoclonus, and seizures, can occur in chronic treatments at high doses.

Opioids can rarely induce other AEs such as (Suga et al., 2019):

- *Renal:* Urinary retention is caused by an increase in the tone of the bladder sphincter together with an inhibition of the reflex response to urination.
- *Dermatological:* Itching can occur due to the release of histamine, and may be due to intolerance of excipients. Sweating is also usually a common AE.
- *Reproductive system:* Opioids have been seen to significantly reduce plasma testosterone levels in people who use heroin, methadone, and opioids for CP. Thus, a decrease in libido and amenorrhea has also been observed.
- *Cardiovascular and respiratory:* Respiratory depression, although rare (less than 1%). At the cardiovascular level, opioids produce vasodilation by direct activity and by the release of histamine. In addition, administered at high doses, they are capable of causing some myocardial depression. Tiredness and loss of appetite are also usually among the most frequent AEs after opioid prescription.

Potential drug-drug interactions

Pharmacological management of pain requires a combination of drugs (polypharmacy) that are usually prescribed at high doses. Thus, drug-drug interactions and possible competitive metabolic destiny should be analyzed in each patient individually. In turn, it is known that a large number of CYP2D6 enzyme substrates have the capacity to inhibit it. Comedication with these inhibitors could cause severe drug-drug interactions, many of which have been observed in clinical practice. Even more, because this pharmacological profiles are usually ignored or simplified in clinical trials (He, Chen, Zhou, & Zhou, 2015). Next are more prevalent combinations drugs:

- *Antiepileptics or anticonvulsants:* pregabalin and gabapentin are anticonvulsants that have shown efficacy in monotherapy but not in combined therapy in pain management. They are included in most neuropathic pain guidelines as the first line of treatment at the same level as antidepressants. Both are active drugs on the central nervous system (CNS) that in their own right occupy a prominent place in the treatment of pain, particularly in the neuropathic type of CP, although not exclusively (Moulin et al., 2014). Although they are chemically and pharmacologically similar, they present a series of differences that should be taken into account when choosing one or the other in a given patient or in a specific situation. The anticonvulsant potency of pregabalin is 3–10 times greater than gabapentin and a longer duration of action with a lower dose range than gabapentin. Pregabalin reaches an optimal dose from the beginning and presents greater therapeutic effectiveness with a better cost-effectiveness and cost-utility profile (Senderovich & Jeyapragasan, 2018).

The most representative AEs of these two anticonvulsants are sedation and drowsiness and, in some cases, pregabalin is associated with edema. In general, both pregabalin and gabapentin are well-tolerated drugs and with little capacity to produce drug interactions with other analgesics or psychoactive drugs in general. According to a study that compared mortality in patients with moderate-to-severe CP, treated long term with opioids or antidepressant and anticonvulsant drugs, an increased risk of mortality from any cause was associated in patients treated with opioids in the long term (Ray, Chung, Murray, Hall, & Stein, 2016).

Polypharmacy of pain management

The CP is usually accompanied by various comorbidities that feed to each other, such as anxiety, depression, or insomnia. In patients with CP, a high prevalence of anxiety and depression has been observed, which occurs in 50%–60% of cases, highlighting the interaction of pain with the psychological sphere, and highlighting the importance of addressing the patient with pain of an interdisciplinary way. CNS active drugs such as tricyclic antidepressants can increase the magnitude and size of all opioid effects. In addition, these antidepressants will induce, at medium or high doses, sedation, drowsiness, urinary retention, and orthostatic hypotension. Taking these considerations into account, the evidence of the relationship between anxiety-depression and pain, justifies the need for mental disorders to be evaluated in patients with CP (Uguz et al., 2010).

Sex influence: Understanding our differences

In addition to the different frequency of CP between sexes, it has been suggested that the response to opioid analgesics may vary between sexes. Based on clinical studies, a greater analgesic effect in women has been suggested globally (Fillingim & Gear, 2004). A compendium of physiological, genetic, hormonal factors and a differential epigenetic regulation based on a hormonal issue could determine sex differences. However, these data are usually based on studies with a low number of subjects and sometimes, of very determined ethnicities, so there are no proven data that can provide clear evidence (Stephens et al., 2017).

Our research group found certain sex differences with reference to the safety profile of opioids prescribed for CP (Muriel et al., 2019). In patients with chronic low back pain, a greater presence of nausea (52% vs 31%) and pruritus (42% vs 12%) and lower headache (10% vs 27%) were observed in women (Salzman et al., 1999). In contrast, sexual dysfunction were significant more prevalent in men (33% affected) and significantly associated with the opioid dose (Ajo et al., 2016). It is important to take in account the sex variable for analysis in patients with CP who have developed any AE to understand analgesic safety profile.

Fentanyl misuse and abuse

There is still controversy in the use of long-term opioid treatment for CP, as fentanyl users, due to issues related to long-term effectiveness and safety, particularly because of the risk of tolerance, dependence, or abuse. In fact, alarming epidemiological medicine reports, accumulated mainly during the last two decades, point to a growing increase in illicit use of fentanyl. Abuse of fentanyl initially appeared in the 1970s and has increased in recent years. Until 2013, sporadic outbreaks of fentanyl and fentanyl analogues contaminating the heroin supply caused some deaths in heroin users in the United States. Since then, fentanyl has caused deaths being completely infiltrated the North American heroin supply. The drug can be obtained by diverting from legitimate medical supplies, or it can be manufactured in illegal laboratories. Synthetic opioids are an increasingly major public health in US threat requiring vigilance from multiple fields. However, in Europe, the situation is different due to (a) the different regulation of prescription drugs, (b) a less aggressive and nonconsumer pharmaceutical advertising, (c) different attitudes about medical and nonmedical use, and (d) the culture in terms of prescription and patient expectations, among others. In addition, opioid crisis coexists with another situation of an under-prescription in other countries, such as Latin America or Africa, showing the existence of a huge gap between the opioid abuses epidemic in the United States, and the shortage of them in many other countries. For these reasons, although the risk in Europe seems rather limited, this situation clearly supports the need for long-term studies of opioid use in CP (Armenian, Vo, Barr-Walker, & Lynch, 2018).

Risk of dependence

As with any opiate, there is a risk of dependency, tolerance, abuse, and addiction with fentanyl use. According to a recent metaanalysis, dependency ratios are estimated between 8% and 12% of patients with CP (Vowles et al., 2015). This has led to the fact that, since 1999, overdose mortality associated with prescription opioids has quadrupled. In addition, although sales of prescription opioids in the United States have increased 300% since 1999, the pain referred by patients has remained unchanged, an even more worrying fact (Vadivelu et al., 2018). This pattern of opioid misuse has caused, in the United States, greater control of the maximum opioid doses, prescribed per person. Since 2010, the maximum prescribed dose has been gradually decreasing every year until 2015, without improving control of pain disease. Risk factors for misuse or addiction to opioids include current or prior abuse of other substances, untreated psychiatric disorders, young ages, and social or family environments that promote their misuse. Health professionals should be able to identify factors that predispose certain individuals to misuse or overdose of prescription opioids (Cragg et al., 2017).

Physical dependence results in withdrawal symptoms when individuals abruptly stop taking the drug. Thus, withdrawal symptoms usually start within 12 h of the last dose of fentanyl and can last 1 week or more. Opioid disruption can trigger a characteristic opioid withdrawal syndrome, which may include tremor, sweating, insomnia, or lack of appetite, among many others. That is why the opioid withdrawal should be carried out slowly and progressively minimizing its symptoms opioid rotation, discontinuation of therapy, prevention of AE, and close supervision of these patients (Just, Mücke, & Bleckwenn, 2016).

Deprescription program

The term deprescription is the cessation of a long-term treatment under the supervision of a medical professional, reconsidering the prescription of the drugs from the beginning. It should be applied whenever medications are accumulated in the therapeutic regimen, before relevant clinical changes or when the patient's perception and attitude toward the medications have been altered, as is the case with fentanyl dependence (Lumish, Goga, & Brandt, 2018). Occasionally, they could include opioid rotation with a better safety profile. The possibility of using opioid maintenance agonists such as transdermal buprenorphine has been suggested, due to its pharmacokinetic profile, its greater tolerability, and comfort of administration, having demonstrated its effectiveness in several studies, and a higher adherence and treatment compliance. Also, prescription opioid detoxification generally includes structured therapies with opioid agonists (Gowing, Ali, & White, 2009).

Potential pharmacogenetics markers in fentanyl pain management

The field of pharmacogenomics (PGx) seeks to understand how this individual's unique gene sequence can affect their response to certain drugs. It is particularly relevant in anesthesia when the interindividual response to pain medication is essential. This is part of personalized medicine which has the objective of setting the optimal therapies in the light of the functional biochemical asset and of the comorbidities of the individual patient, in order to obtain the best clinical response (Gray, Adhikary, & Janicki, 2018).

In the field of pain, the most studied genes are the ones that encode opioid receptors (*OPRM1, OPRD1*), membrane transporters (*ABCB1*), molecules involved in the transmission of the painful impulse (*COMT*) and enzymes responsible for opioid metabolism (such as different isoforms of CYP P450 due to fentanyl as *CYP3A4* or concomitant drugs due to *CYP2D6*) (Yoshida et al., 2018) (Table 1).

TABLE 1 Pharmacogenetic markers in opioid pain management.

Gene	Variants	Clinical effect
OPRM1 Opioid receptor μ	A118G *rs1799971*	Loss of an *N*-glycosylation site in the receptor. GG—higher morphine dose requirement
OPRD1 Opioid receptor δ	T921C *rs4985*	Related to dependency vulnerability
ABCB1 P-glicoproteína	C3435T *rs1045642*	Decreased expression of p-glycoprotein, influencing drug distribution
COMT Catecol-*O*-metiltransferasa	G472A *rs4985*	Reduction of COMT enzyme activity. Association with minor analgesic requirements
ARRB2 ß2-arrestina	C8622T *rs1045280*	Possible contribution to interindividual variability in the response to opioid disabling
CYP2D6 Citocromo P450	Lento Extensivo Ultra-rápido	Influence on the metabolism and plasma concentration of several opioid drugs

Translation effects of gene variants most frequent used in opioid used for pain management.

Opioids reach their place of action by crossing the blood brain barrier and activating the mu receptor encoded by the *OPRM1* gene. Transporters such as the P-glycoprotein encoded by the *ABCB1* gene regulate the passage of BHE. The C3435T single nucleotide polymorphism (SNP) of this gene has been associated with an alteration of the expression of p-glycoprotein, altering the distribution of the drug. The activation of the receptors inhibits the transmission of pain through an intracellular signaling cascade that includes interaction with other systems such as the catecholaminergic. The enzyme catechol-*O*-methyltransferase, regulated by the *COMT* gene, plays a fundamental role in the degradation of catecholamines. Variants as G472A, whose variant leads to a reduction in enzyme activity was associated with a decrease in the response of the opioid system to pain and an increase in the density of receptors, suggesting an improvement in the efficacy of opioid analgesics, needing less dose of opioids to relieve pain (Kambur & Männistö, 2010).

Candidate genes

OPRM1 *gene*

The μ1 opioid receptor (OPRM1) is the main target of opioid molecules. The most studied SNP of the gene that encodes it is A118G (rs1799971) which leads to the loss of an *N*-glycosylation site in the extracellular region of the receptor, implying a greater affinity of the receptor for β-endorphin (Kapur, Lala, & Shaw, 2014). This SNP, not synonymous, is located in exon 1 of the gene and consists in the substitution of an A for a G, which once transcribed becomes the replacement of the amino acid asparagine (Asn) by aspartic acid (Asp) in the position 40 of the protein (Asn40Asp) with the consequent loss of *N*-glycosylation in the extracellular region of the receptor (Mura et al., 2013).

The functional association of SNP A118G with pain is biologically reasonable because it affects the most important target of opioids, the opioid receptor. Some studies indicated that carriers of the allelic variant G required higher doses of morphine or fentanyl to achieve analgesia (Zhang et al., 2011). Even more, the effect of the OPRM1 A118G on the regulation of pain, placebo analgesia, and stress-related personality traits has been described. These findings point to a potentially significant role in illnesses where vulnerability to physical and emotional stressors represents a core feature, which include persistent pain conditions, the addictions and the mood disorders, as well as their comorbidities (Pecina, Love, Stohler, Goldman, & Zubieta, 2015). On the other hand, the *OPRD1* gene encodes the δ opioid receptor, whose main endogenous ligand is the endorphins. It has been suggested that an increase in its expression could increase the reward effects of substances of abuse, through a greater release of dopamine in the nucleus accumbens, thus increasing vulnerability to dependence on opioids (Levran et al., 2008).

Even more, the presence of the allelic G variant of SNP A118G is associated with a reduction in the expression of the opioid receptor in the brain or with a decrease in the effectiveness of the opioid receptor signal in the secondary somatosensory area (Oertel et al., 2009). Thus, several studies have shown that individuals carrying the 118G allele require more doses of opioid to achieve the analgesic effect (Webster & Belfer, 2016). In fact, fentanyl consumption increased in accordance with the number of 118G alleles (Chou et al., 2006). In the obstetric population, women with SNP A118G showed, regardless of being homozygous or heterozygous for the G allele, pain thresholds were higher (Fillingim et al., 2005), and the intensity of pain secondary to migraines (Menon et al., 2012), fibromyalgia (Finan et al., 2010), or postoperative pain (Zhang et al., 2010). In addition, the SNP was found to have a specific effect according to sex, with evidence of female carriers of two 118G alleles who experienced twice as much pain as male carriers (Olsen et al., 2012).

Regarding its influence on the opioid safety profile, several studies found no association between AE and A118G SNP related with nausea or vomiting (Laugsand et al., 2015). The data, today, reveal that more studies are needed to establish this relationship.

COMT *gene*

The *COMT* (*catechol-O-methyltransferase*) gene, located on the long arm of chromosome 22, encodes an enzyme responsible for the inactivation of catecholamine-type neurotransmitters (dopamine, norepinephrine, and adrenaline). Catecholamines modify the perception of pain as they modulate the levels of endogenous opioids and the density of mu opioid receptors in different CNS regions (Zubieta et al., 2003).

One of the best-characterized SNP in the *COMT* gene is the G472A polymorphism (dbSNP rs4680), also known as Val158Met. The ancestral allele G codes for the amino acid valine, while the allelic variant A codes for methionine in codon 158; the change of the allele G for A in position 472 implies a variation of a functional type. The existence of the A allele implies a lower activity rate of the enzyme together with a higher degradation rate for the *COMT* enzyme compared to that considered normal basal activity (Diatchenko et al., 2006). Therefore, the allelic variant A, and especially the existence of the two AA polymorphic alleles, implies a lower capacity to metabolize monoamine neurotransmitters and, therefore, higher levels of dopamine, and a lower pain threshold that implies a 1.5 times increased sensitivity to pain compared to homozygous GG subjects, with an increase in postoperative fentanyl consumption in some studies (Zhang et al., 2015). In general, the presence of the allelic variant A seems to identify individuals who are more likely to respond positively to dopaminergic drugs (Schacht, 2016). In contrast, patients with GG genotype have a higher enzyme activity, lower levels of dopamine that imply lower sensitivity to pain (higher pain threshold) and better ability to recover from stress (Cargnin et al., 2013).

Regarding safety, SNP G472A is associated with an increase in AE at least in chronic cancer pain (Kambur & Männistö, 2010). The main idea is that SNPs can influence nausea and vomiting since the COMT enzyme modulates neurotransmission through the metabolism of dopamine. The blockade of central D2 dopaminergic receptors can inhibit the medullary chemoreceptor trigger zone by blocking emetic impulses to the vomiting center. However, no differences were found in the incidence of AE such as nausea, vomiting, and dizziness in a postoperative acute pain model treated with fentanyl in which four SNPs (rs0269, rs4633, rs4818, and rs4680) and their haplotypes were analyzed. In other study, differences were found in the presence of AE such as dizziness, confusion, and hallucinations associated with certain *COMT* SNPs, which influenced the tolerability of morphine (Ross et al., 2005). Also, the Val158Met SNP can be a genetic biomarker to predict individual differences in sleep physiology, resulting from sleep loss in a healthy and ethnic diverse adult population of men and women (Goel, Banks, Lin, Mignot, & Dinges, 2011).

CYP450

The CYP450 system is of major significance in the metabolism of codeine, fentanyl, methadone, oxycodone, and oxymorphone. Due to genetic variation in *CYP450* and to the potential for drug interactions, these later drugs can have significant risks for wide serum level variation. Researchers have discovered more than 50 CYP450 enzymes, but the CYP1A2, CYP2C9, CYP2D6, CYP3A4, and CYP3A5 enzymes account for the metabolism of up to 90% of drugs that undergo this type of biotransformation (DePriest, Puet, Holt, Roberts, & Cone, 2015).

CYP3A4/5 polymorphisms

Fentanyl is a high extraction drug and therefore, theoretically, relatively insensitive to changes in hepatic intrinsic clearance caused by altered CYP3A4/5 activity. The role of *CYP3A5*3* on treatment toxicity with fentanyl in chronic cancer pain patients, there was a greater incidence of central adverse effects in **3* homozygotes than in **1* carriers implying a correlation with the higher absorption rate (Takashina et al., 2012). Some results indicated that the *ABCB/CYP3A4/OPRM1* gene polymorphisms influenced the fentanyl consumption and the physiological effects of intravenous analgesia in the Chinese women who received lower segment caesarean section surgeries (Zhang et al., 2018). Also, *CYP3A4*1G* polymorphism could decreases the metabolism of fentanyl with a positive correlation between CYP3A4 mRNA level and its metabolism (Yuan et al., 2015).

CYP2D6 phenotypes

CYP2D6 gene interacts with a large number of drugs; either in their metabolism or that influences their catalytic activity. Therefore, the study or genotyping of said gene must be considered, since it can influence adjuvant medication against pain. The CYP2D6 enzyme, a member of the cytochrome P450 family of enzymes, is responsible for the metabolism of a large number of drugs including several opioids (Smith, 2009). From the presence of certain genetic variants in the *CYP2D6* gene, it is possible to estimate the resulting enzyme activity, which could be increased or decreased. Therefore, from the analysis of the CYP2D6 genotype, the metabolizing profile (phenotype) could be estimated, being classified as a slow (ML), extensive (ME) or ultra-fast (UM) metabolizer.

Some drugs need metabolism for activation and trigger their pharmacological effect. In several studies, it has been observed that individuals with a slower metabolism (ML) had a higher rate of therapeutic failure due to lack of efficacy, while MUs have been associated with greater opioid-related toxicity (Gaedigk et al., 2008). Clinical recommendations exist to guide drug therapy based on *CYP2D6* genotype for codeine, tramadol, oxycodone, and hydrocodone, although the level of supporting evidence differs by drug. Limited evidence also supports the use of genetic data to guide other medications in chronic pain therapy, including tricyclic antidepressants and celecoxib.

However, in addition to the genotypic configuration, the activity of the CYP2D6 enzyme is regulated by a large number of physiological factors (menstrual cycle, pregnancy, among others) pathological (liver disease, inflammation, among others), or environmental (tobacco, alcohol, among others) (Tanaka, 1999). Thus, the expression and therefore the activity of this enzyme may be influenced by epigenetic factors, which refer to genomic modifications that can condition expression and phenotypes without changing the DNA sequence. Hypermethylation and histone modifications inactivate the transcription of the *CYP2D6* gene, repressing its expression.

Other genes variants

Next to enzymes, also the efflux drug transporter ABCB1 (P-glycoprotein) was studied. This protein is responsible for the transport of fentanyl through the blood-brain barrier. SLCO1B1 is another protein responsible for transport. Variations in genes coding for these proteins may therefore influence fentanyl pharmacokinetics. However, both the *ABCB1* T1236C and *SLCO1B1*1a* and *15 polymorphisms were not found to influence fentanyl pharmacokinetics (Takashina et al., 2012). More research is needed before implementing genotyping in clinical practice.

Barriers to implementation

Despite the benefits that could be gained by understanding a patient's genomic profile and potential to metabolize opioids and other pain medications, there are still many barriers to more widespread pharmacogenetics testing:

- Differences between labs, testing, and variability in reporting.
- Insurance coverage: DNA sequencing costs continue to decline, but lack of consistent coverage and other regulatory restrictions can discourage patients.
- Lack of data: Some evidence is often limited to retrospective and observational studies.
- Accessibility: Future developments will more consistently include genomic results in the patient's electronic medical record, but right now, results are not always accessible to every physician providing patient care.
- Clinical implementation: Issues of validity (testing to predict the presence or absence of the drug-related phenotype or disease of interest) vs utility (likelihood the test results will alter clinical outcomes or patient treatment decisions).
- Ethical concerns, including potential discrimination based on genomic results.
- Variability in interpretation, stemming from lack of clear, curated guidelines, which translate laboratory results into actionable decisions for specific medications.

Education and awareness of pharmacogenetics has been growing over the last few years and will continue to do so. To advance the practice, it would require the creation of multidisciplinary team, bringing together physician, pharmacist, medical geneticist, laboratory personnel, and nursing staff, to establish pharmacogenomics testing as a part of daily patient care.

Conclusions

Fentanyl is one of the most frequently used opioids for CP. However, large interindividual differences are found in the daily dosages needed to achieve pain control. Some of this variation in pain sensitivity, opioid dose requirements, and opioid-induced adverse effects could be explained by pharmacogenetic variants, which also vary largely between individuals. Gene variation caused by genotypes in receptor signaling *OPRM1*, pain modulators as *COMT* genes or CYP450 liver enzymes, may also play an important role that is not completely clear. These different mechanisms suggest that baseline pharmacogenetic screening, before initiation of fentanyl treatment, may in a future serve as a helpful diagnostic tool to predict the individual response. Due to the different causes and characteristics of patients with CP, under fentanyl prescription, further investigation into the effectiveness and safety of this medication is required, which may target specific phenotypic subgroups rather than comprehensive analysis across the entire patient pool.

Applications to other areas

Pharmacogenetics testing for fentanyl is important in reducing AEs.

Pharmacovigilance is warranted to curb addiction of fentanyl abuse.

Pharmacogenetics testing can be extended to other drugs used for cancer-derived pain, to minimize AEs.

Fentanyl as opioid is one of the most commonly prescribed medicines for chronic pain. The objective of this review was to identify the role of genetic polymorphisms on patient treatment parameters related to adverse effects of fentanyl. Trials included in this review showed that a part of the variation in fentanyl concentration can be explained by the *OPRM1, COMT, and* CYP3A5*3 polymorphisms and thus, postoperative nausea, vomiting, and dizziness. Preoperative genotyping could provide the basis to modulate individual opioid dosage to achieve satisfactory pain control while reducing potential side effects. This genotype may therefore serve as a potential biomarker of vulnerability for those conditions and interindividual variations in responses to their treatments. However, not all studies showed the same effect, and the effects of the investigated polymorphisms in relation to fentanyl pharmacokinetics were small and do not support, nowadays, a routine genotyping in clinical pain practice.

Other agents of interest

In this chapter, we have described detailed mechanisms of how gene polymorphism can influence on fentanyl safety pattern profile. Evidence from recent studies show that several polymorphisms in *ABCB1, OPRM1,* and *COMT* have been significantly associated with opioid dosing and could be useful in determining opioid dose for patients.

Morphine
Tramadol
Oxycodone

Interestingly, most studies that showed statistically significant associations focused on patients with CP associated with malignancy, likely reflecting the fewer studies investigating chronic nonmalignant pain. Important differences in allele frequencies between ethnic and gender groups should also be taken into consideration when designing pharmacogenomics studies. For example, *CYP2D6* plays a vital role in the metabolism of many of the opioids, but it is also the gene that shows great interethnic variability. The prevalence of PMs and UMs range from 7% to 10% and 4.3% in Caucasians and about 0% to 1% in Asians, respectively. This difference in allele distribution may explain the lack of reproducible results across different ethnicities.

Mini-dictionary of terms

Adverse event is any undesirable event experienced by a patient, regardless of whether or not the medication administered is suspected.

Candidate genes is a gene or variation on a gene that may relate to the construct of interest, given the role of the gene in a distinct biological pathway or findings from previous studies.

Deprescription program is the cessation of a long-term treatment under the supervision of a medical professional, reconsidering the prescription of the drugs from the beginning.

Drug-drug interactions refer to the influence of one drug upon another, e.g., stimulates/inhibits catabolism, thereby reducing/increasing efficacy (or toxicity).

Gender-sex: Sex is the property or quality by which organisms are classified as female or male on the basis of their reproductive organs and functions, while gender is expressed in terms of masculinity and femininity.
Opioid use disorder is defined as a problematic pattern of opioid use that leads to serious impairment or distress.
Polypharmacy has come to mean the use of several (usually five or more) medications on a daily basis, with the possibility that these may not all be clinically necessary.

Key facts of fentanyl

Fentanyl is one of the most frequently used opioids used in pain management, more potent than morphine.
Fentanyl is extensively metabolized by CYP3A. However, little is known about the identity and formation of human fentanyl metabolites.
Most common adverse events are gastrointestinal (nausea 21%, vomiting 10%, constipation 15%), nervous system (dizziness 14%, confusion 14%), or dermatological (pruritus 13%).
There is still controversy in the use of long-term opioid treatment for CP, as fentanyl users, due to issues related to long-term effectiveness and safety, particularly because of the risk of tolerance, dependence, or abuse.
Dependency ratios related to general opioids are estimated between 8% and 12% of patients with chronic pain, even more for lozenges oral fentanyl absorption.
Most studied genes are the ones that encode opioid receptors (*OPRM1*), molecules involved in the transmission of the painful impulse (*COMT*) and enzymes responsible for opioid metabolism (*CYP3A4*/5 due to fentanyl or *CYP2D6 due to* concomitant drugs).

Summary points

Opioid fentanyl is used to manage pain during surgery and to treat moderate-to-severe of several chronic pain syndromes and to manage pain during surgery.
Fentanyl is full agonists of varying potencies at the μ-opioid receptor, leading to typical clinical effects of miosis and respiratory and central nervous system depression.
Fentanyl is well tolerated in patients receiving opioids chronically, although some of patients have nausea (frequent but transient), topical skin reactions, dry mouth, asthenia, sweating, and drowsiness can appear.
Abuse of fentanyl initially appeared in the 1970s and has increased in recent years. Until 2013, sporadic outbreaks of fentanyl and fentanyl analogues contaminating the heroin supply caused some deaths in heroin users in the United States.
Gene variation caused by genotypes in receptor signaling *OPRM1*, pain modulators as *COMT* genes or CYP3A4 liver enzymes, can affect pain sensitivity and opioid-induced effects.
Opioids reach their place of action by crossing the blood-brain barrier and activating the mu receptor encoded by the *OPRM1* gene, one of the most candidate and actionable genes nowadays in pain medicine.
COMT gene plays a fundamental role in the degradation of catecholamines. G472A variant, whose variant leads to a reduction in enzyme activity, was associated with a decrease in the response of the opioid system to pain and an increase in the density of receptors, suggesting an improvement in the efficacy of opioid analgesics.
Variants in CYP3A4/5 gene could condition a differential pattern of fentanyl consumption and physiological effects.

References

Ahn, J. S., Lin, J., Ogawa, S., Yuan, C., O'Brien, T., Le, B. H., … Ganapathi, A. (2017). Transdermal buprenorphine and fentanyl patches in cancer pain: A network systematic review. *Journal of Pain Research*, *10*, 1963.

Ajo, R., Segura, A., Inda, M. M., Planelles, B., Martínez, L., Ferrández, G., … Peiró, A.-M. (2016). Opioids increase sexual dysfunction in patients with non-cancer pain. *Journal of Sexual Medicine*, *13*(9), 1377–1386.

Armenian, P., Vo, K. T., Barr-Walker, J., & Lynch, K. L. (2018). Fentanyl, fentanyl analogs and novel synthetic opioids: A comprehensive review. *Neuropharmacology*, *134*, 121–132.

Camilleri, M., Lembo, A., & Katzka, D. A. (2017). Opioids in gastroenterology: Treating adverse effects and creating therapeutic benefits. *Clinical Gastroenterology and Hepatology*, *15*(9), 1338–1349.

Cargnin, S., Magnani, F., Viana, M., Tassorelli, C., Mittino, D., Cantello, R., … Genazzani, A. A. (2013). An opposite-direction modulation of the COMT Val158Met polymorphism on the clinical response to intrathecal morphine and triptans. *Journal of Pain*, *14*(10), 1097–1106.

Chou, W.-Y., Wang, C.-H., Liu, P.-H., Liu, C.-C., Tseng, C.-C., & Jawan, B. (2006). Human opioid receptor A118G polymorphism affects intravenous patient-controlled analgesia morphine consumption after total abdominal hysterectomy. *Anesthesiology: The Journal of the American Society of Anesthesiologists*, *105*(2), 334–337.

Cragg, A., Hau, J. P., Woo, S. A., Liu, C., Doyle-Waters, M. M., & Hohl, C. M. (2017). Risk factors for addiction among patients receiving prescribed opioids: A systematic review protocol. *Systematic Reviews*, *6*(1), 265.

DePriest, A. Z., Puet, B. L., Holt, A. C., Roberts, A., & Cone, E. J. (2015). Metabolism and disposition of prescription opioids: A review. *Forensic Science Review*, *27*(2), 115–145.

Diatchenko, L., Nackley, A. G., Slade, G. D., Bhalang, K., Belfer, I., Max, M. B., … Maixner, W. (2006). Catechol-O-methyltransferase gene polymorphisms are associated with multiple pain-evoking stimuli. *Pain*, *125*(3), 216–224.

Fillingim, R. B., & Gear, R. W. (2004). Sex differences in opioid analgesia: Clinical and experimental findings. *European Journal of Pain*, *8*(5), 413–425.

Fillingim, R. B., Kaplan, L., Staud, R., Ness, T. J., Glover, T. L., Campbell, C. M., … Wallace, M. R. (2005). The A118G single nucleotide polymorphism of the μ-opioid receptor gene (OPRM1) is associated with pressure pain sensitivity in humans. *Journal of Pain*, *6*(3), 159–167.

Finan, P. H., Zautra, A. J., Davis, M. C., Lemery-Chalfant, K., Covault, J., & Tennen, H. (2010). Genetic influences on the dynamics of pain and affect in fibromyalgia. Correction to Finan et al. (2010). *Health Psychology*, *29*, 134–142.

Gaedigk, A., Simon, S., Pearce, R., Bradford, L., Kennedy, M., & Leeder, J. (2008). The CYP2D6 activity score: Translating genotype information into a qualitative measure of phenotype. *Clinical Pharmacology & Therapeutics*, *83*(2), 234–242.

Geist, M. J., Ziesenitz, V. C., Bardenheuer, H. J., Burhenne, J., Skopp, G., & Mikus, G. (2019). Minor contribution of cytochrome P450 3A activity on fentanyl exposure in palliative care cancer patients. *Scientific Reports*, *9*(1), 1–6.

Goel, N., Banks, S., Lin, L., Mignot, E., & Dinges, D. F. (2011). Catechol-O-methyltransferase Val158Met polymorphism associates with individual differences in sleep physiologic responses to chronic sleep loss. *PLoS One*, *6*(12), e29283.

Gowing, L., Ali, R., & White, J. M. (2009). Buprenorphine for the management of opioid withdrawal. *Cochrane Database of Systematic Reviews*, (3).

Gray, K., Adhikary, S. D., & Janicki, P. (2018). Pharmacogenomics of analgesics in anesthesia practice: A current update of literature. *Journal of Anaesthesiology Clinical Pharmacology*, *34*(2), 155.

He, Z.-X., Chen, X.-W., Zhou, Z.-W., & Zhou, S.-F. (2015). Impact of physiological, pathological and environmental factors on the expression and activity of human cytochrome P450 2D6 and implications in precision medicine. *Drug Metabolism Reviews*, *47*(4), 470–519.

Hooman Khademi, M., Farin Kamangar, M., Paul Brennan, M., & Reza Malekzadeh, M. (2016). Opioid therapy and its side effects: A review. *Archives of Iranian Medicine*, *19*(12), 870.

Just, J., Mücke, M., & Bleckwenn, M. (2016). Dependence on prescription opioids: Prevention, diagnosis and treatment. *Deutsches Ärzteblatt International*, *113*(13), 213.

Kambur, O., & Männistö, P. T. (2010). Catechol-O-methyltransferase and pain. In *Vol. 95. International review of neurobiology* (pp. 227–279). Elsevier.

Kapur, B. M., Lala, P. K., & Shaw, J. L. (2014). Pharmacogenetics of chronic pain management. *Clinical Biochemistry*, *47*(13–14), 1169–1187.

Kuczynska, K., Grzonkowski, P., Kacprzak, L., & Zawilska, J. B. (2018). Abuse of fentanyl: An emerging problem to face. *Forensic Science International*, *289*, 207–214. https://doi.org/10.1016/j.forsciint.2018.05.042.

Labroo, R. B., Paine, M. F., Thummel, K. E., & Kharasch, E. D. (1997). Fentanyl metabolism by human hepatic and intestinal cytochrome P450 3A4: Implications for interindividual variability in disposition, efficacy, and drug interactions. *Drug Metabolism and Disposition*, *25*(9), 1072–1080. Retrieved from http://www.ncbi.nlm.nih.gov/pubmed/9311623.

Laugsand, E. A., Skorpen, F., Kaasa, S., Sabatowski, R., Strasser, F., Fayers, P., & Klepstad, P. (2015). Genetic and non-genetic factors associated with constipation in cancer patients receiving opioids. *Clinical and Translational Gastroenterology*, *6*(6), e90.

Levran, O., Londono, D., O'hara, K., Nielsen, D., Peles, E., Rotrosen, J., … Ott, J. (2008). Genetic susceptibility to heroin addiction: A candidate gene association study. *Genes, Brain and Behavior*, *7*(7), 720–729.

Lumish, R., Goga, J. K., & Brandt, N. J. (2018). Optimizing pain management through opioid deprescribing. *Journal of Gerontological Nursing*, *44*(1), 9–14.

Manchikanti, L., Kaye, A. M., Knezevic, N. N., McAnally, H., Slavin, K., Trescot, A. M., … Hirsch, J. A. (2017). Responsible, safe, and effective prescription of opioids for chronic non-cancer pain: American Society of Interventional Pain Physicians (ASIPP) guidelines. *Pain Physician*, *20*(2S), S3–S92. Retrieved from http://www.ncbi.nlm.nih.gov/pubmed/28226332.

Menon, S., Lea, R. A., Roy, B., Hanna, M., Wee, S., Haupt, L. M., & Griffiths, L. R. (2012). The human μ-opioid receptor gene polymorphism (A118G) is associated with head pain severity in a clinical cohort of female migraine with aura patients. *Journal of Headache and Pain*, *13*(7), 513–519.

Moulin, D., Boulanger, A., Clark, A., Clarke, H., Dao, T., Finley, G., … Morley-Forster, P. (2014). Pharmacological management of chronic neuropathic pain: Revised consensus statement from the Canadian Pain Society. *Pain Research & Management*, *19*(6), 328–335.

Mura, E., Govoni, S., Racchi, M., Carossa, V., Ranzani, G. N., Allegri, M., & van Schaik, R. H. (2013). Consequences of the 118A > G polymorphism in the OPRM1 gene: Translation from bench to bedside? *Journal of Pain Research*, *6*, 331.

Muriel, J., Margarit, C., Barrachina, J., Ballester, P., Flor, A., Morales, D., … Peiró, A. M. (2019). Pharmacogenetics and prediction of adverse events in prescription opioid use disorder patients. *Basic & Clinical Pharmacology & Toxicology*, *124*(4), 439–448.

Narumol, J., Arunrot, P., & Krska, J. (2015). Survey of patients' experiences and their certainty of suspected adverse drug reactions. *International Journal of Clinical Pharmacy*, *37*(1), 168–174.

Nelson, L., & Schwaner, R. (2009). Transdermal fentanyl: Pharmacology and toxicology. *Journal of Medical Toxicology*, *5*(4), 230.

Oertel, B. G., Kettner, M., Scholich, K., Renné, C., Roskam, B., Geisslinger, G., … Lötsch, J. (2009). A common human μ-opioid receptor genetic variant diminishes the receptor signaling efficacy in brain regions processing the sensory information of pain. *Journal of Biological Chemistry*, *284*(10), 6530–6535.

Olsen, M. B., Jacobsen, L. M., Schistad, E. I., Pedersen, L. M., Rygh, L. J., Røe, C., & Gjerstad, J. (2012). Pain intensity the first year after lumbar disc herniation is associated with the A118G polymorphism in the opioid receptor mu 1 gene: Evidence of a sex and genotype interaction. *Journal of Neuroscience*, *32*(29), 9831–9834.

Pecina, M., Love, T., Stohler, C. S., Goldman, D., & Zubieta, J.-K. (2015). Effects of the Mu opioid receptor polymorphism (OPRM1 A118G) on pain regulation, placebo effects and associated personality trait measures. *Neuropsychopharmacology*, *40*(4), 957–965.

Ray, W. A., Chung, C. P., Murray, K. T., Hall, K., & Stein, C. M. (2016). Prescription of long-acting opioids and mortality in patients with chronic noncancer pain. *JAMA*, *315*(22), 2415–2423.

Ross, J., Rutter, D., Welsh, K., Joel, S., Goller, K., Wells, A., … Riley, J. (2005). Clinical response to morphine in cancer patients and genetic variation in candidate genes. *Pharmacogenomics Journal*, *5*(5), 324–336.

Salzman, R. T., Roberts, M. S., Wild, J., Fabian, C., Reder, R. F., & Goldenheim, P. D. (1999). Can a controlled-release oral dose form of oxycodone be used as readily as an immediate-release form for the purpose of titrating to stable pain control? *Journal of Pain and Symptom Management*, *18*(4), 271–279.

Schacht, J. P. (2016). COMT val158met moderation of dopaminergic drug effects on cognitive function: A critical review. *Pharmacogenomics Journal*, *16*(5), 430–438.

Senderovich, H., & Jeyapragasan, G. (2018). Is there a role for combined use of gabapentin and pregabalin in pain control? Too good to be true? *Current Medical Research and Opinion*, *34*(4), 677–682.

Smith, H. S. (2009). Opioid metabolism. In *Paper presented at the Mayo Clinic Proceedings*.

Stephens, K. E., Levine, J. D., Aouizerat, B. E., Paul, S. M., Abrams, G., Conley, Y. P., & Miaskowski, C. (2017). Associations between genetic and epigenetic variations in cytokine genes and mild persistent breast pain in women following breast cancer surgery. *Cytokine*, *99*, 203–213.

Suga, Y., Uchida, M., Suzuki, S., Sugawara, H., Torigoe, K., Futamura, A., … Takase, H. (2019). Current status of adverse events related with opioid analgesics in Japan: Assessment based on Japanese adverse drug event report database. *Biological and Pharmaceutical Bulletin*, *42*(5), 801–806.

Takashina, Y., Naito, T., Mino, Y., Yagi, T., Ohnishi, K., & Kawakami, J. (2012). Impact of CYP3A5 and ABCB1 gene polymorphisms on fentanyl pharmacokinetics and clinical responses in cancer patients undergoing conversion to a transdermal system. *Drug Metabolism and Pharmacokinetics*, *27*(4), 414–421.

Tanaka, E. (1999). Gender-related differences in pharmacokinetics and their clinical significance. *Journal of Clinical Pharmacy and Therapeutics*, *24*(5), 339–346.

Uguz, F., Çiçek, E., Salli, A., Karahan, A. Y., Albayrak, İ., Kaya, N., & Uğurlu, H. (2010). Axis I and Axis II psychiatric disorders in patients with fibromyalgia. *General Hospital Psychiatry*, *32*(1), 105–107.

Vadivelu, N., Kai, A. M., Kodumudi, G., Haddad, D., Kodumudi, V., Kuruvilla, N., … Urman, R. D. (2018). Recommendations for substance abuse and pain control in patients with chronic pain. *Current Pain and Headache Reports*, *22*(4), 25.

Vowles, K. E., McEntee, M. L., Julnes, P. S., Frohe, T., Ney, J. P., & van der Goes, D. N. (2015). Rates of opioid misuse, abuse, and addiction in chronic pain: A systematic review and data synthesis. *Pain*, *156*(4), 569–576.

Webster, L. R., & Belfer, I. (2016). Pharmacogenetics and personalized medicine in pain management. *Clinics in Laboratory Medicine*, *36*(3), 493–506.

Woodcock, J. (2009). A difficult balance—Pain management, drug safety, and the FDA. *New England Journal of Medicine*, *361*(22), 2105–2107. https://doi.org/10.1056/NEJMp0908913.

Xu, Y., & Johnson, A. (2013). Opioid therapy pharmacogenomics for noncancer pain: Efficacy, adverse events, and costs. *Pain Research and Treatment*, *2013*, 943014.

Yoshida, K., Nishizawa, D., Ide, S., Ichinohe, T., Fukuda, K.i., & Ikeda, K. (2018). A pharmacogenetics approach to pain management. *Neuropsychopharmacology Reports*, *38*(1), 2–8.

Yuan, J.-J., Hou, J.-K., Zhang, W., Chang, Y.-Z., Li, Z.-S., Wang, Z.-Y., … Kan, Q.-C. (2015). CYP3A4* 1G genetic polymorphism influences metabolism of fentanyl in human liver microsomes in Chinese patients. *Pharmacology*, *96*(1–2), 55–60.

Zhang, W., Chang, Y., Kan, Q., Zhang, L., Lu, H., Chu, Q., … Zhang, J. (2010). Association of human μ-opioid receptor gene polymorphism A118G with fentanyl analgesia consumption in Chinese gynaecological patients. *Anaesthesia*, *65*(2), 130–135.

Zhang, F., Tong, J., Hu, J., Zhang, H., Ouyang, W., Huang, D., … Liao, Q. (2015). COMT gene haplotypes are closely associated with postoperative fentanyl dose in patients. *Anesthesia & Analgesia*, *120*(4), 933–940.

Zhang, W., Yuan, J., Kan, Q., Zhang, L., Chang, Y., & Wang, Z. (2011). Study of the OPRM1 A118G genetic polymorphism associated with postoperative nausea and vomiting induced by fentanyl intravenous analgesia. *Minerva Anestesiologica*, *77*(1), 33–39.

Zhang, J., Zhang, L., Zhao, X., Shen, S., Luo, X., & Zhang, Y. (2018). Association between MDR1/CYP3A4/OPRM1 gene polymorphisms and the post-caesarean fentanyl analgesic effect on Chinese women. *Gene*, *661*, 78–84.

Zubieta, J.-K., Heitzeg, M. M., Smith, Y. R., Bueller, J. A., Xu, K., Xu, Y., … Goldman, D. (2003). COMT val158met genotype affects μ-opioid neurotransmitter responses to a pain stressor. *Science*, *299*(5610), 1240–1243.

Chapter 6

Propofol anesthesia and molecular changes in the brain

Desanka Milanović, Željko Pavković, and Vesna Pešić
Department of Neurobiology, Institute for Biological Research "Siniša Stanković"—National Institute of Republic of Serbia, University of Belgrade, Belgrade, Serbia

Abbreviations

BDNF	brain-derived neurotrophic factor
Ca^{2+}	calcium
CaMKII	Ca^{2+}/calmodulin (CaM)-dependent protein kinase II
CB1	cannabinoid receptors type 1
DARPP-32	dopamine- and cyclic adenosine monophosphate-regulated phosphoprotein with a Mr 32 kDa
Egr-1	early growth response protein 1
ERK1/2	extracellular signal regulated kinases 1 and 2
FAAH	fatty acid amide hydrolase
FosB	Finkel-Biskis-Jinkins murine osteosarcoma viral oncogene homolog B
GABA	γ-aminobutyric acid
LORR	loss of righting reflex
mRNA	messenger ribonucleic acid
Na^{+}	sodium
NGF	nerve growth factor
NMDA	*N*-methyl-D-aspartate
$p75^{NTR}$	the p75 neurotrophin receptor
PI3K/Akt	phosphatidylinositol 3-kinase
PSD-95	postsynaptic density protein 95
TNFα	tumor necrosis factor α
TrkB	tropomyosin receptor kinase B
VTA	ventral tegmental area
XIAP	X-linked inhibitor of apoptosis protein

Introduction

Discovery of substances with anesthetic properties has endorsed development of anesthesiology, which rate of advance has surpassed most other branches of medicine enabling expansion of diagnostic procedures and surgery. General anesthesia is a drug-induced, reversible condition that includes specific behavioral and physiological traits—unconsciousness, amnesia, analgesia, and akinesia—with concomitant stability of the autonomic, cardiovascular, respiratory, and thermoregulatory systems (Brown, Lydic, & Schiff, 2010). Ideal properties of an intravenous (i.v.) anesthetic induction agent include rapid onset of action, rapid recovery from the drug's effects, and minimal depression of cardiorespiratory function, which are the properties that propofol has. Propofol (2,6-diisopropylphenol) is a versatile, short-acting, intravenous sedative-hypnotic agent initially marketed as an anesthetic, and now also widely used for the sedation of patients in the intensive care unit and in those undergoing brief ambulatory surgery. It was discovered by John (Iain) Glen and Roger James in 1977 (Glen & James, 1977). John Glen received the 2018 Lasker-DeBakey Clinical Medical Research Award for this discovering and the formula refinement.

The Neurobiology, Physiology, and Psychology of Pain. https://doi.org/10.1016/B978-0-12-820589-1.00006-3

Formulation matter

Propofol is formulated as an oil-in-water emulsion for intravenous application (1% propofol in 10% soybean oil emulsion, a vehicle very similar to 10% intralipid), being the most lipophilic anesthetic in clinical use (Egan, 2010). It rapidly crosses the blood-brain barrier resulting in rapid action, short duration of action, and minimal side effects that have made it the most widely used agent for induction of anesthesia. Despite its success, current propofol formulation has a number of undesirable properties, with the most common being pain on injection, hyperlipidemia associated with prolonged propofol administration in the intensive care unit, allergic reactions, and postoperative sepsis as intralipid formulation supports rapid microbial growth despite antimicrobial additives (Egan, 2010). Importantly, it was shown that propofol administered in lipid emulsion was more potent and rapid acting than equivalent doses of propofol administered by a method involving a lipid-free vehicle (Dutta & Ebling, 1997). As it was excellently defined by Egan (2010), for propofol, the formulation strategy is not just a "vehicle" to deliver the drug but integral part of the drug.

Neural and molecular targets of propofol

It is now known that different groups of general anesthetics, which represent the only drugs used to produce unconsciousness as a therapeutic goal, create this state through the production of a neuronal hyperpolarization at the level of the thalamus, thalamocortical, and corticocortical (fronto-parietal) reverberant loops, which have been recognized as extremely important in the neurobiology of consciousness (Alkire, Hudetz, & Tononi, 2008; Brown et al., 2010). Induction of general anesthesia also inhibits the excitatory arousal pathways originating in the brain stem and pons, which are involved in maintaining cortical arousal and higher-order integration in the cortex, indicating that the anesthetic cascade may appear as a top-down failure really driven by bottom-up effects (Hudetz, 2012).

At the molecular level, the $GABA_A$ receptor has a key role in hypnotic effect of propofol that, at clinically relevant concentrations, markedly potentiates the effects of the inhibitory neurotransmitter GABA and enhances the GABA-induced chloride current while at higher concentrations directly activates the channel opening (reviewed in Sahinovic, Struys, & Absalom, 2018). The point mutation in the β_3 subunit (β3N265M) of $GABA_A$ receptor greatly reduces the ability of propofol to produce LORR in animals, a surrogate for loss of consciousness (Jurd et al., 2003).

A model of intravenous anesthetic agents-induced affected glutamate release (Buggy, Nicol, Rowbotham, & Lambert, 2000) assumes that $GABA_A$ receptor activation of glutamatergic neurons by propofol results in hyperpolarization, indirectly reducing the amount of glutamate released. This model explains why ketamine anesthesia, which is induced by reduced glutamate release via an action of ketamine at the NMDA receptor, is characterized by "dissociative anesthesia," with analgesia, sedation, and detachment, while propofol is noted for profound depression of consciousness and lack of analgesic effects (discussed in Buggy et al., 2000). Behavioral experiments performed on mice showed that the NMDA receptor antagonist MK-801, as well as non-NMDA receptor agonist kainate, augmented propofol anesthesia suggesting that it is mediated, in addition to $GABA_A$ receptors, by excitatory amino acid receptors (Irifune et al., 2003).

The voltage-gated Na^+ channels, which conduct the Na^+ current inward and are responsible for controlling the initiation and propagation of action potentials in excitable cells, also play a critical role in the disruption of synaptic transmission by general anesthetics including propofol (Frenkel & Urban, 1991). Recent research (Wang et al., 2018) showed that propofol has the net inhibitory action on these channels by facilitating the channel activation process and potently promoting activation-coupled inactivation, thus shifting the channel population from the resting to the inactivated states. Moreover, previous experiments provided biochemical evidence for an action of propofol on neuronal Na^+ channels in isolated presynaptic nerve endings, resulting in presynaptic inhibition of Na^+ channel-dependent glutamate release (Ratnakumari & Hemmings, 1997).

The most recent findings suggest that in addition to presynaptic Na^+ channels, propofol can inhibit the presynaptic Ca^{2+} channels (the P/Q-type calcium channel may be the main target, without affecting their characteristics), thus producing substantial decline in the release of glutamate, impairing both rapid and slow endocytosis and reducing synaptic strength (Liu et al., 2019). Previous experiments showed that propofol inhibits the T-type and L-type components of the Ca^{2+} current, which may play a role in the cardiovascular side effects that have been clinically observed including a decrease in peripheral resistance and hypotension (Olcese, Usai, Maestrone, & Nobile, 1994), which may have clinical implications for hypertensive patients.

Suggestions that sedative properties of propofol may be caused by the activation of CB1 receptors should be mentioned as well. Actually, it has been shown that propofol increases the whole-brain content of anandamide (that binds and activates CB1 receptors) by inhibiting FAAH, which catalyzes the degradation of anandamide (Patel et al., 2003). The authors showed that pretreatment of mice with the CB1 receptor antagonist significantly reduced the number of mice that lost their righting reflex in response to propofol, while pretreatment of mice with the CB1 receptor agonist significantly potentiated the LORR produced by propofol (Patel et al., 2003). This mechanism may also contribute to the unusual antiemetic properties of propofol.

Propofol-induced brain molecular changes during postanesthesia period

There was an opinion that the general anesthesia is completely reversible and that the central nervous system returns to its original state after the anesthetic elimination. However, general anesthetics are major pharmacological modulators of neuronal activity, very potent and effective in transiently inhibiting neuronal communication (Hudetz, 2012), thus influencing a plethora of activity-dependent molecular events and processes that result in long-term morphological and functional alterations especially at the extremes of age (reviewed in Vutskits & Xie, 2016). Propofol exposure also affects activity-dependent development of adult-born neurons (involved in cognitive plasticity), indicating that neuronal immaturity rather than the age of the whole organism is important for the impact of anesthesia (Krzisch et al., 2013).

Molecular changes that reflect alterations in neuronal activity at the peak of brain growth spurt

Changes in the expression of activity-dependent genes/proteins have been examined in the context of molecular alterations in the brain induced by general anesthesia starting with the report that anesthetics with a clinical application (e.g., midazolam, nitrous oxide, isoflurane) induce apoptotic damage in the developing brain at the peak of brain growth spurt, at least in part, through the BDNF-modulated apoptotic cascade (Lu, Yon, Carter, & Jevtovic-Todorovic, 2006). The BDNF and the NGF expression is activity-dependent and the balance between the activities of the glutamatergic and GABAergic systems controls the physiological level of these neurotrophins in neurons in vitro and in vivo (Zafra, Castren, Thoenen, & Lindholm, 1991).

Our group was the first that examined changes in the expression of neurotrophins and c-Fos (an indirect marker for neuronal activity) due to single propofol anesthesia exposure at the peak of brain growth spurt (Pesic et al., 2009, 2015) but see also (Karen et al., 2013; Ponten, Fredriksson, Gordh, Eriksson, & Viberg, 2011). Comparative analysis of mRNA/protein expression showed that changes in BDNF, NGF, and c-Fos in immature rat brain after single exposure to anesthetic dose of propofol are time- and brain region–specific and that alterations at mRNA and protein level are not straight forward (Pesic et al., 2009, 2015). In the frontal cortex, a decrease in NGF was accompanied by an increase in BDNF (as well as phosphorylated/activated form of its receptor TrkB) and relatively unchanged level of c-Fos while in the thalamus the level of all three proteins was decreased 2–4 h after the treatment. These data accentuate that at the peak of brain growth spurt, propofol anesthesia produces fewer disturbances in activity-regulated proteins the frontal cortex than in the thalamus, which may explain early increase (4 h after the treatment) in the active subunit of the caspase-3 (an executioner caspase in apoptosis) in the thalamus but not in the cortex (Pesic et al., 2009). Nevertheless, increased expression of caspase-3 and an increased number of disseminated Fluoro-Jade B positive cells was observed throughout the cortex and in thalamic nuclei of P7 rats 24 h after the propofol treatment along with an increase in the TNFα expression. Overall, these experimental evidences indicate that in the developing brain, at the peak of brain growth spurt period, short-term propofol anesthesia exposure could be related to the apoptotic neurodegeneration that is induced in a neurotrophic- and extrinsic apoptotic pathway-dependent manner (Pesic et al., 2009, 2015). Changes in the expression of different activity-regulated molecules under the same anesthetic treatment are limited and considering those rare (Karen et al., 2013; Lu et al., 2006; Ponten et al., 2011) as well as our findings (Pesic et al., 2009, 2015) it could be speculated that an increase in BDNF protein in the cortex and its decrease in the thalamus of developing brain (P7) is a response to anesthesia exposure in general rather than to propofol exposure per se.

Age-dependent peculiarities in the expression of neurotrophins and their downstream signaling pathways

Considering that vulnerability of developing brain to general anesthesia is less prominent after the peak of brain growth spurt period (in rodents postnatal (P) days P7–P10; Yon, Daniel-Johnson, Carter, & Jevtovic-Todorovic, 2005) in addition to P7 rats, a series of experiments was done on P14 rats (Popic et al., 2012, 2015). In contrast to above-discussed findings in P7 rats, in P14 rats, we observed downregulation of mature BDNF protein in the cortex and an increase in the thalamus while the level of NGF was downregulated in both examined brain structures (Popic et al., 2012). We detected brain-region changes in the activation of the PI3K/Akt (in the thalamus) and the ERK1/2 (in the cortex), with the balanced expression of these prosurvival kinases in the same brain region as important endogenous defense mechanisms that protect the developing

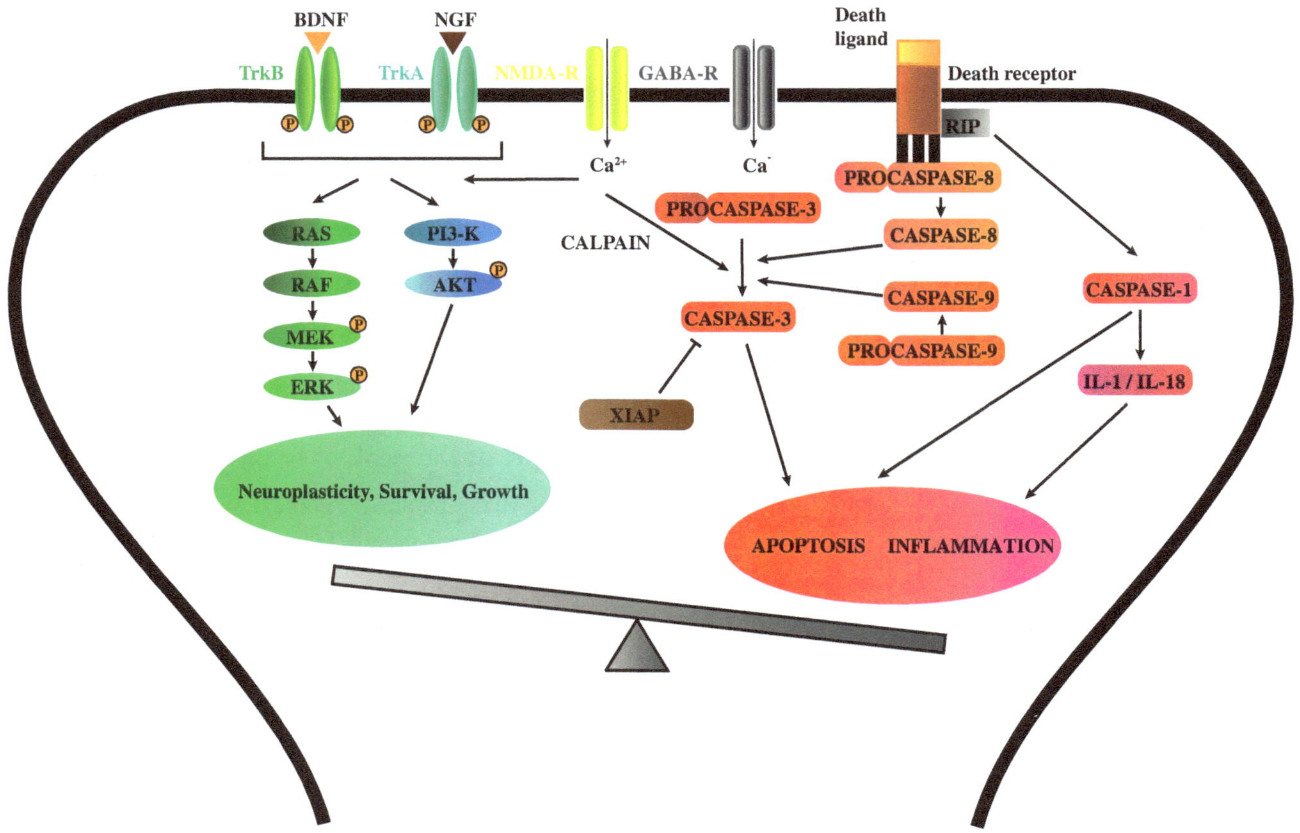

FIG. 1 Molecular players involved in the delicate balance between the *survival* and synaptic integration, at the one side, and apoptotic and *inflammatory response*, at the other side, in the developing brain after propofol anesthesia exposure.

brain from potential damage (Fluoro-Jade B staining did not reveal neuronal degeneration despite the observed transient increase in the expression of the caspase-3 active fragment in both examined brain regions) (Popic et al., 2012). Moreover, in P14 rats, propofol anesthesia exposure promoted both proapoptotic (an increase in TNFα, a proinflammatory cytokine able to induce apoptosis, and caspase-8, an initiator protease in apoptosis mediated by death receptors) and antiapoptotic signals (the nuclear factor-κB (Bhakar et al., 2002) and the XIAP) especially in the thalamus (Popic et al., 2015). Overall, these data showed that in healthy developing brain, propofol anesthesia–induced neurotoxicity is determined by age-dependent peculiarities in the balance between competing pro- and antiapoptotic signals (Fig. 1).

Neuronal activity and synaptic plasticity

There is a suggestion that during early development, attenuation of neuronal activity could be a signal per se for alterations in both neuronal viability and morphology, and synaptic remodeling has been added to the list of potential mechanisms through which anesthetics could induce long-lasting changes in the developing brain (De Roo et al., 2009; Lu et al., 2006).

The research by Briner et al. (2011) proved in vivo morphologic evidence that even a single-bolus exposure to propofol can rapidly induce lasting structural modifications of developing cortical circuitry in a developmental stage-dependent manner, i.e., decreasing spine density in prefrontal cortical pyramidal neurons when administered to rat pups until P10 and increasing the number of synaptic contacts if animals are older (P15 to P30). Our findings indicate that single exposure to propofol anesthesia in neonatal rat brain could be related with the nonfunctional synaptic contacts, i.e., decreased expression of presynaptic protein synaptophysin and increased expression of drebrin (a postsynaptic protein particularly concentrated in dendritic spines) specifically in the thalamus, during a period of diminished neuronal activity in this brain region (2–4 h after anesthesia exposure), while such changes were not observed in the cortex that, actually, did not show negative changes in neuronal activity (Pesic et al., 2015). Overall, these data indicate that decreased neuronal activity could favor excitatory synaptic contacts with the goal of helping to revive/activate the cells (homeostatic synaptic scaling; Turrigiano, 2012).

Molecular autograph of longer exposures to propofol at the peak of the brain growth spurt

Our findings also showed that longer exposures (up to 6 h) to propofol upregulate NGF precursor (proNGF) and $p75^{NTR}$ protein expression, especially in the thalamus, elevate expression of Fas and TNF1 receptors (members of TNF-death receptor family) and their ligands FasL and TNFα, which are part of extrinsic apoptotic pathway (Milanovic et al., 2014, 2016). Upregulation of downstream caspase-8 lead to caspase-3 activation directly or indirectly, by converging to mitochondrial apoptotic pathway. Propofol decreased transcription of mitochondrial antiapoptotic Bcl-2 and Bcl-x genes what might be sufficient to misbalance in favor of proapoptotic signals carried by unchanged Bax and Bad levels (Milanovic et al., 2016). General anesthetics were reported to be particular damaging to mitochondria and endoplasmic reticulum, the main stores of Ca^{2+} in the cell (Jevtovic-Todorovic, 2018). Longer exposures to propofol at the peak of the brain growth spurt activate calpains, Ca^{2+}-dependent cysteine proteases, what we confirmed by generation of α-II-spectin fragment of specific molecular weight representing the signature product of calpain proteolytic activity (Milanovic et al., 2010). Beside changing the spectrin, an actin-binding protein in the cytoskeleton, propofol promotes generation of fractin, a marker of caspase-mediated proteolytic degradation of actin (Creeley et al., 2013). Changes in cytoskeleton inhibit axonal elongation or induce growth cone collapse eventually resulting in apoptosis (Pearn et al., 2012). Along with the activation of proapoptotic pathways, we also observed in elevation of prosurvival molecules Akt and XIAP what, to a certain extent, can explain absence of cell death despite caspase-3 activation in P7 rats (Milanovic et al., 2014). On the other side, this can indicate regulated participation of caspase-3 in nonapoptotic cellular events in synapses and spines that contributes to sublethal morphogenic changes, which may interfere with neuronal plasticity (with an important influence of brain region examined and anesthesia duration), circuitry organization and functional connectivity in the developing brain, i.e., brain functions overall (Milanovic et al., 2017).

At the initial studies, the research was oriented mainly on detrimental effects of anesthesia on neuronal cells but later studies questioned whether glial cells also suffer. Thus, it was reported (Creeley et al., 2013) that 5 h-long propofol anesthesia increases apoptotic degeneration in both neurons and oligodendrocytes in P6 nonhumane primates but to a lesser extent than was previously reported with isoflurane (Brambrink et al., 2012). However, other study found that propofol did not harm other cells (astrocytes, oligodendrocytes, neural stem cells) than neurons in P7 mice (Yan et al., 2017). The controversial findings may be due to vulnerability of neural cells from different species due to different internal signaling (regulation of oligodendrocyte differentiation and myelination), doses and exposure time to propofol (anesthetic or subanesthetic), and the exact developmental age across different species.

Propofol and age-related brain pathology

Majority of studies focused on an anesthetics' ability to induce neuronal apoptosis and to potentiate Alzheimer's disease–associated pathology, including defective amyloid protein pathway, tau protein phosphorylation, and neuroinflammation (Wu, Zhao, Weng, & Ma, 2019). In contrast to isofluran and desflurane that promote amyloid-beta oligomerization even at clinically relevant concentration in transgenic animals, it was found that propofol enhanced oligomerization at very high concentration only (Mandal & Fodale, 2009). Propofol has also been shown to promote phosphorylation of tau protein, with repeated exposure resulting in hippocampal hyperphosphorylation and significant impairment of spatial memory (Whittington et al., 2011). However, in animal studies conducted to examine the deleterious effect of general anesthesia in the absence of surgical insult, it was reported that propofol preserved long-term spatial memory and learning in aged, 18-month-old rats (Lee et al., 2008). Also, clinical studies reported that propofol-based anesthesia, as compared to isoflurane, decreased the incidence of delayed neurocognitive recovery, a form of postoperative cognitive dysfunction in older adults (Zhang et al., 2018). Generally, exposure of the aged brain to general anesthesia can be a risk of the long-lasting cognitive impairment, with certain anesthetics being more favorable, though neuroprotective property of these agents in brain injury is also increasingly recognized (Wu et al., 2019).

Applications to other areas

Molecular and functional consequences of propofol anesthesia usage have been widely addressed with respect to the extremes of age while the outcomes in adolescence, regardless the fact that it is second only to infancy with respect to the neural changes that occur (Fuhrmann, Knoll, & Blakemore, 2015), have been first accentuated in a few papers of our group (Pavkovic et al., 2017; Pavkovic, Milanovic, Ruzdijic, Kanazir, & Pesic, 2018; Pavkovic, Potrebic, Kanazir, & Pesic, 2020). Using peripubertal rats as a rodent model of adolescence, our studies showed that propofol anesthesia produces transient and brain region–specific changes in the expression/phosphorylation of biochemical markers already described as essential in the molecular pathways underlying drug addiction and development of behavioral sensitization (CaMKIIα, DARPP-32, FosB; Pavkovic et al., 2017), and alters the expression of NMDA receptor subunits NR2A and NR2B, PSD-95, and N-cadherin (Pavkovic et al., 2020), thus leaving transient synaptic trace recognized as early form of synaptic plasticity related to passive drug exposure (Hopf, 2017; van Huijstee & Mansvelder, 2014). Behavioral testing showed that propofol-exposed peripubertal rats after awaking favors risk-taking and prefer novel/intense stimuli thus repairing otherwise present motivational deficiency (Pavkovic et al., 2020). The treatment also influenced (a) activity-dependent processes in the dorsal hippocampus including BDNF synthesis, the TrkB activation, and the level of Egr-1, which are important for storage and availability of long-term memory; (b) ERK1/2 activity, which is important for retrieval performance in a test session; and (c) promoted the activity of CaMK IIα (phosphorylated at threonine 286) that is related to memory destabilization following retrieval and increases the expression of FosB/ΔFosB, overall affecting elementary learning forms in the spatial and nonspatial nonaversive learning tasks within days after the treatment (Pavkovic et al., 2018).

It is of importance to accentuate these findings because, due to favorable recovery profile, propofol holds a central place in day-case anesthesia for diagnostic and therapeutic procedures in children (Millar et al., 2014). Notably, recent work of another research group showed that isoflurane exposure in adolescent rats elicited atypical dendritic spine morphology in the prefrontal cortex and hippocampus, hypermaturation of extrasynaptic $GABA_A$ receptor inhibition, immediate and persistent anxiety- and impulsive-like responding and delayed cognitive impairment into adulthood (Landin et al., 2019), as well as exacerbation of adolescent-typical alcohol responding (Landin, Gore-Langton, Varlinskaya, Spear, & Werner, 2020). Overall, all the above-mentioned findings suggest that the developmental window for susceptibility to the functional impairments/malfunctioning after general anesthesia exposure may be longer than previously thought, highlighting the area of adolescent health (Fig. 2).

Other agents of interest

Postanesthesia-related molecular events in the brain, in addition to direct responses to the treatment, include molecular traces of the awakening process that need not be a mirror image of anesthetic induction (reviewed in Kelz, Garcia, Mashour, & Solt, 2019). The role of dopamine in promoting wakefulness has been underappreciated until recently when it was demonstrated that for propofol and isoflurane ($GABA_A$ agent, one of the most widely used inhalational volatile anesthetic) the dopaminergic neurons that originate in the VTA are the important source of arousal-promoting dopamine, suggesting that activating this pathway at the end of surgery may provide a novel approach to accelerate recovery from general anesthesia (Solt et al., 2014). Considering the role of mesolimbic dopamine in mood and motivation, this can explain pleasant mood during recovery from the anesthetic doses of propofol in rats (Pain, Oberling, Sandner, & Di Scala, 1996) and in the patients (Brechmann et al., 2018), as well as addictive potential of propofol (Li, Xiao, Xiong, Delphin, & Ye, 2008; Zacny, Lichtor, Thompson, & Apfelbaum, 1993). On the other side, these facts have not been reported for isoflurane, which is in agreement with findings that lesions of dopaminergic neurons in the VTA do not contribute significantly to emergence from isoflurane but prolonged the recovery time of propofol without affecting its onset time (Zhou et al., 2015). Recent study by Dr Hemmings' group (Torturo, Zhou, Ryan, & Hemmings, 2019) found that isoflurane inhibits synaptic vesicle exocytosis from midbrain dopamine neurons providing a support for the role for dopaminergic VTA neurons in modulating isoflurane-induced unconsciousness. Overall, these findings indicate that the mesocorticolimbic dopamine system is not affected at the same way by anesthetics acting as $GABA_A$ receptor agents, while for some of them, it could be involved in the induction of anesthesia state for others; activation of this system could be particularly important for the emergence. Considering wide usage of propofol and isoflurane in clinical practice, it is of significance to decipher their influence on mesocorticolimbic dopamine system, especially taking into account novel findings that both could have a role in a treatment of depression (Tadler & Mickey, 2018), a disease that is associated with altered cellular function in the VTA.

Mini-dictionary of terms

Intralipid. A source of calories and essential fatty acids for intravenous nutrition available as intralipid 10%, intralipid 20%, and intralipid 30%, with the percentage indicating the relative amount of lipids.

MK-801. An extremely potent noncompetitive antagonist of the NMDA receptor channel when it is in the open state, as the binding site is within the ion channel itself.

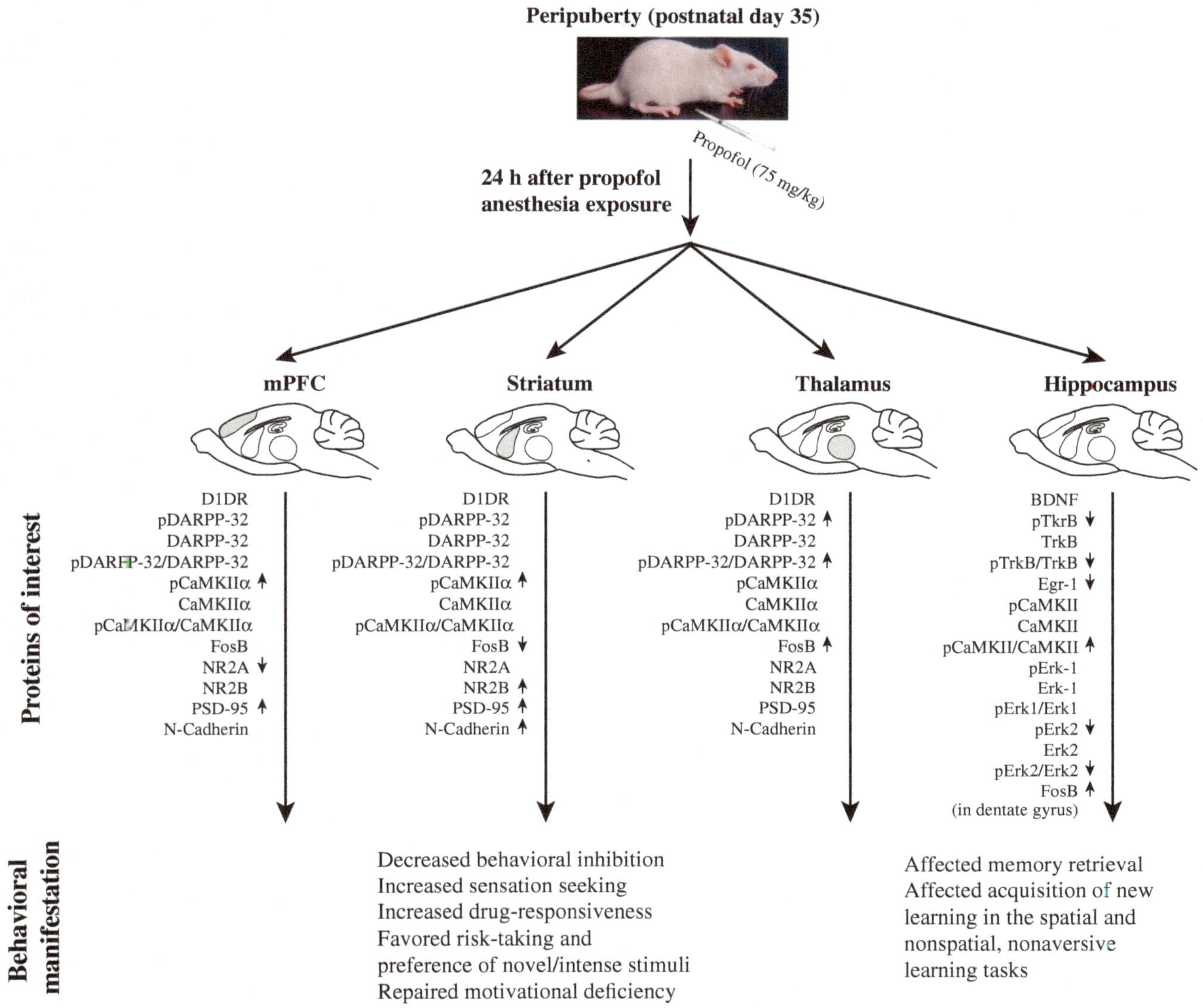

FIG. 2 Brain molecular changes and behavioral manifestations induced by propofol anesthesia exposure in peripubertal rats.

ΔFosB. The truncated splice variant of FosB and a member of the Fos family of transcription factors. Posttranslational modifications increase the stability of ΔFosB, which is implicated in addiction.
Fluoro-Jade B. An anionic fluorescein derivative used for histological staining of dead or dying neurons.
Dendritic spines. Small membranous protrusions in neuronal dendrites and the major postsynaptic targets of excitatory synapses in the central nervous system. They are extremely rich in the actin cytoskeleton.
Behavioral sensitization. A greater behavioral response to one drug due to previous exposure to the same or different drug (behavioral cross-sensitization). It is implicated in drug abuse liability.
Memory retrieval. The use of learned information. The fundamental component of memory processing.

Key facts of propofol anesthesia and molecular changes in the brain

- Propofol rapidly crosses the blood-brain barrier resulting in rapid action, short duration of action, and minimal side effects that have made it the most widely used anesthetic agent.
- The $GABA_A$ receptor has a key role in hypnosis and immobilization under propofol.
- By promoting the response of inhibitory synapses, propofol modulates neuronal activity and neural functions.

- Propofol could induce developmental neurotoxicity and affect dendrite morphology and synaptic plasticity.
- Exposure of the aged brain to propofol anesthesia could be a risk of the long-lasting cognitive impairment.
- Adolescence, a critical period of development, is underestimated regarding brain molecular changes associated with anesthesia exposure.
- Dopamine released by the VTA plays an important role in restoring consciousness after propofol anesthesia.

Summary points

- Propofol influences activity-dependent molecular events and processes that result in long-term morphological and functional alterations especially at the extremes of age.
- Affected crosstalk between growth factor signaling pathways and the apoptotic pathways is important during the critical periods of brain development when apoptosis plays a fundamental role in regulating cell fate.
- Age-dependent peculiarities in the balance between competing pro- and antiapoptotic signals determine propofol-induced neurotoxicity.
- Neuronal age rather than the age of the whole organism is important for the impact of propofol on activity-dependent development.
- Due to a more favorable outcome in terms of molecular and cellular changes, propofol anesthesia may be associated with better cognitive outcomes in the older vulnerable brain compared to inhaled anesthesia.
- In adolescent brain, propofol anesthesia produces transient and brain region–specific alterations in the molecular pathways and synaptic changes related to drug exposure.
- The involvement of the VTA dopamine neurons in the emergence from propofol anesthesia might contribute to its anti-depressant properties, overall supporting additional research on context-dependent influence of general anesthetics on neural networks and brain functions.

References

Alkire, M. T., Hudetz, A. G., & Tononi, G. (2008). Consciousness and anesthesia. *Science*, *322*, 876–880.

Bhakar, A. L., Tannis, L. L., Zeindler, C., Russo, M. P., Jobin, C., Park, D. S., et al. (2002). Constitutive nuclear factor-kappa B activity is required for central neuron survival. *Journal of Neuroscience*, *22*, 8466–8475.

Brambrink, A. M., Back, S. A., Riddle, A., Gong, X., Moravec, M. D., Dissen, G. A., et al. (2012). Isoflurane-induced apoptosis of oligodendrocytes in the neonatal primate brain. *Annals of Neurology*, *72*, 525–535.

Brechmann, T., Maier, C., Kaisler, M., Vollert, J., Schmiegel, W., Pak, S., et al. (2018). Propofol sedation during gastrointestinal endoscopy arouses euphoria in a large subset of patients. *United European Gastroenterology Journal*, *6*, 536–546.

Briner, A., Nikonenko, I., De Roo, M., Dayer, A., Muller, D., & Vutskits, L. (2011). Developmental stage-dependent persistent impact of propofol anesthesia on dendritic spines in the rat medial prefrontal cortex. *Anesthesiology*, *115*, 282–293.

Brown, E. N., Lydic, R., & Schiff, N. D. (2010). General anesthesia, sleep, and coma. *New England Journal of Medicine*, *363*, 2638–2650.

Buggy, D. J., Nicol, B., Rowbotham, D. J., & Lambert, D. G. (2000). Effects of intravenous anesthetic agents on glutamate release: A role for GABAA receptor-mediated inhibition. *Anesthesiology*, *92*, 1067–1073.

Creeley, C., Dikranian, K., Dissen, G., Martin, L., Olney, J., & Brambrink, A. (2013). Propofol-induced apoptosis of neurones and oligodendrocytes in fetal and neonatal rhesus macaque brain. *British Journal of Anaesthesia*, *110*(Suppl. 1), i29–i38.

De Roo, M., Klauser, P., Briner, A., Nikonenko, I., Mendez, P., Dayer, A., et al. (2009). Anesthetics rapidly promote synaptogenesis during a critical period of brain development. *PLoS One*, *4*, e7043.

Dutta, S., & Ebling, W. F. (1997). Emulsion formulation reduces propofol's dose requirements and enhances safety. *Anesthesiology*, *87*, 1394–1405.

Egan, T. D. (2010). Exploring the frontiers of propofol formulation strategy: Is there life beyond the milky way? *British Journal of Anaesthesia*, *104*, 533–535.

Frenkel, C., & Urban, B. W. (1991). Human brain sodium channels as one of the molecular target sites for the new intravenous anaesthetic propofol (2,6-diisopropylphenol). *European Journal of Pharmacology*, *208*, 75–79.

Fuhrmann, D., Knoll, L. J., & Blakemore, S. J. (2015). Adolescence as a sensitive period of brain development. *Trends in Cognitive Sciences*, *19*, 558–566.

Glen, J. B., & James, R. (1977). *2,6-Diisopropylphenol as an anaesthetic agent* (pp. 1–10). London: United States Patent and Trademark Office.

Hopf, F. W. (2017). Do specific NMDA receptor subunits act as gateways for addictive behaviors? *Genes, Brain, and Behavior*, *16*, 118–138.

Hudetz, A. G. (2012). General anesthesia and human brain connectivity. *Brain Connectivity*, *2*, 291–302.

Irifune, M., Takarada, T., Shimizu, Y., Endo, C., Katayama, S., Dohi, T., et al. (2003). Propofol-induced anesthesia in mice is mediated by gamma-aminobutyric acid-A and excitatory amino acid receptors. *Anesthésie et Analgésie*, *97*, 424–429. table of contents.

Jevtovic-Todorovic, V. (2018). Exposure of developing brain to general anesthesia: What is the animal evidence? *Anesthesiology*, *128*, 832–839.

Jurd, R., Arras, M., Lambert, S., Drexler, B., Siegwart, R., Crestani, F., et al. (2003). General anesthetic actions in vivo strongly attenuated by a point mutation in the GABA(A) receptor beta3 subunit. *FASEB Journal*, *17*, 250–252.

Karen, T., Schlager, G. W., Bendix, I., Sifringer, M., Herrmann, R., Pantazis, C., et al. (2013). Effect of propofol in the immature rat brain on short- and long-term neurodevelopmental outcome. *PLoS One*, *8*, e64480.

Kelz, M. B., Garcia, P. S., Mashour, G. A., & Solt, K. (2019). Escape from oblivion: Neural mechanisms of emergence from general anesthesia. *Anesthesia and Analgesia*, *128*, 726–736.

Krzisch, M., Sultan, S., Sandell, J., Demeter, K., Vutskits, L., & Toni, N. (2013). Propofol anesthesia impairs the maturation and survival of adult-born hippocampal neurons. *Anesthesiology*, *118*, 602–610.

Landin, J. D., Gore-Langton, J. K., Varlinskaya, E. I., Spear, L. P., & Werner, D. F. (2020). General anesthetic exposure during early adolescence persistently alters ethanol responses. *Alcoholism, Clinical and Experimental Research*, *44*, 611–619.

Landin, J. D., Palac, M., Carter, J. M., Dzumaga, Y., Santerre-Anderson, J. L., Fernandez, G. M., et al. (2019). General anesthetic exposure in adolescent rats causes persistent maladaptations in cognitive and affective behaviors and neuroplasticity. *Neuropharmacology*, *150*, 153–163.

Lee, I. H., Culley, D. J., Baxter, M. G., Xie, Z., Tanzi, R. E., & Crosby, G. (2008). Spatial memory is intact in aged rats after propofol anesthesia. *Anesthesia and Analgesia*, *107*, 1211–1215.

Li, K. Y., Xiao, C., Xiong, M., Delphin, E., & Ye, J. H. (2008). Nanomolar propofol stimulates glutamate transmission to dopamine neurons: A possible mechanism of abuse potential? *Journal of Pharmacology and Experimental Therapeutics*, *325*, 165–174.

Liu, Q.-Z., Hao, M., Zhou, Z.-Y., Ge, J.-L., Wu, Y.-C., Zhao, L.-L., et al. (2019). Propofol reduces synaptic strength by inhibiting sodium and calcium channels at nerve terminals. *Protein & Cell*, *10*, 688–693.

Lu, L. X., Yon, J. H., Carter, L. B., & Jevtovic-Todorovic, V. (2006). General anesthesia activates BDNF-dependent neuroapoptosis in the developing rat brain. *Apoptosis*, *11*, 1603–1615.

Mandal, P. K., & Fodale, V. (2009). Smaller molecular-sized anaesthetics oligomerize Abeta peptide simulating Alzheimer's disease: A relevant issue. *European Journal of Anaesthesiology*, *26*, 805–806.

Milanovic, D., Pesic, V., Loncarevic-Vasiljkovic, N., Avramovic, V., Tesic, V., Jevtovic-Todorovic, V., et al. (2017). Neonatal propofol anesthesia changes expression of synaptic plasticity proteins and increases stereotypic and anxyolitic behavior in adult rats. *Neurotoxicity Research*, *32*, 247–263.

Milanovic, D., Pesic, V., Loncarevic-Vasiljkovic, N., Pavkovic, Z., Popic, J., Kanazir, S., et al. (2016). The Fas ligand/Fas death receptor pathways contribute to propofol-induced apoptosis and neuroinflammation in the brain of neonatal rats. *Neurotoxicity Research*, *30*, 434–452.

Milanovic, D., Pesic, V., Popic, J., Tanic, N., Kanazir, S., Jevtovic-Todorovic, V., et al. (2014). Propofol anesthesia induces proapoptotic tumor necrosis factor-alpha and pro-nerve growth factor signaling and prosurvival Akt and XIAP expression in neonatal rat brain. *Journal of Neuroscience Research*, *92*, 1362–1373.

Milanovic, D., Popic, J., Pesic, V., Loncarevic-Vasiljkovic, N., Kanazir, S., Jevtovic-Todorovic, V., et al. (2010). Regional and temporal profiles of calpain and caspase-3 activities in postnatal rat brain following repeated propofol administration. *Developmental Neuroscience*, *32*, 288–301.

Millar, K., Bowman, A. W., Burns, D., McLaughlin, P., Moores, T., Morton, N. S., et al. (2014). Children's cognitive recovery after day-case general anesthesia: A randomized trial of propofol or isoflurane for dental procedures. *Paediatric Anaesthesia*, *24*, 201–207.

Olcese, R., Usai, C., Maestrone, E., & Nobile, M. (1994). The general anesthetic propofol inhibits transmembrane calcium current in chick sensory neurons. *Anesthesia and Analgesia*, *78*, 955–960.

Pain, L., Oberling, P., Sandner, G., & Di Scala, G. (1996). Effect of propofol on affective state as assessed by place conditioning paradigm in rats. *Anesthesiology*, *85*, 121–128.

Patel, S., Wohlfeil, E. R., Rademacher, D. J., Carrier, E. J., Perry, L. J., Kundu, A., et al. (2003). The general anesthetic propofol increases brain N-arachidonylethanolamine (anandamide) content and inhibits fatty acid amide hydrolase. *British Journal of Pharmacology*, *139*, 1005–1013.

Pavkovic, Z., Milanovic, D., Ruzdijic, S., Kanazir, S., & Pesic, V. (2018). The influence of propofol anesthesia exposure on nonaversive memory retrieval and expression of molecules involved in memory process in the dorsal hippocampus in peripubertal rats. *Paediatric Anaesthesia*, *28*, 537–546.

Pavkovic, Z., Potrebic, M., Kanazir, S., & Pesic, V. (2020). Motivation, risk-taking and sensation seeking behavior in propofol anesthesia exposed peripubertal rats. *Progress in Neuro-Psychopharmacology & Biological Psychiatry*, *96*, 109733.

Pavkovic, Z., Smiljanic, K., Kanazir, S., Milanovic, D., Pesic, V., & Ruzdijic, S. (2017). Brain molecular changes and behavioral alterations induced by propofol anesthesia exposure in peripubertal rats. *Paediatric Anaesthesia*, *27*, 962–972.

Pearn, M. L., Hu, Y., Niesman, I. R., Patel, H. H., Drummond, J. C., Roth, D. M., et al. (2012). Propofol neurotoxicity is mediated by p75 neurotrophin receptor activation. *Anesthesiology*, *116*, 352–361.

Pesic, V., Milanovic, D., Popic, J., Smiljanic, K., Tesic, V., Kanazir, S., et al. (2015). Neonatal propofol anesthesia modifies activity-dependent processes and induces transient hyperlocomotor response to d-amphetamine during adolescence in rats. *International Journal of Developmental Neuroscience*, *47*, 266–277.

Pesic, V., Milanovic, D., Tanic, N., Popic, J., Kanazir, S., Jevtovic-Todorovic, V., et al. (2009). Potential mechanism of cell death in the developing rat brain induced by propofol anesthesia. *International Journal of Developmental Neuroscience*, *27*, 279–287.

Ponten, E., Fredriksson, A., Gordh, T., Eriksson, P., & Viberg, H. (2011). Neonatal exposure to propofol affects BDNF but not CaMKII, GAP-43, synaptophysin and tau in the neonatal brain and causes an altered behavioural response to diazepam in the adult mouse brain. *Behavioural Brain Research*, *223*, 75–80.

Popic, J., Pesic, V., Milanovic, D., Loncarevic-Vasiljkovic, N., Smiljanic, K., Kanazir, S., et al. (2015). Induction of TNF-alpha signaling cascade in neonatal rat brain during propofol anesthesia. *International Journal of Developmental Neuroscience*, *44*, 22–32.

Popic, J., Pesic, V., Milanovic, D., Todorovic, S., Kanazir, S., Jevtovic-Todorovic, V., et al. (2012). Propofol-induced changes in neurotrophic signaling in the developing nervous system in vivo. *PLoS One*, *7*, e34396.

Ratnakumari, L., & Hemmings, H. C., Jr. (1997). Effects of propofol on sodium channel-dependent sodium influx and glutamate release in rat cerebrocortical synaptosomes. *Anesthesiology*, *86*, 428–439.

Sahinovic, M. M., Struys, M., & Absalom, A. R. (2018). Clinical pharmacokinetics and pharmacodynamics of propofol. *Clinical Pharmacokinetics*, *57*, 1539–1558.

Solt, K., Van Dort, C. J., Chemali, J. J., Taylor, N. E., Kenny, J. D., & Brown, E. N. (2014). Electrical stimulation of the ventral tegmental area induces reanimation from general anesthesia. *Anesthesiology*, *121*, 311–319.

Tadler, S. C., & Mickey, B. J. (2018). Emerging evidence for antidepressant actions of anesthetic agents. *Current Opinion in Anaesthesiology*, *31*, 439–445.

Torturo, C. L., Zhou, Z. Y., Ryan, T. A., & Hemmings, H. C. (2019). Isoflurane inhibits dopaminergic synaptic vesicle exocytosis coupled to $Ca_V2.1$ and $Ca_V2.2$ in rat midbrain neurons. *eNeuro*, *6*(1). https://doi.org/10.1523/ENEURO.0278-18.2018. ENEURO.0278-18.2018.

Turrigiano, G. (2012). Homeostatic synaptic plasticity: Local and global mechanisms for stabilizing neuronal function. *Cold Spring Harbor Perspectives in Biology*, *4*, a005736.

van Huijstee, A. N., & Mansvelder, H. D. (2014). Glutamatergic synaptic plasticity in the mesocorticolimbic system in addiction. *Frontiers in Cellular Neuroscience*, *8*, 466.

Vutskits, L., & Xie, Z. (2016). Lasting impact of general anaesthesia on the brain: Mechanisms and relevance. *Nature Reviews. Neuroscience*, *17*, 705–717.

Wang, Y., Yang, E., Wells, M. M., Bondarenko, V., Woll, K., Carnevale, V., et al. (2018). Propofol inhibits the voltage-gated sodium channel NaChBac at multiple sites. *Journal of General Physiology*, *150*, 1317–1331.

Whittington, R. A., Virag, L., Marcouiller, F., Papon, M. A., El Khoury, N. B., Julien, C., et al. (2011). Propofol directly increases tau phosphorylation. *PLoS One*, *6*, e16648.

Wu, L., Zhao, H., Weng, H., & Ma, D. (2019). Lasting effects of general anesthetics on the brain in the young and elderly: "Mixed picture" of neurotoxicity, neuroprotection and cognitive impairment. *Journal of Anesthesia*, *33*, 321–335.

Yan, Y., Qiao, S., Kikuchi, C., Zaja, I., Logan, S., Jiang, C., et al. (2017). Propofol induces apoptosis of neurons but not astrocytes, oligodendrocytes, or neural stem cells in the neonatal mouse hippocampus. *Brain Sciences*, *7*(10), 130. https://doi.org/10.3390/brainsci7100130.

Yon, J. H., Daniel-Johnson, J., Carter, L. B., & Jevtovic-Todorovic, V. (2005). Anesthesia induces neuronal cell death in the developing rat brain via the intrinsic and extrinsic apoptotic pathways. *Neuroscience*, *135*, 815–827.

Zacny, J. P., Lichtor, J. L., Thompson, W., & Apfelbaum, J. L. (1993). Propofol at a subanesthetic dose may have abuse potential in healthy volunteers. *Anesthesia and Analgesia*, *77*, 544–552.

Zafra, F., Castren, E., Thoenen, H., & Lindholm, D. (1991). Interplay between glutamate and gamma-aminobutyric acid transmitter systems in the physiological regulation of brain-derived neurotrophic factor and nerve growth factor synthesis in hippocampal neurons. *Proceedings of the National Academy of Sciences of the United States of America*, *88*, 10037–10041.

Zhang, Y., Shan, G. J., Zhang, Y. X., Cao, S. J., Zhu, S. N., Li, H. J., et al. (2018). Propofol compared with sevoflurane general anaesthesia is associated with decreased delayed neurocognitive recovery in older adults. *British Journal of Anaesthesia*, *121*, 595–604.

Zhou, X., Wang, Y., Zhang, C., Wang, M., Zhang, M., Yu, L., et al. (2015). The role of dopaminergic VTA neurons in general anesthesia. *PLoS One*, *10*, e0138187.

Chapter 7

Protein kinase G is a molecular switch for pain

Ying-Ju Sung and Richard T. Ambron
Department of Pathology, Anatomy, and Cell Biology, Vagelos College of Physicians and Surgeons, Columbia University, New York, NY, United States

Abbreviations

cGMP	cyclic guanosine monophosphate
CNS	central nervous system
CPIP	chronic postischemia pain
CRPS1	type I complex regional pain syndrome
DRG	dorsal root ganglion
LTH	long-term hyperexcitability
LTP	long-term potentiation
PKG	protein kinase G

Introduction

The perception of pain reaches consciousness when neuronal circuits in the brain are activated by inputs from peripheral nociceptive pathways that convey information from the site of an injury or inflammation. Pain is both necessary and informative, but chronic pain is a pathological condition that is mentally and physically debilitating and adversely affects the quality of life. More than 20 million people in the United States suffer from chronic pain and efforts to alleviate the pain often fail (Boyer, 2012). The main recourse is analgesics containing opiates that result in sedation, tolerance, and potential addiction. This has created an epidemic that has resulted in thousands of deaths and an alarming number of emergency room visits.

The ideal solution would be to develop an analgesic to chronic pain that targets an essential and unique nonopioid component in the pain pathway. This now appears possible as we begin to identify the proteins responsible for the transition of pain from a normal process that informs us of a lesion to the condition where the pain has become chronic. However, the primary focus to date has been on components within the central nervous system (CNS), which vastly complicates drug development given the presence of the blood-brain barrier and the possibility of severe side effects from alterations in brain chemistry. Largely overlooked are processes in the periphery that contribute to chronic pain that are more accessible to treatment. Below is a discussion of how our understanding of these processes has led to the development of a promising new class of analgesics to treat chronic pain.

Pain is perceived via inputs from nociceptive circuits that are adaptive

Nociception

The brain receives information about an injury or inflammation from the skin or internal organs via dedicated nociceptive (pain) pathways. Each pathway consists of first-order (primary) nociceptive neurons whose cell bodies reside either in a dorsal root ganglion (DRG) or a ganglion associated with certain cranial nerves. The single axon that emerges from each cell body divides into a long peripheral and short central branch. Some peripheral branches course through nerves that terminate in the skin and associated structures, whereas others travel within nerves to internal organs. The central branches enter the CNS where they form excitatory synapses on second-order neurons located in laminae within the dorsal horn in the

The Neurobiology, Physiology, and Psychology of Pain. https://doi.org/10.1016/B978-0-12-820589-1.00007-5

spinal cord or the spinal nucleus in the brainstem. These neurons disseminate the inputs to centers in the CNS, such as the thalamus and somatosensory cortex, that contribute to the perception of the pain and an awareness of its origin.

Whereas the primary function of the first- and second-order neurons is to inform the central circuits of a lesion, these neurons adapt to the nature and severity of the lesion, as pain is necessary for survival. In fact, pain occurs in phases, each arising from a unique adaptive response. Consequently, a guiding principle in the development of analgesics is to identify targets that will prevent the first- and second-order neurons from communicating with the higher pain centers. This requires an understanding of the molecular events that mediate these adaptations.

The acute perception of pain

An injury or other noxious stimulus elicits a transient barrage of action potentials in the peripheral branches of the first-order neurons. The potentials propagate along the central branches of the first-order neurons to synapses on the second-order neurons. The number and duration of the initial lesion-induced action potential barrage indicates the extent of the lesion. The action potentials elicited in the second-order neurons then propagate along their axons to the thalamus where they result in an acute perception of pain that alerts us to the lesion in the periphery. Information is subsequently disseminated to areas of the CNS that regulate attentiveness and the emotional consequences of pain.

The prolonged perception of pain

If the injury is evaluated as being of little consequence, the pain diminishes and is soon forgotten. With a more serious injury, however, the pain must be prolonged to ensure a continued awareness of the lesion so as to protect the site from further injury. The prolongation of pain is affected by changes in the efficacy of the synapse between the first- and second-order neurons such that any subsequent inputs from the site are enhanced. Thus, the arrival at the synapse of action potentials that would ordinarily cause only a mild pain now produces more severe pain (i.e., hyperalgesia) whereas the arrival of only a few action potentials that would not normally result in pain is perceived as hurtful (i.e., allodynia). Both hyperalgesia and allodynia are forms of potentiation that result from alterations in the electrophysiological properties of the synapse. Two terms refer to this adaptive potentiation. Central sensitization includes all changes in nociceptive neurons without distinguishing between those that directly elicit the perception of pain from those that are extrinsic to the pathway (Woolf, 2011). The other term, long-term potentiation (LTP), refers to an increase in synaptic strength that, in the context of pain, refers only to the changes that occur at the synapses between the first- and second-order neurons. LTP is the term that will be used here.

Long-term potentiation

LTP is manifest as a rapidly induced increase in synaptic strength that persists for at least 30 min after the cessation of the action potential barrage from the lesion site (Sandkuhler, 2010). LTP at synapses between the first- and second-order neurons is essential to the expression of pain in animal models and the perception of pain in humans (Sandkuhler, 2007). Consequently, identifying the molecular components responsible for LTP should provide targets for drugs to alleviate pain.

Several classes of first-order nociceptive neurons provide inputs resulting in LTP and these can be distinguished by their morphology, electrophysiological properties, molecular composition, and by the degree to which their axons are myelinated. LTP is best understood at the synapse between small, lightly myelinated, slowly conducting axons of first-order C-type neurons, and second-order neurons located in lamina I of the dorsal horn. The essential role of these neurons in promoting pain is supported by their absence in humans who have a congenital insensitivity to pain (Bowsher et al., 2009). The importance of C-fiber inputs is also underscored by studies showing that they are responsible for many types of hyperalgesia (Ikeda et al., 2006). LTP is induced by noxious stimulation, e.g., by the subcutaneous administration of capsaicin, or by noxious heat, pinching, or nerve injuries (Sandkuhler & Liu, 1998).

There are two phases of LTP that differ in mechanism and duration. The early phase appears when the initial barrage of action potentials from the lesion is prolonged, indicating that the injury is serious and requires attention. This results in the activation of kinases that phosphorylate proteins at the synapse, altering their function to increase excitability and pain intensity (Ji, Kohno, Moore, & Woolf, 2003). Key events in the development of early phase LTP include the coactivation of NK1 and NK2 receptors, opening of AMPA sodium channels and NMDA voltage-gated calcium channels, and the activation of metabotropic glutamate receptors (Sandkuhler & Gruber-Schoffnegger, 2012). These sensitize the synapse resulting in hyperalgesia and allodynia. Because these changes are posttranslational modifications of existing proteins, they can be readily reversed and the early phase of LTP lasts minutes to hours, depending on the severity of the lesion.

When the injury or inflammation is more extensive, the pain must continue to sustain the awareness of the lesion. This pain is mediated by the late phase of LTP that involves changes in protein transcription in the second-order neurons. Because the phenotype is altered, these changes can persist for a long time. As is the case with the early phase LTP, the late phase changes are activity driven, meaning that they are responses to continued inputs from the periphery (Sandkuhler & Gruber-Schoffnegger, 2012). Several proteins responsible for the increase in excitability have been identified with the hope that one or more will be targets for the development of drugs to alleviate chronic pain (Chen et al., 2014; Marchand et al., 2011; Yang et al., 2004; Zhou et al., 2008). However, there are several problems with this approach. First, the molecular processes responsible for the genesis of the late phase LTP occur in the highly complex CNS environment where other neuronal circuits are also responding to the lesion (Gruber-Schoffnegger et al., 2013). Identifying potential targets is therefore challenging. Second, a target molecule has to be unique to the pain pathway because inhibiting a target common to other CNS pathways will likely result in severe side effects. Third, the drug must enter the CNS, which complicates its design since it must bypass the blood-brain barrier. An even more fundamental issue is that the duration of LTP under controlled experimental conditions does not correlate with the duration of pain that persists for a long time. For example, when human volunteers were subjected to a relatively mild conditioning stimulation the hyperalgesia that was elicited lasted for about a day (Klein, Magerl, & Treede, 2006). Thus, the two phases of LTP ensure that the duration of pain is commensurate with the seriousness of the initiating lesion, but only up to a point.

These experiments showed that the pain perceived via the late phase LTP decrements over time, yet experience tells us that pain resulting from a severe injury or inflammation can last for weeks. It is therefore significant that severe lesions have been found to result in yet another alteration in the nociceptive pathway, one that results in persistent inputs from the lesion site that prolongs the duration of pain, perhaps by extending the late phase of LTP. This alteration is a long-term hyperexcitability (LTH) that appears in the cell bodies of first-order neurons in peripheral ganglia, such as the DRG. Thus, unlike the events responsible for LTP, LTH occurs in neurons in the periphery with accessible targets to alleviate persistent pain (Sung, Walters, & Ambron, 2004).

LTH and persistent pain

LTH is an evolutionarily conserved phenomenon that mediates defensive behaviors in response to serious nerve injury. It is a lesion-induced change in the phenotype of the affected neurons that lowers the threshold for the generation of action potentials. Like LTP at the synapse, action potentials from the lesion site that reach the cell body of these neurons will generate multiple action potentials that will then propagate along the central process to the synapse on the second-order neurons and then to the thalamus. Because LTH is a phenotypic change it can, in theory, last indefinitely. Since the LTH can sustain both allodynia and hyperalgesia, even a light touch or mild heat applied to the area of the lesion will be painful. This type of hypersensitivity is characteristic of many types of chronic pain, and Table 1 shows that LTH is associated with many conditions that produce persistent, as well as chronic pain.

TABLE 1 Chronic and persistent pain conditions that are accompanied by LTH.

1	Axotomy (neuroma)	Abdulla and Smith (2001)
2	Nerve constriction	Zhang, Song, and LaMotte (1999)
3	Colitis	Beyak, Ramji, Krol, Kawaja, and Vanner (2004)
4	Cystitis	Sculptoreanu, Yoshimura, and de Groat (2004), Sculptoreanu, de Groat, Buffington, and Birder (2005)
5	Osteoarthritis	Morales-Aza, Chillingworth, Payne, and Donaldson (2004)
6	Nerve inflammation	Bennett (2004)
7	Diabetes mellitus neuropathies	Dobretsov, Hastings, Romanovsky, Stimers, and Zhang (2003)
8	Metastatic bone cancer	Peters et al. (2005)
9	Ischemia	Bulka, Hao, and Wiesenfeld-Hallin (2002)

Distinction between persistent and chronic pain

However, even persistent pain will disappear when the cause is no longer present. Thus, pain from an inflammation will last only as long as the inflammatory agents are present. Consequently, there must be a mechanism in the first-order neurons that downregulates the LTH. In contrast, chronic pain continues without an obvious external cause because there is a fault in the intrinsic mechanism that shuts down the LTH. For example, neuropathic pain caused by cutting a nerve is chronic because the LTH is maintained and the pain can arise spontaneously. Any effort to attenuate persistent and chronic pain depends on identifying the proteins responsible for the induction of the LTH.

LTH is induced by protein kinase G, a positive injury signal in nociceptive neurons

Identifying the molecular events responsible for the LTH would be difficult, if not impossible, in vertebrates due to the complexity of the nervous system. Given that the LTH is evolutionarily conserved, it was possible to take a reductionist approach and use the relatively simple nervous system of the marine mollusk *Aplysia californica* (Sung & Ambron, 2004). Aplysia has a cluster of sensory neurons that have properties similar to vertebrate first-order nociceptive neurons and exhibit a transcription-dependent LTH when their axons are injured in vivo (Clatworthy, Castro, Budelmann, & Walters, 1994; Walters, Alizadeh, & Castro, 1991) and in vitro (Gunstream, Castro, & Walters, 1995) (Fig. 1).

The finding was that the LTH appears only after a delay indicated that it might be induced by a positive injury signal, i.e., an axonal protein that is activated at the site of an injury and retrogradely transported to the cell soma where it alters transcription (Ambron, Dulin, Zhang, Schmied, & Walters, 1995; Ambron, Zhang, Gunstream, Povelones, & Walters, 1996). The delay would therefore be due to the transit time for the transport. Direct evidence to support this idea was obtained by demonstrating that LTH could be induced by injecting noninjured nerves with axoplasm enriched in retrogradely transported proteins from injured nerves (Ambron et al., 1995). Subsequent studies determined that the retrogradely transported signal in the axon responsible for inducing the LTH was the Aplysia homologue of vertebrate protein kinase G-1α (PKG-1α) (Sung et al., 2004).

PKG-1α is a nociceptive positive injury signal for LTH in rats

Subsequent studies in rats revealed that PKG-1α is present in several classes of DRG neurons but is markedly enriched in the axons of non- or lightly myelinated axons of the C-type nociceptors that exhibit both the late phase LTP and LTH (Sung, Chiu, & Ambron, 2006). PKG-1α has low basal activity in these neurons, but is activated not only by nerve crush but also by an inflammation (Sung et al., 2006). Moreover, PKG-1α was not detected in motor axons or glia, meaning that a PKG-1α antagonist will not affect motor systems.

The idea of PKG-1α as a nociceptive-specific signal is supported by many other studies of pain in rodents. Thus, Chu et al. (2005) reported that thermal hyperalgesia in rats was attenuated by an inhibitor of NOS, the upstream activator of

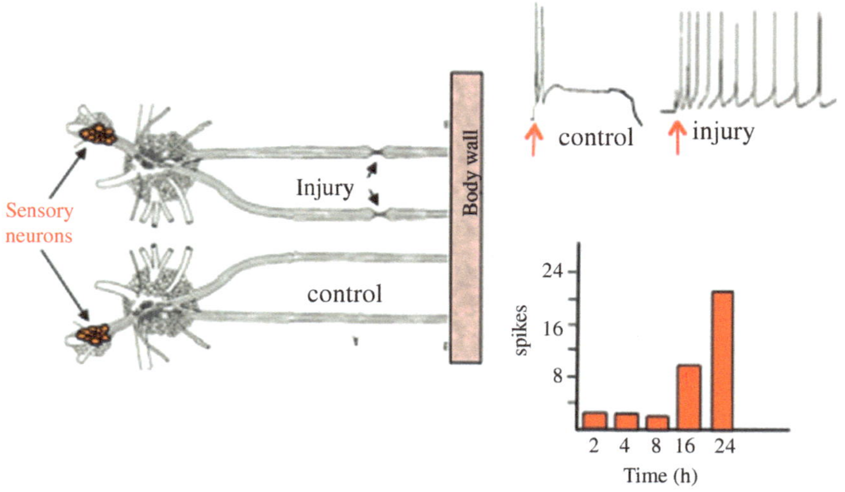

FIG. 1 The development of LTH after nerve injury in the isolated nervous system of *Aplysia californica*. (Left) Part of the nervous system in isolation showing the location of the bilateral cluster of sensory neurons. To induce LTH, the nerves to the body wall that contain axons of the sensory neurons were crushed on one side. The nerves on the other side were untouched. (Top right) A representative result 24 h later showing that stimulus just above threshold to a cell body on the noninjured side resulted in a few action potentials whereas a normalized stimulus to a neuron on the injured resulted in a barrage of action potentials. (Bottom right) A time course study shows that the increase in excitability appeared only after a 16 h delay, confirming the expression of LTH.

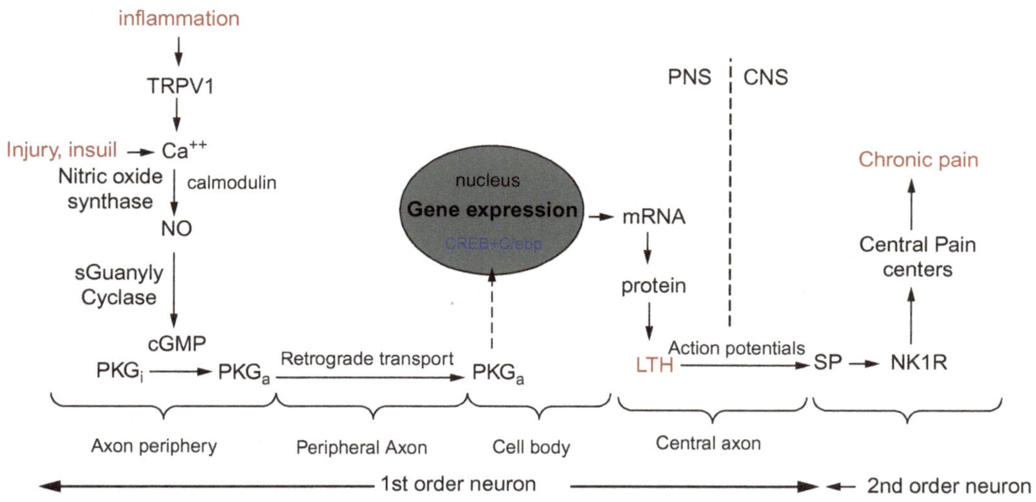

FIG. 2 The induction of LTH caused by an injury or inflammation. Algogenic agents from the inflammation, or physical damage to the axon, initiates a cascade that results in the activation of PKG-1α and its subsequent transport retrogradely to the cell bodies of the affected nociceptive neurons where it is required for the induction of LTH. The LTH amplifies action potential discharge that results in an increase in the release of neurotransmitters, such as Substance P, that bind to receptors second-order neurons in the dorsal horn of the spinal cord. Signals then propagate to pain centers.

PKG-1α, and Morita, Kitayama, Morioka, and Dohi (2008) demonstrated that activating PKG elicits thermal hyperalgesia and allodynia in mice. Finally, a direct connection between PKG, LTH, and pain was obtained by Song et al. (Song, Wang, Gan, & Walters, 2006) who used a rat model of human sciatica, and Liu et al. (Liu, Zhang, Chen, Liu, & Liu, 2014) who used a rat model of bone cancer. Taken together, these findings offer compelling evidence that LTH and the activation of PKG-1α are essential elements in the maintenance of both persistent and chronic pain. Consequently, a drug that can inhibit PKG-1α will block the LTH and alleviate many forms of chronic pain.

Fig. 2 summarizes the role of PKG-1α after a lesion. PKG-1α is present in axoplasm in an inactive form. Following an injury or inflammation, the influx of Ca^{2+} into the axons at the lesion site activates the nitric oxide synthase cascade. This results in the production of cyclic guanosine monophosphate (cGMP) and the activation of PKG-1α. The activation exposes a peptide sequence in PKG-1α that provides access to the retrograde transport machinery and the transport of the kinase back to the cell body where it induces the proteins responsible for the increase in excitability. Since there is a finite time during which the PKG-1α is being transported (Sung et al., 2006) PKG-1α, and by extrapolation, LTH, are not associated with acute pain. Rather, as mentioned earlier, PKG-1α signals the presence of a serious lesion and that the induction of the LTH involves major alterations in the first-order neurons that are designed to maintain an awareness of the injured site during the expected long period of healing. It also implies that the persistence of the pain is dependent upon active PKG-1α. We tested this idea using two animal models of chronic pain.

Persistent activity of PKG-1α in sensory neurons after nerve injury in rats

Painful neuromas are formed when a nerve is severed. Sensitized axon terminals in the neuroma respond to stimuli that cause chronic burning sensations, hot and cold hyperalgesia and chronic tactile allodynia. Neuroma pain is particularly resistant to standard treatments. Significantly, transecting a nerve in a rat leads to the development of LTH (Abdulla & Smith, 2001). To examine a role for PKG-1α, we severed the common fibular nerve in rats in vivo, waited 30d, by which time a neuroma had formed, and then assayed the DRG. There was a significant elevation of PKG-1α activity in the ganglia relative to the control (Fig. 3A).

Type I complex regional pain syndrome (CRPS-1), which develops after fractures, tissue trauma, or surgery, is accompanied by chronic sensory hypersensitivity, allodynia, and uncontrollable spontaneous pain that is alleviated most effectively by opiate analgesics. There is no direct nerve injury, but rather a lesion due to ischemia and/or inflammation. Recently, a chronic postischemia pain (CPIP) model in rats was developed that evokes a mechanical allodynia comparable to the inflammatory/ischemic neuropathic pain in CRPS-I (Coderre, Xanthos, Francis, & Bennett, 2004). The pain, which is induced by applying a tight ligature around the distal portion of the rat's hind limb, lasts for weeks, and is readily quantified using a graded series of von Frey fibers. Inhibitors of nitric oxide synthase attenuated the pain, strongly suggesting that PKG is involved (Xanthos, Bennett, & Coderre, 2008). To test this hypothesis, we followed the CPIP protocol and observed the

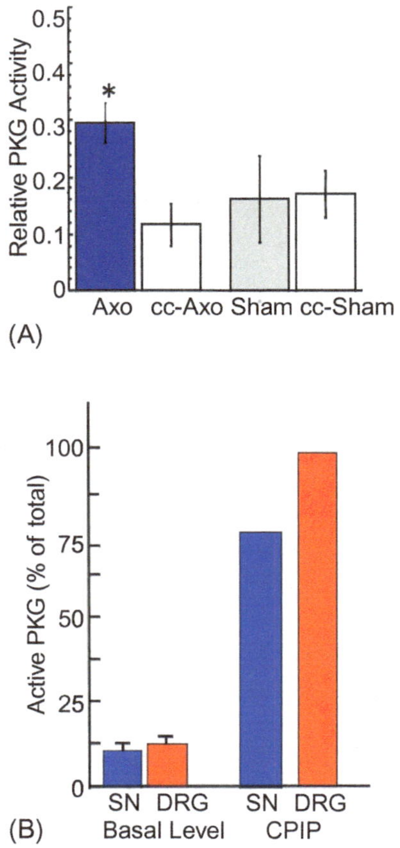

FIG. 3 PKG activation. (A) Activation of PKG in the injured DRG. The right common fibular nerve was isolated and ligated with 6-0 silk and transected. Thirty days later, the L4/L5 DRGs were collected and assayed for PKG activity. Significant PKG activity was present in DRGs removed from the axotomized nerve (Axo, *blue bar*) compared to the contra-lateral uninjured DRG (cc-Axo), $P = .0265$. In contrast, there was no change in the PKG activity in DRGs from the sham-operated animals *(gray bar)* vs the contra-lateral uninjured DRG (cc-sham). $N = 4$ for each group. (B) PKG is robustly activated in the CPIP model of complex regional pain syndrome. The hind paw of three rats was ligated according to the CPIP protocol. Two weeks after the onset of the pain, the sural nerve (SN) and L4/L5 DRG were collected for PKG activity assays. The right panel shows that a high percentage of total PKG is active in the SN *(blue bar)* and in the L4/L5 DRG *(red bar)* in the CPIP model. The left panel shows the baseline PKG activity in the control SN *(blue bar)* and in the control L4/L5 DRG *(red bar)*.

onset of tactile allodynia in the distal hind paw. We then assayed the percentage of total PKG that is active in the sural nerve, which provides nociceptive information from this region, and in the L4/L5 DRG, which contain the cell bodies of the affected neurons. In control rats, 8% of the total PKG in the sural nerve is active and 9% is active in these DRG (Fig. 3B). In striking contrast, assays of 2-week CPIP rats revealed that 80% of the total PKG in the sural nerve and 100% of the PKG in the DRG were in the active form.

Development of a novel potent PKG inhibitor

There are many challenges to developing a protein kinase inhibitor that is both potent and selective. PKG-1α belongs to the AGC family of serine/threonine-specific kinases that includes PKA, PKB, and PKC. PKG-1α is most homologous to PKA and hence selectivity for PKG-1α over PKA is important. PKG inhibitors generally target either the ATP-binding site or the cGMP-binding site. Interrupting the binding of cGMP prevents the activation of PKG-1α whereas blocking the binding of ATP prevents PKG-1α from activating its targets. Commercially available PKG inhibitors, such as KT5823, which acts at the ATP-binding site, or Rp-8-pCPT-cGMPS and the DT-oligopeptides act at the cGMP-binding site. However, Rp-8-pCPT-cGMPS and KT5823 have only micromolar potency ($K_i = 0.5$ μM and 0.234 μM, respectively), are accompanied by nonspecific effects, and have poor pharmacokinetic profiles (Fouty et al., 1998). The oligopeptide PKG inhibitors D and DT-2 are potent and highly specific for PKG in vitro, but lose specificity in cell homogenates (Gambaryan et al., 2012). Thus, suitable PKG inhibitors are lacking for drug development or as a research tool.

The amino acid sequence within the ATP-binding site of PKG differs from that of PKA, PKB, and PKC (Niwa, 2006) and using the X-ray crystallography of the site and computer modeling, we designed and synthesized a novel class of potent, specific inhibitors of PKG-1α (Sung et al., 2017). The effort began with (−)-balanol, a natural product of the fungus *Verticillium balanoides*. Balanol inhibits Ser/Thr kinases in the 1–4 nM range by competing for the ATP-binding site (Koide et al., 1995). It is therefore potent, but not selective. Two of its structural domains are also linked by an ester moiety so that balanol is not stable in serum. Given that the ATP-binding site in PKG-1α differs from that in these other kinases, it was possible to modify the structure of balanol to fit selectively into the PKG site, but not the others (Fig. 4A and B). This resulted in compound N46 (Fig. 4C) that inhibited PKG-1α with an IC_{50} of 7 nM vs 5000 nM for PKA and higher values for other members of the family. About 274 other kinases were also assayed and only 12 were inhibited in the range of

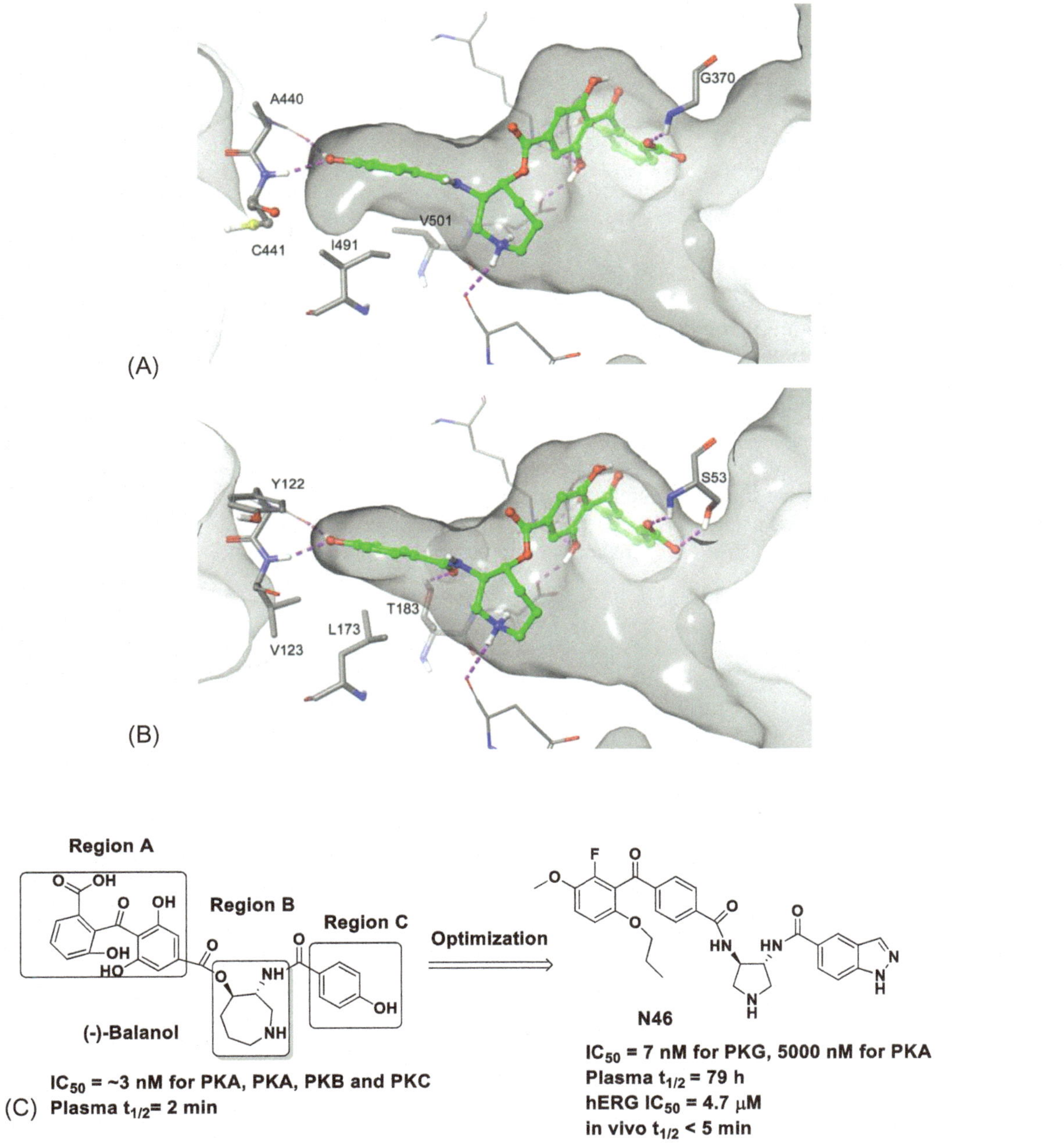

FIG. 4 Binding of balanol to PKA and PKG. (A) Balanol bound in the active site of the PKA Cα—subunit (PDB ID: 1BX6). (B) PKG-1α bound with balanol based on a homology model. Key active site residues that are different between PKG and PKA are shown. (C) Modification of (−)-balanol to produce N46.

PKG-1α, making N46 a very selective inhibitor (Sung et al., 2017). N46 not only prevents the activation of PKG-1α but also inhibits the already-activated kinase, which means that PKG-1α should prevent the onset of pain and also diminish ongoing pain. The half-life of N46 in serum was markedly improved by changing the ester linkages in balanol to amide bridges (Sung et al., 2017).

N46 effectively alleviates chronic osteoarthritic and inflammation-induced pain

The direct correlation between the activation of PKG-1α, the induction of LTH, and the onset of chronic pain strongly suggests that selective inhibition of PKG-1α using N46 would effectively attenuate any pain associated with LTH. We therefore tested this hypothesis in two rat models of LTH-mediated chronic pain and found that intravenous injection of N46 alleviates chronic inflammation-induced thermal hyperalgesia and that intraarticular injection relieves the pain of osteoarthritis. Moreover, this occurred even weeks after the pain was well established (Sung et al., 2017). The latter reflects that N46 not only inhibits the activation of PKG-1α but also reverses the effect of the previously activated kinase.

Side effects and the fate of N46 in vivo

PKG-1α is not present in motor axons (Sung et al., 2006), which explains why rats injected with N46 showed no impairment of motor function or any observable effects on behavior (Sung et al., 2017). We also examined the distribution of N46 after injection into the tail vein. Within 10 min, N46 is widely distributed among the tissues of the body (Table 2) (Sung et al., 2017). By 20 h, much of it was secreted in urine. Significantly, N46 was below the levels of detection in the brain and spinal cord at both times, indicating that N46 does not cross the blood-brain barrier and will not interfere with higher brain function. Indeed, rats injected with N46 in our experiments showed no signs of sedation, which is a highly significant departure from the effects of opiates. N46 was found in the DRG at both times, which is important because these ganglia house the neurons that exhibit the LTH. N46 accumulates in the DRG because these ganglia have an extensive network of fenestrated capillaries (Jimenez-Andrade et al., 2008). N46 was present in the heart at 10 min but was below detection at the longer time point. When we examined several heart parameters after N46 injection, there were no unwanted side effects (Sung et al., 2017). Finally, constipation, which is often associated with opiates, was not observed in rats injected with N46. This is consistent with studies using a gut preparation in vitro, where it was found that N46 had an IC_{50} of 60 μM for the inhibition of peristalsis, which is orders of magnitude higher than its inhibition of PKG-1 (Li et al., 2013).

TABLE 2 Tissue distribution of N46 in vivo.

Tissue	10 min	16–20 h
Plasma	0.47 μg/mL	5.38 μg/mL
Urine	26.3 μg/mL	0.05 μg/mL
Heart	6.55 μg/g	BDL
Liver	28.05 μg/g	9.09 μg/g
Lung	13.25 μg/g	3.15 μg/g
Kidney	82.8 μg/g	0.97 μg/g
DRG	13.75 μg/g	9 μg/g
Brain	BDL	BDL
Spinal cord	BDL	BDL
Cerebellum	0.95 μg/g	BDL

N46 (559 μg = 1 μmol, in 1% DMSO/normal saline) was injected into the tail vein of two rats. Tissue samples from each rat, collected at the times indicated, were combined, homogenized on ice, and the amount of N46 was determined by LC/MS. Baseline values, determined from the control, were subtracted. *BDL*, below detection limit.

Conclusions

PKG-1α is a positive injury signal in peripheral neurons that is activated by nerve injury or inflammation and is necessary for the induction of a persistent pain state. Therefore, PKG-1α is a major concomitant of pain. A measure of our success has been the development of compound N46 that blocks both the activation of PKG-1α and the active kinase. N46 is a highly selective and potent inhibitor of PKG-1α that effectively attenuates two types of LTH-mediated pain without sedation or other severe side effects. More importantly, we demonstrated that PKG-1α is an excellent target for the development of nonopiate analgesics for the treatment of chronic pain. Such analgesics will benefit millions who suffer from neuropathic and inflammatory chronic pain without fear of addiction.

Applications to other areas

N46, an inhibitor of PKG, alleviates chronic pain in animal models. N46 acts at the cell bodies of first-order nociceptive neurons in peripheral sensory ganglia. It does not enter the CNS, has no effect on motor neurons or glia, and is not addictive. As such, it is a potential replacement for opiates in the treatment of chronic pain.

Other agents of interest

We have synthesized N46, a novel, potent inhibitor of PKG that alleviates chronic pain in animal models, even after the pain is established. N46 acts at the cell bodies of first-order nociceptive neurons in peripheral sensory ganglia: it does not enter the CNS, has no effect on motor neurons or glia, and is not addictive. As such, it is a potential replacement for opiates in the treatment of chronic pain.

Mini-dictionary of terms

Long-term potentiation (LTP): An alteration in the electrophysiological properties of a synapse that amplifies responses to a stimulus and extends the duration of pain.

Long-term hyperexcitability (LTH): An alteration in the electrophysiological properties of first-order nociceptive neurons that amplifies responses to a lesion and is associated with many types of chronic pain.

Protein kinase G (PKG): PKG is a molecular switch for pain that is activated by a lesion and transported to the cell bodies of nociceptive neurons where it initiates events that result in LTH.

Positive injury signal: A protein, such as PKG, that is activated by a lesion and transported back to the cell bodies of the affected neurons where it alters the phenotype.

Complex regional pain syndrome type-1: A chronic condition that develops after fractures, tissue trauma, or surgery that is accompanied by chronic sensory hypersensitivity, allodynia, and uncontrollable spontaneous pain.

Key facts

1. First-order neurons in the nociceptive pathway communicate information about a peripheral lesion to second-order neurons in the CNS.
2. Axons of second-order neurons transmit the information to the thalamus and postcentral gyrus, resulting in the perception of the pain and the location of the lesion.
3. Chronic pain is associated with the presence of a long-term hyperexcitability (LTH) in the cell bodies of first-order nociceptive neurons. The LTH amplifies signals from the site lesion to the CNS contributing to allodynia and hyperalgesia.
4. Induction of LTH requires the activation of protein kinase G at the injury site and its transport to the cell bodies of the affected neurons where it alters the phenotype resulting in LTH.
5. N46, a novel, highly potent and selective inhibitor of PKG, blocks the induction of the LTH and alleviates chronic pain in several animal models, even after the pain has been established.

Summary points

1. Chronic pain is associated with the presence of a long-term hyperexcitability (LTH) in the cell bodies of first-order nociceptive neurons.

2. LTH amplifies signals from the lesion site to the CNS contributing to allodynia and hyperalgesia.
3. The induction of LTH requires the activation of protein kinase G (PKG) at the injury site and its transport to the cell bodies of the affected neurons where it initiates events resulting in LTH.
4. LTH is the result of a phenotypic change and can therefore last indefinitely.
5. Inhibiting PKG with N46, a novel, highly potent and selective inhibitor of PKG, both blocks the induction of the LTH and alleviates chronic pain in several animal models, even after the pain has been established.

References

Abdulla, F. A., & Smith, P. A. (2001). Axotomy- and autotomy-induced changes in the excitability of rat dorsal root ganglion neurons. *Journal of Neurophysiology*, *85*(2), 630–643. https://doi.org/10.1152/jn.2001.85.2.630.

Ambron, R. T., Dulin, M. F., Zhang, X. P., Schmied, R., & Walters, E. T. (1995). Axoplasm enriched in a protein mobilized by nerve injury induces memory-like alterations in Aplysia neurons. *Journal of Neuroscience*, *15*(5 Pt 1), 3440–3446.

Ambron, R. T., Zhang, X. P., Gunstream, J. D., Povelones, M., & Walters, E. T. (1996). Intrinsic injury signals enhance growth, survival, and excitability of Aplysia neurons. *Journal of Neuroscience*, *16*(23), 7469–7477.

Bennett, G. J. (2004). Neuropathic pain in the orofacial region: Clinical and research challenges. *Journal of Orofacial Pain*, *18*(4), 281–286.

Beyak, M. J., Ramji, N., Krol, K. M., Kawaja, M. D., & Vanner, S. J. (2004). Two TTX-resistant Na^+ currents in mouse colonic dorsal root ganglia neurons and their role in colitis-induced hyperexcitability. *The American Journal of Physiology-Gastrointestinal and Liver Physiology*, *287*(4), G845–G855.

Bowsher, D., Geoffrey Woods, C., Nicholas, A. K., Carvalho, O. M., Haggett, C. E., Tedman, B., … Rice, F. L. (2009). Absence of pain with hyperhidrosis: A new syndrome where vascular afferents may mediate cutaneous sensation. *Pain*, *147*(1–3), 287–298. https://doi.org/10.1016/j.pain.2009.09.007.

Boyer, E. W. (2012). Management of opioid analgesic overdose. *New England Journal of Medicine*, *367*(2), 146–155. https://doi.org/10.1056/NEJMra1202561.

Bulka, A., Hao, J. X., & Wiesenfeld-Hallin, Z. (2002). Response characteristics of cutaneous mechanoreceptors in neuropathic rats. *Neuroscience Letters*, *317*(2), 89–92. https://doi.org/10.1016/s0304-3940(01)02439-9.

Chen, T., O'Den, G., Song, Q., Koga, K., Zhang, M. M., & Zhuo, M. (2014). Adenylyl cyclase subtype 1 is essential for late-phase long term potentiation and spatial propagation of synaptic responses in the anterior cingulate cortex of adult mice. *Molecular Pain*, *10*, 65. https://doi.org/10.1186/1744-8069-10-65.

Chu, Y. C., Guan, Y., Skinner, J., Raja, S. N., Johns, R. A., & Tao, Y. X. (2005). Effect of genetic knockout or pharmacologic inhibition of neuronal nitric oxide synthase on complete Freund's adjuvant-induced persistent pain. *Pain*, *119*(1–3), 113–123. https://doi.org/10.1016/j.pain.2005.09.024.

Clatworthy, A. L., Castro, G. A., Budelmann, B. U., & Walters, E. T. (1994). Induction of a cellular defense reaction is accompanied by an increase in sensory neuron excitability in Aplysia. *Journal of Neuroscience*, *14*(5 Pt 2), 3263–3270.

Coderre, T. J., Xanthos, D. N., Francis, L., & Bennett, G. J. (2004). Chronic post-ischemia pain (CPIP): A novel animal model of complex regional pain syndrome-type I (CRPS-I; reflex sympathetic dystrophy) produced by prolonged hindpaw ischemia and reperfusion in the rat. *Pain*, *112*(1–2), 94–105. https://doi.org/10.1016/j.pain.2004.08.001.

Dobretsov, M., Hastings, S. L., Romanovsky, D., Stimers, J. R., & Zhang, J. M. (2003). Mechanical hyperalgesia in rat models of systemic and local hyperglycemia. *Brain Research*, *960*(1–2), 174–183. https://doi.org/10.1016/s0006-8993(02)03828-3.

Fouty, B., Komalavilas, P., Muramatsu, M., Cohen, A., McMurtry, I. F., Lincoln, T. M., & Rodman, D. M. (1998). Protein kinase G is not essential to NO-cGMP modulation of basal tone in rat pulmonary circulation. *American Journal of Physiology*, *274*(2), H672–H678. https://doi.org/10.1152/ajpheart.1998.274.2.H672.

Gambaryan, S., Butt, E., Kobsar, A., Geiger, J., Rukoyatkina, N., Parnova, R., … Walter, U. (2012). The oligopeptide DT-2 is a specific PKG I inhibitor only in vitro, not in living cells. *British Journal of Pharmacology*, *167*(4), 826–838. https://doi.org/10.1111/j.1476-5381.2012.02044.x.

Gruber-Schoffnegger, D., Drdla-Schutting, R., Honigsperger, C., Wunderbaldinger, G., Gassner, M., & Sandkuhler, J. (2013). Induction of thermal hyperalgesia and synaptic long-term potentiation in the spinal cord lamina I by TNF-alpha and IL-1beta is mediated by glial cells. *Journal of Neuroscience*, *33*(15), 6540–6551. https://doi.org/10.1523/JNEUROSCI.5087-12.2013.

Gunstream, J. D., Castro, G. A., & Walters, E. T. (1995). Retrograde transport of plasticity signals in Aplysia sensory neurons following axonal injury. *Journal of Neuroscience*, *15*(1 Pt 1), 439–448.

Ikeda, H., Stark, J., Fischer, H., Wagner, M., Drdla, R., Jager, T., & Sandkuhler, J. (2006). Synaptic amplifier of inflammatory pain in the spinal dorsal horn. *Science*, *312*(5780), 1659–1662. https://doi.org/10.1126/science.1127233.

Ji, R. R., Kohno, T., Moore, K. A., & Woolf, C. J. (2003). Central sensitization and LTP: Do pain and memory share similar mechanisms? *Trends in Neurosciences*, *26*(12), 696–705. https://doi.org/10.1016/j.tins.2003.09.017.

Jimenez-Andrade, J. M., Herrera, M. B., Ghilardi, J. R., Vardanyan, M., Melemedjian, O. K., & Mantyh, P. W. (2008). Vascularization of the dorsal root ganglia and peripheral nerve of the mouse: Implications for chemical-induced peripheral sensory neuropathies. *Molecular Pain*, *4*, 10. https://doi.org/10.1186/1744-8069-4-10.

Klein, T., Magerl, W., & Treede, R. D. (2006). Perceptual correlate of nociceptive long-term potentiation (LTP) in humans shares the time course of early-LTP. *Journal of Neurophysiology*, *96*(6), 3551–3555. https://doi.org/10.1152/jn.00755.2006.

Koide, K., Bunnage, M. E., Gomez Paloma, L., Kanter, J. R., Taylor, S. S., Brunton, L. L., & Nicolaou, K. C. (1995). Molecular design and biological activity of potent and selective protein kinase inhibitors related to balanol. *Chemistry & Biology*, *2*(9), 601–608. https://doi.org/10.1016/1074-5521(95)90124-8.

Li, Z., Ambron, R. T., Margolis, K. G., Sung, Y.-J., Welch, M. G., & Gershon, M. D. (2013). The α isoform of cGMP-dependent protein kinase 1 (PKG1α) is selectively expressed in and critical for the function of intrinsic primary afferent neurons (IPANs) of the guinea pig enteric nervous system (ENS). *Gastroenterology*, *144*(5). https://doi.org/10.1016/S0016-5085(13)60387-X. S-106.

Liu, S., Zhang, M. Y., Chen, L. P., Liu, Y. P., & Liu, G. J. (2014). cGMP and cGMP-dependent protein kinase I pathway in dorsal root ganglia contributes to bone cancer pain in rats. *Spine (Phila Pa 1976)*, *39*(19), 1533–1541. https://doi.org/10.1097/BRS.0000000000000456.

Marchand, F. D'Mello, R., Yip, P. K., Calvo, M., Muller, E., Pezet, S., … McMahon, S. B. (2011). Specific involvement of atypical PKCzeta/PKMzeta in spinal persistent nociceptive processing following peripheral inflammation in rat. *Molecular Pain*, *7*, 86. https://doi.org/10.1186/1744-8069-7-86.

Morales-Aza, B. M., Chillingworth, N. L., Payne, J. A., & Donaldson, L. F. (2004). Inflammation alters cation chloride cotransporter expression in sensory neurons. *Neurobiology of Disease*, *17*(1), 62–69.

Morita, K., Kitayama, T., Morioka, N., & Dohi, T. (2008). Glycinergic mediation of tactile allodynia induced by platelet-activating factor (PAF) through glutamate-NO-cyclic GMP signalling in spinal cord in mice. *Pain*, *138*(3), 525–536. https://doi.org/10.1016/j.pain.2008.01.030.

Niwa, T. (2006). Elucidation of characteristic structural features of ligand binding sites of protein kinases: A neural network approach. *Journal of Chemical Information and Modeling*, *46*(5), 2158–2166. https://doi.org/10.1021/ci050528t.

Peters, C. M., Ghilardi, J. R., Keyser, C. P., Kubota, K., Lindsay, T. H., Luger, N. M., … Mantyh, P. W. (2005). Tumor-induced injury of primary afferent sensory nerve fibers in bone cancer pain. *Experimental Neurology*, *193*(1), 85–100.

Sandkuhler, J. (2007). Understanding LTP in pain pathways. *Molecular Pain*, *3*, 9. https://doi.org/10.1186/1744-8069-3-9.

Sandkuhler, J. (2010). Central sensitization versus synaptic long-term potentiation (LTP): A critical comment. *Journal of Pain*, *11*(8), 798–800. https://doi.org/10.1016/j.jpain.2010.05.002.

Sandkuhler, J., & Gruber-Schoffnegger, D. (2012). Hyperalgesia by synaptic long-term potentiation (LTP): An update. *Current Opinion in Pharmacology*, *12*(1), 18–27. https://doi.org/10.1016/j.coph.2011.10.018.

Sandkuhler, J., & Liu, X. (1998). Induction of long-term potentiation at spinal synapses by noxious stimulation or nerve injury. *European Journal of Neuroscience*, *10*(7), 2476–2480. https://doi.org/10.1046/j.1460-9568.1998.00278.x.

Sculptoreanu, A., de Groat, W. C., Buffington, C. A., & Birder, L. A. (2005). Abnormal excitability in capsaicin-responsive DRG neurons from cats with feline interstitial cystitis. *Experimental Neurology*, *193*(2), 437–443.

Sculptoreanu, A., Yoshimura, N., & de Groat, W. C. (2004). KW-7158 [(2S)-(+)-3,3,3-trifluoro-2-hydroxy-2-methyl-N-(5,5,10-trioxo-4,10-dihydro thieno[3,2-c][1]benzothiepin-9-yl)propanamide] enhances A-type K+ currents in neurons of the dorsal root ganglion of the adult rat. *Journal of Pharmacology and Experimental Therapeutics*, *310*(1), 159–168.

Song, X. J., Wang, Z. B., Gan, Q., & Walters, E. T. (2006). cAMP and cGMP contribute to sensory neuron hyperexcitability and hyperalgesia in rats with dorsal root ganglia compression. *Journal of Neurophysiology*, *95*(1), 479–492. https://doi.org/10.1152/jn.00503.2005.

Sung, Y. J., & Ambron, R. T. (2004). Pathways that elicit long-term changes in gene expression in nociceptive neurons following nerve injury: Contributions to neuropathic pain. *Neurological Research*, *26*(2), 195–203. https://doi.org/10.1179/016164104225013761.

Sung, Y. J., Chiu, D. T., & Ambron, R. T. (2006). Activation and retrograde transport of protein kinase G in rat nociceptive neurons after nerve injury and inflammation. *Neuroscience*, *141*(2), 697–709. https://doi.org/10.1016/j.neuroscience.2006.04.033.

Sung, Y. J., Sofoluke, N., Nkamany, M., Deng, S., Xie, Y., Greenwood, J., … Ambron, R. T. (2017). A novel inhibitor of active protein kinase G attenuates chronic inflammatory and osteoarthritic pain. *Pain*, *158*(5), 822–832. https://doi.org/10.1097/j.pain.0000000000000832.

Sung, Y. J., Walters, E. T., & Ambron, R. T. (2004). A neuronal isoform of protein kinase G couples mitogen-activated protein kinase nuclear import to axotomy-induced long-term hyperexcitability in Aplysia sensory neurons. *Journal of Neuroscience*, *24*(34), 7583–7595. https://doi.org/10.1523/JNEUROSCI.1445-04.2004.

Walters, E. T., Alizadeh, H., & Castro, G. A. (1991). Similar neuronal alterations induced by axonal injury and learning in Aplysia. *Science*, *253*(5021), 797–799. https://doi.org/10.1126/science.1652154.

Woolf, C. J. (2011). Central sensitization: Implications for the diagnosis and treatment of pain. *Pain*, *152*(3 Suppl), S2–15. https://doi.org/10.1016/j.pain.2010.09.030.

Xanthos, D. N., Bennett, G. J., & Coderre, T. J. (2008). Norepinephrine-induced nociception and vasoconstrictor hypersensitivity in rats with chronic post-ischemia pain. *Pain*, *137*(3), 640–651. pii: S0304-3959(07)00648-3 https://doi.org/10.1016/j.pain.2007.10.031.

Yang, H. W., Hu, X. D., Zhang, H. M., Xin, W. J., Li, M. T., Zhang, T., … Liu, X. G. (2004). Roles of CaMKII, PKA, and PKC in the induction and maintenance of LTP of C-fiber-evoked field potentials in rat spinal dorsal horn. *Journal of Neurophysiology*, *91*(3), 1122–1133. https://doi.org/10.1152/jn.00735.2003.

Zhang, J. M., Song, X. J., & LaMotte, R. H. (1999). Enhanced excitability of sensory neurons in rats with cutaneous hyperalgesia produced by chronic compression of the dorsal root ganglion. *Journal of Neurophysiology*, *82*(6), 3359–3366.

Zhou, L. J., Zhong, Y., Ren, W. J., Li, Y. Y., Zhang, T., & Liu, X. G. (2008). BDNF induces late-phase LTP of C-fiber evoked field potentials in rat spinal dorsal horn. *Experimental Neurology*, *212*(2), 507–514. https://doi.org/10.1016/j.expneurol.2008.04.034.

Chapter 8

Adrenergic agonists and antagonists enhance opioid receptor activity

Robert Root-Bernstein
Professor of Physiology, Michigan State University, East Lansing, MI, United States

Abbreviations

B2AR	beta-2 adrenergic receptor
GDP	guanine-diphosphate
GPCR	G-protein–coupled receptor
GTP	guanine-triphosphate
OPR	opioid receptor

Introduction

Aira, Barrenetxea, Buesa, Gómez-Esteban, and Azkue (2014) have proposed that, "A sound strategy for improving the clinical efficacy of opioids involves exploiting positive interactions with drugs directed at other targets in pain pathways." In that vein, this chapter explores a range of opioid-adrenergic interactions and ways in which these have been exploited to improve pain control and may lead to new approaches.

The opioid and adrenergic systems both mediate analgesia and nociception and can do so in a synergistic manner. Natural selection to link the two systems makes evolutionary sense since the adrenergic system prepares the body for fight-or-flight responses, while the opioid system mediates the experience of pain, a common trigger for, and result of, both fighting and fleeing. The resulting Interactions between adrenergic and opioid compounds span the range from cellular organization of neuronal pathways and mutual enhancement of each other's receptors, to heterodimerization of those receptors, cross-talk between their second messenger pathways, and even to direct binding between the compounds themselves. Some of the effects of lidocaine (also known as lignocaine or xylocaine), including its opiate-sparing applications, appear to be mediated through this adrenergic-opioid axis. Tramadol and tapentadol interact with both the adrenergic and opioid systems as does vitamin C (ascorbic acid). This many-layered set of interactions makes for a complex web of mutual effects that pose both opportunities and pitfalls in the use of opioid-adrenergic drug combinations for treating pain.

Anatomical and cellular codistribution of opioid and adrenergic functions

Anatomically, the adrenergic and opioid systems are integrated through mutual binding of their ligands and receptors and by colocalization in cell membranes as well as in tissues and organs. While a few tissues sensitive to adrenergic activation are not sensitive to opioid stimulation (e.g., vascular smooth muscle endothelium), it appears that every tissue sensitive to opioids is also sensitive to adrenergic compounds. The specific combinations of adrenergic and opioid receptor subtypes present in different tissues vary so that alpha and beta adrenergic agonists or antagonists may or may not combine in any given tissue to enhance or synergize with any particular mu, delta, or kappa opioid receptor agonist or antagonist. Understanding the tissue distributions of the receptor pairs is therefore important for the proper application of the types of adrenergic-opioid interactions described later. Space does not permit, however, an adequate description of this complex subject here. The important point is that systemic delivery of opioids and adrenergic drugs together will necessarily have broad effects beyond managing nociception. Localized delivery will limit these systemic effects.

In many tissues, the colocalization of adrenergic and opioid sensitivity occurs within individual cells. For example, mu opioid receptors are colocalized on neurons also expressing noradrenergic receptors in the anterior cingulate cortex, a brain region involved in pain perception (Vogt, Sim-Selley, Childers, Wiley, & Vogt, 2001). The shared hypotensive actions of

The Neurobiology, Physiology, and Psychology of Pain. https://doi.org/10.1016/B978-0-12-820589-1.00008-7

catecholamines and central opioids are mediated by medullary neurons that costore or cotransmit both compounds and share common synaptic junctions (Milner, Pickel, & Reis, 1989). Enkephalins are costored and cotransmitted with norepinephrine (NE) in many tissues and with epinephrine (EPI) in adrenal chromaffin cells. Examples could easily be multiplied. This costorage and cotransmission may have evolved because adrenergic compounds bind directly to opioids with sufficient affinity and specificity to form complexes stabilizing their costorage within vesicles and mediating their combined activity at both sets of receptors simultaneously (Dillon, Root-Bernstein, & Lieder, 2006b; Root-Bernstein, 1987).

Each receptor type has evolved to be capable of binding the ligands of the other so that opioid receptors have an allosteric binding site for adrenergic compounds and adrenergic receptors have an allosteric binding site for opioid compounds. In cells expressing both adrenergic and opioid receptors, these receptors can form heterodimers that mediate cross-talk between their activity (Samoshkin et al., 2015). These interactions are essential for understanding how opioids and adrenergics interact to control pain.

Mechanisms of adrenergic receptor-opioid receptor cross-talk

Adrenergic and opioid receptors are both seven-transmembrane G-coupled protein receptors. These receptors undergo allosteric conformational changes upon binding their ligands that activate the receptors, initiating a reaction cascade in which G-proteins bound to the receptors in their inactive state are released. G-proteins are so-called because they bind guanine-triphosphate (GTP). The release of the G-proteins is accompanied by conversion of GTP into guanine diphosphate (GDP), which further activates a second messenger system to produce the specific function associated with that receptor. The release of the G-proteins also reveals intracellular receptor sites that can be phosphorylated by phosphodiesterases. This phosphorylation returns the receptor to an inactive state by causing it to release its ligand and begins a process of receptor internalization. Some of the internalized receptors will be returned to the cell membrane while some will be degraded depending in large part on whether the receptors have dimerized or not. Dimerization, in turn, reflects the degree of receptor activation occurring in any particular cell. Both aminergic and opioid receptors are known to homodimerize leading some investigators to argue that their active conformations involve receptor pairs. Heterodimeric receptor pairs are also known to form among the various classes of aminergic receptors (Breit, Lagacé, & Bouvier, 2004; Chen & Minneman, 2005) and the same kind of mixed-class heterodimer formation is found among opioid receptors (Fujita, Gomes, & Devi, 2014).

Three main types of interactions directly mediate cross-talk between adrenergic receptors and opioid receptors. One involves opioids binding to adrenergic receptors; the second, adrenergic compounds binding to opioid receptors; and the third results from the formation of heterodimer receptor complexes composed of pairs of adrenergic and opioid receptors.

Opioids bind to adrenergic receptors

Opioid compounds bind promiscuously to some nonopioid receptors including adrenergic receptors which bind morphine (a mu opioid agonist), kappa opioid antagonists and endomorphin (Lengyel et al., 2016; Munro et al., 2013). Opioids also bind to serotonin receptors, a topic not covered in this chapter.

Opioid binding to adrenergic receptors results in enhancement of adrenergic receptor activity. Enhancement is a type of interaction in which one compound has no effect on a system but significantly increases the activity of a second compound such that the dose response curve is shifted to the left. Morphine enhances norepinephrine, epinephrine, and isoproterenol activity in a variety of tissues without itself activating the tissue response. This enhancement is not be blocked by naloxone arguing against an opioid-receptor–mediated mechanism. In fact, the opioid antagonists naloxone and naltrexone are themselves able to enhance catecholamine activity (reviewed in Root-Bernstein & Dillon, 2014).

The binding site of opioids on aminergic G-protein–coupled receptors is localized to regions on the first and second extracellular loops of the receptor (Fig. 1 and Table 1) consisting of cysteine- and methionine-rich residues involved in the maintenance of its high-affinity, active conformation (Dillon et al., 2010; Root-Bernstein et al., 2018, 2019, 2020). Opioids binding cause allosteric conformational changes in the receptor, inhibiting phosphodiesterase (PDE) phosphorylation of the receptor, thereby preventing receptor desensitization and internalization (He, Molnar, & Barraclough, 1993; Moffat, Patterson, Curry, & Owen, 1972; Park et al., 2009).

Adrenergic compounds bind to opioid receptors

The binding of opioid compounds to adrenergic receptors is reciprocated by the binding of adrenergic compounds to opioid receptors (Table 1). The alpha-1 adrenergic blocker phentolamine binds to mu, delta, and kappa opioid receptors, competing

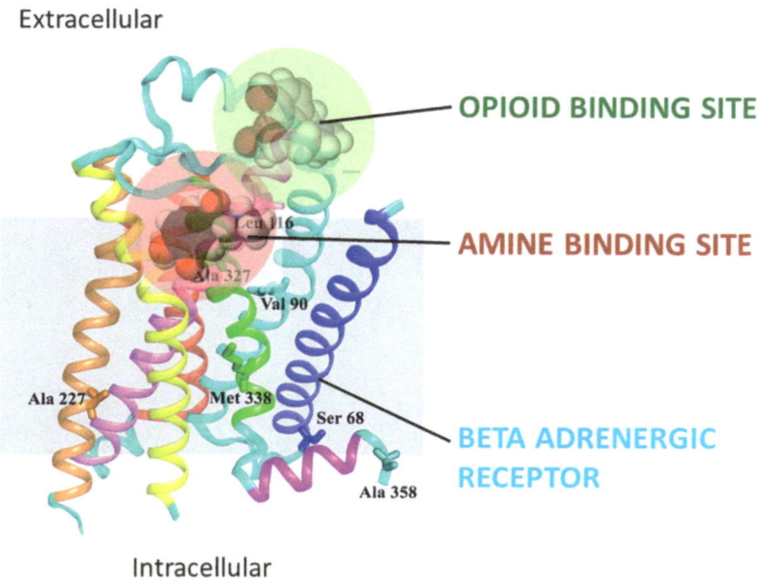

FIG. 1 Ribbon model of beta 2 adrenergic receptor showing the locations of the amine binding site at the center of the seven transmembrane helices and the opioid binding site at the junction of the first and second extracellular loops. The opioid binding site is highly conserved across aminergic receptors (Dillon, Root-Bernstein, & Lieder, 2006a; Dillon, Root-Bernstein, Robinson, Abraham, & Berney, 2010; Root-Bernstein, Churchill, & Turke, 2020; Root-Bernstein, Churchill, Turke, Subhramanyam, & Labahn, 2019; Root-Bernstein, Turke, Subhramanyam, Churchill, & Labahn, 2018).

with opioid agonists; prazosin, another alpha-1 blocker, binds to mu and delta but not kappa opioid receptors; while clonidine, an alpha-2 agonist, is specific for kappa opioid receptors (Monroe, Perschke, Crisp, & Smith, 1991). Xylamine, another alpha-2 agonist that additionally blocks norepinephrine and dopamine reuptake, competes with naloxone for opioid binding sites as does phenoxybenzamine, a nonselective alpha-adrenergic receptor antagonist (Wilkinson, Jacobson, & Wilkinson, 1985). Once again, the binding site for adrenergic compounds on opioid receptors has been localized to the second extracellular loop (Fig. 2) (Root-Bernstein et al., 2018, 2019, 2020).

Just as opioids enhance catecholamine receptor activity, binding of adrenergic compounds to opioid receptors enhances opioid activity (Root-Bernstein et al., 2018, 2019, 2020). Once again, the mechanism of enhancement appears to involve allosteric modifications of the receptor structure that interfere with PDE activity, thereby preventing downregulation and internalization of receptor function (Puri, Cochin, & Volicer, 1975).

Opioid receptor-adrenergic receptor heterodimerization

Finally, just as opioid receptors and adrenergic receptors can homodimerize, opioid receptors and adrenergic receptors can also *heterodimerize with each other* (Chabot-Doré, Schuster, Stone, & Wilcox, 2015; Fujita et al., 2014). This heterodimerization mediates some of the antinociceptive effects of adrenergics through two mechanisms. One involves adrenergics binding directly to the opioid receptor to enhance concentrations of endogenous opioids that are otherwise too low to produce an effect on their own while the other involves simultaneously binding to the adrenergic component of the heterodimer pair to produce an indirect, allosterically mediated enhancement of the opioid receptor. Both aspects of the mechanism inhibit PDE-mediated phosphorylation of the receptor pair thus providing long-term activation of adrenergic and opioid receptor activity simultaneously.

A model of adrenoceptor-opioid receptor cross-talk

The following model (Fig. 3) describes most of the known sets of interactions that have thus far been characterized with regard to opioid-adrenergic interactions.

Briefly, binding of a ligand to its receptor activates a series of steps that produce an intracellular message. The top set of models in Fig. 3 illustrates the normal activation process. In the first step (A), an opioid approaches the receptor. It initially binds (B) nonspecifically to the extracellular loop region of the receptor and is then pulled into the high affinity site at the

TABLE 1 Binding constants of various drugs and neurotransmitters to the extracellular loop regions of the beta 2 adrenergic receptor (B2AR) and the mu opioid receptor (Mu OPR).

Kd (micromolar)	B2AR 97–103	B2AR 105–108	B2AR 183–185	Mu OPR 38–51	Mu OPR 111–122	Mu OPR 121–131	Mu OPR 132–143	Mu OPR 211–226
Morphine	1	30	>1000	35	50	900	35	30
Methadone	300	30	>1000	70	70	>1000	50	150
Naloxone	6	30	>1000	0.5/35	0.5/38	>1000	0.5/42	1.0/45
Met-Enkephalin	130	30	>1000	1.2/35	0.33/80	3.5/90	0.4/70	1.0/65
Phenylephrine	>1000	35	30	1.0/30	150	>1000	60	150
Propranolol	>1000	40	22	25	70	>1000	90	200
Amphetamine	130	20	2.3	1.3/90	1.3/100	>1000	1.1/85	1.2/90
Epinephrine	120	35	25	1.2/35	1.3/40	>1000	1.4/35	1.2/45
Norepinephrine	600	35	12	1.4/45	1.3/40	>1000	1.4/40	1.3/45
Dopamine	>1000	30	35	60	65	>1000	60	65
L-DOPA	200	40	25	80	60	>1000	70	150
Histamine	>1000	50	>1000	>1000	>1000	>1000	>1000	>1000
Serotonin	430	45	>1000	100	100	350	100	90
Melatonin	600	25	>1000	50	400	>1000	60	130
Acetylcholine	>1000	>1000	>1000	>1000	>1000	>1000	>1000	>1000
Glutamate	>1000	>1000	>1000	>1000	>1000	>1000	>1000	>1000
Glycine	>1000	>1000	>1000	>1000	>1000	>1000	>1000	>1000
Glucose	>1000	>1000	>1000	>1000	>1000	>1000	>1000	>1000

Sequence numbers from UniProt identify each receptor region. Binding constants are expressed as Kd in micromolar concentrations. Some ligands bound in both high affinity and low affinity modes that are expressed as a small number followed by a slash and the larger number. The smaller the binding constant, the higher the affinity for the receptor peptide (assembled from Root-Bernstein et al., 2018, 2019, 2020).

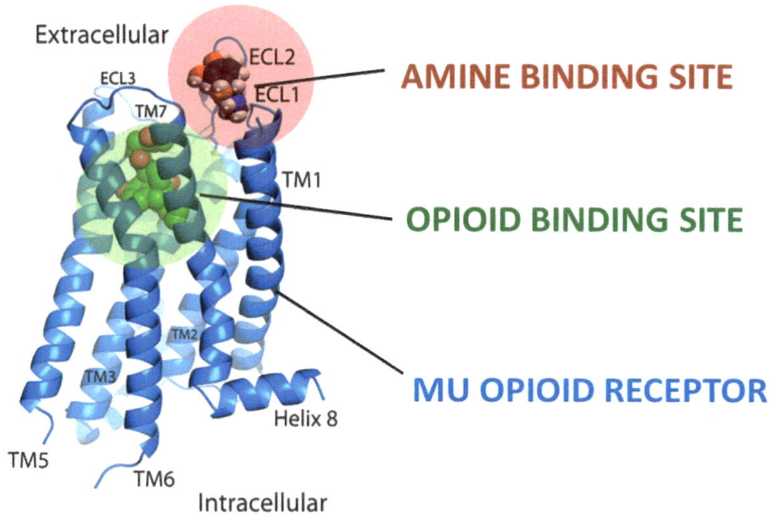

FIG. 2 Ribbon model of the mu opioid receptor showing the opioid binding site at the center of the seven transmembrane helices and the aminergic binding site at the intersection of the first and second extracellular loops (ECL1 and ECL2). The aminergic binding site is highly conserved across opioid receptor types (Root-Bernstein et al., 2018, 2019, 2020).

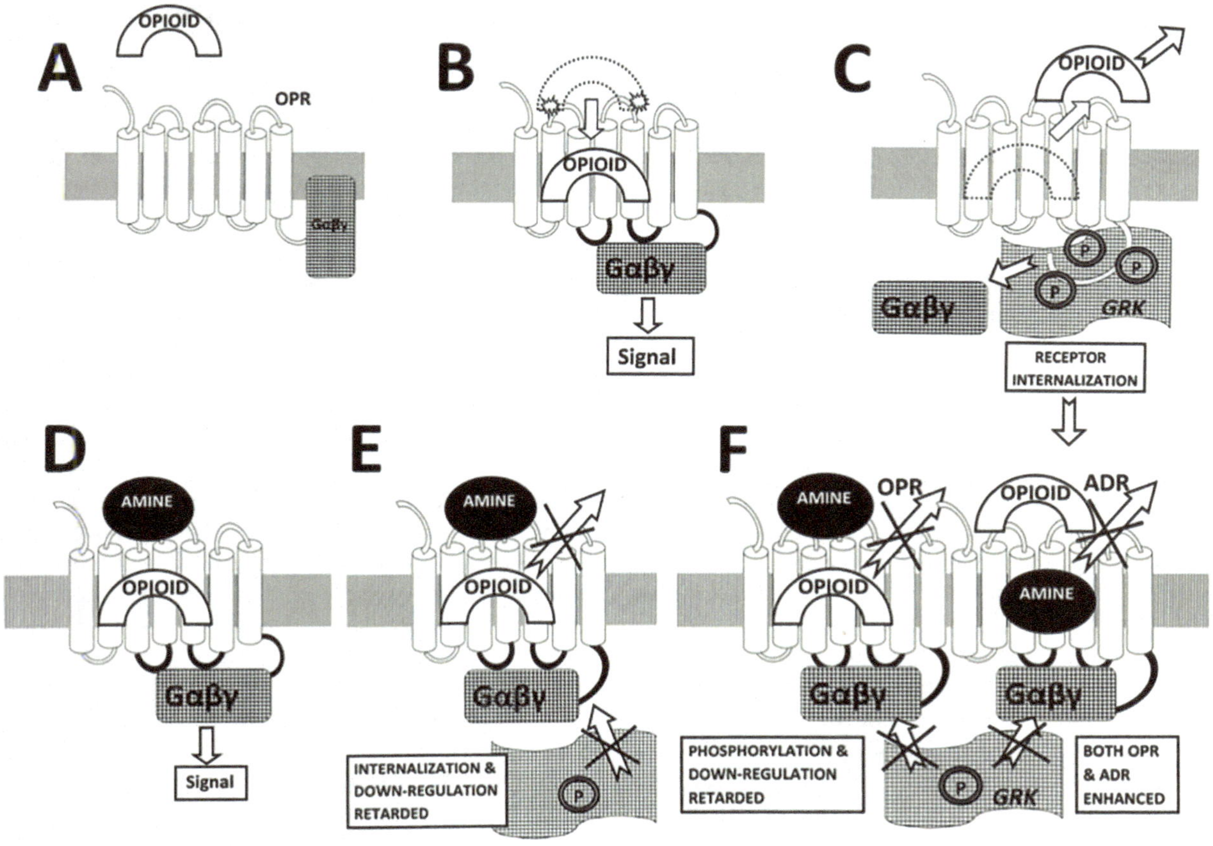

FIG. 3 Graphic model of opioid receptor signaling and desensitization and the effects of aminergic enhancement and heterodimerization with aminergic receptors. See text for a description of the processes illustrated. The activity produced by opioid binding is enhanced in the presence of adrenergic (and serotoninergic) compounds. The opioid binds as in (A) but is accompanied by amine binding to the extracellular loops of the receptor (D). Activation of signal transduction occurs as when the opioid is alone but can be achieved at far lower concentrations of the opioid. This enhanced opioid activity may be due (E) to "capping" of the receptor by the amine, trapping the opioid in the receptor and maintaining its activation, as well as by allosteric changes to the receptor that prevent GRK displacement of the G-proteins (Gαβγ). The same enhancement by the same mechanisms can be produced by opioids binding to adrenergic receptors. Because adrenergic and opioid receptors can heterodimerize (F), activation of one receptor can result in allosteric enhancement of the activity of the other (derived from Root-Bernstein et al., 2018, 2019).

receptor core. This binding causes G-protein binding (Gαβγ) to the intracellular loops of the receptor, activating the signal transduction pathway. A G-coupled receptor kinase (GRK) then displaces the G-proteins (C), phosphorylating the intracellular loops of the receptor. This phosphorylation allosterically modifies the receptor, releasing the opioid, desensitizing it to further opioid binding, and initiating internalization of the receptor. Exactly the same process characterizes the activation of aminergic G-protein–coupled receptors (GPCR) (not shown).

The bottom row of models in Fig. 3 illustrates how the normal function of GPCR is modified and results in receptor cross-talk by binding an enhancer. The activity produced by opioid binding is enhanced in the presence of adrenergic (and serotoninergic) compounds. The opioid binds as in (A) but is accompanied by amine binding to the extracellular loops of the receptor (D). Activation of signal transduction occurs as when the opioid is alone but can be achieved at far lower concentrations of the opioid. This enhanced opioid activity may be due (E) to "capping" of the receptor by the amine, trapping the opioid in the receptor and maintaining its activation, as well as by allosteric changes to the receptor that prevent GRK displacement of the G-proteins (Gαβγ). The same enhancement by the same mechanisms can be produced by opioids binding to adrenergic receptors. Because adrenergic and opioid receptors can heterodimerize (F), activation of one receptor can result in allosteric enhancement of the activity of the other (Dillon et al., 2004; Root-Bernstein et al., 2018, 2019).

This model explains the otherwise puzzling observation that adrenergic receptors mediate analgesia such that their inactivation also impairs opioid-mediated analgesia (Auclair, Drouin, Cotecchia, Glowinski, & Tassin, 2004; Roerig, Lei, Kitto, Hylden, & Wilcox, 1992; Stone, MacMillan, Kitto, Limbird, & Wilcox, 1997). Maximal pain inhibition at minimal dosages requires activation of both receptors simultaneously.

Synergistic effects of adrenergic-opioid receptor cross-talk in the treatment of pain

Processing of pain signals via the spinal cord occurs by means of descending control pathways that can be affected jointly by opioids and neurotransmitters such as norepinephrine and serotonin in the dorsal horn of the spinal cord (Bannister & Dickenson, 2016). The consequence of colocalization of opioid and aminergic receptors within these neurons and subsequent coactivation and heterodimerization is to inextricably integrate adrenergic and opioid function. Purposefully using this integration can improve pain management (Aira et al., 2014; Guidolin, Marcoli, Tortorella, Maura, & Agnati, 2019; Root-Bernstein et al., 2018, 2019).

For example, Meert and De Kock (1994) and Unal et al. (2013) demonstrated that doses of the opioids sufentanil and fentanyl that were too small to produce any antinociceptive effect exerted significant pain tolerance and analgesia when combined with doses of alpha-2 agonists that also had no antinociceptive effect. The order of this adrenergic enhancement was medetomidine > dexmedetomidine > xylazine > clonidine > detomidine. The enhancing effect was observed whether the opioid and adrenergic agonists were delivered simultaneously as subcutaneous injections, as a combination of intraperitoneal (alpha-2 agonist) plus subcutaneous (opioid), intravenous (alpha-2 agonist) plus epidural (opioid), or simultaneous epidural or intrathecal administrations. Clonidine and other alpha-2 agonists such as dexmedetomidine (Keating, 2015) are often used to potentiate opioids following surgery and for the treatment of chronic pain (Prabhakar et al., 2019) resulting in opioid sparing (Tonner, 2017). Several of the cited studies have also documented an increased duration of analgesia produced by these adrenergic-opioid combinations. Moreover, adrenergic agonists have been used to enhance opioid anesthesia and analgesia in opioid-tolerant individuals (Stromer, Michaeli, & Sanders-Kiesling, 2013).

Similar enhancement of opioid-induced antinociception and analgesia has been reported for adrenergic beta blockers such as propranolol (Afify & Andijani, 2017) resulting in their widespread use during general anesthesia to extend its duration and reduce postoperative pain (Härkänen, Halonen, Selander, & Kokki, 2015). The beta blocker esmolol, while exhibiting no anesthetic or analgesic properties itself, has been also found to be an effective opioid enhancer resulting in opioid sparing (Bahr & Williams, 2018). Dopamine D2 agonists also enhance, in a superadditive manner, the activity of opioids in treating peripheral nerve injury (Aira et al., 2014).

Beyond these well-established opioid-adrenergic combinations, there appears to be a huge range of alpha- and beta-adrenergic agonists and antagonists still to be explored for their opioid enhancing and sparing effects.

Prevention of desensitization and mutual resensitization of opioid and adrenergic receptor function by each other's ligands

Perhaps the most surprising and clinically important observation regarding opioid-adrenergic cross-talk is that each class of compounds is able to resensitize the other's receptor function after fade, tachyphylaxis, or development of tolerance, greatly increasing the duration of drug activity and preventing the development of drug dependence. Fade is the acute diminution of activity caused by a single dose of a drug and involving inactivation of receptors by phosphorylation. Tachyphylaxis is the decreased response of a tissue to repeated doses of a drug due to internalization of receptors following their phosphorylation. Chronic use of a drug results in tolerance due to the decrease in the number of activatable receptors on the cell surface. As can be predicted from the model described earlier (Fig. 3), adrenergics can prevent or reverse fade, tachyphylaxis, and tolerization to opioids and vice-versa. In both cases, the presumed mechanism is interference with phosphodiesterase activity and thereby receptor inactivation and internalization.

Many experiments confirm that cells or tissues tolerized to either an opioid or an alpha-2-adrenergic receptor agonist and then deprived of that compound can have their tolerance reversed by treatment with the other compound (Lee, Rosenberg, & Musacchio, 1988). Adrenergic downregulation is reversed by opioids such as morphine and enkephalins (Deyo, Swift, & Miller, 1979; Park et al., 2009; Puri et al., 1975). As predicted by the model in Fig. 3, the reversal is mediated by increased cGMP production and concomitant inhibition of the G-protein–mediated phosphorylation. Conversely, alpha-2 adrenoceptor agonists (West, Dykstra, & Lysle, 1999), alpha-2 adrenoceptor antagonists (Milne, Sutak, Cahill, & Jhamandas, 2008), and beta-2 adrenoceptor agonists (Satarian, Javan, & Fathollahi, 2008) prevent opioid receptor desensitization. Tolerance to opioids such as fentanyl is prevented or reversed by both adrenergic alpha-2-agonists (De Kockm & Meert, 1995) and antagonists (Boronat, Olmos, & García-Sevilla, 1998) with the result that the development of opioid tolerance is inhibited and reversible (De Kockm & Meert, 1995; Milne et al., 2008; Satarian et al., 2008; West et al., 1999).

The utility of delivering appropriate combinations of adrenergic and opioid drugs in order to lower the doses (and thus, side effects) of each and to simultaneously inhibit the development of tolerance to these drugs should be evident. Notably, however, there appear to be no *clinical* studies of resensitization of a faded or tachyphylactic opioid dose but the animal studies cited above strongly suggest that small doses of an adrenergic compound might reconstitute opioid analgesia after it has waned. As noted above, clinically inactive doses of adrenergic agonists and antagonists are able to enhance opioids. However, it must be cautioned that such enhancement also confers risks since the effects of coadministering opioids and adrenergics together are superadditive.

An even better approach to improving opioid analgesia and antinociception might be to deliver low doses of opioids in combinations with adrenergics so as to enhance and prolong their activity. Such an approach would drop the dose of opioid required to achieve any given degree of anesthesia in the first place; lower doses of opioids required to maintain analgesia; decrease risk of adverse side effects; and to prevent desensitization and the development of tolerance to both the opioid and adrenergic.

Applications to other areas: Local anesthetic enhancement and opioid sparing uses

The mechanisms underlying opioid-adrenergic enhancement and modulation may also explain the use of intravenous lidocaine as an opioid-sparing therapy (Beaussier, Delbos, Maurice-Szamburski, Ecoffey, & Mercada, 2018; Li, Wang, Xu, Ding, & Yu, 2018; MacFater et al., 2018) as well as the synergy of this class of local anesthetics with adrenergic and opioid drugs.

Epinephrine has been used for decades to increase the duration of action of local anesthetics such as lidocaine. The putative mechanism for this effect is a reduction in local blood flow due to the epinephrine. However, Sinnott, Cogswel III, Johnson, and Strichartz (2003) and Palmer et al. (2002) have argued that, in Palmer's words, "mechanisms other than local vasoconstriction may contribute to the prolongation of lidocaine nerve blocks by epinephrine." These mechanisms include direct actions on opioid and adrenergic receptors.

Amino-amide anesthetics, including lidocaine, procaine, prilocaine, tetracaine, and bupivacaine, bind directly to kappa and mu opioid receptors at clinically relevant concentrations (Cohen & Mao, 2003; Hirota, Okawa, Appadu, Grandy, & Lambert, 2000) and the anesthetic effects of lidocaine are enhanced in an concentration-dependent fashion by catecholamines such as epinephrine, dopamine, clonidine, and tizanidine (Ohkado, Ichinohe, & Kaneko, 2001) exactly as are opioids.

Lidocaine also competes with adrenergic drugs for adrenergic receptors. Mepivacaine, ropivacaine, bupivacaine, lidocaine, prilocaine, etidocaine, procaine, and tetracaine bind to beta1 and beta2 adrenergic receptors (Butterworth, James, & Grimes, 1997; Li, Cheng, & Guo, 1997) and naloxone can enhance their effects (Chen, Shieh, Liu, Wang, & Hung, 2017) just as it enhances catecholamine effects. Lidocaine binding to adrenergic receptors might explain the cardiovascular inhibition observed in overdoses with this class of drugs.

Other agents of interest: Ascorbic acid, tramadol and tapentadol, and ketamine

Vitamin C (ascorbic acid) binds to adrenergic receptors at the same sites identified for opioid binding, enhancing the effect of any submaximal dose of the adrenergic drug (Dillon et al., 2010; Dillon, Root-Bernstein, & Lieder, 2004; Root-Bernstein et al., 2019, 2020; Root-Bernstein & Dillon, 2002). In light of the adrenergic-opioid heterodimerization model presented above, this observation suggests that vitamin C should have opioid-like activity. In fact, vitamin C has demonstrated analgesic properties at high doses (Chaitanya et al., 2018; Chen, Roffey, Dion, Arab, & Wai, 2016; Jain, Dar, Kumar, Yadav, & Kearns, 2019); it has been used as an adjunctive peri- and postoperative therapy for opiate-sparing (reviewed in Carr & McCall, 2017; Chen et al., 2016; Laflı Tunay et al., 2020); and it can prevent development of morphine tolerance in mice (Khanna & Sharma, 1983). Notably, vitamin C also potentiates local anesthetics such as lidocaine (Tian & Li, 2016) and tramadol (Zeraati, Araghchian, & Farjoo, 2014). This body of evidence suggests that adding vitamin C to morphine drips might both decrease self-administered dose and prevent the development of opioid dependence.

Tramadol and tapentadol are also interesting drugs to consider from the perspective of adrenergic-opioid interactions. Often classified as atypical opioid analgesics, tramadol and tapentadol also acts as adrenergic reuptake inhibitors, thereby stimulating the opioid receptor directly and also enhancing their own effects by increasing the concentration of local adrenergics. Additionally, these drugs produce antinociceptive effects by binding directly to adrenoceptors (Höcker et al., 2008; Li, Chen, Chen, Huang, & Liu, 2012; Torres-Sanchez, Borges, Mico, & Berrocoso, 2018) and chronic use results in down-regulation of these receptors (Faron-Górecka et al., 2004). This class of drugs therefore points toward the efficacy of designing drugs with dual adrenergic-opioid binding capacity.

Finally, ketamine has also been found to enhance opioid receptor function (Gupta, Devi, & Gomes, 2011) through an alpha-2-adrenoceptor mechanism (Cabral, Kapusta, Kenigs, & Varner, 1998) reversible by yohimbine, an alpha-2-adrenergic antagonist (Kilander & Williams, 1992). This mechanism may explain the synergistic antinociceptive effects of combining ketamine with opioids (Pelissier, Laurido, Kramer, Hernández, & Paeile, 2003).

Mini-dictionary of terms

GPCR: G-protein–coupled receptor.
Receptor homodimer: Complex formed by a pair of identical receptors.
Receptor heterodimer: Complex formed by a pair of different classes of receptors.
Oligomerization: Another word for dimerization.
Allostery: Regulation of a receptor by binding an effector molecule at a site other than the receptor's active site.
Enhancement: Increase in receptor activity due the allosteric binding of a compound that is not able to activate the receptor by itself.
Synergy: Superadditive activity produced by a combination of two compounds.
Fade: Acute loss of receptor activity following a single dose of a drug.
Tachyphylaxis: Loss of receptor activity following repeated doses of a drug.
Tolerance: Loss of drug activity due to decreased numbers of receptors.

Key facts

- Pain control using opioids can be improved by combining them with adrenergic drugs in order enhance their efficacy in a superadditive manner that retards the development of tolerance and increases their duration of action.
- This enhanced activity is mediated by cross-talk between adrenergic and opioid receptors involving multiple layers of interaction including adrenergic drugs allosterically binding to opioid receptors, opioids binding allosterically to adrenergic receptors, and heterodimerization between opioid and adrenergic receptors.
- Enhancement results in greater opioid activity at lower doses; increased duration of activity; decreased side effects; and resistance to development of opioid tolerance.

Summary points

- Adrenergic drugs bind to opioid receptors enhancing opioid activity.
- Opioids bind to adrenergic receptors enhancing adrenergic activity.
- Adrenergic receptors dimerize with opioid receptors integrating their functional cross-talk.
- Adrenergic drugs enhance or synergize with opioids producing superadditive effects.
- Enhancement and synergy lower the dose of opioids necessary to produce analgesia and antinociception.
- Adrenergic enhancement extends the duration of activity of opioids and vice-versa.
- Adrenergic enhancement prevents development of opioid tolerance and vice-versa.
- Ascorbic acid can also enhance the adrenergic-opioid system preventing development of tolerance.
- The opiate-sparing effects of lidocaine and tramadol are mediated through the combined adrenergic-opioid receptor system.

References

Afify, E. A., & Andijani, N. M. (2017). Potentiation of morphine-induced antinociception by propranolol: The involvement of dopamine and GABA systems. *Frontiers in Pharmacology*, *8*, 794. https://doi.org/10.3389/fphar.2017.00794.

Aira, Z., Barrenetxea, T., Buesa, I., Gómez-Esteban, J. C., & Azkue, J. J. (2014). Synaptic upregulation and superadditive interaction of dopamine D2- and μ-opioid receptors after peripheral nerve injury. *Pain*, *155*(12), 2526–2533. https://doi.org/10.1016/j.pain.2014.09.012.

Auclair, A., Drouin, C., Cotecchia, S., Glowinski, J., & Tassin, J. P. (2004). 5-HT2A and alpha1b-adrenergic receptors entirely mediate dopamine release, locomotor response and behavioural sensitization to opiates and psychostimulants. *European Journal of Neuroscience*, *20*(11), 3073–3084.

Bahr, M. P., & Williams, B. A. (2018). Esmolol, antinociception, and its potential opioid-sparing role in routine anesthesia care. *Regional Anesthesia and Pain Medicine*, *43*(8), 815–818. https://doi.org/10.1097/AAP.0000000000000873.

Bannister, K., & Dickenson, A. H. (2016). What do monoamines do in pain modulation? *Current Opinion in Supportive and Palliative Care*, *10*(2), 143–148. https://doi.org/10.1097/SPC.0000000000000207.

Beaussier, M., Delbos, A., Maurice-Szamburski, A., Ecoffey, C., & Mercada, L. (2018). Perioperative use of intravenous lidocaine. *Drugs*, *78*(12), 1229–1246. https://doi.org/10.1007/s40265-018-0955-x.

Boronat, M. A., Olmos, G., & García-Sevilla, J. A. (1998). Attenuation of tolerance to opioid-induced antinociception and protection against morphine-induced decrease of neurofilament proteins by idazoxan and other I2-imidazoline ligands. *British Journal of Pharmacology*, *125*(1), 175–185.

Breit, A., Lagacé, M., & Bouvier, M. (2004). Hetero-oligomerization between beta2- and beta3-adrenergic receptors generates a beta-adrenergic signaling unit with distinct functional properties. *Journal of Biological Chemistry*, *279*(27), 28756–28765.

Butterworth, J., James, R. L., & Grimes, J. (1997). Structure-affinity relationships and stereospecificity of several homologous series of local anesthetics for the beta2-adrenergic receptor. *Anesthesia and Analgesia*, *85*(2), 336–342.

Cabral, A. D., Kapusta, D. R., Kenigs, V. A., & Varner, K. J. (1998). Central alpha2-receptor mechanisms contribute to enhanced renal responses during ketamine-xylazine anesthesia. *American Journal of Physiology*, *275*(6), R1867–R1874. https://doi.org/10.1152/ajpregu.1998.275.6.R1867.

Carr, A. C., & McCall, C. (2017). The role of vitamin C in the treatment of pain, new insights. *Journal of Translational Medicine*, *15*(1), 77. https://doi.org/10.1186/s12967-017-1179-7.

Chabot-Doré, A. J., Schuster, D. J., Stone, L. S., & Wilcox, G. L. (2015). Analgesic synergy between opioid and α2-adrenoceptors. *British Journal of Pharmacology*, *172*(2), 388–402. https://doi.org/10.1111/bph.12695.

Chaitanya, N. C., Muthukrishnan, A., Krishnaprasad, C. M. S., Sanjuprasanna, G., Pillay, P., & Mounika, B. (2018). An insight and update on the analgesic properties of vitamin C. *Journal of Pharmacy and Bioallied Sciences*, *10*(3), 119–125. https://doi.org/10.4103/jpbs.JPBS_12_18.

Chen, Z. J., & Minneman, K. P. (2005). Recent progress in alpha1-adrenergic receptor research. *Acta Pharmacologica Sinica*, *26*(11), 1281–1287.

Chen, S., Roffey, D. M., Dion, C.-A., Arab, A., & Wai, E. K. (2016). Effect of perioperative vitamin C supplementation on postoperative pain and the incidence of chronic regional pain syndrome. *Clinical Journal of Pain*, *32*(2), 179–185.

Chen, Y. W., Shieh, J. P., Liu, K. S., Wang, J. J., & Hung, C. H. (2017). Naloxone prolongs cutaneous nociceptive block by lidocaine in rats. *Fundamental and Clinical Pharmacology*, *31*(6), 636–642. https://doi.org/10.1111/fcp.12302.

Cohen, S. P., & Mao, J. (2003). Is the analgesic effect of systemic lidocaine mediated through opioid receptors? *Acta Anaesthesiologica Scandinavica*, *47*(7). https://doi.org/10.1034/j.1399-6576.2003.00163.x. 910-1.

De Kockm, M., & Meert, T. F. (1995). Can alpha 2-adrenoceptor agonists reverse or prevent tolerance to the antinociceptive activity of opioids in rats? *Acta Anaesthesiologica Belgica*, *46*(3–4), 141–152.

Deyo, S. N., Swift, R. M., & Miller, R. J. (1979). Morphine and endorphins modulate dopamine turnover in rat median eminence. *Proceedings of the National Academy of Sciences of the United States of America*, *76*, 3006–3009.

Dillon, P. F., Root-Bernstein, R. S., & Lieder, C. M. (2004). Antioxidant-independent ascorbate enhancement of catecholamine-induced contractions of vascular smooth muscle. *American Journal of Physiology—Heart and Circulatory Physiology*, *286*(6), H2353–H2360.

Dillon, P. F., Root-Bernstein, R. S., & Lieder, C. M. (2006a). Ascorbate enhancement of H1 histamine receptor sensitivity coincides with ascorbate oxidation inhibition by histamine receptors. *American Journal of Physiology—Cell Physiology*, *291*(5), C977–C984.

Dillon, P. F., Root-Bernstein, R. S., & Lieder, C. M. (2006b). Molecular shielding of electric field complex dissociation. *Biophysical Journal*, *90*(4), 1432–1438.

Dillon, P. F., Root-Bernstein, R., Robinson, N. E., Abraham, W. M., & Berney, C. (2010). Receptor-mediated enhancement of beta adrenergic drug activity by ascorbate in vitro and in vivo. *PLoS One*, *5*(12), e15130. https://doi.org/10.1371/journal.pone.0015130.

Faron-Górecka, A., Kuśmider, M., Inan, S. Y., Siwanowicz, J., Piwowarczyk, T., & Dziedzicka-Wasylewska, M. (2004). Long-term exposure of rats to tramadol alters brain dopamine and alpha 1-adrenoceptor function that may be related to antidepressant potency. *European Journal of Pharmacology*, *501*(1–3), 103–110.

Fujita, W., Gomes, I., & Devi, L. A. (2014). Revolution in GPCR signalling, opioid receptor heteromers as novel therapeutic targets, IUPHAR review 10. *British Journal of Pharmacology*, *171*(18), 4155–4176. https://doi.org/10.1111/bph.12798.

Guidolin, D., Marcoli, M., Tortorella, C., Maura, G., & Agnati, L. F. (2019). Receptor-receptor interactions as a widespread phenomenon, novel targets for drug development? *Frontiers in Endocrinology (Lausanne)*, *10*, 53. https://doi.org/10.3389/fendo.2019.00053.

Gupta, A., Devi, L. A., & Gomes, I. (2011). Potentiation of μ-opioid receptor-mediated signaling by ketamine. *Journal of Neurochemistry*, *119*(2), 294–302. https://doi.org/10.1111/j.1471-4159.2011.07361.x.

Härkänen, L., Halonen, J., Selander, T., & Kokki, H. (2015). Beta-adrenergic antagonists during general anesthesia reduced postoperative pain, a systematic review and a meta-analysis of randomized controlled trials. *Journal of Anesthesia*, *29*(6), 934–943. https://doi.org/10.1007/s00540-015-2041-9.

He, J.-R., Molnar, J., & Barraclough, C. A. (1993). Morphine amplifies norepinephrine, NE, -induced LH release but blocks NE-stimulated increases in LHRH mRNA levels, comparison of responses obtained in ovariectomized, estrogen-treated normal and androgen-sterilized rats. *Molecular Brain Research*, *20*, 71–78.

Hirota, K., Okawa, H., Appadu, B. L., Grandy, D. K., & Lambert, D. G. (2000). Interaction of local anaesthetics with recombinant mu, kappa, and delta-opioid receptors expressed in Chinese hamster ovary cells. *British Journal of Anaesthesia*, *85*(5), 740–746.

Höcker, J., Weber, R. B., Tonner, P. H., Scholz, J., Brand, P. A., Ohnesorge, H., et al. (2008). Meperidine, remifentanil and tramadol but not sufentanil interact with alpha(2)-adrenoceptors in alpha(2A)-, alpha(2B)- and alpha(2C)-adrenoceptor knock out mice brain. *European Journal of Pharmacology*, *582*(1–3), 70–77. https://doi.org/10.1016/j.ejphar.2007.12.022.

Jain, S. K., Dar, M. Y., Kumar, S., Yadav, A., & Kearns, S. R. (2019). Role of anti-oxidant (vitamin-C) in post-operative pain relief in foot and ankle trauma surgery: A prospective randomized trial. *Foot and Ankle Surgery*, *25*(4), 542–545. https://doi.org/10.1016/j.fas.2018.05.001.

Keating, G. M. (2015). Dexmedetomidine: A review of its use for sedation in the intensive care setting. *Drugs*, *75*(10), 1119–1130. https://doi.org/10.1007/s40265-015-0419-5.

Khanna, N. C., & Sharma, S. K. (1983). Megadoses of vitamin C prevent the development of tolerance and physical dependence on morphine in mice. *Life Sciences*, *33*(Suppl. 1), 401–404.

Kilander, K., & Williams, H. (1992). Yohimbine reduces neuropathology induced by ketamine/xylazine anesthesia. *Physiology & Behavior*, *51*(3), 657–659.

Laflı Tunay, D., Türkeün Ilgınel, M., Ünlügenç, H., Tunay, M., Karacaer, F., & Biricik, E. (2020). Comparison of the effects of preoperative melatonin or vitamin C administration on postoperative analgesia. *Bosnian Journal of Basic Medical Sciences*, *20*(1), 117–124. https://doi.org/10.17305/bjbms.2019.4379.

Lee, S., Rosenberg, C. R., & Musacchio, J. M. (1988). Cross-dependence to opioid and alpha 2-adrenergic receptor agonists in NG108-15 cells. *The FASEB Journal*, *2*(1), 52–55.

Lengyel, I., Toth, F., Biyashev, D., Szatmari, I., Monory, K., Tomboly, C., et al. (2016). A novel non-opioid binding site for endomorphin-1. *Journal of Physiology and Pharmacology*, *67*(4), 605–616.

Li, Y., Cheng, S., & Guo, G. (1997). The effects of lidocaine on lymphocyte beta-adrenergic receptor at different pH. [Article in Chinese]. *Zhongguo Ying Yong Sheng Li Xue Za Zhi*, *13*(2), 168–170.

Li, C., Chen, S. Q., Chen, B. X., Huang, W. Q., & Liu, K. X. (2012). The antinociceptive effect of intrathecal tramadol in rats: The role of alpha 2-adrenoceptors in the spinal cord. *Journal of Anesthesia*, *26*(2), 230–235. https://doi.org/10.1007/s00540-011-1267-4.

Li, J., Wang, G., Xu, W., Ding, M., & Yu, W. (2018). Efficacy of intravenous lidocaine on pain relief in patients undergoing laparoscopic cholecystectomy, a meta-analysis from randomized controlled trials. *International Journal of Surgery*, *50*, 137–145.

MacFater, W. S., Xia, W., Barazanchi, A., Su'a, B., Svirskis, D., & Hill, A. G. (2018). Intravenous local anaesthetic compared with intraperitoneal local anaesthetic in abdominal surgery: A systematic review. *World Journal of Surgery*, *42*(10), 3112–3119. https://doi.org/10.1007/s00268-018-4623-9.

Meert, T. F., & De Kock, M. (1994). Potentiation of the analgesic properties of fentanyl-like opioids with alpha 2-adrenoceptor agonists in rats. *Anesthesiology*, *81*(3), 677–688.

Milne, B., Sutak, M., Cahill, C. M., & Jhamandas, K. (2008). Low doses of alpha 2-adrenoceptor antagonists augment spinal morphine analgesia and inhibit development of acute and chronic tolerance. *British Journal of Pharmacology*, *155*(8), 1264–1278. https://doi.org/10.1038/bjp.2008.353.

Milner, T. A., Pickel, V. M., & Reis, D. J. (1989). Ultrastructural basis for interactions between central opioids and catecholamines: I. Rostral ventrolateral medulla. *Journal of Neuroscience*, *9*(6), 2114–2130.

Moffat, A. C., Patterson, D. A., Curry, A. S., & Owen, P. (1972). Inhibition in vitro of cyclic adenosine 3′4′-monophosphate phosphodiesterase activity by drugs. *European Journal of Toxicology*, *5*, 160–162.

Monroe, P. J., Perschke, S. E., Crisp, T., & Smith, D. J. (1991). Evaluation of the interactions of serotonergic and adrenergic drugs with mu, delta, and kappa opioid binding sites. *Neuroscience Letters*, *133*(2), 229–232.

Munro, T. A., Huang, X.-P., Inglese, C., Perrone, M. G., Van't Veer, A., et al. (2013). Selective k opioid antagonists nor-BNI, GNTI and JDTic have low affinities for non-opioid receptors and transporters. *PLoS One*, *8*(8), e70701. https://doi.org/10.1371/journal.pone.0070701.

Ohkado, S., Ichinohe, T., & Kaneko, Y. (2001). Comparative study on anesthetic potency depending on concentrations of lidocaine and epinephrine, assessment of dental local anesthetics using the jaw-opening reflex. *Anesthesia Progress*, *48*(1), 16–20.

Palmer, G. M., Cairns, B. E., Berkes, S. L., Dunning, P. S., Taylor, G. A., & Berde, C. B. (2002). The effects of lidocaine and adrenergic agonists on rat sciatic nerve and skeletal muscle blood flow in vivo. *Anesthesia and Analgesia*, *95*(4), 1080–1086.

Park, W. K., Chang, C. H., Chae, J. E., Kim, M. H., Cho, Y. L., & Ahn, D. S. (2009). Phosphodiesterase inhibition by naloxone augments the inotropic actions of beta-adrenergic stimulation. *Acta Anaesthesiologica Scandinavica*, *53*, 1043–1051.

Pelissier, T., Laurido, C., Kramer, V., Hernández, A., & Paeile, C. (2003). Antinociceptive interactions of ketamine with morphine or methadone in mononeuropathic rats. *European Journal of Pharmacology*, *477*(1), 23–28.

Prabhakar, A., Lambert, T., Kaye, R. J., Gaignard, S. M., Ragusa, J., Wheat, S., et al. (2019). Adjuvants in clinical regional anesthesia practice: A comprehensive review. *Best Practice and Research: Clinical Anaesthesiology*, *33*(4), 415–423. https://doi.org/10.1016/j.bpa.2019.06.001.

Puri, S. K., Cochin, J., & Volicer, L. (1975). Effect of morphine sulfate on adenylate cyclase and phosphodiesterase activities in rat corpus striatum. *Life Sciences*, *16*(5), 759–767.

Roerig, S. C., Lei, S., Kitto, K., Hylden, J. K., & Wilcox, G. L. (1992). Spinal interactions between opioid and noradrenergic agonists in mice: Multiplicativity involves delta and alpha-2 receptors. *Journal of Pharmacology and Experimental Therapeutics*, *262*(1), 365–374.

Root-Bernstein, R. S. (1987). Catecholamines bind to enkephalins, morphiceptin, and morphine. *Brain Research Bulletin*, *18*(4), 509–532.

Root-Bernstein, R., Churchill, B., & Turke, M. (2020). Glutathione and glutathione-like sequences of aminergic receptors bind ascorbic acid, adrenergic and opioid drugs mediating anti-oxidant function: Relevance for anesthesia and addiction. *International Journal of Molecular Sciences*, *21*(17). https://doi.org/10.3390/ijms21176230. E6230.

Root-Bernstein, R., Churchill, B., Turke, M., Subhramanyam, U. K. T., & Labahn, J. (2019). Mutual enhancement of opioid and adrenergic receptors by combinations of opioids and adrenergic ligands is reflected in molecular complementarity of ligands: Drug development possibilities. *International Journal of Molecular Sciences*, *20*(17). https://doi.org/10.3390/ijms20174137. pii: E4137.

Root-Bernstein, R. S., & Dillon, P. F. (2002). Fostering adventure research. A case study of the discovery that ascorbic acid enhances adrenergic drug activity. *Drug Development Research*, *57*, 58–74.

Root-Bernstein, R., & Dillon, P. F. (2014). A common molecular motif characterizes extracellular allosteric enhancers of GPCR aminergic receptors and suggests enhancer mechanism of action. *Current Medicinal Chemistry*, *21*(32), 3673–3686.

Root-Bernstein, R., Turke, M., Subhramanyam, U. K. T., Churchill, B., & Labahn, J. (2018). Adrenergic agonists bind to adrenergic-receptor-like regions of the mu opioid receptor, enhancing morphine and methionine-enkephalin binding: A new approach to "biased opioids"? *International Journal of Molecular Sciences*, *19*(1). https://doi.org/10.3390/ijms19010272. pii: E272.

Samoshkin, A., Convertino, M., Viet, C. T., Wieskopf, J. S., Kambur, O., Marcovitz, J., et al. (2015). Structural and functional interactions between six-transmembrane μ-opioid receptors and β2-adrenoreceptors modulate opioid signaling. *Scientific Reports*, *5*, 181–198. https://doi.org/10.1038/srep18198.

Satarian, L., Javan, M., & Fathollahi, Y. (2008). Epinephrine inhibits analgesic tolerance to intrathecal administrated morphine and increases the expression of calcium-calmodulin-dependent protein kinase IIalpha. *Neuroscience Letters*, *430*(3), 213–217.

Sinnott, C. J., Cogswel, L. P., III, Johnson, A., & Strichartz, G. R. (2003). On the mechanism by which epinephrine potentiates lidocaine's peripheral nerve block. *Anesthesiology*, *98*, 181–188. https://doi.org/10.1097/0000542-200301000-00028.

Stone, L. S., MacMillan, L. B., Kitto, K. F., Limbird, L. E., & Wilcox, G. L. (1997). The alpha2a adrenergic receptor subtype mediates spinal analgesia evoked by alpha2 agonists and is necessary for spinal adrenergic-opioid synergy. *Journal of Neuroscience*, *17*(18), 7157–7165.

Stromer, W., Michaeli, K., & Sanders-Kiesling, A. (2013). Perioperative pain therapy in opioid abuse. *European Journal of Anaesthesiology*, *30*, 55–64.

Tian, J., & Li, Y. (2016). Comparative effects of vitamin C on the effects of local anesthetics ropivacaine, bupivacaine, and lidocaine on human chondrocytes. *Brazilian Journal of Anesthesiology*, *66*(1), 29–36. https://doi.org/10.1016/j.bjane.2015.01.006.

Tonner, P. H. (2017). Additives used to reduce perioperative opioid consumption 1: Alpha2-agonists. *Best Practice and Research: Clinical Anaesthesiology*, *31*(4), 505–512. https://doi.org/10.1016/j.bpa.2017.10.004.

Torres-Sanchez, S., Borges, G. D. S., Mico, J. A., & Berrocoso, E. (2018). Opioid and noradrenergic contributions of tapentadol to the inhibition of locus coeruleus neurons in the streptozotocin rat model of polyneuropathic pain. *Neuropharmacology*, *135*, 202–210. https://doi.org/10.1016/j.neuropharm.2018.03.014.

Unal, M., Gursoy, S., Altun, A., Duger, C., Kol, I. O., Kaygusuz, K., et al. (2013). Ineffective doses of dexmedetomidine potentiates the antinociception induced by morphine and fentanyl in acute pain model. *Korean Journal of Physiology and Pharmacology*, *17*(5), 417–422. https://doi.org/10.4196/kjpp.2013.17.5.417.

Vogt, L. J., Sim-Selley, L. J., Childers, S. R., Wiley, R. G., & Vogt, B. A. (2001). Colocalization of mu-opioid receptors and activated G-proteins in rat cingulate cortex. *Journal of Pharmacology and Experimental Therapeutics*, *299*(3), 840–848.

West, J. P., Dykstra, L. A., & Lysle, D. T. (1999). Immunomodulatory effects of morphine withdrawal in the rat are time dependent and reversible by clonidine. *Psychopharmacology*, *146*(3), 320–327.

Wilkinson, M., Jacobson, W., & Wilkinson, D. A. (1985). The noradrenergic neurotoxins DSP4 and xylamine bind to opiate receptors. *Brain Research Bulletin*, *14*(5), 493–495.

Zeraati, F., Araghchian, M., & Farjoo, M. H. (2014). Ascorbic acid interaction with analgesic effect of morphine and tramadol in mice. *Anesthesia and Pain Medicine*, *4*(3). https://doi.org/10.5812/aapm.19529, e19529.

Chapter 9

Inflammatory and neuropathic pain impact on the opioid function in the mesocorticolimbic system

Yolanda Campos-Jurado, Javier Cuitavi, Natalia Landsberg, Jesús D. Lorente, and Lucia Hipólito
Department of Pharmacy and Pharmaceutical Technology and Parasitology, School of Pharmacy, University of Valencia, Burjassot, Valencia, Spain

Abbreviations

ACC	anterior cingulate cortex
BLA	basolateral amygdala
BNST	bed nucleus of the stria terminalis
CCP	conditioned place preference
CFA	complete Freund adjuvant
CNS	central nervous system
DA	dopamine
DAMGO	D-Ala^2, *N*-MePhe^4, Gly-ol]-enkephalin
DOR(s)	delta opioid receptor(s)
Dyn	dynorphyn
Dyn/KOR	dynorphyn/kappa opioid receptor system
GABA	gamma aminobutyric acid
IL	infralimbic cortex
KOR(s)	kappa opioid receptor
MAPK	mitogen-activated protein kinases
MCLS	mesocorticolimbic system
MOR(s)	mu opioid receptor(s)
NAc	nucleus accumbens
OIH	opioid-induced hyperalgesia
OLR1(s)	opioid like receptor(s)-1
OR(s)	opioid receptor(s)
PET	positron emission tomography
PFC	prefrontal cortex
PL	prelimbic cortex
VP	ventral pallidum
VTA	ventral tegmental area

Introduction

Chronic pain greatly impacts a considerable part of the world's population and around 30% of people suffer from it, burdening health care systems worldwide (Nijs et al., 2020). The way pain is processed by the central nervous system (CNS) has been widely studied, and the pain afferent pathways are well known (Basbaum, Bautista, Scherrer, & Julius, 2009). Nonetheless, pain can also modify the normal functioning of other neuronal pathways such as the mesocorticolimbic system (MCLS) through the action of opioid receptors (ORs).

The MCLS processes rewarding and aversive stimuli by controlling dopamine (DA) release in the different areas of the system (Darcq & Kieffer, 2018). In this sense, studies on ORs within this system have shown their key role in pain-related

The Neurobiology, Physiology, and Psychology of Pain. https://doi.org/10.1016/B978-0-12-820589-1.00045-2

comorbidities such as negative affect, anxiety, and addiction. In this chapter, we will address the impact of pain on the MCLS and its effects as a driver for these comorbidities while focusing mainly on pain-induced addiction disorders.

A short introduction to the opioid receptors

The endogenous opioidergic system was discovered in the 70s and since then the understanding of how it works has widely increased. This system comprises four ORs and several endogenous opioid peptides. ORs, namely μ (MOR), δ (DOR), κ (KOR), and opioid receptor like-1 (ORL1), are widespread rhodopsin-like G protein–coupled receptors that work as neuromodulators to produce analgesia. They share the 60% of their structure and possess 7-transmembraned domains; however, they are encoded by different genes (*OPRM1, OPRK1, OPRD1,* and *OPRL1*) (Wei & Loh, 2011). Moreover, ORs participate not only in pain processing but also in many other important mechanisms such as euphoria, miosis, sedation, truncation rigidity, addiction, nausea, and respiration depression (Al-Hasani & Bruchas, 2011). Nonetheless, not all ORs induce the same behavioral effects since, for example, KORs and DORs activation induces dysphoria whereas MORs activation leads to euphoria (Darcq & Kieffer, 2018). Furthermore, endogenous opioids, such as endorphins, dynorphins (Dyn), enkephalins, endomorphins, and orphanin FQ/nociceptin, act as neurotransmitters by binding ORs.

The neuromodulatory function of ORs derives from the intracellular pathway triggered by their activation. This cascade starts with G protein recruitment followed by its dissociation into the Gα and Gβγ subunits. This prevents Ca^{2+} from entering the cell and promotes the exit of K^+, which results in neuron hyperpolarization and neuronal firing inhibition. ORs activation also initiates MAPK and β-arrestin cascades and inhibits cAMP production (Fig. 1A).

The endogenous opioid system: A key component of analgesia, reward, and aversion

Under normal conditions, ORs in periaqueductal gray and the dorsal horn of the spinal cord and dorsal root ganglia act to calm the painful feeling when activated, by inhibiting neuronal firing (Nestler, Hyman, & Malenka, 2001). A recent study in humans showed that approximately 50% of dorsal root ganglia neurons contain MORs, and DORs are present in nearly the same neurons. Only 44% and 40% contained KORs and ORL1s, respectively. Interestingly, 83% of neurons contain KORs and MORs at the same time (Moy et al., 2020). MORs in the spinal cord are located on second-order nociceptive neurons among which we find lamina I projection neurons of the anterolateral tract and excitatory interneurons (Aicher, Punnoose, & Goldberg, 2000; Spike et al., 2002). DORs are widely expressed in the spinal cord and overlap with MORs only in some projection neurons of the anterolateral tract (Wang et al., 2018). Finally, KORs are on inhibitory interneurons (Xu, Bruchas, Ippolito, Gendron, & Chavkin, 2007) and OPRL1s are on laminae I–III (Ozawa et al., 2015). The action of the receptors on those specific neurons shapes the endogenous system of analgesia. Interestingly, exogenous opiates can also activate ORs to promote analgesia. However, the repeated use of those drugs derives in the tolerance and the opioid-induced hyperalgesia (OIH) phenomena (Lavand'homme & Steyaert, 2017).

As aforementioned, ORs are also present in the MCLS. This system is principally composed by DA neurons from the ventral tegmental area (VTA) that send direct projections to the nucleus accumbens (NAc) and the prefrontal cortex (PFC) and all the nuclei that are connecting with these areas, including amygdala basolateral (BLA) and central, thalamus, hippocampus, insula, dorsal striatum, bed nucleus of the stria terminalis (BNST), and habenula (Volkow & Morales, 2015). This system processes rewarding and aversive stimuli by mainly controlling DA release in the different areas. Interestingly, ORs, particularly MORs and KORs, play a key role in the normal functioning of the MCLS. In fact, DA release, controlled in part by local MORs and KORs, accounts for the activation of the system, which in turn encodes for reinforcement (natural and drug-induced) and aversion (Darcq & Kieffer, 2018) (Fig. 1B). Morphine, fentanyl, oxycodone, alcohol, and other synthetic opioids can interact with the opioid system within the MCLS and alter its normal functioning, which in turn derivates in addiction. Interestingly, some ORs exogenous ligands that act as antagonists have been used as parts of addiction therapies, such as naloxone, naltrexone, norbinaltorphimine, naltrindole, and β-funaltrexamine (Stein, 2016). A list of the different OR ligands and their selectivity to the different types of ORs can be found in Table 1.

Chronic pain alters brain function and connectivity recruiting motivational and emotional regions

Brain neuroimaging techniques have been classically selected to directly or indirectly measure the function and connectivity of brain regions. In this way, several differences have been reported between the brain regions involved in acute versus chronic pain, in both human and animal studies. Thereby, during acute pain, brain regions involved in the ascending

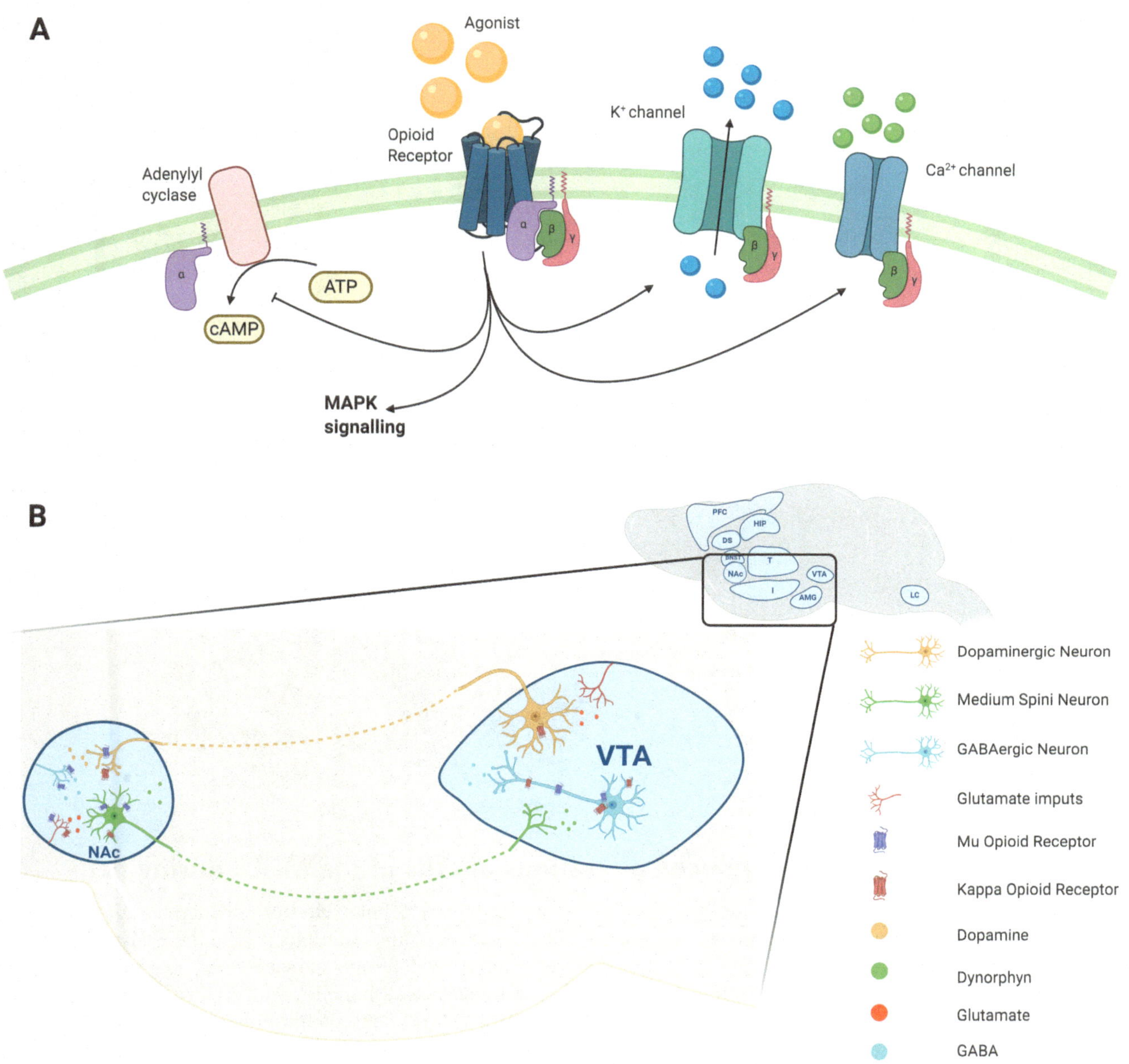

FIG. 1 OR from their cellular signaling to their location in the MCLS to modulate DA release. (A) When an agonist binds the OR, the G protein that is activated alongside the OR splits into the α subunit and the βγ subunit. The former inhibits cAMP production whereas the latter produces the hyperpolarization of the cell by inhibiting the entry of Ca^{2+} and promoting the exit of K^+. OR activation also promotes MAPK cascade activation. (B) Simplified diagram of VTA to NAc dopaminergic pathway of the MCLS. MORs and KORs differentially control DA release in the NAc. The activation of VTA MORs inhibits GABA interneurons and, as a consequence, activate local DA neurons increasing DA release in the NAc. The activation of MORs in the NAc can either increase or decrease local DA release, probably derived from MORs different cellular location in NAc subregions (Hipólito, Sánchez-Catalán, Zanolini, Polache, & Granero, 2008). The activation of VTA KORs inhibits DA neurons and decreases DA release in the NAc. Locally in the NAc, the activation of KORs, which are expressed in DA terminals, glutamatergic terminals, and medium spiny neurons, also leads to a decrease of DA release (Darcq & Kieffer, 2018). *(Figure created with biorender.com.)*

and descending nociceptive pathways are commonly activated (DaSilva & Seminowicz, 2019; Morton, Sandhu, & Jones, 2016). However, later stages of pain seem to specifically correlate with changes in brain structures involved in affection and motivation/reward. Actually, preclinical and clinical studies have reported enhanced activity of PFC, anterior cingulate cortex (ACC), hippocampus, amygdala, basal ganglia, and NAc in pain-suffering subjects (Bushnell, Ceko, & Low, 2013; DaSilva & Seminowicz, 2019).

TABLE 1 OR ligands and their selectivity to the different types of ORs.

Substance	Origin	Selectivity			
		MOR	KOR	DOR	OPRL1
β-Endorphins	Endogenous	+++	+++	+++	−
Enkephalins	Endogenous	++	−	+++	−
Dynorphins	Endogenous	++	+++	+	+
Endomorphins	Endogenous	+++	−	−	−
Orphanin FQ	Endogenous	−	−	−	+++
Fentanyl	Synthetic	+++	+	+	−
DAMGO	Synthetic	+++	−	−	−
Codeine	Synthetic	+++	+	+	−
Oxycodone	Synthetic	+++	+++	+	−
Morphine	Synthetic	+++	+	−	−
Methadone	Synthetic	++	−	−	−
Buprenorphine	Synthetic	++	+	−	−
Pentazocine	Synthetic	−	++	−	−
Naloxone	Synthetic	+++	++	++	−
Naltrexone	Synthetic	+++	++	++	−
Norbinaltorphimine	Synthetic	+	+++	+	−
Naltrindole	Synthetic	+	+	+++	−
β-Funaltrexamine	Synthetic	+++	−	−	−

PET imaging confirms pain-induced alterations of ORs in the MCLS during pain

In clinical studies, the positron emission tomography (PET) is a widely used technique to study changes in OR with the help of a radiotracer that can specifically bind one or all of the OR. Therefore, as the radiotracers bind the available ORs, PET results showing higher levels of signal indicate higher levels of receptor availability, and vice-versa. Regarding ORs in the MCLS in pain-suffering patients, two different types of radiotracers have been used. First, the [11C]diprenorphine has been selected as a nonspecific radiotracer for all the ORs. In Maarrawi et al. (2007), neuropathic pain patients showed a decrease in OR availability in the insula and PFC, suggesting a loss of OR expression or increased occupancy by the endogenous ligands. However, in another study with arthritic patients, higher levels of chronic pain correlated with greater OR availability in the basal ganglia, insula, caudate, and NAc (Brown et al., 2015). It is important to note that [11C]diprenorphine binds all the OR types, and, therefore, these results cannot be specifically attributed to changes in MORs function or expression. Second, [11C]carfentil has also been used as a specific radiotracer for MORs. Studies using [11C]carfentil showed a decrease of MORs availability in the amygdala, NAc, and ACC in patients suffering trigeminal neuropathic pain and fibromyalgia (Fig. 2) (DosSantos et al., 2012; Harris et al., 2007). These findings suggest a possible decrease in MORs function in the rewarding system in chronic pain patients. However, in human cross-sectional studies, findings cannot be directly attributed to the effect of the pain condition. In turn, the observed reduction in opioid binding could also be caused by the pain opioidergic treatment itself or even represent an intrinsic brain difference in people who might be prone to develop chronic pain.

Preclinical studies show altered MOR density or function derived from the presence of pain

A recent preclinical study has also shown interesting PET results in rats under chronic neuropathic pain. In this case, a lower binding of the OR tracer [18F]FDPN was found in the insula and caudate-putamen of chronic neuropathic pain rats

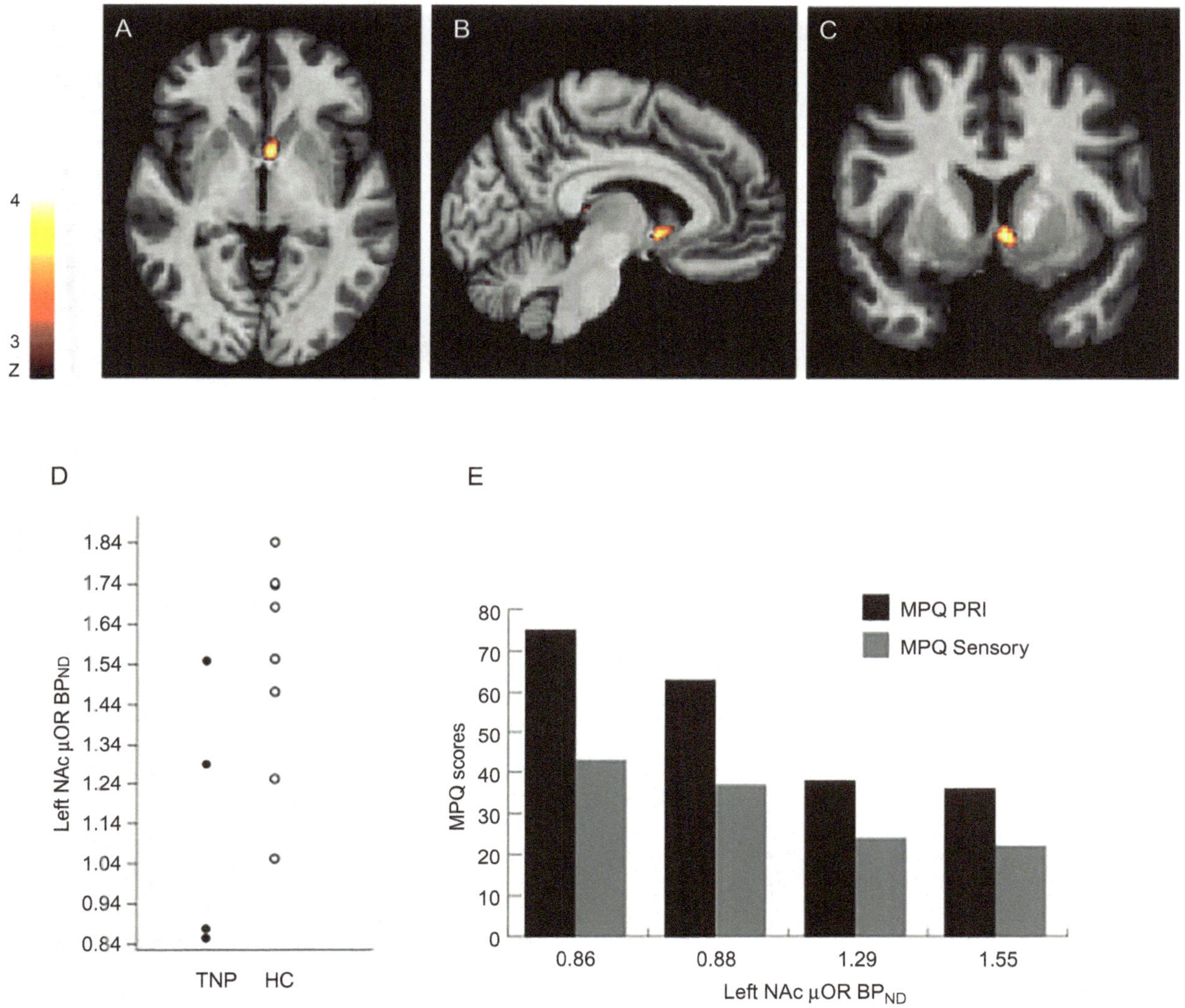

FIG. 2 MOR availability in the NAc in trigeminal neuropathic pain patients. (A–C) Decreased MOR nondisplaceable binding potential (BP_{ND}) in the left NAc in axial, sagittal and coronal planes ($T = 3.2$) respectively. (D) Representation of left NAc individual values of MOR BP_{ND} (*black circles* represent pain patients and *empty circles* represent healthy subjects). (E) Correlations between MacGuill Pain Questionnaire scores (PRI and Sensory) and MOR BP_{ND} in the left NAc of pain patients. *(From DosSantos, M. F., Martikainen, I. K., Nascimento, T. D., Love, T. M., Deboer, M. D., Maslowski, E. C., … DaSilva, A. F. (2012). Reduced basal ganglia µ-opioid receptor availability in trigeminal neuropathic pain: A pilot study.* Molecular Pain, 8, *74. https://doi.org/10.1186/1744-8069-8-74, with permission.)*

(see section "Mini-dictionary of terms), when compared to the sham group (Thompson et al., 2018). Very interestingly, these results correlated with reductions in MORs immunoreactivity in the same areas when analyzing the brain tissue from those rats. Moreover, the authors reported nonsignificant reductions in enkephalin content or in the number of neurons, likely indicating a reduced MORs density.

Apart from the PET results, several preclinical studies have analyzed the effects that different models of pain elicit in the function of MORs in the MCLS. Thus, some authors have demonstrated that pain induces a desensitization of the local MORs in the VTA. Concretely, Campos-Jurado et al. (2019), evaluated the neural activation of different VTA-projecting regions elicited by the administration of the MOR selective agonist, DAMGO, into the VTA. For that, cFos, an immediate early gene traditionally used as an indicator of neural activation, was measured by an immunohistochemical procedure in the following VTA-projecting regions: ACC, prelimbic cortex (PL), infralimbic cortex (IL), NAc, BNST, ventral pallidum (VP), and BLA. The effect of two different intra-VTA doses of DAMGO (7 and 14 ng) on the aforementioned regions was studied in rats under inflammatory pain, induced by the subcutaneous administration of the complete Freund adjuvant (CFA) into the hindpaw 48 h prior to the VTA treatment, or in control animals, subcutaneously injected with saline. Results from this study showed that, in pain-free rats, local MOR agonism in the VTA increased cFos expression in the majority of

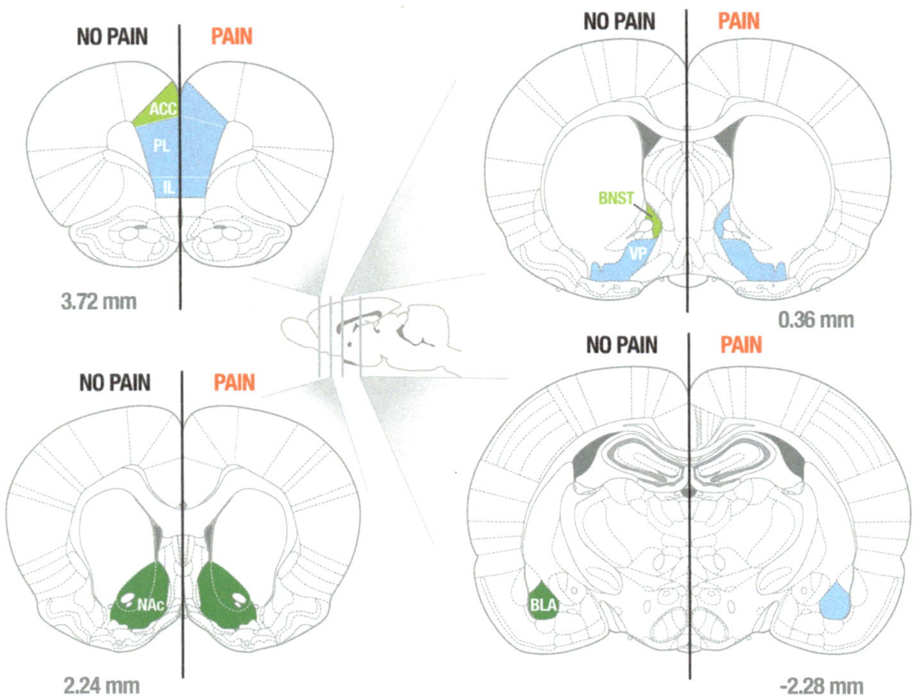

FIG. 3 Intra-VTA DAMGO-mediated activation (as c-Fos expression) of the MCLS of control and inflammatory pain rats. Colored areas indicate significant increase of c-Fos immunoreactive counting induced by intra-VTA administration of DAMGO 7 ng *(light green)* or DAMGO 14 ng *(dark green)*, compared to vehicle, or no significant changes *(blue)*. Left part of the coronal sections represents results for control rats and right part for inflammatory pain rats. Numbers indicate distance from bregma. *(Adapted from Paxinos, G., & Whatson, C. (2007).* The rat brain in stereotaxic coordinates *(6th ed.). London: Academic Press.)*

its projection areas (NAc, BLA, ACC, and BNST), with the exception of PL, IL, and VP (Fig. 3). However, when rats were under an inflammatory pain condition (CFA rats), this response was significantly diminished in most regions (BLA, ACC, and BNST).

Additionally, it is widely accepted that dopaminergic activity in the VTA-NAc pathway is tightly regulated by MORs (Fig. 1) (Darcq & Kieffer, 2018). Therefore, the possible effect of pain in MORs-evoked DA release in that pathway has also been explored. In fact, Ozaki et al. (2002), by combining the microdialysis and GTPγS techniques, concluded that, in the presence of neuropathic pain, rats showed a loss of morphine-induced DA release in the NAc due to reduced MORs function in the VTA. Also, Hipólito et al. (2015), showed that the local administration of DAMGO into the VTA failed to increase DA levels in rats under inflammatory pain only in the case of the lower DAMGO dose (7 ng), but not when a higher dose (67 ng) was administered (Fig. 4), further supporting a possible loss of MORs function in pain-suffering subjects. This fact is also supported by electrophysiological studies showing that inflammatory pain alters MOR-mediated GABA transmission in the VTA. Concretely, in neurons from CFA-treated rats, the lower DAMGO dose was unable to decrease GABA release in the VTA, whereas a higher dose was able to overcome that MOR desensitization. These results suggested a shift to the right in the dose-response curve for the MOR agonists; however, it is not completely clear if this shift is derived from a decreased expression of the MORs or a decreased function of the already expressed MORs (Hipólito et al., 2015).

Animal models of pain showed altered DA-related behaviors derived from the pharmacological activation of MORs

There are also behavioral studies showing that either neuropathic or inflammatory pain induce changes in the ability of MORs agonists to normally evoke their behavioral response that is mediated by the modulation of the DA activity in the MCLS. The MORs-mediated activation of the MCLS can result in the development of a preference for a specific environment, when the administration of the drug is performed in association with that context. These context-drug associations can be studied with the conditioned place preference (CPP) paradigm. In this way, Narita et al. (2005), showed an impairment of morphine-induced CPP (8 mg/kg, i.p.) in rats suffering from inflammatory pain. Moreover, a decrease

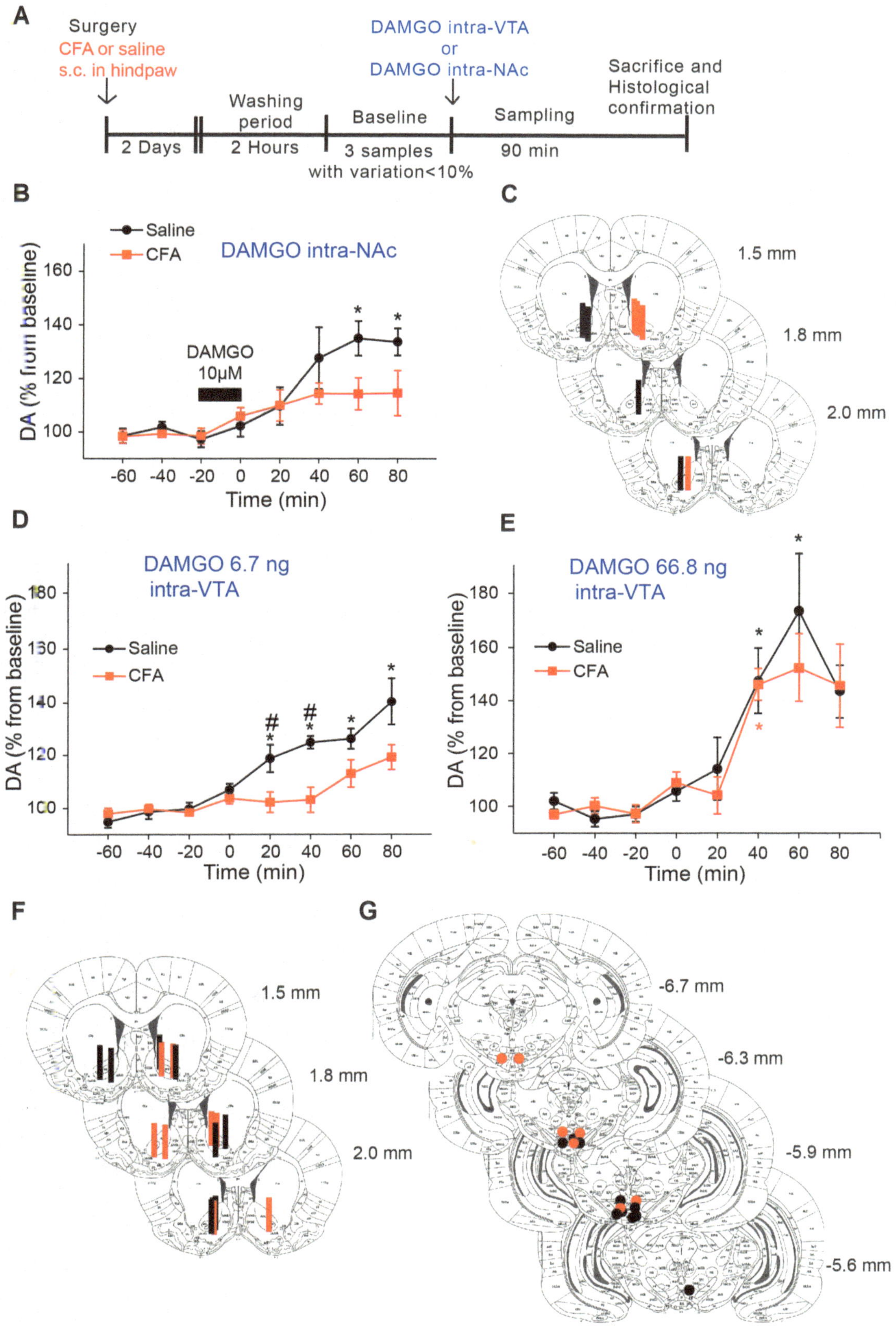

FIG. 4 Accumbal DA release (% from baseline) induced by DAMGO intra-VTA in control and inflammatory pain rats. (A) Experimental design. Effect of DAMGO applied in the NAc (B) or microinjected in the VTA on accumbal DA extracellular levels (D, E) of control *(black dots)* or inflammatory pain rats *(red squares)*. Data are mean ± SEM of DA in % from baseline ($n = 5$–6/group). $^{\#}P < .05$, significant differences between groups. $^{*}P < .05$, significant differences in the within-subjects effect for animals treated with saline or CFA (Bonferroni multiple comparisons). (C and F) Placement of microdialysis probes in the NAc for each experiment and (G) placements of the microinjections in VTA. *Bars and dots* could represent more than one probe or cannula placement. *(From Hipólito, L., Wilson-Poe, A., Campos-Jurado, Y., Zhong, E., Gonzalez-Romero, J., Virag, L., … Bruchas, M. R. (2015). Inflammatory pain promotes increased opioid self-administration: Role of dysregulated ventral tegmental area μ opioid receptors.* Journal of Neuroscience, 35 *(35), 12217–12231, with permission.)*

in morphine-evoked DA release in the NAc was reported to correlate with this behavior. In this study, none of the morphine doses administered (2, 4, or 8 mg/kg) were able to induce morphine place preference in the inflammatory pain group. However, another CPP study also using morphine administration reported a shift in the dose-response of the morphine CPP (Wu et al., 2014), as shown by previous reports aforementioned. In this case, the development of neuropathic pain in rats only blocked morphine-induced CPP at the lower dose (3.5 mg/kg, s.c.), but not at higher doses (5 or 7 mg/kg, s.c.). Furthermore, these differences in morphine-induced CPP seem to be unrelated to differences in its analgesic effect, as no differences in analgesia were found after 30 min of morphine administration (which was the duration of the conditioning sessions). Unfortunately, the fact that these studies were performed using systemic administration impairs us to attribute the observed changes to a loss of MORs function in specific regions of the MCLS. However, in another CPP study, Taylor et al. (2015), did actually administered DAMGO into the VTA. Concretely, in animals under neuropathic pain, the CPP elicited by intra-VTA administration of DAMGO was not developed when a lower dose was selected (1 ng). Contrarily, a higher DAMGO dose of 25 ng was able to overcome this impairment, and rats showed a robust preference for the DAMGO-paired compartment. Thus, an abnormal function of MORs in the VTA of pain animals appears to be the responsible for the shift in drugs dose-response for CPP.

Inflammatory pain promotes increased opioid self-administration

This shift in the dose-response of opioids, probably mediated by a dysregulation of the MORs, has also been reported in self-administration studies. Concretely, rats under inflammatory pain showed an altered heroin self-administration pattern, by not adapting their response to the different doses (i.v.) (Hipólito et al., 2015). In self-administration studies, animals are trained to perform a response (i.e., press a lever) to receive one administration of a drug at a specific dose. Therefore, under normal conditions, animals adapt the number of lever presses depending on the drug dosage, to obtain similar plasmatic concentrations and similar pharmacological effects. However, in Hipólito et al. (2015), rats under inflammatory pain did not readjust the number of lever presses according to the different heroin concentrations. Consequently, pain-suffering animals significantly increased their total heroin intake compared to the control group only when a high dose was available, even when both the lower and the highest dose presented similar antinociceptive properties. This finding suggests that the effect of pain on MORs function in the MCLS may cause a shift in the dose-response of opioidergic drugs, what may be translated into a higher dose drug consumption in pain-suffering patients.

Dynorfin/KOR system (Dyn/KOR) of the MCLS: A new key player in pain field

Recent preclinical reports have pointed to other OR located in MCLS areas (ACC, NAc, and amygdala) as a key mediator of pain-derived negative affect. Although KORs have classically been involved in stress-mediated behaviors and negative affect, its role in pain-induced negative affect has been recently uncovered (Lorente, Cuitavi, Campos-Jurado, & Hipólito, 2020). In fact, pain has shown to induce Dyn/KOR hyperactivity in amygdala and NAc (Massaly et al., 2019; Navratilova et al., 2019) and modify prodynorphin and OPRK1 mRNA levels in ACC, PFC, and NAc (Palmisano, Caputi, Mercatelli, Romualdi, & Candeletti, 2019). Indeed, the blockade of KORs in the central nucleus of the amygdala in neuropathic pain rats prevented gabapentin-induced CPP suggesting that these receptors might be involved in pain unpleasantness (Navratilova et al., 2019). Additionally, a very elegant study using a combination of behavioral, electrophysiology, immunohistochemistry, chemogenetics, and PET in rodents revealed that hyperactivity of the Dyn/KOR in the NAc medial shell was responsible of the pain-induced negative affect and decrease in motivation (Massaly et al., 2019). However, the specific mechanisms involved in how pain produces this Dyn/KOR hyperactivity and if it translated to human pain patients remain to elucidate (Fig. 5).

Applications to other areas

The United States of America are currently facing the opioid epidemic, which was created by a repetitive opioid misuse from medical prescriptions and irregular sources. In fact, 130 people die every day due to an opioid overdose (Nolan, Socias, & Wood, 2018). Nonetheless, the opioid crisis also affects countries and regions such as Canada where in 2017, they reported an increase of the 34% of the opioid overdose death rate in just 1 year (Lisa & Jessica, 2018), and central Europe and Australia with a progressive rise in opioid misuse (Berterame et al., 2016). Interestingly, in the United States, 80% of heroin users first misused prescribed opioids (Muhuri, Gfroerer, & Davies, 2013). Furthermore, recent investigations relate pain to drug addiction, giving hints of the role of pain as a risk factor toward drug addiction (Campos-Jurado et al., 2020; Hipólito et al., 2015). As shown in here, pain induces MORs and KORs alterations in the MCLS which, very

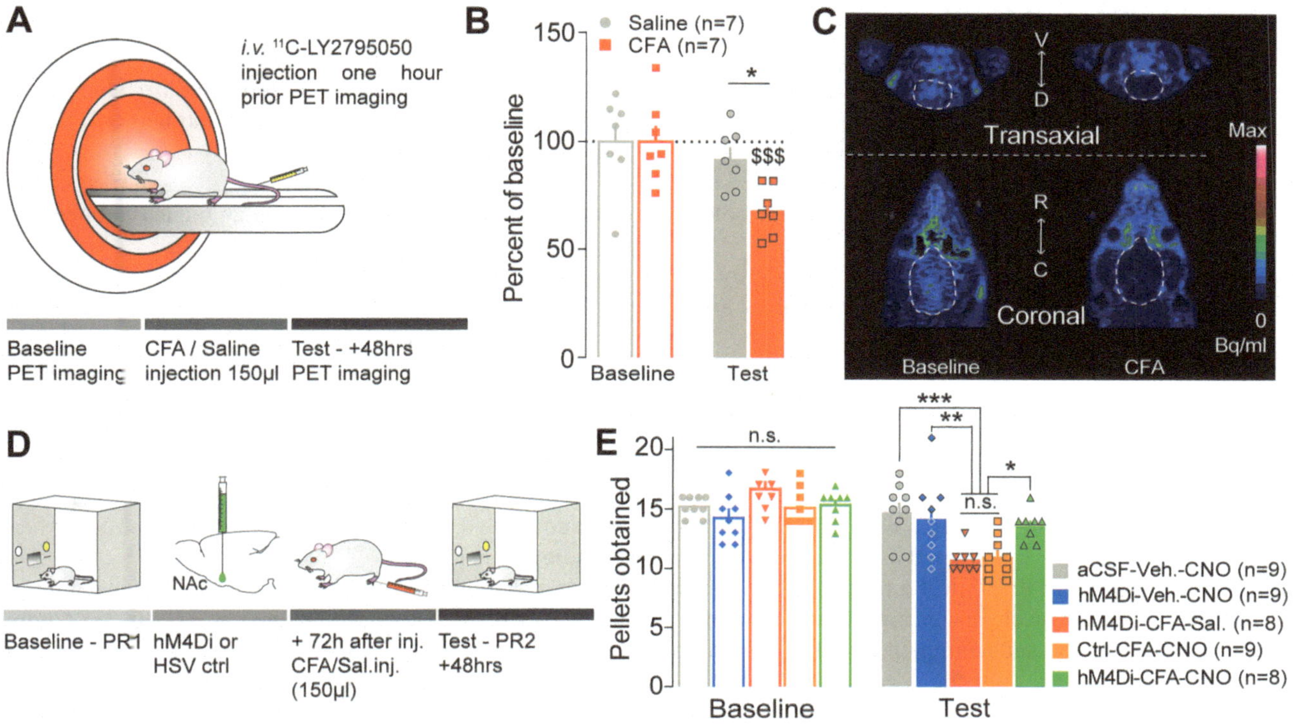

FIG. 5 Inflammatory pain mediates negative affective states through recruitment of dynorphin-containing neurons in the NAc. (A) Schematic representation of the PET imaging methodology in rats. (B) Distribution volume of 11C-LY2795050 in rat's brain (Sidak posthoc test: $*P < .05$, $$$ $P < .0001$, $n = 7$). (C) Representative images of PET imaging before (left) and after (right) CFA-injection. (D) Schematic representation of the behavioral methodology. (E) Silencing dynorphin neurons in the NAc prevented the inflammatory pain-induced decrease in motivation for sucrose self-administration (Sidak posthoc test: $*P < .05$, $**P < .01$, $***P < .001$). *(From Massaly, N., Copits, B. A., Wilson-Poe, A. R., Hipólito, L., Markovic, T., Yoon, H. J., ..., Morón, J. A. (2019). Pain-induced negative affect is mediated via recruitment of the nucleus accumbens kappa opioid system.* Neuron, 102*(3), 564–573.e6, with permission.)*

interestingly, are similar to those produced by opioid chronic consumption and withdrawal. Indeed, opioid addiction is characterized by MORs tolerance resulting in reduced opioid reward, and hyperactivity of the Dyn/KOR system during opioid withdrawal leading to negative affect. These two mechanisms are postulated to drive opioid compulsive consumption, dose escalation, and relapse (Darcq & Kieffer, 2018; Koob & Volkow, 2016).

Notably, some studies suggest that the outcome derived from ORs activation is sex-dependent, which implies that there might also be differences in the pain-induced changes in the ORs of the MCLS (Becker & Chartoff, 2019). In fact, a recent preclinical study shows that CFA injection, during the alcohol abstinence, induced higher KORs expression only in the NAc of female rats (Lorente et al., 2021). Therefore, it is essential to delve into sex-dependent changes when doing research related to pain and its comorbidities, including addiction, since this would help to address more personalized treatments.

Moreover, the study of the relationship between the CNS and the immune system has provided a new approach to many pathological conditions among which we find addiction and pain. In fact, the understanding of neuroimmune and, more concretely, neuroinflammatory mechanisms has unraveled many aspects of the noxious effects derived from drug exposure in the MCLS (Cantacorps et al., 2017). Interestingly, glial activation and neuroinflammation work alongside ORs and, therefore, play a definite role in addiction. Some studies relate the release of reactive oxygen species and proinflammatory cytokines with an upregulation of the MOR mRNA in different cell types (Börner, Kraus, Schröder, Ammer, & Höllt, 2004; Byrne, Peng, Sarkar, & Chang, 2012; Kraus, Börner, Giannini, & Höllt, 2003; Langsdorf, Mao, & Chang, 2011). Likewise, MOR activation enhances the production of inflammatory mediators, proinflammatory cytokines, and chemokines (Gessi et al., 2016; Shrivastava et al., 2017). Furthermore, Cahill and Taylor (2017) gathered up-to-date articles that show how neuroinflammation is partially responsible of linking chronic pain and addiction disorders. Nonetheless, the information available is still reduced and further research is required, but the confirmation of neuroinflammatory mechanisms would allow the development of new therapeutic approaches for both pain and its comorbidities including addiction.

Mini-dictionary of terms

Negative affective state: Is a psychophysical construct that refers to a subjective negative emotional state.

Pain-induced negative affective state: Is defined classically by the presence of unpleasant feelings derived from the painful stimulus, although current researches have broadened this definition to negative emotional and motivational consequences of chronic pain suffering. In this chapter, we will use this latest definition.

Reward: A stimulus that produces positive hedonic interoceptive responses.

Aversion: A tendency to extinguish a behavior or to avoid a situation when is associated with an unpleasant negative stimulus.

Reinforcement: Process that increases the probability of response. It is defined as ***positive*** when the stimulus is pleasant (e.g., food) and ***negative*** when the removal of an aversive stimulus (e.g., footshock) increases this probability of response.

Drug use disorder: "A chronically relapsing disorder, characterized by compulsion to seek and take the drug, loss of control in limiting intake, and emergence of a negative emotional state when access to the drug is prevented" (Koob & Volkow, 2016).

Conditioned place preference/aversion: Paradigm to study drug-induced context associations where animals show preference or aversion to the compartment previously paired to the drug exposition.

Drug self-administration: Paradigm used for study addiction under controlled conditions using laboratory animals where the dose of the drug is delivered as a consequence of a response of the animal.

Animal models of pain: Preclinical models of pain developed to mimic the clinical condition. These models are helpful to study pain-related mechanisms and potential therapeutic strategies (for a review, see Gregory et al., 2013). Among those, inflammatory and neuropathic pain models are two of the most commonly used. In **inflammatory pain models**, several irritant agents (capsaicin, carrageenan) or CFA can be injected in different organs to induce acute or chronic inflammation. Direct injury of a peripheral nerve has been used as a **model of neuropathic pain**, which includes the ligation of the spinal nerves, the ligation of the sciatic nerve (chronic constriction injury), and the ligation of distal branches (peroneal, tibial) of the sciatic nerve (spared nerve injury).

Key facts of pain-induced changes in OR function

- Pain impacts MOR function in the MCLS leading to a shift to the right of the dose-response curve and, consequently, impacting motivated behaviors.
- Dyn/KOR hyperactivity in the NAc mediates pain-induced negative affect.

Summary points

- The opioid system is crucial in the modulation of the analgesia and in the regulation of motivated behaviors.
- Inflammatory and neuropathic pain have shown to alter the functioning of the opioid system in other CNS areas than the canonical pain pathway.
- MRI studies revealed an enhanced activity of PFC, ACC, hippocampus, amygdala, and NAc in pain patients.
- PET imaging shows an increase of ORs availability and a specific decrease of MORs availability in the MCLS suggesting different pain-induced adaptations of the ORs subtypes.
- Preclinical studies have demonstrated, by using animal models of pain and a variety of techniques, that pain impacts functioning or density of MORs and KORs in the MCLS. Whereas pain induces a decrease in MORs function, the opposite effect has been observed in the case of KORs. Both pain-induced effects result in altered dopaminergic transmission that might underlie pain comorbidities, including negative affect, and drug addiction.
- There are still many unknown mechanisms that modulate pain impact on the opioid function in the MCLS. The thorough study of these mechanisms might help to clarify preventive actions to fight against opioid overdose epidemic.

References

Aicher, S. A., Punnoose, A., & Goldberg, A. (2000). μ-Opioid receptors often colocalize with the substance P receptor (NK1) in the trigeminal dorsal horn. *Journal of Neuroscience*, *20*(11), 4345–4354.

Al-Hasani, R., & Bruchas, M. R. (2011). Molecular mechanisms of opioid receptor-dependent signaling and behavior. *Anesthesiology*, *115*(6), 1363–1381.

Basbaum, A. I., Bautista, D. M., Scherrer, G., & Julius, D. (2009). Cellular and molecular mechanisms of pain. *Cell*, *139*(2), 267–284.

Becker, J. B., & Chartoff, E. (2019). Sex differences in neural mechanisms mediating reward and addiction. *Neuropsychopharmacology*, *44*(1), 166–183.

Berterame, S., Erthal, J., Thomas, J., Fellner, S., Vosse, B., Clare, P., … Samak, A. K. E. (2016). Use of and barriers to access to opioid analgesics: A worldwide, regional, and national study. *The Lancet*, *387*(10028), 1644–1656.

Börner, C., Kraus, J., Schröder, H., Ammer, H., & Höllt, V. (2004). Transcriptional regulation of the human μ-opioid receptor gene by interleukin-6. *Molecular Pharmacology*, *66*(6), 1719–1726.

Brown, C. A., Matthews, J., Fairclough, M., McMahon, A., Barnett, E., Al-Kaysi, A., … Jones, A. K. (2015). Striatal opioid receptor availability is related to acute and chronic pain perception in arthritis: Does opioid adaptation increase resilience to chronic pain? *Pain*, *156*(11), 2267–2275.

Bushnell, M. C., Ceko, M., & Low, L. A. (2013). Cognitive and emotional control of pain and its disruption in chronic pain. *Nature Reviews Neuroscience*, *14*(7), 502–511.

Byrne, L. S. Peng, J., Sarkar, S., & Chang, S. L. (2012). Interleukin-1 beta-induced up-regulation of opioid receptors in the untreated and morphine-desensitized U87 MG human astrocytoma cells. *Journal of Neuroinflammation*, *9*(1), 1–12.

Cahill, C. M., & Taylor, A. M. (2017). Neuroinflammation—A co-occurring phenomenon linking chronic pain and opioid dependence. *Current Opinion in Behavioral Sciences*, *13*, 171–177.

Campos-Jurado, Y., Igual-López, M., Padilla, F., Zornoza, T., Granero, L., Polache, A., … Hipólito, L. (2019). Activation of MORs in the VTA induces changes on cFos expression in different projecting regions: Effect of inflammatory pain. *Neurochemistry International*, *131*, 104521.

Campos-Jurado, Y., Lorente, J. D., González-Romero, J. L., Granero, L., Polache, A., & Hipólito, L. (2020). Impaired alcohol-induced dopamine release in the nucleus accumbens in an inflammatory pain model: Behavioural implications in male rats. *Pain*, *161*, 2203–2211. Online ahead of print.

Cantacorps, L., Alfonso-Loeches, S., Moscoso-Castro, M., Cuitavi, J., Gracia-Rubio, I., López-Arnau, R., … Valverde, O. (2017). Maternal alcohol binge drinking induces persistent neuroinflammation associated with myelin damage and behavioural dysfunctions in offspring mice. *Neuropharmacology*, *123*, 368–384.

Darcq, E., & Kieffer, B. L. (2018). Opioid receptors: Drivers to addiction? *Nature Reviews Neuroscience*, *19*(8), 499–514.

DaSilva, J. T., & Seminowicz, D. A. (2019). Neuroimaging of pain in animal models: A review of recent literature. *Pain Reports*, *4*(4), e732.

DosSantos, M. F., Martikainen, I. K., Nascimento, T. D., Love, T. M., Deboer, M. D., Maslowski, E. C., … DaSilva, A. F. (2012). Reduced basal ganglia μ-opioid receptor availability in trigeminal neuropathic pain: A pilot study. *Molecular Pain*, *8*, 74. https://doi.org/10.1186/1744-8069-8-74.

Gessi, S., Borea, P., Bencivenni, A., Fazzi, S., Varani, D., & K. & Merighi, S. (2016). The activation of μ-opioid receptor potentiates LPS-induced NF-kB promoting an inflammatory phenotype in microglia. *FEBS Letters*, *590*(17), 2813–2826.

Gregory, N. S., Harris, A. L., Robinson, C. R., Dougherty, P. M., Fuchs, P. N., & Sluka, K. A. (2013). An overview of animal models of pain: Disease models and outcome measures. *Journal of Pain*, *14*(11), 1255–1269. https://doi.org/10.1016/j.jpain.2013.06.008.

Harris, R. E., Clauw, D. J., Scott, D. J., McLean, S. A., Gracely, R. H., & Zubieta, J. K. (2007). Decreased central mu-opioid receptor availability in fibromyalgia. *Journal of Neuroscience*, *27*(37), 10000–10006. https://doi.org/10.1523/JNEUROSCI.2849-07.2007.

Hipólito, L., Sánchez-Catalán, M. J., Zanolini, I., Polache, A., & Granero, L. (2008). Shell/core differences in mu- and delta-opioid receptor modulation of dopamine efflux in nucleus accumbens. *Neuropharmacology*, *55*(2), 183–189. https://doi.org/10.1016/j.neuropharm.2008.05.012.

Hipólito, L., Wilson-Poe, A., Campos-Jurado, Y., Zhong, E., Gonzalez-Romero, J., Virag, L., … Bruchas, M. R. (2015). Inflammatory pain promotes increased opioid self-administration: Role of dysregulated ventral tegmental area μ opioid receptors. *Journal of Neuroscience*, *35*(35), 12217–12231.

Koob, G. F., & Volkow, N. D. (2016). Neurobiology of addiction: A neurocircuitry analysis. *Lancet Psychiatry*, *3*(8), 760–773.

Kraus, J., Börner, C., Giannini, E., & Höllt, V. (2003). The role of nuclear factor κB in tumor necrosis factor-regulated transcription of the human μ-opioid receptor gene. *Molecular Pharmacology*, *64*(4), 876–884.

Langsdorf, E. F., Mao, X., & Chang, S. L. (2011). A role for reactive oxygen species in endotoxin-induced elevation of MOR expression in the nervous and immune systems. *Journal of Neuroimmunology*, *236*(1–2), 57–64.

Lavand'homme, P., & Steyaert, A. (2017). Opioid-free anesthesia opioid side effects: Tolerance and hyperalgesia. *Best Practice and Research Clinical Anaesthesiology*, *31*(4), 487–498.

Lisa, B., & Jessica, H. (2018). Evidence synthesis—The opioid crisis in Canada: A national perspective. *Health Promotion and Chronic Disease Prevention in Canada: Research, Policy and Practice*, *38*(6), 224.

Lorente, J. D., Cuitavi, J., Campos-Jurado, Y., & Hipólito, L. (2020). Pain-induced alterations in the dynorphinergic system within the mesocorticolimbic pathway: Implication for alcohol addiction. *Journal of Neuroscience Research*. https://doi.org/10.1002/jnr.24703. Advance online publication.

Lorente, J. D., Cuitavi, J., Campos-Jurado, Y., Montón-Molina, R., González-Romero, J. L., & Hipólito, L. (2021). Kappa opioid receptor blockade in the nucleus accembens shell prevents sex-dependent alcohol deprivation effect induced by inflammatory pain. *PAIN*, 1–11. https://doi.org/10.1097/j.pain.0000000000002332. In press.

Maarrawi, J., Peyron, R., Mertens, P., Costes, N., Magnin, M., Sindou, M., … Garcia-Larrea, L. (2007). Differential brain opioid receptor availability in central and peripheral neuropathic pain. *Pain*, *127*(1–2), 183–194.

Massaly, N., Copits, B. A., Wilson-Poe, A. R., Hipólito, L., Markovic, T., Yoon, H. J., … Morón, J. A. (2019). Pain-induced negative affect is mediated via recruitment of the nucleus accumbens kappa opioid system. *Neuron*, *102*(3), 564–573. e6.

Morton, D. L., Sandhu, J. S., & Jones, A. K. (2016). Brain imaging of pain: State of the art. *Journal of Pain Research*, *9*, 613–624.

Moy, J. K., Hartung, J. E., Duque, M. G., Friedman, R., Nagarajan, V., Loeza-Alcocer, E., … Gold, M. S. (2020). Distribution of functional opioid receptors in human dorsal root ganglion neurons. *Pain*, *161*, 1636–1649. Online ahead of printing.

Muhuri, P. K., Gfroerer, J. C., & Davies, M. C. (2013). *CBHSQ data review: Associations of nonmedical pain reliever use and initiation of heroin use in the United States*. Center for Behavioral Health Statistics and Quality, Substance Abuse and Mental Health Services Administration.

Narita, M., Kishimoto, Y., Ise, Y., Yajima, Y., Misawa, K., & Suzuki, T. (2005). Direct evidence for the involvement of the mesolimbic kappa-opioid system in the morphine-induced rewarding effect under an inflammatory pain-like state. *Neuropsychopharmacology*, *30*(1), 111–118.

Navratilova, E., Ji, G., Phelps, C., Qu, C., Hein, M., Yakhnitsa, V., … Porreca, F. (2019). Kappa opioid signaling in the central nucleus of the amygdala promotes disinhibition and aversiveness of chronic neuropathic pain. *Pain*, *160*(4), 824–832.

Nestler, E. J., Hyman, S. E., & Malenka, R. C. (2001). *Molecular neuropharmacology: A foundation for clinical neuroscience*. McGraw-Hill Medical.

Nijs, J., D'Hondt, E., Clarys, P., Deliens, T., Polli, A., Malfliet, A., … Ickmans, K. (2020). Lifestyle and chronic pain across the lifespan: An inconvenient truth? *PM&R*, *12*(4), 410–419.

Nolan, S., Socias, M. E., & Wood, E. (2018). The threat of an international opioid crisis. *Current Addiction Reports*, *5*(4), 473–477.

Ozaki, S., Narita, M., Narita, M., Iino, M., Sugita, J., Matsumura, Y., & Suzuki, T. (2002). Suppression of the morphine-induced rewarding effect in the rat with neuropathic pain: Implication of the reduction in mu-opioid receptor functions in the ventral tegmental area. *Journal of Neurochemistry*, *82*(5), 1192–1198.

Ozawa, A., Brunori, G., Mercatelli, D., Wu, J., Cippitelli, A., Zou, B., … Scherrer, G. (2015). Knock-in mice with NOP-eGFP receptors identify receptor cellular and regional localization. *Journal of Neuroscience*, *35*(33), 11682–11693.

Palmisano, M., Caputi, F. F., Mercatelli, D., Romualdi, P., & Candeletti, S. (2019). Dynorphinergic system alterations in the corticostriatal circuitry of neuropathic mice support its role in the negative affective component of pain. *Genes, Brain, and Behavior*, *18*(6). https://doi.org/10.1111/gbb.12467, e12467.

Shrivastava, P., Cabrera, M. A., Chastain, L. G., Boyadjieva, N. I., Jabbar, S., Franklin, T., & Sarkar, D. K. (2017). Mu-opioid receptor and delta-opioid receptor differentially regulate microglial inflammatory response to control proopiomelanocortin neuronal apoptosis in the hypothalamus: Effects of neonatal alcohol. *Journal of Neuroinflammation*, *14*(1), 83.

Spike, R. C., Puskar, Z., Sakamoto, H., Stewart, W., Watt, C., & Todd, A. J. (2002). MOR-1-immunoreactive neurons in the dorsal horn of the rat spinal cord: Evidence for nonsynaptic innervation by substance P-containing primary afferents and for selective activation by noxious thermal stimuli. *European Journal of Neuroscience*, *15*(8), 1306–1316.

Stein, C. (2016). Opioid receptors. *Annual Review of Medicine*, *67*, 433–451.

Taylor, A. M., Castonguay, A., Taylor, A. J., Murphy, N. P., Ghogha, A., Cook, C., … Cahill, C. M. (2015). Microglia disrupt mesolimbic reward circuitry in chronic pain. *Journal of Neuroscience*, *35*(22), 8442–8450.

Thompson, S. J., Pitcher, M. H., Stone, L. S., Tarum, F., Niu, G., Chen, X., … Bushnell, M. C. (2018). Chronic neuropathic pain reduces opioid receptor availability with associated anhedonia in rat. *Pain*, *159*(9), 1856–1866.

Volkow, N. D., & Morales, M. (2015). The brain on drugs: From reward to addiction. *Cell*, *162*(4), 712–725.

Wang, D., Tawfik, V. L., Corder, G., Low, S. A., François, A., Basbaum, A. I., & Scherrer, G. (2018). Functional divergence of delta and mu opioid receptor organization in CNS pain circuits. *Neuron*, *98*(1), 90–108.e5. https://doi.org/10.1016/j.neuron.2018.03.002.

Wei, L. N., & Loh, H. H. (2011). Transcriptional and epigenetic regulation of opioid receptor genes: Present and future. *Annual Review of Pharmacology and Toxicology*, *51*, 75–97.

Wu, Y., Na, X., Zang, Y., Cui, Y., Xin, W., Pang, R., … Liu, X. (2014). Upregulation of tumor necrosis factor-alpha in nucleus accumbens attenuates morphine-induced rewarding in a neuropathic pain model. *Biochemical and Biophysical Research Communications*, *449*(4), 502–507.

Xu, M., Bruchas, M. R., Ippolito, D. L., Gendron, L., & Chavkin, C. (2007). Sciatic nerve ligation-induced proliferation of spinal cord astrocytes is mediated by κ opioid activation of p38 mitogen-activated protein kinase. *Journal of Neuroscience*, *27*(10), 2570–2581.

Chapter 10

Clinacanthus nutans L., analgesia, and the L-arginine/nitric oxide-mediated/cyclic-guanosine monophosphate-independent pathway

Zainul Amiruddin Zakaria

Department of Biomedical Sciences, Faculty of Medicine and Health Sciences, Universiti Putra Malaysia, Serdang, Selangor, Malaysia

Abbreviations

ACT	abdominal constriction test
ASA	acetylsalicylic acid
ATP	adenosine triphosphate
CAM	complementary and alternative medicine
CNS	central nervous system
COX	cyclooxygenase
DMSO	dimethyl sulfoxide
DW	distilled water
DWP	distilled water partition
EA	ethyl acetate
EAP	ethyl acetate partition
FT	formalin-induced paw licking test
GC-MS	gas chromatography-mass spectrometry
HPLC	high performance liquid chromatography
HPT	hot plate test
L-NAME	Nω-nitro-L-arginine methyl ester hydrochloride
MCNL	methanol extract of *C. nutans* leaves
MOR	morphine
mRNA	mitochondrial ribonucleic acid
NMDA	*N*-methyl-D-aspartate
NO/cGMP	nitric oxide/cyclic guanosine monophosphate
NSAID	nonsteroidal antiinflammatory drugs
ODQ	1H-[1,2,4]oxadiazole[4,3-*a*]quinoxalin-1-one
$ONOO^-$	peroxynitrite
p38 MAPK	p38 mitogen-activated protein kinases
PE	petroleum ether
PEP	petroleum ether partition
PGI2	prostacyclin
PGs	prostaglandins
PKC	protein kinase C
PMA	phorbol 12-myristate 13-acetate
PNS	peripheral nervous system
RNS	reactive nitrogen species
TRPV1	transient receptor potential vanilloid type 1
TXA2	thromboxane A2
UHPLC-ESI-MS	ultrahigh performance liquid chromatography-electrospray ionization-mass spectrometry

The Neurobiology, Physiology, and Psychology of Pain. https://doi.org/10.1016/B978-0-12-820589-1.00010-5

Introduction

Herbal remedies for the treatment of pain

Pain, affected 20% of adults globally, is associated with deficit in quality of life and has become a universal public health problem (Goldberg & McGee, 2011). Despite their effectiveness in curing pain, prolonged usage of currently available analgesic drugs either alone or in combination has been associated with various unwanted side effects (Palos, Mendoza, Cantor, Aday, & Cleeland, 2004). For example, the risk from nonsteroidal antiinflammatory drugs (NSAIDs) usage includes gastrointestinal bleeding and ulceration, potential for myocardial infarction, stroke, and Stevens-Johnson syndrome, whereas the side effects of opioids usage include sedation, nausea, confusion, delirium, tolerance, and dependence. Moreover, extensive efforts and investments from academia and the pharmaceutical industry failed to provide sufficient numbers of new and effective medications to treat pain at clinical phase (Stolz et al., 2014). Consequently, pain treatment remains to be contingent on therapeutic modalities that offer limited efficacy and/or significant adverse effect leading to an increase number of patients to turn to herbal remedies, as part of the complementary and alternative medicine (CAM), as an alternative source for the treatment of pain (Ekor, 2014). Americans, in particular, spend billions of dollars annually to find a more "natural" or holistic treatment with effective pain relief and few side effects, from herbal therapies (Wirth, Hudgins, & Paice, 2005). Considering the above issues, the need to find new analgesic agents with high efficacy, less side effects and reduce costs that helps to improve quality of life have become one of the primary objectives in analgesic drug discovery (Manjiani, Paul, Kunnumpurath, Kaye, & Vadivelu, 2014) with medicinal plants utilized as pain-relieving agents in traditional ethnomedicine being explore and investigate in the search for new and safe analgesic drugs due to their high chemical diversity (Lautié, Russo, Ducrot, & Boutin, 2020). Accordingly, one of the plants that have been studied and proven to possess antinociceptive activity is *Clinacanthus nutans* L.

Clinacanthus nutans L.

Clinacanthus nutans (Burm. f.) Lindau (Fig. 1), a plant belonging to the family Acanthaceae, is a shrub native to the tropical Southeast Asian countries. It has gained a medicinal status in the Southeast Asian region, including Malaysia, where it is locally known as "*Pokok Belalai Gajah*." The plants have been traditionally used in many Southeast Asian countries like Indonesia, Malaysia, and Thailand to treat various ailments (Table 1) while scientifically the plant has been reviewed by numerous authors to exert various pharmacological activities (i.e., antinociceptive, antiinflammatory, antioxidant, immunomodulatory, neuroprotective and neuromodulatory, antidiabetic, antioxidant, antiviral, antibacterial, antifungal, anticancer, wound healing, genoprotective, antihyperlipidemia, antimucositis, and antistomatistis) and to contain numerous bioactive compounds (Alam et al., 2016; Kamarudin, Sarker, Kadir, & Ming, 2017; Zulkipli, Rajabalaya, Idris, Sulaiman, & David, 2017). The antinociceptive potential of *C. nutans*, in particular, its methanol extract of the leaves (MCNL), was profiled (Abdul Rahim et al., 2016) and later the mechanisms of action involved were elucidated (Zakaria et al., 2018). These were later followed by determination of the antinociceptive profile of petroleum ether partition (PEP) derived from MCNL and investigation on the possible mechanisms of antinociception involved (Zakaria, Abdul Rahim, Mohd Sani, et al., 2019; Zakaria, Abdul Rahim, Roosli, et al., 2019). In the next sections, the findings of each report will be described in detail, and a conclusion from each finding will be drawn.

Report on findings related to the antinociceptive activity of *C. nutans*

Antinociceptive profile of MCNL and the possible mechanisms of antinociception

In our first attempt to determine the antinociceptive potential of *C. nutans* (Abdul Rahim et al., 2016), MCNL (100, 250, and 500 mg/kg; administered orally) attenuated nociceptive response when assessed using the acetic acid-induced abdominal constriction (ACT) ($ED_{50} \approx 273$ mg/kg; 30%–70% antinociception), hot plate (HPT), and formalin-induced paw licking (FT) tests. The 100 mg/kg acetylsalicylic acid (ASA)-induced antinociceptive intensity was comparable to that of 250 mg/kg MCNL when assessed using the ACT. With regard to the HPT, MCNL exerted significant antinociceptive activity only at the dose of 500 mg/kg when compared to the vehicle-treated control (10% dimethyl sulfoxide (DMSO)). The antinociceptive activity of MCNL lasted until the end of the experiment (210 min) and the antinociceptive intensity was comparable to 5 mg/kg morphine (MOR). Assessment using the FT showed that only 250 and 500 mg/kg MCNL attenuate nociception in the early/neurogenic phase (0%–40% antinociception; $ED_{50} \approx > 500$ mg/kg) whereas all doses significantly attenuated nociceptive response in the inflammatory (second) phase (40%–70% antinociception; $ED_{50} \approx 228$ mg/kg). In comparison, ASA (a standard peripheral analgesic) only attenuated nociceptive response in the late

FIG. 1 *Clinacanthus nutans* L. (A) The whole tree. (B) The leaves of *C. nutans* L. (C) The leave of *C. nutans* L. *(Photos were courtesy of www. NatureLoveYou.sg.)*

phase (61% antinociception) while MOR (a standard central analgesic) deduced the nociceptive response in both phases with the recorded 77% and 96% antinociception, respectively. Further analyses on the involvement of opioidergic and nitric oxide/cyclic guanosine monophosphate (NO/cGMP) systems in the modulation of MCNL-induced antinociception were performed using the ACT. The findings revealed that 5 mg/kg naloxone (intraperitoneal; i.p.), a nonselective opioid antagonist, completely antagonized MCNL-induced antinociception, whereas i.p. administration of L-arginine, a precursor of NO, Nω-nitro-L-arginine methyl ester hydrochloride (L-NAME), an inhibitor of NO synthase, or their combination thereof caused partial attenuation of the extract's antinociceptive activity. In addition, pretreatment (i.p.) with an inhibitor of soluble guanylyl cyclase, 1H-[1,2,4]oxadiazole[4,3-*a*]quinoxalin-1-one (ODQ), failed to attenuate the extract's antinociception.

Later, Zakaria et al. (2018) further investigated the possible mechanisms of antinociception modulated by MCNL (100–500 mg/kg; orally) found that the extract significantly attenuated nociceptive response when assessed using various nociceptive assays (i.e., capsaicin-, glutamate-, phorbol 12-myristate 13-acetate (PMA)-, bradykinin-induced nociception models). Furthermore, in separate studies assessed using the ACT: (i) pretreatment with nonselective antagonist of opioid

TABLE 1 Traditional usages of *C. nutans* as medicine among local communities in Asian countries.

Country	Part use	Mode of use	Traditional uses	References
Malaysia	Fresh leaves are boiled with water	Orally	Consumed as herbal tea	Shim, Aziana, and Khoo (2013) and Alam et al. (2016)
Thailand	Alcoholic extract of fresh leaves	Used externally	To treat skin rashes, snake and insect bite, herpes simplex virus, and varicella zoster virus lesions	Shim et al. (2013) and Alam et al. (2016)
	Fresh leaves are chewed for internal consumption	Orally	• Consumed raw as vegetables • Used as antisnake venom	Yahaya et al. (2015) and Alam et al. (2016)
	Fresh leaves are chewed and pounded		• Applied to the site of snake bite • Used to treat scorpion bites and nettle rash	Yahaya et al. (2015)
	Fresh leaves mixed with apple juice, sugarcane or green tea	Orally	• Provide as fresh drink and as antisnake venom • Used as antisnake venom • Used to treat scorpion bites and nettle rash	Alam et al. (2016)
	Decoction of dried leaves	Orally	• To treat herpes infection	Yahaya et al. (2015)
	Infusion or decoction of dried leaves and stem	Orally	• To treat hepatitis infection	Yahaya et al. (2015)
China	Whole plant	Not specified	• To treat inflammatory conditions like hematoma, contusion, strains and sprains of injuries and rheumatism • To treat gastrointestinal complications	Shim et al. (2013) and Alam et al. (2016)
	Not specified	Not specified	To control menstrual function, to relieve pain, anemia, jaundice and fractured bones	Alam et al. (2016)
Indonesia	Fresh leaves were boiled in water (decoction)	Orally	To treat dysentery, cancer, diabetes, dysuria, fever	Shim et al. (2013) and Alam et al. (2016)

The leaves were consumed fresh or prepared in various form prior to application to treat diseases.

receptor subtypes, namely β-funaltrexamine (μ-type), naltrindole (δ-type), or nor-binaltorphimine (κ-type), significantly attenuated the antinociception exerts by 500 mg/kg MCNL; (ii) pretreatment with antagonist of the respective non-opioid receptor system, namely 0.15 mg/kg yohimbine (α2-noradrenergic), 1 mg/kg pindolol (β-adrenergic), 3 mg/kg caffeine (adenosinergic), 0.2 mg/kg haloperidol (dopaminergic), or 10 mg/kg atropine (muscarinic cholinergic), significantly but partially inhibited the antinociception exhibited by 500 mg/kg MCNL; and (iii) pretreatment (i.p.) with the respective inhibitor of K^+ channels, namely 10 mg/kg glibenclamide (ATP-sensitive), 0.04 mg/kg apamin (small conductance Ca^{2+}-activated-), 0.02 mg/kg charybdotoxin (large conductance Ca^{2+}-activated-), or 4 mg/kg tetraethylammonium chloride (nonselective voltage-activated-K^+ channel), significantly but partially blocked the antinociception exhibited by 500 mg/kg MCNL.

Antinociceptive profile of PEP and the possible mechanisms of antinociception

In an attempt to further determine the polarity of bioactive compounds responsible for MCNL-induced antinociception, the extract was partitioned according to their polarity using distilled water (DW), petroleum ether (PE) and ethyl acetate (EA) to obtain the respective polar (DWP), nonpolar (PEP) and intermediate polar (EAP) extracts (Zakaria, Abdul Rahim, Mohd Sani, et al., 2019). Each partition (100, 250, 500 mg/kg; orally) was then subjected to the ACT whereby PEP was demonstrated to be the most effective fraction with 64%–77% antinociception and selected for further investigation.

Subsequently, all doses of PEP attenuated nociceptive response when assessed using the HPT whereas assessment using the FT showed that only 250 and 500 mg/kg PEP significantly reduced the nociceptive response in the early phase (10%–39% antinociception), whereas all doses of PEP inhibited the late phase (34%–68% antinociception). In comparison, ASA and MOR produced approximately 3% and 78% antinociception in the early phase and, 61% and 97% antinociception in the late phase, respectively. In separated studies assessed using the ACT, (i) pretreatment with 10 mg/kg β-funaltrexamine, 1 mg/kg, naltrindole or 1 mg/kg nor-binaltorphimine also significantly attenuated 500 mg/kg PEP-exerted antinociception whereas; (ii) pretreatment with L-arginine, but not ODQ, significantly reversed the antinociception produced by PEP. Moreover, PEP when subjected to the Rota rod test did not affect the locomotor activity of treated mice.

Further investigation by Zakaria, Abdul Rahim, Roosli, et al. (2019) revealed that PEP (100–500 mg/kg; orally) significantly attenuated the nociceptive response induced by capsaicin-, glutamate-, PMA-, and bradykinin-induced nociception model. Moreover, in separated investigations, PEP-induced antinociception was significantly inhibited by pretreatment (i.p.) with several nonopioid receptor system (i.e., yohimbine, pindolol, caffeine, haloperidol, or atropine) when assessed using the ACT. The inhibition of PEP-induced antinociception occurs partially with pindolol, yohimbine, and haloperidol but fully with caffeine and atropine. Then, pretreatment with the respective blockers of several K^+ channels (i.e., glibenclamide, apamin, charybdotoxin, or tetraethylammonium chloride) significantly attenuated the antinociception of PEP. Apamin and charybdotoxin caused full reversal whereas glibenclamide and tetraethylammonium chloride caused partial inhibition of the extract-induced antinociception.

Phytoconstituents of MCNL and PEP

Phytoconstituents of MCNL was determined through qualitative phytochemical screening (Table 2) followed by the ultrahigh-performance liquid chromatography-electrospray ionization-mass spectrometry (UHPLC-ESI-MS) (Table 3) and gas chromatography-mass spectrometry (GC-MS) (Table 5) analyses (Abdul Rahim et al., 2016; Zakaria et al., 2018). On the other hand, the phytoconstituents of PEP, determined through qualitative phytochemical screening, and high-performance liquid chromatography (HPLC) and GC-MS analyses were shown in Table 2, Table 4, and Table 5, respectively (Zakaria, Abdul Rahim, Mohd Sani, et al., 2019; Zakaria, Abdul Rahim, Roosli, et al., 2019).

Conclusion based on the reported antinociceptive activity of MCNL and PEP

According to Olesen, Andresen, Staahl, and Drewes (2012), pain, in physiological conditions, is basically a complex spectacle consequential of the modulation of several peripheral and central mechanisms, which constitutes an alarm that has the supporting role of preserving the organism, reducing whatever is activating the pain and, consequently, restricting the damaging consequences. Typically, polymodal nociceptors can be stimulated by several stimuli, such as thermal, mechanical, and toxic molecules or inflammatory mediators (Basbaum, Bautista, Scherrer, & Julius, 2009). Numerous receptors and ion channels are expressed in nociceptors, and these receptors/channels are significant for the discovery of several noxious stimuli and the generation of action potentials (Mickle, Shepherd, & Mohapatra, 2016).

TABLE 2 Classes of phytoconstituents presence in MCNL and PEP.

		Class of bioactive compounds detected	
No.	Class of bioactive compounds	MCNL	PEP
1	Flavonoids	+	+
2	Saponins	+	+
3	Triterpenes	+	+
4	Tannins	−	−
5	Alkaloids	−	−
6	Steroids	+	+

Phytochemical screening was performed to determine the presence of flavonoids, saponins, triterpenes, tannins, alkaloids, and steroids.

TABLE 3 Phytoconstituents of MCNL detected using the UHPLC-ESI-MS.

No.	Bioactive compounds detected MCNL	Compound with antinociceptive potential (√)
1	Gallic acid	√
2	4-Hydroxybenzoic acid	
3	Caffeic acid	√
4	Coumaric acid	
5	Schaftoside	
6	Ferulic acid	√
7	Vitexin	√
8	Orientin	
9	Forsythoside H	
10	Isoorientin	√
11	Forsythoside I	
12	Isovitexin	
13	Diosmetin glycoside	
14	Luteolin	√
15	Apigenin	√

Various phenolic compounds were detected using the UHPLC-ESI-MS.
[a]PEP was not analyzed using the UHPLC-ESI-MS (Zakaria, Abdul Rahim, Mohd Sani, et al., 2019; Zakaria, Abdul Rahim, Roosli, et al., 2019).

TABLE 4 Phytoconstituents of PEP as detected using the HPLC.

No.	Bioactive compounds detected PEP	Compound with antinociceptive potential (√)
1	Cinnamic acid	√

Comparison between chromatogram of PEP against the respective chromatogram of several flavonoids.
[a]Chromatogram of PEP was compared against chromatogram of 15 pure compounds namely cinnamic acid, caffeic acid, coumaric acid, vanillic acid, ferulic acid, gallic acid, epicatechin gallate, catechin, rutin, myricetin, epicatechin, 3,4-hydrobenzoic acid, ellagic acid, protocatechuic acid, and epigallocatechin gallate (Zakaria, Abdul Rahim, Mohd Sani, et al., 2019; Zakaria, Abdul Rahim, Roosli, et al., 2019).

Involvement of peripheral and central antinociceptive mechanisms

Based on the antinociceptive findings related to the application of MCNL (Abdul Rahim et al., 2016; Zakaria et al., 2018) and PEP (Zakaria, Abdul Rahim, Mohd Sani, et al., 2019; Zakaria, Abdul Rahim, Roosli, et al., 2019), several conclusions can be made. The ability of MCNL and PEP to reduce nociceptive response as seen with the chemically induced models (ACT and FT) and thermally induced model (HPT) is concurrent with the previous claim made by Khatun, Imam, and Rana

TABLE 5 Major volatile phytoconstituents of MCNL and PEP.

	Volatile bioactive compounds detected			
No.	MCNL	% of presence	PEP	% of presence
1	2-Ethyl-oxetane	16.6	Lupeol trimethylsilyl ether (TMS)	4.08
2	9,12,15-Octadecatrienoic acid	7.6	Stigmasta-3,5-diene	3.10
3	2,3-Dimethylpyridine	6.4	Cyclononasiloxane, octadecamethyl-	2.51
4	3-Deoxy-D-mannoic lactone	5.7	γ-Tocopherol TMS derivative	2.20
5	Neophytadiene	5.4	Hexasiloxane, tetradecamethyl-	2.19
6	Phytol	5.3	Stigmast-5-ene, 3β-(trimethyl siloxy)-, (24*S*)-	2.18
7	*n*-Hexadecanoic acid	4.6	Cyclodecasiloxane, eicosamethyl-	1.99
8	2,3-Dihydrobenzofuran	4.5	2H-1-Benzopyran-2-one, 7-(5-butoxy-6-methyl-2H-benzotriazol-2-yl)-3-phenyl-	1.93

Volatile compounds presence in both samples were detected using the GC-MS.
[a]Only eight major volatile compounds are shown for each sample (Abdul Rahim et al., 2016; Zakaria, Abdul Rahim, Roosli et al., 2019).

(2015) and Rezaee-Asl, Sabour, Nikoui, Ostadhadi, and Bakhtiarian (2014), which suggests that both compounds exerted peripheral and central antinociception.

In addition, the ability of MCNL and PEP to attenuate different models of nociception suggests the involvement of different mechanisms of antinociception. ACT, considered as a model for peripheral inflammatory pain based on the ability of administered acetic acid to indirectly induces the release of endogenous inflammatory mediators (i.e., cyclooxygenase (COX), prostaglandins (PGs), histamine, serotonin, bradykinin, etc.), which then stimulate the peripheral nociceptive neurons that enter the dorsal horn of the central nervous system (CNS), is sensitive to NSAIDs, opioids, and other centrally acting drugs (Dzoyem, McGaw, Kuete, & Bakowsky, 2017). Therefore, the ability of these compounds to attenuate acetic acid-induced nociception indicates their ability to attenuate the action of inflammatory mediators and impeded pain transduction. On the other hand, HPT, considered to be a supraspinally organized response and is specifically affected only by the centrally acting drugs (i.e., opioid analgesics), is explicitly used to examine the involvement of centrally mediated antinociception of any extract/compound (Dzoyem et al., 2017). Hence, MCNL attenuation of thermal-induced nociception only at the highest dose tested suggests its mild central antinociception, whereas PEP attenuation of the same stimuli at all doses tested suggests its strong central antinociception. In contrast, FT, a popular chemical assay of injury-produced inflammatory pain, is usually used to examine the ability of new compound to modulate the peripheral or central nociceptive pathways due to its characteristic biphasic nociception, known as early/neurogenic- and late/inflammatory-phase. The early phase (between 0 and 5 min), observed immediately after the administration of formalin as a response to the direct action of formalin on nociceptors in the subplantar region, represents neurogenic pain whereas the late phase (between 15 and 30 min), resulted from a tonic response due to the release of inflammatory mediators, represents inflammatory-mediated pain (Dzoyem et al., 2017). It is widely known that centrally acting analgesics inhibit both phases whereas peripherally acting analgesics inhibit only the late phase (Zakaria, Abdul Rahim, Mohd Sani, et al., 2019; Zakaria, Abdul Rahim, Roosli, et al., 2019). Therefore, MCNL and PEP ability to inhibit both phases of the FT suggests that both compounds possess a characteristic of the centrally acting analgesics (i.e., MOR).

Involvement of various mechanisms of antinociception

Role of general opioidergic system

With regards to the mechanisms of antinociception exerted by MCNL and PEP, several conclusions can be made based on the findings described earlier. The ability of naloxone, a nonselective opioid antagonist, to reverse MCNL- and PEP-induced antinociception in all models of nociception (i.e., ACT, HPT, and FT) indicates a strong involvement of peripheral and central opioid receptor system in the compounds-induced antinociception.

Involvement of specific opioidergic system subtypes

Further investigation shows that the antinociceptive activity of 500 mg/kg MCNL and PEP was significantly inhibited by the respective antagonists of μ-, δ-, and κ-opioid receptors, namely β-funaltrexamine, naltrindole, and nor-binaltorphimine; thus, suggesting these compounds to possess a characteristic of a nonselective opioid action (Zakaria, Abdul Rahim, Mohd Sani, et al., 2019; Zakaria et al., 2018). The importance of opioids and opioid receptors in mediating antinociceptive effect at various levels of the neuraxis has been discussed in detail by Herz and Millan (1990). According to Holtsman and Fishman (2005), these opioid subtypes receptors are widely distributed in the brain, spinal cord, and peripheral nociceptors and their roles in nociceptive transmission depends on the inducer used. For examples, μ-opioid receptors influenced responses to chemical, thermal, and mechanical nociception at a supraspinal level; κ-opioid receptors modulated spinally mediated thermal nociception and chemical visceral pain while δ-opioid receptors modulated mechanical nociception and inflammatory pain. This explanation justifies MCNL and PEP ability to exert opioid-mediated antinociception in different nociceptive models and, thus, considered to be a nonselective opioid-acting agent.

Role of nonopioidergic systems

Further investigations also revealed the involvement of various nonopioid systems in the modulation of MCNL- and PEP-induced antinociception. Both compounds significantly attenuated the nociceptive response induced by capsaicin, glutamate, PMA, and bradykinin, which further indicates their capability to modulate a diverse number of pain pathways depending on the stimuli involve. The interaction and link between these pain pathways has been discussed in detailed by Zakaria, Abdul Rahim, Mohd Sani, et al. (2019) and briefly mentioned below to support findings related to *C. nutans*.

Involvement of transient receptor potential vanilloid type 1 (TRPV1) receptors, glutamatergic system, protein kinase C (PKC)-mediated pathway, and bradykininergic system

Capsaicin binds to TRPV1 receptors leading to the activation of pain response. Phosphorylation of TRPV1 receptors by PKC in particular is link to a cascade of event involving bradykinin and positively regulates the receptors activity inside the cell including activation of different modalities of pain sensation (i.e., chemical-, thermal-, mechanical-induced nociception) and tissue injury-induced thermal hyperalgesia (Rosenbaum & Simon, 2007). On the other hand, glutamate, a main excitatory neurotransmitter, is colocalized with its receptors in areas of the brain, spinal cord, and periphery that involved in the modulation of pain (Bardoni, 2013). According to Bardoni (2013), the peripheral, spinal, and supraspinal modulation of nociceptive neurotransmission via the glutamatergic system occurs through a complex cascade of event involving among others the *N*-methyl-D-aspartate (NMDA) receptors, NO, and PKC. Instead, PKC, localized in both peripheral nervous system (PNS) and CNS sites that process pain, is an enzyme with significant regulatory functions including in the modulation of peripheral and central pain transmission (Souza et al., 2002). PKC activation via interaction with intracellular lipid second messengers leads to the phosphorylation of many cellular components that serve as key regulatory components in signal transduction pathways of nociceptor excitation and sensitization (i.e., TRPV1, glutamate, and bradykinin receptors) (Bardoni, 2013; Rosenbaum & Simon, 2007). Moreover, bradykinin is an inflammatory peptide messenger and is among the major pain mediators during inflammation. The link between bradykinin with some of the nociceptors modulating factors such as TRPV1 receptors PKC- and glutamatergic-mediated systems, as well as the voltage-gated and Ca^{2+}-activated K^+ channels, have been extensively reviewed (Rosenbaum & Simon, 2007; Wang et al., 2005). Taking all these reports into consideration, the link between capsaicin, TRPV1, PKC, glutamatergic, and bradykininergic systems could be used to explain the ability of MCNL and PEP to inhibit capsaicin-, PMA-, glutamate-, and bradykinin-induced nociception.

Involvement of α2-adrenergic, β-adrenergic, adenosinergic, dopaminergic, or muscarinic cholinergic receptor systems

With regard to the nonopioid receptor systems' involvement, MCNL- and PEP-induced antinociception was only partially inhibited by the antagonist of the respective α2-adrenergic, β-adrenergic, adenosinergic, dopaminergic, or muscarinic cholinergic receptor systems; hence, suggesting to work as a partial agonist at the respective receptor. Various nonopioid receptor systems have been proven to play role in pain modulation. Evidences have shown that activation of α2-adrenergic receptors generate impressive analgesic effects (Thomas, Bantel, Stone, & Wilcox, 2013), whereas Abbas (2019) suggested that β-adrenergic receptors could be a novel and viable therapeutic target for pain reduction. Martikainen et al. (2015), on the other hand, suggested that activation of dopamine D_2/D_3 receptors on striatal partly contribute to top-down pain regulation in humans. Other than that, adenosinergic system involvement in the modulation of antinociception has also been cited by Marcondes Sari et al. (2014) while Mojtahedin, Tamaddonfard, and Zanbouri (2009) reported on the involvement of muscarinic cholinergic system in the central and peripheral pain modulation. These reports might help to partly explain MCNL ability to partially reverse nociceptive response mediated via those receptors.

Involvement of various types of K^+ channels

The importance of different subtypes of K^+ channels, namely adenosine triphosphate (ATP)-sensitive-, small conductance-, and large-conductance-Ca^{2+}-activated, and nonselective voltage-dependent-K^+ channels, in pain modulation, has been described by Galeotti, Ghelardini, Vinci, and Bartolini (1999). Moreover, their roles in the modulation of antinociceptive effect of drugs or compounds have also been reported elsewhere (Granados-Soto, Argüelles, & Ortiz, 2002; Landa-Juárez, Ortiz, Castañeda-Hernández, & Chávez-Piña, 2016; Ortiz et al., 2002). Based on the findings, MCNL-induced antinociception was only partially reversed by the blocker of each K^+ channels; thus, suggesting that the extract acts as a partial agonist in the opening of different subtypes of K^+ channels resulting in the observed antinociceptive effect. On the other hand, PEP-induced antinociception was partially reversed by the blocker of ATP-sensitive and nonselective voltage-dependent K^+ channels, but fully inhibited by the blocker of small- and large-conductance Ca^{2+}-activated K^+ channels. Briefly, the link between some of the K+ channel subtypes with some of the nociceptor modulating factors affected by MCNL or PEP have also been reported such as: (i) capsaicin inhibitory effect against voltage-gated K^+ channel occurs through TRPV1-dependent or TRPV1-independent pathways (Yang, Xiong, Liu, & Liu, 2014); (ii) PKC phosphorylation has been reported to inhibit voltage-gated K^+ channels (Du & Gamper, 2013); (iii) ATP-sensitive K^+ channel activators antagonized the bradykinin-induced nociceptive response, as well as the thermal- and mechanical-induced nociception (Du, Wang, & Zhang, 2011); (iv) voltage-gated K+ channel have a significant impact on glutamatergic synaptic transmission in the dorsal horn (Muqeem, Ghosh, Pinto, Lepore, & Covarrubias, 2018); (v) large-conductance Ca^{2+}-activated K^+ channels have been linked to opioid (Song & Marvizon, 2005), PKC, glutamate receptors (Carrasquel-Ursulaez, Lorenzo, Echeverria, & Latorre, 2018) to name a few. These links among the nociceptor modulating factors might be used to support the ability of MCNL and PEP to modulate nociceptive response via multiple pathways.

Role of NO-mediated pathways

NO has been implicated in the central and peripheral modulation of pain through multiple mechanisms that may underlie its distinctive roles in different pain states. Other than through the cGMP pathway, NO-mediated pain modulation can also occur via the cGMP-independent pathways (Seabra, 2017), particularly via *S*-nitrosylation, which is referred to the covalent and reversible attachment of NO to a reactive cysteine thiol. Moreover, the cGMP-independent modulation of pain transmission by NO occurs via the involvement of opioidergic-, adenosinergic-, muscarinic cholinergic- (Kokkola, Savinainen, Mönkkönen, Retamal, & Laitinen, 2005), glutamatergic- (Choi et al., 2000), dopaminergic- (Lu et al., 2009), and TRPV1-receptors (Miyamoto, Dubin, Petrus, & Patapoutian, 2009) and K^+ channels, especially the ATP-dependent- and large conductance Ca^{2+}-activated K^+ channels (Park, Lee, Semjid, Kim, & Chun, 2014). Although the NO/cGMP-dependent pathway has been widely acknowledged to regulate peripheral and central antinociception of various analgesic drugs, including MOR (Kampa et al., 2001), based on the findings that only L-arginine (a precursor of NO), and L-NAME (an inhibitor of NO synthase), or their combinations thereof, but not ODQ (an inhibitor of soluble guanylyl cyclase), significantly reversed the antinociceptive activity of MCNL, it is plausible to proposed the involvement of NO-mediated, cGMP-independent pathway.

Applications to other areas

In this chapter, the antinociceptive potential of *C. nutans* and possible mechanisms of action have been reviewed. The role of NO-mediated, cGMP-independent pathway in antinociception and its contribution to *C. nutans* action were highlighted. In the following section, we describe how the cGMP-independent effects of NO play roles in modulating various physiological activities in the body. Under physiological conditions, NO regulates vasculature activities (i.e., blood pressure, cytoprotection, and antiadhesion), inflammatory responses, and cell proliferation of which failure of this regulatory pathway has been linked to atherosclerosis and other forms of vascular dysfunction. Moreover, NO coordinates a highly integrated program of cell cycle arrest that regulates a large number of genes in human via a mechanism that is independent of cGMP, which contradicted the same process in rodents. Utilizing the transcriptome-wide approach, NO was shown to exert broad control over the cell cycle through p38 mitogen-activated protein kinases (MAPK) activation and mitochondrial ribonucleic acid (mRNA) stabilization.

The important of cGMP-independent pathway, particularly the *S*-nitrosylation, wherein NO interacts with a thiol group, in protecting the physiological balance of the body can be seen in the cardiovascular cells. Peroxynitrite ($ONOO^-$), a key metabolite of NO formed endogenously in cardiovascular cells by the reaction of NO with O_2^-, activates prostaglandin endoperoxide H2 synthase and leads to increased production of prostanoid under physiologic condition. However, increase in the concentration of $ONOO^-$ inhibits prostacyclin (PGI2)-synthase formation and consequently switches the production of PGI2 to that of thromboxane A2 (TXA2). These reactive nitrogen species (RNS) if not neutralized via *S*-nitrosylation process may cause irreversible damage to the respiratory chain, inhibition of ATP synthesis, cytochrome *c* release, and induction of caspase-dependent apoptosis. RNS at high concentrations may lead to lipid peroxidation, protein oxidation and nitration, enzyme inactivation, and eventually cell necrosis.

Other agents of interest

In this chapter, we have highlighted the mechanisms of antinociception modulated by *C. nutans*. Several prescription analgesia medications currently available in the market are:

Morphine
Codeine
Buprenorphine
Fentanyl
Hydromorphone
Tramadol
Tapentadol
Oxycodone
Oxymorphone
Butorphanol

The most common modern use of opioid analgesics is to treat acute pain, but an increase in their usage to treat chronic pain was seen in the 1990s. Long-term opioid analgesics treatment raised various issues related to their safety and efficacy, and has been overshadowed by various unwanted side effects. Although the short-term medical use of opioid analgesics rarely leads to an opioid misuse or addiction, but their regular or longer-term use can lead to drug dependence, tolerance, hyperalgesia, and, in some cases, addiction. Moreover, their action on the brain regions involved in reward can induce euphoria, which underlies the potential for misuse and substance use disorder, if taken at a higher-than-prescribed dose or administered in other ways than intended. Furthermore, opioid analgesics misuse can cause overdose leading to suppress breathing or suffocation if not properly treated due to their interaction with parts of the brain stem that control breathing. For these reasons, the clinicians should be vigilant and attentive to their patients, and need to be careful in prescribing the medications to prevent the escalation of a patient's misuse to a substance use disorder. Moreover, the need to find replacement for opioid analgesics has grown tremendously in the last decade wherein the development of effective, non-addicting pain medications has become a public health priority. Researchers working in the field of drug discovery are exploring various alternative treatment approaches including the use of herbal medicine that target other pain signaling systems in the body.

Mini-dictionary of terms

Antinociception. The action or process of blocking the detection of a painful or injurious stimulus by sensory neurons.
Ethnomedicine. The comparative study of how different cultures view disease and how they treat or prevent it.
Herbal remedy. Medicine made from a plant or plants.
Herbal medicine. A plant or plant part or an extract or mixture of these used in herbal remedy.
Alternative medicine. Any of various systems of healing or treating disease (such as homeopathy) not included in the traditional medical curricula of the United States and Britain.
Complementary medicine. Any of the practices (include natural and herbal remedies) of alternative medicine accepted and utilized by mainstream medical practitioners.

Opioids. A natural, semisynthetic, or synthetic substance that typically binds to the same cell receptors as opium and produces similar narcotic effects.

***S*-Nitrosylation**. The covalent attachment of a nitric oxide group (-NO) to cysteine thiol within a protein to form an *S*-nitrosothiol (SNO).

Key facts

Key facts of analgesia

- Opioid analgesics have been demonstrated to be effective medications for various acute and chronic pain conditions.
- The abuse of opioid analgesics increased dramatically in the last decade, resulting in significant public health problems.
- Opioid analgesics account for 38.2% of the 22,400 drug overdose deaths in the United States in 2005.
- Chronic pain treatment using opioid analgesics can cause dependence, tolerance, and addiction possibly due to strong action of the drugs on the brain.
- The development of effective, nonaddicting analgesics is a public health priority.
- Researchers are exploring alternative treatment approaches that target other signaling systems in the body that is also involved in pain.

Summary points

- Pain management using analgesics have been associated with various unwanted side effects including dependence, tolerance, and addiction; thus, increase the demand for new analgesic agents that are effective and safe.
- Researchers have been driven to explore new sources of analgesic compounds including plants and *Clinacanthus nutans*, a well-known traditional herb in the Asian region, has been identified as a potential candidate.
- Investigation revealed that the methanol extract of *C. nutans* (MCNL) exerted peripheral and central antinociception indicated by its ability to attenuate chemical- and thermal-induced nociceptive responses.
- MCNL exerted antinociception via activation of μ, κ, and δ opioid receptors indicating it possessed a characteristic of a nonselective opioid analgesic (i.e., MOR).
- MCNL inhibited nociception mediated via TRPV1 receptors and, PKC-, glutamate-, and bradykinin-mediated systems and exerted antinociception via activation of α2-adrenergic-, β-adrenergic-, adenosinergic-, dopaminergic-, and muscarinic cholinergic-receptor systems, various K^+ channels, and modulation of the NO-mediated, cGMP-independent pathway.
- Interestingly PEP, partitioned from MCNL, also demonstrated the same pharmacological profile as MCNL suggesting in part that the lipid-soluble compounds play roles in the activity.
- Phytochemical analyses revealed the presence of several antinociceptive compounds; their synergistic action is thought to contribute to the modulation of *C. nutans* antinociceptive activity via various nociceptive pathways.

References

Abbas, M. A. (2019). Analgesic effect of *Allium ampeloprasum*: Evidence for the involvement of beta-adrenergic system. *Journal of Functional Foods*, *57*, 1–6.

Abdul Rahim, M. H., Zakaria, Z. A., Mohd Sani, M. H., Omar, M. H., Yakob, Y., Cheema, M. S., et al. (2016). Methanolic extract of *Clinacanthus nutans* exerts antinociceptive activity via the opioid/nitric oxide-mediated, but cGMP-independent, pathways. *Evidence-based Complementary and Alternative Medicine*, *2016*, 1494981. https://doi.org/10.1155/2016/1494981.

Alam, A., Ferdosh, S., Ghafoor, K., Hakim, A., Juraimi, A. S., Khatib, A., et al. (2016). *Clinacanthus nutans*: A review of the medicinal uses, pharmacology and phytochemistry. *Asian Pacific Journal of Tropical Medicine*, *9*, 402–409. https://doi.org/10.1016/j.apjtm.2016.03.011.

Bardoni, R. (2013). Role of presynaptic glutamate receptors in pain transmission at the spinal cord level. *Current Neuropharmacology*, *11*, 477–483.

Basbaum, A. I., Bautista, D. M., Scherrer, G., & Julius, D. (2009). Cellular and molecular mechanisms of pain. *Cell*, *139*, 267–284.

Carrasquel-Ursulaez, W., Lorenzo, Y., Echeverria, F., & Latorre, R. (2018). Large conductance potassium channels in the nervous system. In A. Bhattacharjee (Ed.), *The Oxford handbook of neuronal ion channels* (1st ed., pp. 1–74). NY, USA: Oxford University Press. https://doi.org/10.1093/oxfordhb/9780190669164.013.11.

Choi, Y. B., Tenneti, L., Le, D. A., Ortiz, J., Bai, G., Chen, H. S., et al. (2000). Molecular basis of NMDA receptor-coupled ion channel modulation by S-nitrosylation. *Nature Neuroscience*, *3*, 15–21. https://doi.org/10.1038/71090.

Du, X., & Gamper, N. (2013). Potassium channels in peripheral pain pathways: Expression, function and therapeutic potential. *Current Neuropharmacology*, *11*, 621–640.

Du, X., Wang, C., & Zhang, H. (2011). Activation of ATP-sensitive potassium channels antagonize nociceptive behavior and hyperexcitability of DRG neurons from rats. *Molecular Pain*, *7*, 35. https://doi.org/10.1186/1744-8069-7-35.

Dzoyem, J. P., McGaw, L. J., Kuete, V., & Bakowsky, U. (2017). Anti-inflammatory and anti-nociceptive activities of African medicinal spices and vegetables. In C. Kuete (Ed.), *Medicinal spices and vegetables from Africa: Therapeutic potential against metabolic, inflammatory, infectious and systemic diseases* (1st ed., pp. 239–270). MA, USA: Elsevier Academic Press (Chapter 9).

Ekor, M. (2014). The growing use of herbal medicines: Issues relating to adverse reactions and challenges in monitoring safety. *Frontiers in Pharmacology*, *4*, 177. https://doi.org/10.3389/fphar.2013.00177.

Galeotti, N., Ghelardini, C., Vinci, M. C., & Bartolini, A. (1999). Role of potassium channels in the antinociception induced by agonists of alpha2-adrenoceptors. *British Journal of Pharmacology*, *126*, 1214–1220.

Goldberg, D. S., & McGee, S. J. (2011). Pain as a global public health priority. *BMC Public Health*, *11*, 770. https://doi.org/10.1186/1471-2458-11-770.

Granados-Soto, V., Argüelles, C. F., & Ortiz, M. I. (2002). The peripheral antinociceptive effect of resveratrol is associated with activation of potassium channels. *Neuropharmacology*, *43*, 917–923.

Herz, A., & Millan, M. J. (1990). Opioids and opioid receptors mediating antinociception at various levels of the neuraxis. *Physiologia Bohemoslovaca*, *39*, 395–401.

Holtsman, M., & Fishman, S. M. (2005). Opioid receptors. In H. T. Benzon, S. N. Raja, R. E. Molloy, S. S. Liu, & S. M. Fishman (Eds.), *Essentials of pain medicine and regional anesthesia* (2nd ed.). PA, USA: Elsevier Churchill Livingstone. 752 p.

Kamarudin, M., Sarker, M., Kadir, H. A., & Ming, L. C. (2017). Ethnopharmacological uses, phytochemistry, biological activities, and therapeutic applications of Clinacanthus nutans (Burm. f.) Lindau: A comprehensive review. *Journal of Ethnopharmacology*, *206*, 245–266. https://doi.org/10.1016/j.jep.2017.05.007.

Kampa, M., Hatzoglou, A., Notas, G., Niniraki, M., Kouroumalis, E., & Castanas, E. (2001). Opioids are non-competitive inhibitors of nitric oxide synthase in T47D human breast cancer cells. *Cell Death and Differentiation*, *8*, 943–952.

Khatun, A., Imam, M. Z., & Rana, M. S. (2015). Antinociceptive effect of methanol extract of leaves of *Persicaria hydropiper* in mice. *BMC Complementary and Alternative Medicine*, *15*, 63. https://doi.org/10.1186/s12906-015-0558-y.

Kokkola, T., Savinainen, J. R., Mönkkönen, K. S., Retamal, M. D., & Laitinen, J. T. (2005). S-nitrosothiols modulate G protein-coupled receptor signaling in a reversible and highly receptor-specific manner. *BMC Molecular and Cellular Biology*, *6*, 21. https://doi.org/10.1186/1471-2121-6-21.

Landa-Juárez, A. Y., Ortiz, M. I., Castañeda-Hernández, G., & Chávez-Piña, A. E. (2016). Participation of potassium channels in the antinociceptive effect of docosahexaenoic acid in the rat formalin test. *European Journal of Pharmacology*, *793*, 95–100.

Lautié, E., Russo, O., Ducrot, P., & Boutin, J. A. (2020). Unraveling plant natural chemical diversity for drug discovery purposes. *Frontiers in Pharmacology*, *11*, 397. https://doi.org/10.3389/fphar.2020.00397.

Lu, J., Katano, T., Okuda-Ashitaka, E., Oishi, Y., Urade, Y., & Ito, S. (2009). Involvement of S-nitrosylation of actin in inhibition of neurotransmitter release by nitric oxide. *Molecular Pain*, *5*, 58. https://doi.org/10.1186/1744-8069-5-58.

Manjiani, D., Paul, D. B., Kunnumpurath, S., Kaye, A. D., & Vadivelu, N. (2014). Availability and utilization of opioids for pain management: Global issues. *The Ochsner Journal*, *14*, 208–215.

Marcondes Sari, M. H., Guerra Souza, A. C., Gonçalves Rosa, S., Souza, D., Dorneles Rodrigues, O. E., & Wayne Nogueira, C. (2014). Contribution of dopaminergic and adenosinergic systems in the antinociceptive effect of p-chloro-selenosteroid. *European Journal of Pharmacology*, *725*, 79–86.

Martikainen, I. K., Nuechterlein, E. B., Peciña, M., Love, T. M., Cummiford, C. M., Green, C. R., et al. (2015). Chronic back pain is associated with alterations in dopamine neurotransmission in the ventral striatum. *Journal of Neuroscience*, *35*, 9957–9965.

Mickle, A. D., Shepherd, A. J., & Mohapatra, D. P. (2016). Nociceptive TRP channels: Sensory detectors and transducers in multiple pain pathologies. *Pharmaceuticals*, *9*, 72. https://doi.org/10.3390/ph9040072.

Miyamoto, T., Dubin, A. E., Petrus, M. J., & Patapoutian, A. (2009). TRPV1 and TRPA1 mediate peripheral nitric oxide-induced nociception in mice. *PLoS One*, *4*(10). https://doi.org/10.1371/journal.pone.0007596, e7596.

Mojtahedin, A., Tamaddonfard, E., & Zanbouri, A. (2009). Role of central muscarinic cholinergic receptors in the formalin-induced pain in rats. *Indian Journal of Pharmacology*, *41*, 144–147.

Muqeem, T., Ghosh, B., Pinto, V., Lepore, A. C., & Covarrubias, M. (2018). Regulation of nociceptive glutamatergic signaling by presynaptic Kv3.4 channels in the rat spinal dorsal horn. *Journal of Neuroscience*, *38*, 3729–3740.

Olesen, A. E., Andresen, T., Staahl, C., & Drewes, A. M. (2012). Human experimental pain models for assessing the therapeutic efficacy of analgesic drugs. *Pharmacological Reviews*, *64*, 722–779. https://doi.org/10.1124/pr.111.005447.

Ortiz, M. I., Torres-López, J. E., Castañeda-Hernández, G., Rosas, R., Vidal-Cantú, G. C., & Granados-Soto, V. (2002). Pharmacological evidence for the activation of K+ channels by diclofenac. *European Journal of Pharmacology*, *438*, 85–91.

Palos, G. R., Mendoza, T. R., Cantor, S. B., Aday, L. A., & Cleeland, C. S. (2004). Perceptions of analgesic use and side effects: What the public values in pain management. *Journal of Pain and Symptom Management*, *28*, 460–473.

Park, A. R., Lee, H. I., Semjid, D., Kim, D. K., & Chun, S. W. (2014). Dual effect of exogenous nitric oxide on neuronal excitability in rat substantia gelatinosa neurons. *Neural Plasticity*, *2014*, 628531. https://doi.org/10.1155/2014/628531.

Rezaee-Asl, M., Sabour, M., Nikoui, V., Ostadhadi, S., & Bakhtiarian, A. (2014). The study of analgesic effects of *Leonurus cardiaca* L. in mice by formalin, tail flick and hot plate tests. *International Scholarly Research Notices*, *2014*, 687697. https://doi.org/10.1155/2014/687697.

Rosenbaum, T., & Simon, S. A. (2007). TRPV1 receptors and signal transduction. In W. B. Liedtke, et al. (Eds.), *TRP ion channel function in sensory transduction and cellular signaling cascades* CRC Press/Taylor & Francis.

Seabra, A. B. (2017). *Nitric oxide donors: Novel biomedical applications and perspectives*. London, UK: Elsevier Academic Press.

Shim, S. Y., Aziana, I., & Khoo, B. Y. (2013). Perspective and insight on *Clinacanthus nutans* Lindau in traditional medicine. *International Journal of Integrative Biology*, *14*, 7–9.

Song, B., & Marvizon, J. C. (2005). N-methyl-D-aspartate receptors and large conductance calcium-sensitive potassium channels inhibit the release of opioid peptides that induce mu-opioid receptor internalization in the rat spinal cord. *Neuroscience*, *136*, 549–562.

Souza, A. L., Moreira, F. A., Almeida, K. R., Bertollo, C. M., Costa, K. A., & Coelho, M. M. (2002). In vivo evidence for a role of protein kinase C in peripheral nociceptive processing. *British Journal of Pharmacology*, *135*, 239–247.

Stolz, E. D., Müller, L. G., Trojan-Rodrigues, M., Baumhardt, E., Ritter, M. R., & Rates, S. M. K. (2014). Survey of plants popularly used for pain relief in Rio Grande do Sul, southern Brazil. *Revista Brasileira de Farmacognosia*, *24*, 185–196.

Thomas, B., Bantel, C., Stone, L., & Wilcox, G. L. (2013). α2-adrenergic agonists in pain treatment. In G. F. Gebhart, & R. F. Schmidt (Eds.), *Encyclopedia of pain*. Berlin, Heidelberg: Springer. 9 p.

Wang, H., Kohno, T., Amaya, F., Brenner, G. J., Ito, N., Allchorne, A., et al. (2005). Bradykinin produces pain hypersensitivity by potentiating spinal cord glutamatergic synaptic transmission. *Journal of Neuroscience*, *25*, 7986–7992.

Wirth, J. H., Hudgins, J. C., & Paice, J. A. (2005). Use of herbal therapies to relieve pain: A review of efficacy and adverse effects. *Pain Management Nursing*, *6*, 145–167. https://doi.org/10.1016/j.pmn.2005.08.003.

Yahaya, R., Kumar Dash, G., Abdullah, M. S., & Mathews. (2015). *Clinacanthus nutans* (burm. F.) Lindau: An useful medicinal plant of Southeast Asia. *International Journal of Pharmacognosy and Phytochemical Research*, *7*, 1244–1250.

Yang, R., Xiong, Z., Liu, C., & Liu, L. (2014). Inhibitory effects of capsaicin on voltage-gated potassium channels by TRPV1-independent pathway. *Cellular and Molecular Neurobiology*, *34*, 565–576.

Zakaria, Z. A., Abdul Rahim, M. H., Mohd Sani, M. H., Omar, M. H., Ching, S. M., Abdul Kadir, A., et al. (2019). Antinociceptive activity of petroleum ether fraction obtained from methanolic extract of Clinacanthus nutans leaves involves the activation of opioid receptors and NO-mediated/cGMP-independent pathway. *BMC Complementary and Alternative Medicine*, *19*, 79. https://doi.org/10.1186/s12906-019-2486-8.

Zakaria, Z. A., Abdul Rahim, M. H., Roosli, R. A. J., Mohd Sani, M. H., Marmaya, N. H., Omar, M. H., et al. (2019). Antinociceptive activity of petroleum ether fraction of *Clinacanthus nutans* leaves methanolic extract: Roles of nonopioid pain modulatory systems and potassium channels. *BioMed Research International*, *2019*, 6593125. https://doi.org/10.1155/2019/6593125.

Zakaria, Z. A., Abdul Rahim, M. H., Roosli, R. A. J., Mohd Sani, M. H., Omar, M. H., Mohd Tohid, S. F., et al. (2018). Antinociceptive activity of methanolic extract of *Clinacanthus nutans* leaves: Possible mechanisms of action involved. *Pain Research & Management*, 9536406. https://doi.org/10.1155/2018/9536406.

Zulkipli, I. N., Rajabalaya, R., Idris, A., Sulaiman, N. A., & David, S. R. (2017). Clinacanthus nutans: A review on ethnomedicinal uses, chemical constituents and pharmacological properties. *Pharmaceutical Biology*, *55*(1), 1093–1113. https://doi.org/10.1080/13880209.2017.1288749.

Chapter 11

The orally bioavailable imidazodiazepine, KRM-II-81, is a novel potentiator of α2/3-containing $GABA_A$ receptors with analgesic efficacy

Rok Cerne[a,b], Jodi L. Smith[b], Janet L. Fisher[c], Lalit K. Golani[d], Daniel E. Knutson[d], James M. Cook[d], and Jeffrey M. Witkin[b,d,e]

[a]*Institute of Pathophysiology, Faculty of Medicine, University of Ljubljana, Ljubljana, Slovenia,* [b]*Laboratory of Antiepileptic Drug Discovery, Peyton Manning Hospital for Children, Ascension St. Vincent, Indianapolis, IN, United States,* [c]*Department of Pharmacology, Physiology & Neuroscience, School of Medicine, University of South Carolina, Columbia, SC, United States,* [d]*Department of Chemistry and Biochemistry, University of Wisconsin-Milwaukee, Milwaukee, WI, United States,* [e]*Departments of Neuroscience and Trauma Research, Ascension St. Vincent Hospital, Indianapolis, IN, United States*

Abbreviations

GABA	γ-aminobutyric acid
KRM-II-81	5-(8-ethynyl-6-(pyridin-2-yl)-4H-benzo[f] imidazo[1,5-a][1,4]diazepin-3-yl)oxazole
PAM	positive allosteric modulator

Introduction

Despite advances, the treatment of chronic pain remains a substantial unmet medical need. The relief of pain by currently available medications is, in most cases, partial and accompanied with side effects that include constipation, tolerance, physical dependence, abuse, and death. Therefore, substantial effort has been dedicated to the discovery and development of analgesic drugs with novel mechanisms of action exhibiting improved efficacy and safety profiles. A host of different targets is currently being pursued (Woolf, 2020), and targeting ionotropic GABA ($GABA_A$) receptors has emerged as one valid proposition.

GABA receptors

γ-Amino-butyric acid (GABA) is the primary inhibitory neurotransmitter in the mammalian nervous system, and GABAergic neurotransmission has long been considered to contribute to the regulation of pain. For example, intrathecally administered muscimol blocked pain behaviors induced by formalin injections in rats and these effects were attenuated by the $GABA_A$ receptor antagonist bicuculline (Dirig & Yaksh, 1995).

$GABA_A$ receptors are found on multiple relays of pain pathway and are crucial for controlling the gain and fidelity of pain neurotransmission. Two such relays have gained particular attention in recent years—dorsal root ganglion (Du et al., 2017) and spinal dorsal horn (Melzack, 1996). For example, Weir et al. (Weir et al., 2017) utilized viral delivery of modified glutamate activated chloride channels (designer receptor) to dorsal root ganglion neurons to promote long-lasting pain relief by increasing hyperpolarizing GABA currents after ivermectin (designer drug) administration. An example of modulation at the spinal cord level comes from experiments in which immature telencephalic GABAergic interneurons from the mouse medial ganglionic eminence were transplanted into the adult mouse spinal cord to completely reverse the mechanical hypersensitivity produced by peripheral nerve injury (Etlin et al., 2016).

The Neurobiology, Physiology, and Psychology of Pain. https://doi.org/10.1016/B978-0-12-820589-1.00011-7

$GABA_A$ receptor potentiating benzodiazepines and pain

Benzodiazepines were first developed in the late 1950s, and diazepam (Fig. 1) is still widely used. Currently, no benzodiazepine is approved as a primary therapy for chronic pain. While initial studies demonstrated activity of benzodiazepines in the treatment of lower back pain (Salzmann, Pforringer, Paal, & Gierend, 1992), tic douloureux (Court & Kase, 1976), and cranial neuralgia (Smirne & Scarlato, 1977), subsequent examinations failed to confirm efficacy (Chou et al., 2017). It is likely that the lack of efficacy is due to inadequate dosing due to sedation at subtherapeutic doses (Knabl, Zeilhofer, Crestani, Rudolph, & Zeilhofer, 2009). Consistent with this hypothesis are data showing weak antihyperalgesic activity at mildly sedating doses of clobazam (Besson et al., 2015) and the analgesic efficacy of intrathecally administered benzodiazepines in animal models (Knabl et al., 2008).

Despite their shortcomings, prescriptions for benzodiazepines for chronic pain have increased over the last two decades (Agarwal & Landon, 2019). Increased use might be due to their anxiolytic activity as it has been shown that anxiety increases the unpleasantness of pain, muscle tension, and sympathetic outflow (Dellemijn & Fields, 1994).

Efforts directed at creating improved $GABA_A$ receptor positive allosteric modulators (PAMs) or GABAkines derives from the discovery of the benzodiazepine receptor and its role in potentiation of GABA currents. Promising ligands were evaluated in animal models for efficacy and reduced side effects (reviewed in Skolnick, 2012). Based upon the ability of some compounds to produce anxiolytic-like effects without sedation in animal models (e.g., CL218-872), it was early hypothesized that distinct benzodiazepine receptors might exist that mediate anxiolytic versus sedative effects (Klepner, Lippa, Benson, Sano, & Beer, 1979). The advent of molecular biology enabled further refinement in the search for drugs with therapeutic specificity.

Alpha 2/3-containing $GABA_A$ receptors

The $GABA_A$ receptor is a pentameric ligand-gated ion channel, allowing for various combinations of five different subunits (Olsen & Sieghart, 2009). A typically benzodiazepine-sensitive $GABA_A$ receptor is composed of two α-, two β-, and one γ-subunit; α1,2,3, and 5 create diazepam-sensitive receptors while α4 and α6 are diazepam-insensitive (Sigel & Ernst, 2018). Studies demonstrated that the His→Arg point mutation in the benzodiazepine binding site drastically reduced the affinity for benzodiazepine binding (Wieland, Lüddens, & Seeburg, 1992). Selective use of point mutations has been used to probe

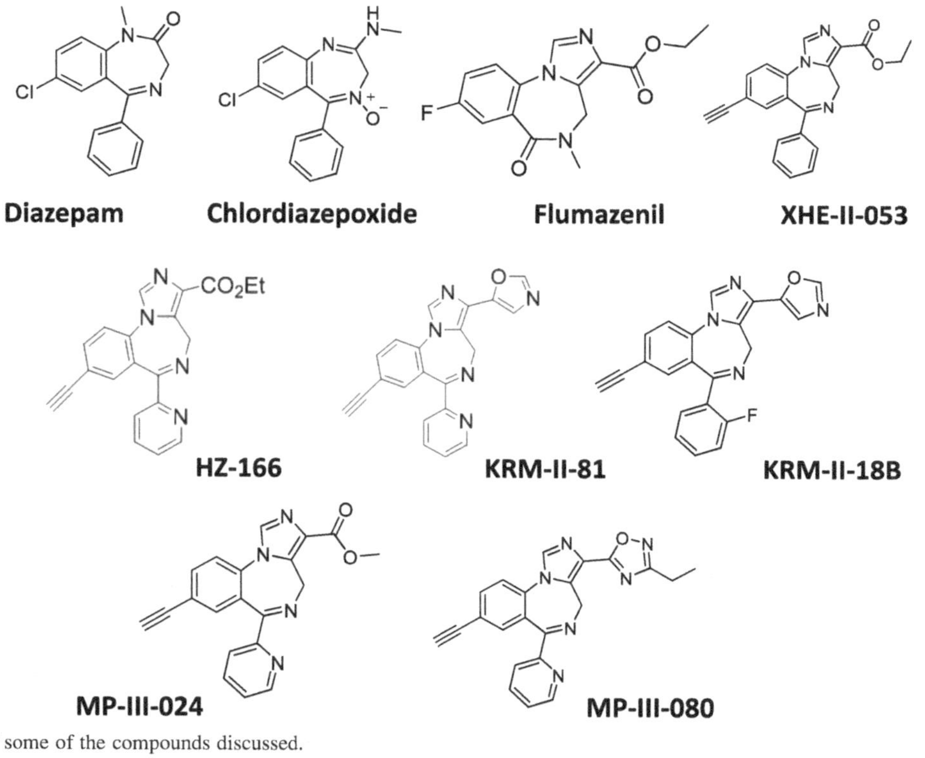

FIG. 1 Structures of some of the compounds discussed.

the contributions of a particular α-subunit to the receptor's pharmacological properties. Such data have suggested that α1-subtype-containing $GABA_A$ receptors preferentially mediate the sedative, amnestic, ataxic, and dependence-producing effects of ligands (Ator, Atack, Hargreaves, Burns, & Dawson, 2010; Licata, Platt, Cook, Van Linn, & Rowlett, 2009; McKernan et al., 2000; Rudolph et al., 1999), whereas α2- and α3-subtypes mediate anxiolytic (Löw et al., 2000; Rudolph & Möhler, 2014) and analgesic effects. When the α1-subtype was rendered insensitive to benzodiazepines, the therapeutic window of diazepam was markedly increased (Löw et al., 2000), while the anxiolytic efficacy of diazepam was retained (McKernan et al., 2000). Analgesic efficacy, not previously observed with diazepam, was uncovered due to the decreased sedation and motor impairment that resulted from the deletion of its interactions with α1-containing $GABA_A$ receptors (Ralvenius, Benke, Acuña, Rudolph, & Zeilhofer, 2015). Work over the last 15 years has been directed at the identification and development of $GABA_A$ receptor PAM analgesics with preference for α2- and α3- over α1-containing $GABA_A$ receptors (Rudolph & Knoflach, 2011) (Atack, 2011; Cerne et al., 2019).

Some of the evidence linking α2/3-selective $GABA_A$ receptors to the regulation of pain is summarized in Table 1. Genetic inactivation studies have shown that when the sedative/motor-impairing effects of nonselective $GABA_A$ PAMs like diazepam are reduced, antihyperalgesic efficacy can be revealed. In studies using pain behaviors in rodent models, several compounds with various degrees of selectivity for α2/3-containing $GABA_A$ receptors reduce pain behaviors, including studies with HZ-166 and KRM-II-81 (Fig. 1). Limited data in humans have also been consistent with the α2/3 hypothesis.

TABLE 1 **Data relating α2/3-containing $GABA_A$ receptors to pain.**

Study	Evidence	References
Genetic models	Sedative, amnesic and anticonvulsant actions of diazepam are reduced in α1(H101R) mice[a]	McKernan et al. (2000) and Rudolph et al. (1999)
	Analgesic effects of spinal diazepam are reduced in α2 (H101R) and to lesser extent in α3(H126R) mice[a]	Knabl et al. (2008)
	Analgesia by systemic diazepam is unmasked in α1(H101R) mice[a]	Knabl et al. (2009)
	Intact α2 GABAAR is sufficient for mediation of diazepam-mediated antihyperalgesia[b]	Ralvenius et al. (2015)
	Intact α1 GABAAR is sufficient for mediation of diazepam-mediated sedation[b]	Ralvenius et al. (2015)
Preclinical models	L-838417 is analgesic in inflammatory and neuropathic pain	Hofmann et al. (2012), Knabl et al. (2008), and Nickolls et al. (2011)
	NS11394 is analgesic in inflammatory and neuropathic pain	Munro, Ahring, and Mirza (2009) and Munro et al. (2008)
	NS16085 is analgesic in inflammatory and neuropathic pain	de Lucas et al. (2015)
	TPA023 is analgesic in inflammatory and neuropathic pain	Nickolls et al. (2011)
	PF-06372865 is analgesic in neuropathic pain	Owen et al. (2019)
	HZ-166 is analgesic in inflammatory and neuropathic pain without sedation or tolerance	Di Lio et al. (2011)
	Antihyperalgesia by HZ166 is reduced when α2-GABAARs are made benzodiazepine insensitive	Paul et al. (2014) and Ralvenius et al. (2015)
	MP-III-024 is analgesic in an inflammatory pain model without motor impairment	Fischer et al. (2017)
	KRM-II-81 is analgesic in inflammatory and neuropathic pain models without motor impairment	Lewter et al. (2017) and Witkin, Cerne, et al. (2019)
	NDMC (α2 preferring clobazam metabolite) has an improved therapeutic window for antihyperalgesia	Ralvenius et al. (2016)

Continued

TABLE 1 Data relating α2/3-containing $GABA_A$ receptors to pain—cont'd

Study	Evidence	References
Human Studies	Clobazam decreased the area of secondary hyperalgesia in human volunteers	Besson et al. (2015) and Vuilleumier, Besson, Desmeules, Arendt-Nielsen, and Curatolo (2013)
	NDMC failed to reduce hyperalgesia in human volunteers	Matthey et al. (2020)
	Hypnotic effects of zolpidem are driven by preference for α1-GABAARs	Hevers and Lüddens (1998) and Wafford, Whiting, and Kemp (1993)
	PF-06372865 displayed PD biomarker engagement in a Ph-1 clinical study	Nickolls et al. (2018)
	PF-06372865 suppressed evoked pain (electrical, pressure, heat, cold, inflammatory)	van Amerongen et al. (2019)
	PF-06372865 failed to demonstrate efficacy in Ph-2 for chronic low back pain (no sedation)	Gurrell et al. (2018)

[a]*Studies used point-mutated mice in which α-subunits were rendered diazepam insensitive (α1(H101R), α2(H101R), α3(H126R) and α5(H105R)).*
[b]*Studies using triple mutant mice in which only one α-subunit is diazepam sensitive.*

PF-06372865 is the most advanced compound currently in clinical development. This compound is selective for α2/3 over α1 but, unlike KRM-II-81, PF-06372865 also potentiates α5-containing $GABA_A$ receptors (Owen et al., 2019).

KRM-II-81

KRM-II-81 was developed in a progression of stages with the goal of creating a $GABA_A$ receptor PAM that retains the efficacy of diazepam while reducing side effect burden. The laboratory of James M. Cook had the goal of developing novel anxiolytic drugs using a privileged structure instead of high throughput screening. The privileged structural framework was the hybrid structure of diazepam and flumazenil (Fig. 1). These new structures deviated from the benzodiazepine scaffold and are termed imidazodiazepines. Structural modifications enabled the discovery of XHE-II-053 (Cook et al., 2006; Cook, Zhou, Huang, Sarma, & Zhang, 2009; Fischer et al., 2010) (Fig. 1). XHE-II-053 was shown to be a nonsedating anxiolytic, without untoward effects or toxicity (Cook et al., 2006, 2009). However, XHE-II-053 was metabolized rapidly in humans (Namjoshi et al., 2013). An analog, HZ-166, was subsequently developed (Cook et al., 2009, 2010; Fischer et al., 2010; Rivas et al., 2009) (Fig. 1). HZ-166 was longer lived than XHE-II-053 and was a nonsedating anxiolytic in mice and primates (Cook et al., 2009, 2010; Fischer et al., 2010; Namjoshi et al., 2013). HZ-166 was also anticonvulsant and antinociceptive (Cook et al., 2009, 2010; Di Lio et al., 2011; Rivas et al., 2009). Importantly, HZ-166 did not display tolerance to its anticonvulsant (Cook et al., 2009, 2010) or antihyperalgesic effects (Di Lio et al., 2011). However, HZ-166 had metabolic limitations arising from its labile ester functionality (Poe et al., 2016).

KRM-II-81 (Fig. 1) was designed to increase the metabolic stability of HZ-166. Construction of KRM-II-81 through bioisosteric substitution was first described by Poe et al. (2016) where the C3 ester function of HZ-166 was replaced with a 1,3-oxazole. The resulting compound, KRM-II-81, demonstrated the desired $GABA_A$ receptor subtype selectivity (Fig. 2, top panel) comparable to HZ-166. KRM-II-81 does not significantly impact α1 or α5-containing $GABA_A$ receptors, whereas diazepam nondiscriminately interacts with these configurations (Fig. 2, top panel). The ability of KRM-II-81 to function as a positive allosteric modulator of $GABA_A$ receptors was demonstrated in electrophysiological studies (Poe et al., 2016). In addition, pharmacokinetic experiments revealed a marked improvement over that of HZ-166: orally bioavailability, long exposure duration, and excellent brain penetrability (Poe et al., 2016). Translation of these in vitro and pharmacokinetic properties into in vivo activity was also demonstrated with the production, by KRM-II-81 and structural analogs like MP-II-080 (Fig. 1), of anxiolytic-like effects in rodents with reduced motor-impairing effects (Poe et al., 2016; Witkin et al., 2017).

KRM-II-81 and pain

Electrophysiological studies were conducted to link the GABA modulatory effects of KRM-II-81 with pain pathways. The inhibitory neurotransmission of GABA was enhanced in isolated rat dorsal root ganglion neurons (Witkin, Cerne, et al., 2019). Fig. 2 shows a representative trace illustrating the enhancement of GABA inhibition in these neurons where a nonmaximal concentration of GABA was augmented by application of KRM-II-81. The resulting concentration effect curve for

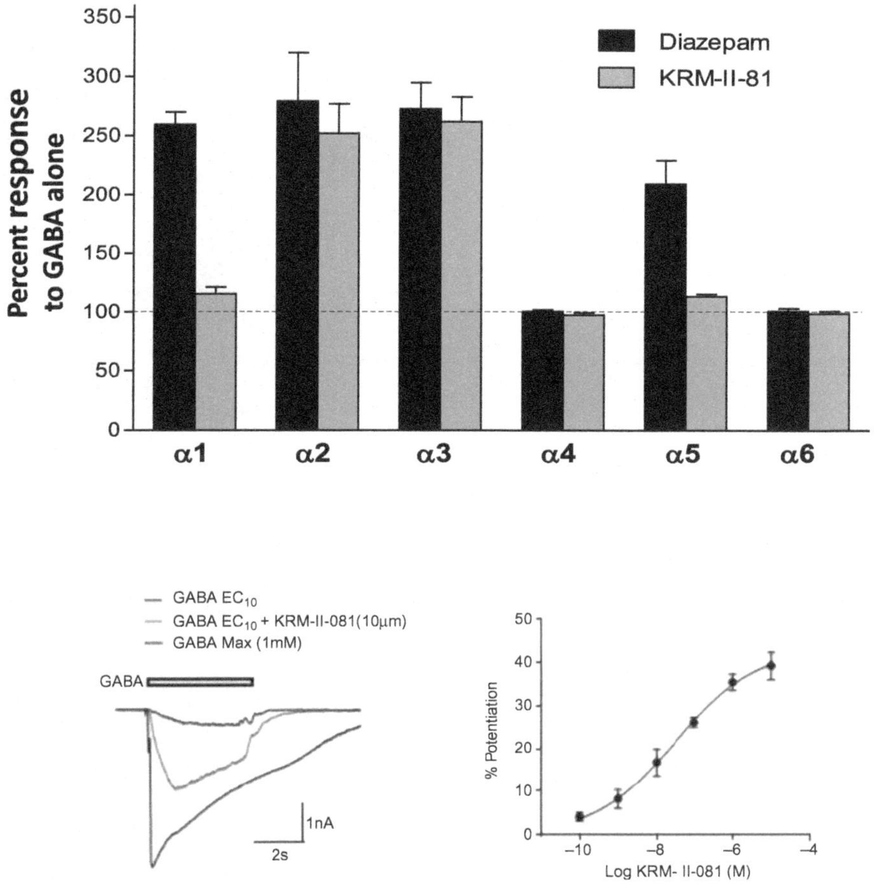

FIG. 2 *Top:* Average enhancement of the current evoked to GABA EC_3 by 0.1 μM (α1, α2, α3, α5) or 1 μM (α4, α6) of either diazepam or KRM-II-81. Data for KRM-II-81 and diazepam are replotted from Lewter et al. (2017). *Bottom:* Effects of KRM-II-81 as a potentiator of GABA currents in dorsal root ganglia preparations from rats. *Left:* Manual patch clamp recording of currents evoked by saturating GABA concentration (1 mM) and by an EC_{10} concentration of GABA (5 μM). GABA current evoked with 5 μM GABA was potentiated by KRM-II-081(10 μM). *Right:* Concentration response curve of KRM-II-81 effects on currents evoked with an EC_{10} concentration of GABA. The potentiation is expressed as a percent increase in the amplitude of partially activated GABA current (GABA EC_{10}). *(Data are from Witkin, J. M., Cerne, R., Davis, P. G., Freeman, K. B., do Carmo, J. M., Rowlett, J. K., et al. (2019). The α2,3-selective potentiator of GABAA receptors, KRM-II-81, reduces nociceptive-associated behaviors induced by formalin and spinal nerve ligation in rats.* Pharmacology, Biochemistry, and Behavior, 180, *22–31. doi:10.1016/j.pbb.2019.02.013 and Witkin, J. M., Li, G., Golani, L. K., Xiong, W., Smith, J. L., Ping, X., et al. (2019). The positive allosteric modulator of α2/3-containing GABAA receptors, KRM-II-81, is active in pharmaco-resistant models of epilepsy and reduces hyperexcitability after traumatic brain injury.* The Journal of Pharmacology and Experimental Therapeutics, *https://doi.org/10.1124/jpet.119.260968 with permission.)*

KRM-II-81 to potentiate GABA inhibition in the dorsal root ganglion neurons (Fig. 2, bottom panel) generated an EC_{50} of 32 nM.

KRM-II-81 was shown to be effective in reducing pain behaviors induced by acute pain provocation in rodents (Table 2). KRM-II-81 reduced acid-induced changes in writhing (Lewter et al., 2017), nest-building (Lewter et al., 2017), and intracranial self-stimulation behaviors (Moerke et al., 2019), but not locomotion for which acetaminophen was active (Lewter et al., 2017). A structural analog, KRM-II-18b (Fig. 1), was also active (Lewter et al., 2017).

Inflammatory pain is also reduced by KRM-II-81 and analogs, including neonatal inflammatory bladder pain (Table 2). Formalin-induced inflammatory pain (late phase of the formalin response) is used here as an illustration of these effects in comparison to diazepam. Diazepam did not attenuate nociceptive behaviors in formalin-treated rats up to a dose of 10 mg/kg, p.o. (Fig. 3, left panel). In contrast, orally administered KRM-II-81 suppressed pain behaviors with equivalent or greater efficacy than the positive control tramadol (Fig. 3, right panel).

KRM-II-81 is also active in models of chronic pain (Table 2). For example, the chronic hyperalgesia induced by ligation of spinal nerves L5/6 in rats is significantly attenuated by KRM-II-81(i.p.) (Fig. 4, left panel). In a variation of this assay where baseline hyperalgesia scores are greater (Fig. 4, right panel), KRM-II-81 (p.o.) reduced the chronic tactile hyperalgesia induced by spinal nerve ligation. The effects of KRM-II-81 were often greater than that of the neuropathic pain medicine gabapentin (Fig. 4, right panel).

TABLE 2 Effects of KRM-II-81 and structural analogs in rodent pain models.

Compound	Pain model	Species	Comparators	References
KRM-II-81	Acetic and lactic-acid-induced writhing, nesting and locomotion	ICR mice	Morphine	Lewter et al. (2017)
KRM-II-18B	Acetic and lactic-acid-induced writhing, nesting and locomotion	ICR mice	Morphine	Lewter et al. (2017)
KRM-II-81	Lactic-acid and ICSS behavior	Sprague-Dawley rats	Ketorolac and diazepam	Moerke, Li, Golani, Cook, and Negus (2019)
MP-III-024	Zymosin A-induced mechanical hyperalgesia	C57BL/6 mice		Fischer et al. (2017)
KRM-II-81	Formalin-induced tactile hyperalgesia	Sprague-Dawley rats	Tramadol and diazepam	Witkin, Cerne, et al. (2019)
KRM-II-81	L5/6 nerve ligation—induced tactile hyperalgesia	Sprague-Dawley rats	Gabapentin	Witkin, Cerne, et al. (2019)
KRM-II-81	L5/6 nerve ligation—sensitization training-induced tactile hyperalgesia	Sprague-Dawley rats	Gabapentin	Witkin, Cerne, et al. (2019)
KRM-II-81	Chemotherapy-induced thermal hyperalgesia	C57BL/6 mice	Gabapentin	Biggerstaff et al. (2020)
KRM-II-81	Chemotherapy-induced tactile hyperalgesia	C57BL/6 mice	Gabapentin	Biggerstaff et al. (2020)
HZ-166	Inflammatory bladder pain	Neonatal Sprague-Dawley rats		Kannampalli et al. (2017)

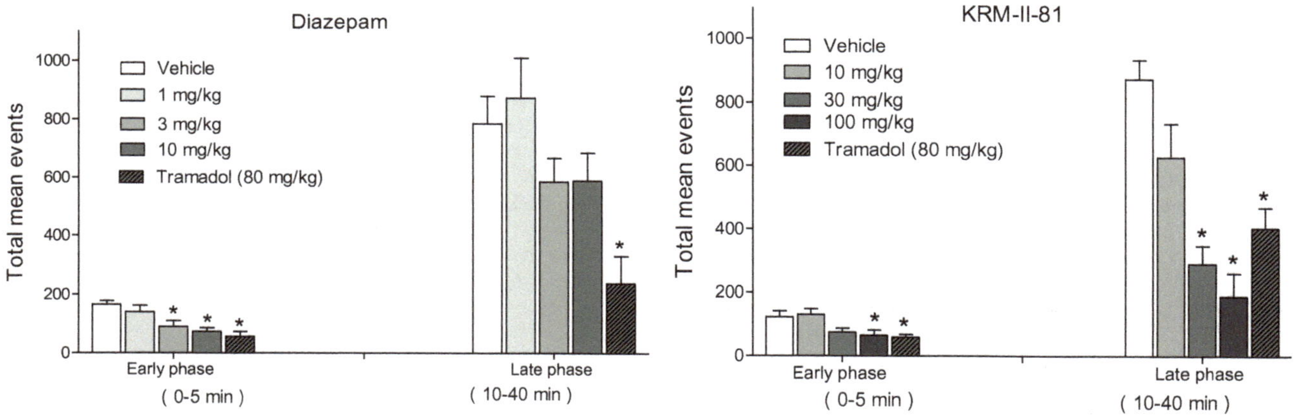

FIG. 3 *Left:* Diazepam (i.p., 30 min prior) did not reduce paw-withdrawal thresholds in the formalin assay during the late phase but decreased early phase behavioral events. *Right:* KRM-II-81 (i.p., 30 min prior) significantly decreased nociceptive responses in the late phase of the formalin assay. Tramadol (80 mg/kg, i.p., 30 min prior) was active as a positive control in both studies. $P < .05$ compared to vehicle control values. *(Data are from Witkin, J. M., Cerne, R., Davis, P. G., Freeman, K. B., do Carmo, J. M., Rowlett, J. K., et al. (2019). The α2,3-selective potentiator of GABAA receptors, KRM-II-81, reduces nociceptive-associated behaviors induced by formalin and spinal nerve ligation in rats.* Pharmacology, Biochemistry, and Behavior, 180, *22–31. doi:10.1016/j.pbb.2019.02.013 and Witkin, J. M., Li, G., Golani, L. K., Xiong, W., Smith, J. L., Ping, X., et al. (2019). The positive allosteric modulator of α2/3-containing GABAA receptors, KRM-II-81, is active in pharmaco-resistant models of epilepsy and reduces hyperexcitability after traumatic brain injury.* The Journal of Pharmacology and Experimental Therapeutics, *https://doi.org/10.1124/jpet.119.260968 with permission.)*

Side effect burden

KRM-II-81 was specifically designed to provide the efficacy of a $GABA_A$ receptor PAM with reduced motor-impairing effects. The reduced impact of KRM-II-81 on motor performances has been demonstrated (Poe et al., 2016; Witkin et al., 2018; Witkin, Li, et al., 2019). Motor impairment can limit the doses that can be given to patients and thereby limit drug

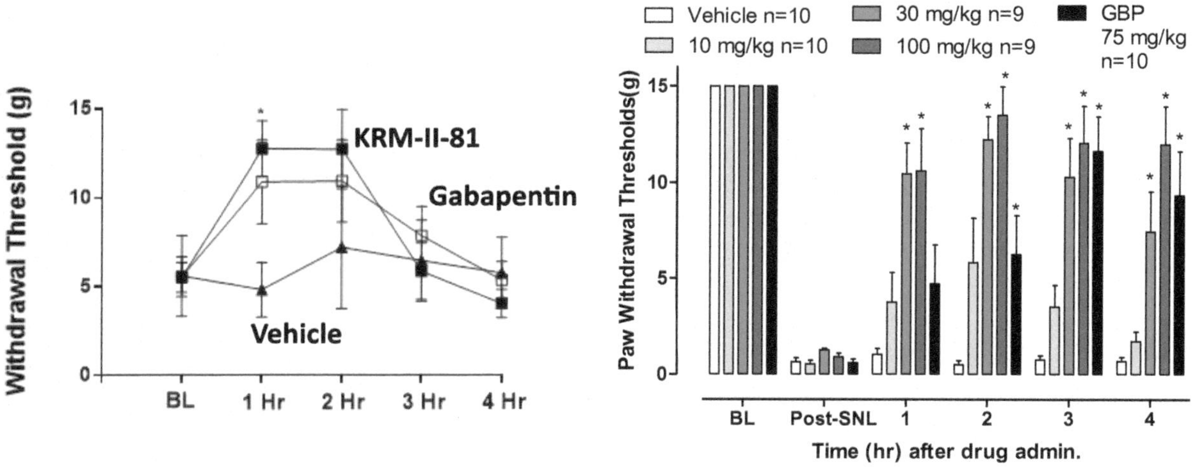

FIG. 4 *Left:* KRM-II-81 (i.p.) produced analgesic effects in rats undergoing chronic nociceptive hypersensitivity induced by spinal nerve ligation (SNL). KRM-II-81 (50 mg/kg); Gabapentin (75 mg/kg). $^*P < .05$ compared to vehicle control values. BL: baseline paw withdrawal latencies prior to drug administration. *Right:* KRM-II-81 (p.o.) produced analgesic effects in rats after chronic SNL in a sensitized model with lower baseline values after SNL (Post-SNL). BL: baseline paw withdrawal latencies prior to spinal nerve ligation (SNL). Post-SNL: paw withdrawal latencies after SNL. *(Data are from Witkin, J. M., Cerne, R., Davis, P. G., Freeman, K. B., do Carmo, J. M., Rowlett, J. K., et al. (2019). The α2,3-selective potentiator of GABAA receptors, KRM-II-81, reduces nociceptive-associated behaviors induced by formalin and spinal nerve ligation in rats.* Pharmacology, Biochemistry, and Behavior, *180, 22–31. doi:10.1016/j.pbb.2019.02.013 and Witkin, J. M., Li, G., Golani, L. K., Xiong, W., Smith, J. L., Ping, X., et al. (2019). The positive allosteric modulator of α2/3-containing GABAA receptors, KRM-II-81, is active in pharmaco-resistant models of epilepsy and reduces hyperexcitability after traumatic brain injury.* The Journal of Pharmacology and Experimental Therapeutics, *https://doi.org/10.1124/jpet.119.260968 with permission.)*

efficacy. Studies with genetically modified mice have shown that manipulations that reduce sedative and motor-impairing effects of diazepam are able to reveal antihyperalgesic efficacy that is otherwise not observed (Table 1). The structural basis for the sparing of the motor-impairing effects of KRM-II-81 was recently disclosed (Witkin, Li, et al., 2019).

Compared to alprazolam, KRM-II-81 produced a lower propensity for respiratory depression (Witkin et al., 2018). Studies on the analgesic effects of HZ-166 have shown no tolerance after 9 days of dosing (Di Lio et al., 2011). KRM-II-81 was studied with subchronic dosing for 22 days in a chemotherapy (paclitaxel)-induced chronic pain assay in mice. No tolerance developed during this dosing period. In contrast, there was evidence of tolerance development to the effects of gabapentin (Biggerstaff et al., 2020). It has been suggested that α1- and α5-containing $GABA_A$ receptors play a role in the development of tolerance (van Rijnsoever et al., 2004; Vinkers & Olivier, 2012). The absence of positive allosterism at these specific $GABA_A$ receptors might account for the differential tolerance profile of KRM-II-81. This possibility is further supported by the general lack of tolerance development to the administration of PAMs selective for α2/3-containing $GABA_A$ receptors (Vinkers & Olivier, 2012). In a study utilizing behavior of rats self-stimulating reinforcing brain areas (ICSS), Moerke et al. (2019) showed that whereas diazepam produced behavioral changes indicative of abuse liability, KRM-II-81 did not.

Applications to other therapeutic areas

In addition to its potential for pain management, KRM-II-81 also produces robust anxiolytic, anticonvulsant, and antidepressant activity in rodent models (Cerne et al., 2019; Methuku et al., 2018; Poe et al., 2016; Witkin et al., 2017, 2018; Witkin, Ping, et al., 2019). Pain is associated with anxiety and depression and, along with other disturbances like sleep, these psychiatric disorders are comorbid in patients suffering from chronic pain. Taken as a whole, the potential for KRM-II-81 to help moderate anxiety and depression symptoms could be an additional therapeutic benefit. Some medicines used to treat pain are also antiepileptics. Epilepsy and pain are also considered to be comorbid, and epilepsy might contribute to increased pain sensitivity (c.f., Velioglu, Gedikli, Yıldırım, & Ayar, 2017). Again, given the combined antinociceptive efficacy of KRM-II-81 and its broad anticonvulsant effects, KRM-II-81 could additionally help to fill this medical niche.

Other agents of interest

Structural analogs of KRM-II-81 (Fig. 1) have been discussed throughout. Discussion of other compounds with selectivity for α2/3-containing $GABA_A$ receptors can be found in Atack (2011) and Cerne et al. (2019).

Mini-dictionary of terms

- $GABA_A$ receptors—one class ligand-gated ion channels that regulate neuronal inhibition through the gating of ions upon binding of γ-aminobutyric acid (GABA) to the orthosteric binding domain of the receptor and through the binding of various allosteric modulators.
- Positive allosteric modulator (PAM)—a compound that enhances the activity of a receptor by binding to an allosteric domain (distinct from the orthosteric or neurotransmitter site) and augmenting the activity of the orthosteric ligand.
- Hyperalgesia—a state of heighted pain sensitivity.
- ICSS—Intracranial self-administration, a method for assessing brain reward thresholds and used to gauge abuse liability.

Key facts

- KRM-II-81 is a selective potentiator of α2/3-containing $GABA_A$ receptors.
- KRM-II-81 is less motor-impairing than anxiolytic benzodiazepines like diazepam.
- KRM-II-81 reduces pain behaviors in acute and chronic pain models in rodents.

Summary points

- Drugs that dampen excitatory neurotransmission, and positive allosteric modulators of $GABA_A$ receptors (PAMs) in particular, have been in use in psychiatry and neurology for over half a century (Cerne et al., 2019).
- Although GABA is an integral mechanism regulating pain in the central nervous system, the exploitation of this mechanism for the treatment of pain has been essentially nonexistent.
- KRM-II-81 is a novel, orally bioavailable compound with efficacy across a range of pain states as assessed by rodent models.
- KRM-II-81 displays a relatively mild side effect profile in preclinical models and has not shown tolerance development in the studies conducted to date.
- KRM-II-81 is also a robust anxiolytic and anticonvulsant in rodent models and demonstrates antidepressant-like activity.
- The potential of KRM-II-81 to additionally manage comorbid anxiety and depression in pain patients creates added value.

Acknowledgments

We are grateful for the generous research support of John and Nancy Peterson. We thank the following granting agencies for support: DA-043204 and NS-076517 and the National Science Foundation, Division of Chemistry [Grant CHE-1625735]. We also acknowledge the UW-Milwaukee Shimadzu Laboratory for Advanced and Applied Analytical Chemistry and support from the Milwaukee Institute of Drug Discovery and the University of Wisconsin-Milwaukee Research Foundation.

Conflict of interest

Authors report no conflict of interest. James M. Cook is a patent holder for the invention of KRM-II-81. The University of Wisconsin-Milwaukee owns the patent with a license agreement to RespireRx Pharmaceticals Inc.

References

Agarwal, S. D., & Landon, B. E. (2019). Patterns in outpatient benzodiazepine prescribing in the United States. *JAMA Network Open*, *2*(1), e187399. https://doi.org/10.1001/jamanetworkopen.2018.7399. Erratum in: JAMA Network Open 2019;2(3):e191203.

Atack, J. R. (2011). GABAA receptor subtype-selective modulators. I. α2/α3-selective agonists as non-sedating anxiolytics. *Current Topics in Medicinal Chemistry*, *11*(9), 1176–1202.

Ator, N. A., Atack, J. R., Hargreaves, R. J., Burns, H. D., & Dawson, G. R. (2010). Reducing abuse liability of GABAA/benzodiazepine ligands via selective partial agonist efficacy at α1 and α2/3 subtypes. *Journal of Pharmacology and Experimental Therapeutics*, *332*(1), 4–16. https://doi.org/10.1124/jpet.109.158303.

Besson, M., Matthey, A., Daali, Y., Poncet, A., Vuilleumier, P., Vuillemier, P., et al. (2015). GABAergic modulation in central sensitization in humans: A randomized placebo-controlled pharmacokinetic-pharmacodynamic study comparing clobazam with clonazepam in healthy volunteers. *Pain*, *156*(3), 397–404. https://doi.org/10.1097/01.j.pain.0000460331.33385.e8.

Biggerstaff, A., Kivell, B., Smith, J. L., Mian, Y., Golani, L. K., Rashid, F., et al. (2020). The α2,3-selective potentiators of GABAA receptors, KRM-II-81 and MP-III-80, produce anxiolytic-like effects and block chemotherapy-induced hyperalgesia in mice without tolerance development. *Pharmacology, Biochemistry and Behavior*, 24. *Submitted for publication.*

Cerne, R., Fisher, J. L., Siemian, J. N., Smith, J. L., Golani, L. K., Knutson, D. E., et al. (2019). Improvements in the pharmacological profile of diazepam by KRM-II-81, an imidazodiazepine positive allosteric modulator of α 2/3-containing GABAA receptors: Preclinical data predict enhanced efficacy for epilepsy, chronic pain, anxiety, and depression. *Pharmaceutical Sciences and Biomedical Analysis*, 14. *Special Issue Article "Diazepam"*.

Chou, R., Deyo, R., Friedly, J., Skelly, A., Weimer, M., Fu, R., et al. (2017). Systemic pharmacologic therapies for low Back pain: A systematic review for an American College of Physicians Clinical Practice Guideline. *Annals of Internal Medicine*, *166*(7), 480. https://doi.org/10.7326/M16-2458.

Cook, J. M., Edwankar, R. V., Edwankar, C. R., Huang, S., Jain, H. D., Yang, J., et al. (2010). *Selective anticonvulsant agents and their uses* (United States Patent No. US20100261711A1) https://patents.google.com/patent/US20100261711A1/en.

Cook, J. M., Huang, Q., He, X., Li, X., Yu, J., Han, D., et al. (2006). *Anxiolytic agents with reduced sedative and ataxic effects* (United States Patent No. US7119196B2) https://patents.google.com/patent/US7119196B2/en?oq=Anxiolytic+agents+with+reduced+sedative+and+ataxic+effects.++US+Patent+7%2c119%2c196.

Cook, J. M., Zhou, H., Huang, S., Sarma, P. V. V. S., & Zhang, C. (2009). *Stereospecific anxiolytic and anticonvulsant agents with reduced muscle-relaxant, sedative-hypnotic and ataxic effects* (United States Patent No. US7618958B2) https://patents.google.com/patent/US7618958B2/en.

Court, J. E., & Kase, C. S. (1976). Treatment of tic douloureux with a new anticonvulsant (clonazepam). *Journal of Neurology, Neurosurgery, and Psychiatry*, *39*(3), 297–299.

de Lucas, A. G., Ahring, P. K., Larsen, J. S., Rivera-Arconada, I., Lopez-Garcia, J. A., Mirza, N. R., et al. (2015). GABAA α5 subunit-containing receptors do not contribute to reversal of inflammatory-induced spinal sensitization as indicated by the unique selectivity profile of the GABAA receptor allosteric modulator NS16085. *Biochemical Pharmacology*, *93*(3), 370–379. https://doi.org/10.1016/j.bcp.2014.12.010.

Dellemijn, P. L., & Fields, H. L. (1994). Do benzodiazepines have a role in chronic pain management? *Pain*, *57*(2), 137–152. https://doi.org/10.1016/0304-3959(94)90217-8.

Di Lio, A., Benke, D., Besson, M., Desmeules, J., Daali, Y., Wang, Z., et al. (2011). HZ166, a novel GABAA receptor subtype-selective benzodiazepine site ligand, is antihyperalgesic in mouse models of inflammatory and neuropathic pain. *Neuropharmacology*, *60*(4), 626–632. https://doi.org/10.1016/j.neuropharm.2010.11.026.

Dirig, D. M., & Yaksh, T. L. (1995). Intrathecal baclofen and muscimol, but not midazolam, are antinociceptive using the rat-formalin model. *Journal of Pharmacology and Experimental Therapeutics*, *275*, 219–227.

Du, X., Hao, H., Yang, Y., Huang, S., Wang, C., Gigout, S., et al. (2017). Local GABAergic signaling within sensory ganglia controls peripheral nociceptive transmission. *The Journal of Clinical Investigation*, *127*(5), 1741–1756. https://doi.org/10.1172/JCI86812.

Etlin, A., Bráz, J. M., Kuhn, J. A., Wang, X., Hamel, K. A., Llewellyn-Smith, I. J., et al. (2016). Functional synaptic integration of forebrain GABAergic precursors into the adult spinal cord. *The Journal of Neuroscience: The Official Journal of the Society for Neuroscience*, *36*(46), 11634–11645. https://doi.org/10.1523/JNEUROSCI.2301-16.2016.

Fischer, B. D., Licata, S. C., Edwankar, R. V., Wang, Z.-J., Huang, S., He, X., et al. (2010). Anxiolytic-like effects of 8-acetylene imidazobenzodiazepines in a rhesus monkey conflict procedure. *Neuropharmacology*, *59*(7–8), 612–618. https://doi.org/10.1016/j.neuropharm.2010.08.011.

Fischer, B. D., Schlitt, R. J., Hamade, B. Z., Rehman, S., Ernst, M., Poe, M. M., et al. (2017). Pharmacological and antihyperalgesic properties of the novel α2/3 preferring GABAA receptor ligand MP-III-024. *Brain Research Bulletin*, *131*, 62–69. https://doi.org/10.1016/j.brainresbull.2017.03.001.

Gurrell, R., Dua, P., Feng, G., Sudworth, M., Whitlock, M., Reynolds, D. S., et al. (2018). A randomised, placebo-controlled clinical trial with the α2/3/5 subunit selective GABAA positive allosteric modulator PF-06372865 in patients with chronic low back pain. *Pain*, *159*(9), 1742–1751. https://doi.org/10.1097/j.pain.0000000000001267.

Hevers, W., & Lüddens, H. (1998). The diversity of $GABA_A$ receptors. Pharmacological and electrophysiological properties of $GABA_A$ channel subtypes. *Molecular Neurobiology*, *18*, 35–86.

Hofmann, M., Kordás, K. S., Gravius, A., Bölcskei, K., Parsons, C. G., Dekundy, A., et al. (2012). Assessment of the effects of NS11394 and L-838417, α2/3 subunit-selective GABA(A) [corrected] receptor-positive allosteric modulators, in tests for pain, anxiety, memory and motor function. *Behavioural Pharmacology*. https://doi.org/10.1097/FBP.0b013e32835a7c7e.

Kannampalli, P., Babygirija, R., Zhang, J., Poe, M. M., Li, G., Cook, J. M., et al. (2017). Neonatal bladder inflammation induces long-term visceral pain and altered responses of spinal neurons in adult rats. *Neuroscience*, *346*, 349–364. https://doi.org/10.1016/j.neuroscience.2017.01.021.

Klepner, C. A., Lippa, A. S., Benson, D. I., Sano, M. C., & Beer, B. (1979). Resolution of two biochemically and pharmacologically distinct benzodiazepine receptors. *Pharmacology, Biochemistry, and Behavior*, *11*(4), 457–462. https://doi.org/10.1016/0091-3057(79)90125-4.

Knabl, J., Witschi, R., Hösl, K., Reinold, H., Zeilhofer, U. B., Ahmadi, S., et al. (2008). Reversal of pathological pain through specific spinal GABAA receptor subtypes. *Nature*, *451*(7176), 330–334. https://doi.org/10.1038/nature06493.

Knabl, J., Zeilhofer, U. B., Crestani, F., Rudolph, U., & Zeilhofer, H. U. (2009). Genuine antihyperalgesia by systemic diazepam revealed by experiments in GABAA receptor point-mutated mice. *Pain*, *141*(3), 233–238. https://doi.org/10.1016/j.pain.2008.10.015.

Lewter, L. A., Fisher, J. L., Siemian, J. N., Methuku, K. R., Poe, M. M., Cook, J. M., et al. (2017). Antinociceptive effects of a novel α2/α3 subtype selective GABAA receptor positive allosteric modulator. *ACS Chemical Neuroscience*, *8*(6), 1305–1312. https://doi.org/10.1021/acschemneuro.6b00447.

Licata, S. C., Platt, D. M., Cook, J. M., Van Linn, M. L., & Rowlett, J. K. (2009). Contribution of alpha1 subunit-containing gamma-aminobutyric acidA (GABAA) receptors to motor-impairing effects of benzodiazepines in squirrel monkeys. *Psychopharmacology*, *203*(3), 539–546. https://doi.org/10.1007/s00213-008-1401-7.

Löw, K., Crestani, F., Keist, R., Benke, D., Brünig, I., Benson, J. A., et al. (2000). Molecular and neuronal substrate for the selective attenuation of anxiety. *Science (New York, N.Y.)*, *290*(5489), 131–134. https://doi.org/10.1126/science.290.5489.131.

Matthey, A., Daali, Y., Curtin, F., Poncet, A., Desmeules, J., & Besson, M. (2020). GABAergic modulation of secondary hyperalgesia: A randomized controlled 4-way crossover trial with the α2-subunit preferring GABA positive allosteric modulator, N-desmethyl-clobazam in healthy volunteers. *European Journal of Pain*. https://doi.org/10.1002/ejp.1554.

McKernan, R. M., Rosahl, T. W., Reynolds, D. S., Sur, C., Wafford, K. A., Atack, J. R., et al. (2000). Sedative but not anxiolytic properties of benzodiazepines are mediated by the GABA(A) receptor alpha1 subtype. *Nature Neuroscience*, *3*(6), 587–592. https://doi.org/10.1038/75761.

Melzack, R. (1996). Gate control theory: On the evolution of pain concepts. *The Journal of Pain*, *5*(2), 128–138. https://doi.org/10.1016/S1082-3174(96)80050-X.

Methuku, K. R., Li, X., Cerne, R., Gleason, S. D., Schkeryantz, J. M., Tiruveedhula, V. V. N. P. B., et al. (2018). An antidepressant-related pharmacological signature for positive allosteric modulators of α2/3-containing GABAA receptors. *Pharmacology Biochemistry and Behavior*, *170*, 9–13. https://doi.org/10.1016/j.pbb.2018.04.009.

Moerke, M. J., Li, G., Golani, L. K., Cook, J., & Negus, S. S. (2019). Effects of the α2/α3-subtype-selective GABAA receptor positive allosteric modulator KRM-II-81 on pain-depressed behavior in rats: Comparison with ketorolac and diazepam. *Behavioural Pharmacology*, *30*(5), 452–461. https://doi.org/10.1097/FBP.0000000000000464.

Munro, G., Ahring, P. K., & Mirza, N. R. (2009). Developing analgesics by enhancing spinal inhibition after injury: GABAA receptor subtypes as novel targets. *Trends in Pharmacological Sciences*, *30*(9), 453–459. https://doi.org/10.1016/j.tips.2009.06.004.

Munro, G., Lopez-Garcia, J. A., Rivera-Arconada, I., Erichsen, H. K., Nielsen, E.Ø., Larsen, J. S., et al. (2008). Comparison of the novel subtype-selective GABAA receptor-positive allosteric modulator NS11394 [3′-[5-(1-hydroxy-1-methyl-ethyl)-benzoimidazol-1-yl]-biphenyl-2-carbonitrile] with diazepam, zolpidem, bretazenil, and gaboxadol in rat models of inflammatory and neuropathic pain. *The Journal of Pharmacology and Experimental Therapeutics*, *327*(3), 969–981. https://doi.org/10.1124/jpet.108.144568.

Namjoshi, O. A., Wang, Z., Rallapalli, S. K., Johnson, E. M., Johnson, Y.-T., Ng, H., et al. (2013). Search for $\alpha 3\beta_{2/3}\gamma 2$ subtype selective ligands that are stable on human liver microsomes. *Bioorganic & Medicinal Chemistry*, *21*(1), 93–101. https://doi.org/10.1016/j.bmc.2012.10.057.

Nickolls, S. A., Gurrell, R., van Amerongen, G., Kammonen, J., Cao, L., Brown, A. R., et al. (2018). Pharmacology in translation: The preclinical and early clinical profile of the novel α2/3 functionally selective GABAA receptor positive allosteric modulator PF-06372865. *British Journal of Pharmacology*, *175*(4), 708–725. https://doi.org/10.1111/bph.14119.

Nickolls, S. A., Mace, H., Fish, R., Edye, M., Gurrell, R., Ivarsson, M., et al. (2011). A comparison of the α2/3/5 selective positive allosteric modulators L-838,417 and TPA023 in preclinical models of inflammatory and neuropathic pain. *Advances in Pharmacological Sciences*, *2011*, 608912. https://doi.org/10.1155/2011/608912.

Olsen, R. W., & Sieghart, W. (2009). $GABA_A$ receptors: Subtypes provide diversity of function and pharmacology. *Neuropharmacology*, *56*, 141–148.

Owen, R. M., Blakemore, D., Cao, L., Flanagan, N., Fish, R., Gibson, K. R., et al. (2019). Design and identification of a novel, functionally subtype selective GABAA positive allosteric modulator (PF-06372865). *Journal of Medicinal Chemistry*, *62*(12), 5773–5796. https://doi.org/10.1021/acs.jmedchem.9b00322.

Paul, J., Yévenes, G. E., Benke, D., Lio, A. D., Ralvenius, W. T., Witschi, R., et al. (2014). Antihyperalgesia by α2-GABAA receptors occurs via a genuine spinal action and does not involve supraspinal sites. *Neuropsychopharmacology*, *39*(2), 477–487. https://doi.org/10.1038/npp.2013.221.

Poe, M. M., Methuku, K. R., Li, G., Verma, A. R., Teske, K. A., Stafford, D. C., et al. (2016). Synthesis and characterization of a novel gamma-aminobutyric acid type A (GABAA) receptor ligand that combines outstanding metabolic stability, pharmacokinetics, and anxiolytic efficacy. *Journal of Medicinal Chemistry*, *59*(23), 10800–10806. https://doi.org/10.1021/acs.jmedchem.6b01332.

Ralvenius, W. T., Acuña, M. A., Benke, D., Matthey, A., Daali, Y., Rudolph, U., et al. (2016). The clobazam metabolite N-desmethyl clobazam is an α2 preferring benzodiazepine with an improved therapeutic window for antihyperalgesia. *Neuropharmacology*, *109*, 366–375. https://doi.org/10.1016/j.neuropharm.2016.07.004.

Ralvenius, W. T., Benke, D., Acuña, M. A., Rudolph, U., & Zeilhofer, H. U. (2015). Analgesia and unwanted benzodiazepine effects in point-mutated mice expressing only one benzodiazepine-sensitive GABAA receptor subtype. *Nature Communications*, *6*, 6803. https://doi.org/10.1038/ncomms7803.

Rivas, F. M., Stables, J. P., Murphree, L., Edwankar, R. V., Edwankar, C. R., Huang, S., et al. (2009). Antiseizure activity of novel gamma-aminobutyric acid (A) receptor subtype-selective benzodiazepine analogues in mice and rat models. *Journal of Medicinal Chemistry*, *52*(7), 1795–1798. https://doi.org/10.1021/jm801652d.

Rudolph, U., Crestani, F., Benke, D., Brünig, I., Benson, J. A., Fritschy, J. M., et al. (1999). Benzodiazepine actions mediated by specific gamma-aminobutyric acid(A) receptor subtypes. *Nature*, *401*(6755), 796–800. https://doi.org/10.1038/44579.

Rudolph, U., & Knoflach, F. (2011). Beyond classical benzodiazepines: Novel therapeutic potential of GABA A receptor subtypes. *Nature Reviews Drug Discovery*, *10*(9), 685–697. https://doi.org/10.1038/nrd3502.

Rudolph, U., & Möhler, H. (2014). GABAA receptor subtypes: Therapeutic potential in down syndrome, affective disorders, schizophrenia, and autism. *Annual Review of Pharmacology and Toxicology*, *54*, 483–507. https://doi.org/10.1146/annurev-pharmtox-011613-135947.

Salzmann, E., Pforringer, W., Paal, G., & Gierend, M. (1992). Treatment of chronic low-back syndrome with tetrazepam in a placebo controlled double-blind trial. *Journal of Drug Development*. https://eurekamag.com/research/010/020/010020799.php.

Sigel, E., & Ernst, M. (2018). The benzodiazepine binding sites of GABAA receptors. *Trends in Pharmacological Sciences*, *39*(7), 659–671. https://doi.org/10.1016/j.tips.2018.03.006.

Skolnick, P. (2012). Anxioselective anxiolytics: On a quest for the Holy Grail. *Trends in Pharmacological Sciences*, *33*(11), 611–620. https://doi.org/10.1016/j.tips.2012.08.003.

Smirne, S., & Scarlato, G. (1977). Clonazepam in cranial neuralgias. *The Medical Journal of Australia*, *1*(4), 93–94.

van Amerongen, G., Siebenga, P. S., Gurrell, R., Dua, P., Whitlock, M., Gorman, D., et al. (2019). Analgesic potential of PF-06372865, an α2/α3/α5 subtype-selective GABAA partial agonist, in humans. *British Journal of Anaesthesia*, *123*(2), e194–e203. https://doi.org/10.1016/j.bja.2018.12.006.

van Rijnsoever, C., Täuber, M., Choulli, M. K., Keist, R., Rudolph, U., Mohler, H., et al. (2004). Requirement of alpha5-GABAA receptors for the development of tolerance to the sedative action of diazepam in mice. *The Journal of Neuroscience: The Official Journal of the Society for Neuroscience*, *24*(30), 6785–6790. https://doi.org/10.1523/JNEUROSCI.1067-04.2004.

Velioglu, S. K., Gedikli, O., Yıldırım, M., & Ayar, A. (2017). Epilepsy may cause increased pain sensitivity: Evidence from absence epileptic WAG/Rij rats. *Epilepsy & Behavior*, *75*, 146–150. https://doi.org/10.1016/j.yebeh.2017.07.007.

Vinkers, C. H., & Olivier, B. (2012). Mechanisms underlying tolerance after long-term benzodiazepine use: A future for subtype-selective GABA(A) receptor modulators? *Advances in Pharmacological Sciences*, *2012*, 416864. https://doi.org/10.1155/2012/416864.

Vuilleumier, P. H., Besson, M., Desmeules, J., Arendt-Nielsen, L., & Curatolo, M. (2013). Evaluation of anti-hyperalgesic and analgesic effects of two benzodiazepines in human experimental pain: A randomized placebo-controlled study. *PLoS One*, *8*(3). https://doi.org/10.1371/journal.pone.0043896.

Wafford, K. A., Whiting, P. J., & Kemp, J. A. (1993). Differences in affinity and efficacy of benzodiazepine receptor ligands at recombinant gamma-aminobutyric acidA receptor subtypes. *Molecular Pharmacology*, *43*(2), 240–244.

Weir, G. A., Middleton, S. J., Clark, A. J., Daniel, T., Khovanov, N., McMahon, S. B., et al. (2017). Using an engineered glutamate-gated chloride channel to silence sensory neurons and treat neuropathic pain at the source. *Brain: A Journal of Neurology*, *140*(10), 2570–2585. https://doi.org/10.1093/brain/awx201.

Wieland, H. A., Lüddens, H., & Seeburg, P. H. (1992). A single histidine in GABAA receptors is essential for benzodiazepine agonist binding. *The Journal of Biological Chemistry*, *267*(3), 1426–1429.

Witkin, J. M., Cerne, R., Davis, P. G., Freeman, K. B., do Carmo, J. M., Rowlett, J. K., et al. (2019). The α2,3-selective potentiator of GABAA receptors, KRM-II-81, reduces nociceptive-associated behaviors induced by formalin and spinal nerve ligation in rats. *Pharmacology, Biochemistry, and Behavior*, *180*, 22–31. https://doi.org/10.1016/j.pbb.2019.02.013.

Witkin, J. M., Cerne, R., Wakulchik, M., S., J., Gleason, S. D., Jones, T. M., et al. (2017). Further evaluation of the potential anxiolytic activity of imidazo[1,5-a][1,4]diazepin agents selective for α2/3-containing GABAA receptors. *Pharmacology Biochemistry and Behavior*, *157*, 35–40. https://doi.org/10.1016/j.pbb.2017.04.009.

Witkin, J. M., Li, G., Golani, L. K., Xiong, W., Smith, J. L., Ping, X., et al. (2019). The positive allosteric modulator of α2/3-containing GABAA receptors, KRM-II-81, is active in pharmaco-resistant models of epilepsy and reduces hyperexcitability after traumatic brain injury. *The Journal of Pharmacology and Experimental Therapeutics*. https://doi.org/10.1124/jpet.119.260968.

Witkin, J. M., Ping, X., Mouser, C., Jin, X., Hobbs, J., Tiruveedhula, V. V. N. P. B., … Smith, J. L. (2019). The value of human epileptic tissue in the discovery of novel antiepileptic drugs: The example of CERC-611 and KRM-II-81. *Brain Research*, *1722*, 146356.

Witkin, J. M., Smith, J. L., Ping, X., Gleason, S. D., Poe, M. M., Li, G., et al. (2018). Bioisosteres of ethyl 8-ethynyl-6-(pyridin-2-yl)-4H-benzo[f]imidazo[1,5-a][1,4]diazepine-3-carboxylate (HZ-166) as novel alpha 2,3 selective potentiators of GABAA receptors: Improved bioavailability enhances anticonvulsant efficacy. *Neuropharmacology*, *137*, 332–343. https://doi.org/10.1016/j.neuropharm.2018.05.006.

Woolf, C. J. (2020). Capturing novel non-opioid pain targets. *Biological Psychiatry*, *87*(1), 74–81. https://doi.org/10.1016/j.biopsych.2019.06.017.

Further reading

Benyamin, R., Trescot, A. M., Datta, S., Buenaventura, R., Adlaka, R., Sehgal, N., et al. (2008). Opioid complications and side effects. *Pain Physician*, *11*(2 Suppl), S105–S120.

Derry, S., Wiffen, P. J., Kalso, E. A., Bell, R. F., Aldington, D., Phillips, T., et al. (2017). Topical analgesics for acute and chronic pain in adults—An overview of cochrane reviews. *The Cochrane Database of Systematic Reviews*, *5*. https://doi.org/10.1002/14651858.CD008609.pub2, CD008609.

Haefely, W. E. (1989). Pharmacology of the benzodiazepine receptor. *European Archives of Psychiatry and Neurological Sciences*, *238*(5), 294–301. https://doi.org/10.1007/BF00449811.

Hill, R. C., Maurer, R., Buescher, H. H., & Roemer, D. (1981). Analgesic properties of the GABA-mimetic THIP. *European Journal of Pharmacology*, *69*(2), 221–224. https://doi.org/10.1016/0014-2999(81)90419-2.

Jeffreys, D. (2005). *Aspirin: The remarkable story of a wonder Drug (New Ed edition)*. Bloomsbury Pub Ltd.

Krebs, E. E., Gravely, A., Nugent, S., Jensen, A. C., DeRonne, B., Goldsmith, E. S., et al. (2018). Effect of opioid vs nonopioid medications on pain-related function in patients with chronic back pain or hip or knee osteoarthritis pain: The SPACE randomized clinical trial. *JAMA*, *319*(9), 872–882. https://doi.org/10.1001/jama.2018.0899.

Krogsgaard-Larsen, P., Johnston, G. A. R., Lodge, D., & Curtis, D. R. (1977). A new class of GABA agonist. *Nature*, *268*(5615), 53–55. https://doi.org/10.1038/268053a0.

Melzack, R., & Wall, P. D. (1965). Pain mechanisms: A new theory. *Science (New York, N.Y.)*, *150*(3699), 971–979. https://doi.org/10.1126/science.150.3699.971.

Musch, B., Morselli, P. L., & Priore, P. (1988). Clinical studies with the new anxiolytic alpidem in anxious patients: An overview of the European experiences. *Pharmacology, Biochemistry, and Behavior*, *29*(4), 803–806. https://doi.org/10.1016/0091-3057(88)90211-0.

Pohl, M., & Smith, L. (2012). Chronic pain and addiction: Challenging co-occurring disorders. *Journal of Psychoactive Drugs*, *44*(2), 119–124. https://doi.org/10.1080/02791072.2012.684621.

Rømsing, J., & Møiniche, S. (2004). A systematic review of COX-2 inhibitors compared with traditional NSAIDs, or different COX-2 inhibitors for post-operative pain. *Acta Anaesthesiologica Scandinavica*, *48*(5), 525–546. https://doi.org/10.1111/j.0001-5172.2004.00379.x.

Chapter 12

Extrasynaptic α_5GABA$_A$ receptors and their role in nociception and pathological pain

Úrzula Franco-Enzástiga[a], Yarim E. De la Luz-Cuellar[a], Luis Eduardo Hernández-Reyes[b], Guadalupe Raya-Tafolla[b], Jorge E. Torres-López[c,d], Janet Murbartián[e], Vinicio Granados-Soto[a], and Rodolfo Delgado-Lezama[b]

[a]*Neurobiology of Pain Laboratory, Departamento de Farmacobiología, Cinvestav, South Campus, Mexico City, Mexico,* [b]*Departamento de Fisiología, Biofísica y Neurociencias, Cinvestav, Zacatenco, Mexico City, Mexico,* [c]*Mechanisms of Pain Laboratory, División Académica de Ciencias de la Salud, Universidad Juárez Autónoma de Tabasco, Villahermosa, Tabasco, Mexico,* [d]*Hospital Regional de Alta Especialidad "Dr. Juan Graham Casasús", Villahermosa, Tabasco, Mexico,* [e]*Departamento de Farmacobiología, Cinvestav, South Campus, Mexico City, Mexico*

Abbreviations

cAP	compound action potential
DRG	dorsal root ganglion
DRR	dorsal root reflex
EPSP	excitatory postsynaptic potential
GABA$_A$	γ-aminobutyric acid type A receptor
HR	Hoffmann reflex
KCC2	K^+-Cl^- cotransporter 2
MSR	monosynaptic reflex
NAM	negative allosteric modulator
PAD	primary afferent depolarization
PAM	positive allosteric modulator
RDD	rate-dependent depression
siRNA	small interfering RNA

Introduction

Primary afferent fibers convey nociceptive information to the spinal cord. These fibers synapse with inhibitory and excitatory interneurons, as well as projection neurons located in the spinal cord dorsal horn (Todd, 2010). In normal conditions, projection neurons receive excitatory inputs from Aδ and C nociceptive fibers which are subjected to a strong regulation by presynaptic inhibition. In contrast, in pathological conditions, this inhibitory mechanism is lost opening the gate to excitatory inputs coming from low-threshold afferents. Accordingly, intrathecal bicuculline, antagonist of GABA$_A$ (γ-aminobutyric acid type A) receptors, increases hypersensitivity to noxious (hyperalgesia) and innocuous stimuli (allodynia) (Roberts, Beyer, & Komisaruk, 1986). Evidence showing the importance of GABAergic inhibition has generated the interest of different research groups in the pain field to find drugs to reestablish the inhibition through facilitation of fast inhibitory GABAergic neurotransmission (Zeilhofer, Benke, & Yevenes, 2012). The aim is to facilitate the synaptic activity of GABA$_A$ receptors by novel drugs able to bind the benzodiazepine site. A study, based in the use of GABA$_A$ receptor point-mutated knock-in mice, shows that analgesia can be induced by targeting spinal GABA$_A$ receptors containing the α_2 and/or α_3 subunits (Knabl et al., 2008). In parallel, the use of point-mutated mice has shown that the ablation of the α_2GABA$_A$R has a strong antihyperalgesic effect with reduced side effects in different pain models (Ralvenius, Benke, Rudolph, & Zeilhofer, 2015). Unfortunately, all the approaches have been behavioral, but not functional. Recently, our group found a role for extrasynaptic GABA$_A$ receptors in pain. In this chapter, we describe the current knowledge about the role of spinal extrasynaptic α_5 subunit-containing GABA$_A$ (α_5GABA$_A$) receptors in pain and propose a hypothesis to explain these actions in several models of chronic pain.

The Neurobiology, Physiology, and Psychology of Pain. https://doi.org/10.1016/B978-0-12-820589-1.00012-9

$GABA_A$ receptors

$GABA_A$ receptor is an ionotropic ligand-gated ion channel. This receptor is a pentamer assembled out of five subunits that form the central ion channel permeable to chloride and bicarbonate. This receptor mediates two kinds of inhibition, synaptic and tonic. The first one has been recorded in almost all neurons of the CNS and is mediated mainly by receptors containing $\alpha_{1,2,3}$ subunits in combination with β_2 and γ_2 (Farrant & Nusser, 2005). These receptors show fast inactivation and desensitize in the presence of GABA. On the other hand, extrasynaptic $GABA_A$ receptors mediate a tonic current, with slow inactivation and low level of desensitization. These receptors contain mainly $\alpha_{4,5,6}$ subunits in combination with β_2 and γ_2 and, in some neurons, the last one is substituted by δ, ε, or θ subunit (Farrant & Nusser, 2005).

Expression of α_5GABA_A receptors at spinal cord and DRG

Distribution of α_5GABA_A receptor mRNA or protein has been explored using different experimental approaches (Table 1). α_5GABA_A receptor has been found at the spinal cord, specifically in small neurons in the *substantia gelatinosa* (layer II) of the dorsal horn during development (Ma et al., 1993). This receptor has also been detected in layer III–VI of the dorsal horn in adult rats (Bohlhalter et al., 1996). Another study showed that α_5GABA_A receptor is found in laminae II–V of the spinal cord colocalizing with glutamatergic and GABAergic neurons of laminae II–III (Knabl et al., 2008; Paul et al., 2012; Perez-Sanchez et al., 2017). Two-thirds of the α_5GABA_A receptors might correspond to extrasynaptic locations in neurons at the dorsal spinal cord (Perez-Sanchez et al., 2017). The most profuse subunits along the laminae of the spinal cord are $\alpha_3 > \alpha_2 > \alpha_5 > \alpha_{4/6}$ in combination with γ_2 and $\beta_{2/3}$ (Alvarez, Taylor-Blake, Fyffe, De Blas, & Light, 1996; Bohlhalter et al., 1996; Paul et al., 2012). Although there are some discrepancies regarding to the exact localization of α_5GABA_A receptors in the spinal cord layers, there is a consensus in that they are localized in layers I–III of dorsal spinal cord in extrasynaptic sites (Table 1). α_5GABA_A receptors have also been found in DRG neurons (Table 1). The first report analyzing this issue described that *Gabra5* expression is abundant in embryonic DRG in adult rats (Ma et al., 1993). Later, α_5GABA_A receptors were found in primary afferent terminals colocalizing with peptidergic terminals in lamina IIo, non-peptidergic terminals in lamina IIi, and with myelinated terminals in lamina III (Paul et al., 2012). α_5GABA_A receptor mRNA and protein are present in sciatic nerve, dorsal root, and DRG neurons in turtles (Loeza-Alcocer et al., 2013). Interestingly, α_5GABA_A receptor mRNA has also been found in human embryonic and adult DRG cultures (Maddox et al., 2004). These findings support the expression of α_5GABA_A receptors in neurons involved in pain processing information.

GABAergic tonic current in the spinal cord

The GABAergic tonic current was described for the first time in ventral horn neurons from isolated spinal cord of chick embryo. In this preparation, the GABAergic current was uncovered by bicuculline (Chub & O'Donovan, 2001). A few years later, the GABAergic tonic current was characterized in motoneurons and premotor interneurons using an in vitro preparation of spinal cord from adult turtles (Castro, Aguilar, Andrés, et al., 2011; Castro, Aguilar, González-Ramírez, et al., 2011). In this preparation, muscimol ($GABA_A$ receptor agonist) decreased the input resistance and the excitatory postsynaptic potential (EPSP) evoked in motoneurons. Accordingly, bicuculline increased input resistance and facilitated EPSP resulting in higher firing activity and changed the holding current of motoneurons, suggesting that spinal extrasynaptic $GABA_A$ receptors are tonically activated by endogenous GABA. Thus, spinal $GABA_A$ receptors exert a strong control on motoneuron excitability (Castro, Aguilar, Andrés, et al., 2011; Andrés et al., 2014).

Like bicuculline, L-655,708 (α_5GABA_A receptor inverse agonist) increased the input resistance and excitability of motoneurons of spinal cord from turtles (Canto-Bustos et al., 2017). In addition, L-655,708 facilitated the monosynaptic reflex (MSR) evoked by electrical stimulation of primary afferents, suggesting that extrasynaptic α_5GABA_A receptors are involved in motor control (Canto-Bustos et al., 2017).

Synaptic and extrasynaptic $GABA_A$ receptors participate in the presynaptic modulation of MSR. It has been described that presynaptic inhibition of primary afferent fibers associated with the primary afferent depolarization (PAD) depresses the MSR (Rudomin & Schmidt, 1999). Interestingly, picrotoxin (20 μM) abolishes PAD, mediated by fast synaptic $\alpha_{2/3}GABA_A$ receptors (Witschi et al., 2011), reestablishing the MSR, while a greater concentration blocks extrasynaptic $GABA_A$ receptors facilitating MSR (Bautista, Aguilar, Loeza-Alcocer, & Delgado-Lezama, 2010; Farrant & Nusser, 2005). Indeed, L-655,708 facilitates MSR and increases tonic excitability of primary afferent fibers without affecting PAD (Loeza-Alcocer et al., 2013).

Finally, it has been demonstrated that formalin- or capsaicin-induced secondary hyperalgesia is associated with activation of antidromic action potentials in sensory neurons, phenomenon known as dorsal root reflex (DRR) (Willis, 1999). In

TABLE 1 Summarize of the studies showing α_5GABA$_A$ distribution in DRG and spinal cord.

Species	Tissues/ samples	Localization/distribution	Condition	Methods	References
SD rats	Cervical spinal cord and DRG	E11: Null expression at spinal cord E13: Expression in motoneurons E16–E17: Expression at ventral and dorsal column E15–E20: Expression at DRG P0: Null expression at spinal cord Adult: Low expression at spinal cord and DRG	Embryonic, postnatal and adult	In situ HZ and PCR	Ma, Saunders, Somogyi, Poulter, and Barker (1993)
Male SD rats	Cervical spinal cord	Layer I and II: Null expression Layer III: Expressed Layer IV–VI and VIII: Expressed Layer IX: Expression at preganglionic motoneurons	Adult (150–200 g)	IHC	Bohlhalter, Weinmann, Mohler, and Fritschy (1996)
Human	DRG cultures	Expression at 2 of 4 embryonic DRG cultures and adult DRG cultures	Embryonic and adult DRG cultures	PCR	Maddox et al. (2004)
Mice	Spinal cord dorsal horn	Expression at projection neurons	Adult (7–12 weeks)	IF	Knabl et al. (2008)
Turtles	Spinal cord	Expression at ventral spinal cord	Adult	PCR and IF	Castro, Aguilar, Andrés, Felix, and Delgado-Lezama (2011) and Castro, Aguilar, González-Ramírez, et al. (2011)
C57BL6 Mice	Lumbar spinal dorsal horn	Lamina II: Colocalization with Vglut2+ (~10%), GAD67+ (~15%) and GlyT2+ (~15%) Laminae IIo-IIi: Colocalization CGRP+ (15%), IB4+ (3%) and NF200+ (44%) Lamina III: Colocalization with Vglut2+ (~40%), GAD67+ (~60%) and GlyT2+ (~20%) Laminae IV–V: Moderate expression	Adult (6–8 weeks)	IHC	Paul, Zeilhofer, and Fritschy (2012)
Turtle	Dorsal root and DRG	Expression mRNA and protein are confined in the dorsal root and DRG	Adult	PCR, WB and IF	Loeza-Alcocer et al. (2013)
Male C57BL6 Mice	Lumbar dorsal spinal cord	Lamina I: Colocalized with CGRP + (~23%) and IB4+ (~3%) Lamina II: Colocalized with gephyrin+ (~10%), CGRP+ (~13%) and IB4+ (~20%)	Adult (>3 months old)	IF	Perez-Sanchez et al. (2017)

HZ, hybridization; *IHC*, immunohistochemistry; *IF*, immunofluorescence; *SD*, *Sprague-Dawley*; *WB*, Western blot.

the turtle, L-655,708 depresses DRRs without affecting PAD, suggesting that spinal extrasynaptic α_5GABA$_A$ receptors are tonically depolarizing the primary afferents to reach the threshold to activate DRRs (Loeza-Alcocer et al., 2013). At the same time, spinal α_5GABA$_A$ receptors, activated by endogenous GABA, are shunting tonically the primary afferents (Loeza-Alcocer et al., 2013). Also, there is evidence of the tonic function of α_5GABA$_A$ receptors in premotor interneurons in the spinal cord of turtles (Castro, Aguilar, González-Ramírez, et al., 2011). These results constitute the first evidence showing that spinal α_5GABA$_A$ receptors mediate a tonic current that controls the excitability of motoneurons and primary afferents.

GABAergic tonic current in spinal cord in mammals

The network engaged in the processing of nociceptive information involves glutamatergic, GABAergic, and glycinergic interneurons that control the excitability of projection neurons located at the laminae I, II, and III (Todd, 2010). It is well known that projection neurons receive excitatory inputs from Aδ and C primary afferents (Todd, 2010). Projection neurons show a wide repertoire of receptors including synaptic and extrasynaptic $GABA_A$ receptors. Blockade of $GABA_A$ receptors produces a behavior quite similar to that observed in neuropathic pain models (Roberts et al., 1986). Based on this, synaptic $GABA_A$ receptors are considered the main pharmacological target to relief pain (Munro et al., 2008; Munro, Erichsen, Rae, & Mirza, 2011; Zeilhofer et al., 2012). However, slight attention has been given to the tonic current mediated by extrasynaptic $GABA_A$ receptors. It is already known that charge transfer carried by tonic GABAergic inhibitory currents is about six times of the total charge transferred by both GABAergic and glycinergic inhibitory postsynaptic currents (Ataka & Gu, 2006). Thus, GABAergic tonic current may have a profound effect on neuronal excitability and be involved in diverse physiological events at the level of both individual neurons and local neuronal networks (Farrant & Nusser, 2005; Walker, 2008). In this context, GABAergic tonic current has been recorded in many dorsal horn neurons, but there is still a lack of convincing evidence showing its importance in chronic pain-induced hypersensitivity and in physiological conditions. For instance, GABAergic tonic currents have been recorded in neurons of the *substantia gelatinosa* in rats (Ataka & Gu, 2006; Takahashi, Mashimo, & Uchida, 2006), considered crucial elements for transmission and modification of noxious stimuli (Todd, 2010; Zeilhofer et al., 2012). There is evidence that $GABA_A$ receptors mediating tonic current may contain $\alpha_5\beta_x\gamma_2$, $\alpha_x\beta_x\varepsilon$ and $\alpha_4\beta_x\delta$ subunits (Takahashi et al., 2006). More recently, the presence of a tonic current mediated by $\delta GABA_A$ receptors in neurons of lamina II was recorded. The role of these receptors in pain is unknown, as a $\delta GABA_A$ receptor agonist reduces excitability of spinal neurons, but it does not affect formalin test (Bonin et al., 2011). Interestingly, there is evidence that $\alpha_5 GABA_A$ receptors are expressed in laminae I and II neurons at the dorsal horn, mediating a tonic current (Perez-Sanchez et al., 2017). It seems that $\alpha_5 GABA_A$ receptors do not modulate acute nociception, but they play a pronociceptive role at late stages of inflammatory pain (Perez-Sanchez et al., 2017). These authors did not evaluate the participation of $\alpha_5 GABA_A$ receptors in neuropathic pain. Since peripheral nerve injury switches GABAergic action from inhibition to excitation in neurons within laminae I and II, it is likely that tonic activation of $\delta GABA_A$ and $\alpha_5 GABA_A$ receptors by endogenous GABA will lead to allodynia (Coull et al., 2003).

Extrasynaptic $\alpha_5 GABA_A$ receptors in the rate-dependent depression (RDD) of the Hoffmann reflex (HR)

The property of the HR known as RDD has been suggested that could be due to a presynaptic effect, different to the classical presynaptic inhibition mediated by synaptic $GABA_A$ receptors (Hultborn et al., 1996; Rudomin & Schmidt, 1999). The RDD is impaired in rats and humans with diabetic neuropathy; therefore, it has been proposed as a biomarker to detect the presence of neuropathic pain (Lee-Kubli & Calcutt, 2014; Marshall et al., 2017). These authors suggested that disruption of GABAergic inhibition mediated by $GABA_A$ receptors induces the impairment of this mechanism in rats as it was restored by bicuculline (Jolivalt, Lee, Ramos, & Calcutt, 2008; Lee-Kubli & Calcutt, 2014).

The impairment of RDD in diabetic rats was parallel to the reduction of KCC2 expression in the dorsal horn. Based on this, reduced spinal KCC2 expression in diabetic rats might contribute to allodynia and hyperalgesia by shifting the $GABA_A$ receptor-mediated inhibition to excitation as happened in neuropathic pain (Coull et al., 2003; Jolivalt et al., 2008; Lee-Kubli & Calcutt, 2014).

Remarkably, we have shown that L-655,708 restores RDD and reduces tactile allodynia in neuropathic diabetic rats (Hernández-Reyes et al., 2019). This evidence suggests that restoration of RDD and alleviation of tactile allodynia in diabetic rats by bicuculline might be due to the blockade of $\alpha_5 GABA_A$ receptors. Likewise, the induction of neuropathic-like pain in control rats, by bicuculline, might be induced by the blockade of spinal $\alpha_5 GABA_A$ receptors. Both results are contrary to other studies using a positive allosteric modulator (PAM) of $GABA_A$ receptors with selectivity for α_2, α_3, and α_5 subunits, which produce analgesic effects (Knabl et al., 2008; Munro et al., 2008, 2011; Zeilhofer et al., 2012). Antinociceptive and pronociceptive actions of $\alpha_5 GABA_A$ receptors in control and neuropathic rats, respectively, might be due to the tonic current they mediate in projection neurons, which is inhibitory in naïve and might be excitatory in diabetic rats as occurs in neuropathic pain induced by nerve injury (Coull et al., 2003). In addition, blockade of $\alpha_5 GABA_A$ receptors in pain conditions induced by chronic inflammation restores RDD (Bravo-Hernández et al., 2016).

Function of extrasynaptic α_5GABA$_A$ receptors in primary afferent fibers

Although PAD inhibits neurotransmitter release, when strong enough it can reach the threshold for the DRR, which will propagate retrogradely to the peripheral terminal where they will evoke release of inflammatory neuropeptides (substance P and CGRP) (Willis, 1999) contributing to hyperalgesia (Cervero, Laird, & García-Nicas, 2003). This mechanism depends on GABA$_A$ receptors. However, the specific subunit was unknown. Data from our laboratory have shown that α_5GABA$_A$ receptors are tonically depolarizing and shunting the primary afferents of turtles and rats (Hernández-Reyes et al., 2019; Loeza-Alcocer et al., 2013). Thus, blockade of α_5GABA$_A$ receptors depressed the DRR without affecting the PAD (Loeza-Alcocer et al., 2013). Accordingly, L-655,708 induces compound action potential (cAP) facilitation in turtles and rats, suggesting that α_5GABA$_A$ receptors shunt the primary afferent fibers tonically active by ambient GABA (Hernández-Reyes et al., 2019; Loeza-Alcocer et al., 2013). Consequently, α_5GABA$_A$ receptors might be a pharmacological target to prevent DRR generation, peripheral/central sensitization, and chronic pain (Hernández-Reyes et al., 2019; Loeza-Alcocer et al., 2013).

Extrasynaptic spinal α_5GABA$_A$ receptors in pain

The role of α_5GABA$_A$ receptors in nociception remained undiscovered for a long time. In normal conditions, synaptic GABA$_A$ receptors in primary afferent fibers generate PAD that ultimately inhibits the pain sensation (Rudomin & Schmidt, 1999; Willis, 1999). Data strongly suggest that α_5GABA$_A$ receptors located in the primary afferent neurons contribute tonically to this form of inhibition. Accordingly, L-655,708 enhances synaptic transmission of C-fibers of rats and the magnitude of long-term potentiation (Xue et al., 2017). Moreover, L-655,708 reduces antidromic sodium spikes in primary afferents of rats (Lucas-Osma et al., 2018). These data suggest that peripheral and spinal α_5GABA$_A$ receptors exert a tonic inhibition of excitability. In support of this, intrathecal L-655,708 decreases pain thresholds in naïve mice (Xue et al., 2017) and rats where in addition abolishes RDD (De la Luz-Cuellar et al., 2019; Franco-Enzástiga et al., 2021; Hernández-Reyes et al., 2019). Indeed, treatment with a siRNA against α_5GABA$_A$ receptors induces tactile allodynia in naïve rats and reduces reserpine-induced allodynia in female rats (De la Luz-Cuellar et al., 2019), confirming that spinal α_5GABA$_A$ receptors have an antinociceptive and pronociceptive role in healthy and in chronic pain conditions, respectively.

Currently, there is a controversy about the role of this receptor in chronic pain. On one hand, administration of a PAM (NS11394, affinity $\alpha_5 > \alpha_3 > \alpha_2 > \alpha_1$GABA$_A$ receptors) or L-838,417 (agonist or $\alpha_2/\alpha_3/\alpha_5$ subunits) diminished chronic pain and decreased C-fiber-evoked wind-up suggesting that activation of α_5GABA$_A$ receptors leads to antinociception (Munro et al., 2011; Nickolls et al., 2011). On the other hand, blockade of spinal α_5GABA$_A$ receptors restores chronic pain-induced impairment of RDD and reverses long-lasting nociceptive hypersensitivity in rodents (Bravo-Hernández et al., 2016; De la Luz-Cuellar et al., 2019; Franco-Enzástiga et al., 2021; Hernández-Reyes et al., 2019). In support of this, a selective negative allosteric modulator (NAM) (α5IA-II) reverses chronic pain (Munro et al., 2011) (Table 2).

Extrasynaptic α_5GABA$_A$ receptors regulation

Previous pharmacological studies have reported sex differences in α_5GABA$_A$ receptors function. PAMs of α_5GABA$_A$ receptors reduce stress-induced behaviors in female, but not in male, rodents (Piantadosi et al., 2016). More recently, our group reported sex-dependent effect in the expression and function of α_5GABA$_A$ receptors in chronic pain (De la Luz-Cuellar et al., 2019; Franco-Enzástiga et al., 2021). We found that α_5GABA$_A$ receptor mRNA and protein changes in DRG and spinal cord are modulated in a sex-dependent fashion. Nerve injury increases α_5GABA$_A$ receptor in female DRG and spinal cord in rodents, but not in males. These changes are activated by DNA methylation in the CpG island of *Gabra5* in males, but not females. In contrast, expression and function of spinal α_5GABA$_A$ receptors in females is associated with the presence of 17β-estradiol (Franco-Enzástiga et al., 2021).

Applications to other areas

The development of drugs to treat neuropathic pain represents a great challenge. One important point is that this kind of pathology produces irreversible changes in the nervous system that makes difficult to develop effective drugs. Reestablishing GABAergic inhibition to relief pain without knowing where and how the inhibition is lost represents a difficult issue to be successful. The reversion of allodynia in chronic pain models by blockade of α_5GABA$_A$ receptors indicates that loss of inhibition might be related with a change in the homeostasis of Cl^- in projection neurons where this receptor

TABLE 2 Evidence of the pharmacological role of α_5GABA_A receptor in pain models.

Pain models	Animals	Treatment and results	Conclusion	References
5%-Formalin, capsaicin, CFA, CCI and SNI	Male SD rats	NS11394 (30 mg/kg, po) reverses nociception	Activation of α_5GABA_A receptors diminishes nociception	Munro et al. (2008)
5%-Formalin, carrageenan and CCI	Male SD rats	α5IA-II (3–30 mg/kg, ip) reverses nociception	Inhibition of α_5GABA_A receptors promotes antinociception	Munro et al. (2011)
CFA, CCI, and SNL	Male SD rats	L-838,417 (10–30 mg, po) reverses nociception	Activation of α_5GABA_A receptors induces antinociception	Nickolls et al. (2011)
0.5%-Formalin	Female Wistar rats	L-655,708 (0.15–15 nmol, it) prevented nociception of GABAergic agonists	Blockade of spinal α_5GABA_A receptors reduces nociception	Bravo-Hernández et al. (2014)
1%-Formalin and carrageenan		NS11394 and NS16085 (30 mg/kg, po) reverse nociception and sensitization of C-fibers	The α_5GABA_A receptors do not contribute to antinociception	Garcia de Lucas et al. (2015)
1%-Formalin, CFA and SNL	Female Wistar rats	L-655,708 (0.15–15 nmol, it) prevented and reversed nociception L-655,708 restored the loss of the RDD of the Hoffman reflex	Blockade of spinal α_5GABA_A receptors reduces nociception	Bravo-Hernández et al. (2016)
Tail-flick, hot-plate, capsaicin, CFA and formalin	Male *Gabra5*$^{-/-}$ mice	Tail-flick and hot-plate latencies are not altered in *Gabra5*$^{-/-}$ mice *Gabra5*$^{-/-}$ mice have a slower recovery from nociception	Absence of α_5GABA_A receptors do not affect acute nociception, but delays recovery at late states of inflammatory pain	Perez-Sanchez et al. (2017)
NMDA-dependent LTP	Adult male SD rats and male Kumming mice	L-655,708 (0.025–0.5 pmol, it) decreases pain thresholds of intact mice In vivo application of L-655,708 (20 nmol) on rat spinal cord enhances C-fiber-synaptic transmission and facilitates LTP	Inhibition of α_5GABA_A receptors increases excitability at spinal cord level	Xue et al. (2017)
Reserpine-induced fibromyalgia-type pain	Male and female Wistar rats (200–250 g)	L-655,708 (0.15–15 nmol, it) and TB21007 (1.5–150 nmol, it) reversed hypersensitivity in female and male rats L-655,708 (15 nmol, it) induced tactile allodynia in naïve male and female rats	Blockade of spinal α_5GABA_A receptors induces nociception in naïve rats and reduces pain in reserpinized rats Lower effect of L-655,708 in male rats	De la Luz-Cuellar et al. (2019)
Type 1 diabetes	Female Wistar rats (220–240 g)	L-655,708 (15 nmol, it) reduced allodynia in diabetic rats L-655,708 restored the loss of the RDD of the Hoffman reflex L-655,708 (15 nmol, it) induced allodynia in naïve rats	Blockade of α_5GABA_A receptors promotes nociception in normal conditions and alleviates pain in diabetic rats	Hernández-Reyes et al. (2019)
SNL	Female vs male Wistar rats	L-655,708 (0.15–150 nmol, it) reversed hypersensitivity in neuropathic female rats and, in a lesser extent, in males	Blockade of spinal α_5GABA_A receptors reverses nociception in a sex-dependent manner Lower effect of L-655,708 in male rats	Franco-Enzástiga et al. (2021)

CFA, complete Freud's adjuvant; *CCI*, chronic constriction injury; *SD*, *Sprague-Dawley*; *SNI*, spared nerve injury; *SNL*, spinal nerve ligation; *NMDAR*, *N*-methyl-aspartate receptors; *LTP*, long-term potentiation.

has a relevant role mediating a tonic current. The sex differences in α_5GABA$_A$ receptors function have been identified in neuropathic and functional pain; our data suggest that blockade of spinal α_5GABA$_A$ receptors could be helpful to treat chronic pain conditions mainly in women.

Other agents of interest

L-655,708 has been a pivotal drug to study the participation of spinal α_5GABA$_A$ receptors in chronic pain conditions. Our group has also evaluated the effect of another α_5GABA$_A$ receptor inverse agonist (TB 21007) in fibromyalgia-type pain in rats (De la Luz-Cuellar et al., 2019). Like L-655,708, TB 21007 was able to reduce tactile allodynia and muscle hyperalgesia in fibromyalgic female rats. These data reinforce that blockade of spinal α_5GABA$_A$ receptors leads to antinociception in female rodents. Likewise, there is evidence that the NAM α5IA-II also reverses chronic pain in rodents (Munro et al., 2011). Interestingly, PF-06372865 (a PAM of α_5GABA$_A$ receptors) induces antinociception in healthy humans (van Amerongen et al., 2019) further supporting our hypothesis that activation of α_5GABA$_A$ receptors in naïve conditions leads to antinociception, while blockade of these receptors reduces chronic pain in pathological situations.

Mini-dictionary of terms

Dorsal root ganglion: It contains the cell bodies of nociceptive sensory neurons that bring information from the periphery to the spinal cord.

Dorsal root reflex: Action potentials that originate near the primary afferent terminal in the spinal cord and propagate antidromically (out to the periphery).

Excitatory postsynaptic potential: A change in membrane potential in the direction to the threshold to activate an action potential.

KCC2: A neuron-specific K^+-Cl^- symporter responsible for establishing the chloride ion gradient in central neurons through the maintenance of low intracellular chloride concentrations.

Monosynaptic reflex: A reflex involving only one synapse between the Ia primary afferent fiber and motoneuron.

Primary afferent depolarization: A phenomenon induced through activation of synaptic GABA$_A$ receptors at the primary afferent terminal. Generation of PAD results from axo-axonic synapsis between GABAergic terminals and primary afferents.

Rate-dependent depression: Depression in amplitude of the second and consecutive HR with respect to the first HR evoked at a frequency greater than 0.1 Hz.

Key facts on α_5GABA$_A$ receptors in pain

- GABA$_A$ receptors are antinociceptive and pronociceptive in normal and pathological conditions, respectively.
- α_5GABA$_A$ receptors are expressed in axons of sensory neurons and in the spinal cord at extrasynaptic places.
- α_5GABA$_A$ receptors are tonically activated by ambient GABA while synaptic GABA$_A$ receptors regarding PAD and DRRs are activated by GABA release from axo-axonic synapse.
- α_5GABA$_A$ receptors are tonically depolarizing and shunting primary afferent fibers.
- Extrasynaptic α_5GABA$_A$ receptors are antinociceptive in naïve rodents and pronociceptive in pathological conditions.
- Chronic pain seems to switch extrasynaptic spinal α_5GABA$_A$ receptors function from tonic inhibition to disinhibition or excitation.
- Blockade of spinal α_5GABA$_A$ receptors reduces chronic pain in female rodents and recovers RDD of the HR.

Summary points

- This chapter focuses on the role of spinal α_5GABA$_A$ receptors in pain.
- Spinal α_5GABA$_A$ receptors underlay the antinociceptive effect of GABA in normal conditions.
- Spinal α_5GABA$_A$ receptors underlay the nociceptive effect of GABA in pathological conditions.
- Chronic pain enhances expression of α_5GABA$_A$ receptors in DRG and spinal cord in female rodents.
- Chronic pain reduces expression of α_5GABA$_A$ receptors in DRG of male rodents.
- Blockade of spinal α_5GABA$_A$ receptors reduces chronic pain in a sex-dependent manner.
- Spinal α_5GABA$_A$ receptors could be a target to treat pain in women.

Acknowledgments

Research described here was partially supported by Conacyt PN-2017/5098 (RD-L), CB-2017-2018/A1-S-40015 (VG-S) and SEP-Cinvestav 127 (VG-S) and 269 (JM) grants.

References

Alvarez, F. J., Taylor-Blake, B., Fyffe, R. E., De Blas, A. L., & Light, A. R. (1996). Distribution of immunoreactivity for the β2 and β3 subunits of the $GABA_A$ receptor in the mammalian spinal cord. *Journal of Comparative Neurology*, *365*, 392–412.

Andrés, C., Aguilar, J., González-Ramírez, R., Elias-Viñas, D., Felix, R., & Delgado-Lezama, R. (2014). Extrasynaptic α_6 subunit-containing $GABA_A$ receptors modulate excitability in turtle spinal motoneurons. *PLoS One*, *9*(12), e115378.

Ataka, T., & Gu, J. G. (2006). Relationship between tonic inhibitory currents and phasic inhibitory activity in the spinal cord lamina II region of adult mice. *Molecular Pain*, *2*, 36.

Bautista, W., Aguilar, J., Loeza-Alcocer, J. E., & Delgado-Lezama, R. (2010). Pre- and postsynaptic modulation of monosynaptic reflex by $GABA_A$ receptors on turtle spinal cord. *Journal of Physiology*, *588*, 2621–2631.

Bohlhalter, S., Weinmann, O., Mohler, H., & Fritschy, J. M. (1996). Laminar compartmentalization of $GABA_A$-receptor subtypes in the spinal cord: An immunohistochemical study. *Journal of Neuroscience*, *16*, 283–297.

Bonin, R. P., Labrakakis, C., Eng, D. G., Whissell, P. D., De Koninck, Y., & Orser, B. A. (2011). Pharmacological enhancement of α-subunit-containing $GABA_A$ receptors that generate a tonic inhibitory conductance in spinal neurons attenuates acute nociception in mice. *Pain*, *152*(6), 1317–1326.

Bravo-Hernández, M., Corleto, J. A., Barragán-Iglesias, P., González-Ramírez, R., Pineda-Farias, J. B., Felix, R., et al. (2016). The α_5 subunit containing $GABA_A$ receptors contribute to chronic pain. *Pain*, *157*, 613–626.

Bravo-Hernández, M., Feria-Morales, L. A., Torres-López, J. E., Cervantes-Durán, C., Delgado-Lezama, R., Granados-Soto, V., et al. (2014). Evidence for the participation of peripheral α_5 subunit-containing $GABA_A$ receptors in $GABA_A$ agonists-induced nociception in rats. *European Journal of Pharmacology*, *734*, 91–97.

Canto-Bustos, M., Loeza-Alcocer, E., Cuellar, C. A., Osuna, P., Elias-Viñas, D., Granados-Soto, V., et al. (2017). Tonically active α_5GABA_A receptors reduce motoneuron excitability and decrease the monosynaptic reflex. *Frontiers in Cellular Neuroscience*, *11*, 283.

Castro, A., Aguilar, J., Andrés, C., Felix, R., & Delgado-Lezama, R. (2011). $GABA_A$ receptors mediate motoneuron tonic inhibition in the turtle spinal cord. *Neuroscience*, *192*, 74–80.

Castro, A., Aguilar, J., González-Ramírez, R., Loeza-Alcocer, E., Canto-Bustos, M., Felix, R., et al. (2011). Tonic inhibition in spinal ventral horn interneurons mediated by α_5 subunit-containing $GABA_A$ receptors. *Biochemical and Biophysical Research Communications*, *412*, 26–31.

Cervero, F., Laird, J. M. A., & García-Nicas, E. (2003). Secondary hyperalgesia and presynaptic inhibition: An update. *European Journal of Pain*, *7*, 345–351.

Chub, N., & O'Donovan, M. J. (2001). Post-episode depression of GABAergic transmission in spinal neurons of the chick embryo. *Journal of Neurophysiology*, *85*, 2166–2176.

Coull, J. A., Boudreau, D., Bachand, K., Prescott, S. A., Nault, F., Sik, A., et al. (2003). Trans-synaptic shift in anion gradient in spinal lamina I neurons as a mechanism of neurophatic pain. *Nature*, *424*, 938–942.

De la Luz-Cuellar, Y. E., Rodríguez-Palma, E. J., Franco-Enzástiga, Ú., Salinas-Abarca, A. B., Delgado-Lezama, R., & Granados-Soto, V. (2019). Blockade of α_5GABA_A receptors differentially reduces reserpine-induced fibromyalgia-type pain in female rats. *European Journal of Pharmacology*, *858*, 1–12.

Farrant, M., & Nusser, Z. (2005). Variations on an inhibitory theme: Phasic and tonic activation of $GABA_A$ receptors. *Nature Review Neuroscience*, *6*, 215–229.

Franco-Enzástiga, U., García, G., Murbartián, J., González-Barrios, R., Salinas-Abarca, A. B., Sánchez-Hernández, B., et al. (2021). Sex-dependent pronociceptive role of spinal α_5GABA_A receptor and its epigenetic regulation in neuropathic rodents. *Journal of Neurochemistry*, *156*(6), 897–916.

Garcia de Lucas, A., Ahring, P. K., Larsen, J. S., Rivera-Arconada, I., Lopez-Garcia, J. A., Mirza, N. R., et al. (2015). $GABA_A$ α_5 subunit-containing receptors do not contribute to reversal of inflammatory-induced spinal sensitization as indicated by the unique selectivity profile of the $GABA_A$ receptor allosteric modulator NS16085. *Biochemical Pharmacology*, *93*, 370–379.

Hernández-Reyes, J. E., Salinas-Abarca, A. B., Vidal-Cantú, G. C., Raya-Tafolla, G., Elias-Viñas, D., Granados-Soto, V., et al. (2019). α_5-$GABA_A$ receptors play a pronociceptive role and avoid the rate-dependent depression of the Hoffmann reflex in diabetic neuropathic pain and reduce primary afferent excitability. *Pain*, *160*, 1448–1458.

Hultborn, H., Illert, M., Nielsen, J., Paul, A., Ballegaard, M., & Wiese, H. (1996). On the mechanism of the post-activation depression of the H-reflex in human subjects. *Experimental Brain Research*, *108*(3), 450–462.

Jolivalt, C. G., Lee, C. A., Ramos, K. M., & Calcutt, N. A. (2008). Allodynia and hyperalgesia in diabetic rats are mediated by GABA and depletion of spinal potassium-chloride co-transporters. *Pain*, *140*, 48–57.

Knabl, J., Witschi, R., Hösl, K., Reinold, H., Zeilhofer, U. B., Ahmadi, S., et al. (2008). Reversal of pathological pain through specific spinal $GABA_A$ receptor subtypes. *Nature*, *451*, 330–334.

Lee-Kubli, C. A., & Calcutt, N. A. (2014). Altered rate-dependent depression of the spinal H-reflex as an indicator of spinal disinhibition in models of neuropathic pain. *Pain*, *155*, 250–260.

Loeza-Alcocer, E., Canto-Bustos, M., Aguilar, J., González-Ramírez, R., Felix, R., & Delgado-Lezama, R. (2013). α_5GABA$_A$ receptors mediate primary afferent fiber tonic excitability in the turtle spinal cord. *Journal of Neurophysiology*, *110*, 2175–2184.

Lucas-Osma, A. M., Li, Y., Lin, S., Black, S., Singla, R., Fouad, K., et al. (2018). Extrasynaptic α_5GABA$_A$ receptors on proprioceptive afferents produce a tonic depolarization that modulates sodium channel function in the rat spinal cord. *Journal of Neurophysiology*, *120*, 2953–2974.

Ma, W., Saunders, P. A., Somogyi, R., Poulter, M. O., & Barker, J. L. (1993). Ontogeny of GABA$_A$ receptor subunit mRNAs in rat spinal cord and dorsal root ganglia. *Journal of Comparative Neurology*, *338*, 337–359.

Maddox, F. N., Valeyev, A. Y., Poth, K., Holohean, A. M., Wood, P. M., Davidoff, R. A., et al. (2004). GABA$_A$ receptor subunit mRNA expression in cultured embryonic and adult human dorsal root ganglion neurons. *Brain Research. Developmental Brain Research*, *149*, 143–151.

Marshall, A. G., Lee-Kubli, C., Azmi, S., Zhang, M., Ferdousi, M., Mixcoatl-Zecuatl, T., et al. (2017). Spinal disinhibition in experimental and clinical painful diabetic neuropathy. *Diabetes*, *66*, 1380–1390.

Munro, G., Erichsen, H. K., Rae, M. G., & Mirza, N. R. (2011). A question of balance—Positive versus negative allosteric modulation of GABA$_A$ receptor subtypes as a driver of analgesic efficacy in rat models of inflammatory and neuropathic pain. *Neuropharmacology*, *61*, 121–132.

Munro, G., López-García, J. A., Rivera-Arconada, I., Erichsen, H. K., Nielsen, E.Ø., Larsen, J. S., et al. (2008). Comparison of the novel subtype-selective GABA$_A$ receptor-positive allosteric modulator NS11394 [3′-[5-(hydroxyl-1-methyl-ethyl)-benzoimidazol-1-yl]-biphenyl-2-carbonitrile] with diazepam, zolpidem, bretanezil, and gabodaxol in rat models of inflammatory and neuropathic pain. *Journal of Pharmacology and Experimental Therapeutics*, *327*, 969–981.

Nickolls, S., Mace, H., Fish, R., Edye, M., Gurrell, R., Ivarsson, M., et al. (2011). A comparison of the $\alpha_{2/3/5}$ selective positive allosteric modulators L-838,417 and TPA023 in preclinical models of inflammatory and neuropathic pain. *Advances in Pharmacological Sciences*, *2011*, 608912.

Paul, J., Zeilhofer, H. U., & Fritschy, J. M. (2012). Selective distribution of GABA$_A$ receptor subtypes in mouse spinal dorsal horn neurons and primary afferents. *Journal of Comparative Neurology*, *520*, 3895–3911.

Perez-Sanchez, J., Lorenzo, L. E., Lecker, I., Zurek, A. A., Labrakakis, C., Bridgwater, E. M., et al. (2017). α_5GABA$_A$ receptors mediate tonic inhibition in the spinal cord dorsal horn and contribute to the resolution of hyperalgesia. *Journal of Neuroscience Research*, *95*, 1307–1318.

Piantadosi, S. C., French, B. J., Poe, M. M., Timić, T., Marković, B. D., Pabba, M., et al. (2016). Sex-dependent anti-stress effect of an α_5 subunit containing GABA$_A$ receptor positive allosteric modulator. *Frontiers in Pharmacology*, *7*, 446.

Ralvenius, W. T., Benke, D., Rudolph, U., & Zeilhofer, H. U. (2015). Analgesia and unwanted benzodiazepine effects in point-mutated mice expressing only one benzodiazepine-sensitive GABA$_A$ receptor subtype. *Nature Communications*, *6*, 6803.

Roberts, L. A., Beyer, C., & Komisaruk, B. R. (1986). Nociceptive responses to altered GABAergic activity at the spinal cord. *Life Sciences*, *39*, 1667–1674.

Rudomin, P., & Schmidt, R. F. (1999). Presynaptic inhibition in the vertebrate spinal cord revisited. *Experimental Brain Research*, *129*, 1–37.

Takahashi, A., Mashimo, T., & Uchida, I. (2006). GABAergic tonic inhibition of substantia gelatinosa neurons in mouse spinal cord. *Neuroreport*, *17*(12), 1331–1335.

Todd, A. J. (2010). Neuronal circuitry for pain processing in the dorsal horn. *Nature Review Neuroscience*, *11*, 823–836.

van Amerongen, G., Siebenga, P. S., Gurrell, R., Dua, P., Whitlock, M., Gorman, D., et al. (2019). Analgesic potential of PF-06372865, an $\alpha2/\alpha3/\alpha5$ subtype-selective GABA$_A$ partial agonist, in humans. *British Journal of Anaesthesia*, *123*(2), 194–203.

Walker, M. C. (2008). GABA$_A$ receptor subunit specificity: A tonic for the excited brain. *Journal of Physiology*, *586*(4), 921–922.

Willis, W. D. (1999). Dorsal root potentials and dorsal root reflexes: A double-edged sword. *Experimental Brain Research*, *124*, 395–421.

Witschi, R., Punnakkal, P., Paul, J., Walczak, J. S., Cervero, F., Fritschy, J. M., et al. (2011). Presynaptic α_2-GABA$_A$ receptors in primary afferent depolarization and spinal pain control. *Journal of Neuroscience*, *31*, 8134–8142.

Xue, M., Liu, J. P., Yang, Y. H., Suo, Z. W., Yang, X., & Hu, X. D. (2017). Inhibition of α_5 subunit-containing GABA$_A$ receptors facilitated spinal nociceptive transmission and plasticity. *European Journal of Pain*, *21*, 1061–1071.

Zeilhofer, H. U., Benke, D., & Yevenes, G. E. (2012). Chronic pain states: Pharmacological strategies to restore diminished inhibitory spinal pain control. *Annual Review of Pharmacology and Toxicology*, *52*, 111–133.

Chapter 13

ATP-sensitive potassium channels in pain and analgesia

Taís de Campos Lima[a], Débora de Oliveira Santos[b], and Celina Monteiro da Cruz Lotufo[a]
[a]*Department of Physiology, Biomedical Sciences Institute, Federal University of Uberlândia, Uberlândia, Minas Gerais, Brazil,*
[b]*Bucomaxillofacial Pathology Laboratory, Odontology Faculty, Federal University of Uberlândia, Uberlândia, Minas Gerais, Brazil*

Abbreviations

ADP	adenosine diphosphate
Akt	protein kinase B
ANOVA	analysis of variance
ATP	adenosine triphosphate
Ca^{2+}	calcium
cAMP	cyclic adenosine monophosphate
cGMP	cyclic guanosine monophosphate
$DiBAC_4(3)$	bis-(1,3-dibutylbarbituric acid)trimethineoxonol
DMSO	dimethyl sulfoxide
DRG	dorsal root ganglion
GIRK	G protein-gated inwardly rectifying potassium channel
Glucose-6-P	glucose-6-phosphate
GLUT2	glucose transporter 2
K^+	potassium
K2P	two-pore potassium channel
Katp	ATP-sensitive potassium channel
KCa	calcium-activated potassium channel
Kir.x	inward rectifier potassium channel
KV	voltage-gated potassium channel
Mg	magnesium
NBF	nucleotide binding folds
nNOS	nitric oxide synthases
NO	nitric oxide
Pi3K	phosphatidylinositol-3-kinase
PIP2	phosphatidylinositol-4,5-bisphosphate
PKG	protein kinase G
SEM	standard error of the mean
SUR	sulphonylurea receptor
TEA	tetraethylammonium
TMD	transmembrane domains

ATP-sensitive potassium channels (Katp)

Potassium channels are a family of membrane proteins that selectively conducts potassium ion (K^+) across the cell membrane and play important roles in cellular signaling processes in excitable and nonexcitable cells, such as heart rate, neuronal excitability, neurotransmitters release, insulin secretion, and cell volume regulation (Mannhold, 2004; Tinker, Aziz, Li, & Specterman, 2018).

The Neurobiology, Physiology, and Psychology of Pain. https://doi.org/10.1016/B978-0-12-820589-1.00009-9

Potassium channels are classified according to their electrophysiological properties into voltage-gated (Kv), calcium-activated (KCa), inward rectifier (Kir or IRK), and two-pore (K2P) K^+ channels (Ocaña, Cendán, Cobos, Entrena, & Baeyens, 2004). ATP-sensitive potassium channels (Katp) are included in the Kir subtype of potassium channels. This means that differently from the Kv channels that are mostly involved in the outflow of potassium ions during the repolarization phase of an action potential, Kir channels are specialized in inward potassium currents that regulate the resting membrane potential when it became more negative than the potassium's reversal potential. However, the main characteristic of Katp channels is its regulation by intracellular adenine nucleotide concentrations: it closes as cellular ATP levels increase and opens when the levels fall, causing depolarization or hyperpolarization of the membrane potential. Therefore, they are considered metabolic sensors as they couple cellular metabolism to electrical activity in many tissues. Katp channels were first described in heart (Noma, 1983) but are widely distributed in many cell types, with a variety of physiological roles (Seino & Miki, 2003; Tinker et al., 2018; Velasco, Díaz-García, Larqúe, & Hiriart, 2016).

The physiological role of Katp channels is well characterized in pancreatic β cells, in which they are essential to glucose homeostasis by determining insulin secretion. This mechanism is dependent on plasma glucose levels: as glucose enter the cells through GLUT2 transporters, glucose metabolism leads to an increase in intracellular ATP/ADT ratio causing Katp channels closure and membrane depolarization. This triggers the opening of voltage-dependent calcium channels, inducing insulin exocytosis (Mctaggart, Clark, & Ashcroft, 2010).

The molecular structure of the Katp consists in a heterooctameric complex with four subunits forming a central pore and four regulatory subunits. The pore-forming subunits consist of inward rectifier potassium channel (Kir6.x) and regulatory ones in the sulphonylurea receptors (SUR.x) (Seino & Miki, 2003; Tinker et al., 2018) (Fig. 1).

The inward rectifier potassium channels exist in the Kir6.1 and Kir6.2 isoforms and present two transmembrane domains. ATP closes the Katp channel by binding to Kir6.2. This causes membrane depolarization and modulating functions in many cell types, such as hormonal secretion and neurotransmitter release (Mctaggart et al., 2010). Phospholipids, such as phosphatidylinositol-4,5-bisphosphate (PIP2), can decrease the ATP sensitivity of Katp channels (Baukrowitz & Fakler, 2000; Klein, Lichtenberg, Stephan, & Quast, 2005).

Sulphonylurea receptors exist in three isoforms: SUR1, SUR2A, and SUR2B, which have different sensitivities to drugs and nucleotides. The combination of a particular SUR and Kir 6.x differs among tissues. SURs receptors show three transmembrane domains, TMD0, TMD1, and TMD2, each with five, six, and six transmembrane propellers, respectively, and two nucleotide binding folds (NBF-1 and NBF-2) on the cytoplasmic side. Mg-nucleotide stimulates Katp channels

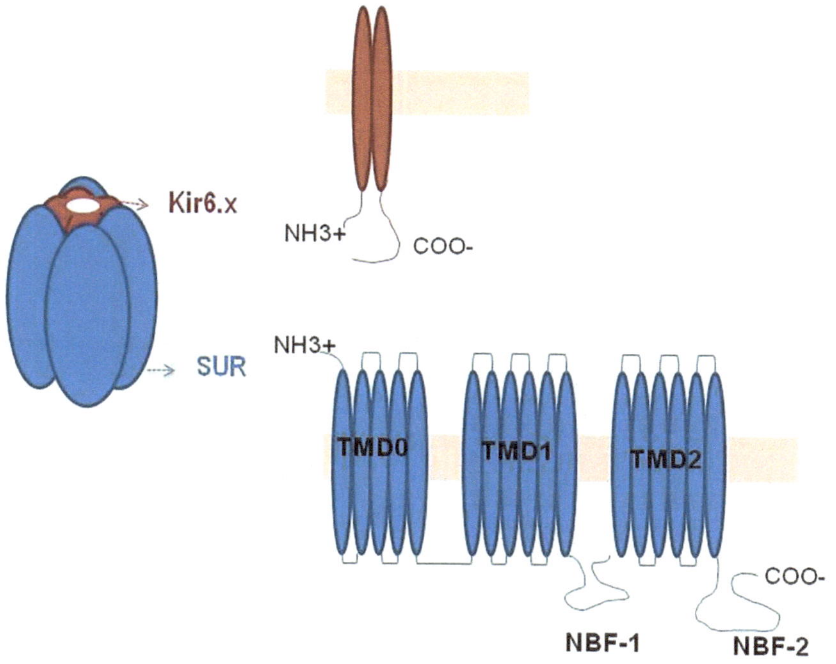

FIG. 1 Molecular structure of ATP-sensitive potassium channels (Katp). A heterooctameric complex with four subunits forming a central pore (inward rectifier potassium channels, Kir6.x) and four regulatory subunits (sulphonylurea receptors, SUR). *TMD*, transmembrane domains; *NBF*, nucleotide binding folds. *Own authorship.*

opening by interacting with the NBFs (Velasco et al., 2016). Furthermore, SUR receptors are regulated by substances that act as specific inhibitors and activators. Drugs of the sulphonylurea class, mainly glibenclamide and tolbutamide, are SURs receptors inhibitors, and widely used for type 2 diabetes mellitus treatment. On the other hand, K^+ channel openers, such as diazoxide and pinacidil among others, cause SURs activation (Edwards & Weston, 1990; Mctaggart et al., 2010; Velasco et al., 2016).

Katp channels are also present in mitochondrial membrane (Lacza et al., 2003; Shimizu et al., 2002). Interestingly, mitochondrial Katp channels are also modulated by openers and blockers, as diazoxide and glibenclamide (Katakam et al., 2016; Lacza et al., 2003; Lutas, Birnbaumer, & Yellen, 2014; Shimizu et al., 2002). Data suggest that diazoxide might be a selective opener of mitochondrial channels where it has protective effects in heart (McCully & Levitsky, 2003) and brain neurons (Shimizu et al., 2002; Teshima et al., 2003).

In the nervous system, Katp channels plays an essential role in regulating resting membrane potential and neuronal excitability where it seems to mediate neuroprotection and control neurotransmitter release (Tinker et al., 2018; Zoga et al., 2010). These channels are widely expressed in the nervous system, being reported in neurons of brain (Seino & Miki, 2003; Tinker et al., 2018), spinal cord (Zhu, Liu, Gao, Cao, & Shen, 2015), primary sensory neurons at the dorsal root ganglia, and also in glial cells (Zoga et al., 2010). Alterations in the expression of these channels in DRG have been linked to pathophysiology of pain after peripheral nerve injury. Katp channel activity was suppressed in neurons from hyperalgesic rats submitted to spinal nerve ligation (Kawano et al., 2009) and it is downregulated at the DRG (Zoga et al., 2010) and also in spinal cord (Zhu et al., 2015) after nerve injury. In the following sections, it will be discussed a possible role for these channels in enhancing nociceptive sensibility during hyperglycemia and also evidence of the importance of Katp channels in the mechanisms of action of central and peripheral analgesics.

Possible role for Katp channels in primary nociceptive neurons during hyperglycemia

Chronic hyperglycemia seems to be the main cause of diabetic neuropathy, wich can in many cases manifest as neuropathic pain. Studies indicate that glucose metabolism in the neuron may be similar to what occurs in pancreatic β cells. The entry of glucose into beta cells is facilitated by the GLUT2 transporter. Once glucose enters pancreatic β cells, it is phosphorylated into Glucose-6-phosphate by glucokinase. Glucose-6-P metabolism in pancreatic β cells increases the proportion of ATP/ADP and induces the closure of ATP-sensitive potassium channels. This results in the depolarization of the cell membrane, which promotes the opening of voltage-gated Ca^{2+} channels, activating the exocytosis of vesicles containing insulin (Velasco et al., 2016).

Based on these data, some studies have shown the importance of a similar mechanism in peripheral sensory neurons. Zhou, Lu, Song, and Owyang (2011) showed that the inhibition of gastric motility induced by hyperglycemia occurs due to the closure of Katp channels present in primary sensory neurons of the nodal ganglion. In this study, it was found that the Katp channels are involved in neuronal depolarization and glucose-induced gastric relaxation.

Results of de Campos Lima, de Oliveira Santos, Lemes, Chiovato, and da Cruz Lotufo (2019) support the idea that hyperglycemia has a direct effect on nociceptive neurons that contribute to a state of hyperalgesia, as also suggested by other authors (Lee, Cox, Mook, & McCarty, 1990; Lee & McCarty, 1990, 1992). Using the fluorescent probe DiBAC4(3) in cultured primary nociceptive neurons, it was observed that glucose alters the resting membrane potential of these cells, causing a membrane depolarization that was reduced in the presence of diazoxide, suggesting the participation of Katp channels (Fig. 2).

As seen in in vitro experiments, behavioral in vivo experiments show that hyperglycemia alters mechanical sensitivity. Chronic administration of glucose solutions at the DRG using osmotic pumps causes hyperalgesia in the hind limb innervated by the perfused ganglion (Dobretsov, Hastings, Stimers, & Zhang, 2001). Hyperglycemic rats after streptozotocin treatment shown decrease in mechanical nociceptive threshold (Romanovsky, Hastings, Stimers, & Dobretsov, 2004; Teixeira et al., 2019). In the study of de Campos Lima et al. (2019), glucose solutions were injected directly at the DRG (right L5) where it was observed an acute increase in mechanical sensitivity. This effect was blocked by coadministration of the Katp channel opener diazoxide, reinforcing the hypothesis that high glucose inhibits Katp channels at primary nociceptive neurons resulting in increased sensitivity (Fig. 3).

Therefore, studies indicate that the same depolarization mechanism induced by glucose in pancreatic β cells may be occurring in some sensory neurons as shown in Fig. 4.

Sulphonylureas such as glibenclamide are commonly prescribed to patients with type 2 diabetes because inhibiting Katp channels at pancreatic β cells cause membrane depolarization and consequent insulin release. However, as Katp channels may be involved on glucose direct effect in primary sensory neurons (de Campos Lima et al., 2019), it might be important to study the impact of sulphonylureas prescription on peripheral neurons. Although sulphonylureas are well tolerated by

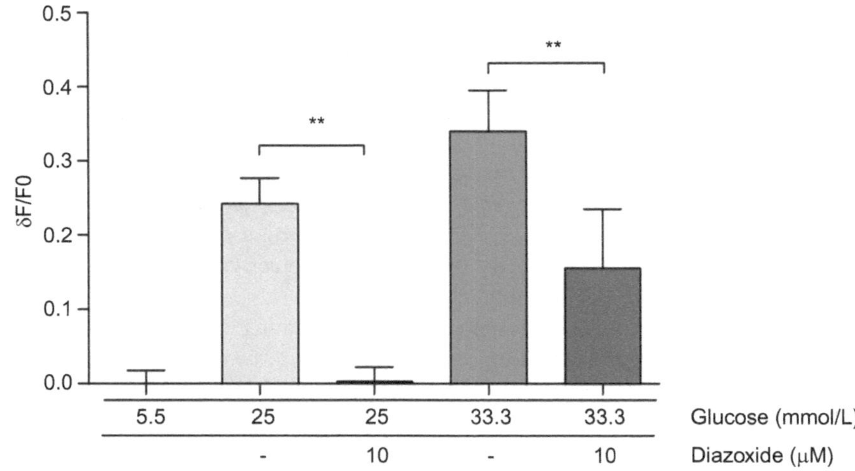

FIG. 2 Effect of glucose on the resting membrane potential of primary nociceptive neurons. Variations in resting membrane potential were accessed by use of the fluorescent indicator DiBAC4(3) in cultured primary nociceptive neurons. An increase in fluorescence indicates membrane depolarization. Neurons fluorescence was tested in Hank's solution (5.5 mmol/L glucose) or after administering glucose (25 or 33.3 mmol/L) or glucose plus diazoxide (10 μM). Only nociceptive neurons that responded to capsaicin (1 μM) were considered for analysis. Data shown as means ± SEM of 20–125 neurons. $^{**}P < .05$. (ANOVA followed by Bonferroni posttest.) *de Campos Lima, T., de Oliveira Santos, D., Lemes, J. B. P., Chiovato, L. M., da Cruz Lotufo, C. M. (2019). Hyperglycemia induces mechanical hyperalgesia and depolarization of the resting membrane potential of primary nociceptive neurons: Role of ATP-sensitive potassium channels.* Journal of the Neurological Sciences, 401, *55–61. https://doi.org/10.1016/j.jns.2019.03.025.*

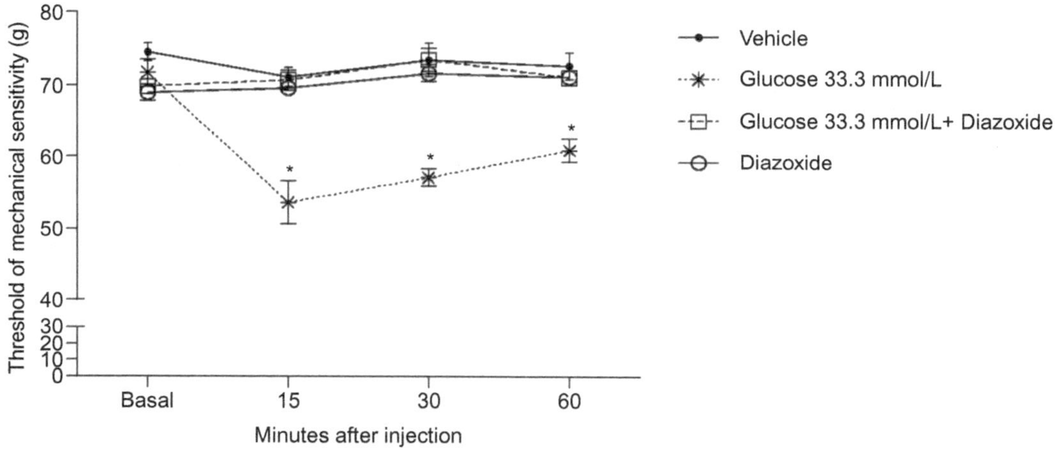

FIG. 3 In vivo effect of glucose injection at rat dorsal root ganglia on mechanical sensitivity. The threshold of mechanical sensitivity (g) of glucose (33.3 mmol/L), glucose plus diazoxide (10 μM), diazoxide alone, or vehicle (1% DMSO in saline) administration is shown. Data shown as means ± SEM 6 animals per group.$^{*}P < .05$. (ANOVA followed by Bonferroni posttest.) *de Campos Lima, T., de Oliveira Santos, D., Lemes, J. B. P., Chiovato, L. M., da Cruz Lotufo, C. M. (2019). Hyperglycemia induces mechanical hyperalgesia and depolarization of the resting membrane potential of primary nociceptive neurons: Role of ATP-sensitive potassium channels.* Journal of the Neurological Sciences, 401, *55–61. https://doi.org/10.1016/j.jns.2019.03.025.*

patients, a study that followed diabetic patients in treatment for 5 years showed that most agents used in the treatment of hyperglycemia can act directly in the development of complications regardless of glucose levels, and treatments that involve an increase in the availability of insulin, such as use of sulphonylureas, tend to have a higher incidence of peripheral diabetic neuropathy than treatments that involve sensitization to the action of insulin (Pop-Busui et al., 2013). Also, a cohort study using binary logistic regression to identify potential risk factors for diabetic peripheral neuropathy showed a strong association with treatment with sulphonylurea and development of diabetic neuropathy in comparison with patients treated with metformin (Katulanda et al., 2012).

Besides the possible direct effect of sulphonylureas treatment at the sensory system, in the next section of this chapter, it will be discussed an increasing amount of evidence indicating that the opening of Katp channels are involved in the

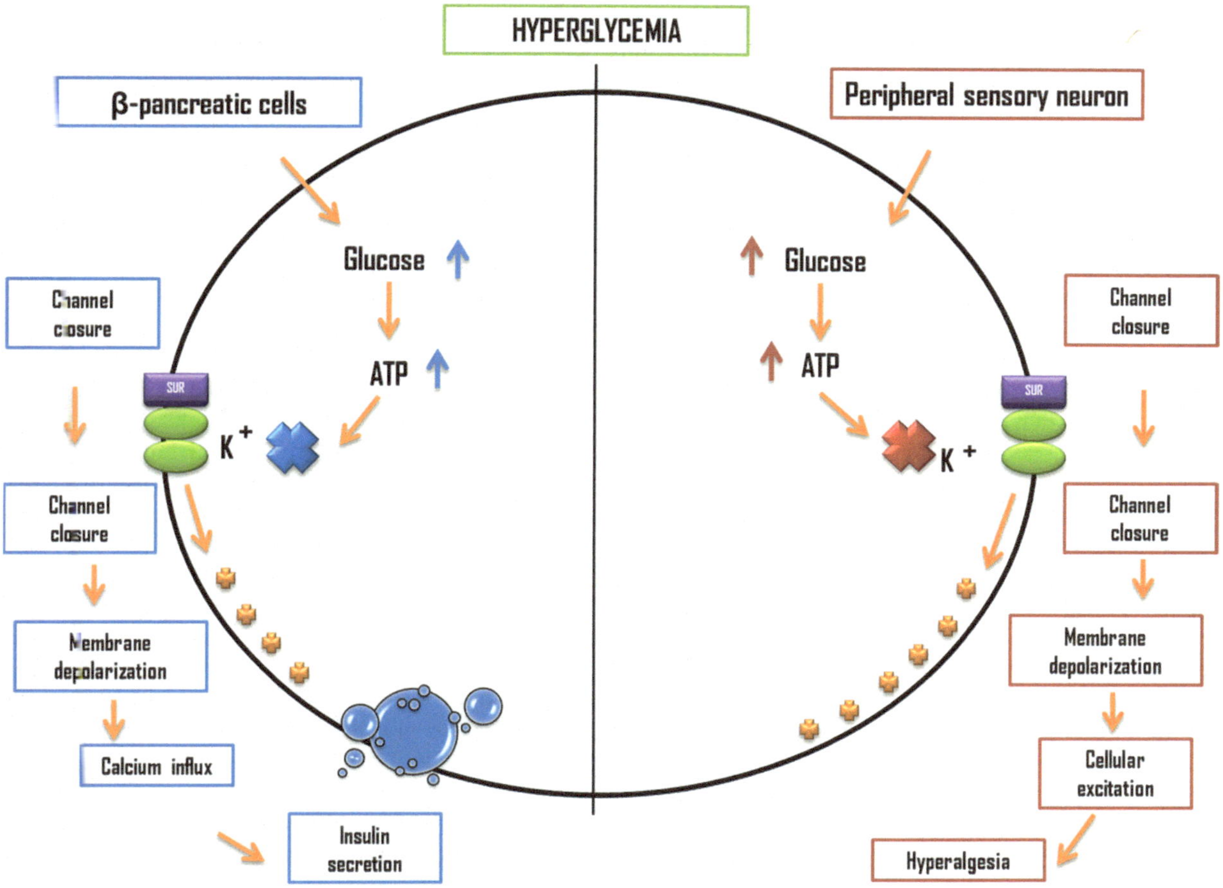

FIG. 4 Katp inhibition by glucose metabolism in β-pancreatic cells and in peripheral sensory neurons. In pancreatic β cells, glucose metabolism causes an increase in ATP levels that closes Katp channels inducing insulin secretion. A similar mechanism seems to occur in primary nociceptive neurons where glucose induces the closure of Katp channels, promoting membrane depolarization and increasing excitability. *Own authorship.*

mechanisms of action of several opioids and nonopioids analgesics, both in central and peripheral systems. Therefore, it might be important to evaluate the efficacy of analgesics in patients using sulphonylureas for diabetes treatment.

Katp involvement in the mechanisms of analgesic substances

The participation of Katp channels in pain was initially investigated as a mechanism for opioid actions at the central nervous system (North & Williams, 1985). It has been observed that opioids inhibit neuronal firing due to hyperpolarization of the resting membrane potential. North and Williams (1985) performed electrophysiological experiments in locus coeruleus neurons and observed that opioids and α_2-adrenoceptor agonists induce potassium currents that are insensitive to tetraethylammonium (TEA) blockade. Later on, Ocaña, Del Pozo, Barrios, Robles, & Baeyens, 1990 showed that the analgesic effect of systemic morphine in the hot plate test in mice was reduced by intracerebroventricular administration of glibenclamide but not by other potassium channel blockers such as TEA and quinine. This was the first indicative of the Katp channel involvement in analgesia. Following experiments suggested that not all opioid receptors act through Katp channels opening at the central nervous system (Wild, Vanderah, Mosberg, & Porreca, 1991), since glibenclamide was able to reduce the analgesic effects of μ and some δ agonists, while it did not affect κ agonists effects. Other studies reinforced the hypothesis of opening Katp channels by opioids in the brain (Ocaña, Del Pozo, & Baeyens, 1993; Ocaña, Del Pozo, Barrios, & Baeyens, 1995; Roane & Boyd, 1993). In all these studies, this assumption was based on the antagonism of the analgesic actions of opioids or other molecules by sulphonylureas. Raffa and Codd (1994) strengthened the hypothesis that opioids induce analgesia by opening Katp channels by showing that glibenclamide did not bind to any of the opioid receptors, excluding the possibility of a competitive antagonism of glibenclamide by blocking opioid receptors. Another interesting study was performed by testing the activity of opioids and several Katp channels modulators using the tail flick test (Welch & Dunlow, 1993). It was observed that Katp channel blockers inhibited the antinociceptive effect of morphine but also that Katp channel openers, as minocidil, lemakalim, and glyburide, produced an antinociceptive effect that was blocked by administration of opioid receptor antagonists. Such relation between Katp openers and opioid receptors was also corroborated by a study that tested the selective knockdown of μ, κ, and δ opioid receptors using oligodeoxinucleotides and again obtained a reduced analgesic effect from Katp channel openers (Lohmann & Welch, 1999). These studies indicated that the relationship between Katp channels and opioid receptors is more complex than opening of Katp channels upon activation of opioid receptors. It was suggested that opening Katp channels might induce the release of endogenous opioids (Lohmann & Welch, 1999). If that was the case, the ability of opioid receptors antagonists to inhibit analgesia induced by Katp channel openers could be explained. This hypothesis was tested by Campbell and Welch (2001) whom did not find any increase in endogenous opioid peptides in spinal perfusate. Also, administration of opioid antagonists did not affect basal endogenous opioids levels. Taking together, data available regarding the Katp potassium channel involvement in opioid-induced analgesia at the central nervous system suggest a relevant relationship. However, the intracellular signaling responsible to Katp channels opening after opioid receptor activation is still not comprehended. Furthermore, there are now convincing evidences that opioid actions at the central nervous system involve another potassium channel the G protein-gated inwardly rectifying potassium channels (GIRK), also known as Kir3 (Ikeda, Kobayashi, Kumanishi, Niki, & Yano, 2000; Marker, Cintora, Roman, Stoffel, & Wickman, 2002; Marker, Luján, Loh, & Wickman, 2005; Nagi & Pineyro, 2014). It is possible that both potassium channels, GIRK/Kir3 and Katp, are involved in opioid mechanism for central analgesia. There are no reports of interactions between GIRK/Kir3 and Katp channels in the literature.

In addition to the involvement of Katp channels at opioid-induced analgesia at the central nervous system, several studies have suggested a role for these channels at peripheral level. Again, the first indicative of Katp participation in peripheral analgesia came from behavioral studies regarding the mechanism of opioid analgesia (Bentley, Newton, & Starr, 1981; Ferreira & Nakamura, 1979). Rodrigues and Duarte (2000) tested the effect of several potassium channel blockers at the antinociceptive effect of morphine at the mechanical hyperalgesia induced by carrageenan injection in the rat paw. Morphine effect was blocked by previous administration of either glibenclamide or tolbutamide, but not other channel blockers. Using similar protocols, it also determined the involvement of the Katp channel in the peripheral analgesia induced by nitric oxide (Soares & Duarte, 2001) and by the permeable analog to cGMP, dibutyryl cGMP. It was then postulated that activation of opioid receptors at peripheral nociceptive neurons would result in activation of nitric oxide synthase with consequent NO synthesis, which in turn would activate guanylyl cyclase to produce cGMP. Activation of protein kinase G (PKG) by cGMP would be the responsible for opening Katp potassium channels leading to hyperpolarization of the neuronal membrane potential. Reinforcing this hypothesis, Chai and Lin (2008) transfected human cell lines with Katp channels and used path clamp recordings to show that activation of the NO/guanylyl cyclase/cGMP/PKG pathway results in opening of the Kir6.2/SUR1 channels. Finally, Cunha et al. (2010), using both in vivo and in vitro approaches, determined that morphine peripheral action in primary nociceptive neurons activates the NO/cGMP/PKG pathway via activation of Pi3Kγ, which phosphorylates the AKT protein, also known as protein kinase B (PKB). This pathway is known to be activated by the $G_{\beta\gamma}$ subunit of the Gi protein (Stephens et al., 1994). It was shown that nNOS-deficient mouse lack peripheral morphine analgesic effect and also that Pi3K and AKT inhibitors reduced NO production in dorsal root ganglia slices. Electrophysiological data from morphine effect on DRG neurons indicate that potassium currents altered by morphine are inhibited by glibenclamide as well as by the Pi3Kγ/AKT inhibitors (Fig. 5).

The peripheral analgesia induced by morphine seems to be related to reducing primary sensory neurons excitability by means of the hyperpolarization caused by the opening of Katp channels (Fig. 5). Alteration in Katp channel is probably due to phosphorylation as a final result from the activation of the nNOS/GMPc/PKG pathway (Fig. 6). At peripheral level, the opening Katp channels seem to be involved in μ (Cunha et al., 2010) and δ opioid receptors (Picolo, Cassola, & Cury, 2003). Although peripheral analgesia induced by opioid κ receptors agonists also activates the PI3K/AKT/NO pathway (Cunha et al., 2012), Picolo et al. (2003) results indicate that Katp channels are involved in δ opioid receptor but not in κ receptors peripheral analgesia.

At peripheral level, several studies have been implicating the Katp channels in the effect of different molecules with analgesic effect (Alves, Tatsuo, Leite, & Duarte, 2004; Chaves et al., 2018; Cunha et al., 2008; Dos Santos et al., 2014; Neves et al., 2018; Picolo et al., 2003). In all of these studies, the assumption of the involvement of the Katp channel is based on inhibition of the analgesic effect by a sulphonylurea, glibenclamide, and/or tolbutamide. Katp channels activation has been implicated in the analgesic mechanisms of Heme oxygenase-1/biliverdin/carbon monoxide in a model of arthritis in the rat temporomandibular joint (Chaves et al., 2018), in the antinociceptive effects of hydrogen sulfide (Cunha et al., 2008), of *Crotalus durissus terrificus* snake venom (Picolo et al., 2003), thymoquinone (Parvardeh et al., 2018), low-intensity laser therapy (Neves et al., 2018) in inflammatory hyperalgesia and also in the peripheral actions of some widely used analgesic drugs such as dipyrone (metamizole) (Dos Santos et al., 2014) and diclofenac (Alves, Tatsuo, et al., 2004).

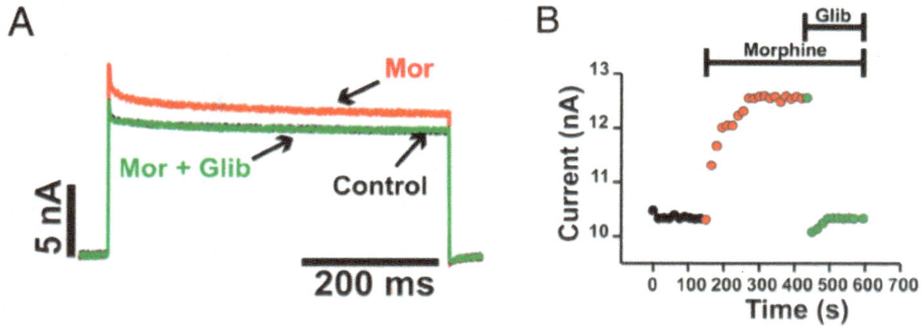

FIG. 5 Morphine-enhanced potassium currents are inhibited by glibenclamide. Whole cell voltage-clamp experiments show that morphine (10 μM, *red*) causes an increase in potassium currents that was abolished by glibenclamide (10 μM, *green*) at primary nociceptive neurons. *Cunha, T. M., Roman-Campos, D., Lotufo, C. M., Duarte, H. L., Souza, G. R., Verri Jr., W. A., et al. (2010). Morphine peripheral analgesia depends on activation of the PI3Kγ/AKT/nNOS/NO/KATP signaling pathway.* Proceedings of the National Academy of Sciences of the United States of America, 107(9). *https://doi.org/10.1073/pnas.0914733107.*

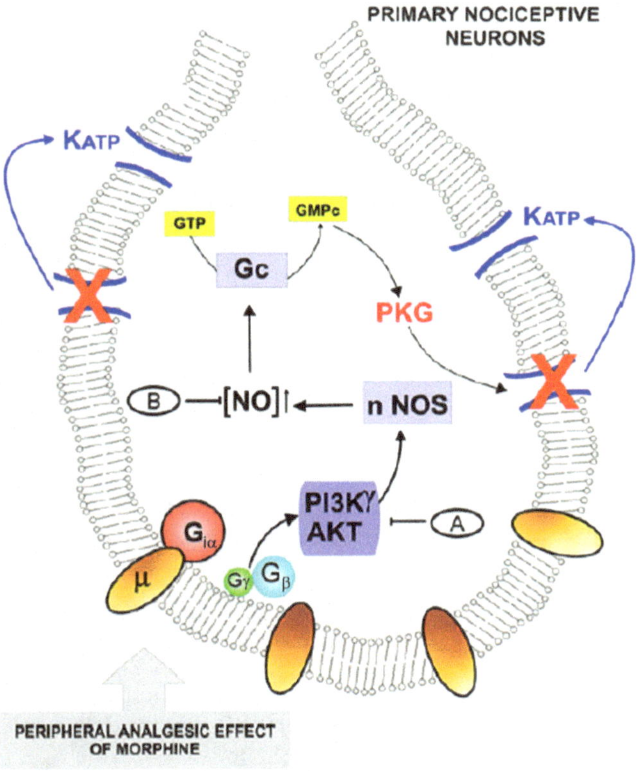

FIG. 6 Schematic representation of the intracellular signaling involved in peripheral analgesia induced by morphine in μ opioid receptors. Protein Gβγ activates PI3Kγ kinase that phosphorylates AKT/PKB protein, which in turn induces nitric oxide production by nNOS. Nitric oxide induces cGMP levels to increase and activate PKG, which possibly cause Katp channel phosphorylation and opening, resulting in membrane hyperpolarization. *Cunha, T. M., Roman-Campos, D., Lotufo, C. M., Duarte, H. L., Souza, G. R., Verri Jr., W. A., et al. (2010). Morphine peripheral analgesia depends on activation of the PI3Kγ/AKT/nNOS/NO/KATP signaling pathway.* Proceedings of the National Academy of Sciences of the United States of America, 107(9). *https://doi.org/10.1073/pnas.0914733107.*

Taking together, data point to an important role for Katp channels in nociceptive regulation at peripheral nociceptive neurons. However, there are few studies regarding the role of such channels at pain signal generation and processing. The physiological role of Katp channels was investigated in a study that tested the effects of Katp activators in pain and in isolated neurons (Du, Wang, & Zhang, 2011). In this study, as well as in others (Alves, Soares, et al., 2004; Picolo et al., 2003), the administration of Katp openers resulted in significant analgesia, reinforcing the importance of Katp channels in pain regulation. On the other hand, administration of the Katp channel blocker glibenclamide did not affect basal or bradykinin-induced nociceptive responses (Du et al., 2011), suggesting that Katp channels might be involved only in downmodulating pain signaling. Data from isolated cells, however, indicate that inhibiting Katp currents with glibenclamide can induce membrane depolarization in nociceptive neurons (de Campos Lima et al., 2019; Du et al., 2011) and these channels participate in controlling resting membrane potential in different neuronal types (Allen & Brown, 2004; Lutas et al., 2014). As sulphonylureas are administered chronically to many diabetic patients, it is important to evaluate the impact of this treatment in pain sensation and in the efficacy of analgesics in this population.

In conclusion, data indicate that ATP-sensitive potassium channels are involved in pain/analgesia by modulating neurons excitability in both central and peripheral neurons. At peripheral nociceptive neurons, the administration of Katp channels blockers reduces the effects of several analgesic substances and the administration of molecules that open Katp channels produce analgesia. More studies on the physiological function of the Katp channels in pain will possibly provide new approaches to pain management and also ensure that long-term use of sulphonylureas by diabetic patients does not impact pain control and the development of peripheral neuropathy.

Application to other areas

ATP-sensitive potassium channels are expressed by many cell types; and besides the well-known involvement in insulin secretion by pancreatic β cells, studies have shown that Katp channels activation has protective effect in ischemic conditions in cardiac cells (Elrod et al., 2008; Flagg, Enkvetchakul, Koster, & Nichols, 2010; Lefer, Nichols, & Coetzee, 2009) and neurons (Szeto, Chen, Feng, & Sun, 2018; Tinker et al., 2018).

Other agents of interest

Mainly due to the protective potential of Katp channel openers in brain and muscle cells, there is an effort to the development of new and more selective molecules (Edwards & Weston, 1990; Mannhold, 2004; Hansen, 2006; Rios-Santos et al., 2007; Sebille et al., 2012). Selective openers of specific subtypes of Katp channels can be investigated as possible agents for pain therapy.

Mini-dictionary of terms

- Diabetic neuropathy: Highly prevalent chronic complication of diabetes, characterized by dysfunction in the peripheral nervous system, with varied clinical manifestation, can manifest as reduced or increased pain sensitivity.
- Diazoxide: A benzothiazine derivative, which opens the ATP-sensitive potassium (Katp) channels, by ligation with its regulatory subunit (SUR). Inhibits insulin release by pancreatic β cells and used as antihypertensive for its action in smooth muscle cells.
- Glibenclamide: A sulphonylurea used as oral hypoglycemiant agent often used to treat diabetes that stimulates the production of insulin by pancreatic β cells.
- Kir/IRK: Inward rectifier potassium channel is a type of potassium channel classified according to their electrophysiological properties, characterized by allowing potassium ions to move more easily into rather than out of the cell.
- Sulphonylurea: A family of molecules used to reduce hyperglycemia for type II diabetic patients. Binds to the regulatory subunit of the Katp channel called sulphonylurea receptors (SUR).
- Opioid: A substance (natural, semisynthetic, or synthetic) that binds specifically to opioid receptors.

Key facts

- Katp channels are members of the inward rectifier (Kir or IRK) potassium channels family involved in regulating resting membrane potential.
- Katp channels are found in plasma and mitochondrial membranes.
- An increase in intracellular ATP levels inhibits Katp channels causing membrane depolarization.
- In pancreatic β cells, Katp channels regulate insulin secretion according to ATP production due to glucose metabolism.
- Katp channels are closed by sulphonylureas, as glibenclamide and tolbutamide and opened by diazoxide, cromakalin, pinacidil, and others.
- Katp channels are expressed by many cell types, including neurons from central and peripheral nervous system.

Summary points

- High-glucose concentration induces resting membrane depolarization in cultured nociceptors and mechanical hyperalgesia in rats. Those effects are reduced in the presence of Katp channel opener diazoxide.
- Katp channels can be involved in the enhanced nociceptive sensitivity caused by hyperglycemia in diabetic patients.
- Central analgesia induced by opioids is reduced by blocking Katp channels.
- Katp channels opening seems to be involved in peripheral analgesia induced by opioids and several other agents, as dipyrone and diclofenac.
- Morphine-induced activation of Katp channels in peripheral nociceptors depends on the Pi3K/AKT/nNOS/cGMP intracellular pathway.

References

Allen, T. G. J., & Brown, D. A. (2004). Modulation of the excitability of cholinergic basal forebrain neurones by KATP channels. *Journal of Physiology*, *554*(2), 353–370. https://doi.org/10.1113/jphysiol.2003.055889.

Alves, D. P., Soares, A. C., Francischi, J. N., Castro, M. S. A., Perez, A. C., & Duarte, I. D. G. (2004). Additive antinociceptive effect of the combination of diazoxide, an activator of ATP-sensitive K+ channels, and sodium nitroprusside and dibutyryl-cGMP. *European Journal of Pharmacology*, *489*(1–2), 59–65. https://doi.org/10.1016/j.ejphar.2004.02.022.

Alves, D. P., Tatsuo, M. A. F., Leite, R., & Duarte, I. D. G. (2004). Diclofenac-induced peripheral antinociception is associated with ATP-sensitive K+ channels activation. *Life Sciences*, *74*(20), 2577–2591. https://doi.org/10.1016/j.lfs.2003.10.012.

Baukrowitz, T., & Fakler, B. (2000). K(ATP) channels: Linker between phospholipid metabolism and excitability. *Biochemical Pharmacology*, *60*(6), 735–740. https://doi.org/10.1016/S0006-2952(00)00267-7.

Bentley, G. A., Newton, S. H., & Starr, J. (1981). Evidence for an action of morphine and the enkephalins on sensory nerve endings in the mouse peritoneum. *British Journal of Pharmacology*, *73*(2), 325–332. https://doi.org/10.1111/j.1476-5381.1981.tb10425.x.

Campbell, V. C., & Welch, S. P. (2001). The role of minoxidil on endogenous opioid peptides in the spinal cord: A putative co-agonist relationship between K-ATP openers and opioids. *European Journal of Pharmacology*, *417*, 91–98. https://doi.org/10.1016/S0014-2999(01)00885-.

Chai, Y., & Lin, Y. F. (2008). Dual regulation of the ATP-sensitive potassium channel by activation of cGMP-dependent protein kinase. *Pflügers Archiv / European Journal of Physiology*, *456*(5), 897–915. https://doi.org/10.1007/s00424-008-0447-z.

Chaves, H. V., do Val, D. R., Ribeiro, K. A., Lemos, J. C., Souza, R. B., Gomes, F. I. F., et al. (2018). Heme oxygenase-1/biliverdin/carbon monoxide pathway downregulates hypernociception in rats by a mechanism dependent on cGMP/ATP-sensitive K+ channels. *Inflammation Research*, *67*(5), 407–422. https://doi.org/10.1007/s00011-018-1133-z.

Cunha, T. M., Dal-Secco, D., Verri, W. A., Jr., Guerrero, A. T., Souza, G. R., Vieira, S. M., et al. (2008). Dual role of hydrogen sulfide in mechanical inflammatory hypernociception. *European Journal of Pharmacology*, *590*(1–3). https://doi.org/10.1016/j.ejphar.2008.05.048.

Cunha, T. M., Roman-Campos, D., Lotufo, C. M., Duarte, H. L., Souza, G. R., Verri, W. A., Jr., et al. (2010). Morphine peripheral analgesia depends on activation of the PI3Kγ/AKT/nNOS/NO/KATP signaling pathway. *Proceedings of the National Academy of Sciences of the United States of America*, *107*(9). https://doi.org/10.1073/pnas.0914733107.

Cunha, T. M., Souza, G. R., Domingues, A. C., Carreira, E. U., Lotufo, C. M., Funez, M. I., et al. (2012). Stimulation of peripheral kappa opioid receptors inhibits inflammatory hyperalgesia via activation of the PI3Kγ/AKT/nNOS/NO signaling pathway. *Molecular Pain*, *8*. https://doi.org/10.1186/1744-8069-8-10.

de Campos Lima, T., de Oliveira Santos, D., Lemes, J. B. P., Chiovato, L. M., & da Cruz Lotufo, C. M. (2019). Hyperglycemia induces mechanical hyperalgesia and depolarization of the resting membrane potential of primary nociceptive neurons: Role of ATP-sensitive potassium channels. *Journal of the Neurological Sciences*, *401*, 55–61. https://doi.org/10.1016/j.jns.2019.03.025.

Dobretsov, M., Hastings, S. L., Stimers, J. R., & Zhang, J. M. (2001). Mechanical hyperalgesia in rats with chronic perfusion of lumbar dorsal root ganglion with hyperglycemic solution. *Journal of Neuroscience Methods*, *110*(1–2), 9–15. https://doi.org/10.1016/S0165-0270(01)00410-1.

Dos Santos, G. G., Dias, E. V., Teixeira, J. M., Athie, M. C. P., Bonet, I. J. M., Tambeli, C. H., et al. (2014). The analgesic effect of dipyrone in peripheral tissue involves two different mechanisms: Neuronal KATPchannel opening and CB1receptor activation. *European Journal of Pharmacology*, *741*, 124–131. https://doi.org/10.1016/j.ejphar.2014.07.019.

Du, X., Wang, C., & Zhang, H. (2011). Activation of ATP-sensitive potassium channels antagonize nociceptive behavior and hyperexcitability of DRG neurons from rats. *Molecular Pain*, *7*, 1–12. https://doi.org/10.1186/1744-8069-7-35.

Edwards, G., & Weston, A. H. (1990). Structure-activity relationships of K+ channel openers. *Trends in Pharmacological Sciences*, *11*(10), 417–422. https://doi.org/10.1016/0165-6147(90)90149-3.

Elrod, J. W., Harrell, M., Flagg, T. P., Gundewar, S., Magnuson, M. A., Nichols, C. G., et al. (2008). Role of sulfonylurea receptor type 1 subunits of ATP-sensitive potassium channels in myocardial ischemia/reperfusion injury. *Circulation*, *117*(11), 1405–1413. https://doi.org/10.1161/CIRCULATIONAHA.107.745539.

Ferreira, S. H., & Nakamura, M. (1979). III—Prostaglandin hyperalgesia: Relevance of the peripheral effect for the analgesic action of opioid-antagonists. *Prostaglandins*, *18*(2), 201–208. https://doi.org/10.1016/0090-6980(79)90105-9.

Flagg, T. P., Enkvetchakul, D., Koster, J. C., & Nichols, C. G. (2010). Muscle KATP channels: Recent insights to energy sensing and myoprotection. *Physiological Reviews*, *90*(3), 799–829. NIH Public Access https://doi.org/10.1152/physrev.00027.2009.

Hansen, J. (2006). Towards selective Kir6.2/SUR1 potassium channel openers, medicinal chemistry and therapeutic perspectives. *Current Medicinal Chemistry*, *13*(4), 361–376. https://doi.org/10.2174/092986706775527947.

Ikeda, K., Kobayashi, T., Kumanishi, T., Niki, H., & Yano, R. (2000). Involvement of G-protein-activated inwardly rectifying K+ (GIRK) channels in opioid-induced analgesia. *Neuroscience Research*, *38*(1), 113–116. https://doi.org/10.1016/S0168-0102(00)00144-9.

Katakam, P. V. G., Dutta, S., Sure, V. N., Grovenburg, S. M., Gordon, A. O., Peterson, N. R., et al. (2016). Depolarization of mitochondria in neurons promotes activation of nitric oxide synthase and generation of nitric oxide. *American Journal of Physiology - Heart and Circulatory Physiology*, *310*(9), 1097–1106. https://doi.org/10.1152/ajpheart.00759.2015.

Katulanda, P., Ranasinghe, P., Jayawardena, R., Constantine, G. R., Sheriff, M. H. R., & Matthews, D. R. (2012). The prevalence, patterns and predictors of diabetic peripheral neuropathy in a developing country. *Diabetology and Metabolic Syndrome*, *4*(1). https://doi.org/10.1186/1758-5996-4-21.

Kawano, T., Zoga, V., Gemes, G., Abram, S., Hogan, Q. H., & Sarantopoulos, C. D. (2009). Atp-sensitive potassium currents in rat primary afferent neurons: Biophysical, pharmacological properties, and alterations by painful nerve injury. *Neuroscience*, *162*, 431–443. https://doi.org/10.1016/j.neuroscience.2009.04.076.

Klein, A., Lichtenberg, J., Stephan, D., & Quast, U. (2005). Lipids modulate ligand binding to sulphonylurea receptors. *British Journal of Pharmacology*, *145*, 907–915. https://doi.org/10.1038/sj.bjp.0706252.

Lacza, Z., Snipes, J. A., Kis, B., Szabó, C., Grover, G., & Busija, D. W. (2003). Investigation of the subunit composition and the pharmacology of the mitochondrial ATP-dependent K+ channel in the brain. *Brain Research*, *994*(1), 27–36. https://doi.org/10.1016/j.brainres.2003.09.046.

Lee, J. H., Cox, D. J., Mook, D. G., & McCarty, R. C. (1990). Effect of hyperglycemia on pain threshold in alloxan-diabetic rats. *Pain*, *40*(1), 105–107. https://doi.org/10.1016/0304-3959(90)91057-P.

Lee, J. H., & McCarty, R. (1990). Glycemic control of pain threshold in diabetic and control rats. *Physiology and Behavior*, *47*(2), 225–230. https://doi.org/10.1016/0031-9384(90)90135-Q.

Lee, J. H., & McCarty, R. (1992). Pain threshold in diabetic rats: Effects of good versus poor diabetic control. *Pain*, *50*(2), 231–236. https://doi.org/10.1016/0304-3959(92)90167-A.

Lefer, D. J., Nichols, C. G., & Coetzee, W. A. (2009). Sulfonylurea receptor 1 subunits of ATP-sensitive potassium channels and myocardial ischemia/reperfusion injury. *Trends in Cardiovascular Medicine*, *19*(2), 61–67. https://doi.org/10.1016/j.tcm.2009.04.008.

Lohmann, A. B., & Welch, S. P. (1999). Antisenses to opioid receptors attenuate ATP-gated K+ channel opener-induced antinociception. *European Journal of Pharmacology*, *384*(2–3), 147–152. https://doi.org/10.1016/S0014-2999(99)00698-6.

Lutas, A., Birnbaumer, L., & Yellen, G. (2014). Metabolism regulates the spontaneous firing of substantia nigra pars reticulata neurons via KATPand nonselective cation channels. *Journal of Neuroscience*, *34*(49), 16336–16347. https://doi.org/10.1523/JNEUROSCI.1357-14.2014.

Mannhold, R. (2004). KATP channel openers: Structure-activity relationships and therapeutic potential. *Medicinal Research Reviews*, *24*(2), 213–266. https://doi.org/10.1002/med.10060.

Marker, C. L., Cintora, S. C., Roman, M. I., Stoffel, M., & Wickman, K. (2002). Hyperalgesia and blunted morphine analgesia in G protein-gated potassium channel subunit knockout mice. *Neuroreport*, *13*(18), 2509–2513. https://doi.org/10.1097/00001756-200212200-00026.

Marker, C. L., Luján, R., Loh, H. H., & Wickman, K. (2005). Spinal G-protein-gated potassium channels contribute in a dose-dependent manner to the analgesic effect of μ- and δ-but not κ-opioids. *Journal of Neuroscience*, *25*(14), 3551–3559. https://doi.org/10.1523/JNEUROSCI.4899-04.2005.

McCully, J. D., & Levitsky, S. (2003). The mitochondrial KATP channel and cardioprotection. *Annals ofThoracic Surgery*, *75*(2). https://doi.org/10.1016/S0003-4975(02)04689-1.

Mctaggart, J. S., Clark, R. H., & Ashcroft, F. M. (2010). The role of the K ATP channel in glucose homeostasis in health and disease: More than meets the islet. *The Journal of Physiology*, *588*, 3201–3209. https://doi.org/10.1113/jphysiol.2010.191767.

Nagi, K., & Pineyro, G. (2014). Kir3 channel signaling complexes : Focus on opioid receptor signaling. *Frontiers in Cellular Neuroscience*, *8*, 1–15. https://doi.org/10.3389/fncel.2014.00186.

Neves, L. M. S., Gonçalves, E. C. D., Cavalli, J., Vieira, G., Laurindo, L. R., Simões, R. R., et al. (2018). Photobiomodulation therapy improves acute inflammatory response in mice: The role of cannabinoid receptors/ATP-sensitive K+ channel/p38-MAPK signalling pathway. *Molecular Neurobiology*, *55*(7), 5580–5593. https://doi.org/10.1007/s12035-017-0792-z.

Noma, A. (1983). ATP-regulated K+ channels in cardiac muscle. *Nature*, *305*(5930), 147–148. https://doi.org/10.1038/305147a0.

North, B. Y. R. A., & Williams, J. T. (1985). On the potassium conductance increased by opioids in rat locus coeruleus neurones. *The Journal of Physiology*, 265–280.

Ocaña, M., Cendán, C. M., Cobos, E. J., Entrena, J. M., & Baeyens, J. M. (2004). Potassium channels and pain: Present realities and future opportunities. *European Journal of Pharmacology*, *500*(1–3), 203–219. https://doi.org/10.1016/j.ejphar.2004.07.026.

Ocaña, M., Del Pozo, E., & Baeyens, J. M. (1993). ATP-dep'endent K+ channel blockers antagonize morphine- but not U-504,88H-induced antinociception. *European Journal of Pharmacology*, *230*(2), 203–207. https://doi.org/10.1016/0014-2999(93)90803-P.

Ocaña, M., Del Pozo, E., Barrios, M., & Baeyens, J. M. (1995). Subgroups among μ-opioid receptor agonists distinguished by ATP-sensitive K+ channel-acting drugs. *British Journal of Pharmacology*, *114*(6), 1296–1302. https://doi.org/10.1111/j.1476-5381.1995.tb13346.x.

Ocaña, M., Del Pozo, E., Barrios, M., Robles, L. I., & Baeyens, J. M. (1990). An ATP-dependent potassium channel blocker antagonizes morphine analgesia. *European Journal of Pharmacology*, *186*(2–3), 377–378. https://doi.org/10.1016/0014-2999(90)90466-J.

Parvardeh, S., Sabetkasaei, M., Moghimi, M., Masoudi, A., Ghafghazi, S., & Mahboobifard, F. (2018). Role of L-arginine/NO/cGMP/KATP channel signaling pathway in the central and peripheral antinociceptive effect of Thymoquinone in rats. *Iranian Journal of Basic Medical Sciences*, *21*(6), 625–633. https://doi.org/10.22038/IJBMS.2018.26255.6438.

Picolo, G., Cassola, A. C., & Cury, Y. (2003). Activation of peripheral ATP-sensitive K+ channels mediates the antinociceptive effect of *Crotalus durissus terrificus* snake venom. *European Journal of Pharmacology*, *469*(1–3), 57–64. https://doi.org/10.1016/S0014-2999(03)01676-5.

Pop-Busui, R., Lu, J., Mori Brooks, M., Albert, S., Althouse, A. D., Escobedo, J., et al. (2013). Impact of glycemic control strategies on the progression of diabetic peripheral neuropathy in the bypass angioplasty revascularization investigation 2 diabetes (BARI 2D) Cohort. *Diabetes Care*, *36*, 3208–3215. https://doi.org/10.2337/dc13-0012.

Raffa, R. B., & Codd, E. E. (1994). Lack of glibenclamide or TEA affinity for opioid receptors : Further evidence for in vivo modulation of antinociception at K + channels. *Brain Research*, *625*, 146–148. https://doi.org/10.1016/0006-8993(94)90217-8.

Rios-Santos, F., Alves-Filho, J. C., Souto, F. O., Spiller, F., Freitas, A., Lotufo, C. M. C., et al. (2007). Down-regulation of CXCR2 on neutrophils in severe sepsis is mediated by inducible nitric oxide synthase-derived nitric oxide. *American Journal of Respiratory and Critical Care Medicine*, *175*(5), 490–497.

Roane, D. S., & Boyd, N. E. (1993). Reduction of food intake and morphine analgesia by central glybenclamide. *Pharmacology, Biochemistry and Behavior*, *46*(1), 205–207. https://doi.org/10.1016/0091-3057(93)90341-P.

Rodrigues, A. R. A., & Duarte, I. D. G. (2000). The peripheral antinociceptive effect induced by morphine is associated with ATP-sensitive K+ channels. *British Journal of Pharmacology*, *129*(1), 110–114. https://doi.org/10.1038/sj.bjp.0703038.

Romanovsky, D., Hastings, S. L., Stimers, J. R., & Dobretsov, M. (2004). Relevance of hyperglycemia to early mechanical hyperalgesia in streptozotocin-induced diabetes. *Journal of the Peripheral Nervous System*, *9*(2), 62–69. https://doi.org/10.1111/j.1085-9489.2004.009204.x.

Sebille, S., Tullio, P.d., Boverie, S., Antoine, M. H., Lebrun, P., & Pirotte, B. (2012). Recent developments in the chemistry of potassium channel activators: The cromakalim analogs. *Current Medicinal Chemistry*, *11*(9), 1213–1222. https://doi.org/10.2174/0929867043365378.

Seino, S., & Miki, T. (2003). Physiological and pathophysiological roles of ATP-sensitive K+ channels. *Progress in Biophysics and Molecular Biology*, *81*(2), 133–176. https://doi.org/10.1016/S0079-6107(02)00053-6.

Shimizu, K., Lacza, Z., Rajapakse, N., Horiguchi, T., Snipes, J., & Busija, D. W. (2002). MitoKATP opener, diazoxide, reduces neuronal damage after middle cerebral artery occlusion in the rat. *American Journal of Physiology - Heart and Circulatory Physiology*, *283*(3), 52–53. https://doi.org/10.1152/ajpheart.00054.2002.

Soares, A. C., & Duarte, I. D. G. (2001). Dibutyryl-cyclic GMP induces peripheral antinociception via activation of ATP-sensitive K+ channels in the rat PGE2-induced hyperalgesic paw. *British Journal of Pharmacology*, *134*(1), 127–131. https://doi.org/10.1038/sj.bjp.0704224.

Stephens, L., Smrcka, A., Cooke, F. T., Jackson, T. R., Sternweis, P. C., & Hawkins, P. T. (1994). A novel phosphoinositide 3 kinase activity in myeloid-derived cells is activated by G protein βγ subunits. *Cell*, *77*(1), 83–93. https://doi.org/10.1016/0092-8674(94)90237-2.

Szeto, V., Chen, N. H., Feng, Z. P., & Sun, H. S. (2018). The role of KATP channels in cerebral ischemic stroke and diabetes. *Acta Pharmacologica Sinica*, *39*(5), 683–694. Nature Publishing Group https://doi.org/10.1038/aps.2018.10.

Teixeira, J. M., dos Santos, G. G., Neves, A. F., Athie, M. C. P., Bonet, I. J. M., Nishijima, C. M., et al. (2019). Diabetes-induced neuropathic mechanical hyperalgesia depends on P2X4 receptor activation in dorsal root ganglia. *Neuroscience*, *398*, 158–170. https://doi.org/10.1016/j.neuroscience.2018.12.003.

Teshima, Y., Akao, M., Li, R. A., Chong, T. H., Baumgartner, W. A., Johnston, M. V., et al. (2003). Mitochondrial ATP-sensitive potassium channel activation protects cerebellar granule neurons from apoptosis induced by oxidative stress. *Stroke*, *34*(7), 1796–1802. https://doi.org/10.1161/01.STR.0000077017.60947.AE.

Tinker, A., Aziz, Q., Li, Y., & Specterman, M. (2018). ATP-sensitive potassium channels and their physiological and pathophysiological roles. *Comprehensive Physiology*, *8*(4), 1463–1511. https://doi.org/10.1002/cphy.c170048.

Velasco, M., Díaz-García, C. M., Larqúe, C., & Hiriart, M. (2016). Modulation of ionic channels and insulin secretion by drugs and hormones in pancreatic beta cells. *Molecular Pharmacology*, *90*(3), 341–357. American Society for Pharmacology and Experimental Therapy https://doi.org/10.1124/mol.116.103861.

Welch, S. P., & Dunlow, L. D. (1993). Antinociceptive activity of intrathecally administered potassium channel openers and opioid agonists: A common mechanism of action? *Journal of Pharmacology and Experimental Therapeutics*, *267*(1), 390–399.

Wild, K. D., Vanderah, T., Mosberg, H. I., & Porreca, F. (1991). Opioid δ receptor subtypes are associated with different potassium channels. *European Journal of Pharmacology*, *193*(1), 135–136. https://doi.org/10.1016/0014-2999(91)90215-C.

Zhou, S.-Y., Lu, Y., Song, I., & Owyang, C. (2011). Inhibition of gastric motility by hyperglycemia is mediated by nodose ganglia K ATP channels. *American Journal of Physiology. Gastrointestinal and Liver Physiology*, *300*(3), 394–400. https://doi.org/10.1152/ajpgi.00493.2010.

Zhu, X., Liu, J., Gao, Y., Cao, S., & Shen, S. (2015). ATP-sensitive potassium channels alleviate postoperative pain through JNK-dependent MCP-1 expression in spinal cord. *International Journal of Molecular Medicine*, *35*(5), 1257–1265. https://doi.org/10.3892/ijmm.2015.2143.

Zoga, V., Kawano, T., Liang, M. Y., Bienengraeber, M., Weihrauch, D., McCallum, B., et al. (2010). KATPchannel subunits in rat dorsal root ganglia: Alterations by painful axotomy. *Molecular Pain*, *6*, 6. https://doi.org/10.1186/1744-8069-6-6.

Chapter 14

Astrocyte-neuron lactate shuttle and pain

Keisuke Miyamoto and Masahiro Ohsawa

Department of Neuropharmacology, Graduate School of Pharmaceutical Sciences, Nagoya City University, Nagoya, Japan

Abbreviations

ACC	anterior cingulate cortex
ANLS	astrocyte-neuron lactate shuttle
ATP	adenosine triphosphate
BDNF	brain-derived neurotrophic factor
BLA	basolateral amygdala
CNS	central nervous system
DREADDs	designer receptors exclusively activated by designer drugs
Fbp2	fructose 1, 6-bisphosphatase 2
L-LDH	L-lactate dehydrogenase
LTM	long-term memory
LTP	long-term potentiation
MCTs	monocarboxylate transporters
NTS	nucleus of the solitary tract
PACAP	pituitary adenylate cyclase activation polypeptide
SDH	spinal dorsal horn

Introduction

Peripheral nociceptive signals are processed and integrated at spinal dorsal horn (SDH), then transmitted to the brain. These pain-transmission pathways are regulated by synaptic transmission, but many nonneuronal cells, such as glial cells or immune cells, are detected as a modulator of neuronal function. In particular, morphological or functional changes of glial cells are related to the enhancement of nociceptive signals in pathophysiological pain. Especially in microglia under the neuropathic pain condition, they are activated in SDH and release inflammatory cytokine or adenosine triphosphate (ATP) to enhance the neurotransmission (Tsuda et al., 2003, 2008). Other study shows brain-derived neurotrophic factor (BDNF) released from SDH microglia contributes to neuropathic pain transmission (Coull et al., 2005). In anterior cingulate cortex (ACC), activated microglia is involved in the development of hyperalgesia by the modulation of GluR1 AMPA receptor function (Miyamoto, Kume, & Ohsawa, 2017). By contrast, the detailed function of astrocytes in the nociceptive transmission remains unknown. However, the activation of SDH astrocytes is associated to be a trigger for hyperalgesia and allodynia in neuropathic pain. Several studies revealed that pharmacological manipulation of spinal astrocytes regulates the neuropathic pain-derived hyperalgesia (Chiang et al., 2007; Ji, Kawasaki, Zhuang, Wen, & Decosterd, 2006; Meller, Dykstra, Grzybycki, Murphy, & Gebhart, 1994; Tsuda et al., 2011). Repeated pharmacological inhibition of astrocytes attenuate mechanical hyperalgesia in neuropathic pain (Raghavendra, Tanga, Rutkowski, & DeLeo, 2003). Nam et al. demonstrated that optogenetic stimulation of spinal astrocytes leads to pain hypersensitivity in mice (Nam et al., 2016). Kohro et al. show that continuous activation of L4 SDH astrocytes is important for maintaining and inducing tactile allodynia (Kohro et al., 2015).

Astrocytes have been considered not to participate in brain function because they do not respond to electric stimulation. However, Ca^{2+} imaging study revealed that astrocytes indicate the Ca^{2+} oscillation in response to various neurotransmitters and they are able to transmit these Ca^{2+} signals to adjacent resting astrocytes as intercellular Ca^{2+} waves (Cornell-Bell & Finkbeiner, 1991). Astrocytes have many process covering synapses and expresses various plasma membrane transporters or receptors for most of the neurotransmitters. These results suggest that astrocytes play a role in CNS information processing. As astrocytes are also in contact with blood vessel, Ca^{2+} wave propagation affects the astrocytes-blood vessel

The Neurobiology, Physiology, and Psychology of Pain. https://doi.org/10.1016/B978-0-12-820589-1.00014-2

interface functions. Increased $[Ca^{2+}]_i$ in astrocyte modifies the diameter of blood vessels (Mulligan & MacVicar, 2004). Dilated blood vessel, that is the increase of blood flow, supports the supply of sufficient glucose to appropriate region. Astrocytes itself also release gliotransmitters, such as glutamate, ATP (Zhang et al., 2003), or D-serine (Henneberger, Papouin, Oliet, & Rusakov, 2010; Zhuang et al., 2010), to activate several receptors and modulate the synaptic transmission (Araque et al., 2014). In particular, D-serine is known as an agonist for NMDA receptor which is involved in synaptic plasticity. These gliotransmitters are suggested to be released in response to the intracellular Ca^{2+} activity that arise from synaptic activity from physiological sensory and motor inputs (Hamilton & Attwell, 2010). Astrocytes also release L-lactate as a product of glycolysis or glycogenolysis. L-Lactate is metabolized to produce ATP through Krebs cycle, and both astrocytes and neurons express the monocarboxylate transporters (MCTs) to transport L-lactate. Magistretti and Pellerin, thus, suggested one hypothesis, astrocyte-neuron lactate shuttle (ANLS), that L-lactate might be an energy source of neuronal activity (Pellerin & Magistretti, 1994).

ANLS hypothesis postulates that astrocytes mainly metabolize glucose to L-lactate during neuronal activation, and the L-lactate is transported to neurons for maintaining and facilitating neurotransmission (Magistretti & Allaman, 2015; Magistretti & Pellerin, 1996; Weber & Barros, 2015). Based on the fact that human brain receives 25% of circulating glucose under physiological conditions (Pellerin, 2010), ATP derived from glucose has been typically considered as the only energy for brain activity. However, Fox et al. revealed that oxygen consumption in visual stimulation-induced neuronal activity shows 5% increase even though the glucose consumption is 50% increase (Fox, Raichle, Mintun, & Dence, 1988). This result suggests that increased neuronal activity promotes anaerobic glycolysis. Other study revealed that visual stimulation transiently increases L-lactate level in human visual cortex (Prichard et al., 1991; Sappey-Marinier et al., 1992). In addition, Magistretti showed the metabolic coupling between neurons and astrocytes (Pierre J. Magistretti, 2011). Astrocytes generally incorporate more glucose than their energy requirements, suggesting that astrocytic glycolysis occurs to maintain higher extracellular L-lactate level for energy requirement of neurons (Mason, 2017). Indeed, the basal L-lactate level of astrocytes is significantly higher than neurons (Mächler et al., 2016). L-Lactate is transported from astrocyte to neurons through MCTs and converted to pyruvate by L-lactate dehydrogenase (L-LDH) as a source of oxidative phosphorylation-derived ATP. Thus, the ANLS is possible to contribute to the persistent synaptic transmission. Illustrated summary of ANLS is shown in Fig. 1. However, it is still debatable whether neurons use L-lactate in preference to glucose as energy source of neuronal activity (Bak & Walls, 2018). Neurons also express glucose transporters and metabolize glucose in an activity-dependent manner (Díaz-García et al., 2017; Lundgaard et al., 2015). Furthermore, the mechanism of the preferential synthesis of L-lactate in astrocytes and consumption in neurons, which is essential for

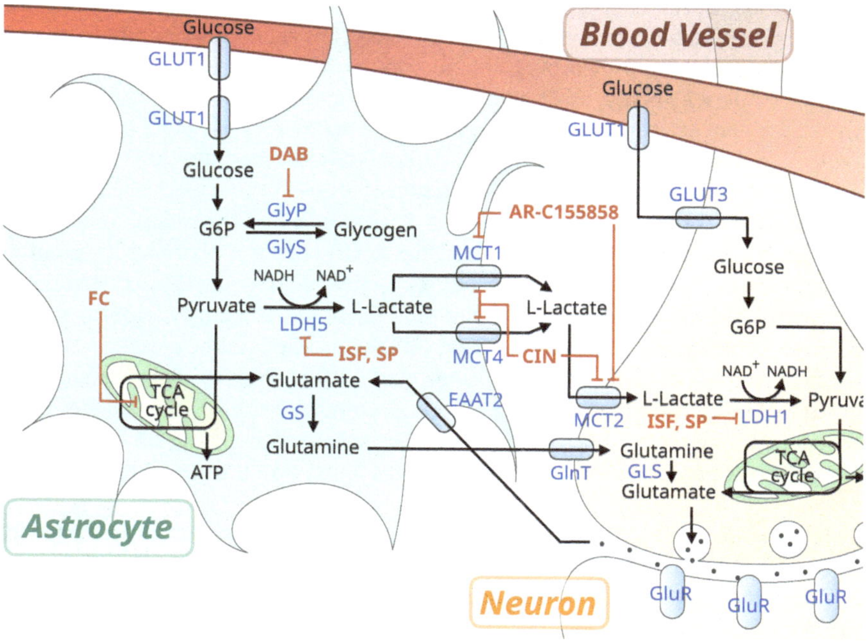

FIG. 1 Intracellular mechanism of L-lactate metabolism and inhibitors of L-lactate shuttle. L-Lactate transported from astrocytes to neurons maintains neurotransmission through the production of ATP or glutamate. *G6P*, glucose-6-phosphate; *GlyP*, glycogen phosphorylase; *GlyS*, glycogen synthase; *GS*, glutamine synthetase; *GLS*, glutaminase; *DAB*, 1,4-dideoxy-1,4-imino-D-arabinitol; *FC*, fluorocitrate; *ISF*, isosafrole; *SP*, stiripentol; *CIN*, α-cyano-4-hydroxycinnamate; *GLUT*, glucose transporter; *EAAT*, excitatory amino acid transporter; *GlnT*, glutamine transporter; *GluR*, glutamate receptor.

demonstration of ANLS hypothesis, is explained by the argument that the kinetic parameters of L-LDH are different in each cell. However, Bak and Schousboe pointed out that distribution of LDH isoenzymes does not form L-lactate flow between the two cellular compartments (Bak & Schousboe, 2017). Based on these discussions, it is premature to define the L-lactate role as constant energy source, but several studies revealed the L-lactate effect of an opportunistic substrate. The L-lactate shuttle is reported to be involved in several functions of central nervous system (CNS), such as long-term memory formation and the maintenance of long-term potentiation (LTP) (Suzuki et al., 2011), dependence formation for cocaine (Boury-Jamot et al., 2016), the basolateral amygdala (BLA)-ACC synchrony during decision-making (Wang et al., 2017) (Table. 1). Thus, it is suggested that the ANLS contributes to the neuronal plasticity in various brain regions.

TABLE 1 Functions regulated by astrocyte-neuron lactate shuttle except for energy supply.

Region	Animal	Effect	References
Hippocampus	Rat	Long-term memory formation	Suzuki et al. (2011)
Hippocampus	Rat	Short-term memory formation	Newman, Korol, and Gold (2011)
Sensory-motor cortex, cell culture	Mouse	Potentiation of NMDA-receptor inward, increase of intracellular calcium, modulation of neuronal redox state	Yang et al. (2014)
Schwann cells, cell culture	Mouse, rat	Myelination	Domènech-Estévez et al. (2015)
Hippocampus	Mouse	Cellular sodium regulation	Karus, Ziemens, and Rose (2015)
Basolateral amygdala	Rat	Reconsolidation of cocaine memory	Zhang et al. (2016)
Basolateral amygdala	Rat	Reconsolidation of cocaine memory	Boury-Jamot et al. (2016)
CA3 of hippocampus	Rat	Ion homeostasis, synaptic signaling	Angamo, Rösner, Liotta, Kovács, and Heinemann (2016)
Lateral hypothalamic area	Mouse	Maintaining normal daily cycle of wakefulness	Clasadonte, Scemes, Wang, Boison, and Haydon (2017)
Anterior cingulate cortex	Rat	Neuronal synchrony, decision making	Wang et al. (2017)
Cell culture, spinal cord	Mouse	Enhancement of nociceptive transmission	Kambe, Yokai, Takasaki, and Takashi Kurihara (2018)
Cell culture	Rat	Neurite outgrowth	Chen et al. (2018)
Spinal cord	Mouse	Enhancement of nociceptive transmission	Miyamoto, Ishikura, Kume, and Ohsawa (2019)
Neonate brain	Rat	Neuroprotection	Roumes et al. (2021)
Somatosensory cortex, CA1 of hippocampus	Mouse	LTP formation	Murphy-Royal et al. (2020)
Cell culture	Mouse	LTP formation	Duda et al. (2020)
Dorsomedial nucleus of the solitary tract	Rat	Maintaining synaptic transmission	Nagase, Takahashi, Watabe, Kubo, and Kato (2014)
Hippocampus	Rat	Long-term memory formation	Gao et al. (2016)
Retina, cell culture	Drosophila, mouse	Lipid synthesis and lipid droplet accumulation	Liu, MacKenzie, Putluri, Maletić-Savatić, and Bellen (2017)
Frontal cortex, hippocampus	Mouse	Spatial memory acquisition	Harris et al. (2019)
cell culture	Mouse	Operation of heart	Gizak, McCubrey, and Rakus (2020)

As a function of activated SDH astrocytes in neuropathic pain, they release approximately double volume of L-lactate and thus enhance the nociceptive transmission through the neuronal L-lactate metabolism (Miyamoto et al., 2019). This report probes that the ANLS also exists in the spinal cord for regulating neuronal plasticity. Recently, pituitary adenylate cyclase activation polypeptide (PACAP)/PAC1 receptor signaling is reported as the trigger of ANLS activation (Kambe et al., 2018). Intrathecal administration of PACAP induces robust activation of spinal dorsal horn astrocytes and long-lasting mechanical allodynia, which is attenuated by the inhibition of astrocytes or MCTs inhibitor. These results also indicate that the enhancement of ANLS triggered by activation of astrocytes is crucial for neuropathic pain onset. Thus, the normalization of ANLS might be a novel target for the treatment of neuropathic pain.

Applications to other areas

This chapter has reviewed the relationship between ANLS and neuropathic pain. It has also described in detail intracellular mechanisms of spinal dorsal horn nociceptive sensitization triggered by the ANLS. The following section focuses on other contribution of the ANLS to neuronal function. Though it is still too early to conclude that the ANLS is universal function, many reports reveal the present of ANLS system in particular brain regions. Suzuki et al. showed that learning leads to the activation of the ANLS and it is essential for long-term memory (LTM) formation and the maintenance of long-term potentiation (LTP) of synaptic strength in hippocampus (Suzuki et al., 2011). LTM is generally believed to be mediated by late-phase LTP (Adams & Dudek, 2005; Ageta et al., 2010) and accompanied by synaptic changes and increase of spine density (Radwanska et al., 2011). Vezzoli et al. demonstrated that the selective inhibition of the astrocytic glycogen metabolism, that is the attenuation of L-lactate production, impaired the LTM-induced increase of spine volume and postsynaptic density surface area (Vezzoli et al., 2020). Recently, Duda et al. demonstrated that fructose 1,6-bisphosphatase 2 (Fbp2), a gluconeogenic enzyme, plays a crucial role in the induction and maintenance of LTP (Duda et al., 2020). As astrocyte-derived L-lactate reduces NAD^+/NADH ratio and increases Fbp2 dimer/tetramer ration, the ANLS might stimulate LTP formation through modifying Fbp2 activity.

All aforementioned ANLS functions are involved in hippocampal neuronal plasticity. However, ANLS is also verified in BLA, neocortex, and nucleus of the solitary tract (NTS). Boury-Jamot et al. demonstrated that the inhibition of astrocytic glycogen metabolism in BLA leads to the deletion of preference for cocaine (Boury-Jamot et al., 2016). Wang et al. indicated that the ANLS facilitates the BLA-anterior cingulate cortex (ACC) synchrony during decision-making (Wang et al., 2017). Murphy-Royal et al. demonstrated that acute stress decreases the astrocyte gap junction and reduces the LTP in neocortex (Murphy-Royal et al., 2020). As the LTP is recovered to nonstressed levels by the supply of L-lactate to astrocytes, the authors concluded that stress-impaired neuronal plasticity is linked with the ANLS dysfunction, induced by the decay in astrocytic gap junction. In dorsomedial NTS, ANLS is verified on the intracellular mechanism level. L-Lactate, transferred from astrocytes to neurons via MCTs, maintains excitatory synaptic transmission through the ATP production (Nagase et al., 2014).

Many previous reports have proposed that synaptic plasticity in CNS can function as a key mechanism perceiving pathological pain (Luo, Kuner, & Kuner, 2014). As described in this section, the energy metabolism depending on the ANLS regulates synaptic plasticity in many regions of CNS. These findings suggest that the ANLS might be an attractive target to treat syndromes related to synaptic plasticity including pathological pain.

Other agents of interest

This section will discuss about the detailed mechanisms of the spinal ANLS regulating neurotransmission based on our latest study (Miyamoto et al., 2019). Local activation of SDH astrocyte using designer receptors exclusively activated by designer drugs (DREADDs) system decreased the mechanical pain threshold and increased the extracellular L-lactate level from 1.0–1.2 mM to 2.0 mM (Fig. 2A–D). This L-lactate level, 2.0 mM, is comparable with that in hippocampus or striatum under supraphysiological condition, such as hypoxia or hyperglycemia (Mosienko, Teschemacher, & Kasparov, 2015). This decreased mechanical threshold derived from activation of SDH astrocytes was completely attenuated by inhibition of L-lactate transporter, MCTs (Fig. 2E). These results indicate that the hypersensitivity after the activation of astrocytes is induced by the excessive production of L-lactate. Moreover, intrathecal administration of L-lactate decreased mechanical threshold, which was attenuated by MCTs inhibitor (Fig. 3A and B). Electrophysiological study also revealed that exogenous L-lactate treatment enhances the spinal dorsal horn neuronal activity to weak filament stimulation to plantar surface of the hind paw without affecting the spontaneous firing or strong filament stimulation evoked firing. Weak filament stimulation means insensitive treatment and strong stimulation is invasive for intact mice. It is important to note that knockdown of MCT2, mainly expressed in neurons, attenuated the L-lactate-induced mechanical hyperalgesia, while

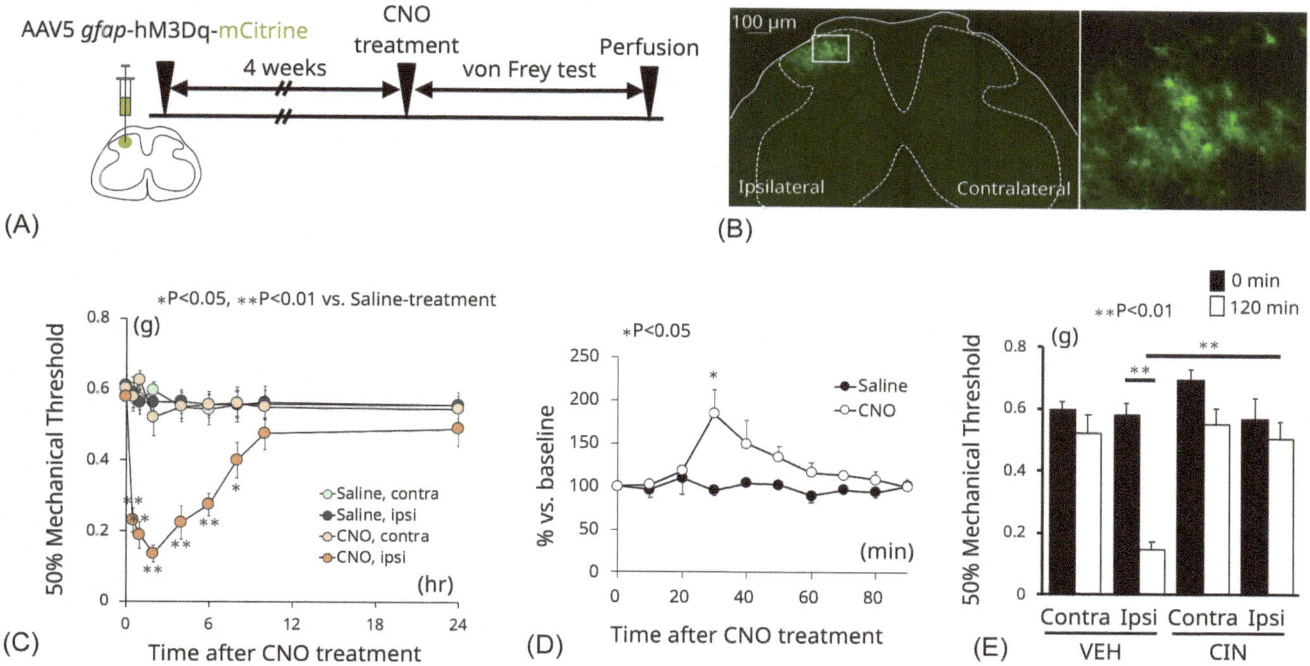

FIG. 2 Effects of the specific activation of SDH astrocytes. (A) A method for SDH astrocytes-specific activation. (B) Immunofluorescence of spinal cord after microinjection of AAV5-*gfap*-EGFP. (C) Effects of the intraperitoneal CNO injection on the mechanical nociceptive threshold. (D) Extracellular L-lactate level after CNO injection. (E) Effects of the MCTs inhibitor CIN on mechanical hypersensitivity following SDH astrocytes-specific activation. Each point and column represent the mean and SEM.

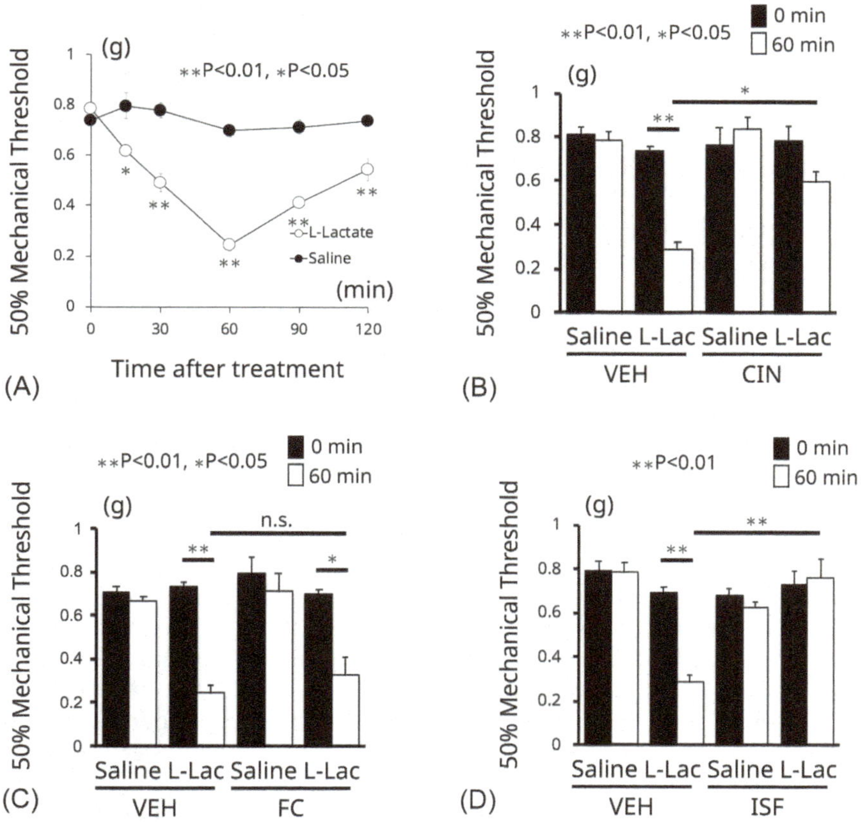

FIG. 3 Effects of intrathecal L-lactate injection and inhibitors related to ANLS. (A) Time course of the effects of intrathecally administered L-lactate on mechanical nociceptive threshold. (B–D) Effects of CIN (B), FC (C), and ISF (D) on L-lactate-induced mechanical hypersensitivity. Each point and column represent the mean and SEM.

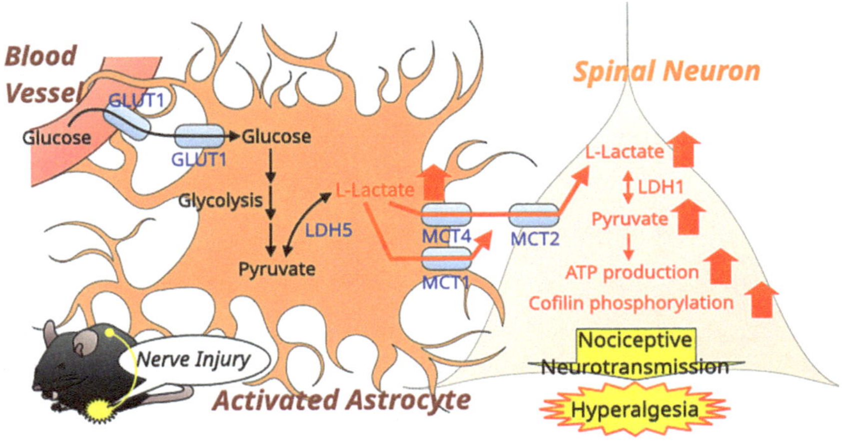

FIG. 4 Summary of spinal ANLS mechanism enhancing nociceptive transmission. Activated SDH astrocytes supply excessive L-lactate to neurons and then increase ATP production and cofilin phosphorylation.

the knockdown of MCT1 or MCT4, mainly expressed in astrocytes, did not affect the mechanical threshold. Pharmacological inhibition of astrocytic metabolism also did not change the mechanical threshold (Fig. 3C). Thus, L-lactate is required to be taken in neurons to enhance the nociceptive transmission.

The intracellular mechanisms of L-lactate-induced mechanical hypersensitivity are illustrated in Fig. 3D. Pharmacological inhibition of L-LDH, converting L-lactate to pyruvate, inhibited L-lactate-induced mechanical hypersensitivity. Synthesized pyruvate supplies ATP as an energy source to neurons through the Krebs cycle and sensitizes nociceptive transmission. In addition, intrathecal administration of L-lactate also increased the phosphorylation of ADF/cofilin, the major actin depolymerizing factor, at Ser 3. ADF/cofilin is inactivated by the phosphorylation of Ser 3, which promote actin polymerization and filament loading (Niwa, Nagata-Ohashi, Takeichi, Mizuno, & Uemura, 2002; Thirone et al., 2009). ADF/cofilin phosphorylation negatively regulates the surface AMPA receptor in the spine (Gu et al., 2010; Yuen, Liu, Kafri, van Praag, & Yan, 2010). Park et al. showed that the proportion of each subtype of spinal dorsal horn AMPA receptor in crude cytosolic fraction and membrane fraction are modified in persistent inflammatory pain model rats (Park et al., 2008). Hence, the L-lactate-induced phosphorylation of cofilin might also be involved in spinal nociceptive sensitization. In conclusion, the enhanced ANLS is supposed to sensitize nociceptive transmission through the excessive production of ATP and inactivation of cofilin (Fig. 4).

Mini-dictionary of terms

L-Lactate: One of the carboxylic acids which is produced by glycolysis. In astrocytes or neurons, L-lactate is metabolized to pyruvate and then used as a material for Krebs cycle or fatty acid synthesis. Most of L-lactate is transported via MCTs. The optical isomer, D-lactate, is also biosynthesized in mammalian cells, but their amounts are drastically tiny compared with that of L-lactate.

Monocarboxylate transporters (MCTs): A transporter which can take monovalent carboxylic acid, such as lactate, ketone body, or acetate, in cells as an energy. Fourteen subtypes have been detected on various cell types (Halestrap, 2012).

Lactate dehydrogenase (LDH): An enzyme which converts lactate to pyruvate and back through the mutual conversion of NAD^+ to NADH. LDH is isomer specific so that metabolism of L-lactate requires L-LDH and D-lactate requires D-LDH. As mammalian cells only contain L-LDH, they utilize the mitochondrial D-2-hydroxyacid-dehydrogenase to catalyze D-lactate to pyruvate (Ewaschuk, Naylor, & Zello, 2005).

DREADD system: Widely used chemogenetic tools for cell type specific regulation using nonendogenous designer receptors, DREADDs, and its ligands with high potency and affinity (Armbruster, Li, Pausch, Herlitze, & Roth, 2007). Some DREADDs are activated and others are inhibited by ligands.

Cofilin: Cofilin binds globular actin (G-actin) and filament actin (F-actin), then cleaves and disassembles F-actin. Cofilin is classified in nonmuscle cofilin, muscle cofilin, and actin depolymerizing factor (ADF)/cofilin. In hippocampal neurons or SDH neurons, ANLS activity regulates the phosphorylation of ADF/cofilin at Ser 3.

Key facts

Key facts of astrocyte-neuron lactate shuttle

- Astrocyte-neuron lactate shuttle (ANLS) hypothesis postulates that astrocytes produce L-lactate through glycolysis during neuronal activation and that the L-lactate is shuttled to neurons for maintaining and facilitating neurotransmission.
- ANLS is associated with several high-order brain functions which are related to neuronal plasticity.
- L-Lactate produced in astrocytes is transported through monocarboxylate transporters (MCTs).
- ANLS contributes the neuronal ATP production through the oxidative phosphorylation.
- Enhancement of ANLS sensitizes nociceptive transmission in spinal dorsal horn.

Summary points

- This chapter focuses on the relationship between neuropathic pain and astrocyte-neuron lactate shuttle (ANLS).
- Functional changes of spinal astrocytes observed in neuropathic pain are considered to play a role in nociceptive hypersensitivity.
- Local activation of spinal astrocytes produces excessive L-lactate and causes transient mechanical hypersensitivity which is attenuated by the inhibition of spinal ANLS.
- Excessive L-lactate supply increases neuronal ATP level, which facilitates the nociceptive neurotransmission.
- Inhibition of spinal ANLS ameliorates hyperalgesia in neuropathic pain.
- Thus, enhancement of spinal ANLS maintains the hyperalgesia in neuropathic pain through the L-lactate overload.

References

Adams, J. P., & Dudek, S. M. (2005). Late-phase long-term potentiation: Getting to the nucleus. *Nature Reviews Neuroscience*, *6*(9), 737–743. https://doi.org/10.1038/nrn1749.

Ageta, H., Ikegami, S., Miura, M., Masuda, M., Migishima, R., Hino, T., … Inokuchi, K. (2010). Activin plays a key role in the maintenance of long-term memory and late-LTP. *Learning and Memory*, *17*(4), 176–185. https://doi.org/10.1101/lm.16659010.

Angamo, E. A., Rösner, J., Liotta, A., Kovács, R., & Heinemann, U. (2016). A neuronal lactate uptake inhibitor slows recovery of extracellular ion concentration changes in the hippocampal CA3 region by affecting energy metabolism. *Journal of Neurophysiology*, *116*(5), 2420–2430. https://doi.org/10.1152/jn.00327.2016.

Araque, A., Carmignoto, G., Haydon, P. G., Oliet, S. H. R., Robitaille, R., & Volterra, A. (2014). Gliotransmitters travel in time and space. *Neuron*, *81*(4), 728–739. https://doi.org/10.1016/j.neuron.2014.02.007.

Armbruster, B. N., Li, X., Pausch, M. H., Herlitze, S., & Roth, B. L. (2007). Evolving the lock to fit the key to create a family of G protein-coupled receptors potently activated by an inert ligand. *Proceedings of the National Academy of Sciences of the United States of America*, *104*(12), 5163–5168. https://doi.org/10.1073/pnas.0700293104.

Bak, L. K., & Schousboe, A. (2017). Misconceptions regarding basic thermodynamics and enzyme kinetics have led to erroneous conclusions regarding the metabolic importance of lactate dehydrogenase isoenzyme expression. *Journal of Neuroscience Research*, *95*(11), 2098–2102. https://doi.org/10.1002/jnr.23994.

Bak, L. K., & Walls, A. B. (2018). CrossTalk opposing view: Lack of evidence supporting an astrocyte-to-neuron lactate shuttle coupling neuronal activity to glucose utilisation in the brain. *Journal of Physiology*, *596*(3), 351–353. https://doi.org/10.1113/JP274945.

Boury-Jamot, B., Carrard, A., Martin, J. L., Halfon, O., Magistretti, P. J., & Boutrel, B. (2016). Disrupting astrocyte-neuron lactate transfer persistently reduces conditioned responses to cocaine. *Molecular Psychiatry*, *21*(8), 1070–1076. https://doi.org/10.1038/mp.2015.157.

Chen, K., Cheng, P., Wang, H., Gao, S., Li, X., Deng, Z., … Sun, X. (2018). Lactate transport facilitates neurite outgrowth. *Bioscience Reports*, *38*(5). https://doi.org/10.1042/BSR20180157.

Chiang, C. Y., Wang, J., Xie, Y. F., Zhang, S., Hu, J. W., Dostrovsky, J. O., & Sessle, B. J. (2007). Astroglial glutamate-glutamine shuttle is involved in central sensitization of nociceptive neurons in rat medullary dorsal horn. *Journal of Neuroscience*, *27*(34), 9068–9076. https://doi.org/10.1523/JNEUROSCI.2260-07.2007.

Clasadonte, J., Scemes, E., Wang, Z., Boison, D., & Haydon, P. G. (2017). Connexin 43-mediated astroglial metabolic networks contribute to the regulation of the sleep-wake cycle. *Neuron*, *95*(6), 1365–1380.e5. https://doi.org/10.1016/j.neuron.2017.08.022.

Cornell-Bell, A. H., & Finkbeiner, S. M. (1991). Ca2+ waves in astrocytes. *Cell Calcium*, *12*(2–3), 185–204. https://doi.org/10.1016/0143-4160(91)90020-F.

Coull, J. A. M., Beggs, S., Boudreau, D., Boivin, D., Tsuda, M., Inoue, K., … De Koninck, Y. (2005). BDNF from microglia causes the shift in neuronal anion gradient underlying neuropathic pain. *Nature*, *438*(7070), 1017–1021. https://doi.org/10.1038/nature04223.

Díaz-García, C. M., Mongeon, R., Lahmann, C., Koveal, D., Zucker, H., & Yellen, G. (2017). Neuronal stimulation triggers neuronal glycolysis and not lactate uptake. *Cell Metabolism*, *26*(2), 361–374.e4. https://doi.org/10.1016/j.cmet.2017.06.021.

Domènech-Estévez, E., Baloui, H., Repond, C., Rosafio, K., Médard, J.-J., Tricaud, N., … Chrast, R. (2015). Distribution of monocarboxylate transporters in the peripheral nervous system suggests putative roles in lactate shuttling and myelination. *The Journal of Neuroscience*, *35*(10), 4151–4156. https://doi.org/10.1523/JNEUROSCI.3534-14.2015.

Duda, P., Wójtowicz, T., Janczara, J., Krowarsch, D., Czyrek, A., Gizak, A., & Rakus, D. (2020). Fructose 1,6-bisphosphatase 2 plays a crucial role in the induction and maintenance of long-term potentiation. *Cell*, *9*(6), 1–22. https://doi.org/10.3390/cells9061375.

Ewaschuk, J. B., Naylor, J. M., & Zello, G. A. (2005). D-lactate in human and ruminant metabolism. *The Journal of Nutrition*, *135*(7), 1619–1625. https://doi.org/10.1093/jn/135.7.1619.

Fox, P. T., Raichle, M. E., Mintun, M. A., & Dence, C. (1988). Nonoxidative glucose consumption during focal physiologic neural activity. *Science*, *241*(4864), 462–464. https://doi.org/10.1126/science.3260686.

Gao, V., Suzuki, A., Magistretti, P. J., Lengacher, S., Pollonini, G., Steinman, M. Q., & Alberini, C. M. (2016). Astrocytic β2-adrenergic receptors mediate hippocampal long-term memory consolidation. *Proceedings of the National Academy of Sciences of the United States of America*, *113*(30), 8526–8531. https://doi.org/10.1073/pnas.1605063113.

Gizak, A., McCubrey, J. A., & Rakus, D. (2020). Cell-to-cell lactate shuttle operates in heart and is important in age-related heart failure. *Aging (Albany NY)*, *12*(4), 3388–3406. https://doi.org/10.18632/aging.102818.

Gu, J., Lee, C. W., Fan, Y., Komlos, D., Tang, X., Sun, C., … Zheng, J. Q. (2010). ADF/cofilin-mediated actin dynamics regulate AMPA receptor trafficking during synaptic plasticity. *Nature Neuroscience*, *13*(10), 1208–1215. JOUR https://doi.org/10.1038/nn.2634.

Halestrap, A. P. (2012). The monocarboxylate transporter family-structure and functional characterization. *IUBMB Life*, *64*(1), 1–9. https://doi.org/10.1002/iub.573.

Hamilton, N. B., & Attwell, D. (2010). Do astrocytes really exocytose neurotransmitters? *Nature Reviews Neuroscience*, *11*(4), 227–238. https://doi.org/10.1038/nrn2803.

Harris, R. A., Lone, A., Lim, H., Martinez, F., Frame, A. K., Scholl, T. J., & Cumming, R. C. (2019). Aerobic glycolysis is required for spatial memory acquisition but not memory retrieval in mice. *eNeuro*, *6*(1). https://doi.org/10.1523/ENEURO.0389-18.2019.

Henneberger, C., Papouin, T., Oliet, S. H. R., & Rusakov, D. A. (2010). Long-term potentiation depends on release of d-serine from astrocytes. *Nature*, *463*(7278), 232–236. https://doi.org/10.1038/nature08673.

Ji, R. R., Kawasaki, Y., Zhuang, Z. Y., Wen, Y. R., & Decosterd, I. (2006). Possible role of spinal astrocytes in maintaining chronic pain sensitization: Review of current evidence with focus on bFGF/JNK pathway. *Neuron Glia Biology*, *2*(4), 259–269. https://doi.org/10.1017/S1740925X07000403.

Kambe, Y., Yokai, M., Takasaki, I., & Takashi Kurihara, A. M. (2018). Spinal astrocyte-neuron lactate shuttle contributes to the PACAP/PAC1 receptor-induced nociceptive behaviors. *BioRxiv*. https://doi.org/10.1101/317156.

Karus, C., Ziemens, D., & Rose, C. R. (2015). Lactate rescues neuronal sodium homeostasis during impaired energy metabolism. *Channels (Austin)*, *9*(4), 200–208. https://doi.org/10.1080/19336950.2015.1050163.

Kohro, Y., Sakaguchi, E., Tashima, R., Tozaki-Saitoh, H., Okano, H., Inoue, K., & Tsuda, M. (2015). A new minimally-invasive method for microinjection into the mouse spinal dorsal horn. *Scientific Reports*, *5*, 1–11. https://doi.org/10.1038/srep14306.

Liu, L., MacKenzie, K. R., Putluri, N., Maletić-Savatić, M., & Bellen, H. J. (2017). The glia-neuron lactate shuttle and elevated ROS promote lipid synthesis in neurons and lipid droplet accumulation in glia via APOE/D. *Cell Metabolism*, *26*(5), 719–737.e6. https://doi.org/10.1016/j.cmet.2017.08.024.

Lundgaard, I., Li, B., Xie, L., Kang, H., Sanggaard, S., Haswell, J. D. R., … Nedergaard, M. (2015). Direct neuronal glucose uptake heralds activity-dependent increases in cerebral metabolism. *Nature Communications*, *6*. https://doi.org/10.1038/ncomms7807.

Luo, C., Kuner, T., & Kuner, R. (2014). Synaptic plasticity in pathological pain. *Trends in Neurosciences*, *37*(6), 343–355. https://doi.org/10.1016/j.tins.2014.04.002.

Mächler, P., Wyss, M. T., Elsayed, M., Stobart, J., Gutierrez, R., Von Faber-Castell, A., … Weber, B. (2016). In vivo evidence for a lactate gradient from astrocytes to neurons. *Cell Metabolism*, *23*(1), 94–102. https://doi.org/10.1016/j.cmet.2015.10.010.

Magistretti, P. J., & Pellerin, L. (1996). Cellular bases of brain energy metabolism and their relevance to functional brain imaging: Evidence for a prominent role of astrocytes. *Cerebral Cortex*, *6*(1), 50–61. https://doi.org/10.1093/cercor/6.1.50.

Magistretti, P. J. (2011). Neuron-glia metabolic coupling and plasticity. *Experimental Physiology*, *96*(4), 407–410. https://doi.org/10.1113/expphysiol.2010.053157.

Magistretti, P. J., & Allaman, I. (2015). A cellular perspective on brain energy metabolism and functional imaging. *Neuron*, *86*(4), 883–901. https://doi.org/10.1016/j.neuron.2015.03.035.

Mason, S. (2017). Lactate shuttles in neuroenergetics-homeostasis, allostasis and beyond. *Frontiers in Neuroscience*, *11*(FEB), 1–15. https://doi.org/10.3389/fnins.2017.00043.

Meller, S. T., Dykstra, C., Grzybycki, D., Murphy, S., & Gebhart, G. F. (1994). The possible role of glia in nociceptive processing and hyperalgesia in the spinal cord of the rat. *Neuropharmacology*, *33*(11), 1471–1478. https://doi.org/10.1016/0028-3908(94)90051-5.

Miyamoto, K., Ishikura, K.-I., Kume, K., & Ohsawa, M. (2019). Astrocyte-neuron lactate shuttle sensitizes nociceptive transmission in the spinal cord. *Glia*, *67*(1), 27–36. https://doi.org/10.1002/glia.23474.

Miyamoto, K., Kume, K., & Ohsawa, M. (2017). Role of microglia in mechanical allodynia in the anterior cingulate cortex. *Journal of Pharmacological Sciences*, *134*(3), 158–165. https://doi.org/10.1016/j.jphs.2017.05.010.

Mosienko, V., Teschemacher, A. G., & Kasparov, S. (2015). Is L-lactate a novel signaling molecule in the brain? *Journal of Cerebral Blood Flow and Metabolism*, *35*(7), 1069–1075. https://doi.org/10.1038/jcbfm.2015.77.

Mulligan, S. J., & MacVicar, B. A. (2004). Calcium transients in astrocyte endfeet cause cerebrovascular constrictions. *Nature*, *431*(7005), 195–199. https://doi.org/10.1038/nature02827.

Murphy-Royal, C., Johnston, A. D., Boyce, A. K. J., Diaz-Castro, B., Institoris, A., Peringod, G., … Gordon, G. R. (2020). Stress gates an astrocytic energy reservoir to impair synaptic plasticity. *Nature Communications*, *11*(1). https://doi.org/10.1038/s41467-020-15778-9.

Nagase, M., Takahashi, Y., Watabe, A. M., Kubo, Y., & Kato, F. (2014). On-site energy supply at synapses through monocarboxylate transporters maintains excitatory synaptic transmission. *Journal of Neuroscience*, *34*(7), 2605–2617. https://doi.org/10.1523/JNEUROSCI.4687-12.2014.

Nam, Y., Kim, J. H., Kim, J. H., Jha, M. K., Jung, J. Y., Lee, M. G., … Suk, K. (2016). Reversible induction of pain hypersensitivity following Optogenetic stimulation of spinal astrocytes. *Cell Reports*, *17*(11), 3049–3061. https://doi.org/10.1016/j.celrep.2016.11.043.

Newman, L. A., Korol, D. L., & Gold, P. E. (2011). Lactate produced by glycogenolysis in astrocytes regulates memory processing. *PLoS One*, *6*(12), e28427. https://doi.org/10.1371/journal.pone.0028427.

Niwa, R., Nagata-Ohashi, K., Takeichi, M., Mizuno, K., & Uemura, T. (2002). Control of actin reorganization by slingshot, a family of phosphatases that dephosphorylate ADF/cofilin. *Cell*, *108*(2), 233–246.

Park, J. S., Yaster, M., Guan, X., Xu, J. T., Shih, M. H., Guan, Y., … Tao, Y. X. (2008). Role of spinal cord alpha-amino-3-hydroxy-5-methyl-4-isoxazolepropionic acid receptors in complete Freund's adjuvant-induced inflammatory pain. *Molecular Pain*, *4*, 1–11. https://doi.org/10.1186/1744-8069-4-67.

Pellerin, L. (2010). Alimenter sa pensée de l'importance du glucose et des autres substrats énergétiques pour soutenir l'activité cérébrale en fionction de son niveau. *Diabetes & Metabolism*, *36*(Suppl. 3), S59–S63. https://doi.org/10.1016/S1262-3636(10)70469-9.

Pellerin, L., & Magistretti, P. J. (1994). Glutamate uptake into astrocytes stimulates aerobic glycolysis: A mechanism coupling neuronal activity to glucose utilization. *Proceedings of the National Academy of Sciences of the United States of America*, *91*(22), 10625–10629. https://doi.org/10.1073/pnas.91.22.10625.

Prichard, J., Rothman, D., Novotny, E., Petroff, O., Kuwabara, T., Avison, M., … Shulman, R. (1991). Lactate rise detected by 1H NMR in human visual cortex during physiologic stimulation. *Proceedings of the National Academy of Sciences of the United States of America*, *88*(13), 5829–5831. https://doi.org/10.1073/pnas.88.13.5829.

Radwanska, K., Medvedev, N. I., Pereira, G. S., Engmann, O., Thiede, N., Moraes, M. F. D., … Giese, K. P. (2011). Mechanism for long-term memory formation when synaptic strengthening is impaired. *Proceedings of the National Academy of Sciences of the United States of America*, *108*(45), 18471–18475. https://doi.org/10.1073/pnas.1109680108.

Raghavendra, V., Tanga, F., Rutkowski, M. D., & DeLeo, J. A. (2003). Anti-hyperalgesic and morphine-sparing actions of propentofylline following peripheral nerve injury in rats: Mechanistic implications of spinal glia and proinflammatory cytokines. *Pain*, *104*(3), 655–664. JOUR.

Roumes, H., Dumont, U., Sanchez, S., Mazuel, L., Blanc, J., Raffard, G., … Bouzier-Sore, A.-K. (2021). Neuroprotective role of lactate in rat neonatal hypoxia-ischemia. *Journal of Cerebral Blood Flow & Metabolism*, *41*(2), 342–358. https://doi.org/10.1177/0271678X20908355.

Sappey-Marinier, D., Calabrese, G., Fein, G., Hugg, J. W., Biggins, C., & Weiner, M. W. (1992). Effect of photic stimulation on human visual cortex lactate and phosphates using 1H and 31P magnetic resonance spectroscopy. *Journal of Cerebral Blood Flow and Metabolism*, *12*(4), 584–592. https://doi.org/10.1038/jcbfm.1992.82.

Suzuki, A., Stern, S. A., Bozdagi, O., Huntley, G. W., Walker, R. H., Magistretti, P. J., & Alberini, C. M. (2011). Astrocyte-neuron lactate transport is required for long-term memory formation. *Cell*, *144*(5), 810–823.

Thirone, A. C. P., Speight, P., Zulys, M., Rotstein, O. D., Szászi, K., Pedersen, S. F., & Kapus, A. (2009). Hyperosmotic stress induces rho/rho kinase/LIM kinase-mediated cofilin phosphorylation in tubular cells: Key role in the osmotically triggered F-actin response. *American Journal of Physiology. Cell Physiology*, *296*(3), C463–C475. https://doi.org/10.1152/ajpcell.00467.2008.

Tsuda, M., Kohro, Y., Yano, T., Tsujikawa, T., Kitano, J., Tozaki-Saitoh, H., … Inoue, K. (2011). JAK-STAT3 pathway regulates spinal astrocyte proliferation and neuropathic pain maintenance in rats. *Brain*, *134*(4), 1127–1139. https://doi.org/10.1093/brain/awr025.

Tsuda, M., Shigemoto-Mogami, Y., Koizumi, S., Mizokoshi, A., Kohsaka, S., Salter, M. W., & Inoue, K. (2003). P2X4 receptors induced in spinal microglia gate tactile allodynia after nerve injury. *Nature*, *424*(6950), 778–783. https://doi.org/10.1038/nature01786.

Tsuda, M., Toyomitsu, E., Komatsu, T., Masuda, T., Kunifusa, E., Nasu-Tada, K., … Inoue, K. (2008). Fibronectin/integrin system is involved in P2X4 receptor upregulation in the spinal cord and neuropathic pain after nerve injury. *Glia*, *56*(5), 579–585. https://doi.org/10.1002/glia.20641.

Vezzoli, E., Calì, C., De Roo, M., Ponzoni, L., Sogne, E., Gagnon, N., … Magistretti, P. J. (2020). Ultrastructural evidence for a role of astrocytes and glycogen-derived lactate in learning-dependent synaptic stabilization. *Cerebral Cortex*, *30*(4), 2114–2127. https://doi.org/10.1093/cercor/bhz226.

Wang, J., Tu, J., Cao, B., Mu, L., Yang, X., Cong, M., … Li, Y. (2017). Astrocytic L-lactate Signaling facilitates amygdala-anterior cingulate cortex synchrony and decision making in rats. *Cell Reports*, *21*(9), 2407–2418. https://doi.org/10.1016/j.celrep.2017.11.012.

Weber, B., & Barros, L. F. (2015). The astrocyte: Powerhouse and recycling center. *Cold Spring Harbor Perspectives in Biology*, *7*(12), 1–16. https://doi.org/10.1101/cshperspect.a020396.

Yang, J., Ruchti, E., Petit, J.-M., Jourdain, P., Grenningloh, G., Allaman, I., & Magistretti, P. J. (2014). Lactate promotes plasticity gene expression by potentiating NMDA signaling in neurons. *Proceedings of the National Academy of Sciences of the United States of America*, *111*(33), 12228–12233. https://doi.org/10.1073/pnas.1322912111.

Yuen, E. Y., Liu, W., Kafri, T., van Praag, H., & Yan, Z. (2010). Regulation of AMPA receptor channels and synaptic plasticity by cofilin phosphatase slingshot in cortical neurons. *Journal of Physiology*, *588*(13), 2361–2371. https://doi.org/10.1113/jphysiol.2009.186353.

Zhang, J. M., Wang, H. K., Ye, C. Q., Ge, W., Chen, Y., Jiang, Z. L., … Duan, S. (2003). ATP released by astrocytes mediates glutamatergic activity-dependent Heterosynaptic suppression. *Neuron*, *40*(5), 971–982. https://doi.org/10.1016/S0896-6273(03)00717-7.

Zhang, Y., Xue, Y., Meng, S., Luo, Y., Liang, J., Li, J., … Shi, J. (2016). Inhibition of lactate transport erases drug memory and prevents drug relapse. *Biological Psychiatry*, *79*(11), 928–939. https://doi.org/10.1016/j.biopsych.2015.07.007.

Zhuang, Z., Yang, B., Theus, M. H., Sick, J. T., Bethea, J. R., Sick, T. J., & Liebl, D. J. (2010). EphrinBs regulate D-serine synthesis and release in astrocytes. *Journal of Neuroscience*, *30*(47), 16015–16024. https://doi.org/10.1523/JNEUROSCI.0481-10.2010.

Chapter 15

Nociception and sweet solutions: Applications to inflammatory pain

Khawla Q. Nuseir[a], Manal Kassab[b,c], and Ahmad Altarifi[d]

[a]*Clinical Pharmacy Department, Faculty of Pharmacy, Jordan University of Science and Technology, Irbid, Jordan* [b]*Department of Maternal and Child Health, Faculty of Nursing, Jordan University of Science and Technology, Irbid, Jordan* [c]*University of Technology, Sydney (UTS), Sydney, NSW, Australia*
[d]*Department of Pharmacology, Faculty of Medicine, Jordan University of Science and Technology, Irbid, Jordan*

Abbreviations

ACC	anterior cingulate cortex
Amg	amygdala
DLPT	dorsolateral pontine nucleus
DRG	dorsal root ganglion
IASP	International Association for the Study of Pain
IL-1β	interleukin-1β
NA	nucleus accumbens
NICU	Neonatal Intensive Care Unit
NSAIDs	nonsteroidal antiinflammatory drugs
PAG	periaqueductal gray
PBN	parabrachial nucleus
PN	primary nociceptors
RMTg	rostromedial tegmentum
RVM	rostroventral medulla
SI	somatosensory cortex I
SII	somatosensory cortex II
VTA	ventral tegmental area

Introduction

As per the IASP, nociception is defined as "The neural process of encoding noxious stimuli," while pain is defined as "An unpleasant sensory and emotional experience associated with, or resembling that associated with, actual or potential tissue damage" (Raja et al., 2020). The distinction between pain and nociception and thus analgesia and antinociception is still not very well defined, particularly, among those nonverbal humans and animals (Anand & Hickey, 1987). Moreover, argument that pain is needed for adaptation of the infant to birth process added to the confusion. Human babies and children who can't express their pain in words are at disadvantage when it comes to treating their pain (Rutter & Doyal, 1998). Premature newborns in particular are exposed to painful procedures with little or without analgesia (Carbajal, 2020), mostly due to myths and misconceptions about pain and analgesia in this vulnerable group (Perry, Tan, Chen, Weidig, & Xu, 2018). Doctors and scientists assumed that premature infants do not feel pain as pain requires consciousness (Anand & Hickey, 1987). The thought was that what premature infants experienced was nociception that resulted in reflexes and autonomic reactions. All this changed with the awareness and knowledge of the development of the nervous system of fetuses and newborns (Anand & Hickey, 1987).

Humans knew pain and had been in search of means to treat it since early history (Chen, 2011; Collier, 2018; Parris & Johnson, 2013). But finding the ideal analgesics proved to be tricky and unattainable. From opioids to NSAIDs, each analgesic suffers from shortcomings (Steinhorn et al., 2015; Tobias, Bhalla, & Shepherd, 2014). Furthermore, for subsets of

The Neurobiology, Physiology, and Psychology of Pain. https://doi.org/10.1016/B978-0-12-820589-1.00015-4

patients such as premature, mature newborns, infants, and young children, the challenges are multifactorial. Thus, sweet solutions use was introduced as analgesics for management of mild-to-moderate painful procedures (De Bernardo et al., 2019). Sweet solutions provided much needed approaches to decrease the impact of painful procedures on short- as well as long-term detrimental effects (Grunau, 2013). On the other hand, sweet solutions analgesic effects on adults were conflicting; either no effect on adults' pain (Mooney, Davies, & Pickering, 2020), or it has an effect (Davies et al., 2019). Also, the use of sweet solutions during infancy was not without proposed adverse effects (Ranger et al., 2019). Even though, the use of sweet solutions for minor procedures in the NICU had become a clinical reality at many hospitals and clinical guidelines had been published to establish their use (Stevens, Yamada, Gy, & Ohlsson, 2013), we are still in need for more evidence regarding the use of sweet solutions. Animal research provides an important insight into this subject.

Inflammatory pain pathways

Pain is a significant and pervasive public health problem. Trauma and inflammation initiate inflammation process by activating the release of different mediators like cytokines, bradykinin, histamine, and Substance P (Bovill, 1997). These stimuli will activate primary afferent nociceptors located in the peripheral nervous system. These pseudounipolar neurons have their cell bodies in the dorsal root ganglion, with a peripheral terminal directed toward the skin, muscles, or visceral tissues, and a central terminal directed to the dorsal horn of the spinal cord. They transmit damage signals from peripheral organs to the centrally located second-order neurons, also called spinothalamic neurons. These neurons have their cell bodies in the spinal cord, and after they cross the midline, they project to thalamic nuclei, where they synapse with tertiary neurons in the thalamus. Axons of the tertiary neurons project to different cortical areas such as somatosensory cortex I (SI), somatosensory cortex II (SII), anterior cingulate cortex (ACC), and insula. Limbic regions receive pain information from two general sources. First, secondary neurons originating in the spinal cord also send bottom-up projections to brainstem areas, such as PAG, RVM, hypothalamus, and parabrachial nucleus (PBN). Some of these areas (e.g., the PBN) send projections to limbic areas such as the VTA (Cliffer, Burstein, & Giesler, 1991; Coizet, Dommett, Klop, Redgrave, & Overton, 2010; Giesler, Menétrey, & Basbaum, 1979). Second, cortical areas such as SII, ACC, and insula send top-down projections to limbic areas, such as VTA, NA, and amygdala. Thus, limbic areas such as VTA, NA, and amygdala receive nociceptive input via two routes: a ventral pathway that is mediated by brain stem areas such as PBN, and a dorsal pathway that is mediated by corticolimbic connections. Moreover, neurons in the PAG send projections to the RVM and dorsolateral pontine tegmentum (DLPT), and both of them send descending projections back to the spinal cord for further modulation of nociceptive information (Basbaum & Fields, 1984; Mantyh & Peschanski, 1982). Fig. 1 summarizes some of these

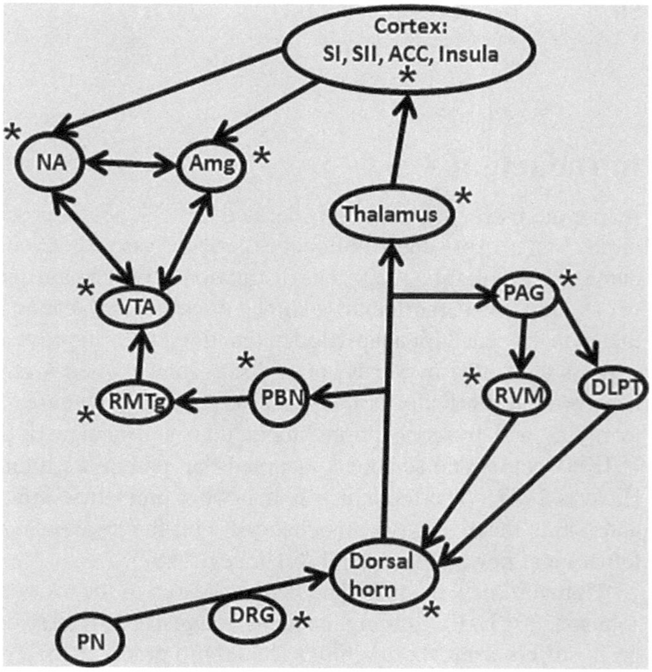

FIG. 1 Simplified cartoon showing major pain pathways and connections between them inside the nervous system. *Arrows* indicate the direction of signal between regions. Some areas have reciprocal connection. *Asterisks* indicate that μ opioid receptors have been localized in that area. *ACC*, anterior cingulate cortex; *Amg*, Amygdala; *DLPT*, dorsolateral pontine nucleus; *DRG*, dorsal root ganglion; *NA*, nucleus accumbens; *PAG*, periaqueductal gray; *PBN*, parabrachial nucleus; *PN*, primary nociceptors; *RMTg*, rostromedial tegmentum; *RVM*, rostroventral medulla; *SI, II*, Sensory area 1 and 2; *VTA*, ventral tegmental area.

TABLE 1 List of causes of inflammatory pain.

Preclinical study of inflammatory pain employs many tools and mediators. The following is a short list of the most widely used compounds (Muley, Krustev, & Mcdougall, 2016)	
Carrageenan	Neugebauer, Han, Adwanikar, Fu, and Ji (2007)
LPS (lipopolysaccharide)	Hudalla et al. (2018)
CFA (complete freund's adjuvant)	Kandasamy, Calsbeek, and Morgan (2016)
H^+ (hydrogen ions)	Eckhardt, Cheplovitz, Lipo, and Govier (1958)
MIA (monoiodoacetate)	Neugebauer et al. (2007)
Mustard oil	Laird, Martinez-Caro, Garcia-Nicas, and Cervero (2001)
Capsaicin	Laird et al. (2001)

connections that are of interest to the current document. These connections have consolidated the view that pain is a complex experience encompassing sensory, affective, and cognitive elements (Neugebauer, Galhardo, Maione, & Mackey, 2009; Zubieta, 2001). Preclinical study of inflammatory pain employs many tools and mediators. Table 1 is a short list of the most widely used compounds.

Pain behaviors and assessment

Pain is essentially a subjective experience, and its existence in humans is typically assessed by verbal report. This report is usually guided by a scale that ranges from 0 to 10, with "0" indicating "no pain" and "10" indicating "worst pain imaginable."

Analgesia is defined as absence of pain in response to stimulation that would normally be painful (IASP, 1994). Despite recent advances in research on pain mechanisms and analgesic drug development, there has been little evolution in the types of drugs used to treat pain. For decades, the most widely used analgesics have included μ opioid agonists such as morphine, antiinflammatory steroids such as cortisone, and nonsteroidal antiinflammatory drugs (NSAIDs) such as aspirin. In preclinical settings, assays of animal behavior have played a key role in research on the neurobiology of pain and development of analgesic drugs (Negus et al., 2006). A noxious stimulus can be chemical, mechanical, thermal, or electrical in modality. In all these cases, application of the noxious stimulus is intended to produce "pain," and changes in behavior produced by the noxious stimulus are interpreted as evidence of "pain." Drugs or other treatments can then be evaluated for their effects on expression of the pain-related behavior. The behavioral outcomes usually fall into one of the following two categories:

(1) Pain-stimulated behaviors: include behaviors that increase in rate, frequency, or intensity after a noxious stimulus. These include withdrawal reflexes of the tail and paw after thermal or chemical irritant induction, respectively.
(2) Pain-depressed behaviors: include behaviors that decrease in rate, frequency, or intensity after a noxious stimulus. Examples of this type of behavior include depression of feeding, locomotion, sleep, and others.

First, the diagnosis of pain in both human and veterinary medicine often relies on measures of pain-depressed behavior (also referred to as "functional impairment"), and restoration of pain-depressed behavior is often a goal of treatment (Cleeland & Ryan, 1994; Dworkin et al., 2005). The utility of these measures in clinical contexts suggests that pain-depressed behaviors may also be useful as endpoints in research. Second, pain-related depression of behavior is often accompanied by comorbid depression of mood in humans (Bair, Robinson, Katon, & Kroenke, 2003; Maletic & Raison, 2009). And preclinical research on pain-depressed behavior may provide insights into the expression, neurobiology, and modulation of the affective dimensions of pain. Thus, assays of pain-depressed behavior would serve as a better tool to evaluate candidate analgesics in the laboratory animals and would produce better translational pain research to the clinical field.

Use of sweet solutions for inflammatory pain

Few researchers used sweet solutions explicitly for management of inflammatory pain. But, with the knowledge that many skin-breaking procedures involve inflammatory (de Carvalho et al., 2019; Ranger et al., 2019) as well as nociceptive component of pain, we can assume that sweet solutions used in the NICU and for immunization pain might have been suppressing both inflammatory and nociceptive pain alike (De Bernardo et al., 2019; Kassab, Almomani, Nuseir, & Alhouary, 2020). Conversely, glucose for procedural pain in preterm neonates failed to alleviate the negative effects on the brain structural development (Schneider et al., 2018). Animal studies using inflammatory nociceptive stimuli specifically suggested the success of sweet solutions for decreasing inflammatory pain in neonatal rats (Nuseir, Altarifi, Tasslaq, Alhusban, & Alzoubi, 2019). On the other hand, neonatal mice treated with sucrose for pinprick pain, which was mild and superficial so (maybe no inflammation), did not affect their performance during adulthood on several behavioral tests (Ranger et al., 2019). All these contradictory results are an argument for the demand to continue research in order to elucidate and shed more light on the use of sweet tasting solutions for procedural pain among infants and young children. As more research is done to clarify the benefits vs the risks of using sweet solutions for pain, more data that will lead to better practices will emerge. However, despite traditionally held beliefs of adults irresponsiveness to sweet-induced analgesia, a recent report showed that sucrose intraorally also had an antinociception in adult rats, albeit through a different mechanism than that of neonatal rats (Davies et al., 2019).

Mechanisms of sweet solutions for inflammatory pain

Endogenous opioid system is probably involved in sweet tasting solutions induced antinociception as evidence from human and animal data (Gibbins & Stevens, 2001). Met-enkephalin and beta-endorphin levels were increased in the serum of rats treated by sucrose solution for pain (Nuseir, Alzoubi, Alabwaini, Khabour, & Kassab, 2015). Opioid antagonist blocked sweet-induced analgesia (Blass, Fitzgerald, & Kehoe, 1987). Alternatively, other researchers showed lack of involvement of the endogenous opioid system in mediating sweet-induced analgesia (Gradin & Schollin, 2005). Other neurotransmitters might also be involved as suggested by research. Monoaminergic neurons of the brain stem were shown to be implicated in this sweet-induced analgesia (Kishi et al., 2006). Descending pain inhibitory brain areas are also involved as evidence by preservation of sweet-induced analgesia after midcollicular transactions (Anseloni, Ren, Dubner, & Ennis, 2005). In Kassab et al. study, it has been shown that sucrose stimulates natural pain killer and decreases levels of excitatory pain neurotransmitters mainly Substance P (Kassab et al., 2020). It has been shown that sucrose led to behavioral stability in the shorter time intervention group as compared to control. Also, Substance P was reduced in the intervention group. The findings of the current study supported that sucrose is effective in reducing pain though increasing natural pain killers and suppression of excitatory neurotransmitters that stimulates pain nociceptors among infants and young children aged 10–18 months (Kassab et al., 2020). As for direct effect on inflammatory mediators, in vivo research is lacking and such effect would, if present, be mediated centrally. In In vitro studies, sucrose blocked IL-1β-induced inflammatory catabolic response via autophagy (Khan, Ansari, & Haqqi, 2017). Adult rats consuming sweet solutions experienced reproducible thermal analgesia following inflammatory insult. This analgesia was not blocked by opioid, dopaminergic, or adrenergic antagonist. However, this analgesia was inhibited by cannabinoid CB1 receptor antagonist (Davies et al., 2019). Thus, analgesia/antinociception induced by sweet solutions of inflammatory pain is probably mediated via several neurotransmitter pathways, and it seems to be dependent on the age of the subjects.

Sweet tasting solutions used for analgesia

Sucrose and glucose are most widely employed compounds to create sweet solutions for use as analgesics/antinociceptives (De Bernardo et al., 2019; Kassab et al., 2020; Schneider et al., 2018). Of the mono and di-saccharides, fructose (Blass & Smith, 1992) and dextrose (Gursul, Hartley, & Slater, 2019) are used but to less extent. The availability of glucose solution at hospitals and clinics makes them accessible and ready to use (Işik, Özek, Bilgen, & Cebeci, 2000). Nonnutritive sweeteners, such as saccharin, provide an alternative (Ahuja, Daga, Gosavi, & Date, 2000), especially when fear from long-term effects of calorie-rich solutions arise (Shankar, Ahuja, & Sriram, 2013). Concentrations which translate into sweetness degree are also studied. Higher concentrations and thus sweeter solutions were more antinociceptive (Segato et al., 2005).

Clinical applications

In this chapter, we have reviewed how sweet solutions were helpful in decreasing neonatal, infant, and young children mild-to-moderate pain. We have described the proposed mechanisms and pathways involved in this action. In the following section, we describe how the application of this nonpharmacological treatment had been exploited clinically and the impact it had on the practice of pain management.

Recently, sweet solutions are known to be effective in reducing procedural pain in neonates and infants (Kassab et al., 2020; Kassab, Roydhouse, Fowler, & Foureur, 2012). Sweet solutions are given orally to infant's tongue using pacifier or syringe (Kassab et al., 2012). Each method has its own benefits. For example, administering sweet solutions using pacifier stimulates continuous nonnutritive sucking reflex in neonates and infants. After 1 year, old sweet solutions are preferable to be given using syringe to prevent suffocation. The appropriate dose of sweet solutions to produce the effect of pain reduction is still debatable. In systematic review, the findings show that a usual single dose of sweet solution was 2 mL of 12%–24% strength (Kassab et al., 2012). Sweet solutions dosing protocol is different from one study to another and needs to be standardized (Campbell, Cleaver, & Davies, 2014; Kassab et al., 2012). Furthermore, no firm-established optimal or minimal effective doses were found for older infants and young children (Mokhnach et al., 2010; Taddio et al., 2015), and the safety of repeated doses requires further investigation (Campbell et al., 2014; Gao et al., 2016). On the other hand, it has been suggested recently that the administration of 10 doses per day, particularly during the first week of life is highly recommended (Kassab et al., 2012). Calming effects may last considerably longer than the analgesic effects as demonstrated by several studies of reduced behavioral distress response during tissue injuries painful procedures. The onset of action is quick (within seconds), the peak effect occurs at 2 min, and the duration of action is up to 10 min (Kassab et al., 2012).

Importance of treatment of neonatal pain

Pain management for infants and children is not just ethical, but also necessary. At this time and age, it is unacceptable to inflict pain (for legitimate reason of course) without taking all possible measures to prevent or at least reduce that pain. That pain during early age has detrimental sequelae later in life is supported by human and animal research. Neonatal rats exposed to repeated nociceptive stimuli were more sensitive to inflammatory pain during adulthood (de Carvalho et al., 2019). Pain during infancy in mice leads to poor short-term memory at adulthood (Ranger et al., 2019). Our own research showed similar results; repetitive neonatal pain leads to poorer long-term memory later in life (Nuseir et al., 2017; Nuseir, Alzoubi, Alabwaini, Khabour, & Kassab, 2015), and higher sensitivity to thermal pain at adulthood (Nuseir et al., 2019).

As far as human studies had shown, immediate beneficial effects of suppressing the pain of procedural pain at the NICU and during immunization include less cry time, lower heart rate, higher oxygen saturation (De Bernardo et al., 2019; Kassab et al., 2020). As well as negative impacts later in life; hypersensitivity to pain, sensory, and cognitive impairments, psychological and social limitation, and overall poor health outcomes (Williams & Lascelles, 2020).

Applications to other areas

Some might argue that the observed "antinociceptive" effect of sweet solutions is merely an extension to sweet solutions calming and soothing effects (Slater et al., 2010). Though the distinction of these effects and antinociceptive effects in neonates and young children is not very clear, others provide evidence from human and animal experiments to the contrary (Blass et al., 1987). Sweet taste has a calming effect that is shown by research; tea sweetened with sugar was calming for humans exposed to acute stress (Samant, Wilkes, Odek, & Seo, 2016). Interestingly, noncaloric sweeteners, such as sucralose or stevia, did not produce this calming effect, but caloric sweeteners (sugar) did as shown by the previous report. Conversely, chronic stress caused an increase in the consumption of sweet food in rats and ingestion of sweet food decreased markers of stress and inflammation (Rho, Kim, Choi, & Lee, 2014). The opioids released in response to sweet taste might be a beneficial mediator of distress relief. Thus, maybe this calming and soothing effect of sweet taste might be an area of importance in clinical practice. Mothers' milk and her touch are known to stop crying babies and to help them calm down. In NICU sometimes, it is not practical to allow mothers presence, so sweet tasting solution might be the "solution."

Other agents of interest

So far use of sweet solutions, despite being effective in certain situations, had been controversary due to proposed adverse effects or even ineffectiveness.

Several nonpharmacological treatments were considered as effective pain management interventions to control pain in neonates and infants including sucrose (Stevens, Yamada, Ohlsson, Haliburton, & Shorkey, 2016), nonnutritive sucking (Shah & Jefferies, 2012), swaddling (Shu, Lee, Hayter, & Wang, 2014), and skin-to-skin care among neonates (Johnston et al., 2017).

Other agents of interest

Water: water was mainly employed as a control for sweet solutions in studying the antianalgesic/antinociceptive effects of oral sweeteners.

Nonnutritive suckling: Nonnutritive sucking is the stimulation of suckling by an infant as a result of a pacifier in his/her mouth without getting any nutrition. The subsequent sucking reflex has been shown to have a soothing and pain-relieving effect on neonates and infants. Sucking is assumed to stimulate the release of serotonin from the brainstem, believed to be involved in modification of the perception of pain. Also, might occur through release of cholecystokinin, an opioid-modulating substance. Mainly, this analgesic effect is bigger as sucking and sucrose are combined than sucking alone (Buscemi, Vandermeer, & Curtis, 2008).

Swaddling: Despite scarcity of reports of employing swaddling alone for analgesia, but it does seem to be effective for decreasing pain behaviors such as crying time, as with other nonpharmacological modalities, the effect is increased when combined with a pacifier (Mangat, Oei, Chen, Quah-Smith, & Schmölzer, 2018).

Skin-to-skin care: Skin-to-skin contact is one of other methods that stimulate sensorial saturation (Bellieni et al., 2007; Moore, Bergman, Anderson, & Medley, 2016). During painful procedure, skin-to-skin contact can be attracting to neonates' attention saturation and stimulates the release through the block of pain fiber transmission (Dubin & Patapoutian, 2010). Skin-to-skin contact is easy to implement intervention and is cost effective. To increase effect of skin-to-skin contact, it is highly suggested to include parental involvement. The effect of skin-to-skin contact is explained through Gate control theory (Sood, 2014) which suggested that touch stimulates traveling ascending pathways inhibiting the nociceptive signals from painful stimuli through various endogenous mechanisms located along spinothalamic tract.

Music therapy: There is some evident that music therapy may be beneficial in relieving procedural pain in young children. A study which included six randomized control trials showed that music was effective in stabilizing neonates during painful procedures such as circumcision and heel lances and it also had a great effect in physiological outcomes (Hartling et al., 2009). However, included studies had several quality issues which could lead to inconsistent reporting of findings across procedures. Another issue is related to the consistency regarding the optimal type or decibel level of the music among various age groups.

Mini-dictionary of terms

Sucrose: The common table sugar, a disaccharide, composed of two monosaccharides glucose and fructose. Produced naturally

Glucose: Monosaccharide, most abundant as it is a component of many polysaccharides

Dextrose: D-glucose, produced from corn, same structure as glucose

Fructose: Fruit sugar, same chemical structure as glucose, the sweetest of all sugars

Saccharin: Artificial sweetener, 300–400 sweeter than sucrose, it provides no calories upon consumption

Sucralose: Artificial sweetener, made from sugar, it provides no calories upon consumption

Stevia: A noncaloric sweetener derived from plant, 30–150 sweeter than sugar

Key facts

Key facts of prematurity (Williams & Lascelles, 2020)

- About 15 million babies are born prematurely every year worldwide.
- Premature birth is any birth before 37 weeks' gestation.
- Prematurity is the leading cause of death for children 5 years old and younger.
- On average, premature babies spend 81 days at NICU, while full-term babies spend on an average 4.9 days.
- Premature babies experience an average of 14 painful procedures daily.

Summary points

- Infants and young children are exposed to many painful procedures after birth and during their stay at NICU, as well as, during immunizations.
- These painful procedures are done with little or no treatment.
- Nonpharmacological managements are available and relatively safe compared to pharmacological treatments.
- Sweet tasting solutions are at the forefront of such nonpharmacological treatment.

References

Ahuja, V. K., Daga, S. R., Gosavi, D. V., & Date, A. M. (2000). Non-sucrose sweetener for pain relief in sick newborns. *Indian Journal of Pediatrics*, *67*(7), 487–489. https://doi.org/10.1007/BF02760473.

Anand, K. J. S., & Hickey, P. R. (1987). Pain and its effects in the human neonate and fetus. *The New England Journal of Medicine*, *317*, 1321–1329.

Anseloni, V. C. Z., Ren, K., Dubner, R., & Ennis, M. (2005). A brainstem substrate for analgesia elicited by intraoral sucrose. *Neuroscience*, *133*(1), 231–243. https://doi.org/10.1016/j.neuroscience.2005.01.055.

Bair, M. J., Robinson, R. L., Katon, W., & Kroenke, K. (2003). Depression and pain comorbidity. *Archives of Internal Medicine*, *163*(20), 2433. https://doi.org/10.1001/archinte.163.20.2433.

Basbaum, A. I., & Fields, H. L. (1984). Endogenous pain control systems: Brainstem spinal pathways and endorphin circuitry. *Annual Review of Neuroscience*, *7*(1), 309–338. https://doi.org/10.1146/annurev.ne.07.030184.001521.

Bellieni, C. V., Cordelli, D. M., Marchi, S., Ceccarelli, S., Perrone, S., Maffei, M., & Buonocore, G. (2007). Sensorial saturation for neonatal analgesia. *The Clinical Journal of Pain*, *23*(3), 219–221. https://doi.org/10.1097/ajp.0b013e31802e3bd7.

Blass, E., Fitzgerald, E., & Kehoe, P. (1987). Interactions between sucrose, pain and isolation distress. *Pharmacology, Biochemistry, and Behavior*, *26*(3), 483–489.

Blass, E. M., & Smith, B. A. (1992). Differential effects of sucrose, fructose, glucose, and lactose on crying in 1- to 3-day-old human infants: Qualitative and quantitative considerations. *Developmental Psychology*, *28*(5), 804–810. Retrieved from http://psycnet.apa.org/index.cfm?fa=buy.optionToBuy&id=1993-00853-001.

Bovill, J. G. (1997). Mechanisms of actions of opioids and non-steroidal anti-inflammatory drugs. *European Journal of Anaesthesiology*, *14*(Suppl. 15), 9–15. https://doi.org/10.1097/00003643-199705001-00003.

Buscemi, N., Vandermeer, B., & Curtis, S. (2008). The Cochrane library and procedural pain in children: An overview of reviews. *Evidence-Based Child Health*, *3*, 260–279. https://doi.org/10.1002/ebch.1864.

Campbell, N., Cleaver, K., & Davies, N. (2014). Oral sucrose as analgesia for neonates: How effective and safe is the sweet solution? A review of the literature. *Journal of Neonatal Nursing*, *20*(6), 274–282. https://doi.org/10.1016/j.jnn.2014.05.008.

Carbajal, R. (2020). Neonatal pain. In E. M. Boyle, & J. Cusack (Eds.), *Emerging topics and controversies in neonatology* (pp. 485–501). Cham: Springer International Publishing. https://doi.org/10.1007/978-3-030-28829-7_28.

Chen, J. (2011). History of pain theories. *Neuroscience Bulletin*, *27*(5), 343–350. https://doi.org/10.1007/s12264-011-0139-0.

Cleeland, C. S., & Ryan, K. M. (1994). Pain assessment: Global use of the brief pain inventory. *Annals of the Academy of Medicine, Singapore*, *23*(2), 129–138. https://doi.org/10.1097/01893697-199513010-00022.

Cliffer, K. D., Burstein, R., & Giesler, G. J. (1991). Distributions of spinothalamic, spinohypothalamic, and spinotelencephalic fibers revealed by anterograde transport of {PHA}-L in rats. *The Journal of Neuroscience*, *11*(3), 852–868. https://doi.org/10.1523/jneurosci.11-03-00852.1991.

Coizet, V., Dommett, E. J., Klop, E. M., Redgrave, P., & Overton, P. G. (2010). The parabrachial nucleus is a critical link in the transmission of short latency nociceptive information to midbrain dopaminergic neurons. *Neuroscience*, *168*(1), 263–272. https://doi.org/10.1016/j.neuroscience.2010.03.049.

Collier, R. (2018). A short history of pain management. *Canadian Medical Association Journal*, E26–E27. https://doi.org/10.1503/cmaj.109-5523.

Davies, A. J., Kim, D., Park, J., Lee, J. Y., Vang, H., Pickering, A. E., & Oh, S. B. (2019). Hedonic drinking engages a supraspinal inhibition of thermal nociception in adult rats. *Pain*, *160*(5), 1059–1069. https://doi.org/10.1097/j.pain.0000000000001482.

De Bernardo, G., Riccitelli, M., Sordino, D., Giordano, M., Piccolo, S., Buonocore, G., & Perrone, S. (2019). Oral 24% sucrose associated with nonnutritive sucking for pain control in healthy term newborns receiving venipuncture beyond the first week of life. *Journal of Pain Research*, *12*, 299–305. https://doi.org/10.2147/JPR.S203526.

de Carvalho, R. C., Prado, L., Rissardo Oliveira, N. C., Vilela Giusti, F. C., Santos Vieira, J., Giusti-Paiva, A., … da Silva, M. L. (2019). Repeated neonatal needle-prick stimulation increases inflammatory mechanical hypersensitivity in adult rats. *International Journal of Developmental Neuroscience*. https://doi.org/10.1016/j.ijdevneu.2019.02.004.

Dubin, A. E., & Patapoutian, A. (2010). Nociceptors: The sensors of the pain pathway. *Journal of Clinical Investigation*, *120*(11), 3760–3772. https://doi.org/10.1172/jci42843.

Dworkin, R. H., Turk, D. C., Farrar, J. T., Haythornthwaite, J. A., Jensen, M. P., Katz, N. P., … Witter, J. (2005). Core outcome measures for chronic pain clinical trials: IMMPACT recommendations. *Pain*. https://doi.org/10.1016/j.pain.2004.09.012. Elsevier B.V.

Eckhardt, E. T., Cheplovitz, F., Lipo, M., & Govier, W. M. (1958). Etiology of chemically induced writhing in mouse and rat. *Proceedings of the Society for Experimental Biology and Medicine*, *98*(1), 186–188. https://doi.org/10.3181/00379727-98-23984.

Gao, H. H., Gao, H. H., Xu, G., Li, M., Du, S., Li, F., … Wang, D. (2016). Efficacy and safety of repeated oral sucrose for repeated procedural pain in neonates: A systematic review. *International Journal of Nursing Studies*, *62*, 118–125. https://doi.org/10.1016/j.ijnurstu.2016.07.015.

Gibbins, S., & Stevens, B. (2001). Mechanisms of sucrose and non-nutritive sucking in procedural pain management in infants. *Pain Research & Management*, *6*(1), 21–28. https://doi.org/10.1155/2001/376819.

Giesler, G. J., Menétrey, D., & Basbaum, A. I. (1979). Differential origins of spinothalamic tract projections to medial and lateral thalamus in the rat. *Journal of Comparative Neurology*, *184*(1), 107–125. https://doi.org/10.1002/cne.901840107.

Gradin, M., & Schollin, J. (2005). The role of endogenous opioids in mediating pain reduction by orally administered glucose among newborns. *Pediatrics*, *115*(4), 1004–1007. https://doi.org/10.1542/peds.2004-1189.

Grunau, R. E. (2013). Neonatal pain in very preterm infants: Long-term effects on brain, neurodevelopment and pain reactivity. *Rambam Maimonides Medical Journal*, *4*(4). https://doi.org/10.5041/rmmj.10132, e0025.

Gursul, D., Hartley, C., & Slater, R. (2019). Nociception and the neonatal brain. *Seminars in Fetal and Neonatal Medicine*, *24*(4), 101016. https://doi.org/10.1016/j.siny.2019.05.008.

Hartling, L., Shaik, M. S., Tjosvold, L., Leicht, R., Liang, Y., & Kumar, M. (2009). Music for medical indications in the neonatal period: A systematic review of randomised controlled trials. *Archives of Disease in Childhood - Fetal and Neonatal Edition*, *94*(5), F349–F354. https://doi.org/10.1136/adc.2008.148411.

Hudalla, H., Karenberg, K., Kuon, R. J., Pöschl, J., Tschada, R., & Frommhold, D. (2018). LPS-induced maternal inflammation promotes fetal leukocyte recruitment and prenatal organ infiltration in mice. *Pediatric Research*, *84*(5), 757–764. https://doi.org/10.1038/s41390-018-0030-z.

IASP. (1994). *Classification of chronic pain* (2nd ed., revised). International Association for the Study of Pain. https://www.iasp-pain.org/PublicationsNews/Content.aspx?ItemNumber=1673&navItemNumber=677.

Işik, U., Özek, E., Bilgen, H., & Cebeci, D. (2000). Comparison of oral glucose and sucrose solutions on pain response in neonates. *The Journal of Pain*, *1*(4), 275–278. https://doi.org/10.1054/jpai.2000.8919.

Johnston, C., Campbell-Yeo, M., Disher, T., Benoit, B., Fernandes, A., Streiner, D., … Zee, R. (2017). Skin-to-skin care for procedural pain in neonates. *Cochrane Database of Systematic Reviews*. https://doi.org/10.1002/14651858.cd008435.pub3.

Kandasamy, R., Calsbeek, J. J., & Morgan, M. M. (2016). Home cage wheel running is an objective and clinically relevant method to assess inflammatory pain in male and female rats. *Journal of Neuroscience Methods*, *1*(263), 115–122. https://doi.org/10.1016/j.gde.2016.03.011.

Kassab, M., Almomani, B., Nuseir, K., & Alhouary, A.a. (2020). Efficacy of sucrose in reducing pain during immunization among 10- to 18-month-old infants and young children: A randomized controlled trial. *Journal of Pediatric Nursing*, *50*, e55–e61. https://doi.org/10.1016/j.pedn.2019.11.010.

Kassab, M. I., Roydhouse, J. K., Fowler, C., & Foureur, M. (2012). The effectiveness of glucose in reducing needle-related procedural pain in infants. *Journal of Pediatric Nursing*, *27*(1), 3–17. https://doi.org/10.1016/j.pedn.2010.10.008.

Khan, N. M., Ansari, M. Y., & Haqqi, T. M. (2017). Sucrose, but not glucose, blocks IL1-β-induced inflammatory response in human chondrocytes by inducing autophagy via AKT/mTOR pathway. *Journal of Cellular Biochemistry*, *118*(3), 629–639. https://doi.org/10.1002/jcb.25750.

Kishi, R., Bongiovanni, R., de Nadai, T. R., Freitas, R. L., de Oliveira, R., Ferreira, C. M. D. R., & Coimbra, N. C. (2006). Dorsal raphe nucleus and locus coeruleus neural networks and the elaboration of the sweet-substance-induced antinociception. *Neuroscience Letters*, *395*(1), 12–17. https://doi.org/10.1016/j.neulet.2005.10.046.

Laird, J. M., Martinez-Caro, L., Garcia-Nicas, E., & Cervero, F. (2001). A new model of visceral pain and referred hyperalgesia in the mouse. *Pain*, *92*(3), 335–342. https://doi.org/10.1016/s0304-3959(01)00275-5.

Maletic, V., & Raison, C. (2009). Neurobiology of depression, fibromyalgia and neuropathic pain. *Frontiers in Bioscience*, *14*(1), 5291. https://doi.org/10.2741/3598.

Mangat, A., Oei, J.-L., Chen, K., Quah-Smith, I., & Schmölzer, G. (2018). A review of non-pharmacological treatments for pain management in newborn infants. *Children*, *5*(10), 130. https://doi.org/10.3390/children5100130.

Mantyh, P. W., & Peschanski, M. (1982). Spinal projections from the periaqueductal grey and dorsal raphe in the rat, cat and monkey. *Neuroscience*, *7*(11), 2769–2776. https://doi.org/10.1016/0306-4522(82)90099-9.

Mokhnach, L., Anderson, M., Glorioso, R., Loeffler, K., Shinabarger, K., Thorngate, L., … Zbirun, I. (2010). NICU procedures are getting sweeter: Development of a sucrose protocol for neonatal procedural pain. *Neonatal Network*, *29*(5), 271–279. https://doi.org/10.1891/0730-0832.29.5.271.

Mooney, E. R., Davies, A. J., & Pickering, A. E. (2020). Sweet taste does not modulate pain perception in adult humans. *Wellcome Open Research*, *5*(43), 1–10.

Moore, E. R., Bergman, N., Anderson, G. C., & Medley, N. (2016). Early skin-to-skin contact for mothers and their healthy newborn infants. *Cochrane Database of Systematic Reviews*, *2016*(11). https://doi.org/10.1002/14651858.CD003519.pub4.

Muley, M. M., Krustev, E., & Mcdougall, J. J. (2016). Preclinical assessment of inflammatory pain. *CNS Neuroscience & Therapeutics*. https://doi.org/10.1111/cns.12486.

Negus, S. S., Vanderah, T. W., Brandt, M. R., Bilsky, E. J., Becerra, L., & Borsook, D. (2006). Preclinical assessment of candidate analgesic drugs: Recent advances and future challenges. *Journal of Pharmacology and Experimental Therapeutics*, *319*(2), 507–514. https://doi.org/10.1124/jpet.106.106377.

Neugebauer, V., Galhardo, V., Maione, S., & Mackey, S. C. (2009). Forebrain pain mechanisms. *Brain Research Reviews*, *60*(1), 226–242. https://doi.org/10.1016/j.brainresrev.2008.12.014.

Neugebauer, V., Han, J. S., Adwanikar, H., Fu, Y., & Ji, G. (2007). Techniques for assessing knee joint pain in arthritis. *Molecular Pain*, *3*, 1–13. https://doi.org/10.1186/1744-8069-3-8.

Nuseir, K. Q., Altarifi, A. Y., Tasslaq, A., Alhusban, A. Y., & Alzoubi, K. H. (2019). Early and late anti nociceptive effects of sucrose on neonatal inflammatory pain in rats: Comparison to a non-steroidal anti-inflammatory drug. *Physiology and Behavior*, *206*(September 2018), 37–42. https://doi.org/10.1016/j.physbeh.2019.03.014.

Nuseir, K. Q., Alzoubi, K. H., Alabwaini, J., Khabour, O. F., & Kassab, M. I. (2015a). Sucrose-induced analgesia during early life modulates adulthood learning and memory formation. *Physiology & Behavior*, *145*, 84–90. https://doi.org/10.1016/j.physbeh.2015.04.002.

Nuseir, K. Q., Alzoubi, K. H., Alhusban, A., Bawaane, A., Al-azzani, M., & Khabour, O. F. (2017). Sucrose and naltrexone prevent increased pain sensitivity and impaired long-term memory induced by repetitive neonatal noxious stimulation : Role of BDNF and β -endorphin. *Physiology & Behavior*, *179*(January), 213–219. https://doi.org/10.1016/j.physbeh.2017.06.015.

Parris, W. C. V., & Johnson, B. W. (2013). The history of pain medicine. In *Practical management of pain (5th ed.).*, 3.e1-12.e1.

Perry, M., Tan, Z., Chen, J., Weidig, T., Xu, W., & Cong, X. S. (2018). Neonatal pain: Perceptions and current practice. *Critical Care Nursing Clinics of North America*, *30*(4), 549–561. https://doi.org/10.1016/j.physbeh.2017.03.040.

Raja, S. N., Carr, D. B., Cohen, M., Finnerup, N. B., Flor, H., Gibson, S., … Vader, K. (2020). The revised International Association for the Study of Pain definition of pain: Concepts, challenges, and compromises. *Pain*. https://doi.org/10.1097/j.pain.0000000000001939.

Ranger, M., Tremblay, S., Chau, C. M. Y., Holsti, L., Grunau, R. E., Goldowitz, D., & Agata, A. L. D. (2019). Adverse behavioral changes in adult mice following neonatal repeated exposure to pain and sucrose. *Frontiers in Psychology*, *9*(January), 1–14. https://doi.org/10.3389/fpsyg.2018.02394.

Rho, S. G., Kim, Y. S., Choi, S. C., & Lee, M. Y. (2014). Sweet food improves chronic stress-induced irritable bowel syndrome-like symptoms in rats. *World Journal of Gastroenterology*, *20*(9), 2365–2373. https://doi.org/10.3748/wjg.v20.i9.2365.

Rutter, N., & Doyal, L. (1998). Neonatal care and management of pain : Historical and ethical issues. *Seminars in Fetal & Neonatal Medicine*, *3*, 297–302.

Samant, S. S., Wilkes, K., Odek, Z., & Seo, H. S. (2016). Tea-induced calmness: Sugar-sweetened tea calms consumers exposed to acute stressor. *Scientific Reports*, *6*(October), 1–8. https://doi.org/10.1038/srep36537.

Schneider, J., Duerden, E. G., Guo, T., Ng, K., Hagmann, P., Bickle, M., … Miller, S. P. (2018). Procedural pain and oral glucose in preterm neonates : Brain development and. *Pain*, *159*(3), 515–525. https://doi.org/10.1097/j.pain.

Segato, E. N., Rebouças, E. C. C. C., Freitas, R. L., Caires, M. P. T. T., Cardoso, A. V., Resende, G. C. C., … Coimbra, N. C. (2005). Effect of chronic intake of sweet substance on nociceptive threshold and feeding behavior of Rattus norvegicus (Rodentia, Muridae). *Nutritional Neuroscience*, *8*(2), 129–140. https://doi.org/10.1080/10284150500069413.

Shah, V., & Jefferies, A. (2012). Preterm infants receiving heel lance procedures have slightly lower pain scores and quicker time to return to baseline heart rate when held in kangaroo care by the mother than by the father. *Evidence-Based Medicine*. https://doi.org/10.1136/ebmed-2012-100525.

Shankar, P., Ahuja, S., & Sriram, K. (2013). Non-nutritive sweeteners: Review and update. *Nutrition*. https://doi.org/10.1016/j.nut.2013.03.024.

Shu, S.-H., Lee, Y.-L., Hayter, M., & Wang, R.-H. (2014). Efficacy of swaddling and heel warming on pain response to heel stick in neonates: A randomised control trial. *Journal of Clinical Nursing*, *23*(21–22), 3107–3114. https://doi.org/10.1111/jocn.12549.

Slater, R., Cornelissedn, L., Fabrizi, L., Patten, D., Yoxen, J., Worley, A., … Fitzgerald, M. (2010). Oral sucrose as an analgesic drug for procedural pain in newborn infants: A randomised controlled trial. *Lancet*, *376*, 1225–1232. https://doi.org/10.1016/S0140-6736(10)61303-7.

Sood, J. (2014). Oxford textbook of Paediatric pain: Patrick J. McGrath, Bonnie J. Stevens, Suellen M. Walker, William T. Zempsky (eds). *The Indian Journal of Pediatrics*, *81*(9), 991. https://doi.org/10.1007/s12098-014-1496-2.

Steinhorn, R., McPherson, C., Anderson, P. J., Neil, J., Doyle, L. W., & Inder, T. (2015). Neonatal morphine exposure in very preterm infants—Cerebral development and outcomes. *Journal of Pediatrics*, *166*(5), 1200–1207.e4. https://doi.org/10.1016/j.jpeds.2015.02.012.

Stevens, B., Yamada, J., Gy, L., & Ohlsson, A. (2013). Sucrose for analgesia in newborn infants undergoing painful procedures—The Cochrane library—Stevens—Wiley online library. *Cochrane Database of Systematic Reviews*, *2013*(1). https://doi.org/10.1002/14651858.CD001069, CD001069.

Stevens, B., Yamada, J., Ohlsson, A., Haliburton, S., & Shorkey, A. (2016). Sucrose for analgesia in newborn infants undergoing painful procedures. *Cochrane Database of Systematic Reviews*. https://doi.org/10.1002/14651858.cd001069.pub5.

Taddio, A., McMurtry, C. M., Shah, V., Riddell, R. P., Chambers, C. T., Noel, M., … Bleeker, E. V. (2015). Reducing pain during vaccine injections: Clinical practice guideline. *CMAJ*, *187*(13), 975–982. https://doi.org/10.1503/cmaj.150391.

Tobias, J., Bhalla, T., & Shepherd, E. (2014). Neonatal pain management. *Saudi Journal of Anaesthesia*, *8*(5), 89–97. https://doi.org/10.4103/1658-354X.144085.

Williams, M. D., & Lascelles, B. D. X. (2020). Early neonatal pain—A review of clinical and experimental implications on painful conditions later in life. *Frontiers in Pediatrics*, *8*(February), 1–18. https://doi.org/10.3389/fped.2020.00030.

Zubieta, J.-K. (2001). Regional mu opioid receptor regulation of sensory and affective dimensions of pain. *Science*, *293*(5528), 311–315. https://doi.org/10.1126/science.1060952.

Chapter 16

Interlinking interleukin-33 (IL-33), neuroinflammation and neuropathic pain

Camila Rodrigues Ferraz[a], Fernanda Soares Rasquel-Oliveira[b], Sergio Marques Borghi[c], Anelise Franciosi[b], Thacyana Teixeira Carvalho[b], Telma Saraiva-Santos[b], Nayara Anitelli Artero[b], Rubia Casagrande[a], and Waldiceu A. Verri, Jr.[b]

[a]*Department of Pharmaceutical Sciences, Centre of Health Sciences, Londrina State University, Londrina, Paraná, Brazil,* [b]*Department of Pathology, Centre of Biological Sciences, Londrina State University, Londrina, Paraná, Brazil,* [c]*Center for Research in Health Sciences, University of Northern Paraná, Londrina, Paraná, Brazil*

Abbreviations

AP-1	activator protein 1
ATP	adenosine triphosphate
CCI	chronic constriction injury
CD11b	cluster of differentiation molecule 11B
CD8⁺ T	cytotoxic T Cell
CFA	complete Freund's adjuvant
CNS	central nervous system
CX3CL1	fractalkine
ERK	extracellular signal-regulated kinase
GFAP	glial fibrillary acidic protein
IL	interleukin
IL-1RAcP	IL-1 receptor accessory protein
ILC2	innate lymphoid cell type 2
JAK2	janus kinase 2
JNK	c-Jun N-terminal kinase
KO	knockout KO
LPA	lysophosphatidic acid
MAPK	mitogen-activated protein kinase
mTOR	mammalian target of rapamycin
MyD88	myeloid differentiation primary response 88
NFκB	nuclear factor κB
p38	p38 mitogen-activated protein kinase
PI3K	phosphatidylinositol 3-kinase
ra	receptor antagonist
rIL-33	recombinant IL-33
rmIL-18	recombinant mouse IL-18
ROS	reactive oxygen species
SIGIRR	single Ig IL-1-related receptor
SNI	spared nerve injury
sST2	small soluble factor
ST2L	longer membrane-bound protein
STAT3	signal transducers and activators of transcription 3
Th	T helper cell
TNF	tumor necrosis factor
Treg	regulatory T cell

The Neurobiology, Physiology, and Psychology of Pain. https://doi.org/10.1016/B978-0-12-820589-1.00016-6

IL-33

Interleukin (IL)-33 is a pleiotropic cytokine involved in innate and adaptive immunity in varied diseases (Drake & Kita, 2017; Molofsky, Savage, & Locksley, 2015). It belongs to the IL-1 family of cytokines as well as IL-1β, IL-1α, IL-18, and IL-36 cytokines and IL-1 and IL-36 receptor antagonists (IL-1ra, IL-36ra) (Molofsky et al., 2015). The secretion of IL-33 is regulated for homeostasis maintenance of the immune system, with the intension of containing its powerful pro-inflammatory effects (Liew, Girard, & Turnquist, 2016). Full-length IL-33 is specifically stored in the nucleus of cells, which is crucial to counteract harmful events to the host related to its dysregulated release (Drake & Kita, 2017; Molofsky et al., 2015). The current understanding is that IL-33 release involves two main mechanisms, a passive release and an active secretion (Drake & Kita, 2017). IL-33 release occurs after irreversible cellular damage leading to necrosis and necroptosis, which renders its alarmin denomination (Molofsky et al., 2015; Zarpelon et al., 2016). Active secretion of IL-33 in living cells can occur after varied cellular insults or inflammatory stimulus that includes mechanical stress such as traumas, infections, allergens, adenosine triphosphate (ATP), which lead to posttranslational modifications of IL-33 structure in the nucleus and its consequent release as a mature form. Full-length IL-33 can be further processed by leukocyte-derived extracellular proteases (Drake & Kita, 2017; Molofsky et al., 2015).

Mechanisms limiting IL-33 activity are also necessary; otherwise, its powerful pro-inflammatory effects could be harmful to the host tissues. Apoptosis-related cleavage by caspases 3 and 7 represents a providential mechanism to limit IL-33 activity. Extracellularly, the role of its soluble receptor (sST2) in sequestrating mature IL-33 is crucial and farther, its quick oxidation and subsequent formation of disulfide bridges also accounts for counteracting its range and duration of activity (Drake & Kita, 2017; Liew et al., 2016). IL-33 is constitutively expressed human and mouse endothelial cells, epithelial cells from respiratory and digestive systems, skin and other organs, fibroblast-like cells in lymphoid organs, skin, intestinal tract and synovia, and other cells types including oligodendrocytes and astrocytes in central nervous system (CNS), smooth muscle cells and hepatocytes, and although its level of expression is high in resting conditions, inflammation can trigger an even greater increase in its production (Liew et al., 2016; Zarpelon et al., 2016). In turn, its receptor is mainly expressed in immune cells, including group 2 innate lymphoid cells (ILC2s), T helper 2 (T_H2) cells, regulatory T (T_{reg}) cells, activated T_H1 cells, cytotoxic $CD8^+$ T cells, mast cells, basophils, neutrophils, macrophages, and dendritic and natural killer cells (Drake & Kita, 2017; Liew et al., 2016).

IL-33 signals through a heterodimer receptor complex (IL-33R) formed by a peculiar subunit named suppressed of tumorigenicity (Fig. 1) (ST2; or alternatively IL-1-recpetor-like 1, IL-1R1) and the IL-1 receptor accessory protein (IL-1RAcP) (Molofsky et al., 2015). ST2 can be a soluble receptor (sST2) or a longer membrane-bound protein (ST2L). The sST2 is constitutively expressed in human serum and serves as a decoy receptor for IL-33. Its neutralizing effects during inflammatory diseases accounts to control IL-33-related detrimental actions (Molofsky et al., 2015). On the other hand, the interaction of IL-33 with ST2L activates a canonical pathway promoting the recruitment of IL-1RAcP that in turn recruits the adapter protein MyD88 (myeloid differentiation primary response 88) (Griesenauer & Paczesny, 2017; Molofsky et al., 2015). This last one recruits IL-1R-associated kinase (IRAK), IRAK-4 and tumor necrosis factor (TNF) receptor associated factor 6 (TRAF6), which induce the activation of mitogen activated proteins (MAP) kinases. These MAP kinases are extracellular signal-regulated kinase (ERK) 1/2, p38 mitogen-activated protein kinases (p38), and c-Jun N-terminal kinase (JNK) that ultimately activate nuclear factor kappa B (NFκB) and activator protein 1 (AP-1) that in turn acting as transcription factors modulate several cellular functions (Griesenauer & Paczesny, 2017; Molofsky et al., 2015).

The activation of IL-33/ST2 signaling drives manifold regulatory effects, for instance, cellular activation and proliferation, cell survival, cytokine secretion, chronic airway inflammation and tissue remodeling, inflammation, and pain (Drake & Kita, 2017; Liew et al., 2016; Molofsky et al., 2015; Verri Jr. et al., 2010; Zarpelon et al., 2016). IL-33 contributes to CNS recovery after spinal cord injury via M2 macrophages activation (Gadani, Walsh, Smirnov, Zheng, & Kipnis, 2015). Besides, negative regulation of IL-33 signaling may be achieved by mechanisms involving the disruption of ST2 by immunoglobulin domain IL-1 receptor-related molecule (SIGIRR), and by ST2 phosphorylation and ubiquitination, resulting in its internalization and degradation (Liew et al., 2016). An interesting aspect is that IL-33 can activate primary sensory neurons mediating itch responses. Of note, IL-33 responsive neurons (67.6% and 85.3%, respectively) were also activated by TRPA1 and TRPV1 agonists indicating that neurons with a nociceptive function were also responsive to IL-33 (Liu et al., 2016). Furthermore, IL-33 mediates inflammatory (Verri Jr. et al., 2008; Zarpelon et al., 2013) and neuropathic pain (Liu et al., 2015; Zarpelon et al., 2016). In the next sections, the role of IL-33 in mediating neuropathic pain will be addressed.

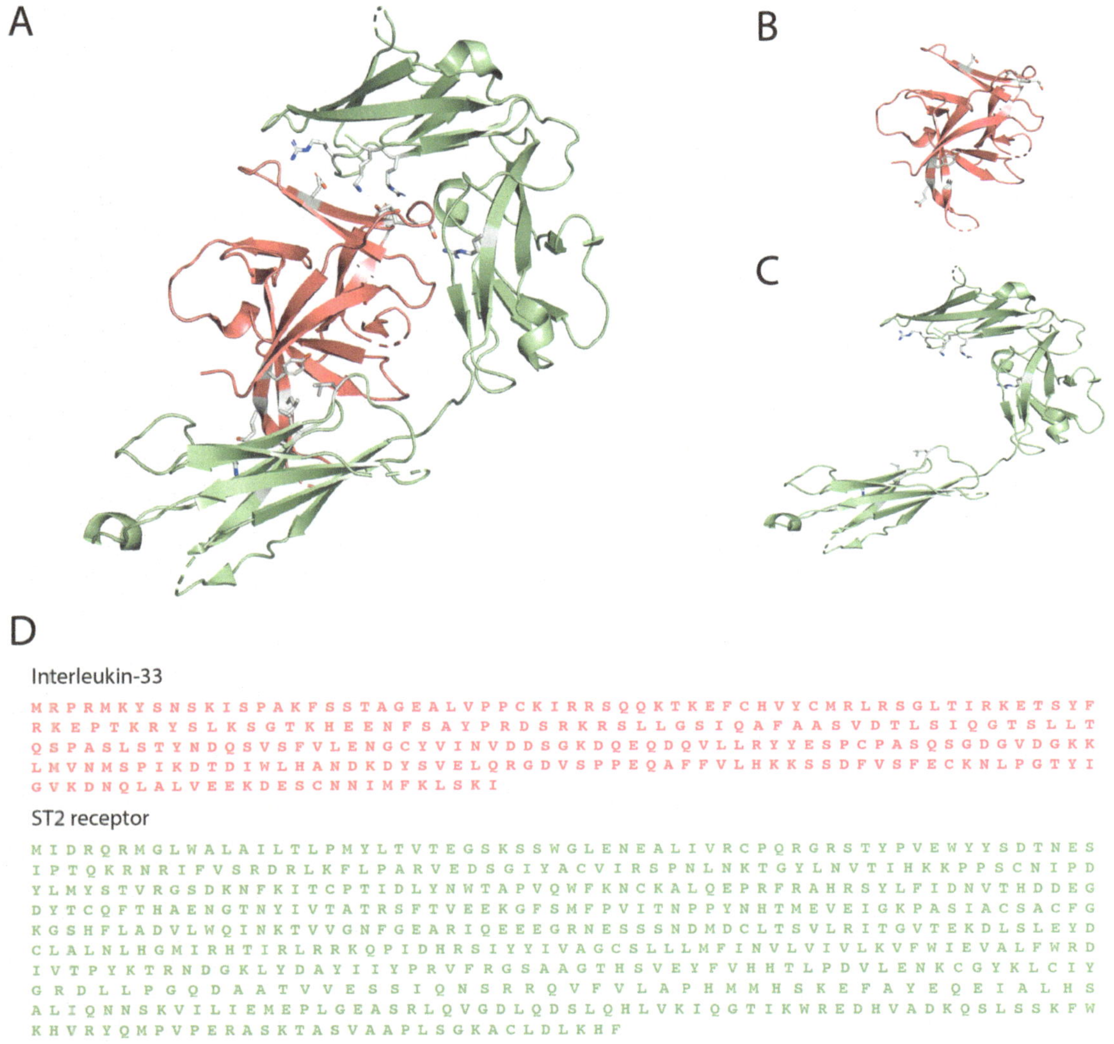

FIG. 1 Tridimensional structure and linear sequence of IL-33 and ST2 receptor. (A) IL-33 bind with ST2 receptor (PDB: 5VI4); (B) IL-33 tridimensional structure (PDB: 5VI4); (C) ST2 tridimensional structure (PDB: 5VI4); (D) linear sequence of IL-33 (UniProt: Q8BVZ5) and ST2 receptor (UniProt: P14719).

Neuropathic pain and glial cells

In sensory neurons of the spinal cord dorsal horn, the predominant events immediately following, for instance, peripheral nerve injury are the production of intense discharge activity from injured and adjacent axons, dramatically increase of extracellular glutamate, pro-inflammatory cytokines, and reactive oxygen species (ROS). These events activate pain-mediating substance transmission in synaptic clefts between neurons and glial cells, and peripheral primary afferent release of neurotransmitters and neuropeptides essential to this communication (Ji, Berta, & Nedergaard, 2013). Glial cells make up over 70% of the total cell population in the CNS, and are classified into oligodendrocytes, microglia, and astrocytes. Glial cells interact with nociceptor sensory neurons regulating neuronal processing in the spinal cord and contributing to the facilitation of pain signaling and nociceptor neuron sensitization (Ji et al., 2013).

In addition to neuronal activation after nerve injury, there is long-lasting surge of extracellular glutamate that is followed by significant increases in pro-inflammatory cytokine secretion, overexpression of membrane-bound receptors and ion channels, and altered expression of transporters in both neurons and activated glial cells (Gwak, Hulsebosch, & Leem, 2017). The glial cells activation is initiated by neuroanatomical and functional changes, followed by increase of pro-inflammatory cytokines, ROS, and chemokines (Gwak, Kang, Unabia, & Hulsebosch, 2012). These mediators initiate activation of neurons and glial cells that stimulate neuronal hyperactivity. Therefore, neuronal-glial interactions initiate and maintain neuroplastic changes that are responsible for acute and chronic pain (Jha, Jeon, & Suk, 2012). Thus, glial hyperactivity is a major factor in the development of neuropathic pain.

IL-33 and glial cells

Growing evidence supports that glial activation contributes to the pro-inflammatory cytokine secretion and other factors crucial for nociceptive transmission (Fattori, Borghi, et al., 2017; Fattori, Hohmann, et al., 2017; Gwak et al., 2017; Zarpelon et al., 2016). Considering that activation of glial cells leads to NFkB activation, this strongly indicates that neuroinflammation promotes pain by releasing inflammatory mediators from activated glial cells. Pharmacological inhibition of glial activation was proposed as an approach to treat neuropathic pain. Inhibiting oligodendrocytes, microglia, and astrocytes can prevent or delay neuropathic pain, indicating the importance of spinal glial cells in the development of pain. MAPK are crucial in intracellular signaling pathways that modulate the nociceptive responses. MAPK may play a key role in production of CNS inflammatory cytokines and activation of NFκB (Ji et al., 2013).

MAPK are activated in oligodendrocytes, microglia, and astrocytes (Ji et al., 2013; Zarpelon et al., 2016). The MAPK activate intracellular events that induce the peripheral and central sensitization that promote alteration at the cellular and behavioral levels (Zarpelon et al., 2016). MAPK exhibit distinct activation patterns in glial cells after painful injuries (Ji et al., 2013). Numerous studies have shown increased phosphorylation of p38 in spinal cord microglia after nerve injury, formalin-induced acute inflammatory pain, and postoperative pain. JNK activation was observed in spinal astrocytes after nerve injury, complete Freund's adjuvant (CFA)-induced persistent inflammatory pain and bone cancer (Ji et al., 2013). ERK phosphorylation, however, is highly dynamic. There is ERK activation in microglia in the early phases and gradually transition to astrocytes in the late phase of neuropathy (Ji et al., 2013). In vitro studies demonstrated elevation of ERK activation in oligodendrocyte-lineage cells (Ishii, Furusho, & Bansal, 2013). The production of IL-33 as well as many other inflammatory mediators is upregulated by MAPK and NFκB in glial cell (Fig. 2).

Oligodendrocytes

Oligodendrocytes function as specialized cells myelinating nervous system axons and the majority of healthy CNS IL-33-positive cells are mature oligodendrocytes (Gadani et al., 2015). A significant contribution of IL-33 signaling to nociceptor sensitization has been described in preclinical models (Borghi, Fattori, Hohmann, & Verri, 2019; Fattori, Borghi, et al., 2017; Fattori, Hohmann, et al., 2017).

The first study demonstrating oligodendrocytes role in a pain showed that IL-33 mediates chronic constriction injury (CCI)-induced neuropathic pain (Zarpelon et al., 2016). Oligodendrocyte-derived IL-33 activates ST2 receptor expressed by other glial cells, including microglia and astrocytes, and triggers intracellular signaling pathways such as JNK, phosphatidylinositol 3-kinase (PI3K), mammalian target of rapamycin (mTOR), MAPK, and NFκB, in addition to cytokines production that enhance central sensitization and IL-33 production (Zarpelon et al., 2016). Corroborating these findings, other reports demonstrated analgesic effects of pharmacological approaches that reduced IL-33 production and inhibited oligodendrocytes activation in the CCI model (Bertozzi et al., 2017; Longhi-Balbinot et al., 2016).

However, in a spinal cord injury (SCI) model, treatment with IL-33 ameliorated spinal cord damage by significantly decreasing tissue loss, demyelination, and astrogliosis. This beneficial role of IL-33 depended on inducing M2 alternatively activated macrophages and reducing pro-inflammatory cytokine production (Pomeshchik et al., 2015). Another beneficial role of IL-33 was observed in CNS injury models of optic nerve crush injury and intraocular glutamate injection in which mice lacking IL-33 had impaired recovery by diminished infiltration of M2-skewed macrophages (Gadani et al., 2015). In vitro and in vivo results also point out to IL-33 influence on recovery and promotion of repair in experimental models of CNS demyelination (Natarajan, Yao, & Sriram, 2016).

The opposing roles of IL-33 described above emphasize the importance of considering the disease context and cytokine milieu in each condition. The type of cell in which IL-33 expression is upregulated is context-dependent as well (Chen, Tsai, Yang, & Li, 2018). The disease outcomes governed by IL-33 might also depend on whether it is released as a cytokine with nociceptive roles in neuropathic pain or as an alarmin by necrotic cells creating an M2 profile (Fattori, Borghi, et al., 2017; Fattori, Hohmann, et al., 2017). Furthermore, knocking down IL-33 expression in oligodendrocytes is not adequate since it results in impaired oligodendrocyte maturation (Sung et al., 2019). In fact, IL-33 seems to control gene expression when stored at the nucleus, which is a role much less understood by now. Thus, it is likely that the cytokine role of IL-33 should be targeted avoiding the downregulation of its nuclear factor activity. In conclusion, growing evidence stress IL-33 and oligodendrocytes as important players in neuropathic pain and that this duo must be further investigated before clinical stage.

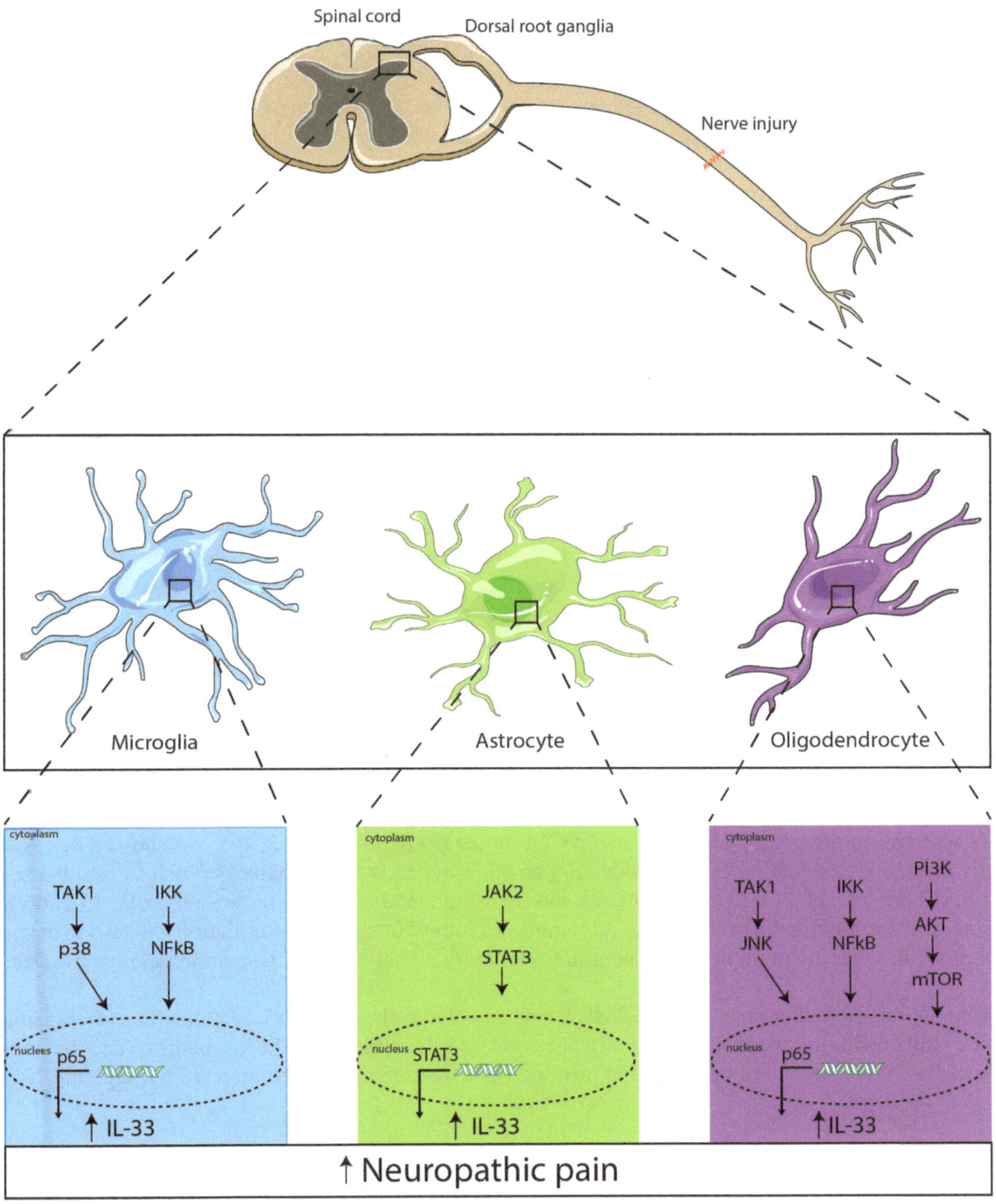

FIG. 2 Molecular mechanism involved in IL-33 production in glial cells. After nerve injury, glial cells (microglia, astrocytes, and oligodendrocytes) are activated in spinal dorsal horn, which release IL-33 and respond to this cytokine contributing to the development and maintenance of neuropathic pain.

Microglia

Microglia are macrophage-like cells in the CNS and represent 10%–15% of all human brain cells (Foster, Talbot, & Woolf, 2015). Microglia are the first line of defense against pathogen invasion, regulate homeostasis in the brain and spinal cord, and play a fundamental role in central sensitization and chronic pain (Ji et al., 2013).

Strong activation of microglia in the spinal cord dorsal horn occurs in neuropathic pain including peripheral nerve injury (Guan et al., 2016; Jin, Zhuang, Woolf, & Ji, 2003; Tsuda, 2016; Tsuda, Inoue, & Salter, 2005), and the injection of minocycline, a microglial inhibitor, into the spinal cord inhibited microglial activation and mechanical hyperalgesia after nerve transection (Raghavendra, Tanga, & DeLeo, 2003). The expression of cluster of differentiation molecule 11b (CD11b) is rapidly increased after nerve injury (Gwak et al., 2012). Peripheral nerve injury results in allodynia and microgliosis (hypertrophy of glial cells) in the spinal cord (Chen, Zhang, Qadri, Serhan, & Ji, 2018). Injured peripheral nerves release fractalkine (CX3CL1) in the spinal cord activating microglial production of TNF-α in a p38 MAPK-dependent manner (Gao & Ji, 2010). The phosphorylation of p38 MAPK in microglia was described in diverse animal models of neuropathic pain and intrathecal injection of p38 inhibitor reduced mechanical allodynia typical of neuropathic pain (Hains & Waxman, 2006; Ito et al., 2007; Jin et al., 2003; Piao et al., 2006; Xu et al., 2007).

Oligodendrocytes are the main source of IL-33 in the spinal cord, but some minor production of IL-33 can also be detected in microglia in CCI-induced neuropathic pain (Zarpelon et al., 2016). IL-33 produced by oligodendrocytes acts on ST2 expressed by microglia (Jiang et al., 2012; Liu et al., 2007; Yasuoka et al., 2011; Zhao et al., 2013). There is an important crosstalk between oligodendrocytes and microglia via IL-33 since this cytokine activates glial cells to produce TNF-α and IL-1β via MAPK, mTOR, and NFkB. TNF-α and IL-1β induce central sensitization essential to chronic pain development (Fattori, Borghi, et al., 2017; Fattori, Hohmann, et al., 2017; Pinho-Ribeiro, Verri Jr., & Chiu, 2017).

Astrocytes

Astrocytes are a class of glial cells present in the CNS that play roles in processes such as maintenance of the blood-brain barrier and antioxidant and metabolic support to neurons (Ben Haim & Rowitch, 2017). These cells express the glial fibrillary acidic protein (GFAP), a marker that distinguishes astrocytes from other glial cells (Ji, Donnelly, & Nedergaard, 2019). In mouse cell cultures, CNS astrocytes express IL-33 and also its receptor ST2 and the IL-1RAcP upon stimulation (Hudson et al., 2008; Yasuoka et al., 2011; Zarpelon et al., 2016).

IL-33 and ST2 in the DRG were found to play an important role in a rat model of neuropathic pain induced by spared nerve injury (SNI) surgery since the injection of IL-33 and ST2 antibodies by the intrathecal route reduced the SNI-induced mechanical allodynia and the expression of the pro-inflammatory cytokines TNF-α and IL-β (Zeng et al., 2020). The use of ST2-neutralizing antibody, recombinant IL-33 or ST2 knockout (KO) mice showed that the mechanical and cold allodynia induced by SNI are IL-33/ST2 signaling pathway-dependent and IL-33 was distributed in astrocytes and ST2 in neurons and astrocytes in the spinal cord (Liu et al., 2015). Also, it was demonstrated that SNI-induced neuropathic pain occurs through the contribution of spinal IL-33/ST2 signaling by upregulating the janus kinase 2/signal transducers and activators of transcription 3 (JAK2/STAT3) cascade in astrocytes (Liu et al., 2015).

According to online publication date, just 1 month before (Liu et al., 2015), we showed that IL-33/ST2 signaling in the spinal cord is essential to the neuropathic pain in the CCI model (Zarpelon et al., 2016). We showed that IL-33 induces astrocyte activation in CCI resulting on the production of other hyperalgesic cytokines, which in turn boost pain transmission and IL-33 production as well. As glial cells and their intracellular signaling depend on IL-33 and vice-versa to continue their activation and production, respectively, this is an important axis in the initiation and also in the perpetuation of chronic neuropathic pain by orchestrating a neuroinflammatory spinal cord mechanism in neuropathic pain (Zarpelon et al., 2016).

Although the inhibition of the spinal IL-33/ST2 signaling and glial cells are potential candidates as therapeutic approach for neuropathic pain conditions, more studies are still necessary to fully understand the mechanisms and the implications of targeting spinal IL-33/ST2 signaling pathway and how astrocyte function would be affected by targeting IL-33.

Applications to other areas

IL-33 has been implicated in many diseases, including inflammation, arthritis, asthma, allergies, cardiovascular, and infectious diseases. IL-33/ST2 is also involved in development of inflammatory pain in several models, which we briefly discuss below (Fattori, Borghi, et al., 2017; Fattori, Hohmann, et al., 2017).

The carrageenin injection increases ST2 and IL-33 mRNA expression and IL-33 production in paw tissues. Carrageenin-induced paw edema and hyperalgesia were reduced in ST2 KO mice compared with WT mice. Carrageenin induced PGE_2, TNF-α, CXCL1, IL-10, and IL-1β production in WT, and these effects were inhibited in ST2 KO mice. IL-33 has a potentiation role in inflammatory pain since doses of carrageenin and IL-33 that did not induce hyperalgesia, and cytokine production per se induced a potentiated effect when administrated together (Zarpelon et al., 2013). In formalin model, the ST2 antibody injection or ST2 deficiency significantly decreased the paw flinch and licking in both phases (Magro et al., 2013). After formalin injection, the duration and the number of ultrasonic vocalizations in the ST2 KO mice were significantly less than in WT mice. rIL-33 injection induced mild spontaneous pain behavior in WT mice and rIL-33 administration significantly exacerbated formalin-induced pain behavior in mice (Han et al., 2013).

Potassium superoxide (KO_2) induces hyperalgesia and spontaneous pain behavior by releasing superoxide anion. There is IL-33 production in this KO_2 model, suggesting the potential involvement of IL-33/ST2 in free radical-triggered pain (Lourenco-Gonzalez et al., 2019). IL-33 contributes to pain in a model of rheumatoid arthritis in mice in which targeting IL-33 with sST2 reduced pain and the production of other cytokines (TNFα, IL-1β and IFNγ), ET-1 and PGE_2. rIL-33 administration in immunized mice could also mimic these painful alterations triggered by antigen challenge in immunized mice (Verri Jr. et al., 2008). In fact, this was the first study to demonstrate that IL-33 is a cytokine involved in pain (Verri Jr. et al., 2008). Using a mouse model of arthritis and neutrophils of RA patients, data revealed that IL-33 accounts to neutrophil recruitment to the joints in arthritis. The neutrophil expression of ST2 is regulated by TNFα, which was observed by differences in expression and function of neutrophils of RA patients treated with methotrexate compared to methotrexate plus infliximab (anti-TNF antibody). Unpublished data from our group (Fattori et al. submitted) also reveal that IL-33 levels are high in the synovial fluid of gout patients compared to osteoarthritis patients, and targeting IL-33 in a mouse model of monosodium urate (MSU) crystals-induced gouty arthritis reduces pain, edema, and neutrophil recruitment. These data suggest IL-33 and its ST2 receptor mediate inflammatory pain, and harnessing this signaling may provide new analgesic therapies.

Other agents of interest

IL-18 is an important pronociceptive cytokine and also a member of the IL-1 family. IL-18 is known to be expressed by different cell types and is upregulated under pathological conditions including injury to the tissue and inflammation. IL-18 signaling is through its receptor, an IL-18Rα chain (IL-18R1, IL-1Rrp) and IL-18Rβ chain (IL-18R accessory protein, IL-1RAcPL) (Yasuda, Nakanishi, & Tsutsui, 2019). IL-18 is an important pronociceptive cytokine with hyperalgesic effect in varied models. Recombinant mouse IL-18 (rmIL-18) administration by intraplantar (Verri Jr. et al., 2004, 2007) and intramuscular (Yoshida et al., 2018) routes induced hyperalgesia, and inhibiting IL-18 actions reduced neuropathic pain (Pilat et al., 2016).

Evidence suggests that IL-18 mediates spinal microglia-astrocyte interactions and plays an important role in the crosstalk between these cells (Daigo, Sakuma, Miyoshi, Noguchi, & Kotani, 2012; Miyoshi, Obata, Kondo, Okamura, & Noguchi, 2008). IL-18 and IL-18R mRNA expression were elevated in the spinal cord after CCI and SNL (Miyoshi et al., 2008; Pilat et al., 2016). In vitro, IL-18 expression was detected in primary cultures of astrocytes and microglia. IL-18 is predominantly expressed in microglia, and IL-18R is primarily expressed on astrocytes in neuropathic model (Miyoshi et al., 2008). Treatment with IL-18-binding protein (IL-18BP) decrease hyperalgesia and microglia and astrocyte activation in CCI model and enhances the analgesic effect of morphine and buprenorphine (Pilat et al., 2016).

IL-18 is known to be upregulated in several human autoimmune and inflammatory diseases, thus this cytokine might represent a novel therapeutic target in inflammatory and neuropathic pain. Therefore, more studies are needed to evaluate the mechanisms that IL-18 orchestrate in pain, in special, neuropathic pain. Table 1 summarizes the major interleukins known to be involved in neuropathic pain and their roles.

Mini-dictionary of terms

Allodynia: "Pain due to a stimulus that does not normally provoke pain" as defined by The International Association for the Study of Pain (IASP).

Glial cells: Nonneuronal cells in the central nervous system.

Hyperalgesia: "Increased pain sensation from a stimulus that normally elicits pain. Hyperalgesia reflects increased pain on suprathreshold stimulation" as defined by IASP.

Neuropathic pain: "Pain caused by a lesion or disease of the somatosensory nervous system" as defined by IASP.

Nociceptor: "A high-threshold sensory neuron of the peripheral somatosensory nervous system that is capable of transducing and encoding noxious stimuli" as defined by IASP.

Key facts of IL-33 in neuropathic pain

- IL-33 and its ST2 receptor mediate the development of neuropathic pain.
- Pharmacological inhibition of IL-33 was proposed as a novel treatment to controlling neuropathic pain.
- Neuron-glia crosstalk plays an important role in the development and maintenance of neuropathic pain.

TABLE 1 Interleukins and their actions on neuropathic pain.

Interleukin	Model	Effect	References
1β	SNI	Nociceptive	Gui et al. (2016)
	SNL		Honore et al. (2006)
4	CCI	Antinociceptive	Sun et al. (2016)
	SNI		Okutani, Yamanaka, Kobayashi, Okubo, and Noguchi (2018)
6	SNI	Nociceptive	Ding et al. (2016) and Ding, Guo, Li, Wang, and Zeng (2018)
	SNL		Liu et al. (2019)
8	CCI	Nociceptive	Khan, Hassun, Zusman, Korczeniewska, and Eliav (2017)
10	CCI	Antinociceptive	Milligan et al. (2006) and Vanderwall et al. (2018)
	CNP		Ledeboer et al. (2007)
17A	SNL	Nociceptive	Sun et al. (2017) and Yao et al. (2016)
	CNP		Luo et al. (2019)
18	SNL	Nociceptive	Miyoshi et al. (2008)
	CCI		Pilat et al. (2016)
27	SNL	Antinociceptive	Fonseca et al. (2019) and Sasaguri et al. (2018)
33	CCI	Nociceptive	Zarpelon et al. (2016)
	SNI		Liu et al. (2015)
35	DNP	Antinociceptive	Jiang, Wang, Li, and Xia (2019, 2020)

SNL, spinal nerve ligation; *CCI*, chronic constriction injury; *SNI*, spared nerve injury; *DNP*, diabetic neuropathic pain induced by streptozotocin; *CNP*, chemotherapy neuropathic pain induced by paclitaxel.

- MAPK signaling pathways are of critical importance in the expression of inflammatory mediator's involved in neuropathic pain.
- NFkB is an important transcriptional factor involved in the expression of inflammatory mediator's involved in neuropathic pain.

Summary points

- IL-33 is member of the IL-1 family.
- IL-33 signals through receptor ST2.
- There is an increase of IL-33 production in the spinal cord after nerve injury.
- Oligodendrocytes are the main glial cells producing IL-33 in the spinal cord.
- Glial cells are an important cell population in the CNS regulated by IL-33.
- Targeting IL-33 is analgesic in neuropathic pain.

Funding

We thank the financial support of Decit/SCTIE/MS intermediated by CNPq with support of the Araucária Foundation and SESA-PR (Brazil; PPSUS Grant agreement041/2017, protocol 48.095); PRONEX grant supported by SETI/Araucária Foundation and MCTI/CNPq; and Paraná State Government (agreement 014/2017, protocol 46.843). FSRO, AF, TSS and NAA thank the Ph.D. scholarship and TTC the postdoc fellowship from CAPES (Brazil, finance code 001). CRF acknowledges a CNPq PDJ postdoc fellowship. SMB acknowledges a FUNADESP research fellowship. WAVJ and RC acknowledge the CNPq Productivity fellowship.

References

Ben Haim, L., & Rowitch, D. H. (2017). Functional diversity of astrocytes in neural circuit regulation. *Nature Reviews. Neuroscience*, *18*(1), 31–41. https://doi.org/10.1038/nrn.2016.159.

Bertozzi, M. M., Rossaneis, A. C., Fattori, V., Longhi-Balbinot, D. T., Freitas, A., Cunha, F. Q., et al. (2017). Diosmin reduces chronic constriction injury-induced neuropathic pain in mice. *Chemico-Biological Interactions*, *273*, 180–189. https://doi.org/10.1016/j.cbi.2017.06.014.

Borghi, S. M., Fattori, V., Hohmann, M. S. N., & Verri, W. A. (2019). Contribution of spinal cord oligodendrocytes to neuroinflammatory diseases and pain. *Current Medicinal Chemistry*, *26*(31), 5781–5810. https://doi.org/10.2174/0929867325666180522112441.

Chen, W. Y., Tsai, T. H., Yang, J. L., & Li, L. C. (2018). Therapeutic strategies for targeting IL-33/ST2 signalling for the treatment of inflammatory diseases. *Cellular Physiology and Biochemistry*, *49*(1), 349–358. https://doi.org/10.1159/000492885.

Chen, G., Zhang, Y. Q., Qadri, Y. J., Serhan, C. N., & Ji, R. R. (2018). Microglia in pain: Detrimental and protective roles in pathogenesis and resolution of pain. *Neuron*, *100*(6), 1292–1311. https://doi.org/10.1016/j.neuron.2018.11.009.

Daigo, E., Sakuma, Y., Miyoshi, K., Noguchi, K., & Kotani, J. (2012). Increased expression of interleukin-18 in the trigeminal spinal subnucleus caudalis after inferior alveolar nerve injury in the rat. *Neuroscience Letters*, *529*(1), 39–44. https://doi.org/10.1016/j.neulet.2012.09.007.

Ding, C. P., Guo, Y. J., Li, H. N., Wang, J. Y., & Zeng, X. Y. (2018). Red nucleus interleukin-6 participates in the maintenance of neuropathic pain through JAK/STAT3 and ERK signaling pathways. *Experimental Neurology*, *300*, 212–221. https://doi.org/10.1016/j.expneurol.2017.11.012.

Ding, C. P., Xue, Y. S., Yu, J., Guo, Y. J., Zeng, X. Y., & Wang, J. Y. (2016). The red nucleus interleukin-6 participates in the maintenance of neuropathic pain induced by spared nerve injury. *Neurochemical Research*, *41*(11), 3042–3051. https://doi.org/10.1007/s11064-016-2023-9.

Drake, L. Y., & Kita, H. (2017). IL-33: Biological properties, functions, and roles in airway disease. *Immunological Reviews*, *278*(1), 173–184. https://doi.org/10.1111/imr.12552.

Fattori, V., Borghi, S. M., Rossaneis, A. C., Bertozzi, M. M., Cunha, T. M., Verri, W. A., et al. (2017). In R. Attaur, & M. I. Choudhary (Eds.), *Vol. 3. Neuroimmune regulation of pain and inflammation: Targeting glial cells and nociceptor sensory neurons interaction* (pp. 146–200). Bentham Science. https://doi.org/10.2174/9781681084435117030l.

Fattori, V., Hohmann, M. S. N., Rossaneis, A. C., Manchope, M. F., Alves-Filho, J. C., Cunha, T. M., et al. (2017). Targeting IL-33/ST2 signaling: Regulation of immune function and analgesia. *Expert Opinion on Therapeutic Targets*, *21*(12), 1141–1152. https://doi.org/10.1080/14728222.2017.1398734.

Fonseca, M. M., Davoli-Ferreira, M., Santa-Cecilia, F., Guimaraes, R. M., Oliveira, F. F. B., Kusuda, R., et al. (2019). IL-27 counteracts neuropathic pain development through induction of IL-10. *Frontiers in Immunology*, *10*, 3059. https://doi.org/10.3389/fimmu.2019.03059.

Foster, S. L., Talbot, S., & Woolf, C. J. (2015). CNS injury: IL-33 sounds the alarm. *Immunity*, *42*(3), 403–405. https://doi.org/10.1016/j.immuni.2015.02.019.

Gadani, S. P., Walsh, J. T., Smirnov, I., Zheng, J., & Kipnis, J. (2015). The glia-derived alarmin IL-33 orchestrates the immune response and promotes recovery following CNS injury. *Neuron*, *85*(4), 703–709. https://doi.org/10.1016/j.neuron.2015.01.013.

Gao, Y. J., & Ji, R. R. (2010). Chemokines, neuronal-glial interactions, and central processing of neuropathic pain. *Pharmacology & Therapeutics*, *126*(1), 56–68. https://doi.org/10.1016/j.pharmthera.2010.01.002.

Griesenauer, B., & Paczesny, S. (2017). The ST2/IL-33 Axis in immune cells during inflammatory diseases. *Frontiers in Immunology*, *8*, 475. https://doi.org/10.3389/fimmu.2017.00475.

Guan, Z., Kuhn, J. A., Wang, X., Colquitt, B., Solorzano, C., Vaman, S., et al. (2016). Injured sensory neuron-derived CSF1 induces microglial proliferation and DAP12-dependent pain. *Nature Neuroscience*, *19*(1), 94–101. https://doi.org/10.1038/nn.4189.

Gui, W. S., Wei, X., Mai, C. L., Murugan, M., Wu, L. J., Xin, W. J., et al. (2016). Interleukin-1beta overproduction is a common cause for neuropathic pain, memory deficit, and depression following peripheral nerve injury in rodents. *Molecular Pain*, *12*. https://doi.org/10.1177/1744806916646784.

Gwak, Y. S., Hulsebosch, C. E., & Leem, J. W. (2017). Neuronal-glial interactions maintain chronic neuropathic pain after spinal cord injury. *Neural Plasticity*, *2017*, 2480689. https://doi.org/10.1155/2017/2480689.

Gwak, Y. S., Kang, J., Unabia, G. C., & Hulsebosch, C. E. (2012). Spatial and temporal activation of spinal glial cells: Role of gliopathy in central neuropathic pain following spinal cord injury in rats. *Experimental Neurology*, *234*(2), 362–372. https://doi.org/10.1016/j.expneurol.2011.10.010.

Hains, B. C., & Waxman, S. G. (2006). Activated microglia contribute to the maintenance of chronic pain after spinal cord injury. *The Journal of Neuroscience*, *26*(16), 4308–4317. https://doi.org/10.1523/JNEUROSCI.0003-06.2006.

Han, P., Zhao, J., Liu, S. B., Yang, C. J., Wang, Y. Q., Wu, G. C., et al. (2013). Interleukin-33 mediates formalin-induced inflammatory pain in mice. *Neuroscience*, *241*, 59–66. https://doi.org/10.1016/j.neuroscience.2013.03.019.

Honore, P., Wade, C. L., Zhong, C., Harris, R. R., Wu, C., Ghayur, T., et al. (2006). Interleukin-1alphabeta gene-deficient mice show reduced nociceptive sensitivity in models of inflammatory and neuropathic pain but not post-operative pain. *Behavioural Brain Research*, *167*(2), 355–364. https://doi.org/10.1016/j.bbr.2005.09.024.

Hudson, C. A., Christophi, G. P., Gruber, R. C., Wilmore, J. R., Lawrence, D. A., & Massa, P. T. (2008). Induction of IL-33 expression and activity in central nervous system glia. *Journal of Leukocyte Biology*, *84*(3), 631–643. https://doi.org/10.1189/jlb.1207830.

Ishii, A., Furusho, M., & Bansal, R. (2013). Sustained activation of ERK1/2 MAPK in oligodendrocytes and schwann cells enhances myelin growth and stimulates oligodendrocyte progenitor expansion. *The Journal of Neuroscience*, *33*(1), 175–186. https://doi.org/10.1523/JNEUROSCI.4403-12.2013.

Ito, T., Ohtori, S., Inoue, G., Koshi, T., Doya, H., Ozawa, T., et al. (2007). Glial phosphorylated p38 MAP kinase mediates pain in a rat model of lumbar disc herniation and induces motor dysfunction in a rat model of lumbar spinal canal stenosis. *Spine (Phila Pa 1976)*, *32*(2), 159–167. https://doi.org/10.1097/01.brs.0000251437.10545.e9.

Jha, M. K., Jeon, S., & Suk, K. (2012). Glia as a link between neuroinflammation and neuropathic pain. *Immune Network*, *12*(2), 41–47. https://doi.org/10.4110/in.2012.12.2.41.

Ji, R. R., Berta, T., & Nedergaard, M. (2013). Glia and pain: Is chronic pain a gliopathy? *Pain*, *154*(Suppl. 1), S10–S28. https://doi.org/10.1016/j.pain.2013.06.022.

Ji, R. R., Donnelly, C. R., & Nedergaard, M. (2019). Astrocytes in chronic pain and itch. *Nature Reviews. Neuroscience*, *20*(11), 667–685. https://doi.org/10.1038/s41583-019-0218-1.

Jiang, H. R., Milovanović, M., Allan, D., Niedbala, W., Besnard, A. G., Fukada, S. Y., et al. (2012). IL-33 attenuates EAE by suppressing IL-17 and IFN-γ production and inducing alternatively activated macrophages. *European Journal of Immunology*, *42*(7), 1804–1814. https://doi.org/10.1002/eji.201141947.

Jiang, Y., Wang, J., Li, H., & Xia, L. (2019). IL-35 alleviates inflammation progression in a rat model of diabetic neuropathic pain via inhibition of JNK signaling. *Journal of Inflammation*, *16*, 19. https://doi.org/10.1186/s12950-019-0217-z.

Jiang, Y., Wang, J., Li, H., & Xia, L. (2020). IL-35 promotes microglial M2 polarization in a rat model of diabetic neuropathic pain. *Archives of Biochemistry and Biophysics*, *685*, 108330. https://doi.org/10.1016/j.abb.2020.108330.

Jin, S. X., Zhuang, Z. Y., Woolf, C. J., & Ji, R. R. (2003). p38 mitogen-activated protein kinase is activated after a spinal nerve ligation in spinal cord microglia and dorsal root ganglion neurons and contributes to the generation of neuropathic pain. *The Journal of Neuroscience*, *23*(10), 4017–4022. https://doi.org/10.1523/jneurosci.23-10-04017.2003.

Khan, J., Hassun, H., Zusman, T., Korczeniewska, O., & Eliav, E. (2017). Interleukin-8 levels in rat models of nerve damage and neuropathic pain. *Neuroscience Letters*, *657*, 106–112. https://doi.org/10.1016/j.neulet.2017.07.049.

Ledeboer, A., Jekich, B. M., Sloane, E. M., Mahoney, J. H., Langer, S. J., Milligan, E. D., et al. (2007). Intrathecal interleukin-10 gene therapy attenuates paclitaxel-induced mechanical allodynia and proinflammatory cytokine expression in dorsal root ganglia in rats. *Brain, Behavior, and Immunity*, *21*(5), 686–698. https://doi.org/10.1016/j.bbi.2006.10.012.

Liew, F. Y., Girard, J. P., & Turnquist, H. R. (2016). Interleukin-33 in health and disease. *Nature Reviews. Immunology*, *16*(11), 676–689. https://doi.org/10.1038/nri.2016.95.

Liu, Q., Chen, W., Fan, X., Wang, J., Fu, S., Cui, S., et al. (2019). Upregulation of interleukin-6 on Cav3.2 T-type calcium channels in dorsal root ganglion neurons contributes to neuropathic pain in rats with spinal nerve ligation. *Experimental Neurology*, *317*, 226–243. https://doi.org/10.1016/j.expneurol.2019.03.005.

Liu, S., Mi, W. L., Li, Q., Zhang, M. T., Han, P., Hu, S., et al. (2015). Spinal IL-33/ST2 Signaling contributes to neuropathic pain via neuronal CaMKII-CREB and Astroglial JAK2-STAT3 cascades in mice. *Anesthesiology*, *123*(5), 1154–1169. https://doi.org/10.1097/ALN.0000000000000850.

Liu, B., Tai, Y., Achanta, S., Kaelberer, M. M., Caceres, A. I., Shao, X., et al. (2016). IL-33/ST2 signaling excites sensory neurons and mediates itch response in a mouse model of poison ivy contact allergy. *Proceedings of the National Academy of Sciences of the United States of America*, *113*(47), E7572–E7579. https://doi.org/10.1073/pnas.1606608113.

Liu, Y. L., Zhou, L. J., Hu, N. W., Xu, J. T., Wu, C. Y., Zhang, T., et al. (2007). Tumor necrosis factor-alpha induces long-term potentiation of C-fiber evoked field potentials in spinal dorsal horn in rats with nerve injury: The role of NF-kappa B, JNK and p38 MAPK. *Neuropharmacology*, *52*(3), 708–715. https://doi.org/10.1016/j.neuropharm.2006.09.011.

Longhi-Balbinot, D. T., Rossaneis, A. C., Pinho-Ribeiro, F. A., Bertozzi, M. M., Cunha, F. Q., Alves-Filho, J. C., et al. (2016). The nitroxyl donor, Angeli's salt, reduces chronic constriction injury-induced neuropathic pain. *Chemico-Biological Interactions*, *256*, 1–8. https://doi.org/10.1016/j.cbi.2016.06.009.

Lourenco-Gonzalez, Y., Fattori, V., Domiciano, T. P., Rossaneis, A. C., Borghi, S. M., Zaninelli, T. H., et al. (2019). Repurposing of the nootropic drug Vinpocetine as an analgesic and anti-inflammatory agent: Evidence in a mouse model of superoxide anion-triggered inflammation. *Mediators of Inflammation*, *2019*, 6481812. https://doi.org/10.1155/2019/6481812.

Luo, H., Liu, H. Z., Zhang, W. W., Matsuda, M., Lv, N., Chen, G., et al. (2019). Interleukin-17 regulates neuron-glial communications, synaptic transmission, and neuropathic pain after chemotherapy. *Cell Reports*, *29*(8), 2384–2397 e2385. https://doi.org/10.1016/j.celrep.2019.10.085.

Magro, D. A., Hohmann, M. S., Mizokami, S. S., Cunha, T. M., Alves-Filho, J. C., Casagrande, R., et al. (2013). An interleukin-33/ST2 signaling deficiency reduces overt pain-like behaviors in mice. *Brazilian Journal of Medical and Biological Research*, *46*(7), 601–606. https://doi.org/10.1590/1414-431X20132894.

Milligan, E. D., Sloane, E. M., Langer, S. J., Hughes, T. S., Jekich, B. M., Frank, M. G., et al. (2006). Repeated intrathecal injections of plasmid DNA encoding interleukin-10 produce prolonged reversal of neuropathic pain. *Pain*, *126*(1–3), 294–308. https://doi.org/10.1016/j.pain.2006.07.009.

Miyoshi, K., Obata, K., Kondo, T., Okamura, H., & Noguchi, K. (2008). Interleukin-18-mediated microglia/astrocyte interaction in the spinal cord enhances neuropathic pain processing after nerve injury. *The Journal of Neuroscience*, *28*(48), 12775–12787. https://doi.org/10.1523/JNEUROSCI.3512-08.2008.

Molofsky, A. B., Savage, A. K., & Locksley, R. M. (2015). Interleukin-33 in tissue homeostasis, injury, and inflammation. *Immunity*, *42*(6), 1005–1019. https://doi.org/10.1016/j.immuni.2015.06.006.

Natarajan, C., Yao, S. Y., & Sriram, S. (2016). TLR3 agonist poly-IC induces IL-33 and promotes myelin repair. *PLoS One*, *11*(3). https://doi.org/10.1371/journal.pone.0152163, e0152163.

Okutani, H., Yamanaka, H., Kobayashi, K., Okubo, M., & Noguchi, K. (2018). Recombinant interleukin-4 alleviates mechanical allodynia via injury-induced interleukin-4 receptor alpha in spinal microglia in a rat model of neuropathic pain. *Glia*, *66*(8), 1775–1787. https://doi.org/10.1002/glia.23340.

Piao, Z. G., Cho, I. H., Park, C. K., Hong, J. P., Choi, S. Y., Lee, S. J., et al. (2006). Activation of glia and microglial p38 MAPK in medullary dorsal horn contributes to tactile hypersensitivity following trigeminal sensory nerve injury. *Pain*, *121*(3), 219–231. https://doi.org/10.1016/j.pain.2005.12.023.

Pilat, D., Piotrowska, A., Rojewska, E., Jurga, A., Slusarczyk, J., Makuch, W., et al. (2016). Blockade of IL-18 signaling diminished neuropathic pain and enhanced the efficacy of morphine and buprenorphine. *Molecular and Cellular Neurosciences*, *71*, 114–124. https://doi.org/10.1016/j.mcn.2015.12.013.

Pinho-Ribeiro, F. A., Verri, W. A., Jr., & Chiu, I. M. (2017). Nociceptor sensory neuron-immune interactions in pain and inflammation. *Trends in Immunology*, *38*(1), 5–19. https://doi.org/10.1016/j.it.2016.10.001.

Pomeshchik, Y., Kidin, I., Korhonen, P., Savchenko, E., Jaronen, M., Lehtonen, S., et al. (2015). Interleukin-33 treatment reduces secondary injury and improves functional recovery after contusion spinal cord injury. *Brain, Behavior, and Immunity*, *44*, 68–81. https://doi.org/10.1016/j.bbi.2014.08.002.

Raghavendra, V., Tanga, F., & DeLeo, J. A. (2003). Inhibition of microglial activation attenuates the development but not existing hypersensitivity in a rat model of neuropathy. *The Journal of Pharmacology and Experimental Therapeutics*, *306*(2), 624–630. https://doi.org/10.1124/jpet.103.052407.

Sasaguri, T., Taguchi, T., Murata, Y., Kobayashi, K., Iizasa, S., Iizasa, E., et al. (2018). Interleukin-27 controls basal pain threshold in physiological and pathological conditions. *Scientific Reports*, *8*(1), 11022. https://doi.org/10.1038/s41598-018-29398-3.

Sun, S., Chen, D., Lin, F., Chen, M., Yu, H., Hou, L., et al. (2016). Role of interleukin-4, the chemokine CCL3 and its receptor CCR5 in neuropathic pain. *Molecular Immunology*, *77*, 184–192. https://doi.org/10.1016/j.molimm.2016.08.006.

Sun, C., Zhang, J., Chen, L., Liu, T., Xu, G., Li, C., et al. (2017). IL-17 contributed to the neuropathic pain following peripheral nerve injury by promoting astrocyte proliferation and secretion of proinflammatory cytokines. *Molecular Medicine Reports*, *15*(1), 89–96. https://doi.org/10.3892/mmr.2016.6018.

Sung, H. Y., Chen, W. Y., Huang, H. T., Wang, C. Y., Chang, S. B., & Tzeng, S. F. (2019). Down-regulation of interleukin-33 expression in oligodendrocyte precursor cells impairs oligodendrocyte lineage progression. *Journal of Neurochemistry*, *150*(6), 691–708. https://doi.org/10.1111/jnc.14788.

Tsuda, M. (2016). Microglia in the spinal cord and neuropathic pain. *Journal of Diabetes Investigation*, *7*(1), 17–26. https://doi.org/10.1111/jdi.12379.

Tsuda, M., Inoue, K., & Salter, M. W. (2005). Neuropathic pain and spinal microglia: A big problem from molecules in "small" glia. *Trends in Neurosciences*, *28*(2), 101–107. https://doi.org/10.1016/j.tins.2004.12.002.

Vanderwall, A. G., Noor, S., Sun, M. S., Sanchez, J. E., Yang, X. O., Jantzie, L. L., et al. (2018). Effects of spinal non-viral interleukin-10 gene therapy formulated with d-mannose in neuropathic interleukin-10 deficient mice: Behavioral characterization, mRNA and protein analysis in pain relevant tissues. *Brain, Behavior, and Immunity*, *69*, 91–112. https://doi.org/10.1016/j.bbi.2017.11.004.

Verri, W. A., Jr., Cunha, T. M., Parada, C. A., Poole, S., Liew, F. Y., Ferreira, S. H., et al. (2007). Antigen-induced inflammatory mechanical hypernociception in mice is mediated by IL-18. *Brain, Behavior, and Immunity*, *21*(5), 535–543. https://doi.org/10.1016/j.bbi.2006.11.005.

Verri, W. A., Jr., Guerrero, A. T., Fukada, S. Y., Valerio, D. A., Cunha, T. M., Xu, D., et al. (2008). IL-33 mediates antigen-induced cutaneous and articular hypernociception in mice. *Proceedings of the National Academy of Sciences of the United States of America*, *105*(7), 2723–2728. https://doi.org/10.1073/pnas.0712116105.

Verri, W. A., Jr., Schivo, I. R., Cunha, T. M., Liew, F. Y., Ferreira, S. H., & Cunha, F. Q. (2004). Interleukin-18 induces mechanical hypernociception in rats via endothelin acting on ETB receptors in a morphine-sensitive manner. *The Journal of Pharmacology and Experimental Therapeutics*, *310*(2), 710–717. https://doi.org/10.1124/jpet.103.063990.

Verri, W. A., Jr., Souto, F. O., Vieira, S. M., Almeida, S. C., Fukada, S. Y., Xu, D., et al. (2010). IL-33 induces neutrophil migration in rheumatoid arthritis and is a target of anti-TNF therapy. *Annals of the Rheumatic Diseases*, *69*(9), 1697–1703. https://doi.org/10.1136/ard.2009.122655.

Xu, J. T., Xin, W. J., Wei, X. H., Wu, C. Y., Ge, Y. X., Liu, Y. L., et al. (2007). p38 activation in uninjured primary afferent neurons and in spinal microglia contributes to the development of neuropathic pain induced by selective motor fiber injury. *Experimental Neurology*, *204*(1), 355–365. https://doi.org/10.1016/j.expneurol.2006.11.016.

Yao, C. Y., Weng, Z. L., Zhang, J. C., Feng, T., Lin, Y., & Yao, S. (2016). Interleukin-17A acts to maintain neuropathic pain through activation of CaMKII/CREB Signaling in spinal neurons. *Molecular Neurobiology*, *53*(6), 3914–3926. https://doi.org/10.1007/s12035-015-9322-z.

Yasuda, K., Nakanishi, K., & Tsutsui, H. (2019). Interleukin-18 in health and disease. *International Journal of Molecular Sciences*, *20*(3). https://doi.org/10.3390/ijms20030649.

Yasuoka, S., Kawanokuchi, J., Parajuli, B., Jin, S., Doi, Y., Noda, M., et al. (2011). Production and functions of IL-33 in the central nervous system. *Brain Research*, *1385*, 8–17. https://doi.org/10.1016/j.brainres.2011.02.045.

Yoshida, S., Hagiwara, Y., Tsuchiya, M., Shinoda, M., Koide, M., Hatakeyama, H., et al. (2018). Involvement of neutrophils and interleukin-18 in nociception in a mouse model of muscle pain. *Molecular Pain*, *14*. https://doi.org/10.1177/1744806918757286, 1744806918757286.

Zarpelon, A. C., Cunha, T. M., Alves-Filho, J. C., Pinto, L. G., Ferreira, S. H., McInnes, I. B., et al. (2013). IL-33/ST2 signalling contributes to carrageenin-induced innate inflammation and inflammatory pain: Role of cytokines, endothelin-1 and prostaglandin E2. *British Journal of Pharmacology*, *169*(1), 90–101. https://doi.org/10.1111/bph.12110.

Zarpelon, A. C., Rodrigues, F. C., Lopes, A. H., Souza, G. R., Carvalho, T. T., Pinto, L. G., et al. (2016). Spinal cord oligodendrocyte-derived alarmin IL-33 mediates neuropathic pain. *The FASEB Journal*, *30*(1), 54–65. https://doi.org/10.1096/fj.14-267146.

Zeng, Y., Shi, Y., Zhan, H., Liu, W., Cai, G., Zhong, H., et al. (2020). Reduction of silent information regulator 1 activates interleukin-33/ST2 Signaling and contributes to neuropathic pain induced by spared nerve injury in rats. *Frontiers in Molecular Neuroscience*, *13*, 17. https://doi.org/10.3389/fnmol.2020.00017.

Zhao, J., Zhang, H., Liu, S. B., Han, P., Hu, S., Li, Q., et al. (2013). Spinal interleukin-33 and its receptor ST2 contribute to bone cancer-induced pain in mice. *Neuroscience*, *253*, 172–182. https://doi.org/10.1016/j.neuroscience.2013.08.026.

Chapter 17

Neurons of the parabrachial nucleus, nociceptive input, and pain pathways

Yosuke Arima, Yoshinori Otani, and Masashi Fujitani
Department of Anatomy and Neuroscience, Faculty of Medicine, Shimane University, Izumo, Shimane, Japan

Abbreviations

AgRP	agouti-related protein
Amy	amygdala
AVPV	anteroventral periventricular nucleus
BLA	basolateral amygdala
BNST	bed nucleus of the stria terminalis
CCK	cholecystokinin
CeA	central nucleus of the amygdala
CGRP	calcitonin gene-related peptide
CM	centromedial thalamic nucleus
CNS	central nervous system
CVLM	caudal ventrolateral medulla
HY	hypothalamus
IC	insular cortex
KF nucleus	Kölliker-Fuse nucleus
LH	lateral hypothalamus
LPB	lateral parabrachial nucleus
MnPO	median preoptic nucleus
MPB	medial parabrachial nucleus
NPY	neuropeptide Y
NST	solitary tract nucleus
PAG	periaqueductal gray
PBel	external lateral subnucleus
PBil	internal lateral subnucleus
PBN	parabrachial nucleus
PBsl	superior lateral subnucleus
PBvl	ventral lateral subnucleus
PNS	peripheral nervous system
Pbcl	central lateral subnucleus
PBcr	lateral crescent subnucleus
PBdl	dorsal lateral subnucleus
PeF	perifornical area
PV	paraventricular thalamic nucleus
RhN	rhomboid nucleus
RN	reticular nuclei
RVM	rostral ventromedial medulla
scp	superior cerebellar peduncle
SNpc	substantia nigra pars compacta
Sp5c	caudal trigeminal spinal nucleus
TH	thalamus
VMH	ventromedial hypothalamic area

The Neurobiology, Physiology, and Psychology of Pain. https://doi.org/10.1016/B978-0-12-820589-1.00017-8

Introduction

Pain information is perceived by a somatic sensory pathway in both the peripheral nervous system (PNS) and central nervous system (CNS), but it also provokes emotional behavior, autonomic reactions, and learning in humans and animals (Baliki & Apkarian, 2015; Bushnell, Ceko, & Low, 2013). To understand the entire pain circuit, it is necessary to understand the precise anatomy of pain circuits from the PNS to the CNS.

Peripheral nociceptors receiving inputs transmit signals to the spinal cord mainly by Aδ and C fibers (Dubin & Patapoutian, 2010). These nociceptive primary afferents enter the dorsal horn of the spinal cord through the dorsal root. Then, secondary nociceptive neurons receive pain information and project to the thalamus via the spinothalamic tract. Subsequently, the primary somatosensory cortex receives the information (Bushnell et al., 2013; Todd & Wang, 2020) (Fig. 1, black arrow). In addition to these direct pathways of pain perception, various regions in the CNS also receive and regulate pain information (Todd & Wang, 2020) (Fig. 1, orange arrow). These distinct pain pathways in humans and animals have been shown to play an important role not only in the perception of pain but also in the manifestation of various physiological phenomena, such as emotional behavior due to pain (Bushnell et al., 2013).

The parabrachial nucleus (PBN) is known as one of the important pontine nuclei that receive various kinds of afferent projections that transmit sensory information such as interoceptive (hunger and satiety) and exteroceptive (gustatory, pain, itch, and thermal sensations) senses (Chiang et al., 2019; Palmiter, 2018; Saper, 2016); it thus serves important cardiopulmonary and autonomic functions such as glucose homeostasis and temperature regulation (Davern, 2014) (Fig. 1).

The PBN is also regulated by other higher brain regions such as the hypothalamus (Fig. 1) and extends efferent projections to the brainstem nuclei to regulate spinal cord neurons (Fig. 2, green arrows). Therefore, these findings support the idea that the PBN is a critical hub region that integrates diverse sensory information, including pain, and provokes emotional behavior, autonomic reactions, and learning (Chiang et al., 2019; Palmiter, 2018).

In recent years, taking advantage of the viral tracing method, genetically modified animals, chemogenetics, and optogenetics, scientists have been able to visualize and control precise and specific neural circuits, including the pain pathway, even those in the PBN (Rodriguez et al., 2017). In this chapter, we review the anatomy of the PBN and the function of neurons in the PBN as well as related pain pathways.

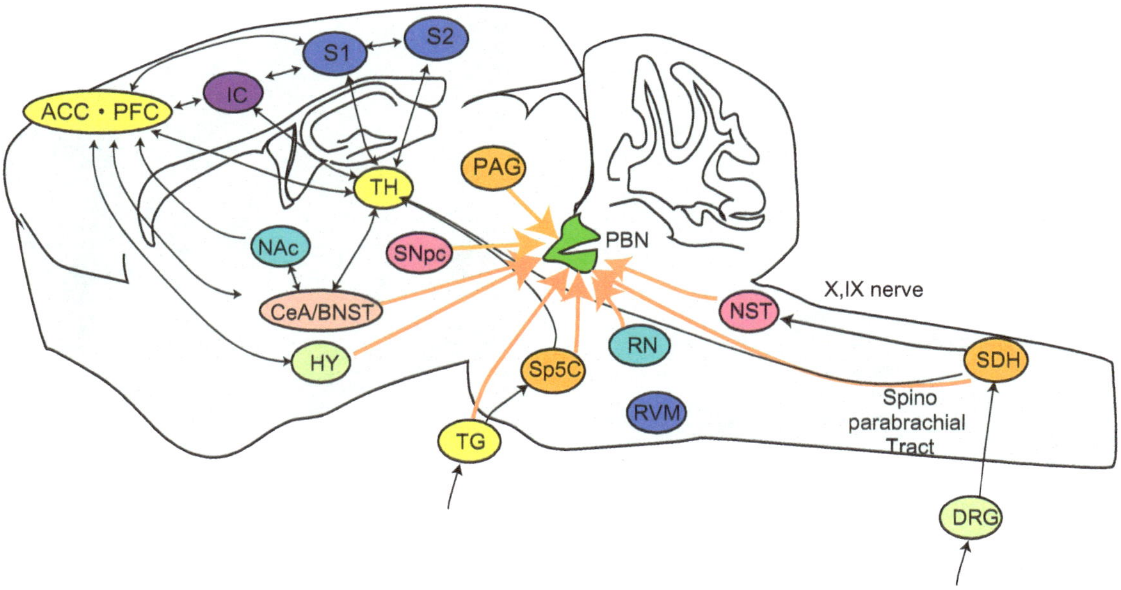

FIG. 1 Schematic summary for the afferents to parabrachial nucleus (PBN), based on the results from Rodriguez et al. (2017). *Orange arrows* represent direct pathways to the PBN. *ACC*, anterior cingulate cortex; *PFC*, prefrontal cortex; *IC*, insular cortex; *S1*, primary somatosensory cortex; *S2*, secondary somatosensory cortex; *TH*, thalamus; *PAG*, periaqueductal gray; *NAc*, nucleus accumbens; *CeA*, central nucleus of amygdala; *BNST*, bed nucleus of the stria terminalis; *SNpc*, substantia nigra pars compacta; *HY*, hypothalamus; *TG*, trigeminal ganglion; *Sp5C*, trigeminal nucleus caudalis; *RN*, reticular nuclei; *RVM*, rostral ventromedial medulla; *NST*, nucleus of solitary tract; *SDH*, spinal dorsal horn; *DRG*, dorsal root ganglion.

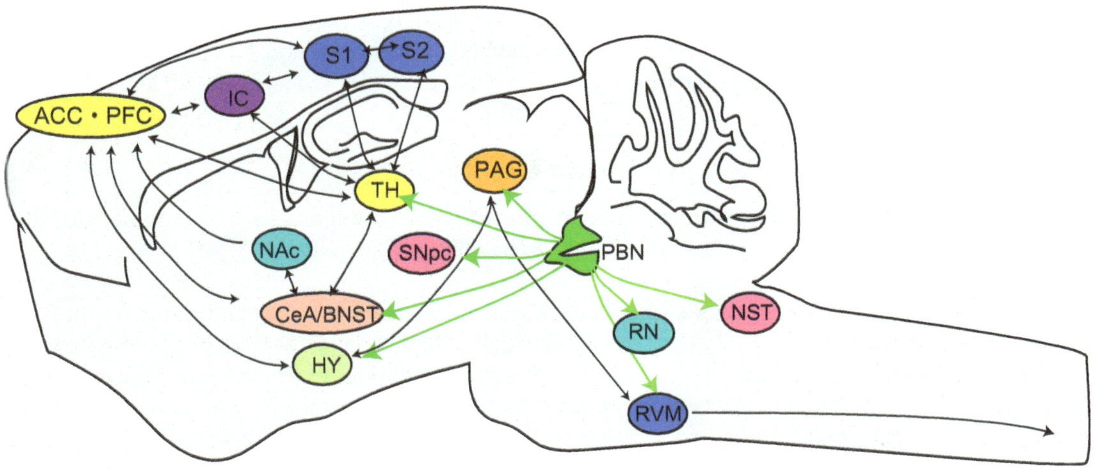

FIG. 2 Schematic summary for the efferents from parabrachial nucleus (PBN), based on the results from Rodriguez et al. (2017). *Green arrows* represent direct pathways from the PBN. *ACC*, anterior cingulate cortex; *PFC*, prefrontal cortex; *IC*, insular cortex; *S1*, primary somatosensory cortex; *S2*, secondary somatosensory cortex; *TH*, thalamus; *PAG*, periaqueductal gray; *NAc*, nucleus accumbens; *CeA*, central nucleus of amygdala; *BNST*, bed nucleus of the stria terminalis; *SNpc*, substantia nigra pars compacta; *HY*, hypothalamus; *TG*, trigeminal ganglion; *Sp5C*, trigeminal nucleus caudalis; *RN*, reticular nuclei; *RVM*, rostral ventromedial medulla; *NST*, nucleus of solitary tract; *SDH*, spinal dorsal horn; *DRG*, dorsal root ganglion.

Anatomy of the PBN

The PBN is the nucleus located around the superior cerebellar peduncle (scp) of the pons (Fig. 3, gray). Three main regions of the PBN have been identified: the medial parabrachial nuclei (MPB) are located in the ventral region of the scp (MPB, blue), the lateral parabrachial nuclei (LPB, yellow) at the dorsal part of the scp, and the Kölliker-Fuse nucleus (KF nucleus, green) is the most laterally located in the LPB (Fulwiler & Saper, 1984; Herbert, Moga, & Saper, 1990). While many distinct subnuclei have been identified in rats (Fulwiler & Saper, 1984; Herbert et al., 1990), only five have been discovered to date in primates (Kitamura, Nagao, Kunimoto, Shirama, & Yamada, 2001; Paxinos, Huang, Petrides, & Toga, 2008). Relative to the rat PB, the human PB is not as obvious (Gioia, Rodella, Petruccioli, & Bianchi, 2000); therefore, in this chapter, we mainly focus on rodent studies.

The PBN can be divided into a dozen regions (Hashimoto, Obata, & Ogawa, 2009) and the LPB comprises seven subnuclei (PBvl, ventral lateral subnucleus; PBil, internal lateral subnucleus; PBsl, superior lateral subnucleus; PBel, external lateral subnucleus; PBcl, central lateral subnucleus; PBcr, lateral crescent subnucleus; PBdl, dorsal lateral subnucleus) (Davern, 2014; Fulwiler & Saper, 1984). Each subnucleus has a characteristic projection pathway to various brain regions (Bernard, Huang, & Besson, 1994; Chiang et al., 2020; Gauriau & Bernard, 2002; Saper & Loewy, 1980). The majority of neurons localized in LPB are glutamatergic, while some are GABAergic (Geerling, Yokota, Rukhadze, Roe, & Chamberlin, 2017; Guthmann, Fritschy, Ottersen, Torp, & Herbert, 1998). A variety of neuropeptides, such as calcitonin

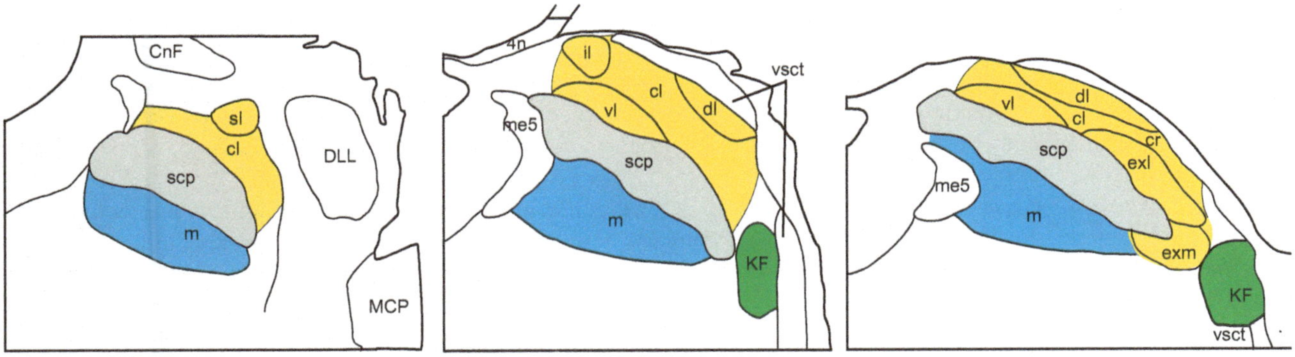

FIG. 3 Schematic summary for the subnuclei of the parabrachial nucleus (PBN). The *yellow area* represents the lateral division of the PBN (LPB), which consists of the following structures: *PBvl*, ventral lateral subnucleus; *PBil*, internal lateral subnucleus; *PBsl*, superior lateral subnucleus; *Pbel*, external lateral subnucleus; *Pbcl*, central lateral subnucleus; *PBcr*, lateral crescent subnucleus; *PBdl*, dorsal lateral subnucleus. The *blue area* represents the medial division of the PBN (MPB). The *gray area* is the superior cerebellar peduncle. The *green area* is the Kölliker-Fuse (KF) nucleus.

gene-related peptide (CGRP), substance P, neurotensin, and dynorphin also colocalize in LPB neurons. It has been revealed that these neurons are involved in the pain pathway, as described below (Barik, Thompson, Seltzer, Ghitani, & Chesler, 2018; Chiang et al., 2019; Han, Soleiman, Soden, Zweifel, & Palmiter, 2015; Okutsu et al., 2017).

Afferent pain transmitting pathways to the LPB

Information on pain sensation is transmitted directly from the dorsal horn of the spinal cord (spinoparabrachial pathway) or the caudal trigeminal spinal nucleus (Sp5c) to the LPB (Cameron et al., 2015; Craig, 1995; Gauriau & Bernard, 2002; Rodriguez et al., 2017) (Fig. 1, orange arrows).

The spinoparabrachial tract is a distinct nociceptive pathway that originates predominantly from lamina I of the dorsal horn (Cechetto, Standaert, & Saper, 1985; Hylden, Anton, & Nahin, 1989; Todd & Wang, 2020). Most targets of the spinoparabrachial tract are contralateral, although some seem to project to both sides of CNS (Spike, Puskar, Andrew, & Todd, 2003; Todd, McGill, & Shehab, 2000). In rodents, approximately 95% of the axonal fibers of lamina I in the lumbar spinal cord are projected to the parabrachial nucleus and caudal ventrolateral medulla (CVLM) (Spike et al., 2003; Todd, 2010). Electrophysiological studies in rats and cats have shown that the majority of spinoparabrachial tract neurons located in lamina I respond to somatic and visceral noxious stimuli (Bester, Beggs, & Woolf, 2000). Importantly, with regard to the target of the spinoparabrachial tract, our earlier study (Asano, Arima, Yokota, & Fujitani, 2019) and another by Chiang et al. (2020) showed that the spinoparabrachial tract is projected only to the dorsal PBN (Pbcl, PBcr, and PBdl) and not to the PBel (Chiang et al., 2020).

A recent transsynaptic viral tracing method also showed that the LPB receives projections from not only Sp5c neurons but also directly from trigeminal ganglion (TG) neurons (Rodriguez et al., 2017) (Fig. 1). Interestingly, TG and Sp5c neurons project Trpv1 positive fibers mainly to the PBel. This anatomical discrepancy could explain the difference in severity between craniofacial and somatic pain (Rodriguez et al., 2017).

In addition, it has been previously shown that the PBN receives various kinds of sensory inputs. The nucleus of the solitary tract (NST) transmits sensory inputs such as taste and other viscerosensory information related to respiratory control, fluid balance, and cardiovascular function (Herbert et al., 1990). MPB receives gustatory information from the rostral NST and visceral sensation from the caudal NST and relays information to the thalamus (Carleton, Accolla, & Simon, 2010). On the other hand, the KF nucleus is involved in the regulation of respiration (Yokota, Kaur, VanderHorst, Saper, & Chamberlin, 2015).

Nociceptive neurons in the LPB and their aversive pathway to amygdala and bed nucleus of the stria terminalis (BNST)

As described in Figs. 1 and 2, based on previous data, the LPB appears to be directly connected to numerous brain regions (Chiang et al., 2020; Gauriau & Bernard, 2002; Rodriguez et al., 2017). For emotional behavior and fear memory formation, the amygdala and the BNST are important brain regions. The amygdala is comprised of multiple nuclei located deep in the temporal lobe. The basolateral amygdala (BLA) receives all sensory inputs concerning the external environment from the sensory thalamus and sensory cortices (Janak & Tye, 2015). The BLA projects to the neighboring central nucleus of the amygdala (CeA), which serves major amygdala output functions. In addition, the lateral part of the CeA has been shown to be important for the pain pathway; thus, it has been termed the "nociceptive amygdala" (Neugebauer, 2015).

On the other hand, the BNST was originally considered a subregion of the "extended amygdala." The BNST is well positioned to integrate information for the limbic system and valence monitoring. The BNST has also become a relevant region of interest in the study of stress-related psychiatric diseases in addition to pain pathway processing (Goode & Maren, 2017; Lebow & Chen, 2016).

In order to determine nociceptive neurons localized in the parabrachial nucleus, a variety of techniques, including retrograde or anterograde tracing techniques, electrophysiology, and calcium imaging have been used (Barik et al., 2018; Campos, Bowen, Roman, & Palmiter, 2018; Chiang et al., 2020; Palmiter, 2018; Uddin et al., 2018). It has been reported that PBel responds to both somatic pain and visceral pain, and projects to the CeA (Bernard et al., 1994; Bernard & Besson, 1990; Gauriau & Bernard, 2002). More importantly, given that we and Chiang et al. (2020) showed that the spinoparabrachial tract did not project to PBel, the connection between the dorsal PBN and PBel should have been determined (Chiang et al., 2020). Chiang et al. (2020) found that dynorphin-expressing neurons in the dorsal PBN transmit pain signals to the PBel. These neurons in the dorsal PBN are required for aversive learning to transmit noxious inputs to the PBel-CeA/BNST pathway.

CGRP-positive neurons in the PBel

On the other hand, CGRP neurons in the PBel projecting to the amygdala are responsible for the transmission of pain (Campos et al., 2018; Campos, Bowen, Schwartz, & Palmiter, 2016; Han et al., 2015; Palmiter, 2018; Saper, 2016). CGRP is a 37-amino acid neuropeptide present in peripheral nerve endings. When CGRP is released, an increased level of intracellular cAMP via a receptor causes vasodilation, swelling, and induces inflammation (Assas, Pennock, & Miyan, 2014). CGRP-positive neurons are known to localize in the PBel of the PBN (Schwaber, Sternini, Brecha, Rogers, & Card, 1988).

There is emerging evidence that CGRP-positive PBN neurons control pain-related aversive behaviors. It is known that direct injection of CGRP into the amygdala induces freezing behavior even without foot shock (unconditioned stimulus [US]) (Kocorowski & Helmstetter, 2001). In addition, CGRP is also involved in synaptic plasticity in the amygdala via CGRP receptors (Han, Li, & Neugebauer, 2005; Shinohara et al., 2017). The neurons expressing CGRP project their axons directly to the CeA (D'Hanis, Linke, & Yilmazer-Hanke, 2007; Gauriau & Bernard, 2002; Schwaber et al., 1988). Furthermore, selective activation or inactivation of CGRP neurons in the PBel showed that they relay the unconditioned stimulus to the CeA (Han et al., 2005).

Recently, Rodriguez and colleagues also showed that craniofacial pain transmitted by the trigeminal nerve and CGRP-positive neurons elicits stronger aversion than pain in the trunk and extremities (Rodriguez et al., 2017). Taken together, CGRP neurons in the PBel are thought to be one of the key neural circuits mediating pain and fear behavior.

More recently, tachykinin-1(Tac1)-positive neurons in PBL have been shown to control escape responses that are interconnected with the reticular formation (Barik et al., 2018). Moreover, Chiang and colleagues showed that the cell bodies of neurons projecting to the CeA/BNST are primarily located in the PBel, whereas VMH (ventromedial hypothalamic area)/PAG (periaqueductal gray)-projecting neurons are concentrated in the dorsal PBN (Chiang et al., 2020). They also found that the PBel-CeA/BNST pathway is responsible for aversive memory and the PBdl-VMH/PAG pathway for escape behavior and analgesia (Chiang et al., 2020).

Nociceptive neurons in the LPB and their autonomic connection with the hypothalamus

The hypothalamus and PAG are deeply involved in the autonomic response due to pain. The hypothalamus plays a central role in the integration of autonomic and endocrine responses necessary for homeostasis. The posterior hypothalamic and lateral hypothalamic areas are related to sleep, autonomic control, and pain modulation (Cortelli, Giannini, Favoni, Cevoli, & Pierangeli, 2013).

Using the retrograde tracing method, Fulwiler and Saper (1984) revealed that LPB densely projects to the hypothalamus (Fulwiler & Saper, 1984). It has been confirmed more recently by the new viral method, but pain-related projections to the hypothalamus were relatively fewer (Rodriguez et al., 2017) than to the CeA or thalamus.

In our recent study, we also revealed that c-fos-positive PBN neurons induced by formalin project to the perifornical area (PeF) in the lateral hypothalamus, mainly localized in the dorsal PBN. They are located in PBcl and PBcr. These results suggest that nociceptive information is transmitted to the PeF via the dorsal PBN, and orexin neurons might be involved in the pain pathway (Asano et al., 2019).

Orexin is a neuropeptide that regulates sleep and arousal (Sakurai et al., 1998). Orexin neurons are known to project efferent fibers to large areas of the brain and play an important role in regulating arousal (Scammell, Arrigoni, & Lipton, 2017) and pain (Razavi & Hosseinzadeh, 2017).

Niu, Yokota, Tsumori, Qin, and Yasui (2010) previously showed that glutamatergic neurons in the LPB establish excitatory synapses with orexin neurons (Niu et al., 2010). In addition to these results, we also showed that LPB neurons send projections via the hypothalamus to brainstem arousal areas, such as the ventral tegmental area (VTA), dorsal raphe nucleus (DR), pedunculopontine tegmental nucleus (PPT), laterodorsal tegmental area (LDT), and locus coeruleus (LC) (Arima, Yokota, & Fujitani, 2019).

Furthermore, it has been shown that agouti-related protein (AgRP) neurons in the arcuate nucleus of the hypothalamus act on the parabrachial nucleus to attenuate pain with neuropeptide Y (NPY). Hunger-sensitive AgRP neurons suppress inflammatory pain in the PBN during hunger (Alhadeff et al., 2018).

Pain-modulating pathways from the LPB

Direct evidence that the LPB is involved in the pain modulatory system has recently emerged. The LPB projects densely to the PAG and stimulation of LPB terminals in the PAG induces analgesia (Chiang et al., 2020). These results are similar to those resulting from the direct stimulation of PAG (Ossipov, Dussor, & Porreca, 2010).

The PAG is located around the middle aqueduct within the tegmentum of the brainstem and is known to play a critical role in responses to analgesia. Electrical or morphine stimulation of the PAG results in immediate analgesia in animals (Ossipov et al., 2010).

The PAG induces analgesia via two pain modulatory systems. One pathway is the ascending inhibitory circuit from the PAG, sending information to the thalamus via the spinothalamic tract (Millan, 2002; Ossipov et al., 2010); another is the descending inhibitory circuit from the PAG, which controls pain pathway in the spinal cord via projections to the rostral ventral medulla (RVM). Thus, the PAG-RVM pathway is an important link in the brainstem pain-modulating circuit (Millan, 2002; Ossipov et al., 2010).

In contrast, the circuit from the LPB to the RVM exerts a pronociceptive effect in persistent pain states (Chen et al., 2017; Chen & Heinricher, 2019). The RVM consists of three nuclei, including the nucleus raphe magnus, nucleus reticularis gigantocellularis, and nucleus reticularis paragigantocellularis (Ossipov et al., 2010). A direct connection from the LPB to RVM pain-modulating neurons allows the LPB to promote or limit the development and maintenance of pathological pain (Chen et al., 2017; Chen & Heinricher, 2019). There are two distinct classes of RVM neurons, "ON-cells" and "OFF-cells," which have pro-nociceptive and antinociceptive effects, respectively (Ossipov et al., 2010).

Chen and Heinricher (2019) recently showed that the contralateral and ipsilateral PBN demonstrate distinct functions when inflammation persists over a period of days. The contralateral PBN relays nociceptive information to the RVM in acute inflammation and contributes to behavioral hyperalgesia, whereas the ipsilateral PBN maintains both the hyperalgesia and responsiveness of RVM neurons in persistent inflammation. These experiments show that the contralateral PBN is required to initiate hyperalgesia, which is then maintained by the ipsilateral PBN, most likely in both cases via the engagement of pain-modulating neurons of the RVM (Chen & Heinricher, 2019).

Conclusion

PBN contributes to distinct behavioral and motivational outputs as a hub of pain information in animals. Based on these observations, researchers are now trying to understand the function of this complex region. However, information is still lacking. First, the internal pain circuit of PBN is elusive. Second, a distinct source and function of each projection connected with PBN have not been fully revealed.

In general, the investigation of neuro-circuit is often based on viral tracing with Cre recombinase-expressing mouse under the transcriptional control of the specific neurotransmitter. However, limitations essentially exist within this powerful method. Multiple neurotransmitters have been revealed to coexist in the same neuron and cotransmit (Granger, Wallace, & Sabatini, 2017). Therefore, it is still unclear whether the PBN has an inhibitory function despite neurons containing glutamate. Further investigation using genetic intersectional method with the entire set of neurotransmitters and their receptors in the rodents may remove the limitations as in the *Drosophila* experiments (Deng et al., 2019).

Last, as mentioned previously, human PBN is not as obvious as animal PBN, highlighting the need to unveil the function of PBN in humans. This information of difference between species may resolve the evolutionary difference of the pain system.

Unveiling the specific neural circuit, neurotransmitter, or neurons in PBNs, especially in humans, will strongly and directly contribute to the development of acute or chronic pain treatment.

Applications to other areas

In the PBN, there are many additional neurons that play important roles in transmitting sensory information besides CGRP neurons, as follows:

The gustatory NST transmits each taste (sweet, sour, bitter, umami, and salty) to the PBN neurons and then to the thalamus and gustatory cortex.

PBN neurons maintain water and salt balance. Salt appetite (salt deprivation) activates LPB neurons via mineralocorticoid-sensitive neurons in the NST.

To mediate satiety, LPB neurons receive input from MC4R-positive paraventricular hypothalamic neurons. In addition, during hypoglycemia, activated glucose sensing LPB neurons that express cholecystokinin (CCK) and the leptin receptor maintain glucose levels. Interestingly, this pathway helps glucose mobilization during noxious stimuli.

Regarding temperature regulation, two distinct populations of neurons in the PBN transmit the behavioral and physiological responses to nonnoxious heat and cold. These neurons project to the preoptic area and their activation stimulates movement to a preferred temperature zone, stimulating heat generation or conservation.

The PBN neurons also modulate respiration and cardiac function. Glutamatergic neurons in the lateral crescent of the PBN and the adjacent KF nucleus are activated by hypercapnia. These neurons send their axons to the medullary regions involved in respiratory control.

Mini-dictionary of terms

Chemogenetics. Chemogenetics is a specific method to control neuronal or glial cell activity, which are most widely used in the neuroscience field. Experiments employ compounds that selectively target a genetically modified G protein-coupled receptor (GPCR) or a chimeric ion channel. The former is generally called DREADD (designer receptor exclusively activated by designer drugs) and the latter is a pharmacologically selective effector molecule (PSAM).

Optogenetics. Optogenetics is also a specific method to excite or inhibit target neurons or glia in vivo at much shorter intervals than chemogenetics, such as millisecond intervals. After expression of chanerhodopsin and halorhodopsin in specific neurons, the irradiation by specific wavelengths of light excites or inhibits target neurons or glia.

AgRP (agouti-related protein). Agouti-related protein (AgRP) is identified as a hypothalamic protein with high sequence homology to Agouti and a natural antagonist of the alpha-melanocyte stimulating hormone action at the melanocortin receptors which are associated with food intake. Also, AgRP is reported that play a role in pain modulation pathway to LPB. Taken together, AgRP is an important modulator of energy balance and pain.

CCK (cholecystokinin). Cholecystokinin (CCK) is a peptide hormone and plays a key role in facilitating digestion within the small intestine. CCK is synthesized and secreted by enteroendocrine cells of the upper small intestine, and also secreted by some neurons in the brain. It acts on two types of receptors found throughout the gut and CNS. CCK have important roles in anxiety, panic disorders, and pain.

Neuropeptide Y (NPY). Neuropeptide Y (NPY) is a 36-amino acid member of the family of peptides comprising NPY, pancreatic polypeptide (PP), and peptide YY (PYY) and is involved in various physiological and homeostatic functions in both CNS and PNS. Recently, AgRP neurons have been observed to suppress neuropeptide Y signaling in the LPB. Therefore, NPY is also modulated to pain pathway with AgRP.

Summary points

- The parabrachial nucleus (PBN) receives various kinds of afferent projections to transmit sensory information, including pain.
- The PBN is comprised of distinct pain pathways that play an important role in the manifestation of emotional behavior, autonomic reaction, and learning due to pain.
- The PBN not only receives sensory information but is also regulated by other higher brain regions.
- The PBN also extends efferent projections to brainstem nuclei to modulate the pain state (analgesia or persistent pain).
- The PBN is a critical hub region that integrates diverse sensory information, including pain, and provokes emotional behavior, autonomic reaction, and learning.

References

Alhadeff, A. L., Su, Z., Hernandez, E., Klima, M. L., Phillips, S. Z., Holland, R. A., et al. (2018). A neural circuit for the suppression of pain by a competing need state. *Cell*, *173*, 140–152.e115.

Arima, Y., Yokota, S., & Fujitani, M. (2019). Lateral parabrachial neurons innervate orexin neurons projecting to brainstem arousal areas in the rat. *Scientific Reports*, *9*, 2830.

Asano, H., Arima, Y., Yokota, S., & Fujitani, M. (2019). New nociceptive circuits to the hypothalamic perifornical area from the spinal cord and spinal trigeminal nucleus via the parabrachial nucleus. *Biochemical and Biophysical Research Communications*, *512*, 705–711.

Assas, B. M., Pennock, J. I., & Miyan, J. A. (2014). Calcitonin gene-related peptide is a key neurotransmitter in the neuro-immune axis. *Frontiers in Neuroscience*, *8*, 23.

Baliki, M. N., & Apkarian, A. V. (2015). Nociception, pain, negative moods, and behavior selection. *Neuron*, *87*, 474–491.

Barik, A., Thompson, J. H., Seltzer, M., Ghitani, N., & Chesler, A. T. (2018). A brainstem-spinal circuit controlling nocifensive behavior. *Neuron*, *100*, 1491–1503.e1493.

Bernard, J. F., & Besson, J. M. (1990). The spino(trigemino)pontoamygdaloid pathway: Electrophysiological evidence for an involvement in pain processes. *Journal of Neurophysiology*, *63*, 473–490.

Bernard, J. F., Huang, G. F., & Besson, J. M. (1994). The parabrachial area: Electrophysiological evidence for an involvement in visceral nociceptive processes. *Journal of Neurophysiology*, *71*, 1646–1660.

Bester, H., Beggs, S., & Woolf, C. J. (2000). Changes in tactile stimuli-induced behavior and c-Fos expression in the superficial dorsal horn and in parabrachial nuclei after sciatic nerve crush. *Journal of Comparative Neurology*, *428*, 45–61.

Bushnell, M. C., Ceko, M., & Low, L. A. (2013). Cognitive and emotional control of pain and its disruption in chronic pain. *Nature Reviews Neuroscience*, *14*, 502–511.

Cameron, D., Polgar, E., Gutierrez-Mecinas, M., Gomez-Lima, M., Watanabe, M., & Todd, A. J. (2015). The organization of spinoparabrachial neurons in the mouse. *Pain*, *156*, 2061–2071.

Campos, C. A., Bowen, A. J., Roman, C. W., & Palmiter, R. D. (2018). Encoding of danger by parabrachial CGRP neurons. *Nature*, *555*, 617–622.

Campos, C. A., Bowen, A. J., Schwartz, M. W., & Palmiter, R. D. (2016). Parabrachial CGRP neurons control meal termination. *Cell Metabolism*, *23*, 811–820.

Carleton, A., Accolla, R., & Simon, S. A. (2010). Coding in the mammalian gustatory system. *Trends in Neurosciences*, *33*, 326–334.

Cechetto, D. F., Standaert, D. G., & Saper, C. B. (1985). Spinal and trigeminal dorsal horn projections to the parabrachial nucleus in the rat. *Journal of Comparative Neurology*, *240*, 153–160.

Chen, Q., & Heinricher, M. M. (2019). Plasticity in the link between pain-transmitting and pain-modulating systems in acute and persistent inflammation. *The Journal of Neuroscience*, *39*, 2065–2079.

Chen, Q., Roeder, Z., Li, M. H., Zhang, Y., Ingram, S. L., & Heinricher, M. M. (2017). Optogenetic evidence for a direct circuit linking nociceptive transmission through the parabrachial complex with pain-modulating neurons of the rostral ventromedial medulla (RVM). *eNeuro*, *4*. ENEURO.0202–17.2017.

Chiang, M. C., Bowen, A., Schier, L. A., Tupone, D., Uddin, O., & Heinricher, M. M. (2019). Parabrachial complex: A hub for pain and aversion. *The Journal of Neuroscience*, *39*, 8225–8230.

Chiang, M. C., Nguyen, E. K., Canto-Bustos, M., Papale, A. E., Oswald, A. M., & Ross, S. E. (2020). Divergent neural pathways emanating from the lateral parabrachial nucleus mediate distinct components of the pain response. *Neuron*, *106*, 927–939.e925.

Cortelli, P., Giannini, G., Favoni, V., Cevoli, S., & Pierangeli, G. (2013). Nociception and autonomic nervous system. *Neurological Sciences*, *34*, S41–S46.

Craig, A. D. (1995). Distribution of brainstem projections from spinal lamina I neurons in the cat and the monkey. *Journal of Comparative Neurology*, *361*, 225–248.

D'Hanis, W., Linke, R., & Yilmazer-Hanke, D. M. (2007). Topography of thalamic and parabrachial calcitonin gene-related peptide (CGRP) immunoreactive neurons projecting to subnuclei of the amygdala and extended amygdala. *Journal of Comparative Neurology*, *505*, 268–291.

Davern, P. J. (2014). A role for the lateral parabrachial nucleus in cardiovascular function and fluid homeostasis. *Frontiers in Physiology*, *5*, 436.

Deng, B., Li, Q., Liu, X., Cao, Y., Li, B., Qian, Y., et al. (2019). Chemoconnectomics: Mapping chemical transmission in Drosophila. *Neuron*, *101*, 876–893.

Dubin, A. E., & Patapoutian, A. (2010). Nociceptors: The sensors of the pain pathway. *Journal of Clinical Investigation*, *120*, 3760–3772.

Fulwiler, C. E., & Saper, C. B. (1984). Subnuclear organization of the efferent connections of the parabrachial nucleus in the rat. *Brain Research*, *319*, 229–259.

Gauriau, C., & Bernard, J. F. (2002). Pain pathways and parabrachial circuits in the rat. *Experimental Physiology*, *87*, 251–258.

Geerling, J. C., Yokota, S., Rukhadze, I., Roe, D., & Chamberlin, N. L. (2017). Kolliker-Fuse GABAergic and glutamatergic neurons project to distinct targets. *Journal of Comparative Neurology*, *525*, 1844–1860.

Gioia, M., Rodella, L., Petruccioli, M. G., & Bianchi, R. (2000). The cytoarchitecture of the adult human parabrachial nucleus: A Nissl and Golgi study. *Archives of Histology and Cytology*, *63*, 411–424.

Goode, T. D., & Maren, S. (2017). Role of the bed nucleus of the stria terminalis in aversive learning and memory. *Learning & Memory*, *24*, 480–491.

Granger, A. J., Wallace, M. L., & Sabatini, B. L. (2017). Multi-transmitter neurons in the mammalian central nervous system. *Current Opinion in Neurobiology*, *45*, 85–91.

Guthmann, A., Fritschy, J. M., Ottersen, O. P., Torp, R., & Herbert, H. (1998). GABA, GABA transporters, GABA(A) receptor subunits, and GAD mRNAs in the rat parabrachial and Kolliker-Fuse nuclei. *Journal of Comparative Neurology*, *400*, 229–243.

Han, J. S., Li, W., & Neugebauer, V. (2005). Critical role of calcitonin gene-related peptide 1 receptors in the amygdala in synaptic plasticity and pain behavior. *The Journal of Neuroscience*, *25*, 10717–10728.

Han, S., Soleiman, M. T., Soden, M. E., Zweifel, L. S., & Palmiter, R. D. (2015). Elucidating an affective pain circuit that creates a threat memory. *Cell*, *162*, 363–374.

Hashimoto, K., Obata, K., & Ogawa, H. (2009). Characterization of parabrachial subnuclei in mice with regard to salt tastants: Possible independence of taste relay from visceral processing. *Chemical Senses*, *34*, 253–267.

Herbert, H., Moga, M. M., & Saper, C. B. (1990). Connections of the parabrachial nucleus with the nucleus of the solitary tract and the medullary reticular formation in the rat. *Journal of Comparative Neurology*, *293*, 540–580.

Hylden, J. L. K., Anton, F., & Nahin, R. L. (1989). Spinal lamina I projection neurons in the rat: Collateral innervation of parabrachial area and thalamus. *Neuroscience*, *28*, 27–37.

Janak, P. H., & Tye, K. M. (2015). From circuits to behavior in the amygdala. *Nature*, *517*, 284–292.

Kitamura, T., Nagao, S., Kunimoto, K., Shirama, K., & Yamada, J. (2001). Cytoarchitectonic subdivisions of the parabrachial nucleus in the Japanese monkey (Macacus fuscatus) with special reference to spinoparabrachial fiber terminals. *Neuroscience Research*, *39*, 95–108.

Kocorowski, L. H., & Helmstetter, F. J. (2001). Calcitonin gene-related peptide released within the amygdala is involved in Pavlovian auditory fear conditioning. *Neurobiology of Learning and Memory*, *75*, 149–163.

Lebow, M. A., & Chen, A. (2016). Overshadowed by the amygdala: The bed nucleus of the stria terminalis emerges as key to psychiatric disorders. *Molecular Psychiatry*, *21*, 450–463.

Millan, M. J. (2002). Descending control of pain. *Progress in Neurobiology*, *66*, 355–474.

Neugebauer, V. (2015). Amygdala pain mechanisms. *Handbook of Experimental Pharmacology*, *227*, 261–284.

Niu, J. G., Yokota, S., Tsumori, T., Qin, Y., & Yasui, Y. (2010). Glutamatergic lateral parabrachial neurons innervate orexin-containing hypothalamic neurons in the rat. *Brain Research*, *1358*, 110–122.

Okutsu, Y., Takahashi, Y., Nagase, M., Shinohara, K., Ikeda, R., & Kato, F. (2017). Potentiation of NMDA receptor-mediated synaptic transmission at the parabrachial-central amygdala synapses by CGRP in mice. *Molecular Pain*, *13*, 1744806917709201.

Ossipov, M. H., Dussor, G. O., & Porreca, F. (2010). Central modulation of pain. *Journal of Clinical Investigation*, *120*, 3779–3787.

Palmiter, R. D. (2018). The parabrachial nucleus: CGRP neurons function as a general alarm. *Trends in Neurosciences*, *41*, 280–293.

Paxinos, G., Huang, X., Petrides, M., & Toga, A. (2008). *The rhesus monkey brain in stereotaxic coordinates* (2nd ed.). New York: Elsevier (Chapter _).

Razavi, B. M., & Hosseinzadeh, H. (2017). A review of the role of orexin system in pain modulation. *Biomedicine & Pharmacotherapy*, *90*, 187–193.

Rodriguez, E., Sakurai, K., Xu, J., Chen, Y., Toda, K., Zhao, S., et al. (2017). A craniofacial-specific monosynaptic circuit enables heightened affective pain. *Nature Neuroscience*, *20*, 1734–1743.

Sakurai, T., Amemiya, A., Ishii, M., Matsuzaki, I., Chemelli, R. M., Tanaka, H., et al. (1998). Orexins and orexin receptors: A family of hypothalamic neuropeptides and G protein-coupled receptors that regulate feeding behavior. *Cell*, *92*, 573–585.

Saper, C. B. (2016). The house alarm. *Cell Metabolism*, *23*, 754–755.

Saper, C. B., & Loewy, A. D. (1980). Efferent connections of the parabrachial nucleus in the rat. *Brain Research*, *197*, 291–317.

Scammell, T. E., Arrigoni, E., & Lipton, J. O. (2017). Neural circuitry of wakefulness and sleep. *Neuron*, *93*, 747–765.

Schwaber, J. S., Sternini, C., Brecha, N. C., Rogers, W. T., & Card, J. P. (1988). Neurons containing calcitonin gene-related peptide in the parabrachial nucleus project to the central nucleus of the amygdala. *Journal of Comparative Neurology*, *270*, 416–426. 398–9.

Shinohara, K., Watabe, A. M., Nagase, M., Okutsu, Y., Takahashi, Y., Kurihara, H., et al. (2017). Essential role of endogenous calcitonin gene-related peptide in pain-associated plasticity in the central amygdala. *European Journal of Neuroscience*, *46*, 2149–2160.

Spike, R. C., Puskar, Z., Andrew, D., & Todd, A. J. (2003). A quantitative and morphological study of projection neurons in lamina I of the rat lumbar spinal cord. *European Journal of Neuroscience*, *18*, 2433–2448.

Todd, A. J. (2010). Neuronal circuitry for pain processing in the dorsal horn. *Nature Reviews Neuroscience*, *11*, 823–836.

Todd, A. J., McGill, M. M., & Shehab, S. A. (2000). Neurokinin 1 receptor expression by neurons in laminae I, III and IV of the rat spinal dorsal horn that project to the brainstem. *European Journal of Neuroscience*, *12*, 689–700.

Todd, A. J., & Wang, F. (2020). Central nervous system pain pathways. In J. N. Wood (Ed.), *The Oxford handbook of the neurobiology of pain* (pp. 414–444). Oxford: Oxford University Press.

Uddin, O., Studlack, P., Akintola, T., Raver, C., Castro, A., Masri, R., et al. (2018). Amplified parabrachial nucleus activity in a rat model of trigeminal neuropathic pain. *Neurobiology of Pain*, *3*, 22–30.

Yokota, S., Kaur, S., VanderHorst, V. G., Saper, C. B., & Chamberlin, N. L. (2015). Respiratory-related outputs of glutamatergic, hypercapnia-responsive parabrachial neurons in mice. *Journal of Comparative Neurology*, *523*, 907–920.

Chapter 18

Anterior cingulate cortex, pain perception, and pathological neuronal plasticity during chronic pain

Fernando Kasanetz[a], Mario A. Acuña[b], and Thomas Nevian[b]

[a]*Universidad de Buenos Aires, CONICET, Instituto de Fisiología y Biofísica (IFIBIO) Bernardo Houssay, Grupo de Neurociencia de Sistemas, Buenos Aires, Argentina,* [b]*Department of Physiology, University of Bern, Bern, Switzerland*

Abbreviations

5-HT7R	serotonin receptor type 7
ACC	anterior cingulate cortex
CPA	conditioned place avoidance
fMRI	functional magnetic resonance
HCN	hyperpolarization-activated cyclic nucleotide-regulated
IC	insular cortex
LTP	long-term potentiation
MCC	medial cingulate cortex
PAG	periaqueductal gray
PET	positron emission tomography

Introduction

The relationship between the intensity of perceived pain and the magnitude of nociceptive inputs is not linear. Like any other sensory modality, pain perception is influenced by a plethora of internal and external factors. The emotional state has an enormous influence on pain, a negative emotional state increases pain, while a positive state reduces pain (Bushnell, Čeko, & Low, 2013). Negative emotional experiences can produce the sensation of pain in the absence of tissue damage, and cognitive and emotional processes such as expectation or anxiety can modulate the efficacy of analgesia (Tracey, 2010). Also, stress, the level of attention, sleep deprivation, or social and cultural variables affect the way pain is perceived.

How does the initial activation of nociceptors in the periphery cause the perception of pain and trigger an adequate behavioral outcome? Taking into account all the variables influencing the experience of pain, it is not hard to imagine that processing nociceptive information in the context of a variety of other signals requires the activity of neurons distributed across extensive brain circuits.

It is considered that sensory-discriminative and affective-motivational domains of pain are processed in separate brain pathways (Treede, Kenshalo, Gracely, & Jones, 1999). Early anatomical and electrophysiological work pointed to lateral thalamic nuclei as mediators of pain localization and medial thalamic nuclei as responsible for affective responses (Albe-Fessar, Berkley, Kruger, Ralston, & Willis, 1985). The advent of human functional imaging allowed the identification of the entire cortical representation of pain perception by measuring brain structures activated by noxious stimuli while people rating a stimulus intensity and unpleasantness. This way, some regions have been related to sensory and discriminatory processing of pain and others presumably important for cognitive-affective-motivational pain processing (Fig. 1). Together, these areas comprise the so-called "pain matrix," and the general activity pattern evoked by a noxious stimulus in these areas constitutes the "neurological signature of physical pain" (Wager et al., 2013).

The Neurobiology, Physiology, and Psychology of Pain. https://doi.org/10.1016/B978-0-12-820589-1.00018-X

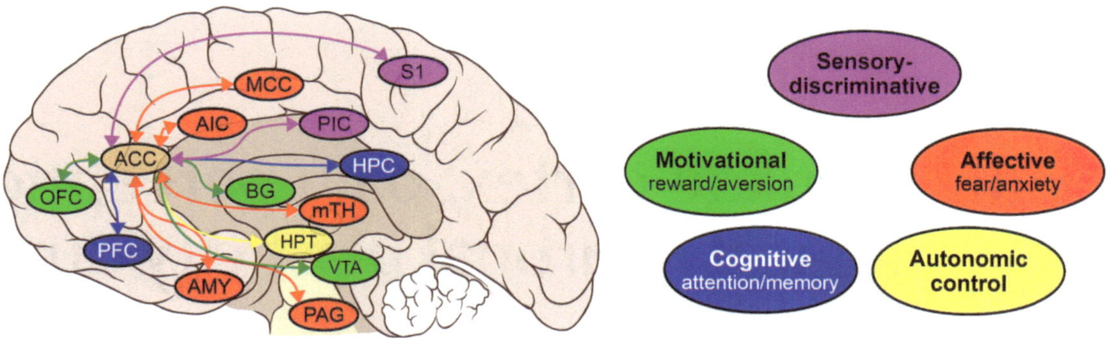

FIG. 1 The ACC is interconnected with brain regions related to the different aspects of pain. *ACC*, anterior cingulate cortex; *AIC*, anterior insular cortex; *AMY*, amygdala; *BG*, basal ganglia; *HPC*, hippocampus; *HPT*, hypothalamus; *MCC*, medial cingulate cortex; *mTH*, medial thalamus; *OFC*, orbitofrontal cortex; *PAG*, periaqueductal gray; *PIC*, posterior insular cortex; *PFC*, prefrontal cortex; *S1*, primary somatosensory cortex; *VTA*, ventral tegmental area.

Despite the tremendous progress made during the last three decades, understanding the precise role of each cortical region and how their neuronal circuits encode the different attributes of the pain experience is still very much elusive. The goal of this chapter is to present an overview of our current understanding on the role of the anterior cingulate cortex (ACC) on the processing of nociceptive information. We will focus on the ACC because it is probably the most profoundly studied cortical region in the context of pain in both humans and preclinical animal models. We will also pay a special attention on how chronic pain conditions affect ACC function and will evaluate a putative role for ACC pathological plasticity as a mechanism for pain chronicity.

ACC is a critical hub linking neuronal circuits for nociception and emotion

The ACC is the more-anterior portion of the cingulate gyrus, surrounding the genu of the corpus callosum. The more caudal part of the cingulate cortex is divided in middle (MCC) and posterior regions. The MCC is involved in several aspects of the pain experience (see below), and the posterior cingulate is associated with cognitive functions like spatial orientation and visual awareness (Stevens, 2011; Vogt, 2005).

Since the pioneer studies on the neurobiological basis of emotions, the ACC was recognized as part of the so-called limbic system. From its extensive interconnections with other brain regions related to fear, anxiety, memory, attention, reward processing, and autonomous control, the ACC is positioned as a critical hub for the integration and regulation of affect (Fig. 1).

The relation between ACC and nociception is also long dated. Initial studies examined the effect of surgical ablation of the ACC and the surrounding cingulate bundle in people suffering persistent intractable pain (Foltz & White, 1962). Most patients reported that they sometimes still felt nociceptive stimuli but that they were less distressing or bothersome. Later, the discovery of dense anatomical projections from medial and intralaminar thalamic nuclei to the ACC identified direct routes by which spinothalamic nociceptive information could be transmitted to this cortical area. More recently, numerous functional brain imaging studies in humans, using positron emission tomography (PET) or functional magnetic resonance imaging (fMRI), consistently reported acute nociceptive responses in cingulate cortex, including ACC and MCC (Apkarian, Bushnell, Treede, & Zubieta, 2005; Jensen et al., 2016; Vogt, 2005). Furthermore, specific patterns of activity in the ACC contribute to a neurologic signature of physical pain that predicts pain intensity at the level of the individual person (Wager et al., 2013).

By no means is neuronal activity in the ACC evoked by nociceptive stimulation selective for pain. Robust ACC responses can be elicited by salient but nonpainful auditory, tactile and visual stimuli, and during conditions that engage sustained attention, memory, emotion not related to pain, reward expectancy or motor activity (Mouraux, Diukova, Lee, Wise, & Iannetti, 2011; Wager et al., 2016). In addition, only a fraction of ACC neurons are modulated by pain (see below). Thus, while the ACC may encode one or several aspect of the multidimensional pain experience, it cannot be considered a "nociceptive cortex."

ACC activation is associated with the affective component of pain

Accumulated evidence from human imaging studies suggests that the ACC may have a role in the affective response to pain. The ACC is activated by both noxious and innocuous thermal stimuli, and the magnitude of the response is correlated with stimulus intensity (Coghill, Sang, Maisog, & Iadarola, 1999; Kwan, Crawley, Mikulis, & Davis, 2000). The feeling of unpleasantness and pain caused by the thermal-grill illusion, in which a combination of nonnoxious heat and cold stimuli are presented to adjacent skin areas, leads to a profound activation of the ACC (Craig, Reiman, Evans, & Bushnell, 1996). Moreover, placebo treatments that decrease pain perception are associated with reduced ACC activation (Wager et al., 2004). A regression analysis between noxious heat-related functional brain signals and experimental parameters reflecting pain perception showed that ACC was related to pain threshold and pain unpleasantness (Tölle et al., 1999) but not to stimulus intensity (Büchel et al., 2002). Furthermore, focusing attention to a painful stimulus enhances its perceived intensity and unpleasantness (Bushnell et al., 2013) and it is associated with increased responses in various brain regions, including the ACC (Peyron et al., 1999). The opposite is also true; distracting attention away from pain decreases its perception with a reduced ACC activity (Frankenstein, Richter, McIntyre, & Rémy, 2001). Although ACC areas recruited by attentional processes seem to not coincide with ACC regions responding to pain (Frankenstein et al., 2001), some studies found increased activity in pain-related areas of the ACC (despite reduced pain perception) when distraction from pain was implemented through a big attention load to another task (Valet et al., 2004). This may reflect the involvement of some ACC subcircuits in top-down cortico-thalamo-PAG pathways that modulate gating of nociceptive signals. In agreement with this, rostral ACC activations have been observed associated to placebo and opioid-mediated analgesia (Petrovic, 2002). Alternatively, ACC recruitment during distraction from pain could be due to the increased negative emotion that commonly accompanies highly demanding attentional tasks (Bushnell et al., 2013).

The perception of pain intensity (how strong we feel it) and unpleasantness (how bothersome it is) are normally coupled, making it difficult to distinguish whether ACC codes for one or the other. Whereas attentional manipulations are able to alter the perception of both pain intensity and unpleasantness, evidences indicate that mood changes selectively alter the affective-reactive response to pain. Manipulations that have a positive effect on mood or emotional state, such as pleasant music, odors or reading, generally reduce pain perception without modifying pain intensity ratings (Bushnell et al., 2013). This way, it was shown that pleasant odors, independent of attentional focus, induced positive mood changes and decreased pain unpleasantness associated with reduced pain-related fMRI signals in the ACC (Bushnell et al., 2013). Furthermore, during hypnotic suggestions that change pain unpleasantness without modifying the perceived intensity of stimuli, ACC activity correlated with people's rating of unpleasantness (Rainville, Duncan, Price, Carrier, & Bushnell, 1997).

A reasonable explanation for the experiments described above would be that ACC is part of the neural circuits that attribute affective meaning to painful events. As pain is the quintessential aversive condition that animals find in nature, this interpretation fits well with the general view of ACC activity as signaling value—or uncertainty—about reward or punishment to solve problems, recognize errors, control emotions, and adapt responses according to ever changing conditions (Davis 2005; Monosov, 2017). On the other hand, the codification of pain in the ACC could be understood rather in terms of using the affective value (codified elsewhere or partially in the ACC) in the context of previously acquired emotional memory to select appropriate behavioral strategies to cope with pain (Vogt, 2005). In both cases, ACC activity would correlate with the perceived unpleasantness of painful events, either attributing the affective value or interpreting it in a proper context.

ACC neuronal activity mediates pain-induced negative affect

Direct evidence of the ACC's instrumental role on encoding the affective meaning of pain was provided by preclinical studies using rodents. This approach allows for invasive interventions that evaluate the necessity and sufficiency of ACC neural activity in nociceptive processing. Nevertheless, an intrinsic pitfall of animal studies lies on our difficulty to estimate precisely the animal's subjective perception of pain unpleasantness (Mogil, 2009). Recently, conditioned place avoidance (CPA) tests were adapted to discriminate the aversive component of nociceptive stimuli. Animals are exposed to a behavioral box with two different compartments, receiving a painful stimulus in one compartment and a neutral stimulus in the other. The time spent in each compartment (measured in real time or during a test session, depending on the protocol used) is quantified and the avoidance of the pain-related context is considered an indication of the unpleasantness of the experience (Fuchs, Peng, Boyette-Davis, & Uhelski, 2014; Johansen, Fields, & Manning, 2001).

Using this strategies, researchers have found that manipulations that reduce or eliminate ACC activity—such as cytotoxic lesions, lidocaine, morphine, or glutamate antagonist microinjection—impaired the expression of CPA to noxious stimuli (Fuchs et al., 2014; Johansen et al., 2001; Johansen & Fields, 2004). Remarkably, none of those manipulations affected mechanical or thermal nociceptive thresholds, indicating that peripheral and spinal nociceptive signaling was intact and that ACC activity was necessary to cope with the unpleasantness of the pain experience. However, these studies do not distinguish whether ACC neurons are necessary to encode the affective quality of pain or for the expression of an avoidance behavior. In a recent report (Zhang et al., 2017), optogenetic manipulations were used to inhibit the activity of

ACC neurons only concomitant to the application of noxious stimuli. This way it was found that ACC neuronal activity in response to noxious events during the conditioning session is necessary to induce CPA 24 h later on a test session. This evidence suggests that the ACC may play a role in mediating or interpreting the aversive quality of pain.

A remaining open question from those results is whether ACC neuronal activity, in addition to be necessary, is sufficient to induce an aversive signal. Zhang and colleges (Zhang et al., 2017) paired optogenetic activation of ACC neurons with a low-intensity noxious stimulus and observed a stronger CPA than the one produced with the stimulus alone. Interestingly, this pain-enhancing effect was not observed when ACC activation was presented with nonnoxious stimuli. Altogether, this suggests that ACC neuronal activity is sufficient to enhance the aversive quality of a noxious stimulus but cannot convert an acute nonnoxious stimulus into a noxious one. This finding is in agreement with observations made in patients undergoing ablation surgeries of the ACC, where electrical stimulation of the ACC failed to produce painful or unpleasant sensations (Hutchison, Davis, Lozano, Tasker, & Dostrovsky, 1999). Conversely, pharmacological prolonged activation of the ACC alone (without pairing with noxious events) was sufficient to produce an aversive teaching signal in rats tested in a CPA paradigm (Johansen & Fields, 2004). Thus, it seems that a sustained activation or transient activation concomitant to noxious (although mild) inputs are required to generate an aversive signal in the ACC.

Nociception-related neurons in the ACC

In addition to the coarse structure of pain processing investigated with PET and fMRI signals, ACC responses to noxious stimuli were also observed at the single neuron level. Electrophysiological recording during cingulotomy procedures for the treatment of psychiatric disease identified neurons in the human ACC that responded selectively to painful thermal and mechanical stimuli (Hutchison et al., 1999). Further experiments in nonhuman primates, rabbits, and rodents have found that ACC neurons fire action potentials in response to nociceptive stimuli (Sikes & Vogt, 1992; Koyama, Kato, Tanaka, & Mikami, 2001; Iwata et al., 2005; Zhang et al., 2017; Fig. 2). While only a fraction of recorded ACC neurons were engaged in nociception (with value ranging from a low 3% up to 40% among the reports), none of them showed specificity for stimulations to a given part of the body, in agreement with a lack of somatotopic representation in the ACC. This suggests that the ACC might generalize nociceptive information independent from its location. In monkeys, ACC neurons fired also

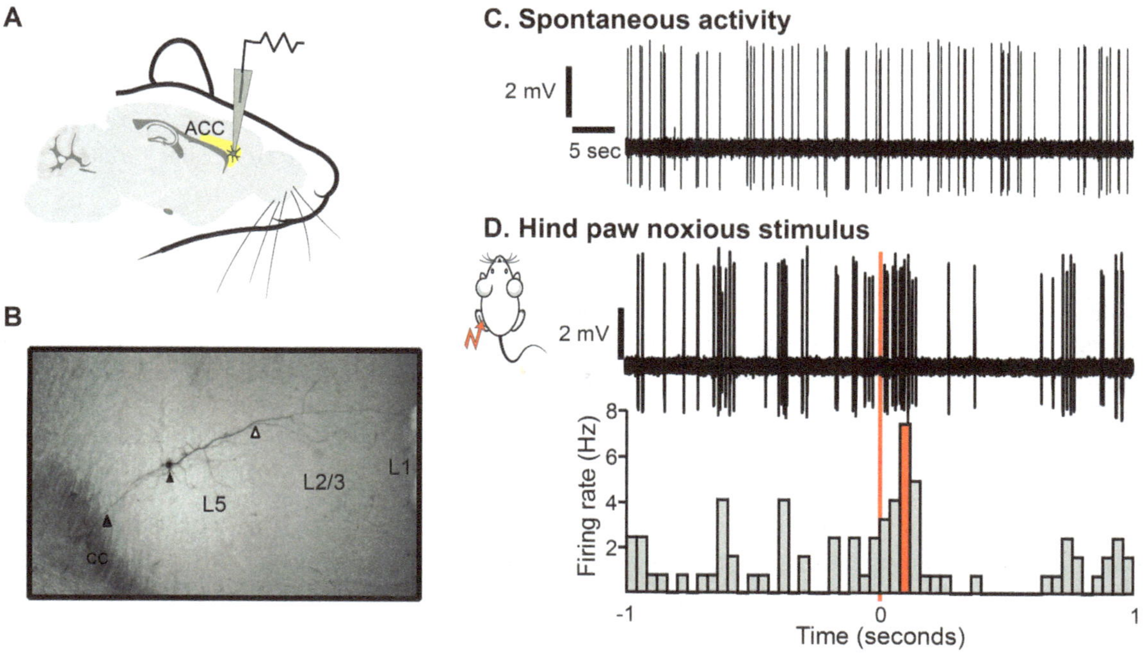

FIG. 2 ACC neurons are activated by noxious stimuli. (A) In vivo juxtacellular electrical recording of ACC neurons. (B) Microphotograph of a layer 5 ACC neuron labeled with biocytin during the electrophysiological recording. (C and D) Representative traces of firing activity of an ACC neuron. Vertical deflections are single action potentials. In D, repeated stimulation trials were aligned at the time of stimulus presentation *(vertical red line)*. *Red bar* in the peri-stimulus histogram represents a significant increase in firing *(red bar)* after stimulus. *ACC*, anterior cingulate cortex; unpublished data from Nevian's lab. *(Mouse cartoon downloaded and modified from Scidraw.io (Federico Claudi).)*

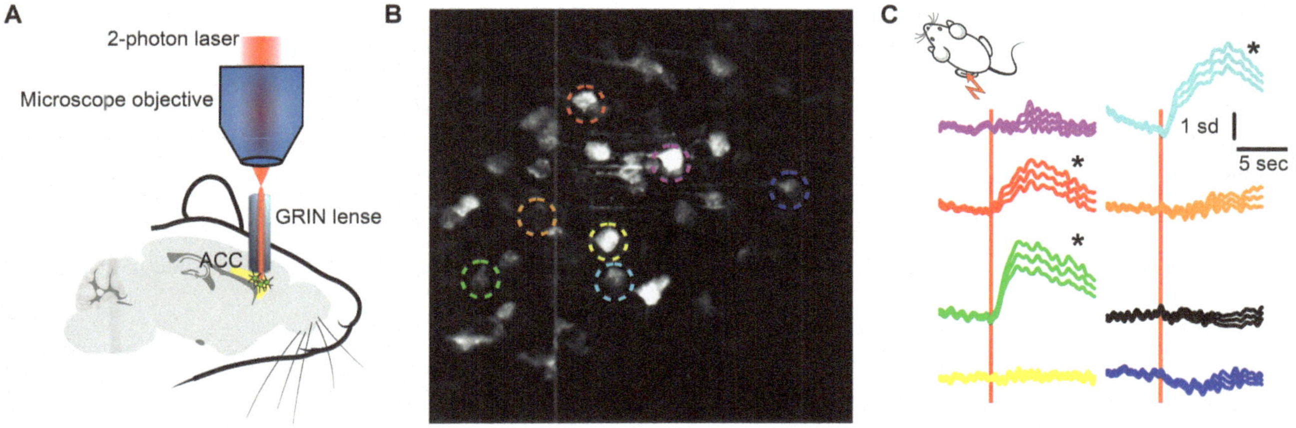

FIG. 3 In vivo calcium imaging of ACC neuronal network with single-cell resolution. (A) ACC neurons are transfected with a genetically encoded calcium indicator and are imaged through a chronically implanted GRIN lens under a two photon microscope. (B) Snapshot image of a group of ACC neurons. (C) Activity traces (mean Z score ± standard errors) in response to noxious stimuli *(vertical red line)* of the neurons encircled in B. *Neurons activated by the stimuli. *sd*, standard deviation; *ACC*, anterior cingulate cortex; *GRIN*, gradient-index; unpublished data from Nevian's lab. *(Mouse cartoon downloaded and modified from Scidraw.io (Federico Claudi).)*

during pain anticipation or response/avoidance, and most neurons activated by noxious thermal stimuli also fired during the behavioral response to cope with it, indicating that single neurons can codify different phases of the nociceptive response (Iwata et al., 2005).

Since the ACC participates in numerous different processes, the discovery of nociceptive neurons raised the possibility that the ACC may contain nociception-selective neuronal assemblies intermixed in a broader multifaceted neuronal network. However, the evidence on whether those neurons are nociception-selective is still controversial. Most of the neurons responding to noxious thermal and mechanical stimuli were not modulated by warmth or innocuous brush (Iwata et al., 2005; Sikes & Vogt, 1992; Zhang et al., 2017). Furthermore, some of the neurons fired in a graded manner to increasing intensities of noxious thermal stimuli. Nevertheless, around half of nociception-related neurons were also activated by an abrupt but innocuous tapping of the skin (Sikes & Vogt, 1992), suggesting that suddenness or salience of the stimulus may be an important attribute for recruiting those cells.

Neuronal networks can process information using rate coding (the magnitude of firing in individual cells) or via the spike-timing coordinated dynamics of neurons within large ensembles. It is currently unknown whether nociception selectivity in the ACC could be attained by such ensemble representation. State-of-the-art in vivo imaging approaches now allow to addressee this question by recording chronic neuronal network dynamics with single-cell resolution (Fig. 3). Future studies using these techniques promise to shed light on the fine structure of pain representation in the brain.

Distinct cortical representation of acute and chronic pain

Despite most chronic pain syndromes originate from peripheral or spinal nociceptive sensitization, pathological mechanisms in pain-related brain regions may contribute to the manifestation of persistence sensory and affective symptoms (Ballantyne & Sullivan, 2015). In fact, functional imaging studies show that experimental pain in normal subjects and chronic pain in patients have distinct but overlapping brain activation patterns, with a decreased sensory processing and enhanced affective, emotional, and cognitive representation during chronic pain (Apkarian et al., 2005).

Based on the prominent role of the ACC in the affective connotation of pain, numerous studies investigated whether neuronal representation of noxious events in the ACC was altered during chronic pain conditions. Enhanced ACC activity in patients compared to controls was observed in a variety of pain conditions (Apkarian et al., 2005; Jensen et al., 2016), including irritable bowel syndrome, migraine attacks, ongoing neuropathic pain, sympathetically mediated chronic pain, phantom limb pain, complex regional pain syndrome, and fibromyalgia. However, other studies have found a diminished ACC activation or no differences at all (Apkarian et al., 2005). Furthermore, a recent metaanalysis of the literature using activation likelihood estimation indicates that although the ACC is commonly activated by pain both in patients and healthy individuals, the likelihood to activate the ACC is decreased in chronic pain patients (Jensen et al., 2016). This controversial data could be related to distinct brain pathways affected by different chronic pain conditions and to the diverse aspects of pain perception in which ACC takes part. Enhanced ACC activity induced by noxious stimuli may be related to a greater discomfort and suffering observed in patients (Gracely, Petzke, Wolf, & Clauw, 2002; Mertz et al., 2000). A failure to increase ACC activity in other patients may be due to a ceiling effect, in which constant noxious input to nociceptive brain regions provokes a maximum level of activity that cannot display additional modulation by noxious stimuli during experimental testing. Alternatively—or even complementary to this explanation—it may occur that ACC networks involved in descending modulation of pain or in triggering appropriate behavioral strategies to cope with pain cannot be properly recruited by nociceptive inputs. This last conclusion is supported by studies investigating conditioned pain modulation, which requires ACC activity, during chronic pain. It has been shown that the ability of conditioning stimuli to reduce pain, such as distractors, is impaired in people with chronic pain (Bushnell et al., 2013). Thus, deficient ACC activation as part of a system that normally inhibits pain may contribute to enhanced pain perception in patients with chronic pain.

Pathological neuronal plasticity in the ACC associated to chronic pain

Neuronal plasticity refers to the ability of the nervous system to change its function in response to previous experience. It is a ubiquitous phenomenon throughout the entire nervous system and is essential for the establishment and refinement of neural networks during development, the formation of memory traces, the acquisition of specific skills, and the storage of information (Bliss, Collingridge, Kaang, & Zhuo, 2016). The cellular correlates underlying these phenomena involve morphological and functional adaptations in synapses, dendrites, and axons. Together, these mechanisms result in a change in the input-output function of the neural network and, therefore, in a modification of information processing. Today we know that numerous diseases are accompanied by profound modifications in cellular and synaptic function (Bliss et al., 2016; Costigan, Scholz, & Woolf, 2009; Kasanetz et al., 2010), suggesting that aberrant neuronal plasticity is a key element in the genesis of mental illness.

It is widely accepted that chronic pain results from a series of pathological mechanisms at all levels of the nociceptive system. Despite initial attention was devoted to changes occurring at primary afferents and spinal cord circuits (Costigan et al., 2009), it is clear now that substantial pathological neuronal plasticity also occurs in brain structures involved in pain perception, including the ACC.

Long-term potentiation (LTP) of glutamatergic synaptic transmission was observed in the ACC of mice models of neuropathic and inflammatory pain (Bliss et al., 2016; Fig. 4). Remarkably, interfering with LTP in the ACC by blocking the intracellular messenger, PKMzeta alleviated neuropathic pain-related allodynia, while the blocker of the hyperpolarization-activated cyclic nucleotide-regulated (HCN) channels, presumably by reversing enhanced glutamate release, interfered with chronic pain-induced anxiety (Bliss et al., 2016).

Neuropathic pain in mice also results in a profound structural modification of the connectivity pattern in layer 5 of the ACC. It was found that neuropathic pain results in a loss of inhibitory synapses onto excitatory pyramidal neurons and a loss of excitatory synapses onto inhibitory fast spiking interneurons (Blom, Pfister, Santello, Senn, & Nevian, 2014; Fig. 4), which altogether may converge to an increased network activity by disinhibition. Furthermore, using the same model of neuropathic pain, a dysfunction of HCN channels in the dendrites of deep layer 5 pyramidal neurons was discovered (Santello & Nevian, 2015). This pain-induced phenotype resulted in enhanced dendritic excitability and increased neuronal firing (Fig. 4). Specific activation of the serotonin receptor type 7 (5-HT7R) in this neurons in vitro alleviated the dendritic pathology by increasing HCN channel function again and restoring normal dendritic integration. Injection of a 5-HT7R agonist specifically into the ACC of neuropathic mice in vivo reduced mechanical pain hypersensitivity (Santello & Nevian, 2015) and pain unpleasantness, tested in a CPA paradigm (Santello et al., 2017).

Finally, in support for a role of ACC activity in the manifestation of chronic pain, recent studies have found that optogenetic inhibition of ACC neurons in rodent models of inflammatory and neuropathic pain reverses chronic pain-related aversion (Zhang et al., 2017) and mood disorders (Sellmeijer et al., 2018), respectably.

Altogether, these findings demonstrate that chronic pain profoundly modifies the structure and function of ACC neurons and that targeting the mechanisms underlying this aberrant neuronal (or network) plasticity may provide new therapeutic approaches for chronic pain.

Applications to other areas

Interfering with pathological cortical plasticity as a therapeutic approach for chronic pain

The discovery that neurological and psychiatric diseases are associated with profound remodeling of neural circuits modified our understanding of brain pathology. A traditional pharmacological view based on the alterations of large modulatory

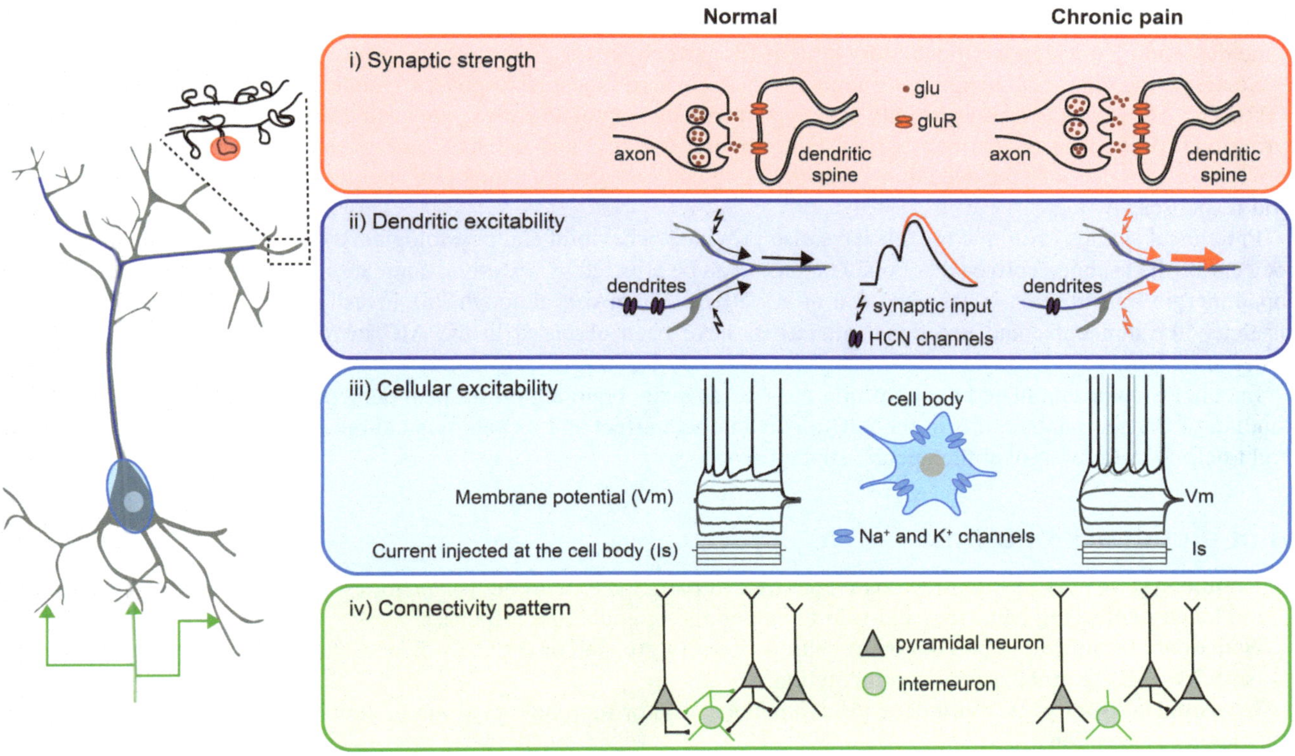

FIG. 4 Schematic representation of the neuronal adaptations observed in the ACC during neuropathic pain. (i) Enhanced synaptic strength involves presynaptic and postsynaptic mechanisms. (ii) Increased dendritic excitability due to functional downregulation of HCN channels facilitates synaptic integration and neuronal firing. (iii) Enhanced somatic excitability results in lower threshold for action potential generation and higher firing frequency. (iv) Altered connectivity pattern results in increased network activity by disinhibition of pyramidal neurons. *HCN*, hyperpolarization-activated cyclic nucleotide-regulated channels. *Na^+*, sodium; *K^+*, potassium. *(Neuron sketch downloaded and modified from Scidraw.io (Federico Claudi).)*

systems, such as dopaminergic, serotoninergic, or stress-related molecules, shifted toward an integrative approach where diseases (or at least some key aspects of it) emerge as a consequence of multiple and diverse synaptic and cellular changes that alter neural circuits function.

Based on this premise, great attention has been directed to interventions that could hijack activity-dependent neuronal plasticity mechanisms to restore function in abnormal brain networks. Thus, noninvasive brain stimulation techniques, such as transcranial magnetic stimulation or transcranial direct current stimulation, have been tested in recent years for numerous brain disorders, including depression, dementia, Parkinson's disease, tinnitus, chronic pain, migraine, and dystonia (Obeso, Oliviero, & Jahanshahi, 2016).

The clinical efficacy of noninvasive brain stimulation techniques for the treatment of chronic pain conditions is still modest (O'Connell, Marston, Spencer, DeSouza, & Wand, 2018), suggesting that further technical refinement is needed to optimize stimulation protocols, select appropriate brain targets, and define suitable clinical symptoms to treat. This should be guided by a mechanistic understanding of the neuronal plasticity that occurs in each of the regions that comprise the pain matrix. For example, the distinction of the aberrant plasticity that is specific to ACC subcircuits involved in top-down inhibition of pain from the changes that take place in ACC ensembles encoding pain unpleasantness may allow to design interventions to enhance activity in the former while tempering responses in the later. Thus, it is expected that a better understanding of the pathological brain alterations associated to chronic pain may lead to a full realization of the therapeutic potential of noninvasive brain stimulation techniques.

Other agents of interest

Anterior insular cortex and pain affect

In addition to the ACC, the anterior IC (AIC) may also participate in encoding pain unpleasantness. The AIC is mostly interconnected with structures of the medial nociceptive pathway and is believed to take part in the emotional and cognitive modulation of pain. Through direct projections to the amygdala, hypothalamic nuclei and PAG, the AIC is also implicated in the descending nociceptive modulatory system (Benarroch, 2019).

Studies using functional imaging in humans have discovered robust IC responses to nociceptive stimuli (Apkarian et al., 2005). The AIC seems to be consistently coactivated with the cingulate cortex, suggesting similar or complementary roles during pain processing (Benarroch, 2019). Furthermore, the AIC and the ACC are considered core regions of a large network for stimulus salience detection, contributing to bodily awareness, directing attention, and engaging proper behavioral responses.

Preclinical studies in animal models have also provided behavioral and physiological evidence implicating the AIC in nociception. Thus, nociceptive behavioral responses can be affected by selective modulation of gabaergic, opiodergic, and dopaminergic transmission in the AIC (Lu et al., 2016). Moreover, although less investigated than the ACC, synaptic plasticity-like molecular and functional alterations have been observed in the AIC in rodent models of chronic pain (Lu et al., 2016).

In conclusion, accumulated data position the AIC as a key brain region for pain perception. Attaining a better understanding of the neuronal circuits in the AIC mediating pain affect and its pathological modifications during chronic pain might help to delineate novel therapeutic strategies.

Mini-dictionary of terms

Limbic system: Set of brain regions recognized to participate in emotion, memories, and arousal. Its activation can modulate reinforcing behaviors and lead to autonomic or endocrine responses.

Neuronal ensembles: A population of neurons, closely grouped or diffusely distributed, capable of firing together to participate in a particular neural computation.

Neuronal plasticity: Activity-dependent functional and/or morphological modifications of neuronal properties and synaptic connections.

Optogenetics: Technique that combines genetic engineering to express light-sensitive ion channels in cells with the use of light stimuli to manipulate the activity of neurons.

Calcium imaging: Microscopy technique using fluorescent molecules that change their fluorescence properties when binding Ca^{2+} ions. In individual neurons, changes in the level of fluorescence correlate with firing activity.

Key facts

Key facts of neuronal plasticity in health and disease

- The idea that functional adaptations in the brain depended on synapses was proposed by Santiago Ramon y Cajal more than a century ago.
- Donald Hebb postulated in 1949 that experience-dependent synaptic modifications would shape the activity of neuronal ensembles.
- It was not until 1973 that the British Timothy Bliss and the Norwegian Terje Lømo first demonstrated in vivo, in mammals, a form of long-term synaptic plasticity.
- It is now widely accepted that the brain is in perpetual reconfiguration, with the capacity to modify its morphology and activity according to the environment, time, and individual experiences.
- Brain diseases hijack neuronal plasticity mechanisms and subvert neuronal circuit functional organization.
- Long-term synaptic plasticity-like mechanisms have been observed in the spinal cord, ACC, insular cortex, amygdala, and other structures in animal models of chronic pain.

Summary points

- Pain is a multidimensional experience engaging—and being influenced by—affect and emotions.
- Pain perception is encoded by the coordinated activity of distributed brain regions.
- Activity in the ACC is related to the affective connotation of pain.
- ACC neuronal activity is necessary for the manifestation of pain-related negative affect in rodents.
- Chronic pain syndromes are associated with altered ACC responses to nociception.
- Aberrant neuronal plasticity in the ACC contributes to the expression of nociceptive sensitization, negative affect, and mood disorders related to chronic pain in animal models.

References

Albe-Fessar, D., Berkley, K. J., Kruger, L., Ralston, H. J., & Willis, W. D. (1985). Diencephalic mechanisms of pain sensation. *Brain Research Reviews*, *9*(3), 217–296. https://doi.org/10.1016/0165-0173(85)90013-X.

Apkarian, A. V., Bushnell, M. C., Treede, R.-D., & Zubieta, J.-K. (2005). Human brain mechanisms of pain perception and regulation in health and disease. *European Journal of Pain*, *9*(4), 463. https://doi.org/10.1016/j.ejpain.2004.11.001.

Ballantyne, J. C., & Sullivan, M. D. (2015). Intensity of chronic pain—The wrong metric? *New England Journal of Medicine*, *373*(22), 2098–2099.

Benarroch, E. E. (2019). Insular cortex: Functional complexity and clinical correlations. *Neurology*, *93*(21), 932–938. https://doi.org/10.1212/WNL.0000000000008525.

Bliss, T. V. P., Collingridge, G. L., Kaang, B.-K., & Zhuo, M. (2016). Synaptic plasticity in the anterior cingulate cortex in acute and chronic pain. *Nature Reviews Neuroscience*, *17*(8), 485–496. https://doi.org/10.1038/nrn.2016.68.

Blom, S. M., Pfister, J.-P., Santello, M., Senn, W., & Nevian, T. (2014). Nerve injury-induced neuropathic pain causes disinhibition of the anterior cingulate cortex. *Journal of Neuroscience*, *34*(17), 5754–5764. https://doi.org/10.1523/JNEUROSCI.3667-13.2014.

Büchel, C., Bornhövd, K., Quante, M., Glauche, V., Bromm, B., & Weiller, C. (2002). Dissociable neural responses related to pain intensity, stimulus intensity, and stimulus awareness within the anterior cingulate cortex: A parametric single-trial laser functional magnetic resonance imaging study. *The Journal of Neuroscience*, *22*(3), 970–976. https://doi.org/10.1523/JNEUROSCI.22-03-00970.2002.

Bushnell, M. C., Čeko, M., & Low, L. A. (2013). Cognitive and emotional control of pain and its disruption in chronic pain. *Nature Reviews Neuroscience*, *14*(7), 502–511. https://doi.org/10.1038/nrn3516.

Coghill, R. C., Sang, C. N., Maisog, J. M., & Iadarola, M. J. (1999). Pain intensity processing within the human brain: A bilateral, distributed mechanism. *Journal of Neurophysiology*, *82*(4), 1934–1943. https://doi.org/10.1152/jn.1999.82.4.1934.

Costigan, M., Scholz, J., & Woolf, C. J. (2009). Neuropathic pain: A maladaptive response of the nervous system to damage. *Annual Review of Neuroscience*, *32*, 1–32.

Craig, A. D., Reiman, E. M., Evans, A., & Bushnell, M. C. (1996). Functional imaging of an illusion of pain. *Nature*, *384*(6606), 258–260. https://doi.org/10.1038/384258a0.

Davis, K. D. (2005). Human anterior cingulate cortex neurons encode cognitive and emotional demands. *Journal of Neuroscience*, *25*(37), 8402–8406. https://doi.org/10.1523/JNEUROSCI.2315-05.2005.

Foltz, E. L., & White, L. E. (1962). Pain "relief" by frontal cingulumotomy. *Journal of Neurosurgery*, *19*, 89–100. https://doi.org/10.3171/jns.1962.19.2.0089.

Frankenstein, U. N., Richter, W., McIntyre, M. C., & Rémy, F. (2001). Distraction modulates anterior cingulate gyrus activations during the cold pressor test. *NeuroImage*, *14*(4), 827–836. https://doi.org/10.1006/nimg.2001.0883.

Fuchs, P. N., Peng, Y. B., Boyette-Davis, J. A., & Uhelski, M. L. (2014). The anterior cingulate cortex and pain processing. *Frontiers in Integrative Neuroscience*, *8*. https://doi.org/10.3389/fnint.2014.00035.

Gracely, R. H., Petzke, F., Wolf, J. M., & Clauw, D. J. (2002). Functional magnetic resonance imaging evidence of augmented pain processing in fibromyalgia. *Arthritis and Rheumatism*, *46*(5), 1333–1343. https://doi.org/10.1002/art.10225.

Hutchison, W. D., Davis, K. D., Lozano, A. M., Tasker, R. R., & Dostrovsky, J. O. (1999). Pain-related neurons in the human cingulate cortex. *Nature Neuroscience*, *2*(5), 403–405. https://doi.org/10.1038/8065.

Iwata, K., Kamo, H., Ogawa, A., Tsuboi, Y., Noma, N., Mitsuhashi, Y., … Kitagawa, J. (2005). Anterior cingulate cortical neuronal activity during perception of noxious thermal stimuli in monkeys. *Journal of Neurophysiology*, *94*(3), 1980–1991. https://doi.org/10.1152/jn.00190.2005.

Jensen, K. B., Regenbogen, C., Ohse, M. C., Frasnelli, J., Freiherr, J., & Lundström, J. N. (2016). Brain activations during pain: A neuroimaging meta-analysis of patients with pain and healthy controls. *Pain*, *157*(6), 1279–1286. https://doi.org/10.1097/j.pain.0000000000000517.

Johansen, J. P., & Fields, H. L. (2004). Glutamatergic activation of anterior cingulate cortex produces an aversive teaching signal. *Nature Neuroscience*, *7*(4), 398–403. https://doi.org/10.1038/nn1207.

Johansen, J. P., Fields, H. L., & Manning, B. H. (2001). The affective component of pain in rodents: Direct evidence for a contribution of the anterior cingulate cortex. *Proceedings of the National Academy of Sciences*, *98*(14), 8077–8082.

Kasanetz, F., Deroche-Gamonet, V., Berson, N., Balado, E., Lafourcade, M., Manzoni, O., & Piazza, P. V. (2010). Transition to addiction is associated with a persistent impairment in synaptic plasticity. *Science (New York, N.Y.)*, *328*(5986), 1709–1712. https://doi.org/10.1126/science.1187801.

Koyama, T., Kato, K., Tanaka, Y. Z., & Mikami, A. (2001). Anterior cingulate activity during pain-avoidance and reward tasks in monkeys. *Neuroscience Research*, *39*(4), 421–430. https://doi.org/10.1016/S0168-0102(01)00197-3.

Kwan, C. L., Crawley, A. P., Mikulis, D. J., & Davis, K. D. (2000). An fMRI study of the anterior cingulate cortex and surrounding medial wall activations evoked by noxious cutaneous heat and cold stimuli. *Pain*, *85*(3), 359–374. https://doi.org/10.1016/S0304-3959(99)00287-0.

Lu, C., Yang, T., Zhao, H., Zhang, M., Meng, F., Fu, H., … Xu, H. (2016). Insular cortex is critical for the perception, modulation, and chronification of pain. *Neuroscience Bulletin*, *32*(2), 191–201. https://doi.org/10.1007/s12264-016-0016-y.

Mertz, H., Morgan, V., Tanner, G., Pickens, D., Price, R., Shyr, Y., & Kessler, R. (2000). Regional cerebral activation in irritable bowel syndrome and control subjects with painful and nonpainful rectal distention. *Gastroenterology*, *118*(5), 842–848. https://doi.org/10.1016/S0016-5085(00)70170-3.

Mogil, J. S. (2009). Animal models of pain: Progress and challenges. *Nature Reviews Neuroscience*, *10*(4), 283–294. https://doi.org/10.1038/nrn2606.

Monosov, I. E. (2017). Anterior cingulate is a source of valence-specific information about value and uncertainty. *Nature Communications*, *8*(1). https://doi.org/10.1038/s41467-017-00072-y.

Mouraux, A., Diukova, A., Lee, M. C., Wise, R. G., & Iannetti, G. D. (2011). A multisensory investigation of the functional significance of the "pain matrix". *NeuroImage*, *54*(3), 2237–2249. https://doi.org/10.1016/j.neuroimage.2010.09.084.

O'Connell, N. E., Marston, L., Spencer, S., DeSouza, L. H., & Wand, B. M. (2018). Non-invasive brain stimulation techniques for chronic pain. *Cochrane Database of Systematic Reviews*. https://doi.org/10.1002/14651858.CD008208.pub5.

Obeso, I., Oliviero, A., & Jahanshahi, M. (2016). Editorial: Non-invasive brain stimulation in neurology and psychiatry. *Frontiers in Neuroscience*, *10*. https://doi.org/10.3389/fnins.2016.00574.

Petrovic, P. (2002). Placebo and opioid analgesia—Imaging a shared neuronal network. *Science*, *295*(5560), 1737–1740. https://doi.org/10.1126/science.1067176.

Peyron, R., García-Larrea, L., Grégoire, M.-C., Costes, N., Convers, P., Lavenne, F., ... Laurent, B. (1999). Haemodynamic brain responses to acute pain in humans. *Brain*, *122*(9), 1765–1780. https://doi.org/10.1093/brain/122.9.1765.

Rainville, P., Duncan, G. H., Price, D. D., Carrier, B., & Bushnell, M. C. (1997). Pain affect encoded in human anterior cingulate but not somatosensory cortex. *Science (New York, N.Y.)*, *277*(5328), 968–971.

Santello, M., Bisco, A., Nevian, N. E., Lacivita, E., Leopoldo, M., & Nevian, T. (2017). The brain-penetrant 5-HT 7 receptor agonist LP-211 reduces the sensory and affective components of neuropathic pain. *Neurobiology of Disease*, *106*, 214–221. https://doi.org/10.1016/j.nbd.2017.07.005.

Santello, M., & Nevian, T. (2015). Dysfunction of cortical dendritic integration in neuropathic pain reversed by serotoninergic neuromodulation. *Neuron*, *86*(1), 233–246. https://doi.org/10.1016/j.neuron.2015.03.003.

Sellmeijer, J., Mathis, V., Hugel, S., Li, X.-H., Song, Q., Chen, Q.-Y., ... Yalcin, I. (2018). Hyperactivity of anterior cingulate cortex areas 24a/24b drives chronic pain-induced anxiodepressive-like consequences. *The Journal of Neuroscience*, *38*(12), 3102–3115. https://doi.org/10.1523/JNEUROSCI.3195-17.2018.

Sikes, R. W., & Vogt, B. A. (1992). Nociceptive neurons in area 24 of rabbit cingulate cortex. *Journal of Neurophysiology*, *68*(5), 1720–1732. https://doi.org/10.1152/jn.1992.68.5.1720.

Stevens, F. L. (2011). Anterior cingulate cortex: Unique role in cognition and emotion. *The Journal of Neuropsychiatry and Clinical Neurosciences*, *6*.

Tölle, T. R., Kaufmann, T., Siessmeier, T., Lautenbacher, S., Berthele, A., Munz, F., ... Bartenstein, P. (1999). Region-specific encoding of sensory and affective components of pain in the human brain: A positron emission tomography correlation analysis. *Annals of Neurology*, *45*(1), 40–47. https://doi.org/10.1002/1531-8249(199901)45:1<40::aid-art8>3.0.co;2-l.

Tracey, I. (2010). Getting the pain you expect: Mechanisms of placebo, nocebo and reappraisal effects in humans. *Nature Medicine*, *16*(11), 1277–1283. https://doi.org/10.1038/nm.2229.

Treede, R. D., Kenshalo, D. R., Gracely, R. H., & Jones, A. K. (1999). The cortical representation of pain. *Pain*, *79*(2–3), 105–111.

Valet, M., Sprenger, T., Boecker, H., Willoch, F., Rummeny, E., Conrad, B., ... Tolle, T. R. (2004). Distraction modulates connectivity of the cingulo-frontal cortex and the midbrain during pain—An fMRI analysis. *Pain*, *109*(3), 399–408. https://doi.org/10.1016/j.pain.2004.02.033.

Vogt, B. A. (2005). Pain and emotion interactions in subregions of the cingulate gyrus. *Nature Reviews Neuroscience*, *6*(7), 533–544. https://doi.org/10.1038/nrn1704.

Wager, T. D., Atlas, L. Y., Botvinick, M. M., Chang, L. J., Coghill, R. C., Davis, K. D., ... Yarkoni, T. (2016). Pain in the ACC? *Proceedings of the National Academy of Sciences*, *113*(18), E2474–E2475. https://doi.org/10.1073/pnas.1600282113.

Wager, T. D., Atlas, L. Y., Lindquist, M. A., Roy, M., Woo, C.-W., & Kross, E. (2013). An fMRI-based neurologic signature of physical pain. *New England Journal of Medicine*, *368*(15), 1388–1397. https://doi.org/10.1056/NEJMoa1204471.

Wager, T. D., Rilling, J. K., Smith, E. E., Sokolik, A., Casey, K. L., Davidson, R. J., ... Cohen, J. D. (2004). Placebo-induced changes in FMRI in the anticipation and experience of pain. *Science (New York, N.Y.)*, *303*(5661), 1162–1167. https://doi.org/10.1126/science.1093065.

Zhang, Q., Manders, T., Tong, A. P., Yang, R., Garg, A., Martinez, E., ... Urien, L. (2017). Chronic pain induces generalized enhancement of aversion. *eLife*, *6*, e25302.

Chapter 19

Sleep deprivation, headache, and Fos immunohistochemistry

Seonghoon Kim and Jeong-wook Park
Department of Neurology, Uijeongbu St. Mary's hospital, College of Medicine, The Catholic University of Korea, Seoul, Republic of Korea

Abbreviations

ARAS	ascending reticular activating system
CB1	cannabinoid receptor agonist
CDS	Cortical spreading depression
DLMO	dim light melatonin onset
LC	locus coeruleus
PAG	periaqueductal gray
REM sleep	rapid eye movement sleep
RVM	rostral ventromedial medulla
SCN	suprachiasmatic nucleus
TAC	trigeminal autonomic cephalalgia
TCC	trigeminocervical complex
TNC	trigeminal nucleus caudalis
vlPAG	ventrolateral periaqueductal gray
VLPO	ventrolateral preoptic nucleus
VPM nucleus	ventral posteromedial (VPM) nucleus

Introduction

Sleep is not passive rest, but an active process that reconstructs information from external stimuli and maintains cognitive and emotional function. Headache and sleep problems are prevalent and burdensome in daily life. They can coincide and aggravate each other. The prevalence of sleep disorder is increased in chronic headache and migraine patients, suggesting a close clinical relationship between them (Sancisi et al., 2010).

Structures from the brainstem to the cerebral cortex are related to sleep and headache modulation. Both central and peripheral mechanisms are involved in the development and transmission of nociceptive pain in headaches. The dural vasculature and trigeminocervical complex are pain-sensitive structures constituting the main contribution to pain in migraine and headache. The periaqueductal gray (PAG), rostral medullar, hypothalamus, and thalamus modulate and deliver nociceptive pain from the periphery to the cerebral cortex (Goadsby, Charbit, Andreou, Akerman, & Holland, 2009). The suprachiasmatic nucleus (SCN) governs circadian rhythm and acts as an internal pacemaker. The ascending reticular activating system (ARAS) mediates arousal and consists of monoaminergic brainstem nuclei. The ventrolateral preoptic nucleus (VLPO) system inhibits the arousal system and promotes sleep (Fuller, Gooley, & Saper, 2006).

Modulation of headache and sleep is facilitated by overlapping networks and mutual effects. In this chapter, we introduce the clinical and pathophysiological relationship between headache and sleep, present an experimental model of sleep deprivation, and discuss Fos immunohistochemistry.

Pain, headache, and sleep deprivation

Sleep deprivation lowers pain thresholds and enhances the pain response. Sleep quality is negatively correlated with pain threshold in fibromyalgia patients (Agargun et al., 1999). In burn patients, poor sleep quality worsens burn-related pain, but severe daytime pain does not worsen sleep quality (Raymond, Nielsen, Lavigne, Manzini, & Choiniere, 2001). In a healthy

The Neurobiology, Physiology, and Psychology of Pain. https://doi.org/10.1016/B978-0-12-820589-1.00019-1

population, sleep deprivation lowered the pain threshold but other somatosensory thresholds were not affected (Kundermann, Spernal, Huber, Krieg, & Lautenbacher, 2004). These results suggest that sleep deprivation has a more selective and direct effect on pain perception than on other sensory inputs.

Headaches are a chronic pain disorder related to sleep disorders such as insomnia, sleep apnea, and restless leg syndrome. In a clinical study of migraine and sleep, chronic migraine patients were found to have more sleep complaints than episodic migraine patients. Short sleepers who slept 6 h per night had more severe and frequent headaches (Kelman & Rains, 2005). Another population-based study reported no differences in actual sleep duration between groups with and without migraine, but the prevalence of insufficient sleep was higher in the migraine group. This suggests that sleep quality has a significant effect on insufficient sleep and sleep duration (Kim et al., 2017). Tension-type headache patients complain more often of insufficient sleep that is correlated with insomnia, poor sleep quality, and short sleep duration (Oh et al., 2018). A short sleep duration of fewer than 6 h exacerbates headache frequency (Song et al., 2018). In a healthy population-based study, 24 h of total sleep deprivation induced mechanical pain sensitization in craniofacial muscles, which was recovered after restoration sleep (Kristiansen et al., 2017). Sleep deprivation induces pain sensitization of facial areas, but slow-wave sleep deprivation did not do so in healthy participants (Arima et al., 2001).

The pathophysiologic relationship between sleep deprivation and headache

Cortical spreading depression (CDS) is the propagation of the depolarization wave of the cerebral cortex, which is related to migraine aura. Cortical dysregulation and hyper-hypo excitation develop during migraine attacks and sleep deprivation. REM sleep deprivation enhances the propagation of CSD, and adequate sleep stabilizes cortical excitability (Amorim, Guedes, Medeiros, da Silva, & Cabral-Filho, 1988; Manganotti, Fuggetta, & Fiaschi, 2004).

The occurrence of headache attacks has circadian variation in migraine patients. A previous study reported that the frequency of migraine attacks peaked in the morning and was reduced at night (Park, Cho, Park, & Chu, 2018). During awakening, the thalamus shows fast rhythm oscillation and enhances sensory transmission via the thalamocortical connection. Sleep induces hyperpolarization of the thalamocortical connection, which decreases peripheral sensory transmission compared to waking (Steriade, McCormick, & Sejnowski, 1993). Therefore, the amount of nociceptive transmission changes depending on the state of sleep.

The SCN is one of the hypothalamic nuclei and acts as a circadian pacemaker. Casein kinase 1δ of SCN modulates the circadian rhythm through the PERIOD protein. Dysfunction of casein kinase 1δ and PERIOD protein is related to the advanced sleep phase, which is characterized by very early sleep onset and offset. Mice with casein kinase 1δ mutations exhibit decreased thresholds for cortical spreading depression, and increased sensitivity to pain after treatment with the migraine trigger nitroglycerin (Brennan et al., 2013). The hypothalamus is connected to the nociceptive transmission pathway and is involved in controlling headaches through the orexin and somatostatin pathways. The hypothalamus is activated during headache attacks in migraine and cluster headache patients, which suggests a relationship between headache modulation and the hypothalamus.

The brainstem nuclei are associated with transmission and modulation of nociceptive pain through the trigeminocervical complex (TCC). These brain stem nuclei have connections to the cerebral cortex, thalamus, and hypothalamus that are associated with modulation of headache and sleep. The PAG is activated during awakening to suppress REM sleep, and electrical stimulation of ventrolateral periaqueductal gray (vlPAG) induces headache (Lu, Jhou, & Saper, 2006). The locus coeruleus (LC) promotes waking and suppresses REM sleep by activating the REM-off cell group. LC receives the nociceptive afferent of TCC and connects to the hypothalamus, which induces intracranial vasoconstriction and CDS (Goadsby, Lambert, & Lance, 1985).

Modulation of headache and sleep-wake cycles

Central nervous structures from the brain stem to the cortex have complex interconnections and interactions that involve modulation of sleep and headaches. Various neuropeptides are associated with the modulation of headache and sleep. Here, we describe the pathophysiological and pharmacological mechanisms related to headache and sleep.

Headache

The stimulation of dural vasculature induces headache and activates the TCC, which includes the trigeminal nucleus caudalis (TNC) and C1, C2 of the cervical cord (Goadsby & Zagami, 1991). The TCC is a major relay neuron of nociception, receiving peripheral nociceptive afferent from the craniocervical area and relaying it to brainstem nuclei such as the vlPAG, areas of the rostral ventromedial medulla (RVM), nucleus raphe magnus, and other pontine nuclei. During migraine attacks, the vlPAG is activated, and stimulation of the vlPAG inhibits nociceptive traffic of TCC (Knight & Goadsby, 2001). Administration of orexin A, or cannabinoid receptor agonist (CB1) to vlPAG reduces nociceptive traffic, as the 5-HT antagonist inhibits these actions (Holland, Akerman, Lasalandra, & Goadsby, 2008). The vlPAG is connected to various brain stem nuclei and inhibits spinal nociceptive processing through RVM. Administration of 5-HT agonist to vlPAG or RVM reduces the nociceptive traffic of TCC (Bartsch, Knight, & Goadsby, 2004). RVM has a direct bidirectional connection between the vlPAG and spinal dorsal horn. RVM has an on-cell and an off-cell group that exert opposite actions on nociceptive transmission. Opioids inhibit the on-cell group and promote the off-cell group, reducing the transmission of pain. The raphe nucleus located in the medulla has serotonergic projection and inhibits pain afferents. The LC is connected to the spinal cord's dorsal horn and suppresses the transmission of pain afferents.

The thalamus relays nociceptive pain information from the TCC to the somatosensory cerebral cortex. During a migraine attack, the thalamus is activated and induces the spread of cutaneous allodynia and hyperalgesia beyond the site of migraine pain (Afridi et al., 2005). The ventral posteromedial (VPM) nucleus is a key structure for nociceptive processing. Propranolol, triptan, valproic acid, and CGRP receptor antagonists have direct action on the VPM nucleus and inhibit trigeminal nociceptive inputs (Andreou, Shields, & Goadsby, 2010; Shields & Goadsby, 2005). The posterior thalamus receives dural vascular nociceptive input and light information from the retinal ganglion cells, and projects to the somatosensory, visual, and associative cortices. The mean firing rate increases 2–4 times with bright light compared to dim light (Noseda et al., 2010).

The hypothalamus is activated during headache attacks such as migraine and trigeminal autonomic cephalalgia (TAC) (Denuelle, Fabre, Payoux, Chollet, & Geraud, 2007; Goadsby, Cittadini, & Cohen, 2010). The hypothalamus modulates nociceptive transmission through direct connections with the TCC and brain stem nuclei. The trigeminohypothalamic tract delivers somatosensory and visceral nociceptive information from the TCC. The hypothalamus secretes orexin and somatostatin, which are related to the modulation of nociceptive transmission. Orexin A has an antinociceptive effect, and orexin B has a pronociceptive effect. When somatostatin antagonist is administered to the posterior hypothalamus, firing of the nociceptive input of the TNC is reduced (Bartsch, Levy, Knight, & Goadsby, 2005).

Sleep

The internal circadian pacemaker and external light stimulation modulate the sleep-wake cycle. The suprachiasmatic nucleus (SCN) is a circadian pacemaker located in the anterior hypothalamus and directly above the optic chiasm. The SCN generates an oscillatory 24-h cycle wake-promoting signal. The ventrolateral part of the SCN receives light signals from retinal ganglion cells via the retinohypothalamic tract and entrains the internal circadian rhythm to the natural external cycle. The dorsomedial SCN does not respond to light stimulation and maintains the endogenous circadian rhythm. The hypothalamus secretes orexin, which is related to maintaining the waking state. Orexin enhances firing of the tuberomammillary nucleus, LC, and dorsal raphe nucleus of ARAS.

The ascending reticular activation system (ARAS) promotes arousal through extensive connections between the brain stem, thalamus, and basal forebrain, and the ventrolateral preoptic nucleus (VLPO)-related network inhibits arousal and induces sleep (Fig. 1C). The ARAS is composed of cholinergic, dopaminergic, serotonergic, and histaminergic neurons. Cholinergic neurons are present in the PPT and LDT nuclei located in the mesopontine tegmentum and send cholinergic projections to the thalamocortical and reticular nucleus. Other cholinergic neurons located in the basal forebrain also project to the cerebral cortex. The monoaminergic systems include the noradrenergic LC, the dopaminergic neurons of the vlPAG, the serotoninergic dorsal and median raphe nuclei, and histaminergic neurons in the tuberomammillary nucleus (Fuller et al., 2006).

The VLPO is located in the anterior hypothalamus and has an inhibitory connection (GABA and galanin) to ARAS, thereby inhibiting arousal (Fig. 1D) (Sherin, Shiromani, McCarley, & Saper, 1996). The core neuron group of VLPO projects to the tuberomammillary nucleus, while others connect to the LC and dorsal and median raphe nuclei (Fuller et al., 2006). The ARAS and orexin pathways inhibit each other when activated, thereby operating like a switch to minimize unnecessary transitions.

Pharmacologic relationships between headache and sleep

Adenosine

Adenosine is related to ATP energy metabolism, neuronal activity, and sleep. The adenosine receptor has four subtypes (A1, A2a, A2b, A3) that sometimes act in opposition. Adenosine inhibits the wake-promoting cholinergic neurons of the basal forebrain through the A1 receptor and activates the sleep-promoting VLPO through the A2a receptor (Huang, Urade,

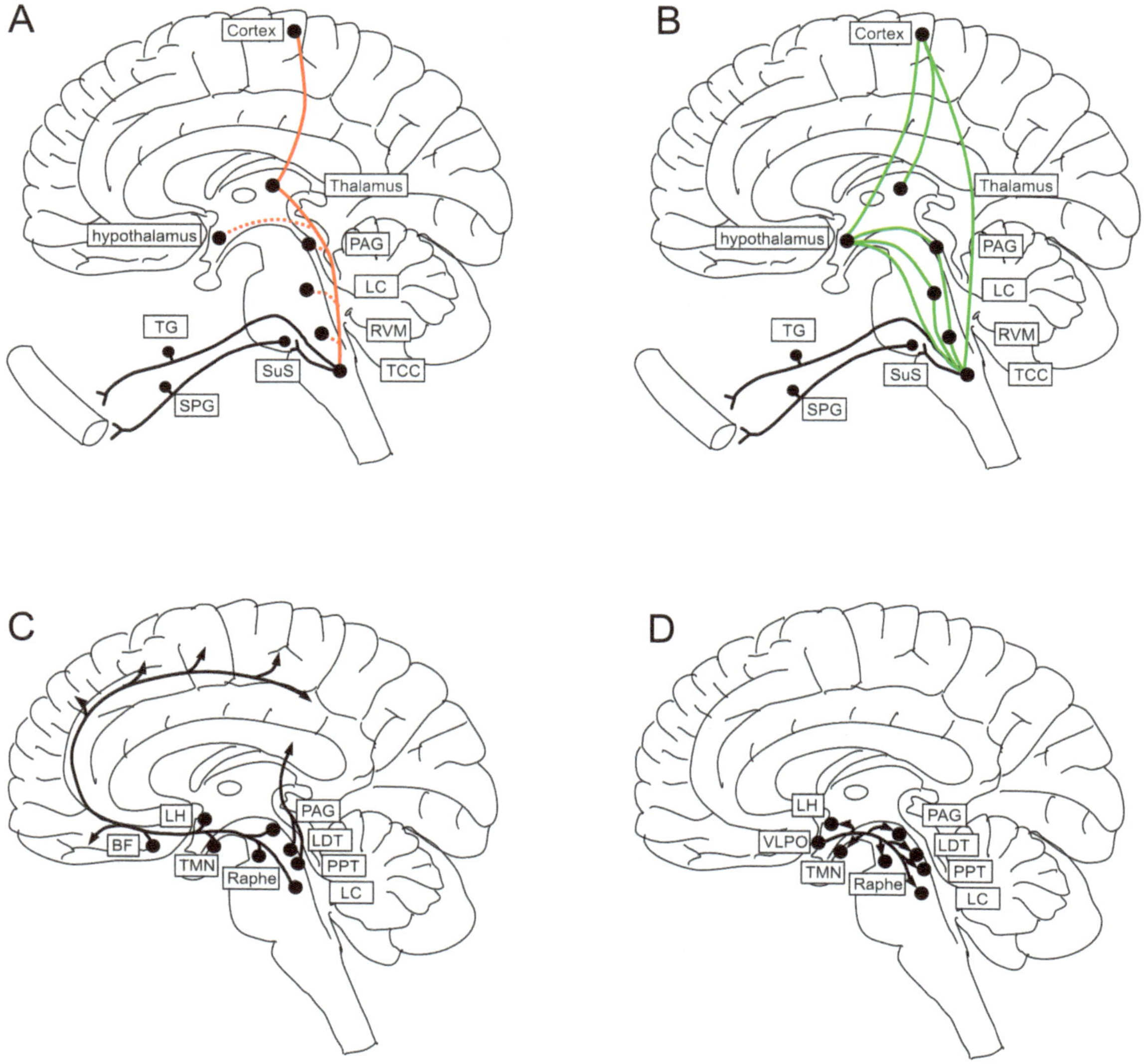

FIG. 1 Headache and sleep modulation pathways. The ascending headache transmission pathway: (A) The pseudo-unipolar neuron of the trigeminal ganglion (TG) synapse in the context of cranial vascular structure and the TCC. The second-order neuron of TCC has a direct connection with the thalamus, which sends nociceptive information to the thalamus. The third-order thalamocortical neuron synapses on the cerebral cortex. There are direct and indirect ascending connections to the LC, PAG, and hypothalamus. The descending headache modulation pathway: (B) The TCC receives the direct projection from the cerebral cortex. There is an indirect connection between cortex and TCC via the thalamus and hypothalamus. The hypothalamus has a direct connection with the TCC. The ascending arousal system enhances waking via two pathways. (C) The major input of the thalamus originates from the cholinergic group of the pedunculopontine (PPT) and laterodorsal tegmental (LDT) nuclei. The second waking pathway arises from monoaminergic cell groups, including the tuberomammillary nucleus (TMN) containing histamine, dorsal, and median raphe nuclei containing serotonin (5-HT), and the LC containing noradrenaline. This pathway also receives contributions from the lateral hypothalamus (LH) containing orexin (ORX), and from basal forebrain (BF). The ventrolateral preoptic nucleus (VLPO) has direct inhibitory connections to the arousal pathway through the LDT, LC, raphe, and orexin neuron group of (LH) (D).

& Hayaishi, 2011). Regarding headache, the A1 receptor of the spinal cord has an antinociceptive effect. A1 receptor knockout mice show increases in pain sensitivity (Wu et al., 2005). In contrast, A2a receptor knock-out mice show hypoalgesia, which suggests the A2a receptor may increase of pain transmission (Ferre et al., 2007). Adenosine inhibits orexinergic neuron groups related to wake-promotion and modulation of nociceptive traffic from the TCC to the brainstem (Liu & Gao, 2007). These results show that adenosine is involved in both sleep and headache control.

Orexin

Orexin is a neuropeptide produced in the hypothalamus that is associated with homeostasis of the sleep-wake cycle and autonomic function. The hypothalamus secretes two types of orexin (orexin A and orexin B), and there are two types of orexin receptor (orexin 1 and orexin 2 receptors). Orexin A has a high affinity with the orexin 1 receptor (OX1R). The orexin 2 receptor (OX2R) has affinity for both orexin A and orexin B. The LC distributes only OX1R, and the tuberomammillary nucleus and RVM distribute only OX2R. The raphe nucleus, amygdala, and cortex have both types of receptors (Scammell & Winrow, 2011). Orexin promotes the waking state through connections to the basal forebrain and monoamine group of the brainstem. When orexin A is administered to the posterior hypothalamus, it decreases the responses to dural electrical stimulation and thermal stimulation of facial skin in the TNC. Conversely, orexin B increases pain responses of the TNC (Bartsch, Levy, Knight, & Goadsby, 2004).

Melatonin

The pineal gland secretes melatonin, which is a sleep-promoting hormone. The secretion of melatonin is related to SCN and has a 24-h cycle, decreasing in the morning, and increasing in the evening. Melatonin plays a vital role in the modulation of circadian rhythm. The point at which melatonin rises in dim light is called the dim light melatonin onset (DLMO). When melatonin is administered before the DLMO, the sleep cycle advances.

Although the role of melatonin in pain control is not clear, it is known to have antinociceptive effects at spinal and supraspinal levels. In rats that have had their pineal glands removed, activation of TCC is enhanced and normalized by melatonin replacement (Tanuri et al., 2009). In migraine patients, plasma melatonin levels are lower than those of healthy controls, while inhibition of melatonin secretion by exposure to light is more severe in migraine sufferers than controls, suggesting an explanation for photosensitivity in migraine sufferers (Claustrat, Brun, Chiquet, Chazot, & Borson-Chazot, 2004). Cluster headache patients show reduced plasma melatonin levels and nonsignificant melatonin rhythms. Melatonin significantly reduces the frequency of cluster headaches and analgesic use during cluster periods (Leone, D'Amico, Moschiano, Fraschini, & Bussone, 1996). These results suggest that melatonin plays roles in sleep control and headache modulation.

Pituitary adenylate cyclase activating peptide (PACAP)

PACAP is a member of the vasoactive intestinal polypeptides and transmits light information from the retina to the SCN via the retinohypothalamic tract. PACAP activates cAMP signaling and entrains the internal circadian rhythm to the external natural cycle (Harrington, Hoque, Hall, Golombek, & Biello, 1999). PACAP is widely distributed in the TNC and crosses the BBB, causing vasodilatation that is associated with headaches (Csati et al., 2012). PACAP is increased during headache attacks and enhances headache intensity (Tuka et al., 2013).

Experimental research to examine headache and sleep deprivation

Animal experiments confirm the behavioral, physiological, and pathological changes caused by sleep deprivation, which cannot be obtained ethically through human research. Adequate animal models for sleep deprivation and headache assessment are needed. The fos immunohistochemistry method is frequently used to evaluate pathophysiology, with special focus on neuronal activation.

Fos-immunohistochemistry

Many studies have used fos-immunohistochemistry to identify and map brain lesions related to various stimuli. The proto-oncogene c-fos is one of the immediate-early genes in neurons that responds to specific situations. After stimulation, the transcription is activated at 5–20 min, and peak of mRNA accumulation occurs at 30–40 min. The c-fos protein can be detected by immunohistochemistry at 20–90-min poststimulation. Fos immunohistochemistry is analyzed according to standardized protocols (Fig. 2).

Fos immunohistochemistry is useful to label polysynaptically activated neurons, thereby enabling researchers to map pathways of neuronal activation at many synapses. Quantitative analysis can be performed by analyzing the population of fos-activated neuronal cells at the specific lesion. However, fos expression can be induced by various nonspecific stimuli and may require stimulation above the physiologic threshold. Fos is not expressed in all activated neurons, so lack of Fos detection does not equate to lack of neuronal activity.

The use of fos immunohistochemistry has the advantage of tracing the pathways of headache and sleep mechanisms in the central nervous system. In headache research, c-fos protein has been observed to be significantly expressed in the trigeminal caudal nucleus and the superficial layer of the upper cervical cord after electrical stimulation of the infusion of inflammatory substances (Offenhauser et al., 2005). Stimulation of the superior sagittal sinus induces headache pain and

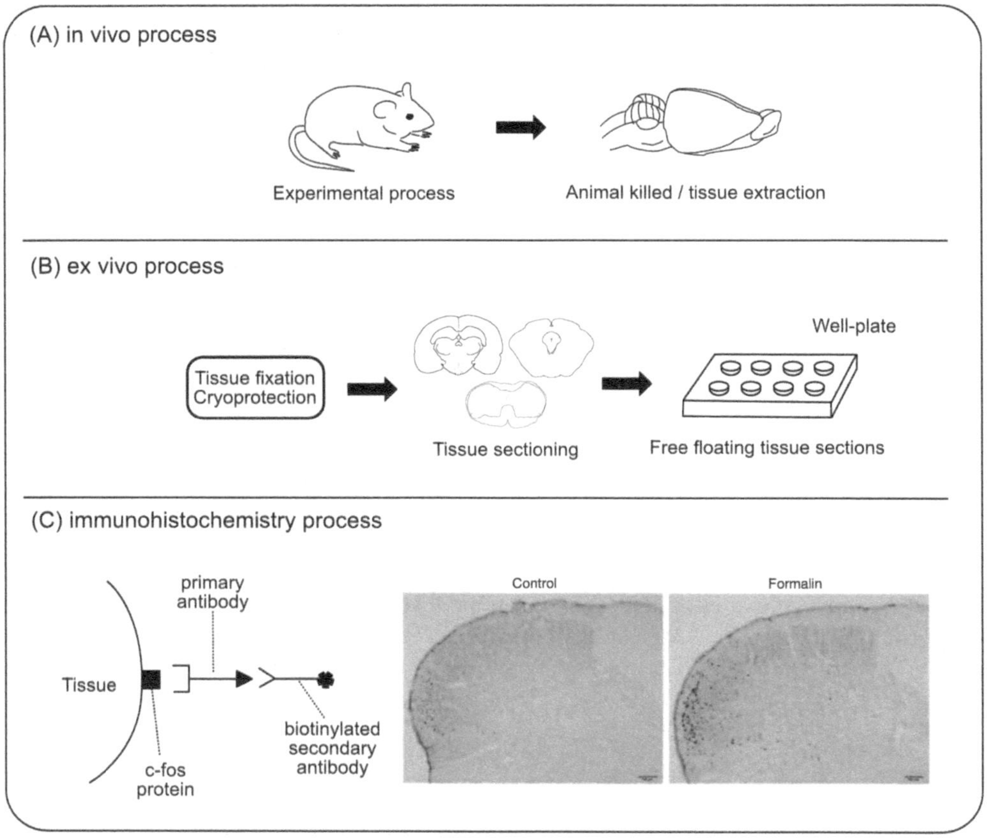

FIG. 2 Tissue preparation and immunohistochemistry. (A) After the experiment, the animal is killed, and CNS tissue is obtained. After tissue fixation and sectioning (B), immunohistochemistry is analyzed by examining primary and biotinylated secondary antibodies. Then, each section is treated with avidin-biotin-peroxidase and stained with 3,3′-diaminobenzidine tetrahydrochloride dihydrate containing nickel. Fos-activated neuronal cells can be observed under the microscope. The formalin-treated group showed more fos-activated cells *(black dot)* than did controls (C) (Park, Suh, Shin, & Park, 2012).

activates the trigeminovascular system. Electrical stimulation of the superior sagittal sinus increases fos activity on the supraoptic nucleus and posterior hypothalamus (Benjamin et al., 2004). Triptan and *N*-methyl-D-aspartate antagonist also suppress nociceptive processing, confirmed by decreased c-fos protein expression in a headache model (Park et al., 2012).

During sleep, fos activation is very low or absent in the brain. However, fos activity of the VLPO increases during sleep, and therefore Fos may be related to sleep induction and suppression of waking (Sherin et al., 1996). Fos expression increases in many portions of the medial and lateral reticular formation of the pons and medulla during REM sleep (Shiromani, Winston, & McCarley, 1996). During spontaneous and forced waking, c-fos is high in most regions of the neocortex and allocortex, including the frontal, motor, parietal, temporal, occipital, cingulate, insular, piriform, and entorhinal areas (Cirelli & Tononi, 2000). In sleep-deprived animals, fos activity is increased in preoptic and lateral areas of the hypothalamus. The fos activity of orexin neurons increases during REM-deprivation, whereas melanin-concentrated hormone neurons are activated only during rebound REM sleep (Arthaud et al., 2015).

Animal model of sleep deprivation

Each sleep deprivation model has advantages and limitations, so researchers select experimental models according to their needs, for example, depending on whether to deprive total sleep, only a portion of sleep, or to adjust for external stimuli or stress. Here, we describe several commonly used animal models.

Gentle handling model

The gentle handling model induces sleep deprivation using external stimuli. When the animal falls asleep, gentle tactile stimuli are given to wake them up (Fig. 3A). Depending on the researcher's purpose, olfactory or visual stimuli can be used

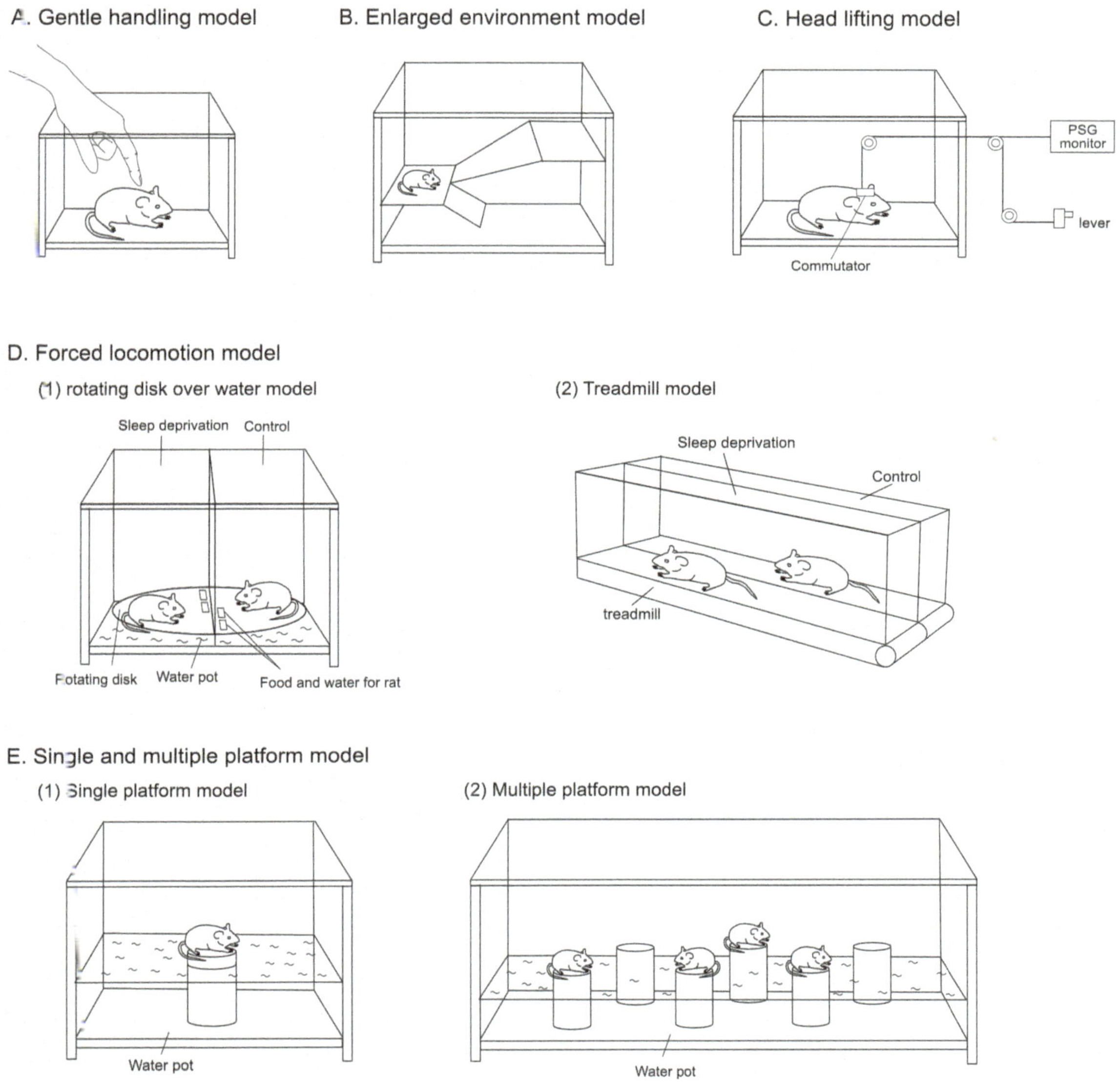

FIG. 3 Animal model of sleep deprivation. (A) Gentle handling model. When animals fall asleep, the researcher gives external stimuli to awaken them. (B) Enlarged environment model. The new environment will cause sleep deprivation because the rat will explore rather than sleep. (C) Head lifting model. The animal's sleep stages are monitored by polysomnography. When the rat falls asleep, the researcher pushes the lever and activates a pulley system that gently lifts the rat's head and awakens it. (D) Forced locomotion model. Rotating disk over water (D1) and treadmill method (D2). In the experimental group, sleep deprivation occurs by forced exercise. The control group is provided with the same environment and exercise but allowed sleep freely. (E) Single or multiple platform model. When a rat enters REM sleep, a part of the body falls into the water, and the rat awakens. The multiple platform model makes up for limitations of single platform models such as social isolation and movement limitation.

instead of tactile stimuli. This method has the advantage of depriving the animal of entire sleep stages or, if necessary, only of certain sleep stages through EEG monitoring, but this method is difficult for long periods of time because the researcher must monitor the animal continuously and provide stimulation. In most cases, it is used for sleep deprivation periods of 2–8 h. When the animals adapt to the stimulation, sleep deprivation no longer occurs. The external stimuli also become a potential confounder by enriching the animal's environment and perhaps affecting subsequent behavioral performance (Alkadhi, Zagaar, Alhaider, Salim, & Aleisa, 2013).

Enlarged environment model

The enlarged environmental model is a method to induce relatively short periods of sleep deprivation of 2–4 h. When the experimental animal falls asleep, the researcher makes environmental changes such as new nesting material and an enlarged cage (Fig. 3B). Compared to other models, it has the advantage of inducing less stress other than sleep deprivation. However, adaptation to environmental change is a confounding factor for the evaluation of behavior tasks or the learning process (Alkadhi et al., 2013).

Head lifting model

The head lifting model is a recently introduced method for selective REM sleep deprivation. This model consists of a spring-action mechanical lever, three pulleys, and light-weight flexible wire. One side of the wire is fixed to the rat's head, and the other side is connected to a mechanical lever through a pully (Fig. 3C). When REM is observed through polysomnography, the researcher pushes the lever and head lifting occurs, causing the rat to wake up. This method has the advantage that it allows for selective REM deprivation, but it requires continuous and careful monitoring by the researcher. Therefore, the head lifting model cannot be used for long durations or for multiple experiments at once (Datta, Mavanji, Ulloor, & Patterson, 2004).

Forced locomotion model

The most representative models of forced locomotion utilize rotating disks over water and treadmills. The rotating disk over water model includes a rotating disk over a water pot with experimental and control animals housed in divided cages. The disk does not rotate while the experimental group is awake. When the experimental animals fall asleep, the disk rotates (3.5 rpm, random direction for 6 s) until the experimental rat awakens (Fig. 3D1). The control group is provided with the same environment on the opposite side of the cage to control for environmental confounders and is free to accumulate sleep while the experimental group is awake because the disk rotates only when the experimental group falls asleep (Campbell, Guinan, & Horowitz, 2002). To ensure total sleep deprivation, the researcher can extend the experimental time and deprive animals of specific sleep stages using EEG monitoring. However, too frequent disk rotation may influence the control group's sleep.

The treadmill model is a commonly used method in which a treadmill activates when the animal falls asleep or enters a specific sleep stage. To control for the confounding factor of physical activity, the experimental group and control group receive the same amount of exercise, but the control group is allowed to sleep freely (Fig. 3D2). For example, the experimental group repeats 3 s of exercise and 12 s of rest in one session to deprive it of sleep, while the control group is able to sleep freely while getting the same amount of exercise due to a schedule of 15 min of exercise and 60 min of rest. However, physical stress is a confounding factor that may affect behavioral performance after the experiment.

Single or multiple platform model

Single or multiple platform models have been developed for REM sleep deprivation. The rat is placed on a small platform over the water. When the rat enters the REM sleep stage, muscle tone decreases, and a part of the body (such as the face) falls into the water, causing REM deprivation. The single platform method uses one animal on one platform, causing stress due to movement restriction and social isolation (Alkadhi et al., 2013). The multiple platform method was developed to address the limitations of single platform models, such as movement restriction. To minimize stress caused by social isolation, the multiple platform model uses a group of rats socialized in the same cage before the experiment. Ten rats are placed on each of 14 narrow platforms (6.5 cm in diameter) in the cage (127.0 cm × 44.0 cm × 45.0 cm, filled with water to 1 cm below the platform). The single and multiple platform models deprive REM sleep without PSG monitoring and expand experiment time without requiring the researcher's close monitoring. However, non-REM sleep can be affected, and this method requires a habituation period (Alkadhi et al., 2013).

Animal model of headache (activation of the trigeminovascular system)

Electrical stimulation of the trigeminal ganglion

Direct electrical stimulation of the trigeminal ganglion activates the TNC and induces headache (Fig. 4a). Electrodes are inserted into both trigeminal ganglia through a 1–2 mm burr hole using a stereotactic frame. In general, 5 ms, 3 mA paired rectangular pulses are used as a single stimulation unit. The exact insertion of electrodes can be confirmed by observing ipsilateral masseter muscle contraction after electrical stimulation. Trigeminal ganglion stimulation increases CGRP and substance P in the external jugular vein and dural plasma protein extravasation. These results indicate that this model reflects neurogenic inflammation, the main mechanism of migraine (Goadsby, Edvinsson, & Ekman, 1988). However,

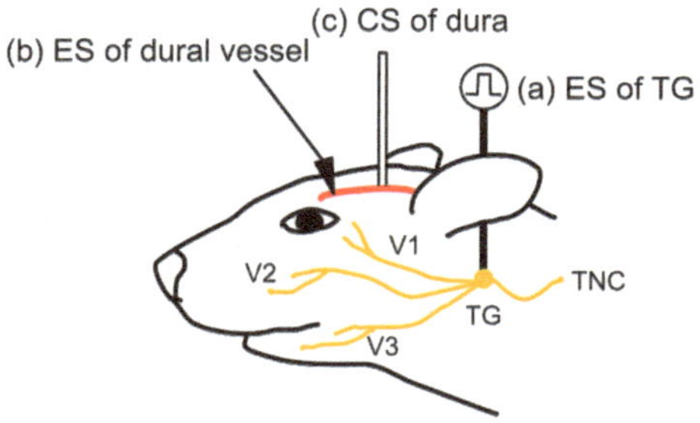

FIG. 4 An animal model of headache. (a) Single pulse electrical stimulation (ES) is applied to the trigeminal ganglion (TG). The superior sagittal sinus or middle meningeal artery may be stimulated by (b) electrical (ES) and (c) chemical (CS) stimulation. These three models activate the trigeminovascular system.

there is a limitation in that stimulation causes unnecessary broad activation of other structures that are innervated by the trigeminal ganglion that are not related to nociceptive processing.

Electrical stimulation of dural vessels

The middle meningeal artery and superior sagittal sinus, which are abundantly innervated by the trigeminal nerve, are mainly used for the electrical stimulation model (Fig. 4b). The skull is removed, the dura is exposed intact, and electrical stimulation is applied to these blood vessels. This model also confirms that the release of pain-related neurotransmitters such as CGRP or VIP increases at the trigeminal nerve terminal (Zagami, Goadsby, & Edvinsson, 1990). Compared to trigeminal ganglion stimulation, it has advantages of more selective pain stimulation and causing definite changes in cerebral blood flow. However, electrical stimulation of vessels is not a physiologic situation.

Chemical stimulation of dural nociceptive field

This method is mainly performed by injecting inflammatory soup or capsaicin around the dura through a catheter into the superior sagittal sinus or the cisterna magna (Fig. 4c). In the superior sagittal sinus model, the catheter is stably positioned on both sides of bregma, and the catheter tip is placed between the skull and the dura mater. Then, the stimulants are injected. In the cisterna magna model, a catheter is inserted by exposing the atlantoaxial membrane after the incision of the cervical spine. Administration of the stimulant enhances the neuron activity of the trigeminal ganglion and reactivity to mechanical stimulation (Burstein, Yamamura, Malick, & Strassman, 1998). The chemical stimulants may induce alterations of the blood-brain barrier (BBB), resulting in direct neural activation rather than synaptic activation via meningeal afferent fibers.

Experimental studies of sleep deprivation and headache pain

There have been some experimental studies of the relationship between sleep deprivation and pain. REM sleep deprivation lowered the pain threshold and persisted for a period of time, even after restoring sleep (Hicks, Coleman, Ferrante, Sahatjian, & Hawkins, 1979; Onen, Alloui, Eschalier, & Dubray, 2000). REM sleep deprivation is known to interfere with opioid and monoamine oxidase-related pain suppression mechanisms (Ukponmwan, Rupreht, & Dzoljic, 1986). Although there are some differences depending on the study, changes of pain thresholds due to REM sleep deprivation are generally restored after sufficient restoring sleep.

One study investigated the effect of sleep deprivation on temporomandibular joint pain using the multiple platform method. Sleep deprivation causes changes of inflammatory cytokines, which affect the occurrence of temporomandibular joint pain sensitization (Wu et al., 2014). Other studies reported that REM sleep deprivation decreases the mechanical pain threshold of the facial area using a single platform method. This study used the chemical stimulation headache model and fos-immunohistochemistry to investigate sleep-headache-related structures. In the REM-deprived group, we observed increased c-fos activity in the posterior and dorsomedial hypothalamus, ventrolateral PAG, and the superficial layer of the TCC (Fig. 5). REM deprivation lowers the headache threshold and prolongs recovery of threshold changes (Fig. 6) (Kim et al., 2019). These results suggest that there are clinical and pathophysiological correlations between sleep deprivation and headache.

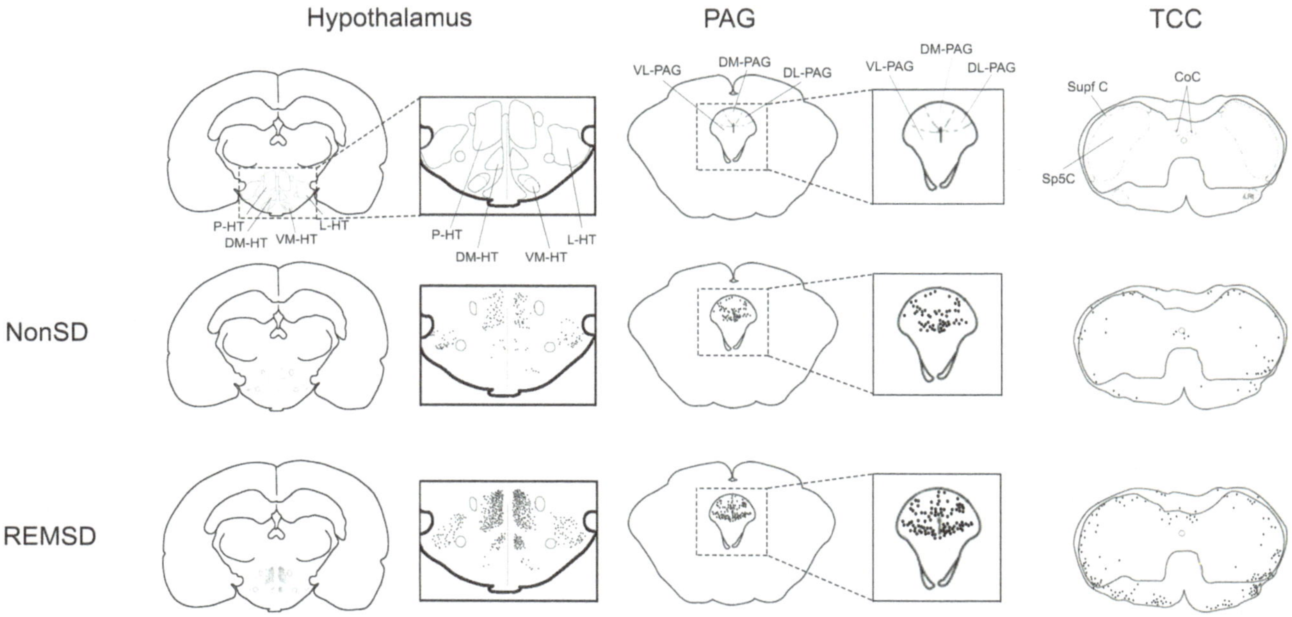

FIG. 5 The distributions of c-fos-positive cells in the hypothalamus, PAG, and TCC after REM sleep deprivation. The numbers of c-fos-positive cells were significantly higher in the P-HT, DM-HT, VL-PAG, and Supf C. *PAG*, periaqueductal gray; *TCC*, trigeminal nucleus caudalis; *P-HT*, posterior hypothalamus; *DM-HT*, dorsomedial hypothalamus; *VL-PAG*, ventrolateral PAG; *Supf C*, superficial layer.

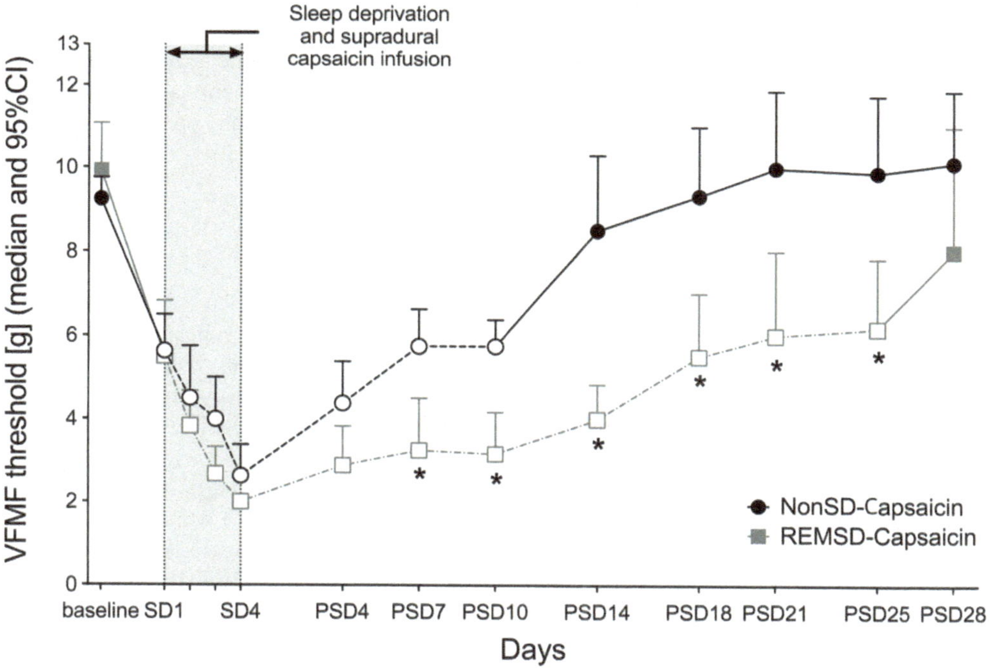

FIG. 6 After REM sleep deprivation, the duration of reduced headache threshold is significantly prolonged. (* indicates $P < .05$ between groups; *blank circles and squares* with a *dotted line* indicate $P < .05$ compared to baseline, *error bar* indicates 95% confidence interval). *SD*, days of sleep deprivation; *PSD*, days of postsleep deprivation; *g*, gram.

Applications to other areas

Clinically, sleep disorders and headaches coexist and interact with each other due to the considerable overlap of structural and pharmacological mechanisms involved in controlling sleep and headaches. Sleep deprivation is especially associated with exacerbation of headaches by lowering the threshold and increasing the sensitivity of headaches. Sleep and headaches are governed by complex mechanisms that require further investigation. It also suggests that sleep deprivation may exacerbate other pain disorders by overlapping mechanisms between sleep and pain.

Mini-dictionary of terms

Sleep deprivation: Sleep deprivation is insufficient duration or quality of sleep. Sleep is essential for maintaining homeostasis. Sleep consists of non-REM sleep and REM sleep. Experimentally, the researcher can deprive the entire sleep or selective sleep stage.

REM sleep: Rapid eye movement sleep is known as paradoxical sleep. REM sleep shows rapid eye movement, decreased muscle tone, and vivid dreams. The flip-flop system modulates the REM on/off, which located in the brain stem.

Nociception: Nociception is neural processing that recognizes the external noxious stimuli. External noxious stimuli activate the peripheral nociceptor that transmits to the nociception center of the central nervous system.

Key facts

Key facts of sleep deprivation and headache

- Sleep deprivation worsens the headache severity and lowers the headache thresholds.
- In pathophysiological aspects in CNS, there is a significant correlation in headache and sleep modulation.
- REM sleep deprivation enhances pain transmission and delays recovery periods of headache in animal research.

Summary points

- Sleep deprivation induces the aggravation of headache severity and reduces the headache threshold.
- Various CNS structures are involved in headache worsening associated with improper sleep quality: cerebral cortex, thalamus, hypothalamus, and brainstem nuclei.
- Adenosine, melatonin, orexin, and PACAP are linked to sleep and headache modulation.
- The pathophysiological mechanism between headache and sleep has been suggested through experimental research.

References

Afridi, S. K., Matharu, M. S., Lee, L., Kaube, H., Friston, K. J., Frackowiak, R. S., & Goadsby, P. J. (2005). A PET study exploring the laterality of brainstem activation in migraine using glyceryl trinitrate. *Brain*, *128*(Pt. 4), 932–939. https://doi.org/10.1093/brain/awh416.

Agargun, M. Y., Tekeoglu, I., Gunes, A., Adak, B., Kara, H., & Ercan, M. (1999). Sleep quality and pain threshold in patients with fibromyalgia. *Comprehensive Psychiatry*, *40*(3), 226–228. Retrieved from https://www.ncbi.nlm.nih.gov/pubmed/10360619.

Alkadhi, K., Zagaar, M., Alhaider, I., Salim, S., & Aleisa, A. (2013). Neurobiological consequences of sleep deprivation. *Current Neuropharmacology*, *11* (3), 231–249. https://doi.org/10.2174/1570159X11311030001.

Amorim, L. F., Guedes, R. C., Medeiros, M. C., da Silva, A. T., & Cabral-Filho, J. E. (1988). Apomorphine does not mimic the effects of REM-sleep deprivation on cortical spreading depression. *Brazilian Journal of Medical and Biological Research*, *21*(3), 611–614. Retrieved from https://www.ncbi.nlm.nih.gov/pubmed/3228646.

Andreou, A. P., Shields, K. G., & Goadsby, P. J. (2010). GABA and valproate modulate trigeminovascular nociceptive transmission in the thalamus. *Neurobiology of Disease*, *37*(2), 314–323. https://doi.org/10.1016/j.nbd.2009.10.007.

Arima, T., Svensson, P., Rasmussen, C., Nielsen, K. D., Drewes, A. M., & Arendt-Nielsen, L. (2001). The relationship between selective sleep deprivation, nocturnal jaw-muscle activity and pain in healthy men. *Journal of Oral Rehabilitation*, *28*(2), 140–148. https://doi.org/10.1046/j.1365-2842.2001.00687.x.

Arthaud, S., Varin, C., Gay, N., Libourel, P. A., Chauveau, F., Fort, P., … Peyron, C. (2015). Paradoxical (REM) sleep deprivation in mice using the small-platforms-over-water method: Polysomnographic analyses and melanin-concentrating hormone and hypocretin/orexin neuronal activation before, during and after deprivation. *Journal of Sleep Research*, *24*(3), 309–319. https://doi.org/10.1111/jsr.12269.

Bartsch, T., Knight, Y. E., & Goadsby, P. J. (2004). Activation of 5-HT(1B/1D) receptor in the periaqueductal gray inhibits nociception. *Annals of Neurology*, *56*(3), 371–381. https://doi.org/10.1002/ana.20193.

Bartsch, T., Levy, M. J., Knight, Y. E., & Goadsby, P. J. (2004). Differential modulation of nociceptive dural input to [hypocretin] orexin a and B receptor activation in the posterior hypothalamic area. *Pain*, *109*(3), 367–378. https://doi.org/10.1016/j.pain.2004.02.005.

Bartsch, T., Levy, M. J., Knight, Y. E., & Goadsby, P. J. (2005). Inhibition of nociceptive dural input in the trigeminal nucleus caudalis by somatostatin receptor blockade in the posterior hypothalamus. *Pain*, *117*(1–2), 30–39. https://doi.org/10.1016/j.pain.2005.05.015.

Benjamin, L., Levy, M. J., Lasalandra, M. P., Knight, Y. E., Akerman, S., Classey, J. D., & Goadsby, P. J. (2004). Hypothalamic activation after stimulation of the superior sagittal sinus in the cat: A Fos study. *Neurobiology of Disease*, *16*(3), 500–505. https://doi.org/10.1016/j.nbd.2004.03.015.

Brennan, K. C., Bates, E. A., Shapiro, R. E., Zyuzin, J., Hallows, W. C., Huang, Y., … Ptacek, L. J. (2013). Casein kinase idelta mutations in familial migraine and advanced sleep phase. *Science Translational Medicine*, *5*(183), 183ra156. 181–111 https://doi.org/10.1126/scitranslmed.3005784.

Burstein, R., Yamamura, H., Malick, A., & Strassman, A. M. (1998). Chemical stimulation of the intracranial dura induces enhanced responses to facial stimulation in brain stem trigeminal neurons. *Journal of Neurophysiology*, *79*(2), 964–982. Retrieved from http://jn.physiology.org/content/jn/79/2/964.full.pdf.

Campbell, I. G., Guinan, M. J., & Horowitz, J. M. (2002). Sleep deprivation impairs long-term potentiation in rat hippocampal slices. *Journal of Neurophysiology*, *88*(2), 1073–1076. https://doi.org/10.1152/jn.2002.88.2.1073.

Cirelli, C., & Tononi, G. (2000). On the functional significance of c-fos induction during the sleep-waking cycle. *Sleep*, *23*(4), 453–469.

Claustrat, B., Brun, J., Chiquet, C., Chazot, G., & Borson-Chazot, F. (2004). Melatonin secretion is supersensitive to light in migraine. *Cephalalgia*, *24*(2), 128–133. https://doi.org/10.1111/j.1468-2982.2004.00645.x.

Csati, A., Tajti, J., Kuris, A., Tuka, B., Edvinsson, L., & Warfvinge, K. (2012). Distribution of vasoactive intestinal peptide, pituitary adenylate cyclase-activating peptide, nitric oxide synthase, and their receptors in human and rat sphenopalatine ganglion. *Neuroscience*, *202*, 158–168. https://doi.org/10.1016/j.neuroscience.2011.10.055.

Datta, S., Mavanji, V., Ulloor, J., & Patterson, E. H. (2004). Activation of phasic pontine-wave generator prevents rapid eye movement sleep deprivation-induced learning impairment in the rat: A mechanism for sleep-dependent plasticity. *The Journal of Neuroscience*, *24*(6), 1416–1427. https://doi.org/10.1523/jneurosci.4111-03.2004.

Denuelle, M., Fabre, N., Payoux, P., Chollet, F., & Geraud, G. (2007). Hypothalamic activation in spontaneous migraine attacks. *Headache*, *47*(10), 1418–1426. https://doi.org/10.1111/j.1526-4610.2007.00776.x.

Ferre, S., Diamond, I., Goldberg, S. R., Yao, L., Hourani, S. M., Huang, Z. L., … Kitchen, I. (2007). Adenosine A2A receptors in ventral striatum, hypothalamus and nociceptive circuitry implications for drug addiction, sleep and pain. *Progress in Neurobiology*, *83*(5), 332–347. https://doi.org/10.1016/j.pneurobio.2007.04.002.

Fuller, P. M., Gooley, J. J., & Saper, C. B. (2006). Neurobiology of the sleep-wake cycle: Sleep architecture, circadian regulation, and regulatory feedback. *Journal of Biological Rhythms*, *21*(6), 482–493. https://doi.org/10.1177/0748730406294627.

Goadsby, P. J., Charbit, A. R., Andreou, A. P., Akerman, S., & Holland, P. R. (2009). Neurobiology of migraine. *Neuroscience*, *161*(2), 327–341. https://doi.org/10.1016/j.neuroscience.2009.03.019.

Goadsby, P. J., Cittadini, E., & Cohen, A. S. (2010). Trigeminal autonomic cephalalgias: Paroxysmal hemicrania, SUNCT/SUNA, and hemicrania continua. *Seminars in Neurology*, *30*(2), 186–191. https://doi.org/10.1055/s-0030-1249227.

Goadsby, P. J., Edvinsson, L., & Ekman, R. (1988). Release of vasoactive peptides in the extracerebral circulation of humans and the cat during activation of the trigeminovascular system. *Annals of Neurology*, *23*(2), 193–196. https://doi.org/10.1002/ana.410230214.

Goadsby, P. J., Lambert, G. A., & Lance, J. W. (1985). The mechanism of cerebrovascular vasoconstriction in response to locus coeruleus stimulation. *Brain Research*, *326*(2), 213–217. https://doi.org/10.1016/0006-8993(85)90030-7.

Goadsby, P. J., & Zagami, A. S. (1991). Stimulation of the superior sagittal sinus increases metabolic activity and blood flow in certain regions of the brainstem and upper cervical spinal cord of the cat. *Brain*, *114*(Pt. 2), 1001–1011. https://doi.org/10.1093/brain/114.2.1001.

Harrington, M. E., Hoque, S., Hall, A., Golombek, D., & Biello, S. (1999). Pituitary adenylate cyclase activating peptide phase shifts circadian rhythms in a manner similar to light. *The Journal of Neuroscience*, *19*(15), 6637–6642. Retrieved from https://www.ncbi.nlm.nih.gov/pubmed/10414992.

Hicks, R. A., Coleman, D. D., Ferrante, F., Sahatjian, M., & Hawkins, J. (1979). Pain thresholds in rats during recovery from REM sleep deprivation. *Perceptual and Motor Skills*, *48*(3 Pt. 1), 687–690. https://doi.org/10.2466/pms.1979.48.3.687.

Holland, P. R., Akerman, S., Lasalandra, M. P., & Goadsby, P. J. (2008). Antinociceptive effects of orexin a in the vlPAG are blocked by 5-HT1B/1D receptor antagonism. *Headache: The Journal of Head and Face Pain*, *48*(s1), S1–S67. https://doi.org/10.1111/j.1526-4610.2008.01172.x.

Huang, Z. L., Urade, Y., & Hayaishi, O. (2011). The role of adenosine in the regulation of sleep. *Current Topics in Medicinal Chemistry*, *11*(8), 1047–1057. https://doi.org/10.2174/156802611795347654.

Kelman, L., & Rains, J. C. (2005). Headache and sleep: Examination of sleep patterns and complaints in a large clinical sample of migraineurs. *Headache*, *45*(7), 904–910. https://doi.org/10.1111/j.1526-4610.2005.05159.x.

Kim, J., Cho, S. J., Kim, W. J., Yang, K. I., Yun, C. H., & Chu, M. K. (2017). Insufficient sleep is prevalent among migraineurs: A population-based study. *The Journal of Headache and Pain*, *18*(1), 50. https://doi.org/10.1186/s10194-017-0756-8.

Kim, S. H., Park, J. Y., Shin, H. E., Lee, S. B., Ryu, D. W., Kim, T. W., & Park, J. W. (2019). The influence of rapid eye movement sleep deprivation on nociceptive transmission and the duration of facial allodynia in rats: A behavioral and Fos immunohistochemical study. *The Journal of Headache and Pain*, *20*(1), 21. https://doi.org/10.1186/s10194-019-0977-0.

Knight, Y. E., & Goadsby, P. J. (2001). The periaqueductal grey matter modulates trigeminovascular input: A role in migraine? *Neuroscience*, *106*(4), 793–800. https://doi.org/10.1016/s0306-4522(01)00303-7.

Kristiansen, E. S., Nielsen, L. S., Christensen, S. S., Botvid, S. H. C., Norgaard Poulsen, J., & Gazerani, P. (2017). Sleep deprivation sensitizes human craniofacial muscles. *Somatosensory & Motor Research*, *34*(2), 116–122. https://doi.org/10.1080/08990220.2017.1317241.

Kundermann, B., Spernal, J., Huber, M. T., Krieg, J. C., & Lautenbacher, S. (2004). Sleep deprivation affects thermal pain thresholds but not somatosensory thresholds in healthy volunteers. *Psychosomatic Medicine*, *66*(6), 932–937. https://doi.org/10.1097/01.psy.0000145912.24553.c0.

Leone, M., D'Amico, D., Moschiano, F., Fraschini, F., & Bussone, G. (1996). Melatonin versus placebo in the prophylaxis of cluster headache: A double-blind pilot study with parallel groups. *Cephalalgia*, *16*(7), 494–496. https://doi.org/10.1046/j.1468-2982.1996.1607494.x.

Liu, Z. W., & Gao, X. B. (2007). Adenosine inhibits activity of hypocretin/orexin neurons by the A1 receptor in the lateral hypothalamus: A possible sleep-promoting effect. *Journal of Neurophysiology*, *97*(1), 837–848. https://doi.org/10.1152/jn.00873.2006.

Lu, J., Jhou, T. C., & Saper, C. B. (2006). Identification of wake-active dopaminergic neurons in the ventral periaqueductal gray matter. *The Journal of Neuroscience*, *26*(1), 193–202. https://doi.org/10.1523/JNEUROSCI.2244-05.2006.

Manganotti, P., Fuggetta, G., & Fiaschi, A. (2004). Changes of motor cortical excitability in human subjects from wakefulness to early stages of sleep: A combined transcranial magnetic stimulation and electroencephalographic study. *Neuroscience Letters*, *362*(1), 31–34. https://doi.org/10.1016/j.neulet.2004.01.081.

Noseda, R., Kainz, V., Jakubowski, M., Gooley, J. J., Saper, C. B., Digre, K., & Burstein, R. (2010). A neural mechanism for exacerbation of headache by light. *Nature Neuroscience*, *13*(2), 239–245. https://doi.org/10.1038/nn.2475.

Offenhauser, N., Zinck, T., Hoffmann, J., Schiemann, K., Schuh-Hofer, S., Rohde, W., … Reuter, U. (2005). CGRP release and c-fos expression within trigeminal nucleus caudalis of the rat following glyceryltrinitrate infusion. *Cephalalgia*, *25*(3), 225–236. https://doi.org/10.1111/j.1468-2982.2004.00845.x.

Oh, J. H., Cho, S. J., Kim, W. J., Yang, K. I., Yun, C. H., & Chu, M. K. (2018). Insufficient sleep in tension-type headache: A population study. *Journal of Clinical Neurology*, *14*(4), 566–573. https://doi.org/10.3988/jcn.2018.14.4.566.

Onen, S. H., Alloui, A., Eschalier, A., & Dubray, C. (2000). Vocalization thresholds related to noxious paw pressure are decreased by paradoxical sleep deprivation and increased after sleep recovery in rat. *Neuroscience Letters*, *291*(1), 25–28. Retrieved from https://www.ncbi.nlm.nih.gov/pubmed/10962145. http://ac.els-cdn.com/S0304394000013835/1-s2.0-S0304394000013835-main.pdf?_tid=5741d678-5bce-11e7-b06e-00000aab0f6c&acdnat=1498632924_5c4eaabca8d4d8b99a35ffeeefddacae.

Park, J. W., Cho, S. J., Park, S. G., & Chu, M. K. (2018). Circadian variations in the clinical presentation of headaches among migraineurs: A study using a smartphone headache diary. *Chronobiology International*, *35*(4), 546–554. https://doi.org/10.1080/07420528.2017.1420076.

Park, J. W., Suh, G. I., Shin, H. E., & Park, G. E. (2012). Influence of memantine on nociceptive responses of the trigeminocervical complex after formalin injection. *Cephalalgia*, *32*(4), 308–316. https://doi.org/10.1177/0333102411435986.

Raymond, I., Nielsen, T. A., Lavigne, G., Manzini, C., & Choiniere, M. (2001). Quality of sleep and its daily relationship to pain intensity in hospitalized adult burn patients. *Pain*, *92*(3), 381–388. Retrieved from https://www.ncbi.nlm.nih.gov/pubmed/11376911. http://ac.els-cdn.com/S0304395901002822/1-s2.0-S0304395901002822-main.pdf?_tid=90e842a0-3918-11e7-a295-00000aacb360&acdnat=1494816513_5793d2ae95ff4dd142ddb7ccaf96104f.

Sancisi, E., Cevoli, S., Vignatelli, L., Nicodemo, M., Pierangeli, G., Zanigni, S., … Montagna, P. (2010). Increased prevalence of sleep disorders in chronic headache: A case-control study. *Headache*, *50*(9), 1464–1472. https://doi.org/10.1111/j.1526-4610.2010.01711.x.

Scammell, T. E., & Winrow, C. J. (2011). Orexin receptors: Pharmacology and therapeutic opportunities. *Annual Review of Pharmacology and Toxicology*, *51*, 243–266. https://doi.org/10.1146/annurev-pharmtox-010510-100528.

Sherin, J. E., Shiromani, P. J., McCarley, R. W., & Saper, C. B. (1996). Activation of ventrolateral preoptic neurons during sleep. *Science*, *271*(5246), 216–219. https://doi.org/10.1126/science.271.5246.216.

Shields, K. G., & Goadsby, P. J. (2005). Propranolol modulates trigeminovascular responses in thalamic ventroposteromedial nucleus: A role in migraine? *Brain*, *128*(Pt. 1), 86–97. https://doi.org/10.1093/brain/awh298.

Shiromani, P. J., Winston, S., & McCarley, R. W. (1996). Pontine cholinergic neurons show Fos-like immunoreactivity associated with cholinergically induced REM sleep. *Brain Research. Molecular Brain Research*, *38*(1), 77–84. https://doi.org/10.1016/0169-328x(95)00325-m.

Song, T. J., Yun, C. H., Cho, S. J., Kim, W. J., Yang, K. I., & Chu, M. K. (2018). Short sleep duration and poor sleep quality among migraineurs: A population-based study. *Cephalalgia*, *38*(5), 855–864. https://doi.org/10.1177/0333102417716936.

Steriade, M., McCormick, D. A., & Sejnowski, T. J. (1993). Thalamocortical oscillations in the sleeping and aroused brain. *Science*, *262*(5134), 679–685. https://doi.org/10.1126/science.8235588.

Tanuri, F. C., de Lima, E., Peres, M. F., Cabral, F. R., da Graca Naffah-Mazzacoratti, M., Cavalheiro, E. A., … Amado, D. (2009). Melatonin treatment decreases c-fos expression in a headache model induced by capsaicin. *The Journal of Headache and Pain*, *10*(2), 105–110. https://doi.org/10.1007/s10194-009-0097-3.

Tuka, B., Helyes, Z., Markovics, A., Bagoly, T., Szolcsanyi, J., Szabo, N., … Tajti, J. (2013). Alterations in PACAP-38-like immunoreactivity in the plasma during ictal and interictal periods of migraine patients. *Cephalalgia*, *33*(13), 1085–1095. https://doi.org/10.1177/0333102413483931.

Ukponmwan, O. E., Rupreht, J., & Dzoljic, M. (1986). An analgesic effect of enkephalinase inhibition is modulated by monoamine oxidase-B and REM sleep deprivations. *Naunyn-Schmiedeberg's Archives of Pharmacology*, *332*(4), 376–379. Retrieved from https://www.ncbi.nlm.nih.gov/pubmed/3090452.

Wu, G., Chen, L., Wei, G., Li, Y., Zhu, G., Zhao, Z., & Huang, F. (2014). Effects of sleep deprivation on pain-related factors in the temporomandibular joint. *The Journal of Surgical Research*, *192*(1), 103–111. https://doi.org/10.1016/j.jss.2014.05.035.

Wu, W. P., Hao, J. X., Halldner, L., Lovdahl, C., DeLander, G. E., Wiesenfeld-Hallin, Z., … Xu, X. J. (2005). Increased nociceptive response in mice lacking the adenosine A1 receptor. *Pain*, *113*(3), 395–404. https://doi.org/10.1016/j.pain.2004.11.020.

Zagami, A. S., Goadsby, P. J., & Edvinsson, L. (1990). Stimulation of the superior sagittal sinus in the cat causes release of vasoactive peptides. *Neuropeptides*, *16*(2), 69–75.

Chapter 20

Antinociceptive glucagon-like peptides

Duygun Altıntaş Aykan
Department of Pharmacology, Kahramanmaras Sutcu Imam University, Faculty of Medicine, Kahramanmaras, Turkey

Abbreviations

5-HT	5-hydroxytryptamine
ACTH	adrenocorticotropic hormone
ADA	adenosine-deaminase
AP-1	activator protein 1
CNS	central nervous system
COX	cyclo-oxygenase
CXCL-10	CXC chemokine ligand-10
DM	diabetes mellitus
DPP4	dipeptidyl peptidase-4
EM1	endomorphine 1
ERK	extracellular signal-regulated kinase
GIP	gastric inhibitor polypeptide
GIS	gastrointestinal system
GLP-1	glucagon-like peptide-1
GLP-1R	GLP-1 receptor
GLP-2	glucagon-like peptide-2
GLP-2R	GLP-2 receptor
IGF-1	insulin-like growth factor-1
IL	interleukin
INS-1E	insulinoma cells
IPI	isoleucin-prolin-isoleucin
JNK-1	c-Jun N-terminal kinases
KGF	keratinocyte growth factor
L-NAME	L-N^G-Nitro arginine methyl ester
LPS	lipopolysaccharide
MAP	mitogen-activated protein
MCP-1	monocyte chemoattractant protein-1
NOS	nitric oxide synthase
NPY	neuropeptide Y
PACAP	pituitary adenylate activating peptide
PCs	prohormone convertase enzymes
PK	protein kinase
SOCS-3/physiological suppressors of cytokine signaling	cytokine signal suppressors pathway
SP	substance P
STAT-3	signal transducer and transcription activation-3
TGF	transforming growth factor
TLR4	Toll-like receptor 4
TNF	tumor necrosis factor
Tpl2	tumor progression locus 2
VIP	vasoactive intestinal peptide

The Neurobiology, Physiology, and Psychology of Pain. https://doi.org/10.1016/B978-0-12-820589-1.00020-8

Introduction to the glucagon-like peptides

Current analgesics for the management of chronic pain, which are induced by tissue damage, inflammation, or existing diseases, are moderately effective and have many adverse effects. Hence, the management of chronic pain is being researched with great curiosity, and the intervention of new drugs to achieve pain control becomes an important target (Király et al., 2018).

Glucagon-like peptides (GLPs) are the peptide molecules secreted from the enteroendocrine L-cells of the intestines. Their precursor protein, proglucagon, is expressed in the pancreatic α-cells, intestinal L-cells, and brain cells. It consists of 160-amino acids and is processed in a tissue-specific manner by prohormone convertase enzymes (PCs). In the pancreatic α-cells, the PC2 enzyme predominates and separates the proglucagon into glucagon and other products, while PC1/3 enzyme is dominated in the gut and brain, and contributes to the formation of glucagon-like peptide-1 (GLP-1), glucagon-like peptide-2 (GLP-2), glycentin, and oxyntomodulin (Aykan, Kesim, Ayan, & Kurt, 2019; Insuela & Carvalho, 2017) (Table 1).

Glucagon-like peptide-1 (GLP-1) is mainly used in the modulation of blood glucose levels, and it is an important choice of treatment for type 2 diabetes mellitus (DM). GLP-1 has exerted antinociceptive effects in peripheric neuropathic pain, cancer-related chronic pain, and inflammation in experimental studies (Aykan et al., 2019). Glucagon-like peptide-2 (GLP-2), a potent growth factor of the gut, is involved in the enteral absorption and mucosal inflammation. GLP-2 is frequently used in the treatment of short bowel syndrome. Activation of GLP-1/GLP-2 signaling pathways has been involved in the neuronal inflammation, oxidative stress, neurotoxicity, and memory impairment (Aykan et al., 2019) (Table 2).

The action of GLPs in the nociceptive response requires importance for the development of new therapeutic strategies in pain, as they can serve as prototypes of the design and synthesis for potential antinociceptive new compounds.

Applications to other areas

Observing that enteral nutrition leads to more insulin secretion than intravenous nutrition, the incretin model has emerged. The first isolated incretin was named as gastric inhibitor polypeptide (GIP) since it caused an inhibition in the gastric acid secretion. Later, the insulinotropic and blood glucose regulating effects of the GIP were strongly defined. After this discovery, the name of the molecule was changed to the glucose-dependent insulinotropic polypeptide. With further studies, more incretins were defined in the gastrointestinal system (GIS) (Lauritsen, Moody, Christensen, & Lindkaer Jensen, 1980).

The most important incretin secreted from the intestinal tract is GLP-1. GLP-1 is synthesized in the jejunal L-cells. The gene that synthesizes it is the same as the glucagon gene in the α-cells of the pancreas. GLP-1 is known as the most potent incretin and is similar to glucagon with 50% homology (Orskov, Holst, Poulsen, & Kirkegaard, 1987).

L-cells secrete equal molar amounts of GLP-1 and GLP-2 as a result of the processing of proglucagon by PC 1/3 in intestinal L-cells. The stimulating signal for GLP secretion is the food intake. Subsequent to the nutrient intake, GLP-1

TABLE 1 Pathways involved in the antiinflammatory and antinociceptive effects of GLP-1.

GLP-1	Decreases or blocks (Insuela & Carvalho, 2017; Kagal et al., 2017)	Increases or activates (Lee & Jun, 2016)
	Inflammation	cAMP/Ca^{2+}, PKA, CAMKKβ, and pAMPK
	Mucus hypersecretion	ERK1/2
	Hyperreactivity	Release of β-endorphins (Fan et al., 2015)
	E-selectin, TNF-α, IL-4, IL-5, IL-13, TGF-β	
	Infiltration of macrophages, lymphocytes	
	LPS-induced permeability	
	NOS induction	
	PKC, NF-κB activation, expression of NLRP3, IL-1β, IL-6, VCAM-1, IFN-γ	
	Chemokines CXCL10, MCP-1	

TABLE 2 Pathways involved in the antiinflammatory effects of GLP-2.

GLP-2	Decreases or blocks	Increases or activates
	TNFα and IFNγ gene expression	Cell proliferation
	Mucosal damage	TGF β and VEGF
	Apoptosis/proteolysis	Wound healing and epithelial restoration
	Expression of i-NOS, COX-2 enzyme, IL-1β and IL-6 mRNA in macrophages (Xie et al., 2014)	CD11b/F4/80 macrophages
	CD4 T cell (Ivory et al., 2008)	IGF-1 production (Ivory et al., 2008)
	STAT-3 (Sigalet et al., 2007)	SOCS-3 (Sigalet et al., 2007)
	ERK1/2, JNK1/2, microglia-mediated nerve damage (Li et al., 2016)	gastric mucosal blood flow (Gulec Suyen et al., 2015)
	Caspase-3 and protein kinase A related Caspase 8 and 9 (Yusta et al., 1999)	cAMP and activator protein 1 (AP-1)
		P70 S6 kinase (Yusta et al., 1999)
		(VIP)—mediated restoration of enteric neural system
		IGF-1, keratinocyte growth factor (KGF), TGFβ (Yusta et al., 1999)

induces insulin secretion in a glucose-dependent manner (Kissow, 2015). GLP-1 also reduces gastric emptying, secretion of gastric acid, and pancreatic enzymes.

GLP-2 is known as a potent growth factor of the intestines attributing to its tropic effects on GIS. It stimulates the proliferation of epithelial cells and inhibits apoptosis (Kissow, 2015). Numerous studies on exogenous administration of GLP-2 have proven its trophic action on the gut, but the exact underlying mechanism is not clear at all. There are studies with GLP-2 receptor (GLP-2R) knock-out mice, in which normal growth and development of the gut was reported, supporting that multiple factors may play role in the bowel adaptation (Lee et al., 2012).

Previous studies have shown that GLP-2 has also several nontrophic effects, such as reducing intestinal motility and gastric emptying, inhibiting gastric acid secretion, and increasing nutrient absorption. GLP-2 is well known as a good choice for intestinal diseases such as inflammatory bowel disease and short bowel syndrome (Meier et al., 2006).

Glucagon-like peptide-1 and its analogs

GLP-1 is a 31-amino acid peptide hormone. Despite its high amino acid sequence homology with glucagon, GLP-1 reacts opposite in glucose modulation (Day et al., 2011). Glucagon defenses the organism against the hypoglycemia by increasing glucose levels in the circulation via glycogenolysis and gluconeogenesis, also inhibits glycolysis and glycogenesis in the liver. As a result of glucagon-induced glucose metabolism, an increase in the hepatic glucose output provides the sufficient amount of glucose to the brain. However, GLP-1 stimulates the glucose-dependent insulin release from pancreatic cells, inhibits glucagon secretion, and increases islet α-cell proliferation (Godoy-Matos, 2014).

GLP-1, commonly indicated in type 2 DM, exerts antiapoptotic, antioxidative, and antiinflammatory properties besides the blood glucose regulation. It is a hormone with neuronal and cardiovascular protection properties (Seufert & Gallwitz, 2014). GLP-1 reduces the activation of CXC chemokine ligand-10 (CXCL-10), signal transducer and transcription activation-3 (STAT-3) Mcp-1, tumor necrosis factor (TNF)-α expression, NFκB activation, thereby reducing the activity of natural killer cells (Campbell & Drucker, 2013). In LPS-induced inflammation models, GLP-1 administration was reported to decrease protein expressions of IL-6, cyclo-oxygenase (COX)-2, nitric oxide synthase (NOS), NF-κB mRNA, TNF-α, and interleukin (IL)-1β (Lin & Lin, 2016).

GLP-1 reduces allergen-induced histamine release and pro-inflammatory cytokines from the mast cells. The inhibitory action of GLP-1 in the lung inflammation was attributed to its ability in decreasing E-selectin, TNF-α, IL-4, IL-5, and

IL-13. GLP-1 also reduces the infiltration of macrophages, lymphocytes, and transforming growth factor (TGF)-β in fibrosis. In addition, GLP-1 reduces lipopolysaccharide (LPS)-induced permeability, NOS induction, and eosinophil-related IL-4, IL-8, and IL-13 production (Insuela & Carvalho, 2017; Kagal, Angadi, & Matule, 2017).

Its receptor, GLP-1R, is a seven-transmembrane G protein-coupled receptor. Although the glucagon receptor binds to only Gs protein, GLP-1R binds to the Gs or Gq protein. GLP-1 leads to increase intracellular cAMP and Ca^{2+} levels by activating adenyl cyclase, protein kinase (PK) A, Epac-2, phospholipase C, and extracellular signal-regulated kinase (ERK) 1/2 pathways (Thompson & Kanamarlapudi, 2015). Activation of GLP-1R by GLP-1 or by exogenous agonists protects β cells from cytokine-induced apoptosis and necrosis via protein kinase B cell signaling pathway (Li, El-Kholy, Rhodes, & Brubaker, 2005). In addition, mitogen-activated protein (MAP) 3 kinase, tumor progression locus 2 (Tpl2) inhibitor, and GLP-1R agonists produce potent antiapoptotic and antiinflammatory effects on insulinoma cells (INS-1E) (Varin et al., 2016).

GLP-1 can function as a neuropeptide. Recently, the functional roles of the GLP-1/GLP-1R signaling pathways have been studied in various CNS disorders. Activation of this signaling axis has been involved in the neuronal inflammation, oxidative stress, neurotoxicity, and memory impairment (Bassil, Fernagut, Bezard, & Meissner, 2014). The regions of GLP-1 and GLP-1R in the CNS are presented in the solitary tract nucleus, dorsal, and ventral parts of the medullary reticular nucleus, in the spinal cord, hypothalamus, thalamus, and cortex (Lee et al., 2018). GLP-1R is upregulated by microglial proliferation after peripheral nerve injury. In the studies with GLP-1 analogs, GLP-1 (7–36) and exenatide reduced the number of paw licking in the tonic phase in formalin test, depending on the dose. The antinociceptive effects were revealed through opioidergic pathway by an induction in the release of spinal β-endorphins (Gong et al., 2014). GLP-1 protects neurons and created analgesia by causing β-endorphine expression in microglial cells (Jia, Gong, Li, Zhu, & Wang, 2015).

Treatments that based on GLP-1 and its analogs have demonstrated antiinflammatory actions in the several systems (Table 3). In the cardiac fibrosis, liraglutide, a GLP-1R agonist, inhibited vascular reactivity, cardiac hypertrophy, fibroblast and collagen accumulation, and monocyte chemoattractant protein-1 (MCP-1) production (Gaspari, Brdar, Lee, et al., 2016). In the CNS, liraglutide decreased the levels of IL-6, IL-1β, IL-12, and NO in the irradiation-induced model (Parthsarathy & Hölscher, 2013). In the diet-induced fatty liver disease model, liraglutide improved insulin resistance and liver function, reduced lipid peroxidation, free fatty acid, and hepatic concentration of TNF-α, and induced the antioxidant enzyme superoxide dismutase. It is observed that the function of liraglutide in liver is modulated by adiponectin and inactivation of c-Jun N-terminal kinases (JNK-1) (Gao et al., 2015). It protects the brain against oxidative stress. In previous studies, liraglutide improved the neurogenesis, cognitive function, and reduced amyloid plaque deposition (Palleria et al., 2017). In addition, it reduced neuron loss after cerebral ischemia due to its antiapoptotic and antioxidant effects. It also exhibited neurotrophic and antiinflammatory effects in rats (Briyal, Shah, & Gulati, 2014). Another study showed that liraglutide improved the kidney damage in diabetic rats by activating the endothelial NOS. This evidence suggests that activating GLP-1 receptors may produce NO in tropic functions (Nozu et al., 2018).

TABLE 3 The most common GLP-based drugs.

Agent	Indication
GLP-1R agonists	
Exenatide (Exendin-4)	DM Vascular disease, nonalcoholic steatohepatitis, nephropathy, neurodegenerative brain disorder
Liraglutide	Diabetes, vascular disease, neurodegenerative brain disorder, nonalcoholic steatohepatitis, nephropathy, asthma, psoriasis
DPP-4 inhibitors	
Sitagliptin	Diabetes, vascular disease, neurodegenerative brain disorder, nephropathy
Linagliptin	Nephropathy, vascular disease
Vildagliptin (PKF-275-055)	Diabetes, nephropathy
GLP-2 analog	
Teduglutide	Short bowel syndrome

Exendin-4, another analog of GLP-1, decreased the synthesis of IL-1 and IL-6 in cardiac remodeling and diastolic dysfunction (Tate, Robinson, Green, McDermott, & Grieve, 2016). GLP-1R agonist, exenatide, inhibited the levels of TNF-α, MCP-1, and F4/80 expression in the testis of mice in an obesity model and demonstrated an improvement in sperm quality (Zhang et al., 2015). In particular, several in vivo evidences have confirmed the significant benefits of exendin-4 in Alzheimer's disease, Parkinson's disease, and amyotrophic lateral sclerosis, which are associated with reduced microglial activation (Darsalia et al., 2012). Exenatide was administered at a dose of 0.1 μg/kg for 14 days to the diabetic mice. It reduced protein synthesis of hippocampal neural cell adhesion molecule, intercellular cell adhesion molecule, and vascular cell adhesion molecule, which are normally increased by pro-inflammatory stimulation (Gumuslu et al., 2016). In another experimental study, time to response pain was shortened after exenatide treatment in mice with type 2 DM (Komsuoglu Celikyurt et al., 2014). WB4-24, another GLP-1R agonist, showed an inhibition in the nociceptive response by stimulating the release of β-endorphins from spinal microglia (Fan et al., 2015).

Glucagon-like peptide-2

GLP-2 is an intestinotropic peptide secreted by enteroendocrine L-cells. It increases epithelial proliferation and inhibits apoptosis. GLP-2 secretion is increased after feeding, which is the same stimulus for GLP-1 secretion (Markovic & Brubaker, 2019).

GLP-2R is placed on the enteroendocrine cells, enteric neurons, and intestinal subepithelial myofibroblasts. The effects of GLP-2 on enterocytes are induced by some downstream mediators. Binding of GLP-2 to its receptor on subepithelial myofibroblasts has been shown to be triggered by insulin-like growth factor-1 (IGF-1), which is required for the proliferation and barrier function of GLP-2 (Austin, Tsang, Chalmers, Maalouf, & Brubaker, 2018).

The action of GLP-2 has been investigated in mice receiving chemotherapy. An increase in survival and a decrease in bacteremia were observed (Boushey, Yusta, & Drucker, 2001). GLP-2 prevents intestinal apoptosis and chemotherapy-induced villus destruction (Kissow et al., 2012). Its antiinflammatory action has been attributed to the inhibition of pro-inflammatory substances mediated by nuclear factor-kB activity and ERK phosphorylation (Xie et al., 2014).

GLP-2 was demonstrated to be protective in the experimental corrosive esophagitis model (Tekin et al., 2015). Dong et al. investigated GLP-2 in irinotecan-induced mucositis. They reported that the effect of GLP-2 was related to IGF-1R, declaring the absence of the beneficial effects of GLP-2 in IGF-1R null mice (Dong et al., 2014).

GLP-2 decreases the expression of i-NOS, COX-2, TNF-α, IL-1β, and IL-6 mRNA in macrophages (Xie et al., 2014). In histopathological studies, it has been reported that GLP-2 administration leads a decrease in CD4 T cells and TNF-α production, an increase in CD11b/F4/80 macrophages and IGF-1 production in fluorescence-active cell analysis of lamina propria of animals with colitis (Ivory, Wallace, McCafferty, & Sigalet, 2008). The antiinflammatory effects of GLP-2 occur via enteric neural pathways. GLP-2 shows antiinflammatory activity in the gut by activation of the cytokine signal suppressors pathway (SOCS-3/physiological suppressors of cytokine signaling) and inhibition of the STAT-3 signaling pathway (Sigalet et al., 2007).

GLP-2 application reduces LPS-induced iNOS, COX-2, IL1-β, IL6, TNF-α production in the brain tissue, associated kinase ERK1/2, JNK1/2, and microglia-mediated nerve damage (Li et al., 2016). In the amygdala of adrenocorticotropic hormone (ACTH)—administered mice, GLP-2 induced 5-hydroxytryptamine (5-HT) levels and presented antidepressant effect (Iwai et al., 2015). NO contributes in the stimulating action of GLP-2 in chylomicron production, and L-N^G-Nitro arginine methyl ester (L-NAME) blocks GLP-2 stimulated apo B48 secretion (Hsieh et al., 2015). GLP-2, which provides an increase in gastric mucosal blood flow, performs this effect through calcitonin gene-related peptide and endogenous prostaglandins, and NO does not play a role in this effect (Gulec Suyen, Isbil-Buyukcoskun, Cam, & Ozluk, 2015).

In addition to the studies related with short bowel syndrome, in which GLP-2 is used most frequently in the clinic, there are many studies on the expansion of the intestinal epithelium and cell proliferation. These studies indicate that GLP-2R signal pathways include the activation of cAMP and activator protein 1 (AP-1)-dependent signal pathways, as in glucagon and GLP-1. The increase in cAMP leads to a rapid synthesized precursor gene activation and cell proliferation other than PKA and AP-1. Another pathway activated by GLP-2 has been shown to be P70 S6 kinase. This pathway is thought to be the basis of the inhibitory effect of GLP-2 in the apoptosis. The contribution of GLP-2 in the cell proliferation is based on the vasoactive intestinal peptide (VIP)-mediated restoration of the enteric neural system. It reduces plasma pro-inflammatory cytokines, markers of tissue oxidative stress, and macrophage inflammation, and subsequent MAPK inhibition (ERK1/2, JNK), PI3Kγ, and Akt phosphorylations. It inhibits more than one apoptotic pathway including Caspase-3 and protein kinase A-related Caspase 8 and 9 activity. GLP-2 increases IGF-1, keratinocyte growth factor (KGF), TGFβ, and decreases TNFα and IFNγ levels (Yusta et al., 1999).

Due to the rapid breakdown of GLP-1 and GLP-2 by dipeptidyl peptidase-4 (DPP4), the eliminating half-life is defined as 2 min for GLP-1 and a few minutes for GLP-2. Hence, circulating endogenous GLP-1 and GLP-2 levels are very low. However, effects of GLP-1 can be seen weeks later. There are studies on the fact that the short half-lives do not reflect their beneficial effects in the long term. In a study, the peak concentration returned to normal levels after 90 min of GLP-1 subcutaneous administration.

The acute phase of exogenous GLP-1 administration corresponds to the first minutes. This period is the phase of glucose-dependent insulinomimetic effect, closure of ATP-dependent K channels, β cell depolarization, opening of voltage-dependent Ca channels and insulin secretion by the cAMP/PKA pathway. In the first hours, GLP-1 activates β cell-specific (insulin, glucose transporter 2, glucokinase) genes, and CNS effects occur in this period. Days after, GLP-1 causes an increase in β cell mass (Hui, Farilla, Merkel, & Perfetti, 2002).

Other agents of interest

DPP4 is a type II integral transmembrane glycoprotein which is expressed in cells and body fluids. Its substrates are incretins (GLP-1 and -2 and glucose-dependent insulinotropic polypeptide), bradykinin, substance P (SP), neuropeptide Y (NPY), VIP, and TNF-α (Mentlein, 1999). It has several binding sites for different extracellular matrix proteins such as adenosine-deaminase (ADA), collagen, and fibronectin (De Meester et al., 1994). DPP4 reacts as a cell surface antigen CD26 on T-lymphocytes (Yu et al., 2011), and as a receptor for Coronaviruses (Lu et al., 2013). Inhibition of DPP4 increases incretin levels and prolongs postprandial insulin effect. Hence, DPP4 has become the main target for type II diabetes. New DPP4 inhibitors are still under investigation for lipid metabolism, myocardial, kidney and liver functions, atherosclerosis, and inflammation.

DPP4 enzyme has many physiological roles in the immunity, metabolism, and CNS (Balogh et al., 2019). The binding of DPP4 with ADA, caveolin-1, CD45, fibronectin, and collagen results in mainly pro-inflammatory events (Klemann, Wagner, Stephan, & von Hörsten, 2016). DPP4 can be involved in the nociceptive transmission since its substrates include pain-related peptides such as endomorphine 1 (EM1) and EM2, SP, NPY, pituitary adenylate activating peptide (PACAP) and GLP1. Among these, SP and PACAP are proallergics, while others are considered analgesic peptides (Gerald et al., 1996).

Despite these, there are only a few reports involving DPP4 or DPP4 inhibitors on the pain pathway. DPP4 expression has recently been reported in spinal dorsal horn neurons, astrocytes, and microglia. The expression of DPP4 is increased after peripheral inflammation in astrocytes and neural damage in microglia (Király et al., 2018). In addition, vincristine-induced neuropathic pain caused an increase in DPP4-like enzymatic activity in the spinal cord (Yang et al., 2014). These reports suggested that the role of DPP4 has become more evident in cases of certain pain conditions. Besides, DPP4 knock-out mice are more sensitive to nociception due to increased SP levels (Guieu et al., 2006).

Although several dipeptides are potential DPP4 substrates, DPP4 cannot break down all peptides in vivo. DPP4 metabolizes substrates with a length of 24 amino acids around the N-terminal position, and is inhibited by gliptins. DPP4 inhibition results in immunosuppression such as asthma, colitis, and rheumatoid arthritis (Iwaya, Fujii, Hagio, Hara, & Ishizuka, 2013). In addition, inhibition of DPP4 suppresses the production of IL-1β, IL-2, IL-6, and TNF-α (Hirakawa et al., 2015).

DPP4 inhibitors are reported to selectively affect hyperalgesia. DPP4 contributes in the modulation of mechanical hyperalgesia in the neuropathy model, but the underlying mechanisms are not clear at all. The contribution of microglia to the development of neuropathic allodynia has been extensively studied (Tsuda, 2016), but its association with DPP4 requires further study. Analgesic and antiinflammatory action of DPP4 inhibitors has also been demonstrated in chronic inflammatory models in rodents. DPP4 inhibitors do not change the nociceptive threshold in the acute nociceptive state (Király et al., 2009). Isoleucin-prolin-isoleucin (IPI) and vildagliptin, DPP4 inhibitors, have been shown to cause a significant decrease in mechanical hyperalgesia (Hamidi et al., 2014; Lee & Jun, 2016).

DPP4 inhibitors suppress Toll-like receptor 4 (TLR4) and mediate the upregulation of pro-inflammatory cytokines. This shows an important interaction between TLR4 and DPP4 activity, which can also be found in CNS and regulate the synthesis and release of inflammatory mediators and endogenous opioids. The possible relationship of the opioidergic pathway and gliptins has been extensively investigated in recent years. Mechanical hyperalgesia is dramatically reduced after administration of spinal DPP4 inhibitors in subacute inflammation. This effect was reversed with the opioid receptor blocker naloxone (Király et al., 2009). Endogenous opioids are secreted from the CNS during acute inflammation, and are degraded rapidly with high enzymatic activity. Selective DPP4 inhibitors, such as vildagliptin, are unlikely to have inhibitory effect on opioid degrading enzymes. Glial cells express opioid receptors and can synthesize endogenous opioids. The interaction between the two systems is mutual because endogenous opioids are effective in the synthesis of inflammatory mediators.

Mini-dictionary of terms

Incretins: Gut peptides that are secreted after nutrient intake and stimulate insulin secretion together with hyperglycemia.

Proglucagon family: The components which are cleaved from proglucagon in the alpha cells of the pancreas and in the intestinal L-cells in the distal ileum and colon. Glicentin (21–89), glicentin-related pancreatic polypeptide (GRPP, 21–50), oxyntomodulin (OXY or OXM, 53–89), glucagon (53–81), glucagon-like peptide 1 (GLP-1, 92–128), and glucagon-like peptide 2 (GLP-2, 146–178).

Tropic hormone: A hormone that has a growth effect, hyperplasia or hypertrophy, on the tissue it is stimulating.

Insulinomimetics: Agents that have been shown to mimic the actions of insulin including promoting the entry of glucose into tissues, activating signal proteins, influencing the expression of genes and regulating metabolic processes.

Gliptins: A novel class of agents that improve beta cell health and suppress glucagon, resulting in improved postprandial and fasting hyperglycemia (Sitagliptin, vildagliptin, saxagliptin, linagliptin, alogliptin).

Neuropeptide: Small protein-like molecules produced and released by neurons through the regulated secretory route and acting on neural substrates.

Key facts of glucagon-like peptides

- GLPs are the eptide molecules secreted from the enteroendocrine L-cells of the intestines.
- GLP-1, indicated in the treatment of type 2 diabetes mellitus.
- GLP-2, indicated in the treatment of short bowel syndrome.
- GLP receptor, a seven-transmembrane G protein-coupled receptor.
- GLP-1/GLP-2 signaling pathways, antinociceptive effects in neuropathic pain, chronic pain, neuronal inflammation, oxidative stress, neurotoxicity, and memory impairment.

Summary points

- GLPs are the peptide molecules secreted from the enteroendocrine L-cells of the intestines after nutrient intake.
- GLP-1 is mainly used in the modulation of blood glucose levels, and is an important treatment for type 2 diabetes mellitus.
- GLP-2, a potent growth factor of the gut, is involved in enteral absorption and is frequently used in the treatment of short bowel syndrome.
- Treatments based on GLPs and analogs have demonstrated antiinflammatory and antinociceptive actions in the several systems.
- The action of GLPs in the nociceptive response requires importance for the development of new therapeutic agents in pain, as they can serve as prototypes of the design and synthesis for potential antinociceptive new compounds.

References

Austin, K., Tsang, D., Chalmers, J. A., Maalouf, M. F., & Brubaker, P. L. (2018). Insulin-like growth factor-binding protein-4 inhibits epithelial growth and proliferation in the rodent intestine. *American Journal of Physiology. Gastrointestinal and Liver Physiology*, *315*(2), G206–G219.

Aykan, D. A., Kesim, M., Ayan, B., & Kurt, A. (2019). Anti-inflammatory and antinociceptive activities of glucagon-like peptides: Evaluation of their actions on serotonergic, nitrergic, and opioidergic systems. *Psychopharmacology*, *236*(6), 1717–1728.

Balogh, M., Varga, B. K., Karádi, D.Á., Riba, P., Puskár, Z., Kozsurek, M., et al. (2019). Similarity and dissimilarity in antinociceptive effects of dipeptidyl-peptidase 4 inhibitors, diprotin a and vildagliptin in rat inflammatory pain models following spinal administration. *Brain Research Bulletin*, *147*, 78–85.

Bassil, F., Fernagut, P. O., Bezard, E., & Meissner, W. G. (2014). Insulin, IGF-1 and GLP-1 signaling in neurodegenerative disorders: Targets for disease modification? *Progress in Neurobiology*, *118*, 1–18.

Boushey, R. P., Yusta, B., & Drucker, D. J. (2001). Glucagon-like peptide (GLP)-2 reduces chemotherapy-associated mortality and enhances cell survival in cells expressing a transfected GLP-2 receptor. *Cancer Research*, *61*(2), 687–693.

Briyal, S., Shah, S., & Gulati, A. (2014). Neuroprotective and anti-apoptotic effects of liraglutide in the rat brain following focal cerebral ischemia. *Neuroscience*, *281*, 269–281.

Campbell, J. E., & Drucker, D. J. (2013). Pharmacology, physiology, and mechanisms of incretin hormone action. *Cell Metabolism*, *17*(6), 819–837.

Darsalia, V., Mansouri, S., Ortsäter, H., Olverling, A., Nozadze, N., Kappe, C., et al. (2012). Glucagon-like peptide-1 receptor activation reduces ischaemic brain damage following stroke in type 2 diabetic rats. *Clinical Science (London, England: 1979)*, *122*(10), 473–483.

Day, J. W., Li, P., Patterson, J. T., Chabenne, J., Chabenne, M. D., Gelfanov, V. M., et al. (2011). Charge inversion at position 68 of the glucagon and glucagon-like peptide-1 receptors supports selectivity in hormone action. *Journal of Peptide Science*, *17*(3), 218–225.

De Meester, I., Vanham, G., Kestens, L., Vanhoof, G., Bosmans, E., Gigase, P., et al. (1994). Binding of adenosine deaminase to the lymphocyte surface via CD26. *European Journal of Immunology*, *24*(3), 566–570.

Dong, C. X., Zhao, W., Solomon, C., Rowland, K. J., Ackerley, C., Robine, S., et al. (2014). The intestinal epithelial insulin-like growth factor-1 receptor links glucagon-like peptide-2 action to gut barrier function. *Endocrinology*, *155*(2), 370–379.

Fan, H., Gong, N., Li, T. F., Ma, A. N., Wu, X. Y., Wang, M. W., et al. (2015). The non-peptide GLP-1 receptor agonist WB4-24 blocks inflammatory nociception by stimulating β-endorphin release from spinal microglia. *British Journal of Pharmacology*, *172*(1), 64–79.

Gao, H., Zeng, Z., Zhang, H., Zhou, X., Guan, L., Deng, W., et al. (2015). The glucagon-like peptide-1 analogue liraglutide inhibits oxidative stress and inflammatory response in the liver of rats with diet-induced non-alcoholic fatty liver disease. *Biological & Pharmaceutical Bulletin*, *38*(5), 694–702.

Gaspari, T., Brdar, M., Lee, H. W., et al. (2016). Molecular and cellular mechanisms of glucagon-like peptide-1 receptor agonist-mediated attenuation of cardiac fibrosis. *Diabetes & Vascular Disease Research*, *13*(1), 56–68.

Gerald, C., Walker, M. W., Criscione, L., Gustafson, E. L., Batzl-Hartmann, C., Smith, K. E., et al. (1996). A receptor subtype involved in neuropeptide-Y-induced food intake. *Nature*, *382*(6587), 168–171.

Godoy-Matos, A. F. (2014). The role of glucagon on type 2 diabetes at a glance. *Diabetology & Metabolic Syndrome*, *6*(1), 91.

Gong, N., Xiao, Q., Zhu, B., Zhang, C. Y., Wang, Y. C., Fan, H., et al. (2014). Activation of spinal glucagon-like peptide-1 receptors specifically suppresses pain hypersensitivity. *The Journal of Neuroscience: The Official Journal of the Society for Neuroscience*, *34*(15), 5322–5334.

Guieu, R., Fenouillet, E., Devaux, C., Fajloun, Z., Carrega, L., Sabatier, J. M., et al. (2006). CD26 modulates nociception in mice via its dipeptidyl-peptidase IV activity. *Behavioural Brain Research*, *166*(2), 230–235.

Gulec Suyen, G., Isbil-Buyukcoskun, N., Cam, B., & Ozluk, K. (2015). Effects of centrally injected glucagon-like peptide-2 on gastric mucosal blood flow in rats: Possible mechanisms. *Peptides*, *64*, 62–66.

Gumuslu, E., Cine, N., Ertan Gökbayrak, M., Mutlu, O., Komsuoglu Celikyurt, I., & Ulak, G. (2016). Exenatide alters gene expression of neural cell adhesion molecule (NCAM), intercellular cell adhesion molecule (ICAM), and vascular cell adhesion molecule (VCAM) in the Hippocampus of type 2 diabetic model mice. *Medical Science Monitor*, *22*, 2664–2669.

Hamidi, G. A., Jafari-Sabet, M., Abed, A., Mesdaghinia, A., Mahlooji, M., & Banafshe, H. R. (2014). Gabapentin enhances anti-nociceptive effects of morphine on heat, cold, and mechanical hyperalgesia in a rat model of neuropathic pain. *Iranian Journal of Basic Medical Sciences*, *17*(10), 753–759.

Hirakawa, H., Zempo, H., Ogawa, M., Watanabe, R., Suzuki, J., Akazawa, H., et al. (2015). A DPP-4 inhibitor suppresses fibrosis and inflammation on experimental autoimmune myocarditis in mice. *PLoS One*, *10*(3), e0119360.

Hsieh, J., Trajcevski, K. E., Farr, S. L., Baker, C. L., Lake, E. J., Taher, J., et al. (2015). Glucagon-like peptide 2 (GLP-2) stimulates postprandial chylomicron production and Postabsorptive release of intestinal triglyceride storage pools via induction of nitric oxide signaling in male hamsters and mice. *Endocrinology*, *156*(10), 3538–3547.

Hui, H., Farilla, L., Merkel, P., & Perfetti, R. (2002). The short half-life of glucagon-like peptide-1 in plasma does not reflect its long-lasting beneficial effects. *European Journal of Endocrinology*, *146*(6), 863–869.

Insuela, D., & Carvalho, V. F. (2017). Glucagon and glucagon-like peptide-1 as novel anti-inflammatory and immunomodulatory compounds. *European Journal of Pharmacology*, *812*, 64–72.

Ivory, C. P., Wallace, L. E., McCafferty, D. M., & Sigalet, D. L. (2008). Interleukin-10-independent anti-inflammatory actions of glucagon-like peptide 2. *American Journal of Physiology. Gastrointestinal and Liver Physiology*, *295*(6), G1202–G1210.

Iwai, T., Jin, K., Ohnuki, T., Sasaki-Hamada, S., Nakamura, M., Saitoh, A., et al. (2015). Glucagon-like peptide-2-induced memory improvement and anxiolytic effects in mice. *Neuropeptides*, *49*, 7–14.

Iwaya, H., Fujii, N., Hagio, M., Hara, H., & Ishizuka, S. (2013). Contribution of dipeptidyl peptidase IV to the severity of dextran sulfate sodium-induced colitis in the early phase. *Bioscience, Biotechnology, and Biochemistry*, *77*(7), 1461–1466.

Jia, Y., Gong, N., Li, T. F., Zhu, B., & Wang, Y. X. (2015). Peptidic exenatide and herbal catalpol mediate neuroprotection via the hippocampal GLP-1 receptor/β-endorphin pathway. *Pharmacological Research*, *102*, 276–285.

Kagal, U. A., Angadi, N. B., & Matule, S. M. (2017). Effect of dipeptidyl peptidase 4 inhibitors on acute and subacute models of inflammation in male Wistar rats: An experimental study. *International Journal of Applied & Basic Medical Research*, *7*(1), 26–31.

Király, K., Kozsurek, M., Lukácsi, E., Barta, B., Alpár, A., Balázsa, T., et al. (2018). Glial cell type-specific changes in spinal dipeptidyl peptidase 4 expression and effects of its inhibitors in inflammatory and neuropatic pain. *Scientific Reports*, *8*, 3490.

Király, K., Szalay, B., Szalai, J., Barna, I., Gyires, K., Verbeken, M., et al. (2009). Intrathecally injected Ile-pro-Ile, an inhibitor of membrane ectoenzyme dipeptidyl peptidase IV, is antihyperalgesic in rats by switching the enzyme from hydrolase to synthase functional mode to generate endomorphin 2. *European Journal of Pharmacology*, *620*(1–3), 21–26.

Kissow, H. (2015). Glucagon-like peptides 1 and 2: Intestinal hormones implicated in the pathophysiology of mucositis. *Current Opinion in Supportive and Palliative Care*, *9*(2), 196–202.

Kissow, H., Viby, N. E., Hartmann, B., Holst, J. J., Timm, M., Thim, L., et al. (2012). Exogenous glucagon-like peptide-2 (GLP-2) prevents chemotherapy-induced mucositis in rat small intestine. *Cancer Chemotherapy and Pharmacology*, *70*(1), 39–48.

Klemann, C., Wagner, L., Stephan, M., & von Hörsten, S. (2016). Cut to the chase: A review of CD26/dipeptidyl peptidase-4's (DPP4) entanglement in the immune system. *Clinical and Experimental Immunology*, *185*(1), 1–21.

Komsuoglu Celikyurt, I., Mutlu, O., Ulak, G., Uyar, E., Bektaş, E., Yildiz Akar, F., et al. (2014). Exenatide treatment exerts anxiolytic- and antidepressant-like effects and reverses neuropathy in a mouse model of type-2 diabetes. *Medical Science Monitor Basic Research*, *20*, 112–117.

Lauritsen, K. B., Moody, A. J., Christensen, K. C., & Lindkaer Jensen, S. (1980). Gastric inhibitory polypeptide (GIP) and insulin release after small-bowel resection in man. *Scandinavian Journal of Gastroenterology*, *15*(7), 833–840.

Lee, C. H., Jeon, S. J., Cho, K. S., Moon, E., Sapkota, A., Jun, H. S., et al. (2018). Activation of glucagon-like peptide-1 receptor promotes neuroprotection in experimental autoimmune encephalomyelitis by reducing neuroinflammatory responses. *Molecular Neurobiology*, *55*(4), 3007–3020.

Lee, Y. S., & Jun, H. S. (2016). Anti-inflammatory effects of GLP-1-based therapies beyond glucose control. *Mediators of Inflammation*, *2016*, 3094642.

Lee, S. J., Lee, J., Li, K. K., Holland, D., Maughan, H., Guttman, D. S., et al. (2012). Disruption of the murine Glp2r impairs Paneth cell function and increases susceptibility to small bowel enteritis. *Endocrinology*, *153*(3), 1141–1151.

Li, L., El-Kholy, W., Rhodes, C. J., & Brubaker, P. L. (2005). Glucagon-like peptide-1 protects beta cells from cytokine-induced apoptosis and necrosis: Role of protein kinase B. *Diabetologia*, *48*(7), 1339–1349.

Li, N., Liu, B. W., Ren, W. Z., Liu, J. X., Li, S. N., Fu, S. P., et al. (2016). GLP-2 attenuates LPS-induced inflammation in BV-2 cells by inhibiting ERK1/2, JNK1/2 and NF-κB signaling pathways. *International Journal of Molecular Sciences*, *17*(2), 190.

Lin, C. H., & Lin, C. C. (2016). Sitagliptin attenuates inflammatory responses in lipopolysaccharide-stimulated cardiomyocytes via nuclear factor-κB pathway inhibition. *Experimental and Therapeutic Medicine*, *11*(6), 2609–2615.

Lu, G., Hu, Y., Wang, Q., Qi, J., Gao, F., Li, Y., et al. (2013). Molecular basis of binding between novel human coronavirus MERS-CoV and its receptor CD26. *Nature*, *500*(7461), 227–231.

Markovic, M. A., & Brubaker, P. L. (2019). The roles of glucagon-like peptide-2 and the intestinal epithelial insulin-like growth factor-1 receptor in regulating microvillus length. *Scientific Reports*, *9*(1), 13010.

Meier, J. J., Nauck, M. A., Pott, A., Heinze, K., Goetze, O., Bulut, K., et al. (2006). Glucagon-like peptide 2 stimulates glucagon secretion, enhances lipid absorption, and inhibits gastric acid secretion in humans. *Gastroenterology*, *130*(1), 44–54.

Mentlein, R. (1999). Dipeptidyl-peptidase IV (CD26)—Role in the inactivation of regulatory peptides. *Regulatory Peptides*, *85*(1), 9–24.

Nozu, T., Miyagishi, S., Kumei, S., Nozu, R., Takakusaki, K., & Okumura, T. (2018). Glucagon-like peptide-1 analog, liraglutide, improves visceral sensation and gut permeability in rats. *Journal of Gastroenterology and Hepatology*, *33*(1), 232–239.

Orskov, C., Holst, J. J., Poulsen, S. S., & Kirkegaard, P. (1987). Pancreatic and intestinal processing of proglucagon in man. *Diabetologia*, *30*(11), 874–881.

Palleria, C., Leo, A., Andreozzi, F., Citraro, R., Iannone, M., Spiga, R., et al. (2017). Liraglutide prevents cognitive decline in a rat model of streptozotocin-induced diabetes independently from its peripheral metabolic effects. *Behavioural Brain Research*, *321*, 157–169.

Parthsarathy, V., & Hölscher, C. (2013). The type 2 diabetes drug liraglutide reduces chronic inflammation induced by irradiation in the mouse brain. *European Journal of Pharmacology*, *700*(1–3), 42–50.

Seufert, J., & Gallwitz, B. (2014). The extra-pancreatic effects of GLP-1 receptor agonists: A focus on the cardiovascular, gastrointestinal and central nervous systems. *Diabetes, Obesity & Metabolism*, *16*(8), 673–688.

Sigalet, D. L., Wallace, L. E., Holst, J. J., Martin, G. R., Kaji, T., Tanaka, H., et al. (2007). Enteric neural pathways mediate the anti-inflammatory actions of glucagon-like peptide 2. *American Journal of Physiology. Gastrointestinal and Liver Physiology*, *293*(1), G211–G221.

Tate, M., Robinson, E., Green, B. D., McDermott, B. J., & Grieve, D. J. (2016). Exendin-4 attenuates adverse cardiac remodelling in streptozocin-induced diabetes via specific actions on infiltrating macrophages. *Basic Research in Cardiology*, *111*(1), 1.

Tekin, M., Topaloğlu, N., Küçük, A., Deniz, M., Yıldırım, Ş., & Erdem, H. (2015). Protective effect of glucagon-like peptide-2 in experimental corrosive esophagitis. *Diseases of the Esophagus: Official Journal of the International Society for Diseases of the Esophagus*, *28*(3), 258–261.

Thompson, A., & Kanamarlapudi, V. (2015). Agonist-induced internalisation of the glucagon-like peptide-1 receptor is mediated by the Gαq pathway. *Biochemical Pharmacology*, *93*(1), 72–84.

Tsuda, M. (2016). Microglia in the spinal cord and neuropathic pain. *Journal of Diabetes Investigation*, *7*(1), 17–26.

Varin, E. M., Wojtusciszyn, A., Broca, C., Muller, D., Ravier, M. A., Ceppo, F., et al. (2016). Inhibition of the MAP3 kinase Tpl2 protects rodent and human β-cells from apoptosis and dysfunction induced by cytokines and enhances anti-inflammatory actions of exendin-4. *Cell Death & Disease*, *7*(1), e2065.

Xie, S., Liu, B., Fu, S., Wang, W., Yin, Y., Li, N., et al. (2014). GLP-2 suppresses LPS-induced inflammation in macrophages by inhibiting ERK phosphorylation and NF-κB activation. *Cellular Physiology and Biochemistry: International Journal of Experimental Cellular Physiology, Biochemistry, and Pharmacology*, *34*(2), 590–602.

Yang, Y., Zhang, Y. G., Lin, G. A., Xie, H. Q., Pan, H. T., Huang, B. Q., et al. (2014). Spinal changes of a newly isolated neuropeptide endomorphin-2 concomitant with vincristine-induced allodynia. *PLoS One*, *9*(2), e89583.

Yu, D. M., Slaitini, L., Gysbers, V., Riekhoff, A. G., Kähne, T., Knott, H. M., et al. (2011). Soluble CD26/dipeptidyl peptidase IV enhances human lymphocyte proliferation in vitro independent of dipeptidyl peptidase enzyme activity and adenosine deaminase binding. *Scandinavian Journal of Immunology*, *73*(2), 102–111.

Yusta, B., Somwar, R., Wang, F., Munroe, D., Grinstein, S., Klip, A., et al. (1999). Identification of glucagon-like peptide-2 (GLP-2)-activated signaling pathways in baby hamster kidney fibroblasts expressing the rat GLP-2 receptor. *The Journal of Biological Chemistry*, *274*(43), 30459–30467.

Zhang, E., Xu, F., Liang, H., Yan, J., Xu, H., Li, Z., et al. (2015). GLP-1 receptor agonist exenatide attenuates the detrimental effects of obesity on inflammatory profile in testis and sperm quality in mice. *American Journal of Reproductive Immunology (New York, N.Y.: 1989)*, *74*(5), 457–466.

Chapter 21

Pain transmission and peripheral group III metabotropic glutamate receptors (mGluRs)

Eui Ho Park and Hee Chul Han
Department of Physiology, College of Medicine and Neuroscience Research Institute, Korea University, Seoul, South Korea

Abbreviations

AC	adenylyl cyclase
AMPA	α-amino-3-hydroxy-5-methyl-4-isoxazolepropionic acid
cAMP	cyclic adenosine monophosphate
Cav	(N-type) voltage-gated Ca^{2+} channels
CFA	complete Freund's adjuvant
CGRP	calcitonin gene-related peptide
CNS	central nervous system
DAG	diacylglycerol
DRG	dorsal root ganglia
EAAT	excitatory amino acid transporter
EC_{50}	half maximal effective concentration
eEPSCs	evoked excitatory postsynaptic currents
EPSPs	excitatory postsynaptic potentials
GDNF	glial-derived neurotrophic factor
GLS	glutaminase
IB4	isolectin B4
iGluRs	ionotropic glutamate receptors
IP3	1,4,5-triphosphate
mEPSC	miniature excitatory postsynaptic current
mGluRs	metabotropic glutamate receptors
mRNA	messenger ribonucleic acid
NAAG	*N*-acetyl-aspartyl-glutamate
NGF	nerve growth factor
NMDA	*N*-methyl-D-aspartate receptor
NTS	nucleus tractus solitarius
PAG	periaqueductal gray
PAM	positive allosteric modulator
PLC	phospholipase C
PNS	peripheral nerve system
RET	rearranged during transfection
SCs	Schwann cells
SGCs	satellite glial cells
SP	substance P
TG	trigeminal ganglia
TrkA	tropomyosin receptor kinase A
TRPV1	transient receptor potential cation channel subfamily V member 1
VGLUTs	vesicular glutamate transporters

The Neurobiology, Physiology, and Psychology of Pain. https://doi.org/10.1016/B978-0-12-820589-1.00021-X

Introduction

Pain is a critical symptom in clinical medicine. Nociceptive pain serves to warn living organism of protecting the body against an actual or potential tissue damage. Since pain is a perception mainly mediated by the nervous system, nociceptors present in terminals of peripheral nerve system (PNS) can transduce noxious stimuli to electrochemical information that is transmitted to central nervous system (CNS).

Small and medium diameter cell bodies of dorsal root ganglia (DRG) or the trigeminal ganglia (TG) neurons exhibit two general types of nociceptors (Basbaum, Bautista, Scherrer, & Julius, 2009). Unmyelinated C fibers nociceptor with slowly conducting velocity mediate a delayed and dull pain, and thinly myelinated Aδ fibers nociceptor with more rapidly conducting velocity mediate a rapid and sharp pain. DRG (for body) and TG (for face) have pseudo-unipolar morphology. Peripheral terminals of axon innervating peripheral tissues are responsible for transducing noxious stimuli and initiating action potential, and central terminals of axon extend into the superficial dorsal horn in the spinal cord and form the first presynaptic circuit. The first relay neurons project to periaqueductal gray (PAG), reticular formation of brain stem and thalamus, finally transferring the nociceptive information to cerebral cortex including the somatosensory cortex, insular cortex, and cingulate cortex. Nociceptive information from the face is relayed into thalamus and somatosensory cortex by the trigeminal system. Most of all, this ascending information of pain delivers into neurons of the rostral ventral medulla and PAG to engage descending pathway to regulate the spinal cord output.

Glutamatergic transmission or signaling play a crucial role in initiating and maintaining pain. Nociceptive Aδ and C fibers are glutamatergic, and several types of glutamate receptors capable of binding with glutamate are abundantly distributed in the pain pathway of nervous system. The hyperexcitability of the glutamatergic system induces the augmented glutamate release and upregulation (or phosphorylation) of glutamate receptor in axon terminal and cell bodies of nervous system. For this reason, the glutamatergic action can be highly associated with the sensitized response of nociceptor in animal and human with acute or chronic pain (Bleakman, Alt, & Nisenbaum, 2006). Multiple studies have identified that the activation of ionotropic or metabotropic glutamate receptors (iGluRs or mGluRs) localized in nervous system can regulate nociceptive processing. Importantly, the role of glutamatergic signaling in PNS is essential in the first phase of nociceptive processing (Carlton, 2001). Accumulated evidences are emerging that NMDA receptor and group I mGluRs in cell bodies and peripheral afferent fibers of PNS can potentiate the nociceptive processing (Bhave, Karim, Carlton, & Gereau, 2001; Cairns et al., 2003; Zhang et al., 2009); however, much attention has been less drawn into the function of peripheral group III mGluRs. Therefore, the present review will target peripheral group III mGluRs of PNS which can provide a better understanding of modulatory role and therapeutic potential in pain processing.

Glutamate metabolism in peripheral nervous system

Plasma glutamate concentration maintain in a low-level range of 20–50 μM. Systemic absorption of glutamate from colonic lumen to portal circulation is poor owing to the first pass metabolism in the splanchnic bed and extensive energy production for lumen function (Julio-Pieper, Flor, Dinan, & Cryan, 2011). Since blood barrier restrict the entry of glutamate and glutamine into PNS and CNS (Hawkins, 2009; Reinhold & Rittner, 2020), glutamate as the main neurotransmitter can be produced (or reuptaked) in primary neuron, transported, and released from peripheral terminals.

The production of endogenous glutamate in PNS is implicated in mitochondrial glutaminase (GLS; the conversion of glutamine into glutamate), excitatory amino acid transporter (EAAT; glutamate reuptake), and the degradation of *N*-acetyl-aspartyl-glutamate (NAAG; *N*-acetyl-aspartyl-glutamate into glutamate and *N*-acetyl-aspartate) in DRG neurons and glia (Miller, Hoffman, Sutharshan, & Schechter, 2011). The synthesized glutamate in DRG cells is packed in vesicular glutamate transporters (VGLUTs) and is transported to peripheral terminals. Intense noxious stimuli, tissue damage, and inflammation causes augmented glutamate to be released from peripheral afferent nerve terminals. Extensive analysis in mice has revealed that mRNA and protein of VGLUT1 are mostly shown in medium-to-large diameter DRG neurons, whereas those of VGLUT2 and VGLUT3 are present in small-to-medium diameter DRG neurons (Malet & Brumovsky, 2015). Especially, VGLUT2 DRG neurons are colocalized with peptidergic calcitonin gene-related peptide (CGRP) and nonpeptidergic isolectin B4 (IB4) (Rogoz, Lagerstrom, Dufour, & Kullander, 2012; Scherrer et al., 2010) and VGLUT2-expressed nerve terminals are abundantly expressed in hindlimb skin and viscera (Brumovsky et al., 2013).

Glutamate release and pain

Glutamatergic transmission is required for both physiological and pathological condition in nervous system. Under a sensitization in the pain pathways, the glutamatergic system is strongly hyperexcitable in response to noxious or innocuous stimuli. Hyperalgesia (increased pain by a stimulus that usually provokes pain) and allodynia (pain by a stimulus that does not usually provoke pain) can be accompanied in animals and patients with inflammatory or chronic pain (Basbaum et al., 2009; Woolf & Ma, 2007). Increased glutamate release in axon terminals of PNS leads to the development of these evoked pain response (Alfredson & Lorentzon, 2002; Schizas et al., 2010).

Besides endogenous glutamate release in pain processing, the exogenous application of glutamate can induce pain response. In a glabrous hindpaw skin of rat, intraplantar application of L-glutamate causes the excitation response of Aδ and C fibers and lowers a temporal threshold of these fibers to heat stimulation (Du, Koltzenburg, & Carlton, 2001). Intraplantar injection of L-glutamate causes both mechanical allodynia and mechanical hyperalgesia (Carlton, Hargett, & Coggeshall, 1995). In a human study, subcutaneous injection of glutamate to forehead reduces pressure pain threshold, causes secondary hyperalgesia, elevates skin temperature by about 1°C (Gazerani, Wang, Cairns, Svensson, & Arendt-Nielsen, 2006). Similar to skin, there are sensitized effects of glutamate on muscle and joint of humans (Castrillon et al., 2008; Svensson et al., 2003).

Glutamate can also play an action into substance P (SP), CGRP, adenosine, and TRPV1 to initiate pain processing. The release of CGRP in bovine dental pulp is found after the treatment of L-glutamate (Jackson & Hargreaves, 1999). Subcutaneous injection of L-glutamate into rat hindlimb dose-dependently evokes adenosine release; however, systemic preadministration of capsaicin to desensitize primary afferents fibers does not induce adenosine release of L-glutamate (Liu, White, & Sawynok, 2002). Subcutaneous application of glutamate with SP into hairy skin of rats increases mean spikes firing of Aδ and C fibers compared to either glutamate or SP alone (Carlton, Zhou, & Coggeshall, 1998).

Glutamate receptor signaling: iGluR and mGluR

Augmented glutamate action is mediated through two different types of receptors: ionotropic glutamate receptors (iGluRs) and metabotropic glutamate receptors (mGluRs). It has been known that iGluRs (NMDA, AMPA, and kainate receptors) play roles to mediate fast excitatory transmission, whereas mGluRs modulate neuronal excitability via G-protein-dependent or G-protein-independent signaling pathways (Neugebauer, 2002; Niswender & Conn, 2010).

The topology of mGluRs comprises seven hydrophobic transmembrane domains with extracellular loops and intracellular loops, a large extracellular N-terminal domain and a cytoplasmic C-terminal domain. Subtypes of mGluRs can be divided into three different groups according to their sequence similarity, pharmacology, and intracellular signal transduction pathways. Group I mGluRs—mGluR1 and 5—are coupled to $G_{aq/11}$proteins and activate phospholipase C (PLC). A signal transduction of PLC leads to activate protein kinase C and mobilize Ca^{2+} from intracellular storage. On the other hand, group II (2 and 3) and group III (4, 6, 7, and 8) mGluRs are negatively coupled to adenylate cyclase and decrease cAMP levels, regulating K^+ and Ca^{2+} channels. This different signal pathways of mGluRs can explain different roles of mGluRs (Fig. 1).

Localization of peripheral group III mGluR

Except for mGluR6 restrictedly localized within the retina, the presence of mGluR4, mGluR7, and mGluR8 is widely identified in PNS associated with nociception (Neugebauer, 2002) (Fig. 2).

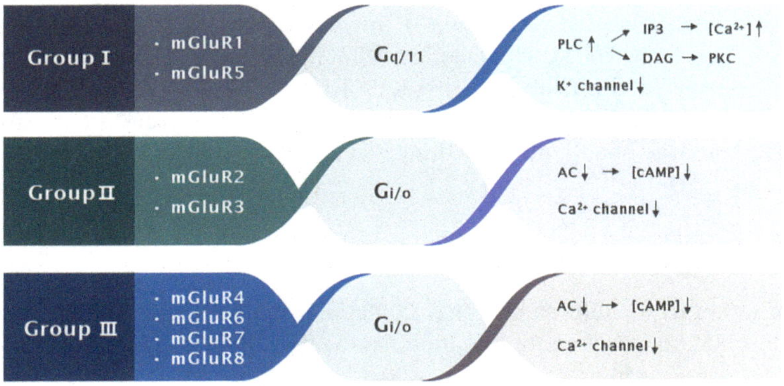

FIG. 1 Classification and signal transduction of mGluRs. mGluRs are classified into three family. Group I induces PLC hydrolysis with formation of IP3 and DAG, whereas groups II and III induce a decrease of intercellular levels of cAMP through negatively coupling to AC.

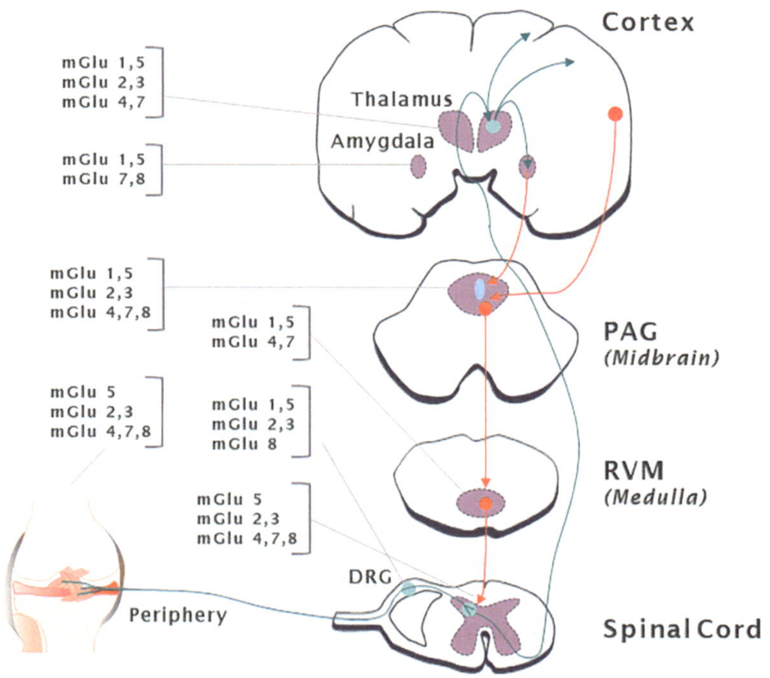

FIG. 2 Localization of mGluRs in pain pathways. Subtypes of mGluRs are widely expressed in pain pathways of nervous system. *Green line* symbolizes ascending pathway and *red line* symbolizes descending pathway.

The immunoreactivity of mGluR4 is found particularly in small-to-medium diameter DRG neurons, and in the axon terminal of laminar II in spinal dorsal horn of rats. In detail, a few studies reveal that mGlu4 are expressed in both the inner lamina II of mice spinal dorsal horn and unmyelinated C afferent fibers expressing VGLUT3 (Azkue et al., 2001; Vilar et al., 2013). A small proportion of mGlu4 receptor staining (<10%) are colocalized with both peptidergic SP/CGRP and nonpeptidergic IB4 (Vilar et al., 2013). It is also reported that mGluR4 mRNA in DRG is downregulated after sciatic nerve ligation of rats (Xiao et al., 2002).

The immunoreactivity of mGluR7 is shown in terminals or cell bodies of DRG and TG neurons of rats (Li et al., 1996, 1997). It is found that mGluR7 is highly immunolabeled in DRG (85%) of rats where the double-labeled proportion with SP, CGRP, and TRPV1 are expressed in 86%, 92%, and 71%, respectively. However, there is a relatively low double-labeling (22%) of mGluR7 with nonpeptidergic IB4 (Li et al., 2012). Interestingly, mGluR7 is accumulated at the proximal site to DRG neurons after sciatic nerve ligation, whereas the expression level of mGluR7 mRNA and protein in DRG neurons is downregulated after this ligation (Li et al., 2012), thus implicating the transportation of this receptor into peripheral and central terminal.

Immunoreactive and stereological study shows that mGluR8 is highly expressed (75%) in small, medium, and large diameter DRG cells, and the double-labeled proportion of mGluR8 with mGluR2/3 have a considerable population (21%) (Carlton, Hargett, & Coggeshall, 2001). The labeling of mGluR8 within digital nerves is shown in unmyelinated (22%) and myelinated (19%) axons, and mGluR8-labeled L4 DRG neurons display double-labeling (25%) for TRPV1 (Govea, Zhou, & Carlton, 2012). In the TG, a high proportion (56%) of the neurons are positive for mGluR8 of which the immunolabeling is distributed as small (6%), medium (41%), and large (53%) diameter neurons (Boye Larsen et al., 2014).

Negative modulation of pain transmission

As with the contribution of group III mGluRs to neuronal excitability in presynaptic location of CNS, these receptors on primary afferent terminal of PNS are also thought to inhibit neuronal excitability by negatively modulating the glutamatergic transmission.

Some behavioral and electrophysiological studies present evidences to identify that group III mGluR located in peripheral terminal of PNS can reduce nociceptive inputs before transmission to the spinal cord (Table 1). Intraplantar

TABLE 1 Summary of the function of peripheral group III mGluRs on pain processing.

Location	Drugs (route)	Receptors	Experimental model	Effects	References
Peripheral terminal of primary DRG neurons	UBP1112 (intraplantar)	Group III antagonist	Capsaicin-injected hindpaw (extracellular recording)	The increased firings of primary afferent C fiber	Carlton, Zhou, Govea, and Du (2011)
	LY341495 (intraplantar)	Group II/III antagonist	Capsaicin- or excess glutamate-injected hindpaw (extracellular recording)	The increased firings of primary afferent C fiber	
				The increased pain behavior (flinching, lifting, licking, and paw withdrawal)	
			Capsaicin-injected primary DRG neurons (calcium imaging)	The increased calcium influx	
	L-AP4 (intraarticular)	Group III agonist	Carrageenan-injected knee joint	The decreased pain behavior (dynamic weight load and paw withdrawal)	Lee, Park, Cho, Kim, and Han (2013)
	L-AP4 (intraarticular)	Group III agonist	CFA-injected knee joint (extracellular recording)	The decreased firings of mechanosensitive afferent fiber	Park et al. (2019)
			CFA-injected knee joint	The decreased pain behavior (dynamic weight load)	
	MSOP (intraplantar)	Group III mGluR antagonist	Capsaicin-injected hindpaw	No effect (paw withdrawal)	Jin et al. (2009)
Central terminal of primary DRG neurons	L-AP4 (intrathecal)	Group III agonist	Spinal nerve-ligated injury	The decreased pain behavior (paw withdrawal and cold water)	Fisher, Lefebvre, and Coderre (2002)
	L-AP4 (intrathecal)	Group III agonist	Capsaicin-injected hindpaw	The decreased pain behavior (paw withdrawal and hot water)	Soliman, Yu, and Coderre (2005)
	L-AP4 (intrathecal)	Group III agonist	Spinal nerve-ligated injury	The decreased pain behavior (paw withdrawal and radiant heat)	Chen and Pan (2005)
			Single unit in dorsal spinal cord (extracellular recording)	The decreased firings of mechanosensitive afferent fiber	
	ACPT-I (intrathecal)	Group III mGluR agonist	Capsaicin-injected hindpaw or CFA-induced ankle	The decreased pain behavior (paw pressure)	Goudet et al. (2008)
			Spinal nerve-ligated injury or Vincristine-injected neuropathy	The decreased pain behavior (paw pressure)	
	PHCCC (intrathecal)	mGlu4 positive allosteric modulator	Capsaicin-injected hindpaw or Spinal nerve-ligated injury	The decreased pain behavior (paw pressure)	

Continued

TABLE 1 Summary of the function of peripheral group III mGluRs on pain processing—cont'd

Location	Drugs (route)	Receptors	Experimental model	Effects	References
	AMN082 (intrathecal)	mGluR7-positive allosteric modulator	Capsaicin-injected or incision hindpaw	The decreased pain behavior (paw withdrawal and radiant heat)	Dolan, Gunn, Biddlestone, and Nolan (2009)
	L-AP4 (intrathecal)	Group III agonist	Spinal nerve-ligated injury	The decreased pain behavior (paw withdrawal)	Wang, Jiang, Yang, and Li (2011)
	VU0155041 (intrathecal)	mGluR4-positive allosteric modulator	Spinal nerve-ligated injury	The decreased pain behavior (paw withdrawal)	
	LSP4-2022 (intrathecal)	mGlu4 agonist	Whole neuron in sliced dorsal spinal cord (intracellular recording)	The inhibition of evoked excitatory postsynaptic current	Vilar et al. (2013)
			Carrageenan-injected hindpaw or Spinal nerve-ligated injury	The decreased pain behavior (paw withdrawal and lifting)	
			Knowdown-mGlu4	The decreased pain behavior (paw lifting and licking, tail withdrawal)	
	AMN082 (intrathecal)	mGluR7-positive allosteric modulator	Spinal nerve-ligated injury	No effect (paw withdrawal)	Wang et al. (2011)

application of UBP1112, selective group III mGluRs antagonist, shows an increased neuronal firings (250%) of primary afferent C fiber in capsaicin-injected hindpaw of rats, and the application of group II and III mGluRs antagonist LY341495 significantly enhance capsaicin-induced calcium influx in primary DRG culture (Carlton et al., 2011). Intraarticular applications of group III mGluR agonist LAP4 dose-dependently inhibit pain behavior (reduced dynamic weight load) and decrease mechanosensitive firings (40%–50% from baseline) of primary afferent fibers in inflamed knee joint of rat (Lee et al., 2013; Park et al., 2019).

In central terminal of PNS, intrathecal pretreatment with L-AP4 increases a mechanical or thermal threshold of capsaicin-injected hindpaw in rats (Soliman et al., 2005). It is also revealed that intrathecal treatments with L-AP4 or mGluR4-positive allosteric modulator VU0155041 reduce mechanical thresholds in rats with neuropathic pain (Fisher et al., 2002; Wang et al., 2011). Intrathecal application with mGluR7 allosteric agonist AMN082 inhibits mechanical and thermal hyperalgesia in rats with inflammatory or incision pain (Dolan et al., 2009). Intrathecal application of LAP4 dose-dependently decreases neuronal firings of dorsal horn neuron evoked by mechanical stimulation (tough, pressure, and pinch) in rats having neuropathic pain, and intrathecal injection of group III mGluR antagonist MAP4 blocks this inhibitory effect of LAP4 (Chen & Pan, 2005). Administration of mGluR4 agonist LSP4-2022 to lumbar spinal slices of mice decreases the amplitude of evoked excitatory postsynaptic current (eEPSC) to an electrical stimulation on the dorsal root, possibly through coupling to Cav2.2. channel (Vilar et al., 2013).

The key comment is that peripheral group III mGluRs can play a modulatory role on glutamatergic neurotransmission through activity-dependent inhibition, rather than simply tonic inhibition. That is to say, the reaction of group III mGluRs is silent to a noxious stimulation in naïve or healthy environment; however, their ability to inhibit excitatory transmission is reinforced when inflammation or tissue damage occur. For instance, intrathecal applications of positive allosteric mGluR4 modulator, PHCCC, dose-dependently increase mechanical thresholds in rats with inflammatory and nerve-injured pain; however, there is no effect of PHCCC on a mechanical threshold of healthy rat (Goudet et al., 2008) (Fig. 3). In addition, depressant effects of LSP4-2022 on eEPSC amplitude of mice lumbar spinal slices remarkably enhance by 68% under an

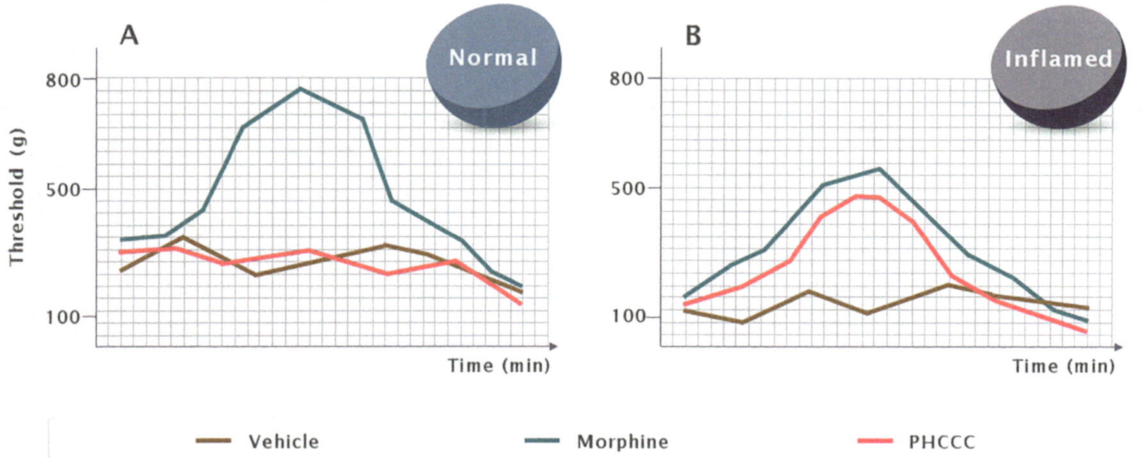

FIG. 3 Negative modulation of peripheral group III mGluRs on pain behavior. (A) Intrathecal application of mGlu4 positive allosteric modulator PHCCC has no effect on mechanical threshold of naïve animals. (B) Intrathecal application of PHCCC leads to an increased mechanical threshold in animals with inflammation or nerve injury. However, intrathecal application of morphine tonically has increased effects on mechanical threshold. *(From Goudet, C., Chapuy, E., Alloui, A., Acher, F., Pin, J. P., Eschalier, A. (2008). Group III metabotropic glutamate receptors inhibit hyperalgesia in animal models of inflammation and neuropathic pain.* Pain, 137(1), *112–124, https://doi.org/10.1016/j.pain.2007.08.020.)*

inflammatory condition, compared to by 53% under a normal condition. In parallel, mGlu4-knockout mice show higher mechanical thresholds and licking/biting behavior in response to noxious mechanical stimulation to tail and formalin injection to hindpaw (Vilar et al., 2013). Intraplantar applications of LY341495, group II/III mGluRs antagonist, produce both enhanced firing rates (400%) and lowered thermal threshold (about 1°C) to heat stimulation of primary afferent C fiber under the presence of excess exogenous glutamate (1 mM) into hindpaw preparation of rat, compared to under the condition of vehicle-control (Carlton et al., 2011). The activated mGluRs via LAP4 have no influence on mechanosensitive responses of primary afferent fibers under saline-injected knee joint (Park et al., 2019) (Fig. 4).

Although both group II and III mGluRs in PNS are known to be responsible for pain modulation, there are actually inconsistent effects of these receptors. Intraplantar applications of MCCG or MSOP, group II or group III mGluR antagonist, in rat have no effects on capsaicin-induced thermal hypersensitivity (Jin et al., 2009). Intrathecal treatment with mGluR7-positive allosteric modulator AMN082 does not reduce a mechanical threshold in rat model of neuropathic pain (Wang et al., 2011). In addition, the activation of group II mGluR via intrathecal treatment with DCG-IV has a pronociceptive effect on the sham control rats, implicating the involvement of NMDARs in spinal cord (Zhou, Chen, Chen, & Pan, 2011).

Group III mGluRs in non-neuronal glial cells

The cell bodies (DRG and TG) and axons fibers of PNS are in close contact with two types of glial cells such as satellite glial cells (SGCs) and myelinating/nonmyelinating Schwann cells (SCc). The morphology of satellite glial cells (SGCs) tightly envelops cell bodies of primary afferent neurons and creates an extracellular space of 15–20 nm. Myelinating Schwann cells wrap around axons of neurons to generate the myelin sheath and shape the axonal activity in peripheral nerves. It can be possible to extensively communicate with neurons and glial cells (Ji, Chamessian, & Zhang, 2016). In fact, glial cells can maintain neuronal homeostasis and glutamate recycling, thus regulating glutamate neurotransmission. However, after nerve or tissue injury SGCs can proliferate and initiate aberrant excitability via the coupling among SGC to SGC or neuron to SGC, therefore, contributing to developing chronic pain (Kim et al., 2016; Takeda, Takahashi, & Matsumoto, 2009).

Glutamate receptors including iGluRs and group I mGluRs in SGCs seem to contribute to glutamatergic signaling (Ferrari et al., 2014; Kung et al., 2013). For example, the application of glutamate into DRG cultures induces calcium influx only in satellite cells, which are suppressed by preadministration of NMDAR antagonist D-AP5 (Ferrari et al., 2014). Most of all, a few evidences implicate possible roles of group III mGluRs of SGCs on pain although there is no precise mechanism of these receptors toward SGCs (Fig. 5). A study reveals that group III mGluR8 is very highly expressed in SGCs of rats, with only minimal expression of group I (mGluR1) and group II (mGluR2/3) (Carlton & Hargett, 2007; Govea et al., 2012). In parallel, mGluR8 and mGluR1α, but not mGluR2/3, are found on SGCs in TG of rats (Boye Larsen et al., 2014).

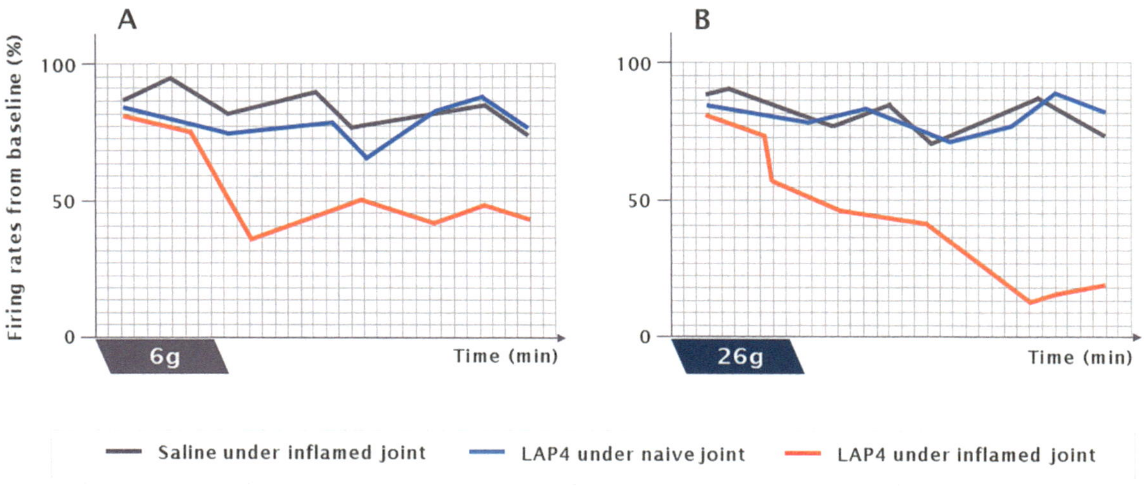

FIG. 4 Negative modulation of peripheral group III mGluRs on excitability of primary afferent fibers. The activation of peripheral group III mGluRs through L-AP4 has inhibitory effect on neuronal excitation evoked by 6 g (A) and 26 g (B) mechanical stimulations under inflamed knee joint. However, there is no effect of these receptors on mechanosensitive neuronal excitation under saline-injected knee joint. *(From Park, E. H., Lee, S. W., Moon, S. W., Suh, H. R., Kim, Y. I., Han, H. C. (2019). Activation of peripheral group III metabotropic glutamate receptors inhibits pain transmission by decreasing neuronal excitability in the CFA-inflamed knee joint.* Neuroscience Letters, 694*, 111–115. https://doi.org/10.1016/j.neulet.2018.11.033.)*

FIG. 5 Peripheral group III mGluRs in satellite glial cells. Non-neuronal glial cells are essential for maintaining neuronal homeostasis and glutamate recycling in peripheral nervous system. Especially, satellite glial cells (SGCs) can contribute to developing chronic pain. Group III mGluR8, as well as NMDA receptor and group I mGluRs, has a possible action for glutamatergic signaling in pain processing.

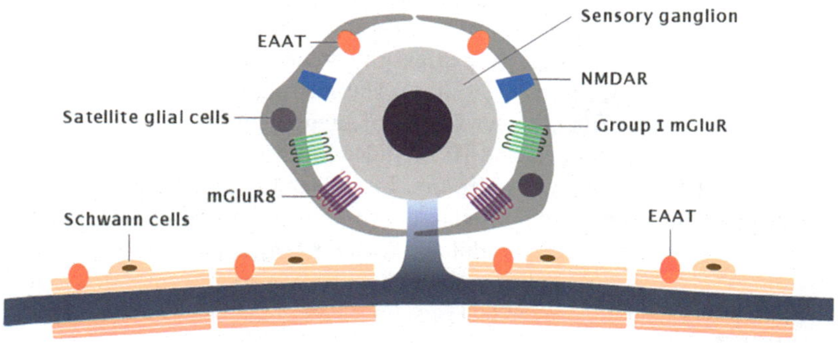

Applications to other areas

Group III mGluR4 in taste sensation. Glutamate deriving from dietary protein and food additive elicits umami taste. An umami-related sensory inputs are delivered to the rostral division of nucleus tractus solitarius (NTS) through facial, glossopharyngeal, and vagus nerves, which are transmitted to thalamus and gustatory cortex. The blockade of mGlu4 with a pharmacological antagonist or mGluR4 knock-out mice reduces the response of the umami taste stimulation (Pal Choudhuri, Delay, & Delay, 2016; Yasumatsu et al., 2015).

Group III mGluR7 and mGluR8 in colonic motor or secretory response. Colonic mGluRs can regulate intestinal mucosal function by acting on enteric neurons or epithelial cells. The activation of mGluR8 using (RS)PPG in the guinea pig colon dose-dependently increases the rate of propulsion (Tong & Kirchgessner, 2003). In addition, selective mGlu7 receptor agonist AMN082 increases water content and remarkably amplifies the bethanechol (acetylcholine analog)-induced electrolyte transport (Julio-Pieper, Hyland, Bravo, Dinan, & Cryan, 2010).

Group III mGluR4 in reward. These glutamatergic inputs of nucleus accumbens originated from the medial prefrontal cortex, hippocampus and amygdala contribute to reward-related perception and memory. The electrophysiological study indicates that group III receptor agonists such as L-SOP and LAP4 decrease the amplitude of cortically evoked

EPSPs in a dose-dependent manner (Pisani, Calabresi, Centonze, & Bernardi, 1997). In addition, a selective mGluR4 agonist LSP1-3081 not only inhibits cortically evoked EPSPs in a dose-dependent manner, but also reduces the frequency of miniature excitatory postsynaptic current (mEPSC) (Cuomo et al., 2009).

Other agents of interest

N-Phenyl-7-(hydroxyamino)cyclopropa[b]chromen-1a-carboxamide (PHCCC). PHCCC is the first selective positive allosteric modulator of mGluR4. Allosteric modulator increases or decreases the affinity of agonists through the binding to allosteric site of target receptor. PHCCC displays a micromolar potency for mGluR4 (EC_{50}, 4.1 μM) (Niswender et al., 2008), and has inhibitory effects on animals with inflammatory and neuropathic pain. However, PHCCC is also known as a partial antagonist (30%) of mGluR1.

Mini-dictionary of terms

Peripheral sensitization. Hyperexcitability or reduced threshold of the nociceptors in peripheral afferent neuron terminals and cell bodies after the occurrence of injury or inflammation.

Peptidergic nociceptor. Peptidergic nociceptor releases the neuropeptide such as substance P and calcitonin-gene-related peptide (CGRP) and also expresses a large proportion of TrkA receptor to bind with nerve growth factor.

Nonpeptidergic nociceptor. Nonpeptidergic nociceptor expresses the c-Ret neurotrophin receptor to respond to glial-derived neurotrophic factor (GDNF), and a large percentage of the c-Ret-positive population also binds with the IB4 isolectin (A lectin from the plant *Griffonia simplicifolia*).

Allosteric modulator. Allosteric modulators positively or negatively regulate receptor activity by binding at a site distinct from the orthosteric pocket of endogenous natural ligands.

Excitatory postsynaptic current. Excitatory neurotransmitters induce the opening of cationic channels in postsynaptic neurons, leading to the depolarizing phase capable of generating an action potential. These electrical events are called as excitatory postsynaptic currents (EPSCs) and excitatory postsynaptic potentials (EPSPs).

Key facts of *activity-dependent inhibition*

- Activity-dependent inhibition mechanism of peripheral group III mGluRs plays a role in negatively regulating glutamatergic neurotransmission.
- It can be recruited in peripheral or central terminals of PNS after the nociceptor is activated through excess glutamate release or inflammation/injury.
- The activation of peripheral group III mGluRs generates both decreased firing rates and increased threshold in primary afferent C fibers of hindpaw to heat stimulation under excess glutamate condition. However, these receptors do not inhibit neuronal excitability evoked by thermal stimulation under vehicle-control condition.
- The activation of peripheral group III mGluRs decreases mechanosensitive neuronal firings of primary afferent fibers innervating inflamed knee joint. However, there are no inhibitory effects of these receptors on mechanosensitivities under saline-injected knee joint.
- In the central terminal of primary afferent neuron, the activation of peripheral mGluR4 produces stronger effects on the reduction of eEPSC amplitude under CFA-induced inflammatory condition, compared to under normal condition.

Summary points

- This review focuses on the localization and function of the peripheral group III mGluRs with regard to glutamatergic transmission.
- After glutamate is produced (or reuptaked) in primary neurons and glia, it is transported through VGLUTs and released from peripheral terminals.
- The increased glutamate release from terminals of PNS involves the sensitization of nociceptor after tissue injury and inflammation occur.
- Glutamatergic transmission can be positively or negatively modulated by peripheral mGluRs.
- The expression and function of mGluR4, mGluR7, and mGluR8 among group III mGluRs are widely identified in PNS associated with nociception.

- Peripheral group III mGluRs of PNS have the ability to play an autoinhibitory role on excitatory neurotransmission under inflammation or tissue injury or excess glutamate.
- Peripheral group III mGluRs in SGCs have a feasible role to communicate with primary neurons for regulating pain.

References

Alfredson, H., & Lorentzon, R. (2002, Feb). Chronic tendon pain: No signs of chemical inflammation but high concentrations of the neurotransmitter glutamate. Implications for treatment? *Current Drug Targets*, *3*(1), 43–54. https://doi.org/10.2174/1389450023348028.

Azkue, J. J., Murga, M., Fernandez-Capetillo, O., Mateos, J. M., Elezgarai, I., Benitez, R., et al. (2001, Feb 19). Immunoreactivity for the group III metabotropic glutamate receptor subtype mGluR4a in the superficial laminae of the rat spinal dorsal horn. *The Journal of Comparative Neurology*, *430*(4), 448–457. https://doi.org/10.1002/1096-9861(20010219)430:4<448::aid-cne1042>3.0.co;2-o.

Basbaum, A. I., Bautista, D. M., Scherrer, G., & Julius, D. (2009). Cellular and molecular mechanisms of pain. *Cell*, *139*(2), 267–284. Oct 16 https://doi.org/10.1016/j.cell.2009.09.028.

Bhave, G., Karim, F., Carlton, S. M., & Gereau, R. W. T. (2001). Peripheral group I metabotropic glutamate receptors modulate nociception in mice. *Nature Neuroscience*, *4*(4), 417–423. Apr https://doi.org/10.1038/86075.

Bleakman, D., Alt, A., & Nisenbaum, E. S. (2006). Glutamate receptors and pain. *Seminars in Cell & Developmental Biology*, *17*(5), 592–604. Oct https://doi.org/10.1016/j.semcdb.2006.10.008.

Boye Larsen, D., Ingemann Kristensen, G., Panchalingam, V., Laursen, J. C., Norgaard Poulsen, J., Skallerup Andersen, M., et al. (2014). Investigating the expression of metabotropic glutamate receptors in trigeminal ganglion neurons and satellite glial cells: Implications for craniofacial pain. *Journal of Receptor and Signal Transduction Research*, *34*(4), 261–269. Aug https://doi.org/10.3109/10799893.2014.885049.

Brumovsky, P. R., Seal, R. P., Lundgren, K. H., Seroogy, K. B., Watanabe, M., & Gebhart, G. F. (2013). Expression of vesicular glutamate transporters in sensory and autonomic neurons innervating the mouse bladder. *The Journal of Urology*, *189*(6), 2342–2349. Jun https://doi.org/10.1016/j.juro.2012.11.046.

Cairns, B. E., Svensson, P., Wang, K., Hupfeld, S., Graven-Nielsen, T., Sessle, B. J., et al. (2003). Activation of peripheral NMDA receptors contributes to human pain and rat afferent discharges evoked by injection of glutamate into the masseter muscle. *Journal of Neurophysiology*, *90*(4), 2098–2105. Oct https://doi.org/10.1152/jn.00353.2003.

Carlton, S. M. (2001). Peripheral excitatory amino acids. *Current Opinion in Pharmacology*, *1*(1), 52–56. Feb https://doi.org/10.1016/s1471-4892(01)00002-9.

Carlton, S. M., & Hargett, G. L. (2007). Colocalization of metabotropic glutamate receptors in rat dorsal root ganglion cells. *The Journal of Comparative Neurology*, *501*(5), 780–789. Apr 10 https://doi.org/10.1002/cne.21285.

Carlton, S. M., Hargett, G. L., & Coggeshall, R. E. (1995). Localization and activation of glutamate receptors in unmyelinated axons of rat glabrous skin. *Neuroscience Letters*, *197*(1), 25–28. Sep 1 https://doi.org/10.1016/0304-3940(95)11889-5.

Carlton, S. M., Hargett, G. L., & Coggeshall, R. E. (2001). Localization of metabotropic glutamate receptors 2/3 on primary afferent axons in the rat. *Neuroscience*, *105*(4), 957–969. https://doi.org/10.1016/s0306-4522(01)00238-x.

Carlton, S. M., Zhou, S., & Coggeshall, R. E. (1998). Evidence for the interaction of glutamate and NK1 receptors in the periphery. *Brain Research*, *790*(1–2), 160–169. Apr 20 https://doi.org/10.1016/s0006-8993(97)01471-6.

Carlton, S. M., Zhou, S., Govea, R., & Du, J. (2011). Group II/III metabotropic glutamate receptors exert endogenous activity-dependent modulation of TRPV1 receptors on peripheral nociceptors. *The Journal of Neuroscience*, *31*(36), 12727–12737. Sep 7 https://doi.org/10.1523/JNEUROSCI.6558-10.2011.

Castrillon, E. E., Cairns, B. E., Ernberg, M., Wang, K., Sessle, B., Arendt-Nielsen, L., et al. (2008). Glutamate-evoked jaw muscle pain as a model of persistent myofascial TMD pain? *Archives of Oral Biology*, *53*(7), 666–676. Jul https://doi.org/10.1016/j.archoralbio.2008.01.008.

Chen, S. R., & Pan, H. L. (2005). Distinct roles of group III metabotropic glutamate receptors in control of nociception and dorsal horn neurons in normal and nerve-injured rats. *The Journal of Pharmacology and Experimental Therapeutics*, *312*(1), 120–126. Jan https://doi.org/10.1124/jpet.104.073817.

Cuomo, D., Martella, G., Barabino, E., Platania, P., Vita, D., Madeo, G., et al. (2009). Metabotropic glutamate receptor subtype 4 selectively modulates both glutamate and GABA transmission in the striatum: Implications for Parkinson's disease treatment. *Journal of Neurochemistry*, *109*(4), 1096–1105. May https://doi.org/10.1111/j.1471-4159.2009.06036.x.

Dolan, S., Gunn, M. D., Biddlestone, L., & Nolan, A. M. (2009). The selective metabotropic glutamate receptor 7 allosteric agonist AMN082 inhibits inflammatory pain-induced and incision-induced hypersensitivity in rat. *Behavioural Pharmacology*, *20*(7), 596–604. Oct https://doi.org/10.1097/FBP.0b013e32832ec5d1.

Du, J., Koltzenburg, M., & Carlton, S. M. (2001). Glutamate-induced excitation and sensitization of nociceptors in rat glabrous skin. *Pain*, *89*(2–3), 187–198. Jan https://doi.org/10.1016/s0304-3959(00)00362-6.

Ferrari, L. F., Lotufo, C. M., Araldi, D., Rodrigues, M. A., Macedo, L. P., Ferreira, S. H., et al. (2014). Inflammatory sensitization of nociceptors depends on activation of NMDA receptors in DRG satellite cells. *Proceedings of the National Academy of Sciences of the United States of America*, *111*(51), 18363–18368. Dec 23 https://doi.org/10.1073/pnas.1420601111.

Fisher, K., Lefebvre, C., & Coderre, T. J. (2002). Antinociceptive effects following intrathecal pretreatment with selective metabotropic glutamate receptor compounds in a rat model of neuropathic pain. *Pharmacology, Biochemistry, and Behavior*, *73*(2), 411–418. Sep https://doi.org/10.1016/s0091-3057(02)00832-8.

Gazerani, P., Wang, K., Cairns, B. E., Svensson, P., & Arendt-Nielsen, L. (2006). Effects of subcutaneous administration of glutamate on pain, sensitization and vasomotor responses in healthy men and women. *Pain*, *124*(3), 338–348. Oct https://doi.org/10.1016/j.pain.2006.06.015.

Goudet, C., Chapuy, E., Alloui, A., Acher, F., Pin, J. P., & Eschalier, A. (2008). Group III metabotropic glutamate receptors inhibit hyperalgesia in animal models of inflammation and neuropathic pain. *Pain*, *137*(1), 112–124. Jul https://doi.org/10.1016/j.pain.2007.08.020.

Govea, R. M., Zhou, S., & Carlton, S. M. (2012). Group III metabotropic glutamate receptors and transient receptor potential vanilloid 1 co-localize and interact on nociceptors. *Neuroscience*, *217*, 130–139. Aug 16 https://doi.org/10.1016/j.neuroscience.2012.05.014.

Hawkins, R. A. (2009). The blood-brain barrier and glutamate. *The American Journal of Clinical Nutrition*, *90*(3), 867S–874S. Sep https://doi.org/10.3945/ajcn.2009.27462BB.

Jackson, D. L., & Hargreaves, K. M. (1999). Activation of excitatory amino acid receptors in bovine dental pulp evokes the release of iCGRP. *Journal of Dental Research*, *78*(1), 54–60. Jan https://doi.org/10.1177/00220345990780010801.

Ji, R. R., Chamessian, A., & Zhang, Y. Q. (2016). Pain regulation by non-neuronal cells and inflammation. *Science*, *354*(6312), 572–577. Nov 4 https://doi.org/10.1126/science.aaf8924.

Jin, Y. H., Yamaki, F., Takemura, M., Koike, Y., Furuyama, A., & Yonehara, N. (2009). Capsaicin-induced glutamate release is implicated in nociceptive processing through activation of ionotropic glutamate receptors and group I metabotropic glutamate receptor in primary afferent fibers. *Journal of Pharmacological Sciences*, *109*(2), 233–241. Feb https://doi.org/10.1254/jphs.08262fp.

Julio-Pieper, M., Flor, P. J., Dinan, T. G., & Cryan, J. F. (2011). Exciting times beyond the brain: Metabotropic glutamate receptors in peripheral and non-neural tissues. *Pharmacological Reviews*, *63*(1), 35–58. Mar https://doi.org/10.1124/pr.110.004036.

Julio-Pieper, M., Hyland, N. P., Bravo, J. A., Dinan, T. G., & Cryan, J. F. (2010). A novel role for the metabotropic glutamate receptor-7: Modulation of faecal water content and colonic electrolyte transport in the mouse. *British Journal of Pharmacology*, *160*(2), 367–375. May https://doi.org/10.1111/j.1476-5381.2010.00713.x.

Kim, Y. S., Anderson, M., Park, K., Zheng, Q., Agarwal, A., Gong, C., et al. (2016). Coupled activation of primary sensory neurons contributes to chronic pain. *Neuron*, *91*(5), 1085–1096. Sep 7 https://doi.org/10.1016/j.neuron.2016.07.044.

Kung, L. H., Gong, K., Adedoyin, M., Ng, J., Bhargava, A., Ohara, P. T., et al. (2013). Evidence for glutamate as a neuroglial transmitter within sensory ganglia. *PLoS One*, *8*(7). https://doi.org/10.1371/journal.pone.0068312, e68312.

Lee, K. S., Park, E. H., Cho, H. Y., Kim, Y. I., & Han, H. C. (2013). Peripheral group II and III metabotropic glutamate receptors in the knee joint attenuate carrageenan-induced nociceptive behavior in rats. *Neuroscience Letters*, *542*, 21–25. May 10 https://doi.org/10.1016/j.neulet.2013.03.006.

Li, J. L., Ohishi, H., Kaneko, T., Shigemoto, R., Neki, A., Nakanishi, S., et al. (1996). Immunohistochemical localization of a metabotropic glutamate receptor, mGluR7, in ganglion neurons of the rat; with special reference to the presence in glutamatergic ganglion neurons. *Neuroscience Letters*, *204*(1–2), 9–12. Feb 2 https://doi.org/10.1016/0304-3940(95)12299-0.

Li, H., Ohishi, H., Kinoshita, A., Shigemoto, R., Nomura, S., & Mizuno, N. (1997). Localization of a metabotropic glutamate receptor, mGluR7, in axon terminals of presumed nociceptive, primary afferent fibers in the superficial layers of the spinal dorsal horn: An electron microscope study in the rat. *Neuroscience Letters*, *223*(3), 153–156. Feb 28 https://doi.org/10.1016/s0304-3940(97)13429-2.

Li, J. Y., Wang, X., Ji, P. T., Li, X. F., Guan, G. H., Jiang, X. S., et al. (2012). Peripheral nerve injury decreases the expression of metabolic glutamate receptor 7 in dorsal root ganglion neurons. *Neuroscience Letters*, *531*(1), 52–56. Nov 30 https://doi.org/10.1016/j.neulet.2012.10.014.

Liu, X. J., White, T. D., & Sawynok, J. (2002). Intraplantar injection of glutamate evokes peripheral adenosine release in the rat hind paw: Involvement of peripheral ionotropic glutamate receptors and capsaicin-sensitive sensory afferents. *Journal of Neurochemistry*, *80*(4), 562–570. Feb https://doi.org/10.1046/j.0022-3042.2001.00721.x.

Malet, M., & Brumovsky, P. R. (2015). VGLUTs and glutamate synthesis-focus on DRG neurons and pain. *Biomolecules*, *5*(4), 3416–3437. Dec 2 https://doi.org/10.3390/biom5043416.

Miller, K. E., Hoffman, E. M., Sutharshan, M., & Schechter, R. (2011). Glutamate pharmacology and metabolism in peripheral primary afferents: Physiological and pathophysiological mechanisms. *Pharmacology & Therapeutics*, *130*(3), 283–309. Jun https://doi.org/10.1016/j.pharmthera.2011.01.005.

Neugebauer, V. (2002). Metabotropic glutamate receptors- -important modulators of nociception and pain behavior. *Pain*, *98*(1–2), 1–8. Jul https://doi.org/10.1016/s0304-3959(02)00140-9.

Niswender, C. M., & Conn, P. J. (2010). Metabotropic glutamate receptors: Physiology, pharmacology, and disease. *Annual Review of Pharmacology and Toxicology*, *50*, 295–322. https://doi.org/10.1146/annurev.pharmtox.011008.145533.

Niswender, C. M., Lebois, E. P., Luo, Q., Kim, K., Muchalski, H., Yin, H., et al. (2008). Positive allosteric modulators of the metabotropic glutamate receptor subtype 4 (mGluR4): Part I. discovery of pyrazolo[3,4-d]pyrimidines as novel mGluR4 positive allosteric modulators. *Bioorganic & Medicinal Chemistry Letters*, *18*(20), 5626–5630. Oct 15 https://doi.org/10.1016/j.bmcl.2008.08.087.

Pal Choudhuri, S., Delay, R. J., & Delay, E. R. (2016). Metabotropic glutamate receptors are involved in the detection of IMP and L-amino acids by mouse taste sensory cells. *Neuroscience*, *316*, 94–108. Mar 1 https://doi.org/10.1016/j.neuroscience.2015.12.008.

Park, E. H., Lee, S. W., Moon, S. W., Suh, H. R., Kim, Y. I., & Han, H. C. (2019). Activation of peripheral group III metabotropic glutamate receptors inhibits pain transmission by decreasing neuronal excitability in the CFA-inflamed knee joint. *Neuroscience Letters*, *694*, 111–115. Feb 16 https://doi.org/10.1016/j.neulet.2018.11.033.

Pisani, A., Calabresi, P., Centonze, D., & Bernardi, G. (1997). Activation of group III metabotropic glutamate receptors depresses glutamatergic transmission at corticostriatal synapse. *Neuropharmacology*, *36*(6), 845–851. Jun https://doi.org/10.1016/s0028-3908(96)00177-3.

Reinhold, A. K., & Rittner, H. L. (2020). Characteristics of the nerve barrier and the blood dorsal root ganglion barrier in health and disease. *Experimental Neurology*, *327*, 113244. May https://doi.org/10.1016/j.expneurol.2020.113244.

Rogoz, K., Lagerstrom, M. C., Dufour, S., & Kullander, K. (2012). VGLUT2-dependent glutamatergic transmission in primary afferents is required for intact nociception in both acute and persistent pain modalities. *Pain*, *153*(7), 1525–1536. Jul https://doi.org/10.1016/j.pain.2012.04.017.

Scherrer, G., Low, S. A., Wang, X., Zhang, J., Yamanaka, H., Urban, R., et al. (2010). VGLUT2 expression in primary afferent neurons is essential for normal acute pain and injury-induced heat hypersensitivity. *Proceedings of the National Academy of Sciences of the United States of America*, *107*(51), 22296–22301. Dec 21 https://doi.org/10.1073/pnas.1013413108.

Schizas, N., Lian, O., Frihagen, F., Engebretsen, L., Bahr, R., & Ackermann, P. W. (2010). Coexistence of up-regulated NMDA receptor 1 and glutamate on nerves, vessels and transformed tenocytes in tendinopathy. *Scandinavian Journal of Medicine & Science in Sports*, *20*(2), 208–215. Apr https://doi.org/10.1111/j.1600-0838.2009.00913.x.

Soliman, A. C., Yu, J. S., & Coderre, T. J. (2005). mGlu and NMDA receptor contributions to capsaicin-induced thermal and mechanical hypersensitivity. *Neuropharmacology*, *48*(3), 325–332. Mar https://doi.org/10.1016/j.neuropharm.2004.10.014.

Svensson, P., Cairns, B. E., Wang, K., Hu, J. W., Graven-Nielsen, T., Arendt-Nielsen, L., et al. (2003). Glutamate-evoked pain and mechanical allodynia in the human masseter muscle. *Pain*, *101*(3), 221–227. Feb https://doi.org/10.1016/s0304-3959(02)00079-9.

Takeda, M., Takahashi, M., & Matsumoto, S. (2009). Contribution of the activation of satellite glia in sensory ganglia to pathological pain. *Neuroscience and Biobehavioral Reviews*, *33*(6), 784–792. Jun https://doi.org/10.1016/j.neubiorev.2008.12.005.

Tong, Q., & Kirchgessner, A. L. (2003). Localization and function of metabotropic glutamate receptor 8 in the enteric nervous system. *American Journal of Physiology. Gastrointestinal and Liver Physiology*, *285*(5), G992–G1003. Nov https://doi.org/10.1152/ajpgi.00118.2003.

Vilar, B., Busserolles, J., Ling, B., Laffray, S., Ulmann, L., Malhaire, F., et al. (2013). Alleviating pain hypersensitivity through activation of type 4 metabotropic glutamate receptor. *The Journal of Neuroscience*, *33*(48), 18951–18965. Nov 27 https://doi.org/10.1523/JNEUROSCI.1221-13.2013.

Wang, H., Jiang, W., Yang, R., & Li, Y. (2011). Spinal metabotropic glutamate receptor 4 is involved in neuropathic pain. *Neuroreport*, *22*(5), 244–248. Mar 30 https://doi.org/10.1097/WNR.0b013e3283453843.

Woolf, C. J., & Ma, Q. (2007). Nociceptors- -noxious stimulus detectors. *Neuron*, *55*(3), 353–364. Aug 2 https://doi.org/10.1016/j.neuron.2007.07.016.

Xiao, H. S., Huang, Q. H., Zhang, F. X., Bao, L., Lu, Y. J., Guo, C., et al. (2002). Identification of gene expression profile of dorsal root ganglion in the rat peripheral axotomy model of neuropathic pain. *Proceedings of the National Academy of Sciences of the United States of America*, *99*(12), 8360–8365. Jun 11 https://doi.org/10.1073/pnas.122231899.

Yasumatsu, K., Manabe, T., Yoshida, R., Iwatsuki, K., Uneyama, H., Takahashi, I., et al. (2015). Involvement of multiple taste receptors in umami taste: Analysis of gustatory nerve responses in metabotropic glutamate receptor 4 knockout mice. *The Journal of Physiology*, *593*(4), 1021–1034. Feb 15 https://doi.org/10.1113/jphysiol.2014.284703.

Zhang, G. H., Min, S. S., Seol, G. H., Lee, K. S., Han, H., Kim, Y. I., et al. (2009). Inhibition of the N-methyl-d-aspartate receptor unmasks the antinociception of endogenous opioids in the periphery. *Pain*, *143*(3), 233–237. Jun https://doi.org/10.1016/j.pain.2009.03.007.

Zhou, H. Y., Chen, S. R., Chen, H., & Pan, H. L. (2011). Functional plasticity of group II metabotropic glutamate receptors in regulating spinal excitatory and inhibitory synaptic input in neuropathic pain. *The Journal of Pharmacology and Experimental Therapeutics*, *336*(1), 254–264. Jan https://doi.org/10.1124/jpet.110.173112.

Chapter 22

TRPM8 receptor and menthol in pain management

Hannu Kokki[a] and Merja Kokki[b]

[a]School of Medicine, University of Eastern Finland, Kuopio, Finland, [b]Department of Anaesthesia and Operative Services, Kuopio University Hospital, Kuopio, Finland

Abbreviations

CIPN	chemotherapy-induced peripheral neuropathy
CRPS	complex regional pain syndrome
DOMS	delayed onset muscle soreness
EC_{50}	effective concentration 50, 50% of a maximal effect
IL1-β	interleukin 1 beta
K_{ow}	octanol-water partition coefficient
LTB4	leukotriene B4
Mg^{++}	magnesium
NMDA	N-methyl-D-aspartate
OGT	oxygenated glycerol triesters
PGE2	prostaglandine E2
SPID8	summed pain intensity difference score through 8 h
TRPA	transient receptor potential ankyrin
TRPC	transient receptor potential canonical
TRPM8	transient receptor potential melastatin 8
TRPML	transient receptor potential mucolipin
TRPN	transient receptor potential NOMPC-like
TRPP	transient receptor potential polycystin
TRPV	transient receptor potential vanilloid
WOMAC	Western Ontario and McMaster Universities osteoarthritis index

Introduction

International Association for the Study of Pain has recently updated the definition for pain; *"An unpleasant sensory and emotional experience associated with, or resembling that associated with, actual or potential tissue damage."* The essential change is, compared to the previous version, that it takes in accounts infants, elderly people, and others who could not verbally articulate their pain (IASP, 2018). Acute and chronic pain conditions are common in all ages, and there is a clinical need for analgesics approaches that can be used in younger children as well as in the elderly.

Apprehension regarding on pharmacotherapies, adverse effect, and interaction and abuse potential has enforced seeking nonpharmacological approaches for effective and safe pain control. Better understanding on the diversity of receptors involved in nociceptive sensory detection, transduction, and modulation has fostered to evaluate some older agent in a new perspective also (Kinnunen, Piirainen, Kokki, Lammi, & Kokki, 2019).

Cool and menthol has been used in nonpharmacological pain management since ancient times and the first scientific publications date back to 1870s (Wright, 1870; MacDonald, 1879). Two decades ago, the receptor, the transient receptor potential melastatin 8 (TRPM8), was identified as the molecular target of cool stimulus. Understanding that sensory afferents having TRPM8 receptors are widely distributed throughout the nociceptive system, in periphery somatosensory neurons, dorsal root ganglia and trigeminal ganglia, and central nervous system has opened a new perspective on menthol use in pain management. New knowledge that the 10%–15% of the somatosensory neurons that express TRPM8 receptors

The Neurobiology, Physiology, and Psychology of Pain. https://doi.org/10.1016/B978-0-12-820589-1.00022-1

are highly sensitive to nociceptive stimuli has led to increased clinical applications on menthol in clinical practice (McKemy, Neuhausser, & Julius, 2002; Julius, 2013).

After identification of TRPM8 channel as the molecular target of cool and menthol in 2002, there has been an increasing interest on TRPM8-mediated analgesia. Menthol has a counterirritant analgesic activity in TRPM8 receptor; it first activates and then desensitizes nociceptors and other structures in the nociceptive system. Pain relief is followed by the initial pleasant cool sensation after menthol application. Cool sensation of menthol is innocuous, compared to that of ice or other direct cold applications that could cause frostbites and other harms in the body. Menthol is well tolerated, it has a low interaction potential, is well absorbed after topical application, and highly effective in divers pain conditions in children, adults, as well in the elderly (McKemy et al., 2002; Peier, Moqrich, Hergarden, et al., 2002).

Menthol is a feasible compound in many acute and chronic pain conditions. Surgical patients are one group who need effective pain management. Patients have often pain before surgery and most have moderate or severe pain after surgery. Preoperative pain and severe postoperative pain are associated with a risk of persistent postoperative pain. In order to ensure calm recovery, pain management should be optimized in these patients (Lahtinen, Kokki, & Hynynen, 2006; Härkänen et al., 2018; Kuronen, Kokki, Nyyssönen, Savolainen, & Kokki, 2018).

Transient receptor potential channels

Transient receptor potential (TRP) channels were first discovered in the fly eye in 1969. Now it is known that these channels have a central role in physiology and diseases, e.g., in the detection of noxious stimuli (Cosens & Manning, 1969; Julius, 2013). TRP channels are grouped into seven subfamilies, TRPA, TRPC, TRPM, TRPML, TRPN, TRPP, and TRPV. TRP channels respond to multiple types of stimuli. In neurons, activation of TRP channels leads increased probability of action potential generation and in nonexcitable cells, other processes such as vasodilatation (Julius, 2013).

Menthol receptor, TRPM8-receptor is an excitatory ion channel receptor. TRPM8-receptor acts as the main mediator of cool and cold sensation in the human body. Menthol receptor is nonselective and polymodal ion channel which opens through chemical agonists, cool and cold temperatures (10–28°C) and voltage. (−)-menthol is one of the most potent agonists that activates TRPM8 channel (McKemy et al., 2002).

TRPM8 belongs to the melastatin TRP subfamily of TRP ion channels that compromise eight putative members. TRPM8 was first identified in prostate (Tsavaler, Shapero, Morkowski, & Laus, 2001), and later described as a cold- and menthol-activated nonselective cation channel expressed in sensory neurons, dorsal root ganglia and trigeminal ganglia, spinal cord, and in central nervous system (McKemy et al., 2002; Peier et al., 2002). TRPM8 is one of the three TRP ion channels; two others are TRPV1 and TRPA1 that are molecular detectors of thermal and chemical stimuli that activate sensory neurons and nociceptive system to produce pain (Julius, 2013).

TRPM8 is expressed by 10%–15% of all somatosensory neurons, small-diameter, unmyelinated C-fibers and lightly myelinated Aδ-fibers. The cell bodies of these neurons from face and head are located in trigeminal ganglia and from rest of the body in dorsal root ganglia. TRPM8 is expressed also in other tissues, e.g., bladder, prostate, and tumor cells; it is assumed to have other functions in the body as well (Julius, 2013). TRPM8 receptors are expressed in endothelium, and after topical application of menthol cutaneous blood flow increases within 5–10 min. In vascular smooth muscles, the actions of menthol depend on previous vasomotor tone. In vasoconstricted vessels, menthol generates relaxation and thus, vasodilatation, increases blood flow and decreases on elevate blood pressure in compromised patients (Silva, 2020).

Recent data indicating that activation of TRPM8 channels with topical menthol activates brown fat tissue and may be used to prevent and treat obesity is interesting also for pain. Obesity is associated with musculoskeletal and joint pain, and low-grade inflammation may contribute several pain conditions (Ma et al., 2012) (Table 1).

Menthol

Menthol, $C_{10}H_{20}O$, 5-methyl-2-propan-2-ylcyclohexan-1-ol, is a covalent organic compound made synthetically or obtained from peppermint or mint oils. Peppermint oil contains 35%–60% menthol. The main form in nature (−)-menthol. It has molecular weight 156.27 g/mol and octanol/water partition coefficient, log Kow 3.2–3.4 and water solubility 456 mg/L at 25°C, i.e., menthol is slightly soluble in water but very soluble in alcohol. Menthol has a characteristic peppermint odor and taste. In room temperature, it is a white crystalline solid and it melts at 43°C. (−)-menthol is threefold more potent agonist in TRPM8 receptor, EC_{50} 4.1 μM, than (+)-menthol, EC_{50} 14.4 μM (Behrendt, Germann, Gillen, Hatt, & Jostock, 2004).

Menthol is primarily metabolized to the glucuronide conjugate and eliminated mostly in the urine within 12–24 h. The plasma half-life of menthol glucuronide averaged 45–60 min (Peat, Frazee, Kearns, & Garg, 2016).

TABLE 1 Transient receptor potential melastatin 8 (TRPM8) expression in the body.

Tissue	Reference
C and Aδ afferent neurons in the skin and mucous membranes	McKemy et al. (2002)
Dorsal root ganglia and trigeminal ganglia	McKemy et al. (2002)
Midbrain and pons	Du, Yang, Zhang, and Zeng (2009)
Ascending and descending pain pathway, dorsal and ventral roots in spinal cord	Facer, Casula, Smith, et al. (2007)
Odontoblasts, tooth pulp	Tazawa, Ikeda, Kawashima, and Okiji (2017)
Nasal epithelial	Keh et al. (2011)
Bronchopulmonary autonomic afferent fibers	Xing, Ling, Chen, et al. (2008)
Keratinocytes, excessive in hypersensitive skin	Facer et al. (2007) and Park et al. (2013)
Lung epithelial cells	Sabnis, Shadid, Yost, and Reilly (2008)
Ganglion cells innervating the upper gut	Zhang, Jones, Brody, Costa, and Brookes (2004)
Gastric fundus	Mustafa and Oriowo (2005)
Colonic sensory neurons	Harrington, Hughes, Martin, et al. (2011)
Bladder urothelium, different tissues in male genital tract—prostate	Tsavaler et al. (2001) and Stein, Santos, Nagatomi, et al. (2004)
Brown adipose tissue	Ma et al. (2012)
White adipose tissue	Rossato, Granzotto, Macchi, et al. (2014)
Articular sensory neurons	da Silva Serra, Husson, Bartlett, and Smith (2016)
Vascular smooth muscle cells and endothelium and smooth muscle	Johnson et al. (2009)
Myocardium	Cheng et al. (2019)
Tumors (e.g., breast, colon, lung, prostate, and skin)	Tsavaler et al. (2001)

Menthol is contraindicated in newborns as glucose-6-phosphate-dehydrogenase-deficiency, and it may cause allergic reactions in individuals having hypersensitive to eucalyptus or menthol.

Menthol is widely used since ancient times in consumer products as well as other uses; antipruritic, dermatologic agents, food additive, disinfectants, and pharmaceutical products among others. Menthol decreases inflammatory mediators (LTB4, PGE2, IL1-β) synthesis, is antiviral, bacteriostatic, and antifungal, and has a substantial antipruritic activity (Bromm, Scharein, Darsow, & Ring, 1995; Park, Na, Choi, Cho, & Lee, 2013; Ludwiczuk, Kieltyka-Dadasiewiczb, Sawicki, Golusd, & Ginalskad, 2016; Suchodolski, Feder-Kubis, & Krasowska, 2017; Taylor, Hamid, Andres, et al., 2020).

Menthol is one of the natural products from medicinal plants that modulate nociceptive input in the pain pathway. Two other well-known pharmacological agents are morphine from the opium poppy and salicylate from the willow bark. Morphine was a key to discovering opioid receptors and salicylate to cyclooxygenases (Julius, 2013).

Topical application of menthol acts as an analgesic in nociceptive, inflammatory, and neuropathic pain conditions. Analgesic action is based on activation of TRPM8 channels in different levels of pain pathways. Other actions may also be involved. Menthol is assumed to act on κ-opioid receptors as analgesic effect can be reversed naloxone. It also has a local analgesic or anesthetic effect (Proudfoot, Garry, Cottrell, et al., 2006; Liu et al., 2013) (Fig. 1 and Table 2).

Multimodal pain therapy

Most patients have moderate or severe pain after injuries, trauma, and surgery. In some pain states, a single approach may be sufficient, but most common approach in pain management is the use of combinations of two or more analgesic methods, a multimodal analgesia. In multimodal analgesia, nonpharmacological interventions and pharmacological techniques with

FIG. 1 *Mentha spicata*. Menthol is an organic compound made synthetically or obtained from mint oils.

TABLE 2 Studies reporting use of menthol in pain management.

Authors and reference	Cohort	Study design	Intervention	Findings
Green and McAuliffe (2000)	12 subjects: 7 male	Open-label trial	Menthol sip before capsaicin application to tongue	The sensory response to capsaicin decreased by menthol
Airaksinen et al. (2003)	74 patients with sport related soft tissue injuries	Randomized, prospective, double-blind trial	Application four times/day for 14 days Topical menthol, $n = 37$ Topical placebo, $n = 37$	Menthol: pain (0–100) decreased from 59 to 30 in 1 week, to 14 in 2 weeks and to 7 at 3 weeks Placebo: from 59 to 45, 26 and 13 respectively.
Kraemer et al. (2005)	28 patients with osteoarthrosis (knee, elbow or wrist), 2 male	Open-label trial	Application of topical menthol cream two times/day for 1 week	WOMAC pain (0–10) decreased from 9.4 to 5.0
Zhang et al. (2008)	36 patients with acute low back pain	Randomized controlled study	Application three times/day for 4 weeks in experimental group ($n = 18$) Control group no topical treatment ($n = 18$)	Experimental group: pain (0–10) decreased from 4.1 to 1.3 Control group: pain increased from 4.4 to 5.2

TABLE 2 Studies reporting use of menthol in pain management—cont'd

Authors and reference	Cohort	Study design	Intervention	Findings
Higashi, Kiuchi, and Furuta (2010)	208 patients with mild to moderate muscle strain	Randomized, double-blind, placebo-controlled study	Application of patch for 8 h Menthol, $n = 105$ Placebo, $n = 103$	Menthol: SPID8 decreased by 182 Placebo: SPID8 decreased by 130
Borhani Haghighi et al. (2010)	35 patients with migraine without aura	Randomized, triple-blind, cross-over study	Menthol: 2 attacks treated with menthol 10% and next 2 with menthol 0.5% ($n = 17$) or vice versa ($n = 18$)	Menthol 10%: pain-free attacks 38% of treated attacks Menthol 0.5%: pain-free attacks 12%
Johar et al. (2012)	16 subjects	Randomized, open-label trial	48 h after DOMS inducing session Topical menthol, $n = 8$ Topical ice, $n = 8$	Menthol: soreness perception (0–10) 1.1 Ice: soreness 3.1
Taylor Jr. et al. (2012)	69 subjects with acute musculoskeletal pain	Double-blind, randomized trial	Application three times/day for 8 days Topical menthol, $n = 34$ Topical menthol + OGT, $n = 35$	Menthol: pain (0–100) decreased from 42 to 29 Menthol + OGT: from 47 to 19
Topp et al. (2013)	20 subjects with osteoarthritis of the knee	Randomized, cross-over, placebo-controlled trial	Application between assessments, first topical menthol and period placebo, $n = 10$, or vice versa, $n = 10$	Menthol: pain decreased during 4/5 functional tests Placebo: no decrease in pain
Sundstrup, Jakobsen, Brandt, et al. (2014)	10 patients with carpal tunnel syndrome patients	Triple-blind, randomized, cross-over, placebo-controlled trial	A single application day 1: topical menthol, day 2 placebo, $n = 5$ or vice versa, $n = 5$	Menthol: pain (0–10) decreased by −1.2 compared to placebo
Fallon et al. (2015)	51 patients with CIPN pain	Open-label, prospective clinical trial	Application twice/day for 4–6 weeks	Pain (0–10) decreased from 4.7 to 3.4, function and mood improved
Keshavarzian and Shahgholian (2017)	105 patients undergoing hemodialysis	Single-blind, placebo-controlled clinical trial	Application three times/day for 3 days Menthol, $n = 35$ Rosemary, $n = 35$ Placebo: $n = 35$	Menthol: pain (0–10) decreased from 7.3 to 4.8 Rosemary: from 7.5 to 4.2 Placebo: from 7.2 to 6.4
Engelhard et al. (2019)	36 patients undergoing anterior cruciate ligament reconstruction	Open-label, randomized, prospective clinical trial	Three times/day for 3 days, and 2 times/day for 4 weeks Menthol-containing bandage, $n = 12$ Ice pack, $n = 12$ No cooling, $n = 12$	Menthol: cross section of vastus medialis at 30 days −3.2%, analgesics 26 tablets Ice pack: −8.8%, 40 tablets No cooling: −7.2% and 35 tablets
Stefanelli et al. (2019)	16 participants with DOMS	Double-blind, placebo-controlled clinical trial	Single application of 2 mg Menthol, $n = 8$ Placebo: $n = 8$	Menthol: Pain pressure threshold before gel 3.1 kg, 45 min after 4.3 kg Placebo: 3.5 kg and 4.1 kg

CIPN, chemotherapy-induced peripheral neuropathy; *DOMS*, delayed onset muscle soreness; *OGT*, oxygenated glycerol triesters; *SPID8*, summed pain intensity difference score through 8 h; *WOMAC*, Western Ontario and McMaster Universities osteoarthritis index.

different sites and mechanism of action are combined to improve analgesia and patient satisfaction, enhance recovery, and reduce/eliminate opioid use and adverse effects (Glare, Aubrey, & Myles, 2019).

Menthol is a feasible component in multimodal analgesia regiments. The topical route of administration is easily available, TRPM8 channel is different to other targets in pain management, menthol can be coadministered with any topical or systemic analgesics with no risk of interactions, has very low risk of adverse effects or overdoses, the onset of action is rapid, and there is no need for dose titration. In acute pain, menthol is often applied four times a day and in persistent pains two times a day.

Posttraumatic and postoperative pain

Since the pivotal study by Airaksinen et al. (2003) where they show that topical menthol gel did not only provide superior pain relief but also enhance recovery in patients with sports-related soft tissue injury, this has been established treatment for musculoskeletal pains and injuries. On the contrary to topical menthol, the effectiveness of ice application on acute injuries has not been proven in any randomized clinical trial and actually, ice can be harmful (Bleakley, McDonough, & MacAuley, 2004).

Cryotherapy has been used in the postoperative management in orthopedics. However, there are no sufficient evidence to support this. Some data show that cryotherapy can have some efficacy as an adjunct analgesic treatment in the early phase of recovery, but not, e.g., in reducing swelling or edema, and does not improve long-term outcome. Actually, cryotherapy has potential risks. Skin can be directly affected by hypothermia and cryotherapy may compromise cutaneous perfusion. Prolonged duration of cryotherapy may cause frostbites, tissue necrosis, and thromboembolic complications, and damage peripheral nerves. On the contrary to ice packs and cryotherapy devices, menthol increases cutaneous blood flow, and the risk for local or underlining tissue damage is minimal.

Topical menthol has substantial analgesic efficacy and could be useful in postoperative pain management; menthol-containing bandage during the rehabilitation after orthopedic surgery prevents muscle cross-section decrease, improves rehabilitation success, and reduces analgesics consumption, thus decreasing the risk of adverse events (Engelhard, Hofer, & Annaheim, 2019).

Complex regional pain syndrome (CRPS) is a severe neuropathic pain condition. It may develop after surgery, trauma, or minor injury. Up to 26% of patients with surgical treatment of traumatic hand injuries may develop CRPS and after elective ankle and foot surgery 5%–6%, respectively (Savaş et al., 2018). CRPS patients have severe pain and after the initial "warm phase" excessive vasoconstriction in the affected limb. Menthol has analgesic efficacy in nociceptive, inflammatory, and neuropathic pain, and it induces vasodilatation (Craighead, McCartney, Tumlinson, & Alexander, 2017). Thus, menthol is assumed to relieve CRSP symptoms by different mechanism, and could be used in multimodal approach in the treatment of refractory CRPS.

Musculoskeletal pain

Knowledge on the feasibility of topical menthol on the management of musculoskeletal pain dates back for centuries (Ringer & Sainsbury, 1897) and since that topical menthol is well-established use for sore muscles and other pain of the musculoskeletal system (EMA, 2020).

The most common musculoskeletal pain is low back pain, the prevalence in the general population 30%–40%, followed by neck and shoulder pain and pain in joints. Fibromyalgia is less common, 2%–5%, but one of the most challenging pain conditions. Evidence is accumulating on involvement TRPM8 receptors, and high efficacy and safety of menthol in these pain conditions (Taylor Jr., Gan, Raffa, et al., 2012).

Low back pain is the leading cause of disability. The lifetime prevalence of nonspecific low back pain is 60%–70%. Acute and recurrent pain is often caused by strain of muscle and ligaments, and degenerative spine conditions can cause more persistent pain. Pain has often neuropathic features radiating to the extremities, while in majority of cases, pain is axial pain in paraspinal muscles, ligaments, and facet joints. Sacroiliac joint pain accounts up to one out of four cases of lower back pain symptoms. Neck pain affects 30%–50% of the general population annually. Both physical and psychological factors contribute to the high prevalence of the symptoms (GBD, 2017).

As most patients have stiff muscles and associated impaired muscle circulation, menthol should relief symptoms as it reliefs pain and increases blood flow, and thus may relax taut muscles. This was supported by a randomized study of adult patients where topical menthol three times a day provided superior pain relief and functional recovery compared to chiropractic adjustment twice a week (Zhang, Enix, Snyder, Giggey, & Tepe, 2008).

Prehabilitation is a process of improving the functional capability of a patient prior to surgery. Cardiorespiratory fitness is a strong and independent predictor of length of stay, postoperative pain, postoperative complications, and all-cause mortality in surgical patients. As functional reserve increases, this enhances patient's capacity to withstand postoperative inactivity and associated decline. Typical exercise regime consists a combination of strength and aerobic exercises four to five times a week for a few weeks before surgery. Despite benefits of exercise, acute bout of physical activity may cause pain associated with exercise (Banugo & Amoako, 2017).

Myalgia after exercise, delayed-onset muscle soreness (DOMS) is a common condition after exercise and physical activity to which one is not customized. After strenuous muscle loading, one may feel muscles tender and sore 24–72 h after exercise. Pain is dull and aching and becomes worse when the muscle is activated, stretched, or palpated. The pathophysiology is assumed to base on inflammatory mediators and other noxious substances in the affected muscles, and the pain modulation in the central nervous system is involved. Acute muscle pain alters motor system also (Johar, Grover, Topp, & Behm, 2012).

Topical application of menthol-based products decreases muscles soreness and pain, and recovers decreased function in subjects having DOMS. Topical menthol is superior to local cold therapy and cold water immersion, and is more feasible mode of therapy (Johar et al., 2012; Stefanelli, Lockyer, Collins, et al., 2019).

Osteoarthrosis is a chronic, disabling, and the most common joint disorder. The prevalence of symptomatic osteoarthrosis in persons aged ≥60 years is >40%. It is the leading cause of disability in older adults. In reflecting the aging population, sedentary lifestyles, and increasing incidence of obesity, the prevalence of osteoarthrosis is likely to increase further (GBD, 2017).

In most subjects, osteoarthrosis is associated with pain, stiffness, and impaired function. However, not all osteoarthritis is associated with pain, and in those with pain, it is fluctuating with temporary flares. Moreover, several pain phenotypes have been identified, and nociceptive, inflammatory and neuropathic pains may occur, and both peripheral and central processes are involved (Kraemer, Ratamess, Maresh, et al., 2005; Topp, Brosky Jr., & Pieschel, 2013).

Nonpharmacological and nonsurgical treatments are preferred in symptomatic osteoarthrosis. Nonsteroidal anti-inflammatory agents are effective, but often contraindicated, and opioid analgesics should be used short time periods if any. Corticosteroid injections relieve pain caused by osteoarthrosis but may enhance the progress of the condition in long term (McAlindon, LaValley, Harvey, et al., 2017).

Topical menthol both decreases pain and improves functioning in patients with osteoarthrosis. Recent systematic reviews confirm that education, the lifestyle interventions of physical activity, exercise, and weight management are the gold standard for managing pain and disability associated with osteoarthrosis. Topical menthol enhances the potential for exercise and physical activity training in this population (Kraemer et al., 2005; Topp et al., 2013).

Neuropathic pain

Persistent postoperative pain has often neuropathic features (Glare et al., 2019). Topical menthol is well-established use for neuropathic pain (EMA, 2020). Neuropathic pain particularly challenging to treat as it involves aberrant neural signaling. Neuropathic pain is associated with unpleasant symptoms, and low health-related quality of life. Patients with neuropathic pain report nonrestorative sleep, daytime tiredness, and physical fatigue. In addition to lesions and diseases of the somatosensory system, many common pains have neuropathic characteristics.

TRPM8 receptors are expressed both at peripheral tissues and central nervous system. TRPM8 receptor expression increases in sensory neurons after nerve injury (Proudfoot et al., 2006). These are the basis on high efficacy on topical menthol in several neuropathic pain state.

The first reports on menthol in neuropathic pain date back to 1870 when peppermint oil was used on neuropathic face pain with a success (Wright, 1870). Since then analgesic efficacy of topical menthol has been shown, e.g., in cold allodynia (Wasner et al., 2008), postherpetic neuralgia (Davies, Harding, & Baranowski, 2002), and chemotherapy-induced peripheral neuropathy (CIPN) (Fallon, Storey, Krishan, et al., 2015) among other.

As life expectancies are increasing, this will increase neuropathic pain prevalence. These patients have also age-related comorbidities and thus, contraindications and interaction potential for analgesics. The life-time risk of herpes zoster is 30%, and most cases and complications, such as post herpetic neuralgia, occur in the elderly. Topical menthol has been used for pain management in postherpetic neuralgia in adjunct to other interventions with a success (Davies et al., 2002).

The elderly surgical population rate is increasing and survival rates of, for example, surgical cancer patients have increased substantially. However, quality of life could be compromised by treatment emergent adverse effects. Certain chemotherapy drugs can cause CIPN. Neuropathy occurs in 60%–80% of patients who receive, e.g., carboplatin and paclitaxel, and one-third of patients have symptoms 6 months after chemotherapy. In many patients, pharmacotherapy of CIPN pain is limited by low efficacy and high incidence and severity of adverse effects (Colvin, 2019).

Topical menthol-cream two times a day is highly effective in the refractory CIPN pain. Menthol can be coadministered with any other intervention, onset of analgesia is relatively rapid and no dose titration is needed. Six weeks' menthol treatment had positive effects on mood, catastrophizing, functional ability to walk, and sensation (Fallon et al., 2015). One of the proposed pathophysiological mechanisms of CIPN is the accumulation of chemotherapy drugs adducts in dorsal root ganglion and spinal cord dorsal horn. As TRPM8 channels are expressed throughout the somatosensory system including dorsal root ganglion and dorsal horn and as the expression of TRPM8 receptors is upregulated following chemotherapy treatment, it is assumed that the high analgesic efficacy of menthol in painful CIPN could be based both on peripheral mechanisms and on decreased neuronal excitability in the dorsal root ganglia and in the central nervous system (Colvin, 2019).

Fibromyalgia affects 2%–5% of people of all ages. Fibromyalgia causes widespread pain and tenderness, fatigue, and sleep problems. The most effective treatment for fibromyalgia is aerobic and muscle strengthening exercise. In patient with fibromyalgia, any acute bout of physical activity may aggravate muscle tenderness and pain. As in any people, topical menthol should decrease muscle soreness and pain, and enhance recovery after intense and unfamiliar exercise, and should be feasible treatment for fibromyalgia patients also (Johar et al., 2012; Stefanelli et al., 2019).

Taken together, these data show that in many patients with peripheral and central neuropathic pain topical, menthol has a substantial analgesic efficacy.

Headache and migraine

One-fifth of surgical patients have postoperative headache and the lifetime prevalence of posttraumatic headache is 2%–5% (Ashina, Porreca, Anderson, et al., 2019). Menthol is well-established use for different types of headache and should be appropriate treatment also for surgical patients (Borhani Haghighi, Motazedian, Rezaii, et al., 2010). In the population, migraine is the sixth highest cause of disability and ranged second in years lived with disability. Tension-type headache is much more common. Cluster headache is the most common trigeminal autonomic headache but the prevalence is low, 0.1% (GBD, 2017).

Nonpharmacological treatments should be the first option for patients with headache. Several pharmacological treatments are effective for different types of headaches but are associated with a risk of adverse outcome. One of the concerns is medication overuse headache. In headache, topical menthol can be combined with any other nonpharmacological and pharmacological treatments, and is assumed to decrease the need for pharmacological approaches.

TRPM8 receptors are expressed in the trigeminal ganglia and contribute in the pathology of headache. Single-nucleotide polymorphisms in the channel gene within or near of genomic loci of TRPM8 contribute features of migraine. Similar, low TRPM8 expression is associated with reduced risk for migraine (Borhani Haghighi et al., 2010; Dussor & Cao, 2016; Schattner & Randerson, 1996).

Topical menthol is a feasible, efficacious, safe, and well-tolerated therapeutic option for different types of headache (Wright, 1870). In migraine, menthol may be applied topically repeatedly as long as the attack subsides (Borhani Haghighi et al., 2010). In tension-type headache, topical menthol is commonly applied four times a day or as a patch that can be leaved on place for 12 h. Menthol may aggravate headache in a few patients, but it alleviates headache in majority of patients.

Pregnant and lactating women

Different types of pain, muscle tension, and associated symptoms are common in pregnant and lactating women. Pain can be associated with pregnancy and some have acute nonobstetric pain-associated minor trauma and injuries, and some may need surgery during pregnancy and breastfeeding. In some women, pain is related to chronic conditions. Low back pain, tension-type headache, other musculoskeletal pain conditions, and migraine are the most common pain states in pregnant and lactating mothers, and some women suffer from neuropathic pain (Black, Khor, Kennedy, et al., 2019).

Fetal and newborn exposure is a concern during pregnancy and breastfeeding. As only 1% of menthol is absorbed into the mother's systemic circulation after skin application, this is not an issue for fetal safety. In lactating women, menthol is used to treat nipple fissures and no safety concerns have been reported in newborns (OECD, 2004).

In pregnant women, nonpharmacological and transdermal treatments should be employed where possible in order to avoid harmful effects of systemic analgesia on the fetus and the pregnancy. Menthol cream can safely be used throughout pregnancy and lactation (Drugs and Lactation Database (LactMed), 2006). Menthol is highly effective in headache and musculoskeletal pains common complaints during pregnancy. In fertile women with neuropathic pain, topical menthol with/without capsaicin is feasible as a sole intervention or as one component of multimodal approach.

Children and adolescent

Both acute and chronic pains are very common in childhood and adolescence, 60% report pain over the last month. Weekly musculoskeletal pain affect 8% and 32% of children and adolescent respectively, 3%–5% of children have migraine and that of tension-type headache is 5- to 10-fold higher. Fibromyalgia affects 2%–6% of school-age children, mostly adolescent girls. The prevalence of pain increases after puberty, and thereafter affects more girls (King, Chambers, Huguet, et al., 2011).

Menthol is well established used in pediatric patients for recurrent abdominal pain, irritable bowel syndrome, nausea, and symptomatic relief of coughs and colds, and oral hygiene (Bardellini, Amadori, & Majorana, 2016; EMA, 2020). Topical menthol is used for relief of different musculoskeletal pains, headaches, and neuralgias. Among other, growing pains are common, 50% of children may have this condition during childhood. As growing pains share the characteristics of DOMS, topical menthol applied one to two times in the evening/bedtime should allow undisturbed sleep. Topical menthol can also be used to ease the pain, pruritus, and urticaria caused by insect bites and vaccination. Topical menthol is well tolerated in children but caution is needed in the children less than 2 years old (EMA, 2020).

Elderly

Pain prevalence increases with increasing age, the prevalence of chronic pain in the community is up to 75% and 25%–30% of the elderly have continuous or daily pain. The prevalence of joint pain, back pain, pain in lower extremities, and neuropathic pain is substantially higher in the elderly. Older patients often experience pain in multiple sites and more often chronic pain than younger individuals (Mäntyselkä, Turunen, Ahonen, & Kumpusalo, 2003).

Older persons have often polypharmacy and comorbid disease that should be taken in account. Elderly patients tolerate opioids, anti-inflammatory drugs, and adjunctive compounds less well than younger people, and the elderly are more prone to develop interactions. Thus, nonpharmacological approaches should be preferred and used in conjunction to pharmacological therapy.

Topical menthol is well-established use for common pain states in the elderly, e.g., pain associated with musculoskeletal system, joints, osteoarthritis, postherpetic neuralgia, and diabetic neuropathy (EMA, 2020). With menthol, there is no interaction potential, contraindications and adverse effects are rare, the onset of analgesic action of menthol is rapid and there is no need for dose titration. Topical application of menthol is a feasible dosage form in the elderly, it is easy to use and patient compliance is often better than that with other administration routes.

Obesity

Obesity rates are increasing in all ages, 40% of adults are overweight, and 15% are obese. Recent data indicate that activation of TRPM8 channels with topical menthol activates brown fat tissue and may be used to prevent and treat obesity. This is interesting also for pain. Obesity is associated with musculoskeletal and joint pain, and low-grade inflammation may contribute several pain conditions (Ma, 2012). Menthol should be feasible compound for obese patients.

Other agents of interest

Menthol enhances transdermal permeation of other substances by interfering with the intracellular lipid structure and via calcium ion interactions (Turunen et al., 2019). This is beneficial property as menthol can be coadministered with other compounds acting in different sites in the pain pathway.

Magnesium is an N-methyl-D-aspartate (NMDA) antagonist. NMDA antagonists are used increasingly for preemptive analgesia and for treating established pain states. Mg^{++} prevents central sensitization and attenuates preexisting pain hypersensitivity. NMDA antagonists have analgesic efficacy in neuropathic pain, such as diabetic neuropathy, postherpetic neuralgia, and CIPN. Mg^{++} is used for headache, migraine, and different musculoskeletal conditions also. Mg^{++} does not permeate skin readily but coadministration with menthol enhanced Mg^{++} permeability substantially. Magnesium and menthol have been used in skin conditions, e.g., in atopic dermatitis, for centuries. Menthol is used as an antiseptic, as it has bacteriostatic, antiviral, and antifungal activity, and it has antipruritic activity also. Moreover, as topical menthol increases skin blood flow and has anti-inflammatory activity, Mg^{++}-levomenthol combination could be a useful combination in skin diseases also (Turunen et al., 2019).

Another interesting combination product is menthol-capsaicin cream. Capsaicin is TRPV1 agonist. Topical application of capsaicin causes transient pain, and as menthol has local anesthetic efficacy, the combination product is better tolerated. Topical application of capsaicin reduced epidermal TRPV1-expressing nociceptive nerve fiber density. This combination is used in peripheral neuropathic pain, e.g., diabetic peripheral neuropathy and postherpetic neuralgia, and in musculoskeletal pain, e.g., backache, osteoarthrosis, sprains, and strains.

Although menthol has anti-inflammatory efficacy itself (Juergens, Stöber, & Vetter, 1998; Park et al., 2013), it is used also in combination products with topical nonsteroidal anti-inflammatory analgesics. These combination products are used in osteoarthrosis and musculoskeletal injuries (Brunner et al., 2011).

Applications to other areas

TRPM8, originally identified in prostate cells, is widely expressed in the human body. Thus, menthol has several applications in areas other than pain management.

Menthol is used as an aroma in many different food products. Several consumer products contain menthol, for example, most dental products, mouthwashes, chewing gums, lip balms, and confectionary. Various nonprescription decongestants and products for mouth irritation, sore throat, and cough contain menthol. Menthol is a common ingredient in many topical antipruritics, which are indicated to reduce itching, rash, and pain caused by insect bites, plants, and minor burns.

One of the most well-known applications of menthols is its use as an additive in cigarettes, both for flavor and to reduce throat and sinus irritation caused by smoking. Menthol enhances nicotine absorption and thereby facilitates smoking initiation and increases addiction to nicotine. In the European Union (EU), menthol cigarettes were smoked by one-fourth of the smokers, whereas in the United States menthol cigarettes account for one-third of the cigarette sales. Younger people are more likely to smoke menthol cigarettes. The EU banned the sale of menthol cigarettes in May 2020.

Recent data indicating that TRPM8 is involved in the pathogenesis of some cancers and that menthol inhibits apoptosis, oxidative stress, and inflammation have increased the interest in the use of menthol in oncology. Menthol has anticancer activity against various human cancers (e.g., glioblastoma, melanoma, and urothelial carcinoma of the bladder).

In gastroenterology, menthol relieves symptoms of irritable bowel syndrome and inflammatory bowel disease, and can possibly be used in pediatric patients to relieve pain and bloating. Menthol is used to treat nipple fissures in women, which is the first exposure to menthol for many neonates. As menthol has antispasmodic properties, it is used in endoscopic gastrointestinal procedures.

Experimental data indicate that menthol-induced vasodilatation could be beneficial in hypertension as well as pulmonary hypertension. Exiting data on the use of menthol in dry eye syndrome, vaginal dryness, and burning mouth syndrome have also been presented. However, confirmatory human data are awaited.

As menthol is antiseptic and has antiviral, bacteriostatic, and antifungal properties, there is an increasing interest in its use for the treatment of viral infections.

Menthol is used in sports to improve physiological performance, but these properties, decreased sense of dyspnea and improved efficacy of exercise therapy, can also be employed in rehabilitation and geriatrics. Enhancement of lung and airway volume may improve aerobic capacity and hasten recovery, and increased blood flow may improve response to muscle strength training in the elderly as well as in younger adults.

Topical menthol is a promising compound for the prevention and treatment of obesity, as menthol increases metabolic rate and induces browning of white adipose tissue. Chronic morphine upregulates TRPM8 channels, and promising results have been presented on menthol use in the management of some withdrawal symptoms of morphine abuse tapering.

Mini-dictionary of terms

Menthol: Menthol is a natural compound, a cyclic terpene alcohol, found in oils of *Mentha* species plants. It is produced also synthetically. Menthol is a counterirritant at TRPM8 receptor, it first excites and subsequently desensitizes TRPM8 channels in pain pathways. Outcome is analgesic effect.

TRPM8: Transient receptor potential melastatin 8, TRPM8, is an excitatory ion channel receptor and a member of the TRP family. Menthol receptor is nonselective and polymodal ion channel which opens through chemical agonists, for example, menthol, cool and cold temperatures (8–28°C), and voltage.

Key facts of TRPM8 receptor and menthol

- TRPM8 (transient receptor potential melastatin 8) receptor, also known as a cool or menthol receptor, is temperature-sensitive receptor in the body.
- TRPM8 receptors are activated, e.g., by cool temperatures (15–28°C) and menthol.
- TRPM8 receptors are highly sensitive to nociceptive stimuli.
- 10%–15% of somatosensory neurons, mainly Aδ- and C-fibers, are activated by menthol.
- At moderate concentrations, <10%, menthol induces first a pleasant cool sensation and subsequently an analgesic effect.
- TRPM8 receptors are expressed also in blood vessels and activation of these receptors will induce vasodilatation. As the blood flow increases, this is beneficial in many pain states.
- Menthol has per se anti-inflammatory and antipruritic activity.
- Menthol is bacteriostatic, antiviral, and antifungal activity, and may thus be used in broken skin also.
- Menthol enhances transdermal permeation of other substances.
- Menthol is well tolerated and has no interaction potential, and thus, is a feasible nonpharmacological compound for pain management in several nociceptive, inflammatory, and neuropathic pain conditions.

Summary points

- Analgesic medicines have contraindications, interaction potential, risk of adverse effects, and abuse potential.
- In multimodal analgesia, interventions with different sites and mechanisms of action are combined to improve analgesia and patient satisfaction, enhance recovery, and reduce/eliminate opioid use and adverse effects.
- Nonpharmacological interventions are highly effective in many pain states.
- TRPM8 receptors are expressed in 10%–15% of somatosensory neurons, mainly Aδ- and C-fibers.
- Menthol is an agonist in TRPM8 receptor.
- At moderate concentrations, <10%, topical menthol has analgesic action, but at high concentrations, >40%, topical application induces cold and mechanical hyperalgesia.
- Topical menthol is a highly effective nonpharmacological intervention for pain management.
- TRPM8-mediated analgesia alone will produce a satisfactory analgesic effect in some pain states.
- In multimodal analgesia, menthol can be combined with other interventions and compounds with low risk of harmful interactions or adverse effects.

References

Airaksinen, O. A., Kyrklund, N., Latvala, K., Kouri, J. P., Grönblad, M., & Kolari, P. (2003). Efficacy of cold gel for soft tissue injuries: A prospective randomized double-blinded trial. *The American Journal of Sports Medicine*, *31*, 680–684.

Ashina, H., Porreca, F., Anderson, T., et al. (2019). Post-traumatic headache: Epidemiology and pathophysiological insights. *Nature Reviews. Neurology*, *15*, 607–617.

Banugo, B., & Amoako, D. (2017). Prehabilitation. *BJA Education*, *17*, 401–405.

Bardellini, E., Amadori, F., & Majorana, A. (2016). Oral hygiene grade and quality of life in children with chemotherapy-related oral mucositis: A randomized study on the impact of a fluoride toothpaste with salivary enzymes, essential oils, proteins and colostrum extract versus a fluoride toothpaste without menthol. *International Journal of Dental Hygiene*, *14*, 314–319.

Behrendt, H. J., Germann, T., Gillen, C., Hatt, H., & Jostock, R. (2004). Characterization of the mouse cold-menthol receptor TRPM8 and vanilloid receptor type-1 VR1 using a fluorometric imaging plate reader (FLIPR) assay. *British Journal of Pharmacology*, *141*, 737–745.

Black, E., Khor, K. E., Kennedy, D., et al. (2019). Medication use and pain management in pregnancy: A critical review. *Pain Practice*, *19*, 875–899.

Bleakley, C., McDonough, S., & MacAuley, D. (2004). The use of ice in the treatment of acute soft-tissue injury: A systematic review of randomized controlled trials. *The American Journal of Sports Medicine*, *32*, 251–261.

Borhani Haghighi, A., Motazedian, S., Rezaii, R., et al. (2010). Cutaneous application of menthol 10% solution as an abortive treatment of migraine without aura: A randomised, double-blind, placebo-controlled, crossed-over study. *International Journal of Clinical Practice*, *64*, 451–456.

Bromm, B., Scharein, E., Darsow, U., & Ring, J. (1995). Effects of menthol and cold on histamine-induced itch and skin reactions in man. *Neuroscience Letters*, *187*, 157–160.

Brunner, M., Davies, D., Martin, W., Leuratti, C., Lackner, E., & Müller, M. (2011). A new topical formulation enhances relative diclofenac bioavailability in healthy male subjects. *British Journal of Clinical Pharmacology*, *71*, 852–859.

Cheng, Q. Y., Yang, M. C., Wu, J., Jia, X. L., Xiao, C., Lian, T., et al. (2019). Reduced cardiac ischemia/reperfusion injury by hypothermic reperfusion via activation of transient receptor potential M8 channel. *Life Sciences*, *232*, 116658.

Colvin, L. A. (2019). Chemotherapy-induced peripheral neuropathy: Where are we now? *Pain*, *160*(Suppl. 1), S1–S10.

Cosens, D. J., & Manning, A. (1969). Abnormal electroretinogram from a Drosophila mutant. *Nature*, *224*, 285–287.

Craighead, D. H., McCartney, N. B., Tumlinson, J. H., & Alexander, L. M. (2017). Mechanisms and time course of menthol-induced cutaneous vasodilation. *Microvascular Research*, *110*, 43–47.

da Silva Serra, I., Husson, Z., Bartlett, J. D., & Smith, E. S. (2016). Characterization of cutaneous and articular sensory neurons. *Molecular Pain*, *12*. 1744806916636387.

Davies, S. J., Harding, L. M., & Baranowski, A. P. (2002). A novel treatment of postherpetic neuralgia using peppermint oil. *The Clinical Journal of Pain*, *18*, 200–202.

Drugs and Lactation Database (LactMed). (2006). *Peppermint*. Bethesda (MD): National Library of Medicine (US) (Internet]. [Read August 30, 2020).

Du, J., Yang, X., Zhang, L., & Zeng, Y. M. (2009). Expression of TRPM8 in the distal cerebrospinal fluid-contacting neurons in the brain mesencephalon of rats. *Cerebrospinal Fluid Research*, *6*, 3.

Dussor, G., & Cao, Y. Q. (2016). TRPM8 and migrane. *Headache*, *56*, 1406–1417.

EMA. (2020). *Assessment report on Mentha x piperita L., folium and aetheroleum*. https://www.ema.europa.eu/en/documents/herbal-report/assessment-report-mentha-x-piperita-l-folium-aetheroleum-revision-1_en.pdf (Read August 30, 2020).

Engelhard, D., Hofer, P., & Annaheim, S. (2019). Evaluation of the effect of cooling strategies on recovery after surgical intervention. *BMJ Open Sport & Exercise Medicine*, *5*, e000527.

Facer, P., Casula, M. A., Smith, G. D., et al. (2007). Differential expression of the capsaicin receptor TRPV1 and related novel receptors TRPV3, TRPV4 and TRPM8 in normal human tissues and changes in traumatic and diabetic neuropathy. *BMC Neurology*, *7*, 11.

Fallon, M. T., Storey, D. J., Krishan, A., et al. (2015). Cancer treatment-related neuropathic pain: Proof of concept study with menthol—A TRPM8 agonist. *Support Care Cancer*, *23*, 2769–2777.

GBD 2016 Disease and Injury Incidence and Prevalence Collaborators. (2017). Global, regional, and national incidence, prevalence, and years lived with disability for 328 diseases and injuries for 195 countries, 1990-2016: A systematic analysis for the global burden of disease study 2016. *Lancet*, *3900*, 1211–1259.

Glare, P., Aubrey, K. R., & Myles, P. S. (2019). Transition from acute to chronic pain after surgery. *Lancet*, *393*, 1537–1546.

Green, B. G., & McAuliffe, B. L. (2000). Menthol desensitization of capsaicin irritation. Evidence of a short-term anti-nociceptive effect. *Physiology & Behavior*, *68*, 631–639.

Härkänen, L., Kokki, M., Huopio, J., Sjövall, S., Auriola, S., Lehtonen, M., et al. (2018). Buprenorphine for pain management in osteoarthritis patients scheduled for arthroplasty—A randomized, double-blind, placebo controlled trial. *Austin Journal of Anesthesia and Analgesia*, *6*, 1065.

Harrington, A. M., Hughes, P. A., Martin, C. M., et al. (2011). A novel role for TRPM8 in visceral afferent function. *Pain*, *152*, 1459–1468.

Higashi, Y., Kiuchi, T., & Furuta, K. (2010). Efficacy and safety profile of a topical methyl salicylate and menthol patch in adult patients with mild to moderate muscle strain: A randomized, double-blind, parallel-group, placebo-controlled, multicenter study. *Clinical Therapeutics*, *32*, 34–43.

IASP. (2018). *IASP terminology*. International Association for the Study of Pain. https://www.iasp-pain.org/Education/Content.aspx?ItemNumber=1698#Peripheralneuropathicpain (Read August 30, 2020).

Johar, P., Grover, V., Topp, R., & Behm, D. G. (2012). A comparison of topical menthol to ice on pain, evoked tetanic and voluntary force during delayed onset muscle soreness. *International Journal of Sports Physical Therapy*, *7*, 314–322.

Johnson, C. D., Melanaphy, D., Purse, A., Stokesberry, S. A., Dickson, P., & Zholos, A. V. (2009). Transient receptor potential melastatin 8 channel involvement in the regulation of vascular tone. *American Journal of Physiology. Heart and Circulatory Physiology*, *296*, H1868–H1877.

Juergens, U. R., Stöber, M., & Vetter, H. (1998). The anti-inflammatory activity of L-menthol compared to mint oil in human monocytes in vitro: A novel perspective for its therapeutic use in inflammatory diseases. *European Journal of Medical Research*, *3*, 539–545.

Julius, D. (2013). TRP channels and pain. *Annual Review of Cell and Developmental Biology*, *29*, 355–384.

Keh, S. M., Facer, P., Yehia, A., Sandhu, G., Saleh, H. A., & Anand, P. (2011). The menthol and cold sensation receptor TRPM8 in normal human nasal mucosa and rhinitis. *Rhinology*, *49*, 453–457.

Keshavarzian, S., & Shahgholian, N. (2017). Comparison of the effect of topical application of rosemary and menthol for musculoskeletal pain in hemodialysis patients. *Iranian Journal of Nursing and Midwifery Research*, *22*, 436–441.

King, S., Chambers, C. T., Huguet, A., et al. (2011). The epidemiology of chronic pain in children and adolescents revisited: A systematic review. *Pain*, *152*, 2729–2738.

Kinnunen, M., Piirainen, P., Kokki, H., Lammi, P., & Kokki, M. (2019). Updated clinical pharmacokinetics and pharmacodynamics of oxycodone. *Clinical Pharmacokinetics*, *58*, 705–725.

Kraemer, W. J., Ratamess, N. A., Maresh, C. M., et al. (2005). Acetylated fatty acid topical cream with menthol reduces pain and improves functional performance in individuals with arthritis. *Journal of Strength and Conditioning Research*, *19*, 475–480.

Kuronen, M., Kokki, H., Nyyssönen, T., Savolainen, S., & Kokki, M. (2018). Life satisfaction and pain interference in spine surgery patients before and after surgery: Comparison between on-opioid and opioid-naïve patients. *Quality of Life Research*, *27*, 3013–3020.

Lahtinen, P., Kokki, H., & Hynynen, M. (2006). Pain after cardiac surgery: A prospective cohort study of one-year incidence and intensity. *Anesthesiology*, *105*, 794–800.

Liu, B., Fan, L., Balakrishna, S., Sui, A., Morris, J. B., & Jordt, S. E. (2013). TRPM8 is the principal mediator of menthol-induced analgesia of acute and inflammatory pain. *Pain*, *154*, 2169–2177.

Ludwiczuk, A., Kieltyka-Dadasiewiczb, A., Sawicki, R., Golusd, J., & Ginalskad, G. (2016). Essential oils of some Mentha species and cultivars, their chemistry and bacteriostatic activity. *Natural Product Communications*, *11*, 1015–1018.

Ma, S., Yu, H., Zhao, Z., Luo, Z., Chen, J., Ni, Y., et al. (2012). Activation of the cold-sensing TRPM8 channel triggers UCP1-dependent thermogenesis and prevents obesity. *Journal of Molecular Cell Biology*, *4*, 88–96.

MacDonald, A. D. (1879). Menthol: An anti-neuralgic. *Edinburgh Medical Journal*, *25*, 286–287.

Mäntyselkä, P. T., Turunen, J. H., Ahonen, R. S., & Kumpusalo, E. A. (2003). Chronic pain and poor self-rated health. *Journal of the American Medical Association*, *290*, 2435–2442.

McAlindon, T. E., LaValley, M. P., Harvey, W. F., et al. (2017). Effect of intra-articular triamcinolone vs saline on knee cartilage volume and pain in patients with knee osteoarthritis: A randomized clinical trial. *Journal of the American Medical Association*, *317*, 1967–1975.

McKemy, D., Neuhausser, W., & Julius, D. (2002). Identification of a cold receptor reveals a general role for TRP channels in thermosensation. *Nature*, *416*, 52–58.

Mustafa, S., & Oriowo, M. (2005). Cooling-induced contraction of the rat gastric fundus: Mediation via transient receptor potential (TRP) cation channel TRPM8 receptor and rho-kinase activation. *Clinical and Experimental Pharmacology & Physiology*, *32*, 832–838.

OECD. (2004). *Organisation for economic co-operation and development*. SIDS.

Park, N. H., Na, Y. J., Choi, H. T., Cho, J. C., & Lee, H. K. (2013). Activation of transient receptor potential melastatin 8 reduces ultraviolet B-induced prostaglandin E2 production in keratinocytes. *The Journal of Dermatology*, *40*, 919–922.

Peat, J., Frazee, C., Kearns, G., & Garg, U. (2016). Determination of menthol in plasma and urine by gas chromatography/mass spectrometry (GC/MS). *Methods in Molecular Biology*, *1383*, 205–211.

Peier, A. M., Moqrich, A., Hergarden, A. C., et al. (2002). A TRP channel that senses cold stimuli and menthol. *Cell*, *108*, 705–715.

Proudfoot, C. J., Garry, E. M., Cottrell, D. F., et al. (2006). Analgesia mediated by the TRPM8 cold receptor in chronic neuropathic pain. *Current Biology*, *16*, 1591–1605.

Ringer, S., & Sainsbury, H. (1897). *A handbook of therapeutics*. New York: William Wood and Company.

Rossato, M., Granzotto, M., Macchi, V., et al. (2014). Human white adipocytes express the cold receptor TRPM8 which activation induces UCP1 expression, mitochondrial activation and heat production. *Molecular and Cellular Endocrinology*, *383*, 137–146.

Sabnis, A. S., Shadid, M., Yost, G. S., & Reilly, C. A. (2008). Human lung epithelial cells express a functional cold-sensing TRPM8 variant. *American Journal of Respiratory Cell and Molecular Biology*, *39*, 466–474.

Savaş, S., İnal, E. E., Yavuz, D. D., Uslusoy, F., Altuntaş, S. H., & Aydın, M. A. (2018). Risk factors for complex regional pain syndrome in patients with surgically treated traumatic injuries attending hand therapy. *Journal of Hand Therapy*, *31*, 250–254.

Silva, H. (2020). Current knowledge on the vascular effects of menthol. *Frontiers in Physiology*, *11*, 298.

Stefanelli, L., Lockyer, E. J., Collins, B. W., et al. (2019). Delayed-onset muscle soreness and topical analgesic alter corticospinal excitability of the biceps brachii. *Medicine and Science in Sports and Exercise*, *51*, 2344–2356.

Stein, R. J., Santos, S., Nagatomi, J., et al. (2004). Cool (TRPM8) and hot (TRPV1) receptors in the bladder and male genital tract. *The Journal of Urology*, *172*, 1175–1178.

Suchodolski, J., Feder-Kubis, J., & Krasowska, A. (2017). Antifungal activity of ionic liquids based on (−)-menthol: A mechanism study. *Microbiological Research*, *197*, 56–64.

Sundstrup, E., Jakobsen, M. D., Brandt, M., et al. (2014). Acute effect of topical menthol on chronic pain in slaughterhouse workers with carpal tunnel syndrome: Triple-blind, randomized placebo-controlled trial. *Rehabilitation Research and Practice*, 310913.

Taylor, D. J. R., Hamid, S. M., Andres, A. M., et al. (2020). Antiviral effects of menthol on coxsackievirus B. *Viruses*, *12*, 373.

Taylor, R., Jr., Gan, T. J., Raffa, R. B., et al. (2012). A randomized, double-blind comparison shows the addition of oxygenated glycerol triesters to topical mentholated cream for the treatment of acute musculoskeletal pain demonstrates incremental benefit over time. *Pain Practice*, *12*, 610–619.

Tazawa, K., Ikeda, H., Kawashima, N., & Okiji, T. (2017). Transient receptor potential melastatin (TRPM) 8 is expressed in freshly isolated native human odontoblasts. *Archives of Oral Biology*, *75*, 55–61.

Topp, R., Brosky, J. A., Jr., & Pieschel, D. (2013). The effect of either topical menthol or a placebo on functioning and knee pain among patients with knee OA. *Journal of Geriatric Physical Therapy (2001)*, *36*, 92–99.

Tsavaler, L., Shapero, M. H., Morkowski, S., & Laus, R. (2001). Trp-p8, a novel prostate-specific gene, is up-regulated in prostate cancer and other malignancies and shares high homology with transient receptor potential calcium channel proteins. *Cancer Research*, *61*, 3760–3769.

Turunen, V., Peräniemi, S., Tolvanen, K., Kokki, M., Ruponen, M., & Kokki, H. (2019). The permeability of magnesium across the skin is enhanced by menthol cream. *Magnesium Research*, *32*, 118–123.

Wasner, G., Naleschinski, D., Binder, A., Schattschneider, J., McLachlan, E. M., & Baron, R. (2008). The effect of menthol on cold allodynia in patients with neuropathic pain. *Pain Medicine*, *9*, 354–358.

Wright, A. (1870). Oil of peppermint as a local anaesthetic. *Lancet*, *2464*, 726.

Xing, H., Ling, J. X., Chen, M., et al. (2008). TRPM8 mechanism of autonomic nerve response to cold in respiratory airway. *Molecular Pain*, *4*, 22.

Zhang, J., Enix, D., Snyder, B., Giggey, K., & Tepe, R. (2008). Effects of biofreeze and chiropractic adjustments on acute low back pain: A pilot study. *Journal of Chiropractic Medicine*, *7*, 59–65.

Zhang, L., Jones, S., Brody, K., Costa, M., & Brookes, S. J. (2004). Thermosensitive transient receptor potential channels in vagal afferent neurons of the mouse. *American Journal of Physiology. Gastrointestinal and Liver Physiology*, *286*, G983–G991.

Further reading

Pan, F., & Jones, G. (2018). Clinical perspective on pain and pain phenotypes in osteoarthritis. *Current Rheumatology Reports*, *20*, 79.

UNEP Publications. (2004). *Initial assessment report on menthols*. http://www.chem.unep.ch/irptc/sids/OECDSIDS/MENTHOLS.pdf. Available at:[Read August 30, 2020].

Part II

Physiology, imaging and physical recordings

Chapter 23

Anesthetic and proconvulsant effects of ketamine on EEG

Shaila Gowda[a] and Charles Akos Szabó[b]

[a]*Department of Neurology, Baylor College of Medicine, Houston, TX, United States,* [b]*Department of Neurology, UT Health, San Antonio, TX, United States*

Abbreviations

cIED	interictal epileptiform discharges
EEG	electroencephalography
GABA	γ-aminobutyric acid
NMDAR	*N*-methyl-D-aspartic acid receptor
PFA	paroxysmal fast activity

Introduction

Ketamine was introduced into clinical practice more than 50 years ago with the purpose of delivering anesthesia and analgesia to people undergoing brief surgical procedures. It is a unique medication as it simultaneously provides hypnosis, analgesia, and amnesia. It can cause bronchodilation and minimal respiratory depression making it a desirable choice for people to avoid intubation during short surgical procedures (Haas & Harper, 1992; Werner, Reeker, Engelhard, Lu, & Kochs, 1997). Ketamine also has a short half-life allowing people to recover quickly, though many experience an emergence phenomenon characterized by euphoria, illusions, hallucinations, and delirium. It is also known to induce seizures in humans predisposed to epilepsy. With introduction of other IV anesthetics and given its proconvulsant and emergence effects, and its abuse potential ketamine became unpopular (Domino & Luby, 2012; Kohrs & Durieux, 1998). However, recent insights into its neuroprotective, antiinflammatory effects, anesthetic mechanisms of action and its expanded use as a treatment for depression and superrefractory status have allowed ketamine to regain its popularity and it continues to remain a subject of immense interest. Over the years, numerous studies including human and animal models have attempted to analyze EEG recordings to understand the influence of anesthetics on neural structure, to describe their unique properties, and demonstrate the dose-effect relationship to quantify the hypnotic component (Shafer, 2004). This chapter will review the anesthetic and proconvulsant effects of ketamine on EEG and compare it with other anesthetics.

Ketamine acts mainly through *N*-methyl-D-aspartic acid receptor (NMDAR) blockade. The NMDAR are located at the cortical, thalamic, limbic, and spinal levels and activated by glutamate which is the most abundant excitatory neurotransmitter in the CNS. Ketamine binds to the PCP binding site in NMDAR and inhibits glutamate activation of the channel in a noncompetitive manner accounting for the majority of its anesthetic and other effects (Chizh, 2007; Lodge, Anis, & Burton, 1982). This blockade is time, dose, and frequency dependent (MacDonald, Miljkovic, & Pennefather, 1987). Unlike many other anesthetics, it does not enhance the effects of inhibitory neurotransmitter γ-aminobutyric acid (GABA) (Antkowiak, 1999). With inhibition of NMDA receptors, a decrease in neuronal output may be expected due to increase in hyperpolarization. However, the release of GABA is prevented by blockade of tonically active channels on inhibitory interneurons in specific regions thereby disinhibiting the downstream excitatory interneurons (Monteggia & Zarate Jr, 2015). Interestingly, ketamine also causes a region-specific increase or decrease in cerebral activity (Cavazzuti, Porro, Biral, Benassi, & Barbieri, 1987; Crosby, Crane, & Sokoloff, 1982) attributed to its action on NMDAR which are ionotropic glutamate receptors allowing calcium influx on synaptic depolarization (Thomson, West, & Lodge, 1985). Its long-term effects via enhancement of non-NMDA glutamatergic receptors such as AMPA and dopaminergic receptors are likely responsible for its beneficial and adverse effects. It is a racemic mixture consisting of two optical enantiomers (Fig. 1) R(−) and S(+);

The Neurobiology, Physiology, and Psychology of Pain. https://doi.org/10.1016/B978-0-12-820589-1.00023-3

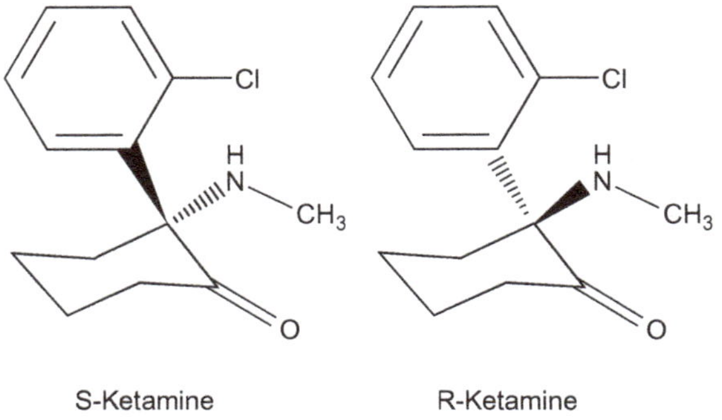

FIG. 1 Structure of ketamine isomers.

S+ isomer being at least 3–4 times more potent than the R-isomer (Sinner & Graf, 2008) due to its higher affinity to PCP binding sites on NMDAR and has beneficial properties such as rapid emergence from anesthesia (Himmelseher & Pfenninger, 1998).

Ketamine effects as anesthetic

It is intriguing that anesthetics can produce a state of brain inactivity and simultaneous hyperexcitability. The anesthetic effects of ketamine are described as dose-dependent CNS depression. At low doses of 1–2 mg/kg, IV or 4–6 mg/kg intramuscularly ketamine produces a dissociative state and at higher doses it produces a state of unconsciousness (Modica, Tempelhoff, & White, 1990). Dissociative anesthesia is characterized by analgesia with changes in perception and vigilance. It is described as a trance-like state, associated eye opening and nystagmus, during which people are unconscious, amnesic, and deeply analgesic. The dissociative state is ascribed to inhibition of thalamocortical pathways with excitation of limbic system (Domino, Chodoff, & Corssen, 1965). However, during emergence from anesthesia, patients have experienced psychomimetic reactions which include hallucinations and vivid dreams which can be unpleasant (Domino, 2010).

Research studies in humans are limited due to increased associated risks with general anesthesia. However, EEG recordings have provided a means to correlate the neurophysiological changes encountered during the routine clinical use with current knowledge of neural mechanisms. The EEG bandwidths recorded during routine studies are shown in Table 1. The EEG changes induced by anesthesia are measured with anesthetic monitors presented as a numerical value after complex processing of the EEG signals for depth of anesthesia assessment. While a whole set of EEG electrode montage recording, and real-time interpretation provides more detailed information, this often not feasible. Routine implementation of EEG recording during clinical anesthesia is confined to the use of predominantly frontocentral channels only.

Schwartz et al. (Schwartz, Virden, & Scott, 1974) studied nine subjects without history of epilepsy. They were given 2 mg/kg of ketamine in two doses at 10-min intervals. Soon after ketamine administration 30–40 Hz, low amplitude paroxysmal fast activity (PFA) emerged frontally on scalp EEG, followed by moderate voltage theta and evolving into intermittent high amplitude delta activity. These EEG changes recurred cyclically at 3–4 s intervals. The frontal PFA persisted even after appearance of the theta and delta activity. A similar pattern was seen with the second injection, however, with quicker onset and shorter interval. No epileptiform discharges were seen. This frontally predominant PFA activity was felt characteristic of CNS stimulant effect of ketamine anesthesia. Higher doses of ketamine (>2 mg/kg) produce a burst

TABLE 1 EEG bandwidth with frequencies.

Brainwave type	Frequency (Hz)	Physiological state
Gamma	30–100	Concentration/higher mental activity
Beta	14–30	Awake/alert
Alpha	8–13	Relaxed wakefulness
Theta	4–7	Drowsiness
Delta	0.5–4	Sleep

suppression pattern. Earlier studies in cats induced with ketamine also reported high voltage beta activity followed by slow waves coined as "ketamine waves" (Celesia, Chen, & Bamforth, 1975) which had a combination of polyspike-and-wave complexes or beta-delta waves. The fast activity alternating with ketamine waves was replaced by burst suppression pattern at higher doses. Varying EEG patterns have been attributed to differences in the two optical isomers of ketamine when used to induce clinical anesthesia, (R) isomer not capable of producing the same degree of EEG changes as (S) isomer.

Ketamine effects as proconvulsant

Ketamine is also known to have proconvulsant properties causing seizures in human and animals with h/o epilepsy and also those without epilepsy when used for sedation (Ferrer-Allado, Brechner, Dymond, Cozen, & Crandall, 1973). An epileptic seizure is characterized by episodes of excessive, abnormal enhanced synchrony in neuronal activity which may originate in one area (focal seizure) or more extensively involve both hemispheres (generalized seizures). The spread of the activity to bilateral hemispheres and or brainstem structures can result in loss of tone and consciousness. On the other hand, myoclonus, described as arrhythmic or rhythmic muscular contractions, can be a nonepileptic phenomenon arising from the brainstem structures or spinal cord (Swanson, Luttrell, & Magladery, 1962). Anesthetic drug-induced clinical seizures or seizure-like activity occur rarely. During normal sleep, spindle oscillation occurs characterized by bursts of 11–14 Hz potentials lasting 1–3 s followed by period of quiescence, generated by GABAergic thalamic reticular neurons transmitted to the thalamocortical relay (TC) neurons (ref). Clinical seizures or seizure-like phenomena have been noted during or soon after induction and during recovery from anesthesia (Voss, Ludbrook, Grant, Sleigh, & Barnard, 2006; Voss, Sleigh, Barnard, & Kirsch, 2008). Synchronous patterns typically occur during sleep and can be detected on the EEG as slow cortical oscillations (<1 Hz) and delta activity (<4 Hz). Similar patterns have been noted during induction and maintenance of anesthesia (Voss et al., 2008). In the nonepileptic patients, soon after induction and with increased dose of ketamine administration at doses of 2 mg/kg/IV or 10–12 mg IM, myoclonic and seizure-like convulsive movements have been reported, felt to be subcortical in origin. This was demonstrated in normal cats chronically implanted with depth electrodes (Bennett, Madsen, Jordan, & Wiser, 1973; Schwartz et al., 1974). EEG recordings showed an intermittent hypersynchronous spiking in limbic system with spread to subcortical and cortical regions which clinically correlated with excitation, catalepsy, and bizarre posturing of cats.

Ketamine is well known to activate epileptogenic foci in people with epilepsy. One of the proposed mechanisms for development of anesthesia-induced seizures is the increased neuronal synchrony in thalamocortical neurons both during anesthesia and sleep. Ferrar-Allado et al. (Ferrer-Allado et al., 1973) in nine patients monitored with intracranial electrodes demonstrated seizure activity arising from the limbic and thalamic structures after intravenous ketamine administration at low doses of 2–4 mg/kg. Subcortical seizure activity was seen even at lower intravenous doses of ketamine (0.1 mg/kg), without any loss of consciousness and with the faster frequency activity similar to EEG pattern demonstrated in people without epilepsy.

On scalp EEG, however, low-dose ketamine (0.5–2 mg/kg, IV) administered in epileptic patients did not show reveal EEG changes on surface EEG nor any clinical seizure activity (Celesia et al., 1975), while higher doses (4–10 mg/kg, IM followed by 1–20 mg/kg, IV, in divided doses) produced clinical convulsions and intermittent bursts of epileptic discharges (Kayama & Iwama, 1972; Mori, Kawamata, Mitani, Yamazaki, & Fujita, 1971). Dose-dependent effects of ketamine were also noted after single intravenous doses of ketamine in patients undergoing temporal lobe epilepsy surgery (Ferrer-Allado et al., 1973). At doses of 0.5 mg/kg, focal IEDs were generated without clinically rendering patients unconscious but 2–4 mg/kg produced subclinical and clinical seizures within 2 min of administration.

EEG analysis of ketamine effects

Understanding the mechanisms of ketamine-induced anesthetic action at neural level and functional connectivity has been a subject of immense interest with use of EEG analysis in anesthesia literature. A disruption of fronto-parietal network connectivity was proposed as a mechanism to explain the state of unconsciousness induced by ketamine in awake patients at subanesthetic doses (Engelhardt, Stahl, Marouche, Hartung, & Dierks, 1994; Lee et al., 2013; Muthukumaraswamy et al., 2015). Akeju et al. attempted to characterize EEG biomarkers associated with ketamine-induced anesthesia (Schultz, Schultz, Zacher, & Pichlmayr, 1990). They analyzed EEG changes with spectral and coherence analyses in 12 people receiving ketamine for induction of anesthesia. They demonstrated a gamma burst pattern consisting of alternating slow delta (0.1–4 Hz) and gamma (27–40 Hz) frequency oscillations, generated by thalamocortical activity reflecting anesthetic effect of ketamine. Similar findings of increased gamma frequency oscillations were noted in other studies (Blain-Moraes,

Lee, Ku, Noh, & Mashour, 2014; Engelhardt et al., 1994; Ferrer-Allado et al., 1973; Schultz et al., 1990; Schwartz et al., 1974). This was ascribed to ketamine's more effective blockade of fast spiking cortical interneurons compared to the pyramidal neurons resulting in downregulation of interneuronal activity with decrease of GABA-related reduction of inhibitory control over pyramidal neurons (Homayoun & Moghaddam, 2007; Seamans, 2008). The disinhibited pyramidal neurons likely produce the gamma oscillations noted on the scalp EEG. In the absence of gamma oscillations, theta frequency oscillations have also been reported, usually when lower doses of ketamine are administered for analgesia or treatment of depression (Kochs, Scharein, Mollenberg, Bromm, & Am Esch, 1996; Muthukumaraswamy et al., 2015). Neurophysiologic correlates and functional connectivity during ketamine at subanesthetic, anesthetic, and recovery phase was further characterized using spectral and connectivity analyses of high-density EEG (Vlisides et al., 2017). At subanesthetic doses, the spectral power decreased from alpha to theta frequencies, but with maintained alpha anterior-to-posterior connectivity. At anesthetic doses, spectral power in the theta bandwidth was increased, along with anterior-to-posterior theta connectivity. At increased doses, the theta frequency activity was replaced by higher gamma power frontally, with reduction of alpha anterior-to-posterior connectivity. They concluded increased theta power correlating with the depth of ketamine anesthesia, similar to the earlier studies of ketamine's pharmacologic effects (Domino et al., 1965). The anterior-to-posterior connectivity was maximally depressed in the alpha bandwidth suggesting a dose-dependent relationship between connectivity and loss of consciousness.

Scalp vs intracranial EEG effects of ketamine in nonhuman primate model of an idiopathic generalized epilepsy

One of the advantages to study ketamine effects in a natural animal model of epilepsy is the ability to evaluate its epileptogenic effects. The epileptic baboon exhibits an idiopathic generalized epilepsy, with spontaneous generalized tonic-clonic seizures as well as absence and myoclonic seizures (Szabó et al., 2005). In particular, ketamine's proconvulsant effects were utilized to characterize epilepsy risk in baboon cohorts and pedigrees housed at the Southwest National Primate Research Center, located in San Antonio, Texas. The other benefit is the ability to compare scalp and intracranial EEG recordings in this model.

The acute effects of intramuscularly administered ketamine (5–6 mg/kg) differed clinically, and by scalp EEG, between baboons with epilepsy and otherwise asymptomatic baboons (Szabó et al., 2005; Szabó, Knape, Leland, & Williams, 2013). Ketamine activated interictal epileptic discharges (IEDs) and clinical seizures within 3 min of the injection. Initially, the baboons exhibited intermittent tonic posturing of their heads, tongue, trunk, extremities last a few seconds followed by eyelid myoclonus or myoclonic jerk in epileptic baboons and a just brief loss of tone in nonepileptic controls. Scalp EEG showed an increase in beta frequency activity of 20–40 Hz with an amplitude of 10–40 mV during the tonic phase, interrupted intermittent generalized high amplitude 1–2 Hz slow ("ketamine") waves during the eyelid myoclonus or atonia; in epileptic baboons, the ketamine waves were often associated with generalized spikes or polyspikes. In later phases, during which the tonic activity decreased, the baboons exhibited vertical nystagmus of 4–6 Hz. The early activation of IEDs were attributed to the use of ketamine's known dose-dependent effects similar to prior studies (Schwartz et al., 1974).

Scalp (Sc) and intracranial (Ic) EEGs were further compared in three epileptic baboons within 10 min of low dose of ketamine administration (Gowda & Szabó, 2019). Sc-EEG showed generalized IED and produced clinically apparent seizures consisting of myoclonic seizures (MS) and generalized tonic-clonic seizure similar to prior studies (Szabó et al., 2005, 2013). Earliest EEG changes noted at 30 s in Sc-EEG and in 15 s in Ic-EEG consisted of PFA of 30–40 Hz frequency. Thirty seconds later, the Ic-EEG 30 s showed slow activity first emerging occipitally and then spreading to frontocentral and orbitofrontal regions. After 2 min, the PFA became more frontally predominant and interrupted by ketamine waves occurring every 3–6 s. Clinically, the baboons exhibited increased muscle tone for 10–15 s correlating with fast activity and brief loss of tone during ketamine waves. A dissociated EEG background between frontocentral and parieto-occipital regions was seen in the Ic-EEG within 2 min. Even while frontocentral fast activity persisted, the parieto-occipital regions showed evolution of burst suppression pattern followed by background suppression. The study highlighted initial changes consisted of slowing and IEDs occurred predominantly in the parieto-occipital and that there was an occipital to frontal propagation of EEG changes suggesting a vulnerability of occipital lobe to ketamine. Early changes in the visual cortices are thought to be due to the relatively low concentration of NMDARs in (Geodes, Cooper, Cotman, Patel, & Meldrum, 1989), leading to their rapid saturation by ketamine. Regional cortical NMDAR concentrations were speculated to mediate both the proconvulsant and anesthetic effects of ketamine.

Effects of other agents on EEG

The main targets of anesthetics are $GABA_A$-receptor, NMDAR, two-pore-domain K+ channels, and several other ion channels and receptors (Alkire, Hudetz, & Tononi, 2008; Franks, 2008). Unlike ketamine, volatile anesthetics, barbiturates, and propofol act by potentiating inhibitory $GABA_A$-receptors. The EEG changes with these agents show decrease in amplitude and increase in frequency of the background activity at low doses, followed by decreases in frequency and increases in amplitude, and appearance of an alpha pattern, at higher doses. Spindle waves occur only as bursts during normal sleep but occur continuously during anesthesia induced by propofol and volatile anesthetics. As the level of anesthesia deepens, theta and delta slowing give way to a burst suppression pattern, which eventually is replaced by a background suppression pattern. Anesthesia shifts the alpha frequency activity generated by the occipital lobes during wakefulness to the frontal regions, characterized as "anteriorization" associated with anesthetics potentiating thalamocortical hypersynchrony due $GABA_A$-mediated effects (Feshchenko, Veselis, & Reinsel, 2004; Mackenzie, Pope, & Willoughby, 2004; Steriade, Gloor, Llinas, Da Silva, & Mesulam, 1990). While EEGs recordings of slow induction and recovery phase from anesthesia with propofol demonstrate maintained symmetry (Doufas, Bakhshandeh, Bjorksten, Shafer, & Sessler, 2004), administration of ketamine following propofol or other ether-based volatile anesthetics has resulted in a shift from alpha to beta or gamma frequencies (Hayashi, Tsuda, Sawa, & Hagihira, 2007). Table 2 summarizes the effects of different anesthetics on EEG, their epileptogenic potential, and mechanisms for mediating proconvulsant actions. The proconvulsant properties have also been noted with volatile agents, opioids, etomidate, propofol, and methohexital. In the opioid group, meperidine is known to have the strongest association with myoclonic and tonic-clonic seizures. Low-to-moderate doses of other agents such as alfentanil, fentanyl, sufentanil, and morphine also have been reported to produce generalized seizures particularly with intrathecal administration. Etomidate produces EEG effects similar to barbiturates. Compared to propofol, people receiving etomidate that were on thiopental and methohexital exhibited more myoclonic and tonic movements. Etomidate enhances the epileptiform activity in patients with focal epilepsy, similar to ketamine at lower doses, followed by suppression with anesthetic doses. Epileptogenic properties of enflurane and etomidate have been used to facilitate in identification of epileptic foci in epilepsy surgery. On the contrary, nitrous oxide when used in conjunction with inhalational anesthetics resulted in attenuation of spikes on electrocorticogram (ECoG) in epileptic patients.

In conclusion, it is evident that specific EEG patterns are associated with particular anesthetic agents, but that the patterns undergo similar changes with increasing doses of the anesthetic. EEG changes associated with lower doses of ketamine, consist of fast activity noted predominantly frontally alternating with ketamine waves, and characteristic of a dissociative state, while the gamma burst pattern evolving into a burst suppression pattern at higher doses, are characteristic for deeper levels of ketamine anesthesia. Specific EEG patterns are associated with particular anesthetic agents, but that the patterns undergo similar changes with increasing doses of the anesthetic. With anesthetics as a group, there is a higher propensity to produce epileptiform discharges than clinical seizures.

TABLE 2 Effects of different anesthetics on EEG.

	Concentration				
Agent	Low	Moderate	High	Epileptic activity	Clinical seizure
Propofol	Increase in fast activity (15–40 Hz)	Gradual increase in theta, delta waves	Burst suppression → Background suppression	+ d/c at low dose Suppressant at high doses	+ during induction and emergence at low doses
PE—Rapid changes of propofol concentration in the brain at the beginning or end of anesthesia, glycine antagonism and specific $GABA_A$ receptor agonism appear to mediate the epileptogenic potential of propofol (San-juan, Chiappa, & Cole, 2010)					
Barbiturates Thiopental Methohexital	Increase in fast activity (15–30 Hz)	Gradual increase in theta, delta waves	Burst Suppression → Background suppression	+ Spiking methohexital in low dose (Loddenkemper, Möddel, Schuele, Wyllie, & Morris III, 2007) Suppresses at high dose	+ Myoclonus, opisthotonus, generalized sz in patients with epilepsy Myoclonus in nonepileptic patients

Continued

TABLE 2 Effects of different anesthetics on EEG—cont'd

Agent	Concentration: Low	Moderate	High	Epileptic activity	Clinical seizure
PE—Thiopental < Methohexital 1 direct excitatory effects on epileptic neurons (Aasly, Blom, Silfvenius, & Zetterlund, 1984)					
Benzodiazepine Diazepam/ lorazepam	Increase in fast activity (12–22 Hz)	Gradual increase in theta, delta waves	Theta/delta slowing	Suppress the d/c	+ Brief episodes of sz reported in LGS (Diazepam)
PE—Increased neuronal hyperexcitability, altered neuronal function					
Etomidate	Increase in alpha amplitude (lack of fast activity)	Theta-delta slowing	Burst suppression	High voltage spikes on EEG with no clinical activity	+ Myoclonic movements with no EEG changes +sz in epileptics
PE—Excessive activation of GABA ergic B2. And effect on KCC2 co-transporter, chloride dominating $GABA_A$ influx currents, and excitatory response to GABA (Hoyer, Kranaster, Janke, & Sartorius, 2014)					
Inhalational agents					
Nitrous oxide	Increase in fast activity Decrease in alpha	Evolution into theta waves 6–8 Hz with short intervals of fast activity (12–16 Hz)	Theta waves 4–7 Hz	Suppression of epileptic activity	+ sz in animal models +human volunteers (hyperbaric N_2O)
Sevoflurane, isoflurane, desflurane, halothane, enflurane, isoflurane	Increase in frequency (12–18 Hz) decrease in amplitude	4–8 Hz theta waves with increase in amplitude	Background suppression MAC > 1.5	Sevoflurane—Increased epileptic activity with MAC >1.5 55 Halothane—Brief, polyspike activity Enflurane- +epileptic d/c in low doses and suppresses at high doses	Sevoflurane- +sz in epileptic patients Desflurane- + transient myoclonus Enflurane- +sz in epileptic + Myoclonic movements in nonepileptic patient
PE—Enflurane > > Sevoflurane > Isoflurane = Desflurane 56 Ability to destabilize the cortex and delay inhibitory circuit (Voss et al., 2006, 2008)					
Opioids	Decrease in frequency, increase in amplitude	4–8 Hz theta slowing	Delta 1–4 Hz slowing	Enhance EEG activity in epileptic patients Remi and alfentanil—used to induce spike activity	+ Myoclonus and generalized seizures in epileptic patients
PE—Neuronal excitation is mediated via selective stimulation of κ and μ opioid receptors in a dose-dependent manner, decreased GABA ergic interneuronal inhibition of pyramidal neurons and inhibition of hyperpolarization-activated K^+ currents (Saboory et al., 2007)					
Local anesthetics	No clear reported EEG changes	Theta slowing	Theta-delta slowing → Background suppression	+ d/c in epileptic patients 59 Suppressant at low doses	+ Seizures in epileptic patients (Kopp, Wynd, Horlocker, Hebl, & Wilson, 2009)
PE—higher doses have the ability to rapidly cross the BBB & decrease seizure threshold in hippocampus, cortex, amygdala (Greensmith & Murray, 2006)					

sz, seizure; *d/c*, discharge; *MAC*, mean alveolar concentration; *PE*, proconvulsant effect; *BBB*, blood-brain barrier; *LGS*, Lennox-Gastaut syndrome.

Key facts of anesthetic and proconvulsant effects of ketamine on EEG

- Ketamine is a dissociative anesthetic that acts by NMDAR receptor blockade with unique side effects including emergence phenomenon.
- It is a proconvulsant at low doses and can cause seizures in humans predisposed to epilepsy.
- At low doses of ketamine, EEG changes consist of frontally predominant paroxysmal fast activity alternating with ketamine waves evolving into burst suppression pattern at higher doses.
- Intracranial studies support initial changes of burst suppression pattern and interictal epileptiform discharges occurring in the parieto-occipital regions suggesting initial anesthetic and proconvulsant effects occurring predominantly in occipital lobes attributed to the regional NMDAR concentrations.

Summary points

- Ketamine is a dissociative anesthetic that acts mainly through noncompetitive *N*-methyl-D-aspartic acid receptor (NMDAR) blockade.
- Unlike many other anesthetics, ketamine does not enhance the effects of inhibitory neurotransmitter γ-aminobutyric acid (GABA).
- Ketamine is known to have proconvulsant properties causing seizures in human and animals with h/o epilepsy and also those without epilepsy when used for sedation.
- EEG changes associated with lower doses of ketamine consist of fast activity noted predominantly frontally alternating with ketamine waves, and characteristic of a dissociative state. The gamma burst pattern evolving into a burst suppression pattern at higher doses is characteristic for deeper levels of ketamine anesthesia.
- Intracranial studies support initial changes of burst suppression pattern and interictal epileptiform discharges occurring in the parieto-occipital regions suggesting a vulnerability of occipital lobes to ketamine.
- Regional NMDAR concentrations are speculated to mediate both the proconvulsant and anesthetic effects of ketamine.
- Specific EEG patterns are associated with particular anesthetic agents, but that the patterns undergo similar changes with increasing doses of the anesthetic.

References

Aasly, J., Blom, S., Silfvenius, H., & Zetterlund, B. (1984). Effects of amobarbital and methohexital on epileptic activity in mesial temporal structures in epileptic patients. An EEG study with depth electrodes. *Acta Neurologica Scandinavica*, *70*(6), 423–431.

Alkire, M. T., Hudetz, A. G., & Tononi, G. (2008). Consciousness and anesthesia. *Science*, *322*(5903), 876–880.

Antkowiak, B. (1999). Different actions of general anesthetics on the firing patterns of neocortical neurons mediated by the GABAAReceptor. *Anesthesiology: The Journal of the American Society of Anesthesiologists*, *91*(2), 500–511.

Bennett, D. R., Madsen, J. A., Jordan, W. S., & Wiser, W. C. (1973). Ketamine anesthesia in brain-damaged epileptics: Electroencephalographic and clinical observations. *Neurology*, *23*(5), 449.

Blain-Moraes, S., Lee, U., Ku, S., Noh, G., & Mashour, G. A. (2014). Electroencephalographic effects of ketamine on power, cross-frequency coupling, and connectivity in the alpha bandwidth. *Frontiers in Systems Neuroscience*, *8*, 114.

Cavazzuti, M., Porro, C. A., Biral, G. P., Benassi, C., & Barbieri, G. C. (1987). Ketamine effects on local cerebral blood flow and metabolism in the rat. *Journal of Cerebral Blood Flow & Metabolism*, *7*(6), 806–811.

Celesia, G. G., Chen, R.-C., & Bamforth, B. J. (1975). Effects of ketamine in epilepsy. *Neurology*, *25*(2), 169.

Chizh, B. A. (2007). Low dose ketamine: A therapeutic and research tool to explore N-methyl-D-aspartate (NMDA) receptor-mediated plasticity in pain pathways. *Journal of Psychopharmacology*, *21*(3), 259–271.

Crosby, G., Crane, A., & Sokoloff, L. (1982). Local changes in cerebral glucose utilization during ketamine anesthesia. *Anesthesiology: The Journal of the American Society of Anesthesiologists*, *56*(6), 437–443.

Domino, E. F. (2010). Taming the ketamine tiger. *Anesthesiology: The Journal of the American Society of Anesthesiologists*, *113*(3), 678–684.

Domino, E. F., Chodoff, P., & Corssen, G. (1965). Pharmacologic effects of CI-581, a new dissociative anesthetic, in man. *Clinical Pharmacology & Therapeutics*, *6*(3), 279–291.

Domino, E. F., & Luby, E. D. (2012). Phencyclidine/schizophrenia: One view toward the past, the other to the future. *Schizophrenia Bulletin*, *38*(5), 914–919.

Doufas, A. G., Bakhshandeh, M., Bjorksten, A. R., Shafer, S. L., & Sessler, D. I. (2004). Induction speed is not a determinant of propofol pharmacodynamics. *Anesthesiology: The Journal of the American Society of Anesthesiologists*, *101*(5), 1112–1121.

Engelhardt, W., Stahl, K., Marouche, A., Hartung, E., & Dierks, T. (1994). Ketamine racemate versus S-(+)-ketamine with or without antagonism with physostigmine. A quantitative EEG study on volunteers. *Der Anaesthesist*, *43*, S76–S82.

Ferrer-Allado, T., Brechner, V. L., Dymond, A., Cozen, H., & Crandall, P. (1973). Ketamine-induced electroconvulsive phenomena in the human limbic and thalamic regions. *Anesthesiology: The Journal of the American Society of Anesthesiologists*, *38*(4), 333–344.

Feshchenko, V. A., Veselis, R. A., & Reinsel, R. A. (2004). Propofol-induced alpha rhythm. *Neuropsychobiology*, *50*(3), 257–266.

Franks, N. P. (2008). General anaesthesia: From molecular targets to neuronal pathways of sleep and arousal. *Nature Reviews Neuroscience*, *9*(5), 370–386.

Geodes, J., Cooper, S., Cotman, C., Patel, S., & Meldrum, B. (1989). N-methyl-d-aspartate receptors in the cortex and hippocampus of baboon (Papio anubis and Papio papio). *Neuroscience*, *32*(1), 39–47.

Gowda, S., & Szabó, C.Á. (2019). Effects of ketamine on EEG in baboons with genetic generalized epilepsy. *Epilepsy Research*, *154*, 50–54.

Greensmith, J. E., & Murray, W. B. (2006). Complications of regional anesthesia. *Current Opinion in Anesthesiology*, *19*(5), 531–537.

Haas, D. A., & Harper, D. G. (1992). Ketamine: A review of its pharmacologic properties and use in ambulatory anesthesia. *Anesthesia Progress*, *39*(3), 61.

Hayashi, K., Tsuda, N., Sawa, T., & Hagihira, S. (2007). Ketamine increases the frequency of electroencephalographic bicoherence peak on the α spindle area induced with propofol. *British Journal of Anaesthesia*, *99*(3), 389–395.

Himmelseher, S., & Pfenninger, E. (1998). The clinical use of S-(+)-ketamine—A determination of its place. *Anasthesiologie, Intensivmedizin, Notfallmedizin, Schmerztherapie*, *33*(12), 764.

Homayoun, H., & Moghaddam, B. (2007). NMDA receptor hypofunction produces opposite effects on prefrontal cortex interneurons and pyramidal neurons. *Journal of Neuroscience*, *27*(43), 11496–11500.

Hoyer, C., Kranaster, L., Janke, C., & Sartorius, A. (2014). Impact of the anesthetic agents ketamine, etomidate, thiopental, and propofol on seizure parameters and seizure quality in electroconvulsive therapy: A retrospective study. *European Archives of Psychiatry and Clinical Neuroscience*, *264*(3), 255–261.

Kayama, Y., & Iwama, K. (1972). The EEG, evoked potentials, and single-unit activity during ketamine anesthesia in cats. *Anesthesiology: The Journal of the American Society of Anesthesiologists*, *36*(4), 316–328.

Kochs, E., Scharein, E., Mollenberg, O., Bromm, B., & Am Esch, J. S. (1996). Analgesic efficacy of low-dose ketamine somatosensory-evoked responses in relation to subjective pain ratings. *Anesthesiology: The Journal of the American Society of Anesthesiologists*, *85*(2), 304–314.

Kohrs, R., & Durieux, M. E. (1998). Ketamine: Teaching an old drug new tricks. *Anesthesia & Analgesia*, *87*(5), 1186–1193.

Kopp, S. L., Wynd, K. P., Horlocker, T. T., Hebl, J. R., & Wilson, J. L. (2009). Regional blockade in patients with a history of a seizure disorder. *Anesthesia & Analgesia*, *109*(1), 272–278.

Lee, U., Ku, S., Noh, G., Baek, S., Choi, B., & Mashour, G. A. (2013). Disruption of frontal–parietal communication by ketamine, propofol, and sevoflurane. *Anesthesiology: The Journal of the American Society of Anesthesiologists*, *118*(6), 1264–1275.

Loddenkemper, T., Möddel, G., Schuele, S. U., Wyllie, E., & Morris, H. H., III. (2007). Seizures during intracarotid methohexital and amobarbital testing. *Epilepsy & Behavior*, *10*(1), 49–54.

Lodge, D., Anis, N., & Burton, N. (1982). Effects of optical isomers of ketamine on excitation of cat and rat spinal neurones by amino acids and acetylcholine. *Neuroscience Letters*, *29*(3), 281–286.

MacDonald, J., Miljkovic, Z., & Pennefather, P. (1987). Use-dependent block of excitatory amino acid currents in cultured neurons by ketamine. *Journal of Neurophysiology*, *58*(2), 251–266.

Mackenzie, L., Pope, K. J., & Willoughby, J. O. (2004). Physiological and pathological spindling phenomena have similar regional EEG power distributions. *Brain Research*, *1008*(1), 92–106.

Modica, P. A., Tempelhoff, R., & White, P. F. (1990). Pro-and anticonvulsant effects of anesthetics (part I). *Anesthesia and Analgesia*, *70*(3), 303–315.

Monteggia, L. M., & Zarate, C., Jr. (2015). Antidepressant actions of ketamine: From molecular mechanisms to clinical practice. *Current Opinion in Neurobiology*, *30*, 139–143.

Mori, K., Kawamata, M., Mitani, H., Yamazaki, Y., & Fujita, M. (1971). A neurophysiologic study of ketamine anesthesia in the cat. *Anesthesiology: The Journal of the American Society of Anesthesiologists*, *35*(4), 373–382.

Muthukumaraswamy, S. D., Shaw, A. D., Jackson, L. E., Hall, J., Moran, R., & Saxena, N. (2015). Evidence that subanesthetic doses of ketamine cause sustained disruptions of NMDA and AMPA-mediated frontoparietal connectivity in humans. *Journal of Neuroscience*, *35*(33), 11694–11706.

Saboory, E., Derchansky, M., Ismaili, M., Jahromi, S. S., Brull, R., Carlen, P. L., et al. (2007). Mechanisms of morphine enhancement of spontaneous seizure activity. *Anesthesia & Analgesia*, *105*(6), 1729–1735.

San-juan, D., Chiappa, K. H., & Cole, A. J. (2010). Propofol and the electroencephalogram. *Clinical Neurophysiology*, *121*(7), 998–1006.

Schultz, A., Schultz, B., Zachen, B., & Pichlmayr, I. (1990). The effects of ketamine on the electroencephalogram- -typical patterns and spectral representations. *Der Anaesthesist*, *39*(4), 222.

Schwartz, M. S., Virden, S., & Scott, D. (1974). Effects of ketamine on the electroencephalograph. *Anaesthesia*, *29*(2), 135–140.

Seamans, J. (2008). Losing inhibition with ketamine. *Nature Chemical Biology*, *4*(2), 91–93.

Shafer, S. L. (2004). Shock values. *Anesthesiology: The Journal of the American Society of Anesthesiologists*, *101*(3), 567–568.

Sinner, B., & Graf, B. (2008). *Ketamine. Modern anesthetics* (pp. 313–333). Springer.

Steriade, M., Gloor, P., Llinas, R. R., Da Silva, F. L., & Mesulam, M.-M. (1990). Basic mechanisms of cerebral rhythmic activities. *Electroencephalography and Clinical Neurophysiology*, *76*(6), 481–508.

Swanson, P. D., Luttrell, C. N., & Magladery, J. W. (1962). Myoclonus—A report of 67 cases and review of the literature. *Medicine*, *41*(4), 339.

Szabó, C.Á., Knape, K. D., Leland, M. M., & Williams, J. T. (2013). Electroclinical phenotypes in a pedigreed baboon colony. *Epilepsy Research*, *105*(1–2), 77–85.

Szabó, C.Á., Leland, M. M., Knape, K., Elliott, J. J., Haines, V., & Williams, J. T. (2005). Clinical and EEG phenotypes of epilepsy in the baboon (*Papio hamadryas* spp.). *Epilepsy Research*, *65*(1–2), 71–80.

Thomson, A. M., West, D. C., & Lodge, D. (1985). An N-methylaspartate receptor-mediated synapse in rat cerebral cortex: A site of action of ketamine? *Nature*, *313*(6002), 479–481.

Vlisides, P. E., Bel-Bahar, T., Lee, U., Li, D., Kim, H., Janke, E., et al. (2017). Neurophysiologic correlates of ketamine sedation and anesthesia. A high-density electroencephalography study in healthy volunteers. *Anesthesiology: The Journal of the American Society of Anesthesiologists*, *127*(1), 58–69.

Voss, L., Ludbrook, G., Grant, C., Sleigh, J., & Barnard, J. (2006). Cerebral cortical effects of desflurane in sheep: Comparison with isoflurane, sevoflurane and enflurane. *Acta Anaesthesiologica Scandinavica*, *50*(3), 313–319.

Voss, L. J., Sleigh, J. W., Barnard, J. P., & Kirsch, H. E. (2008). The howling cortex: Seizures and general anesthetic drugs. *Anesthesia & Analgesia*, *107*(5), 1689–1703.

Werner, C., Reeker, W., Engelhard, K., Lu, H., & Kochs, E. (1997). Ketamine racemate and S-(+)-ketamine. Cerebrovascular effects and neuroprotection following focal ischemia. *Der Anaesthesist*, *46*, S55.

Chapter 24

Electroencephalography and anesthetic doses of ketamine

Logan Voss[a] and Jamie Sleigh[b]

[a]*Department of Anesthesia, Waikato District Health Board, Hamilton, New Zealand;* [b]*Department of Anesthesia, Waikato Clinical Campus, University of Auckland, Hamilton, New Zealand*

Abbreviations

ECoG electrocorticogram
EEG electroencephalogram
GABA gamma-aminobutyric acid
HCN1 hyperpolarization-activated cyclic nucleotide-gated
LOBR loss of behavioral response
NMDA *N*-methyl-D-aspartate
PKPD pharmacokinetic-pharmacodynamic

Introduction

Ketamine has a history dating back to the 1960s and was first trialed as an anesthetic in humans in 1965 (Domino, Chodoff, & Corssen, 1965). Even before its approval for human use by the US Food and Drug Administration in 1970 (Li & Vlisides, 2016), the EEG effects of high doses of ketamine were being investigated in animal models (Miyasaka & Domino, 1968). The first experiments conducted in cats showed that the anesthetized state was characterized by alternating periods of thalamocortical delta wave activity and bursts of cortical high-frequency activity. These results were replicated a few years later in cats and monkeys (Celesia & Chen, 1974). Early on it was noted that these EEG patterns are substantially different from those seen with other classes of anesthetic drugs. Because it is an outlier, understanding the complex effects of ketamine has always been a challenge to the conventional paradigms used to understand the mechanisms of other anesthetic drugs.

Similar EEG changes to those observed in the early animals experiments have now been well described for anesthetic concentrations in humans. A typical example in Fig. 1 shows runs of higher frequency oscillations (in this case around 7 Hz—an example in the red box) separated every few seconds by very large slow delta waves (an example in the blue box). It is of note that the slow waves are commonly over 200 μV in peak to trough amplitude—much larger than those generally seen with propofol anesthesia (usually around 50–100 μV).

Spectral EEG changes

Analyzing the spectral content of the EEG is a convenient way of quantifying how cortical oscillatory activity tracks with time, in response to drug dosing. The time-varying raw EEG waveform is deconstructed into its component frequencies, usually by the Fast Fourier Transformation, to produce a spectrogram of power (amplitude squared) across a continuous frequency range. By convention, the spectral content is reported as power within commonly accepted frequency bands: delta (up to 4 Hz); theta (5–8 Hz); alpha (9–15 Hz); beta (16–32 Hz); and gamma (>32 Hz). This frequency-domain analysis is the cornerstone of EEG-based anesthetic depth monitors, which detect the shift to lower frequencies with agents such as propofol and volatile anesthetics.

The Neurobiology, Physiology, and Psychology of Pain. https://doi.org/10.1016/B978-0-12-820589-1.00024-5

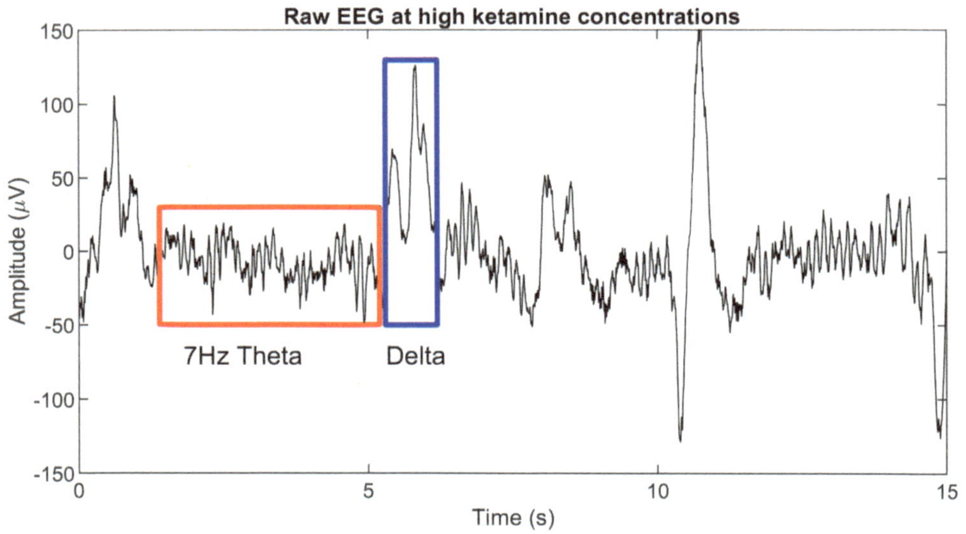

FIG. 1 Raw EEG at high ketamine concentrations. EEG from unresponsive subject anesthetized with ketamine.

Ketamine's unique set of spectral changes are sufficiently distinct from the so-called gamma-aminobutyric acid (GABA)-ergic anesthetics, to trick commercial anesthetic depth monitors into reporting a state of wakefulness when subjects are behaviorally unresponsive. Numerous studies have shown that ketamine induces broad high-frequency activation in the beta-gamma bands, strong theta activation and medio-frontal slow waves in the delta range (Akeju, Song, Hamilos, et al., 2016; Schuttler, Stanski, White, et al., 1987; Sleigh, Pullon, Vlisides, & Warnaby, 2019; Vlisides, Bel-Bahar, Lee, et al., 2017). As described previously (Fig. 1), the high-frequency component occurs as bursts alternating with slow-delta waves (Akeju et al., 2016), which on the spectrogram resembles the burst-suppression pattern seen at deep levels of anesthesia with GABAergic agents. Indeed, the delta-gamma burst pattern predominates at higher ketamine concentrations. As plasma drug levels wain after a bolus dose or infusion, there is progressive loss of the slow-delta component and a shift to predominance of the more stable beta-gamma pattern (Akeju et al., 2016). Schuttler et al. (Schuttler et al., 1987) report a three-stage progression in EEG changes that is consistent with this picture, namely: stage one, alpha power loss (see also (Blain-Moraes, Lee, Ku, Noh, & Mashour, 2014)); stage two, theta and gamma activation; and stage three, superimposed delta and gamma activity. An earlier study identified a similar stepwise progression in EEG changes with increasing concentration: from gamma activation, to a theta dominated waveform and finally, episodic delta waves (Schwartz, Virden, & Scott, 1974). This typical progression is shown in Fig. 2. We see an initial increase in the theta and gamma bands at around 180 s as the estimated brain ketamine concentrations rise. This is followed by a large increase in slow-delta at 200 s, close to the point at which the subject loses responsiveness.

The dose-dependent effects of ketamine on the EEG are important to consider, because they can be related to behavioral endpoints such as loss of consciousness. Typically half of patients are unresponsive when ketamine concentrations are above around 1–2 μg/mL. Because of the complex effects that ketamine has on consciousness, it is an open question as to whether loss of *responsiveness* under ketamine corresponds closely to true loss of *consciousness*. Patients may stop responding to commands because they have a disconnected consciousness, hallucinations and intense analgesia; rather than being truly unaware of their existence. There is a school of thought that true unconsciousness is never achieved, even with huge doses of ketamine.

Few studies have reported specifically on the spectral EEG changes before and after the transition from subhypnotic to hypnotic concentrations. However, there is general agreement that the former is characterized by a pattern of activity consistent with Schuttler's stage-two definition above; predominantly reflecting medial-frontal theta and spatially broad high frequency activity, without significant delta waves (McMillan, Sumner, Forsyth, et al., 2020; Muthukumaraswamy et al., 2015; Vlisides et al., 2017). Dose-response effects can also be approached from the perspective of pharmacokinetic-pharmacodynamic (PKPD) modeling. Typically, PKPD modeling is utilized in pharmacological studies to estimate the time constant describing the movement of a drug from the plasma to the effect-site. For anesthetic drugs, because the effect-site is the brain, the technique involves correlating some predetermined change in the EEG with increasing and decreasing plasma drug concentrations (up to some defined behavioral endpoint). The closer the EEG measure reflects plasma drug levels, the stronger the link to behavioral effect. In this way, PKPD modeling can be used to probe which

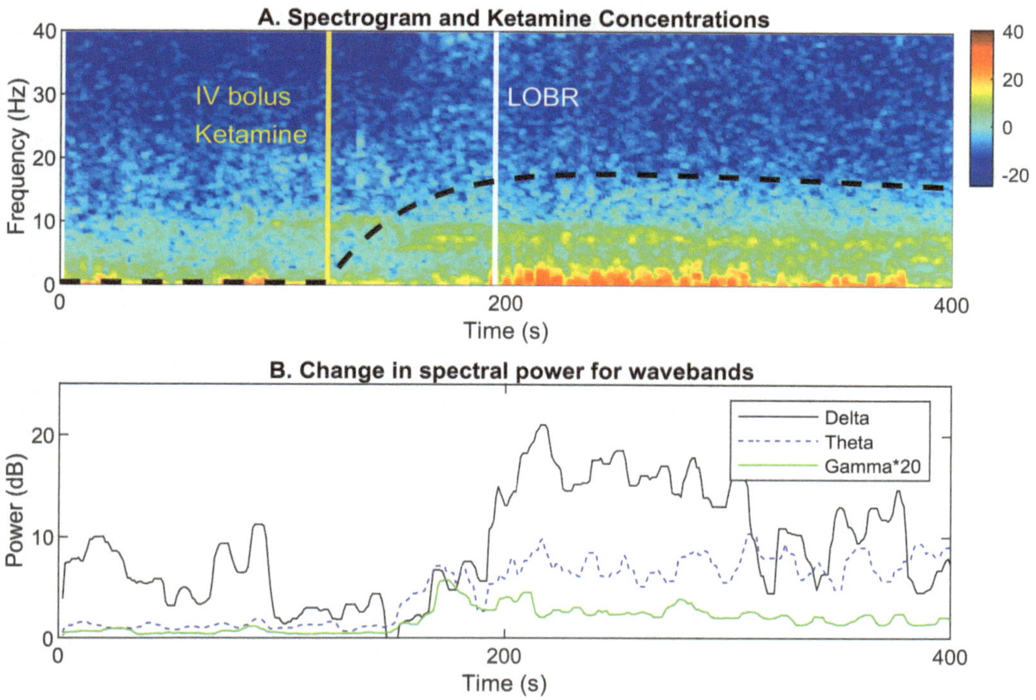

FIG. 2 Changes in EEG spectral power with ketamine anesthesia. An example EEG spectrogram in a subject given 1.5 mg/kg ketamine bolus. *LOBR*, loss of behavioral response. The *dashed black line* in (A) is the estimated brain concentration of ketamine, after the IV induction bolus given at the point of the *vertical solid yellow line*. Power is in decibels *(colorbar)*. In (B), the gamma power is multiplied by a factor of 20, for display purposes. *(Data from Sleigh, J., Pullon, R. M., Vlisides, P. E., Warnaby, C. E. (2019). Electroencephalographic slow wave dynamics and loss of behavioural responsiveness induced by ketamine in human volunteers.* British Journal of Anaesthesia, 123(5), *592–600. https://doi.org/10.1016/j.bja.2019.07.021.)*

EEG parameters correlate most closely with loss of behavioral responsiveness. Of the studies that have investigated the PKPD effects of ketamine, up to loss of consciousness, it is interesting to note that all but one incorporated, either directly or indirectly, a measure of slow wave dominance in the EEG (Schuttler et al., 1987; Sleigh et al., 2019; Voss, Ludbrook, Grant, Upton, & Sleigh, 2007). This suggests that slow wave activity in the EEG is a reasonable spectral indicator of ketamine anesthesia. The exception to these studies is a rodent investigation that quantified very high-frequency oscillations (130–160 Hz range) in the electrocorticogram (ECoG) (Flores, Ching, Hartnack, et al., 2015). These findings are not readily comparable to studies recording the scalp EEG, which is blind to these higher frequencies due to the low-pass filtering effects of the skull. ECoG recordings do, however, enable very precise quantification of cortical activity. In Fig. 3, we see that ketamine concentrations correlate very nicely with the increasing percentage time that the ECoG is in the slow wave pattern. This relationship was used to effectively model brain effects in a ketamine PKPD study (Voss et al., 2007).

Emergence from ketamine anesthesia is not a mirror image of induction. The slow wave EEG activity is lost long before the patients start responding to verbal command. This is presumably because the patient is saturated in ketamine and is almost always in a state of disconnected/dreamlike consciousness for quite some time, as the drug levels drift slowly down from around 200 ng/mL to subpsychomimetic levels of <50 ng/mL. This slow elimination time is driven by the high lipid solubility, and hence volume of distribution, of ketamine.

Connectivity metrics

Spectral analysis of individual EEG channels provides information about the oscillatory activity at specific cortical locations. Neuronal populations, however, do not function in isolation from each other. Coordinated information dissemination between brain regions is the basis for cognition and perception (Bressler & Kelso, 2016; Siegel, Donner, & Engel, 2012)—and oscillatory activity in particular has been hypothesized as the basis by which dynamic information flow occurs (Siegel et al., 2012). This can be quantified statistically under the umbrella of functional connectivity metrics, which capture statistical dependence in signals recorded from spatially distant regions. There is a large repertoire of such metrics and an in-depth discussion on the advantages, disadvantages, and limitations of each is beyond the scope of this review. There are excellent tutorials and reviews elsewhere which cover these details (Bastos & Schoffelen, 2015). However, it is important to bear these in mind when interpreting and comparing functional connectivity findings from different studies.

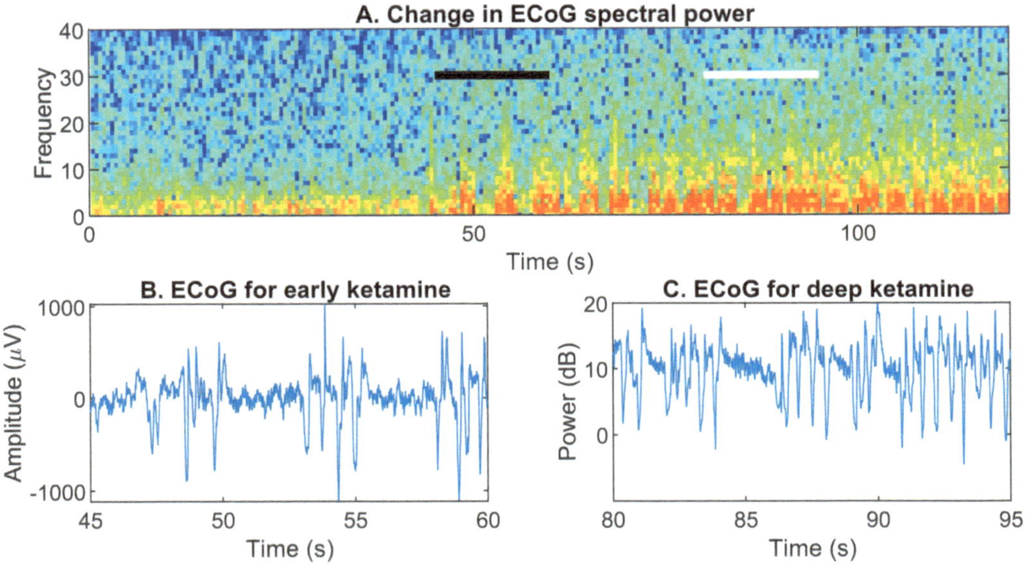

FIG. 3 Changes in spectral power and raw EEG at deepening levels of ketamine anesthesia. An example of ECoG from sheep data. (B) Shows a segment of the ECoG time series from the region of the black line in (A), just as the ketamine brain concentration was rising; and (C) from the region of the *white line* at high concentrations of ketamine. *(Data from Voss, L. J, Ludbrook, G., Grant, C., Upton, R., Sleigh, J. W. (2007). A comparison of pharmacokinetic/pharmacodynamic versus mass-balance measurement of brain concentrations of intravenous anesthetics in sheep.* Anesthesia and Analgesia, 104, *1440–1446.)*

The effect of ketamine on cortical functional connectivity is of interest. In the last decade, there has been burgeoning interest in the so-called "cortical connectivity" hypothesis of anesthetic action (reviewed in Hudetz, 2012). This is a potentially unifying hypothesis to explain the action of different groups of anesthetic drugs which have widely differing molecular mechanisms of action. In essence, it states that anesthetics cause unconsciousness by reducing feedback information flow from the frontal to parietal cortices, usually measured in the alpha frequency band. While the effect of ketamine on functional connectivity has not been widely investigated in human studies, some observations and comparisons can be drawn. At hypnotic ketamine concentrations, a decrease in anteroposterior alpha and increase in posteroanterior theta connectivity has been reported in humans, using the directed phase lag index (Blain-Moraes et al., 2014; Vlisides et al., 2017) and symbolic transfer entropy (Lee, Arbor, Ku, et al., 2013). Interestingly, in one of these studies, prefrontal coherence remained unchanged despite a reduction in directed phase lag index (Blain-Moraes et al., 2014). This would suggest that actual information flow (estimated by the directed phase lag index) is more important than regional brain synchrony (estimated by the coherence). A similar reduction in anteroposterior feedback in the alpha band has also been reported for hypnotic levels of propofol (Jordan, Ilg, Riedl, et al., 2013; Lee, Mashour, Noh, Kim, & Lee, 2013) and sevoflurane (Lee, Arbor, et al., 2013); which are anesthetics that act quite differently, primarily through potentiating GABA at $GABA_A$ receptors (Lee, Mashour, et al., 2013). These findings hint at the possibility of an agent-independent marker of loss of consciousness based on fronto-parietal connectivity metrics. Changes in functional connectivity at subhypnotic ketamine levels have been less well studied, but are equally important for testing the specificity of a putative marker of hypnosis. There is some disagreement here. While no directional changes in alpha were noted in one study (Vlisides et al., 2017), a reduction in fronto-parietal alpha connectivity using dynamic causal modeling has been observed (Muthukumaraswamy et al., 2015). Teasing out discrepancies such as these is not straightforward, given the sheer number of technically complex measures available, their sometimes idiosyncratic implementation and the tendency toward over-interpretation (Bastos & Schoffelen, 2015).

Neurophysiological underpinnings of the EEG effects of ketamine

Heuristic descriptions of how drugs affect EEG patterns are almost meaningless unless these patterns are linked to underlying changes in neurobiology. At present this is difficult, because of our very incomplete understanding of even the basic principles of brain function. How ketamine works as an anesthetic remains an open question in the field of anesthesiology. Conventional wisdom says that its NMDA blocking action is central to causing its anesthetic and analgesic effects, but this is simplistic at best. Recent work on a group of experimental ketamine analogs shows no correlation between the anesthetic potency of these compounds and their affinity for NMDA binding (Dimitrov et al., 2019). Assuming the mechanism of action of the analog compounds is similar to that of ketamine (and behavioral studies support this (Harvey, Sleigh, Voss, et al., 2015a)), the role of NMDA antagonism must be seriously questioned. In support of the idea that NMDA blockade is not necessary for ketamine anesthesia, MK-801 a very specific, and very potent NMDA antagonist, does not have anesthetic properties (Kelland, Soltis, Boldry, & Walters, 1993). Importantly, ketamine has a raft of other neurophysiological effects (Sleigh, Harvey, Voss, & Denny, 2014; Voss, Karalus, Englund, & Sleigh, 2018) and any one or combination of these may be relevant to this discussion.

Can we correlate changes in the EEG pattern with the neurophysiological effects underpinning ketamine's action—and does this tell us anything about how ketamine works as an anesthetic? As we have discussed, one of the earliest changes in the EEG is an increase in high-frequency power in the gamma frequency band. This generally occurs before loss of behavioral responsiveness (Shadli et al., 2018; Sleigh et al., 2019) and is therefore unlikely to be directly related to this behavioral endpoint. Ketamine preferentially inhibits NMDA at the axon initial segment on parvalbumin-containing interneurons (Homayoun & Moghaddam, 2007; Seamans, 2008). The resultant disinhibition of pyramidal cell populations, along with an increase in prefrontal cholinergic tone (Howe, Gritton, Lusk, et al., 2017; Kikuchi, Wang, Shinbori, Sato, & Okumura, 1997), likely contributes to the gamma band EEG excitation (Akeju et al., 2016). The behavioral correlate is likely to be the well-described subhypnotic psychomimetic effects of ketamine, including schizophrenic-like behaviors (Frohlich & Van Horn, 2014; Lahti, Weiler, Tamara Michaelidis, Parwani, & Tamminga, 2001). Schizophrenia itself has been linked to NMDA hypofunction causing glutamatergic disinhibition—and positive schizophrenic symptoms are related to increased gamma-band activity (Frohlich & Van Horn, 2014). Consistent with this picture, the psychomimetic side effects of ketamine appear to be minimized in the ketamine analogs which are void of NMDA action (Dimitrov et al., 2019).

If NMDA is not the primary target for ketamine's anesthesia action, what are the other likely alternatives? Cortically expressed hyperpolarization-activated cyclic nucleotide-gated (HCN) cation channels are a recognized ketamine target (Zhou et al., 2013). Animal experiments using gene knockout technologies suggest that 30%–50% of ketamine's hypnotic effect could be accounted for by HCN1-mediated block of hyperpolarization-activated cation currents (I_h) (Chen, Shu, & Bayliss, 2009; Zhou et al., 2013). Importantly, HCN1 block hyperpolarizes cortical pyramidal neurons, which could be linked to the appearance of cortically generated slow wave activity keen during ketamine anesthesia (Amzica & Steriade, 1998). HCN1 block has also been modeled to reproduce many of the ketamine-induced spectral changes in the EEG (Bojak, Day, & Liley, 2013). The thalamus and the cerebral cortex are usually highly coupled. Thus slow wave EEG activity can also be of thalamocortical origin, occurring when thalamocortical cells are hyperpolarized below the spindle generation zone of −55 to −65 mV (Nunez, Curro Dossi, Contreras, & Steriade, 1992). This mechanism has been proposed for both GABAergic anesthetics (Hagihira, 2015) and for ketamine (Akeju et al., 2016). Perhaps this is a more favorable explanation in light of early cat experiments showing that cortically recorded delta waves during ketamine anesthesia have a similar spatial distribution to thalamically generated spindles (Miyasaka & Domino, 1968). Furthermore, delta wave activity is still recordable in the thamalus after decortication (Miyasaka & Domino, 1968), implying a thalamic origin.

Interactions between ketamine and other hypnotic drugs

In modern clinical anesthetic practice, ketamine is commonly used as an adjunct on top of conventional opioid, and volatile- or propofol-based anesthesia, so as to minimize the deleterious effects of excessive intra- and postoperative opioid usage. It is unclear whether the addition of ketamine produces additive, antagonistic, or supraadditive interactions with the other drugs. This is because ketamine has some actions—such as increasing brain noradrenaline and acetyl choline—which will tend to depolarize the cortex and increase arousal (Garcia & Sleigh, 2017). Conversely, ketamine will also intensely reduce nociceptive arousal input. The question of the ketamine-GABAergic hypnotic interactions is of considerable clinical importance; because the safe use of propofol-based intravenous hypnosis is largely dependent on reliable EEG monitoring to titrate—and detect inadvertent problems with—intravenous infusion delivery. Most commercially available EEG monitoring systems are sensitive to high-frequency power in the EEG and therefore will tend to give a falsely high value with the addition of ketamine. Of course, this is assuming that the ketamine is not antagonistic to the underlying GABAergic hypnotic drug.

There are not any definitive large controlled studies describing how the EEG is changed with the addition of ketamine to conventional ether or propofol-based anesthesia. Small doses of ketamine have minimal effect (Faraoni, Salengros, Engelman, Ickx, & Barvais, 2009), but larger doses (perhaps >0.5 mg/kg) profoundly alter the usual intraoperative EEG, increasing common indices of hypnotic depth of anesthesia (BIS and spectral entropy) by about 15 points (Hans, Dewandre, Brichant, & Bonhomme, 2005). An example is shown in Fig. 4. As the ketamine concentrations increase above

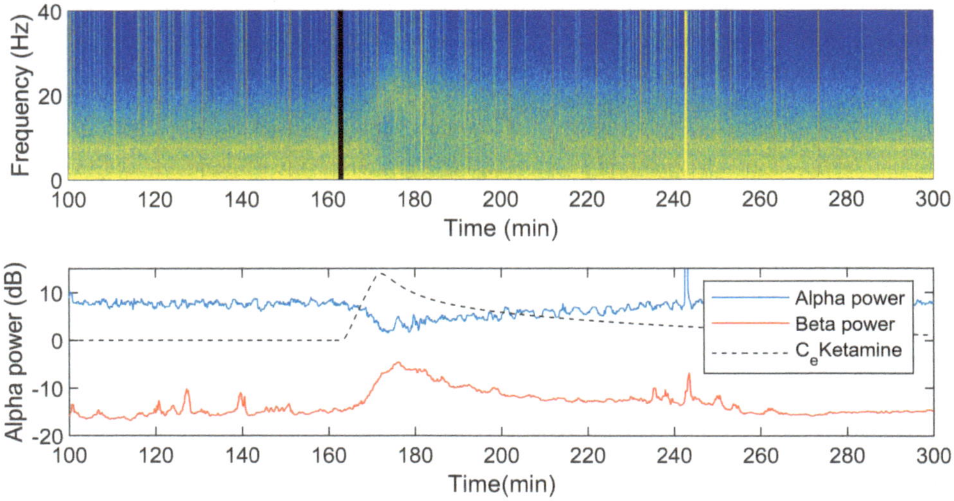

FIG. 4 Changes in EEG spectral power with ketamine interactions. A spectrogram of change in EEG frequencies caused by the infusion of 50 mg of ketamine over 10 min (between the *black* and *yellow bars*); on top of stable sevoflurane and fentanyl anesthesia. C_e*Ketamine*, estimated ketamine effect-site concentrations (ng/mL/5 for display purposes).

about 50 ng/mL, there is a simultaneous 10 dB increase in beta power, and comparable decrease in alpha power, that track the drug concentrations, with a slight delay. The underlying mechanisms have not been investigated thoroughly, but are probably related to cholinergic and aminergic thalamocortical depolarization effects as described previously.

Applications to other areas

The continuous EEG, as discussed in this chapter, records the spontaneous electrical potential activity of the cerebral cortex. A related EEG application involves recording central nervous system responses evoked by stimuli presented to one or other modality of the peripheral nervous system (for example, somatosensory, auditory, or visual). So-called "evoked potentials" are recorded in the EEG as low amplitude responses that can only be resolved by repeated time-locked averaging of multiple stimuli. In the field of anesthesiology, evoked response analysis has been used to investigate the effect of anesthetics on the transmission of sensory signals from the peripheral nervous system to the cerebral cortex. In particular, a reduction in "middle latency" responses (reflecting mainly primary sensory cortical activation) has been linked to adequate anesthesia (Matsushita, Oda, Otaki, Nakane, & Kawamae, 2015). This principle, applied to auditory stimuli, has been incorporated into a commercial depth of anesthesia monitor called the aepEX (auditory evoked potential index) (Cheung, Scoones, Stolker, & Weber, 2020; Matsushita et al., 2015). In keeping with its unique profile and unlike other anesthetics, ketamine has little effect on middle latency evoked responses generated by auditory (Schwender, Klasing, Madler, Poppel, & Peter, 1993) or somatosensory (Schubert, Licina, & Lineberry, 1990) stimuli. This suggests that ketamine affects higher order cortical processing downstream from primary sensory cortices—and may go some way to explaining its dissociative effects at higher concentrations. Interestingly, ketamine has been shown to depress aepEX at loss of consciousness (Matsushita et al., 2015). It is unclear whether the aepEX utilizes proprietary processing algorithms that could account for this apparent contradiction.

Other agents of interest

Other putative NMDA-antagonistic anesthetic drugs have quite different EEG signatures. Nitrous oxide has similar activation effects, but it is not possible to deliver nitrous oxide in high enough concentrations (unless given hyperbarically) to see the same patterns as high-dose ketamine. This gas also shows high-voltage delta activity on sudden changes in concentration. However, these waves are sinusoidal in character, and do not share a similar morphology to those of ketamine. Their significance is unclear. Xenon is also an NMDA blocker—but with a number of other actions—such as potassium channel blockade. It also shows some gamma activation, but unresponsiveness is marked by increased theta power. Experimental ketamine-ester analogs (Harvey et al., 2015b) (discussed above in relation to NMDA binding) have mixed EEG effects (Fig. 5). So-called "R1," the more strongly anesthetic analog, increases delta markedly, with lesser increases in

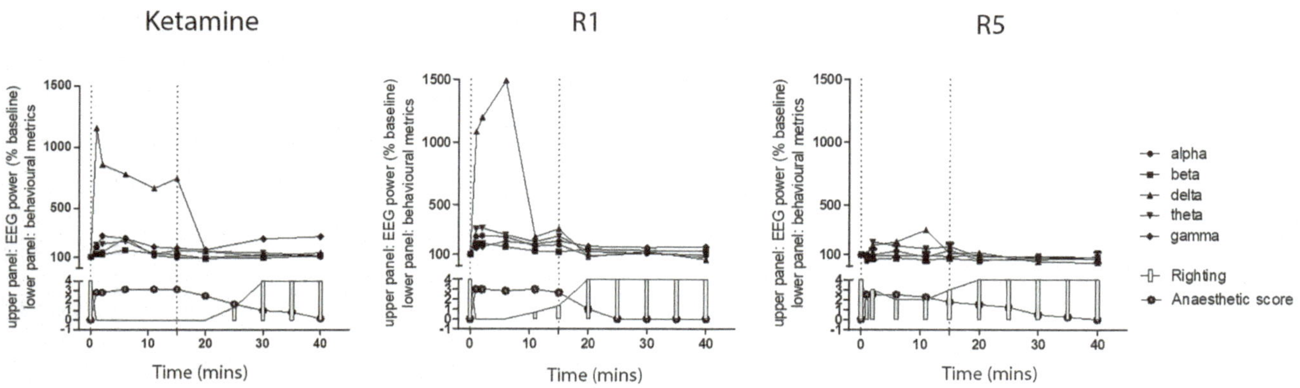

FIG. 5 Effect of ketamine analogs on the EEG power. Time course changes in EEG power recorded in rats anesthetized with ketamine and the experimental ketamine-ester analogs "R1" and "R5" (Harvey, Sleigh, Voss, et al., 2015b). The *dotted vertical lines* represent the period of drug delivery.

higher frequency bandwidths (alpha to gamma). These changes mirror those seen with ketamine. On the other hand, "R5," a more strongly analgesic analog, has very little effect on the EEG.

Mini dictionary of terms

Spectrogram. Graphical representation of the spectral content of an EEG recording, showing time-varying EEG power across the spectrum of frequencies.

Fast Fourier Transformation. Analysis algorithm that converts a time-varying signal into its component frequencies. The computed output is often visualized as a *spectrogram*.

Functional connectivity. A group of analysis metrics that capture statistical dependence in signals recorded from spatially distant regions, providing a measure of dynamic information flow.

Gamma-aminobutyric acid (GABA) receptor. The main inhibitory synaptic receptor in the brain, thought to be the target of most general anesthetics (other than ketamine).

N-Methyl-D-aspartate receptor (NMDA) receptor. The main excitatory synaptic receptor in the brain, thought to be one of the targets of ketamine.

Pharmacokinetic-pharmacodynamic (PKPD) modeling. A mathematical technique that combines pharmacokinetic and pharmacodynamic modeling, giving a description of the time course of drug effect against effect-site concentration.

Key facts

(1) The EEG in ketamine anesthesia is more complex, and unlike that seen with other anesthetic drugs—showing a mixture of simultaneous cortical activation and depression.
(2) Cortical activation by ketamine is probably due to NMDA-mediated disinhibition of pyramidal cell populations.
(3) The appearance of the slow wave component in the EEG is likely to be of thalamocortical origin and may be related to HCN1 antagonism.
(4) Ketamine's NMDA antagonism does not explain its anesthetic properties.
(5) When combined with other hypnotic drugs in general anesthesia, ketamine shows similar EEG changes to when it is used alone.

Summary points

(1) At subhypnotic concentrations, ketamine activates the cortex—as shown by increases in theta, beta, and gamma power, and decreases in alpha power, in the EEG.
(2) The transition to unresponsiveness at higher concentrations of ketamine is marked by the appearance of large slow waves and loss of fronto-parietal connectivity.
(3) Slow wave activity and loss of functional connectivity correlates with loss of behavioral responsiveness and suggests that these may be a common final pathway for detecting unresponsiveness among many classes of anesthetic agents. This has to be confirmed in larger studies.

(4) Loss of slow waves, as the ketamine concentrations fall, occurs some time before the return of wakefulness.
(5) Addition of ketamine to conventional volatile or propofol-based anesthesia—in doses above about 0.25 mg/kg—will reduce alpha power and increase high-frequency EEG power, which will cause an increase in the indices of commercially available monitors of anesthesia, making accurate interpretation of drug dosage difficult.

References

Akeju, O., Song, A. H., Hamilos, A. E., et al. (2016). Electroencephalogram signatures of ketamine anesthesia-induced unconsciousness. *Clinical Neurophysiology*, *127*(6), 2414–2422. https://doi.org/10.1016/j.clinph.2016.03.005.

Amzica, F., & Steriade, M. (1998). Electrophysiological correlates of sleep delta waves. *Electroencephalography and Clinical Neurophysiology*, *107*(2), 69–83.

Bastos, A. M., & Schoffelen, J.-M. (2015). A tutorial review of functional connectivity analysis methods and their interpretational pitfalls. *Frontiers in Systems Neuroscience*, *9*, 175. https://doi.org/10.3389/fnsys.2015.00175.

Blain-Moraes, S., Lee, U., Ku, S., Noh, G., & Mashour, G. A. (2014). Electroencephalographic effects of ketamine on power, cross-frequency coupling, and connectivity in the alpha bandwidth. *Frontiers in Systems Neuroscience*, *8*, 114. https://doi.org/10.3389/fnsys.2014.00114.

Bojak, I., Day, H. C., & Liley, D. T. J. (2013). Ketamine, Propofol, and the EEG: A neural field analysis of HCN1-mediated interactions. *Frontiers in Computational Neuroscience*, *7*, 22. https://doi.org/10.3389/fncom.2013.00022.

Bressler, S. L., & Kelso, J. A. S. (2016). Coordination dynamics in cognitive neuroscience. *Frontiers in Neuroscience*, *10*, 397. https://doi.org/10.3389/fnins.2016.00397.

Celesia, G. G., & Chen, R. C. (1974). Effects of ketamine on EEG activity in cats and monkeys. *Electroencephalography and Clinical Neurophysiology*, *37*(4), 353–354. https://doi.org/10.1016/0013-4694(74)90110-2.

Chen, X., Shu, S., & Bayliss, D. (2009). a. HCN1 channel subunits are a molecular substrate for hypnotic actions of ketamine. *The Journal of Neuroscience*, *29*(3), 600–609. https://doi.org/10.1523/JNEUROSCI.3481-08.2009.

Cheung, Y. M., Scoones, G. P., Stolker, R. J., & Weber, F. (2020). Monitoring depth of hypnosis: Mid-latency auditory evoked potentials derived aepEX in children receiving Desflurane-remifentanil anesthesia. *Anesthesia and Analgesia*, *130*(1), 194–200. https://doi.org/10.1213/ANE.0000000000003546.

Dimitrov, I. V., Harvey, M. G., Voss, L. J., Sleigh, J. W., Bickerdike, M. J., & Denny, W. A. (2019). Ketamine esters and amides as short-acting anaesthetics: Structure-activity relationships for the side-chain. *Bioorganic & Medicinal Chemistry*, *27*(7), 1226–1231. https://doi.org/10.1016/j.bmc.2019.02.010.

Domino, E. F., Chodoff, P., & Corssen, G. (1965). Pharmacologi effects of cl-581, a new dissociative anesthetic, in man. *Clinical Pharmacology and Therapeutics*, *6*, 279–291. https://doi.org/10.1002/cpt196563279.

Faraoni, D., Salengros, J.-C., Engelman, E., Ickx, B., & Barvais, L. (2009). Ketamine has no effect on bispectral index during stable propofol-remifentanil anaesthesia. *British Journal of Anaesthesia*, *102*(3), 336–339. https://doi.org/10.1093/bja/aen403.

Flores, F. J., Ching, S., Hartnack, K., et al. (2015). A PK-PD model of ketamine-induced high-frequency oscillations. *Journal of Neural Engineering*, *12*(5), 56006. https://doi.org/10.1088/1741-2560/12/5/056006.

Frohlich, J., & Van Horn, J. D. (2014). Reviewing the ketamine model for schizophrenia. *Journal of Psychopharmacology*, *28*(4), 287–302. https://doi.org/10.1177/0269881113512909.

Garcia, P., & Sleigh, J. (2017). Ketamine: A drug at war with itself. *Anesthesiology*, *126*(3), 371–372. https://doi.org/10.1097/ALN.0000000000001513.

Hagihira, S. (2015). Changes in the electroencephalogram during anaesthesia and their physiological basis. *British Journal of Anaesthesia*, *115*(Suppl), i27–i31. https://doi.org/10.1093/bja/aev212.

Hans, P., Dewandre, P.-Y., Brichant, J. F., & Bonhomme, V. (2005). Comparative effects of ketamine on bispectral index and spectral entropy of the electroencephalogram under sevoflurane anaesthesia. *British Journal of Anaesthesia*, *94*(3), 336–340. https://doi.org/10.1093/bja/aei047.

Harvey, M., Sleigh, J., Voss, L., et al. (2015a). Determination of the hypnotic potency in rats of the novel ketamine ester analogue SN 35210. *Pharmacology*, *96*(5–6). https://doi.org/10.1159/000439598.

Harvey, M., Sleigh, J., Voss, L., et al. (2015b). Development of rapidly metabolized and ultra-short-acting ketamine analogs. *Anesthesia and Analgesia*, *121*(4), 925–933. https://doi.org/10.1213/ANE.0000000000000719.

Homayoun, H., & Moghaddam, B. (2007). NMDA receptor hypofunction produces opposite effects on prefrontal cortex interneurons and pyramidal neurons. *The Journal of Neuroscience*, *27*(43), 11496–11500. https://doi.org/10.1523/JNEUROSCI.2213-07.2007.

Howe, W. M., Gritton, H. J., Lusk, N. A., et al. (2017). Acetylcholine release in prefrontal cortex promotes gamma oscillations and theta-gamma coupling during Cue detection. *The Journal of Neuroscience*, *37*(12), 3215–3230. https://doi.org/10.1523/JNEUROSCI.2737-16.2017.

Hudetz, A. G. (2012). General anesthesia and human brain connectivity. *Brain Connectivity*, *2*(6), 291–302. https://doi.org/10.1089/brain.2012.0107.

Jordan, D., Ilg, R., Riedl, V., et al. (2013). Simultaneous electroencephalographic and functional magnetic resonance imaging indicate impaired cortical top-down processing in association with anesthetic-induced unconsciousness. *Anesthesiology*, *119*(5), 1031–1042. https://doi.org/10.1097/ALN.0b013e3182a7ca92.

Kelland, M. D., Soltis, R. P., Boldry, R. C., & Walters, J. R. (1993). Behavioral and electrophysiological comparison of ketamine with Dizocilpine in the rat. *Physiology & Behavior*, *54*, 547–554.

Kikuchi, T., Wang, Y., Shinbori, H., Sato, K., & Okumura, F. (1997). Effects of ketamine and pentobarbitone on acetylcholine release from the rat frontal cortex in vivo. *British Journal of Anaesthesia*, *79*, 128–130.

Lahti, A. C., Weiler, M. A., Tamara Michaelidis, B. A., Parwani, A., & Tamminga, C. A. (2001). Effects of ketamine in normal and schizophrenic volunteers. *Neuropsychopharmacology*, *25*(4), 455–467. https://doi.org/10.1016/S0893-133X(01)00243-3.

Lee, U., Arbor, A., Ku, S., et al. (2013). Disruption of fronto-parietal communication by ketamine, propofo, l and sevoflurane. *Anesthesiology*, *118*(6), 1264–1275. https://doi.org/10.1097/ALN.0b013e31829103f5.Disruption.

Lee, H., Mashour, G. A., Noh, G.-J., Kim, S., & Lee, U. (2013). Reconfiguration of network hub structure after propofol-induced unconsciousness. *Anesthesiology*, *119*(6), 1347–1359. https://doi.org/10.1097/ALN.0b013e3182a8ec8c.

Li, L., & Vlisides, P. E. (2016). Ketamine: 50 years of modulating the mind. *Frontiers in Human Neuroscience*, *10*(November), 612. https://doi.org/10.3389/fnhum.2016.00612.

Matsushita, S., Oda, S., Otaki, K., Nakane, M., & Kawamae, K. (2015). Change in auditory evoked potential index and bispectral index during induction of anesthesia with anesthetic drugs. *Journal of Clinical Monitoring and Computing*, *29*(5), 621–626. https://doi.org/10.1007/s10877-014-9643-x.

McMillan, R., Sumner, R., Forsyth, A., et al. (2020). Simultaneous EEG/fMRI recorded during ketamine infusion in patients with major depressive disorder. *Progress in Neuro-Psychopharmacology & Biological Psychiatry*, *99*, 109838. https://doi.org/10.1016/j.pnpbp.2019.109838.

Miyasaka, M., & Domino, E. F. (1968). Neural mechanisms of ketamine-induced anesthesia. *International Journal of Neuropharmacology*, *7*(6), 557–573. https://doi.org/10.1016/0028-3908(68)90067-1.

Muthukumaraswamy, S. D., Shaw, A. D., Jackson, L. E., Hall, J., Moran, R., & Saxena, N. (2015). Evidence that subanesthetic doses of ketamine cause sustained disruptions of NMDA and AMPA-mediated frontoparietal connectivity in humans. *The Journal of Neuroscience*, *35*(33), 11694–11706. https://doi.org/10.1523/JNEUROSCI.0903-15.2015.

Nunez, A., Curro Dossi, R., Contreras, D., & Steriade, M. (1992). Intracellular evidence for incompatibility between spindle and delta oscillations in thalamocortical neurons of cat. *Neuroscience*, *48*(1), 75–85. https://doi.org/10.1016/0306-4522(92)90339-4.

Schubert, A., Licina, M. G., & Lineberry, P. J. (1990). The effect of ketamine on human somatosensory evoked potentials and its modification by nitrous oxide. *Anesthesiology*, *72*(1), 33–39.

Schuttler, J., Stanski, D. R., White, P. F., et al. (1987). Pharmacodynamic modeling of the EEG effects of ketamine and its enantiomers in man. *Journal of Pharmacokinetics and Biopharmaceutics*, *15*(3), 241–253.

Schwartz, M. S., Virden, S., & Scott, D. F. (1974). Effects of ketamine on the electroencephalograph. *Anaesthesia*, *29*(2), 135–140. https://doi.org/10.1111/j.1365-2044.1974.tb00611.x.

Schwender, D., Klasing, S., Madler, C., Poppel, E., & Peter, K. (1993). Mid-latency auditory evoked potentials during ketamine anaesthesia in humans. *British Journal of Anaesthesia*, *71*(5), 629–632.

Seamans, J. (2008). Losing inhibition with ketamine. *Nature Chemical Biology*, *4*, 91–93.

Shadli, S. M., Kawe, T., Martin, D., McNaughton, N., Neehoff, S., & Glue, P. (2018). Ketamine effects on EEG during therapy of treatment-resistant generalized anxiety and social anxiety. *The International Journal of Neuropsychopharmacology*, *21*(8), 717–724. https://doi.org/10.1093/ijnp/pyy032.

Siegel, M., Donner, T. H., & Engel, A. K. (2012). Spectral fingerprints of large-scale neuronal interactions. *Nature Reviews. Neuroscience*, *13*(2), 121–134. https://doi.org/10.1038/nrn3137.

Sleigh, J., Harvey, M., Voss, L., & Denny, B. (2014). Ketamine—More mechanisms of action than just NMDA blockade. *Trends in Anaesthesia and Critical Care*, *4*(2–3), 76–81. https://doi.org/10.1016/j.tacc.2014.03.002.

Sleigh, J., Pullon, R. M., Vlisides, P. E., & Warnaby, C. E. (2019). Electroencephalographic slow wave dynamics and loss of behavioural responsiveness induced by ketamine in human volunteers. *British Journal of Anaesthesia*, *123*(5), 592–600. https://doi.org/10.1016/j.bja.2019.07.021.

Vlisides, P. E., Bel-Bahar, T., Lee, U., et al. (2017). Neurophysiologic correlates of ketamine sedation and anesthesia: A high-density electroencephalography study in healthy volunteers. *Anesthesiology*, *127*(1), 58–69. https://doi.org/10.1097/ALN.0000000000001671.

Voss, L. J., Karalus, S., Englund, V., & Sleigh, J. W. (2018). Ketamine action in the in vitro cortical slice is mitigated by potassium channel blockade. *Anesthesiology*, *128*(6), 1167–1174.

Voss, L. J., Ludbrook, G., Grant, C., Upton, R., & Sleigh, J. W. (2007). A comparison of pharmacokinetic/pharmacodynamic versus mass-balance measurement of brain concentrations of intravenous anesthetics in sheep. *Anesthesia and Analgesia*, *104*, 1440–1446.

Zhou, C., Douglas, J. E., Kumar, N. N., Shu, S., Bayliss, D. A., & Chen, X. (2013). Forebrain HCN1 channels contribute to hypnotic actions of ketamine. *Anesthesiology*, *118*(4), 785–795. https://doi.org/10.1097/ALN.0b013e318287b7c8.

Chapter 25

Linking heart function and analgesia

Giorgia Saltelli[a], Antonella Paladini[b], Martina Rekatsina[c], and Giustino Varrassi[d]

[a]*Department of Anesthesiology and Intensive Care, Sant'Andrea Hospital, "La Sapienza University", Rome, Italy;* [b]*Department MESVA, University of L'Aquila, L'Aquila, Italy;* [c]*Whipps Cross University Hospital, Barts Health NHS Trust, London, United Kingdom;* [d]*Paolo Procacci Foundation, Rome, Italy*

Abbreviations

ADAPT trial	Alzheimer's Disease Anti-Inflammatory Prevention trial
APC trial	Adenoma Prevention with Celecoxib trial
APPROVe trial	Adenomatous Polyp Prevention on Vioxx study
ASA	acetylsalicylic acid
CAD	coronary artery disease
CLASS study	Celecoxib Long-term Arthritis Safety Study
COX	cyclooxygenase
CVD	cardiovascular disease
CVS	cardiovascular system
DOR	δ-opioid receptor
GFR	glomerular filtration rate
KOR	κ-opioid receptor
LAAM	levomethadyl acetate
MI	myocardial infarction
MMPs	matrix metalloproteinases
MOR	μ-opioid receptor
NSAIDs	nonsteroidal antiinflammatory drugs
OPRs	opioid peptides receptors
PGE2	prostaglandin E2
PGI2	prostaglandin I2
PRECISION Trial	Prospective Randomized Evaluation of Celecoxib Integrated Safety Vs Ibuprofen or Naproxen
PreSAP trial	Prevention of Colorectal Sporadic Adenomatous Polyps trial
QTc	corrected QT interval
TdP	Torsades de Pointes
TXA2	thromboxane A2
VIGOR trial	Vioxx Gastrointestinal Outcomes Research

Introduction

The aging of population associated with the improvement of acute and chronic management of cardiovascular diseases, coupled with declines in CVD mortality and with an increasing prevalence of pain diseases. Particularly in industrialized countries, it has resulted in a significant number of patients with, or at high risk for, CVD also requiring analgesic drugs for painful syndromes.

Pain itself affects several different functions of the body, in particular it has important direct and indirect effects on the nervous system and the cardiovascular system. Acute pain is a defensive process that has multiple effects on the whole-body system, both local and systemic. Those effects, normally protective in healthy subjects, could become extremely dangerous in different clinical situation such as multitraumatized or cardiopathic patients and are also related to the entity of tissue damage. The local response to damage is characterized by erythema, vasodilation, local inflammatory response, activation of nociceptors, and neurotransmission of pain information from periphery through the spinal cord to the brain

The Neurobiology, Physiology, and Psychology of Pain. https://doi.org/10.1016/B978-0-12-820589-1.00025-7

(Varrassi, Fusco, Coaccioli, & Paladini, 2015). If the nociceptive stimulus lasts beyond the acute phase, neurotransmitters accumulation in the dorsal column of the spinal cord may occur, resulting in a reduction of pain thresholds and potentially central hypersensitivity (Fusco, Skaper, Coaccioli, Varrassi, & Paladini, 2017). The reaction to acute pain is complex and not only localized, it has also systemic consequences involving nervous, endocrine, immune, and cardiovascular systems: activation of the hypothalamic-pituitary-adrenal axis and therefore sympathetic nervous system. This results in increased heart rate, stroke volume, peripheral resistance, arterial blood pressure, myocardial work, and tissue oxygen demand and in a reduction of peripheral and epicardial blood flow, fibrinolysis and lastly increased risk of venous thrombosis (Beattie, Buckley, et al., 1993; Meissner, Rolf, & VanAken, 1997; Rosenfeld, Beattie, et al., 1993; Stevens, Beardsley, et al., 1995). Consequences of the aforementioned pathophysiological changes are different in healthy or CVD-affected people (Beattie et al., 1993). The sympathetic stimulation of beta and alpha-adrenergic receptors results in an increased myocardial oxygen demand with a reduction of epicardial blood flow, while this phenomenon in a normal patient leads to a coronary artery vasodilatation, in patients with CAD instead, a paradoxical vasoconstriction is often observed (Bruehl & Chung, 2004; Shawaqfeh & Harrington, 2018). An efficacious and early performed analgesia is of paramount importance to reduce the risk of cardiovascular effects due to pain, especially in patients with cardiac disease. On the other hand, it must take into account that analgesics, as any other drug, have many side effects that may involve, among the others, the cardiovascular system (Varrassi, Coaccioli, et al., 2019).

Analgesic drugs

The management of both acute and chronic pain is often challenging, usually managed with a multimodal approach which may include systemic, pharmacological therapy, or interventional analgesic techniques (Varrassi et al., 2020; Varrassi, Hanna, et al., 2017). The analgesics family contains a wide group of drugs; among them the most frequently used, alone or in combination, are opioids, NSAIDs, paracetamol, adjuvants (e.g., antiepileptic agents, etc.), and other agents such as ketamine (Imani & Varrassi, 2019).

Opioids

Opioids are implicated in pain control, and mediate systemic and organ-specific responses to injury and adaptation binding their receptors. OPRs are ubiquitous in the body and are widely represented in the cardiovascular system, particularly in the heart (Khademi, Kamangar, et al., 2016; Treskatsch, Shaqura, et al., 2016; Varrassi, Pergolizzi, Peppin, & Paladini, 2020). The heart expresses very high levels of endogenous opioid peptides and their corresponding receptors OPRs (DORs, KORs, and MORs) which are implicated in adaptive responses to cellular stress (Table 1) (Treskatsch et al., 2016). Myocytes, in fact, possess major sites of storage of genes encoding endogenous opioid precursors (preproenkephalin, prodynorphin, and pro-opiomelanocortin) and are characterized by large opioid peptides synthesis, storage, and release. Those endogenous peptides (endorphins, dynorphins, and enkephalins) activate cellular response by the engagement of MORs, DORs, and KORs (Bolte, Newman, et al., 2009; Chen & Ashburn, 2015; Siong Chan, Cao, & Ng, 2018).

The opioid system plays an important role in the regulation of various cardiac function, including, among others, modulation of beta-adrenoceptor signaling and responses, excitation-contraction coupling, and cardiogenesis (Headrick, See Hoe, et al., 2015). Expression of OPRs in the heart and endogenous opioid production varies in different conditions, both physiological and pathological, e.g., increasing in advanced ages and senescence, or during physical exercise (Bolte et al., 2009; Chen & Ashburn, 2015; Siong Chan et al., 2018). Opioid effects on myocardial performance are complex (Headrick,

TABLE 1 Principal opioid receptors, endogenous ligand, distribution, and their major effects.

Receptor	Endogenous ligand	Distribution	Effects
MOR	Endorphin	Central and peripheral nervous system, gastrointestinal system	Analgesia, euphoria, dependence, respiratory depression, miosis, constipation, vasodilation
DOR	Dynorphin	Central and peripheral nervous system	Analgesia, diuresis, diphoria
KOR	Enkephalin	Central and peripheral nervous system	Analgesia, constipation
NOR	Nociceptin/ orphanin	Central nervous system	Analgesia and hyperalgesia

Pepe, & Peart, 2012). All the OPRs subtypes play a crucial role in local regulation and response to cardiac injury; they are implicated in stress signaling and cardiac protection, myocardial conditioning, and cardioprotective response to different pathophysiological stimuli (Bolte et al., 2009; Chen & Ashburn, 2015; Siong Chan et al., 2018).

Both exogenous and endogenous opioids have a key role in modulating parasympathetic activity which can result in enhanced myocardial performance (Headrick et al., 2012). The cardiovascular response to opioid stimulation can also be accompanied by transient bradycardia and hypotension. A decrease of heart rate and blood pressure is usually seen as opioids cardiodepressive effects, but these manifestations are usually transient and do not reflect long-term hemodynamic or contractile effects (Caffrey, 1999; Hayashi & Tanaka, 2016; Headrick et al., 2012; Jackson, Farias, et al., 2001).

Inotropic and chronotropic effects of opioids are debated. OPRs activity is involved in inotropic myocardial activity, since a dose-dependent reduction of contractility in human atrial trabeculae has been demonstrated with different opioids (e.g., Alfentanil); however, data showing positive inotropic effects have also been reported (Hanouz, Yvon, et al., 2001; Hayashi & Tanaka, 2016; Wu, Fry, & Henry, 1997). Among the analgesics, Remifentanil is one of the major responsible for bradycardia (Hanouz et al., 2001) while is a rare, but possible, effect during chronic Fentanyl administration in analgesic doses (Stanley & Webster, 1978). Hydrocodone and hydromorphone have been associated, in rare cases, with vagally mediated sinus pauses, leading to decreased heart rate even in patients with no previous cardiac conduction disorders (Sudhakaran, Surani, & Surani, 2014). The most frequently observed nonanalgesic effect of these drugs is vasodilatation that usually results in hypotension. The provoked hypotension may also be refractory to treatment and can be associated with arrhythmias in cases of overdoses. As with all opioids, especially when in combinations with benzodiazepines, Fentanyl administration in analgesic doses can cause hypotension but is usually well tolerated even in patients with coexisting cardiac disease (Stanley & Webster, 1978). There is evidence of opioids acting on myocytes but few data suggest that they could have any direct effect on the electrical status of nodal and Purkinje cells (Wedam & Haigney, 2016). Some of opioids receptors have been shown to activate inward rectifier potassium channels and inhibit voltage-dependent calcium channels thus prolonging repolarization and the QT interval (Meents, Juhasz, et al., 2018; Wedam & Haigney, 2016). The arrhythmia most commonly induced by increased QT interval is a type of polymorphic ventricular tachycardia known as Torsades de Pointes (TdP) (Al-Hasani & Bruchas, 2011; Behzadi, Joukar, & Beik, 2018; Isbister, Brown, et al., 2017; Meents et al., 2018; Seewald, Coles Jr, et al., 2017; Tanaka, Kersten, & Riess, 2014; Wedam & Haigney, 2016). All opioids seem capable of affecting cardiac electrophysiology in vitro. Oxycodone and Buprenorphine may have a dose-related effect on QTc but at low doses this effect seems no clinically meaningful. Tapentadol does not impact QT interval. However, potentially responsible for serotonin syndrome, its administration can lead to serious arrhythmias. Clinical data support instead a strong correlation between cardiac arrhythmias and methadone and levomethadyl-acetate (LAAM) which, prolonging QTc, can lead to TdP (Alinejad, Kazemi, et al., 2015; Behzadi et al., 2018; Isbister et al., 2017). On the other hand, morphine, thanks to its strong vagal effect, and fentanyl, that is selective for μ-receptors, can both enhance ventricular fibrillation threshold, partially reducing sympathetic activity and thus reducing the risks of tachycardia (Wedam & Haigney, 2016). However, possible arrhythmogenic effects of opioids on CVS have not been fully understood yet, and this issue would surely benefit from some further research. It is now clear that myocardial preconditioning and postconditioning are two kinds of cardiac protection mechanisms which can be induced both directly and indirectly often via transient heart or other tissues ischemia-reperfusion mechanisms.

Cardiac pre- and postconditioning act involving common cellular pathways such as mitochondrial protection via phosphokinases and reactive oxygen species signals, and appear to be mediated by the OPRs system (Al-Hasani & Bruchas, 2011). It is already well known the benefit of morphine administration during acute coronary syndrome or cardiogenic pulmonary edema and different trials on OPRs agonists have shown cardioprotective benefit in surgical ischemia and angioplasty. There is evidence that OPRs agonists can improve cardioprotection by modulating the responses to physiological and pathological stimuli including ischemic pre- and postconditioning (Al-Hasani & Bruchas, 2011; Seewald et al., 2017; Tanaka et al., 2014). Moreover, opioids play a major role influencing in a beneficial way other factors determining the ischemia-reperfusion outcome, i.e. arrhythmogenesis, contractile dysfunction, inflammation, and apoptosis.

In conclusion, further research is needed to better understand nonanalgesic effects of opioids, particularly on the cardiovascular system. They definitely play an important role in potentially modulating a wide range of cardiovascular responses including conditioning to ischemia. Additional investigation of OPRs role on CVS and how it changes with senescence may pave the way to new applications of opioidergic therapies. At the moment, a correct use of opioids does not appear to be responsible for heart damage or dysfunction.

NSAIDs

NSAIDs are among the most widely used therapeutic class in clinical practice and represent a significant part of a physician's arsenal in managing inflammation and pain (Varrassi, Alon, et al., 2019). NSAIDs use is broadly recognized in current international guidelines for treatment of different pathologies (Sarzi Puttini, Giorgi, et al., 2021), among the other osteoarthritis and ankylosing spondylitis, but they find their application in cancer patients and postsurgical pain as well (Gay-Esconda et al., 2019; Varrassi, Coaccioli, et al., 2019; Varrassi, Pergolizzi, Dowling, & Paladini, 2020) .

Considering that an increasing number of patients use NSAIDs for long periods, the cardiovascular safety of those drugs is of considerable concern, particularly among patients with or at risk for developing cardiovascular disease (Varrassi, Pergolizzi, et al., 2020). Historically, the first representative of this group of drugs was salicylic acid, a natural product converted to acetylsalicylic acid at the end of 19th century and followed by other synthetic versions. NSAIDs are chemically heterogeneous and this structural difference is responsible for their relatively different action and side effects. New drugs are classified for their mode of action on cyclooxygenase (COX) isoforms rather than for the chemical structure, consequently we can divide them in COX-1 selective and nonselective inhibitors and COX-2 selective inhibitors or "coxibs." The nonselective NSAIDs (e.g., diclofenac, naproxen, and ibuprofen) inhibit both COX isomerases, whereas Coxibs inhibit more selectively COX-2 (Fig. 1) minimizing COX-1-mediated side effects (e.g., gastrointestinal and kidney harm) (Table 2).

Although NSAIDS have been in use for more than a century, their mechanism of action is not completely understood. In 1970s, the ability of NSAIDS to block the cyclooxygenase, thus inhibiting arachidonic acid transformation into prostanoids (Fig. 2 and Table 3), was demonstrated (Ferreira, Moncada, & Vane, 1971; Smith & Willis, 1971; Vane, 1971). From then on, the mechanism was made clearer; in the 1990s, in fact, the existence of different isoforms of COX enzyme was identified and defined, being COX-1 a constitutive enzyme while COX-2 an inducible one, inducted by inflammatory processes (Hla & Neilson, 1992; Masferrer, Seibert, et al., 1992). Synthesis and research on new molecules able to inhibit selectively the COX-2 isoenzyme were thus developed (Mitchell, Vane, et al., 1993; Vane, Bakhle, & Botting, 1998). By sparing the COX-1 isoform, which is largely constitutively expressed, Coxibs have met the expectation to reduce the gastrointestinal harm and to minimize the effect on platelet function, whereas other unexpected side effects have occurred with consequent market withdrawal (e.g., valdecoxib, rofecoxib, lumiracoxib) (European Medicines Agency, 2005, 2007; Funk & FitzGerald, 2007; Walker & Biasucci, 2018). The primary concern focused on the apparent excess of cardiovascular harm. It is still not clear which properties of COX inhibition may result in increased CV-risk. Theoretically, selective blocking of COX-2 disrupts homeostasis, reducing PGI2-related vasodilatation with no counterbalancing inhibition of COX-1, thus not discontinuing production of vasoconstrictory thromboxane; these processes could result in an increased thrombogenicity (Mitchell & Warner, 2006; Walker & Biasucci, 2018; Wallace & Devchand, 2005). However, data show both COX-1 and COX-2 inhibitors are associated with cardiovascular harm, suggesting a more complex interaction between cardiovascular risk and NSAIDS that goes beyond the sole selectivity (i.e. drug exposure: dose, duration of action, duration of treatment).

In normal human heart conditions, COX-1 is expressed in endothelial and smooth muscle cells of blood vessels and in endothelial cells of the endocardium. Different animal studies investigated whether static or shear stress could influence COX isoform expression on aortic endothelial cells (Sellers, Radi, & Khan, 2010). Furthermore, the role of aging and senescence on COX-1/COX-2 expression was investigated, demonstrating that aged endothelium has greater propensity to

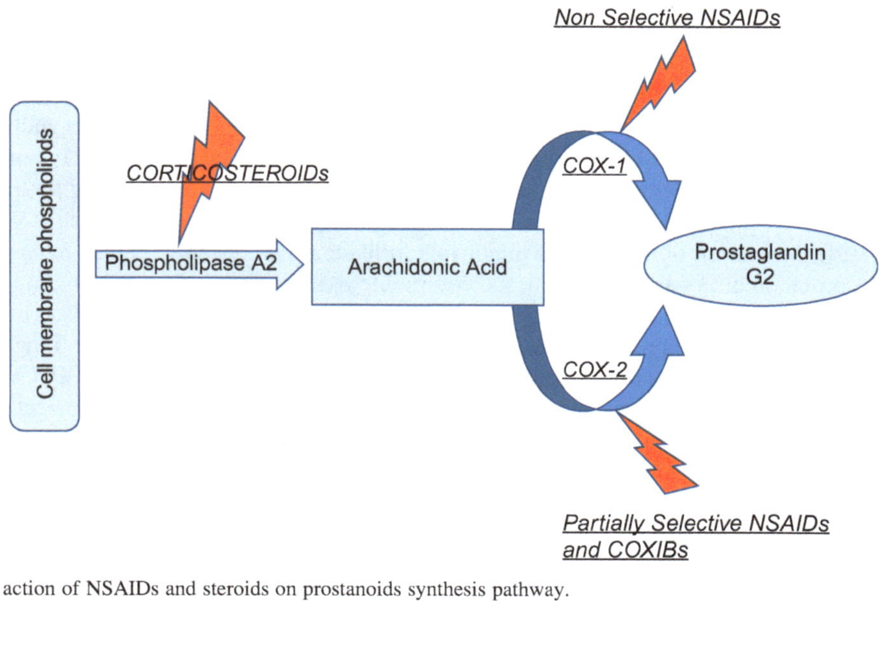

FIG. 1 Mechanism of action of NSAIDs and steroids on prostanoids synthesis pathway.

TABLE 2 NSAIDs chemical classification and COX selectivity.

Chemical class	Nonselective NSAIDs or COX-1-selective	COX-2-selective NSAIDs
Salicylic acid derivates	Acetylsalicylic acid, diflunisal, salicylsalicylic acid, sulfasalazine, olsalazine	
Propionic acid derivates	Dexketoprofen, dexibuprofen, flurbiprofen, ibuprofen, ketoprofen, naproxen, oxaproxin	
Enolic acid derivates	Droxicam, phenylbutazone, piroxicam, tenoxicam	Meloxicam
Acetic acid derivates	Diclofenac, ketorolac, indomethacin, sulindac, tolmetin	Etodolac, lumiracoxib
Fenamic acid derivates	Mefenamic acid, meclofenamic acid, tolfenamic acid	Derivates of meclofenamate
Sulfonamides		Nimesulide, flosulide
COXIBS		Celecoxib, rofecoxib, valdecoxib, parecoxib, etoricoxib

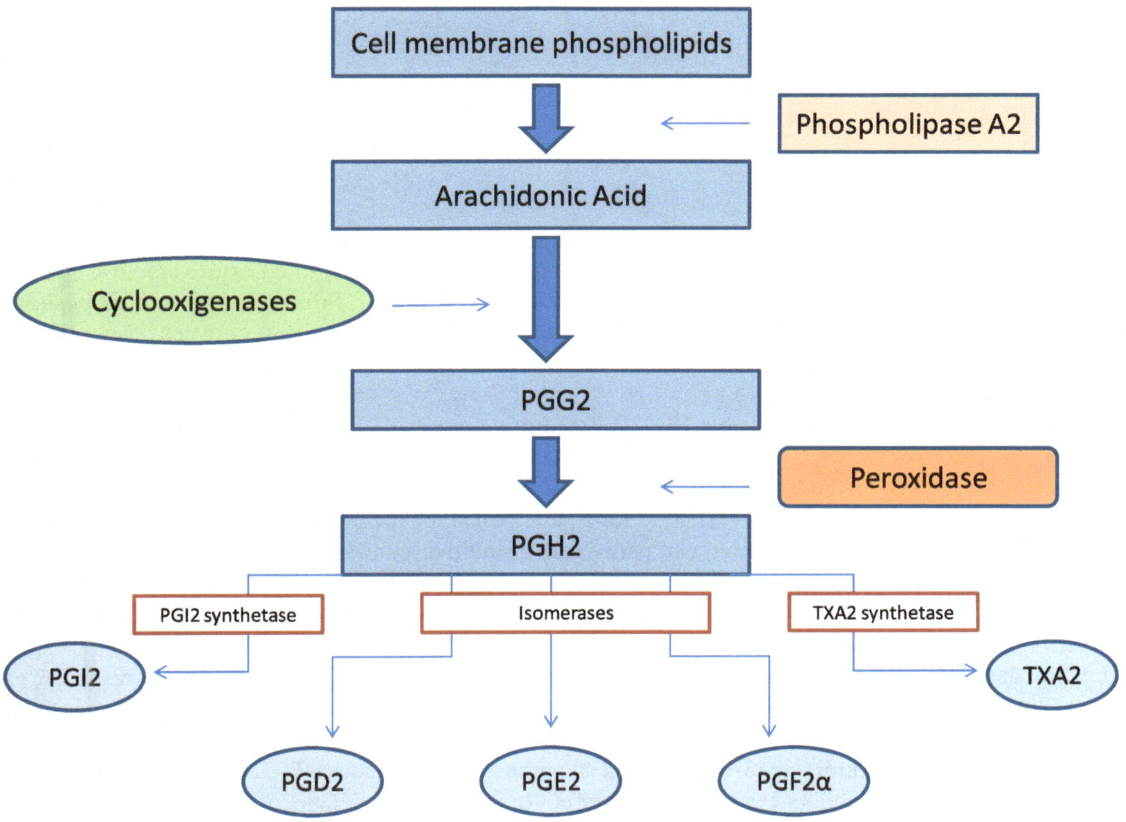

FIG. 2 Synthesis of prostanoids. *PGI2*, prostaglandin I2; *TXA2*, thromboxane A2; *PGE2*, prostaglandin E2; *PGD2*, prostaglandin D2; *PGF2α*, prostaglandin F2α.

release COX-derived vasoconstrictive prostaglandins thus explaining the potential impact of COX inhibition in older patients. It has been assumed that endothelial aging could be correlated to reduced antiplatelet activity and a prothrombotic status. Chenevard et al. showed that humans with coronary artery disease improved endothelium-related vasodilatation with COX-2 inhibition (Chenevard, Hurlimann, et al., 2003; Sellers et al., 2010; Zaher, Radi, & Khan, 2019). While

TABLE 3 Tissue-specific prostanoid functions.

Organ	Prostanoids	Function
Cardiovascular system	TXA2-PGI2	Thrombosis Platelets aggregation
	TXA2	Vascular permeability
	PGE2-PGI2	Arterial vasodilation
	TXA2-PGF2α	Venous vasoconstriction
Kidney	PGE2-PGI2	Renal blood flow regulation Glomerular filtration rate regulation Increased renin secretion
	PGE2	Inhibition of ADH effect
Respiratory system	PGE2	Bronchodilation
	PGF2α-TXA2	Bronchoconstriction
Reproductive system	PGF2α-PGE2	Fertility Uterine contraction
Gastrointestinal system	PGE2-PGI2	Cellular protection
Central nervous system	PGE2	Fever
	PGD2	Sleep
	PGE2-PGI2	Pain

COX-1 is constitutive, COX-2 is usually nonexpressed in normal endothelial and smooth muscular cells, while it is quickly induced in case of vascular damage or inflammatory conditions. In human, only limited expression of COX-2 in normal coronary arteries and nonatheromatous aorta was reported (Zaher et al., 2019). Prostaglandins have opposite effects on cardiac homeostasis and on vascular tone, they play a key role in maintaining the balance between vasoconstriction/vasodilatation and pro/antithrombotic events. In this regard, we must stress that the balance is held by PGI2, TXA2, and PGE2 (Sellers et al., 2010).

Platelets express only COX-1, no inducible COX-2 isoform is present. TXA2 is produced by COX-1 and is mainly platelet-derived, representing the major arachidonic acid metabolite in platelets (Zidar, Odar, et al., 2009). TXA2 causes local vasoconstriction, platelet aggregation, and smooth muscle cells proliferation (Zidar, Dolenc-Strazar, et al., 2007). Experimental studies had shown that PGI2 is essential for counterbalancing the prothrombotic and vasoconstrictive effects of TXA2, while PGE2, causing vasodilatation, can participate in indirectly modulating TXA2-related effects on platelet aggregation, suggesting that other homeostatic mechanism and molecular pathways might play a role in preserving the pro-antithrombotic homeostasis (e.g., nitric oxide) (Chenevard et al., 2003; Rossi, Kuukasjarvi, et al., 1996; Sellers et al., 2010; Zaher et al., 2019).

Experimental data show that prostaglandins promote atherosclerosis by altering the inflammatory response, expression of matrix metalloproteinases (MMPs), and acting as a mitogen for different cell types including vascular muscular smooth cells. The upregulation of MMPs in plaques seems to be mediated by COX-1 and COX-2-derived PGE2 and may contribute to plaque rupture. TXA2 has a central role in development and progression of plaques. PGI2 increases in atherosclerosis and has atheroprotective properties (Sellers et al., 2010; Zaher et al., 2019).

COX-2 is observed in damaged cardiac tissue both in humans and experimental animal models of cardiac disease. In human heart obtained after myocardial infarction, there is evidence of both COX-1and COX-2 expression in inflammatory cells, myofibroblasts and in areas of granulation tissue and fibrosis. In heart failure and MI, COX-2 expression has been detected in cardiomyocytes near to infarcted regions, in inflammatory cells, and vessels associated with regions of cardiac damage (Sellers et al., 2010). In addition, COX-2 expression has been identified in patients with septicemia and myocarditis (Becker, Wang, et al., 2009; Chenevard et al., 2003; McGettigan & Henry, 2011; Pepine & Gurbel, 2017; Rossi et al., 1996; Sellers et al., 2010; Solomon, Wittes, Finn, et al., 2008; Varrassi, Pergolizzi, et al., 2020; Zaher et al., 2019; Zidar et al., 2007, 2009).

Several studies on animal models have been conducted in the last few years evaluating the cardiac effects of COX inhibition on MI or ischemia/reperfusion. Collectively, these studies suggest that COX-1 inhibition may have little impact on short and long effect of MI, while COX-2 inhibition can reduce infarct size and inflammation on the MI site in the acute phase (Becker et al., 2009; Chenevard et al., 2003; McGettigan & Henry, 2011; Pepine & Gurbel, 2017; Rossi et al., 1996; Sellers et al., 2010; Solomon et al., 2008; Varrassi, Pergolizzi, et al., 2020; Zidar et al., 2007, 2009). Acutely the reduction of inflammation could be beneficial but may have adverse long-term consequences, there are data showing a decrease in collagen density, which could result in scar thinning potentially harmful for heart function (Chenevard et al., 2003; Rossi et al., 1996; Sellers et al., 2010; Zidar et al., 2007, 2009). Both Coxibs and nonselective NSAIDs may influence wound healing and clearing of bacterial infections following cardiovascular surgery.

The advantageous effect of ischemic preconditioning is lost when Coxibs are administered prior to surgery whereas the protective preconditioning effect on myocardiocytes is maintained when ASA is administered, thus suggesting the importance of early effects of COX-2 in myocardial preconditioning. However, in animal models, neither COX-1 nor COX-2 block appears to modify the long-term ischemia/reperfusion outcome (Becker et al., 2009; Chenevard et al., 2003; McGettigan & Henry, 2011; Pepine & Gurbel, 2017; Rossi et al., 1996; Sellers et al., 2010; Solomon et al., 2008; Varrassi, Pergolizzi, et al., 2020; Zaher et al., 2019; Zidar et al., 2007, 2009).

Until quite recently, evidence of NSAIDs on cardiac safety was limited to relatively small sample sizes trials. The selective block of the COX-2 has been supposed to be a potential pharmacological target to reduce the side effects of NSAIDs, particularly gastroenteric effects. The clinical trial VIGOR compared rofecoxib with naproxen in a large population of rheumatoid arthritis patients, showing a significant higher incidence of unexpected adverse cardiovascular events in the rofecoxib's group (ADAPT Research Group et al., 2008; Baron, Sandler, Bresalier, et al., 2008; Becker et al., 2009; Bombardier, Laine, et al., 2000; Breitner, Baker, Montine, et al., 2011; Chenevard et al., 2003; FitzGerald & Patrono, 2001; McGettigan & Henry, 2011; Pepine & Gurbel, 2017; Rossi et al., 1996; Sellers et al., 2010; Silverstein, Faich, et al., 2000; Solomon et al., 2008; Zaher et al., 2019; Zidar et al., 2007, 2009). Different possible explanations have been given, among others the antiplatelet effect of naproxen and the low-dose aspirin assumed by high-risk patients. The CLASS study comparing celecoxib with ibuprofen or diclofenac showed no statistically different cardiovascular side effects between the groups (Solomon, McMurray, et al., 2005). Both rofecoxib and celecoxib were trialed in APPROVe, APC, and PreSAP comparing active groups to placebo group in patients with history of colorectal adenoma (Baron et al., 2008; Bombardier et al., 2000; Hernandez-Diaz et al., 2006; Silverstein et al., 2000; Solomon et al., 2005, 2008). In ADAPT trial, celecoxib was compared with naproxen and placebo in patients with family history of Alzheimer's disease. All four trials were interrupted for excessive cardiovascular events. APPROVe and APC demonstrated a significant increase of adverse CV events and increase of systolic blood pressure in the treatment groups (rofecoxib, celecoxib 400 mg/day and celecoxib 800 mg/day) (ADAPT Research Group et al., 2008; Baron et al., 2008; Becker et al., 2009; Bombardier et al., 2000; Breitner et al., 2011; FitzGerald & Patrono, 2001; Silverstein et al., 2000; Solomon et al., 2005, 2008). No significant difference between placebo and celecoxib active group was noticed in the PreSAP and ADAPT trials. In ADAPT, however, a significant increased risk of cardiovascular event was observed in the naproxen group and not in the Coxib group, stressing again the relationship between CV risk and all NSAIDs (ADAPT Research Group et al., 2008; Baron et al., 2008; Breitner et al., 2011; FitzGerald & Patrono, 2001). A metaanalysis evaluating more than 100 trials demonstrated that high doses of diclofenac and ibuprofen are associated with higher risk that cannot be avoided with simultaneous use of low-dose aspirin (Varrassi, Pergolizzi, et al., 2020). Similar results have been obtained with other studies; among the others, a systematic review by Hernandez-Diaz et al. on the risk of MI in patients treated with nonselective NSAIDs and Coxibs has showed an increase of 10% of MI risk in all groups, with some percentage differences between the drugs (Hernandez-Diaz et al., 2006). McGettigan et al. carried out a systematic review showing diclofenac being the NSAID associated to the highest CV risk (McGettigan & Henry, 2011). All these studies were not designed to investigate specifically the cardiovascular side effects with the use of selective and nonselective NSAIDs but to describe only events as complications. Because of the controversy on cardiovascular safety profile of Coxibs, a large randomized, multicenter, double-blind, noninferiority trial including more than 24,000 patients was conducted. Recently, data from the PRECISION (Prospective randomized evaluation of celecoxib integrated safety versus ibuprofen or naproxen) trial in arthritis population with preexisting cardiovascular diseases or at high risk for CVD were published and demonstrated a noninferiority of celecoxib vs ibuprofen and naproxen regarding to CV death, nonfatal MI and stroke. Notably, in contrast with previous metaanalyses, a lower risk of cardiovascular outcomes with naproxen was not demonstrated (Becker et al., 2009; McGettigan & Henry, 2011; Pepine & Gurbel, 2017; Solomon et al., 2008; Varrassi, Pergolizzi, et al., 2020).

All data from preclinical and clinical experience suggest that cardiac safety profile of selective and nonselective NSAIDs is comparable. In such context, the question of which NSAID should safely be recommend to CVD patients and the precise dose for long-term use remains largely unanswered. Nonspecific NSAIDs should be preferred in patients at high risk for cardiovascular issues, in patients with high risk of GI bleeding NSAIDS should be prescribed in association with proton pump inhibitors. In patients with high cardiac risk or in need for secondary prevention, addition of low doses of aspirin is suggested. The best advice would probably be to avoid NSAIDs, especially in patients with CVD. If this is not feasible, these patients should assume these drugs at the lowest possible doses and for the shortest possible period of time, seen that the risk is dose, duration, and recency of exposure dependent (ADAPT Research Group et al., 2008; Becker et al., 2009; Breitner et al., 2011; FitzGerald & Patrono, 2001; McGettigan & Henry, 2011; Pepine & Gurbel, 2017; Rossi et al., 1996; Solomon et al., 2008; Varrassi, Pergolizzi, et al., 2020; Zidar et al., 2007, 2009).

Paracetamol

Paracetamol is one of the most widely used analgesics and antipyretics. It is usually used either alone or in combination with other drugs. When used properly is very safe and effective, while when assumed in excessive dosages or in association with alcohol, hepatotoxicity and liver damage could arise. Hepatotoxicity of paracetamol is a well-known issue. Cardiovascular effects are rare but, in long-term use, elevated blood pressure can occur and, in some cases, may represent the result of kidney injury after chronic intoxication (Ramachandran & Jaeschke, 2018; Roberts, Delgado Nunes, et al., 2016). Laboratory, preclinical, and clinical studies have suggested that the risk for major CV events is increased if paracetamol is taken frequently (>22 days/month) or at high doses. Nonetheless, when assumed at a proper dose, recent investigation revealed that it may work as a potential cardioprotective agent, thanks to antioxidant activity which has beneficial effects also on neuroprotection, blood glucose levels, and on muscle function and structure (Roberts et al., 2016).

Application to other areas: Effects of NSAIDs on the kidney

Both isoforms of the cyclooxygenase, COX-1 and COX-2, are expressed in the kidney (Sellers et al., 2010). In the human kidney, COX-1 is constitutively expressed in glomerular vessels, collecting ducts and interstitial cells, while COX-2 is found mainly in the macula densa, podocytes, and renal vasculature. They both appear to be involved in the regulation of hydroelectrolytic balance and blood pressure: COX-1 regulating mainly local hemodynamics and glomerular filtration rate (GFR), COX-2 modulating salt and water excretion. In this setting, the role of prostaglandin appears fundamental with PGE2 involved in sodium reabsorption in the distal renal tube and, together with PGI2, activating the renin-angiotensin-system stimulating potassium secretion. NSAIDS, inhibiting COX, might be responsible for increases in arterial blood pressure and peripheral edemas lowering GFR through a reduction in renal blood flow and thus potentially leading to electrolytes abnormalities and fluid retention. In nonpathological patients, COX-2 inhibition does not seem to have any consequence on renal function, while in elderly and already CVD-affected patients, Coxibs and nonselective NSAIDs as well could be central in the development of this side effect, especially if associated at hydro-electrolytic imbalance. Further research focusing on whether Coxibs and NSAIDs might be directly responsible for increase of blood pressure would be useful (Khan, Paulson, et al., 2002; Mitchell & Warner, 2006; Walker & Biasucci, 2018; Wallace & Devchand, 2005; Zaher et al., 2019).

Other agents of interest: Adjuvants

Pain is always multifactorial and involves several neurotransmitters and signaling pathways, thus its management is frequently challenging for the clinician. Usually, patients cannot find pain relief by a single drug assumption, whereas a combination of multiple drugs that act through different mechanisms could be more effective. In this context, we have already analyzed in which way the long-term use of powerful analgesics as opioids or NSAIDs could be problematic. For this reason, even powerful analgesic medications may necessitate of a coadministration of other drugs such as paracetamol and the so-called adjuvants (Table 4.). Adjuvants do not find their main application in analgesia, nevertheless, at the same time, they play a very important role in pain management. As any drug, adjuvants have side effects and some of them can affect the cardiovascular system. Multiple medications are part of the "adjuvants group"; herein, we briefly discuss the most frequently used in clinical practice.

Antidepressant, mostly tricyclic agents, is one of the most commonly used class of drugs, especially in chronic neuropathic pain. They have different effects on CVS, particularly on intraventricular conduction prolonging PR, QRS, and QT and so potentially facilitating the onset of arrhythmias (ADAPT Research Group et al., 2008; Baron et al., 2008; Becker et al., 2009; Bombardier et al., 2000; Breitner et al., 2011; FitzGerald & Patrono, 2001; Gillman, 2007; Hernandez-Diaz et al., 2006; McGettigan & Henry, 2011; Pepine & Gurbel, 2017; Silverstein et al., 2000; Solomon et al., 2005, 2008; Varrassi, Pergolizzi, et al., 2020).

TABLE 4 Principal adjuvant coanalgesics.

Class	Most commonly used drugs	Common adverse effects
Tricyclic antidepressants	Amitriptyline	
	Nortriptyline	Anticholinergic
	Desipramine	effects
Anticonvulsants	Gabapentin Pregabalin Clonazepam Carbamazepine	Sedation Dizziness Lower-extremity edema
Corticosteroids	Dexamethasone	Delirium-dyspepsia
	Prednisone	
Local anesthetics	Lidocaine	Tremor-paresthesia
N-Methyl-D-aspartate(NMDA) Receptor agonists	Ketamine	Confusion-Frightening dreams
α2 adrenergic agonists	Clonidine Tizanidine	Hypotension
Bisphosphonates	Pamidronate	Bone/joint/muscle pain-osteonecrosis
Antispasmodic	Baclofen	Muscle weakness
Other	Capsaicin	Erythema

Gabapentinoids are antiepileptic drugs approved for partial seizures but also for several painful neuropathic syndromes. Gabapentin and pregabalin are structural analogs of GABA and selectively modulate voltage-sensitive Ca^{2+} channels in the afferent neurons. They have different pharmacokinetics and pharmacodynamics, explaining the differences in efficacy and adverse events. Mostly gabapentinoids side effects occur in central nervous system. Regarding to vascular system, peripheral edema has been reported by an extensive Cochrane review (Moore et al., 2015; Ramachandran & Jaeschke, 2018; Roberts et al., 2016; Wiffen, Derry, et al., 2017).

Glucocorticoids do not appear responsible for major cardiovascular side effects when used for a short period. Several metaanalyses of RCTs comparing administrations of glucocorticoids with placebo or no treatment, reported neither increased mortality for CV events nor myocardial or pulmonary complications. For these reasons, short-term assumption of glucocorticoids can be effective in reducing inflammation thus supporting the management of painful conditions, without CV complications. On the contrary, long-term use of glucocorticoids should be avoided, if possible, due to the multiples known complications, including cardiovascular ones (Varrassi, Pergolizzi, et al., 2020).

Mini-dictionary of terms

Cardiac preconditioning. Preconditioning is an endogenous adaptive mechanism of cardioprotection. It is an inducible protective response elicited by several stimuli, mostly by brief ischemic insults followed by reperfusion. It is shown that those short-term insults enhance myocardial ability to resist to sustained ischemic injuries, i.e. reducing infarct size and decreasing the incidence of reperfusion-induced arrhythmias.

Cardiac postconditioning. Postconditioning consists in a series of brief mechanical interruptions of reperfusion following a specific algorithm applied at the very onset of reperfusion. The cardioprotective effect is associated with an improvement in endothelial function, reduction in cardiac apoptosis, and decrease in microvascular injury. Different molecular pathways are involved such as adenosine nitric oxide production.

Endogenous opioids peptides and precursors. Endogenous opioids are small peptides, mainly, but not exclusively, produced in the central nervous system which act as hormones and neuromodulators. Endogenous opioids consist in three families derived from protelytic cleavage of protein precursors which are: proopiomelanocortin (POMC)-precursor of β-endorphin, preproenkephalin (PENK)-precursor of Leu- and Met-enkephalins, and preprodynorphin (PDYN)-precursor of dynorphins.

Opioid Receptors. OPRs are G-protein-coupled receptors. G-proteins can be divided based on the pathways activated in the cells: Gs-receptor is stimulatory while Gi-receptor is inhibitory. Both endogenous and exogenous opioids bind these receptors thus mediating several local and systemic function. Several types of receptors have been described: μ-Receptors 1,2,3 (MORs) binding β-endorphins, κ-receptors (KORs) binding dynorphins, δ-receptors (DORs) binding enkephalins, Nociceptin receptors (NORs) binding nociceptin/orphanin FQ and ζ-receptors (ZORs).

Prostanoids. Prostanoids include prostaglandins and thromboxane A2. They derive from arachidonic acid released from the plasma membrane by phospholipase A2 and metabolized by the sequential actions of COX and respective prostaglandins-synthases. The principal bioactive prostanoids generated in vivo are PGE2, PGI2, PGD2, PGF2α, and TXA2.

Summary points

- Analgesics are among the most assumed drugs worldwide
- Clinicians should be aware of the deep interactions between analgesics and CV system
- Heart expresses a high number of opioid receptors
- Opioids have been demonstrated to act on parasympathetic activity, thus widely influencing cardiac function
- Treatment-refractory hypotension and polymorphic ventricular tachycardia seem to represent the most dangerous CV side effects
- COX 2 selective inhibition may lead to an excess in thrombogenicity
- COX-2 inhibition seems to be able to reduce infarct size and inflammation on the MI site in the acute phase
- Nowadays, data from preclinical and clinical experience suggest that cardiac safety profile of selective and nonselective NSAIDs is comparable
- CV effects of paracetamol are rare but sometimes increased blood pressure can occur as a possible result of kidney harm
- Possible CV effects of adjuvant drugs should always be considered
- As a general rule, drugs should always be administered at the lower effective dose and for the shortest period in order to minimize side effects

References

ADAPT Research Group, Martin, B. K., Szekely, C., et al. (2008). Cognitive function over time in the Alzheimer's disease anti-inflammatory prevention trial (ADAPT): Results of a randomized, controlled trial of naproxen and celecoxib. *Archives of Neurology*, *65*(7), 896–905.

Al-Hasani, R., & Bruchas, M. R. (2011). Molecular mechanisms of opioid receptor-dependent signaling and behavior. *Anesthesiology*, *115*(6), 1363–1381.

Alinejad, S., Kazemi, T., et al. (2015). A systematic review of the cardiotoxicity of methadone. *EXCLI Journal*, *14*, 577–600.

Baron, J. A., Sandler, R. S., Bresalier, R. S., et al. (2008). Cardiovascular events associated with rofecoxib: final analysis of the APPROVe trial. *Lancet*, *372* (9651), 1756–1764. 15.

Beattie, W. S., Buckley, D. N., et al. (1993). Epidural morphine reduces the risk of postoperative myocardial ischemia in patients with cardiac risk factors. *Canadian Journal of Anaesthesia*, *40*, 532–538.

Becker, M. C., Wang, T. H., et al. (2009). Rationale, design, and governance of prospective randomized evaluation of celecoxib integrated safety versus ibuprofen or naproxen (PRECISION), a cardiovascular end point trial of nonsteroidal antiinflammatory agents in patients with arthritis. *American Heart Journal*, *157*, 606–612.

Behzadi, M., Joukar, S., & Beik, A. (2018). Opioids and cardiac arrhythmia: A literature review. *Medical Principles and Practice*, *27*(5), 401–414.

Bolte, C., Newman, G., et al. (2009). Kappa and delta opioid receptor signaling is augmented in the failing heart. *Journal of Molecular and Cellular Cardiology*, *47*, 493–503.

Bombardier, C., Laine, L., et al., & VIGOR Study Group. (2000). Comparison of upper gastrointestinal toxicity of rofecoxib and naproxen in patients with rheumatoid arthritis. VIGOR Study Group. *The New England Journal of Medicine*, *343*(21), 1520–1528.

Breitner, J. C., Baker, L. D., Montine, T. J., et al. (2011). Extended results of the Alzheimer's disease anti-inflammatory prevention trial. *Alzheimer's & Dementia*, *7*(4), 402–411.

Bruehl, S., & Chung, O. Y. (2004). Interactions between the cardiovascular and pain regulatory systems: An updated review of mechanisms and possible alterations in chronic pain. *Neuroscience and Biobehavioral Reviews*, *28*, 395–414.

Caffrey, J. L. (1999). Enkephalin inhibits vagal control of heart rate, contractile force and coronary blood flow in the canine heart in vivo. *Journal of the Autonomic Nervous System*, *76*, 75–82.

Chen, A., & Ashburn, M. A. (2015). Cardiac effects of opioid therapy. *Pain Medicine*, *16*, S27–S31.

Chenevard, R., Hurlimann, D., et al. (2003). Selective COX-2 inhibition improves endothelial function in coronary artery disease. *Circulation*, *107*, 405–409.

European Medicines Agency. (2005). *EMEA public statement on the suspension of the marketing authorisation for bextra (valdecoxib) in the European Union.*

European Medicines Agency. (2007). *Questions and answers on the recommendation to withdraw the marketing authorisations for lumiracoxib-containing medicines.*

Ferreira, S. H., Moncada, S., & Vane, J. R. (1971). Indomethacin and aspirin abolish prostaglandin release from the spleen. *Nature: New Biology*, *231*(25), 237–239.

FitzGerald, G. A., & Patrono, C. (2001). The coxibs, selective inhibitors of cyclooxygenase-2. *The New England Journal of Medicine*, *345*, 433–442.

Funk, C. D., & FitzGerald, G. A. (2007). COX-2 inhibitors and cardiovascular risk. *Journal of Cardiovascular Pharmacology*, *50*, 470–479.

Fusco, M., Skaper, S. D., Coaccioli, S., Varrassi, G., & Paladini, A. (2017). Degenerative joint diseases and neuroinflammation. *Pain Practice*, *17*, 522–532. https://doi.org/10.1111/papr.12551.

Gay-Esconda, C., Hanna, M., Montero, A., Dietrich, T., Milleri, S., Giergiel, E., … Varrassi, G. (2019). Tramadol/dexketoprofen (TRAM/DKP) compared with tramadol/paracetamol in moderate to severe acute pain: results of a randomised, double-blind, placebo and active-controlled, parallel group trial in the impacted third molar extraction pain model (DAVID study). *BMJ Open*, *9*(2), e023715. https://doi.org/10.1136/bmjopen-2018-023715.

Gillman, P. K. (2007). Tricyclic antidepressant pharmacology and therapeutic drug interactions updated. *British Journal of Pharmacology*, *151*(6), 737–748.

Hanouz, J. L., Yvon, A., et al. (2001). The in vitro effects of remifentanil, sufentanil, fentanyl, and alfentanil on isolated human right atria. *Anesthesia and Analgesia*, *93*(3), 543–549.

Hayashi, K., & Tanaka, A. (2016). Effect-site concentrations of remifentanil causing bradycardia in hypnotic and non-hypnotic patients. *Journal of Clinical Monitoring and Computing*, *30*(6), 919–924.

Headrick, J. P., Pepe, S., & Peart, J. N. (2012). Non-analgesic effects of opioids: Cardiovascular effects of opioids and their receptor systems. *Current Pharmaceutical Design*, *18*(37), 6090–6100.

Headrick, J. P., See Hoe, L. E., et al. (2015). Opioid receptors and cardioprotection—'Opioidergic conditioning' of the heart. *British Journal of Pharmacology*, *172*(8), 2026–2050.

Hernandez-Diaz, S., et al. (2006). Non-steroidal anti-inflammatory drugs and the risk of acute myocardial infarction. *Basic & Clinical Pharmacology & Toxicology*, *98*(3), 266–274.

Hla, T., & Neilson, K. (1992). Human cyclooxygenase-2 cDNA. *Proceedings of the National Academy of Sciences of the United States of America*, *89*(16), 7384–7388.

Imani, F., & Varrassi, G. (2019). Ketamine as adjuvant for acute pain management. *Anesthesia and Pain Medicine*, *9*(6), e100178. https://doi.org/10.5812/aapm.100178.

Isbister, G. K., Brown, A. L., et al. (2017). QT interval prolongation in opioid agonist treatment: Analysis of continuous 12-lead electrocardiogram recordings. *British Journal of Clinical Pharmacology*, *83*(10), 2274–2282.

Jackson, K. E., Farias, M., et al. (2001). Transient arterial occlusion raises enkephalin in the canine sinoatrial node and improves vagal bradycardia. *Autonomic Neuroscience*, *94*, 84–92.

Khademi, H., Kamangar, F., et al. (2016). Opioid therapy and its side effects: A review. *Archives of Iranian Medicine*, *19*(12), 870–876. 0161912/AIM.0010.

Khan, K. N., Paulson, S. K., et al. (2002). Pharmacology of cyclooxygenase-2 inhibition in the kidney. *Kidney International*, *61*, 1210–1219.

Masferrer, J. L., Seibert, K., et al. (1992). Endogenous glucocorticoids regulate an inducible cyclooxygenase enzyme. *Proceedings of the National Academy of Sciences of the United States of America*, *89*(9), 3917–3921.

McGettigan, P., & Henry, D. (2011). Cardiovascular risk with non-steroidal anti-inflammatory drugs: Systematic review of population-based controlled observational studies. *PLoS Medicine*, *8*(9), e1001098.

Meents, J. E., Juhasz, K., et al. (2018). The opioid oxycodone use-dependently inhibits the cardiac sodium channel NaV 1.5. *British Journal of Pharmacology*, *175*(14), 3007–3020.

Meissner, A., Rolf, N., & VanAken, H. (1997). Thoracic epidural anesthesia and the patient with heart disease: Benefits, risks and controversies. *Anesthesia and Analgesia*, *85*, 517–528.

Mitchell, J. A., Vane, J. R., et al. (1993). Selectivity of nonsteroidal anti-inflammatory drugs as inhibitors of constitutive and inducible cyclooxygenase. *Proceedings of the National Academy of Sciences of the United States of America*, *90*(24), 11693–11697.

Mitchell, J. A., & Warner, T. D. (2006). COX isoforms in the cardiovascular system: Understanding the activities of non-steroidal anti-inflammatory drugs. *Nature Reviews. Drug Discovery*, *5*, 75–86.

Moore, R. A., et al. (2015). Amitriptyline for neuropathic pain in adults. *Cochrane Database of Systematic Reviews*, *2015*(7), CD008242.

Pepine, C. J., & Gurbel, P. A. (2017). Cardiovascular safety of NSAIDs: Additional insights after PRECISION and point of view. *Clinical Cardiology*, *40*(12), 1352–1356.

Ramachandran, A., & Jaeschke, H. (2018). Acetaminophen toxicity: Novel insights into mechanisms and future perspectives. *Gene Expression*, *18*(1), 19–30.

Roberts, E., Delgado Nunes, V., et al. (2016). Paracetamol: Not as safe as we thought? A systematic literature review of observational studies. *Annals of the Rheumatic Diseases*, *75*(3), 552–559.

Rosenfeld, B. A., Beattie, C., et al. (1993). The effects of different anesthetic regimens on fibrinolysis and the development of postoperative arterial thrombosis. PIRAJ Study Group. *Anesthesiology*, *79*, 435–443.

Rossi, P., Kuukasjarvi, P., et al. (1996). Prostacyclin and thromboxane A2 synthesis are increased in acute lower limb ischaemia. *Prostaglandins, Leukotrienes & Essential Fatty Acids*, *55*, 433–436.

Sarzi Puttini, P., Giorgi, V., Atzeni, F., et al. (2021). Fibromyalgia position paper. *Clinical and Experimental Rheumatology*, *suppl. 130*(3), 186–193.

Seewald, M., Coles, J. A., Jr., et al. (2017). Featured article: Pharmacological postconditioning with delta opioid attenuates myocardial reperfusion injury in isolated porcine hearts. *Experimental Biology and Medicine (Maywood, N.J.)*, *242*(9), 986–995.

Sellers, R. S., Radi, Z. A., & Khan, K. N. (2010). Pathophysiology of cyclooxygenase in cardiovascular homeostatis. *Veterinary Pathology*, *47*, 601–613.

Shawaqfeh, M. S., & Harrington, C. (2018). Pain: Systematic review of pharmacy compounding of pain medication. *International Journal of Pharmaceutical Compounding*, *22*(1), 19–24.

Silverstein, F. E., Faich, G., et al. (2000). Gastrointestinal toxicity with celecoxib vs nonsteroidal anti-inflammatory drugs for osteoarthritis and rheumatoid arthritis: The CLASS study: A randomized controlled trial. Celecoxib Long-term Arthritis Safety Study. *JAMA*, *284*(10), 1247–1255.

Siong Chan, D. C., Cao, T. H., & Ng, L. L. (2018). Proenkephalin in heart failure. *Heart Failure Clinics*, *14*(1), 1–11.

Smith, J. B., & Willis, A. L. (1971). Aspirin selectively inhibits prostaglandin production in human platelets. *Nature: New Biology*, *231*(25), 235–237.

Solomon, S. D., McMurray, J. J., et al. (2005). Adenoma prevention with celecoxib (APC) study investigators. Cardiovascular risk associated with celecoxib in a clinical trial for colorectal adenoma prevention. *The New England Journal of Medicine*, *352*(11), 1071–1080.

Solomon, S. D., Wittes, J., Finn, P. V., et al. (2008). Cardiovascular risk of celecoxib in 6 randomized placebo-controlled trials: The cross trial safety analysis. *Circulation*, *117*(16), 2104–2113.

Stanley, T. H., & Webster, L. R. (1978). Anesthetic requirements and cardiovascular effects of fentanyl-oxygen and fentanyl-diazepam-oxygen anesthesia in man. *Anesthesia and Analgesia*, *57*, 411–416.

Stevens, R. A., Beardsley, D., et al. (1995). Does the choice of local anesthetic affect the catecholamine response to stress during epidural anesthesia? *Anesthesiology*, *81*, 1169–1174.

Sudhakaran, S., Surani, S. S., & Surani, S. R. (2014). Prolonged ventricular asystole: A rare adverse effect of hydrocodone use. *American Journal of Case Reports*, *15*, 450–453.

Tanaka, K., Kersten, J. R., & Riess, M. L. (2014). Opioid-induced cardioprotection. *Current Pharmaceutical Design*, *20*(36), 5696–5705.

Treskatsch, S., Shaqura, M., et al. (2016). Evidence for MOR on cell membrane, sarcoplasmatic reticulum and mitochondria in left ventricular myocardium in rats. *Heart and Vessels*, *31*(8), 1380–1388.

Vane, J. R. (1971). Inhibition of prostaglandin synthesis as a mechanism of action for aspirin-like drugs. *Nature: New Biology*, *231*(25), 232–235.

Vane, J. R., Bakhle, Y. S., & Botting, R. M. (1998). Cyclooxygenases1 and 2. *Annual Review of Pharmacology and Toxicology*, *38*, 97–120.

Varrassi, G., Coaccioli, S., De Andres, J., et al. (2019). Expert consensus on clinical use of an orally administered dexketoprofen plus tramadol fixed-dose combination in moderate-to-severe acute pain: A Delphi study. *Advances in Therapy*, *36*(11), 3174–3185. https://doi.org/10.1007/s12325-019-01096-0.

Varrassi, G., Fusco, M., Coaccioli, S., & Paladini, A. (2015). Chronic pain and neurodegenerative process in elderly people. *Pain Practice*, *15*, 1–3. https://doi.org/10.1111/papr.12254.

Varrassi, G., Pergolizzi, J. V., Dowling, P., & Paladini, A. (2020). Ibuprofen safety at the golden anniversary: Are all NSAIDs the same? A narrative review. *Advances in Therapy*, *37*(1), 61–82. https://doi.org/10.1007/s12325-019-01144-9.

Varrassi, G., Pergolizzi, J. V., Peppin, J., & Paladini, A. (2020a). Analgesic drugs and cardiac safety. In S. Govoni, P. Politi, & E. Vanoli (Eds.), *Brain and heart dynamics* (pp. 649–670). Switzerland: Springer Nature. https://doi.org/10.1007/978-3-030-28008-6_43.

Varrassi, G., Yeam, C. T., Rekatsina, M., Pergolizzi, J. V., Zis, P., & Paladini, A. (2020). The expanding role of the COX inhibitor/opioid receptor agonist combination in the management of pain. *Drugs*, *80*(14), 1443–1453. https://doi.org/10.1007/s40265-020-01369-x.

Varrassi, G., Alon, E., Bagnasco, M., et al. (2019). Towards an effective and safe treatment of inflammatory pain: A Delphi-guided expert consensus. *Advances in Therapy*, *36*(10), 2618–2637. https://doi.org/10.1007/s12325-019-01053-x.

Varrassi, G., Hanna, M., Macheras, G., et al. (2017). Multimodal analgesia in moderate-to-severe pain: A role for a new fixed combination of dexketoprofen and tramadol. *Current Medical Research and Opinion*, *33*(6), 1165–1173. https://doi.org/10.1080/03007995.2017.1310092.

Walker, C., & Biasucci, L. M. (2018). Cardiovascular safety of non-steroidal anti-inflammatory drugs revisited. *Postgraduate Medicine*, *130*(1), 55–71.

Wallace, J. L., & Devchand, P. R. (2005). Emerging roles for cyclooxygenase-2 in gastrointestinal mucosal defense. *British Journal of Pharmacology*, *145*, 275–282.

Wedam, E. F., & Haigney, M. C. (2016). The impact of opioids on cardiac electrophysiology. *Current Cardiology Reviews*, *12*(1), 27–36.

Wiffen, P. J., Derry, S., et al. (2017). Gabapentin for chronic neuropathic pain in adults. *Cochrane Database of Systematic Reviews*, *6*(6), CD007938. Published 2017 Jun 9.

Wu, C., Fry, C. H., & Henry, J. (1997). The mode of action of several opioids on cardiac muscle. *Experimental Physiology*, *82*, 261–272.

Zaher, A., Radi, K., & Khan, N. (2019). Cardio-renal safety of non-steroidal anti-inflammatory drugs. *Journal of Toxicological Science*, *44*(6), 373–391.

Zidar, N., Dolenc-Strazar, Z., et al. (2007). Expression of cyclooxigenase-1 and cyclooxigenase-2 in the normal human heart and myocardial infarction. *Cardiovascular Pathology*, *16*, 300–304.

Zidar, N., Odar, K., et al. (2009). Cyclooxigenase in normal human tissue—Is COX-1 really a constitutive isoform, and COX-2 an inducible isoform? J. *Journal of Cellular and Molecular Medicines*, *13*(9B), 3753–3763.

Chapter 26

Computed tomography-guided procedures for epidural injections

Rohit Aiyer[a] **and Semih Gungor**[b]
[a]*Interventional Pain Management and Pain Psychiatry, Henry Ford Health System, Detroit, MI, United States,* [b]*Department of Anesthesiology, Critical Care and Pain Management, Hospital for Special Surgery, Weill Medical School of Cornell University, New York, NY, United States*

Introduction

Epidural steroid injections (ESIs) are among the most commonly performed interventional pain management procedures to treat radiculopathy (Gangi, Dietemann, Mortazavi, et al., 1998). Appropriate image guidance not only allows the interventionalist to avoid vulnerable vascular or neural structures during performance of ESI, but also is critical for appropriate needle placement, for monitoring injectate flow patterns and to ensure injectate is delivered to the correct target (Bogduk, 2013; Rathmell et al., 2015). It is not recommended to perform ESI without appropriate image guidance for treatment of spinal pain as it has been shown to be relatively ineffective, with reported studies indicating as much as 30% failure rate (Chandra, 2013; Vorobeychik et al., 2016). Moreover, the use of image guidance has been shown to reduce the risk of complications during ESI (Bogduk, 2013; Rathmell et al., 2015). Fluoroscopy is utilized more commonly for ESI and allows for accurate needle placement when used in conjunction with contrast agent. However, in certain conditions, CT-guided injections may be preferred over fluoroscopy due its advantage of better visualization of the bony and soft tissues such as the presence of deformities, severe degenerative changes, osteophytes, cysts, vascular structures, and pleura (Gangi et al., 1998). In this chapter, we aim to overview CT-guided ESIs.

General principles and technique common to all types of CT-guided epidural injections

Patient positioning

Appropriate positioning of the patient on the CT table is important. The patient should be positioned as comfortable as possible to prevent the patient movement, thereby to prevent alteration in subsequent CT images during the course of the procedure. The most optimal and stable positions for this purpose are the prone and supine positions. Lateral decubitus position should be avoided as much as possible as it is difficult to maintain stable patient position throughout the procedure (White, Derby, & Wynne, 1980). Increasing the forward flexion of spine with appropriate padding and support will facilitate in opening up the space between the laminae for interlaminar epidural injections.

Three phases of CT-guided interventions

(1) *Survey phase*: A skin grid marker is placed over the target entry site. Initial preliminary axial images are obtained at the target area (Figs. 1 and 2). The interventionalist reviews the images, correlating with the patient's preprocedure diagnostic images and clinical pathology. Correct spinal level is confirmed at this stage. Reviewing these images will allow the interventionalist to visualize and evaluate the relevant bony and soft-tissue structures along the projected needle trajectory. This is one of the advantages of CT-guided interventions over fluoroscopy. Based on this review of these preliminary images, the following parameters are determined to perform the procedure safely:
   (a) Needle entry point
   (b) Needle length requirement
   (c) Needle trajectory
   (d) The degree of needle angulation requirement, if needed.

The Neurobiology, Physiology, and Psychology of Pain. https://doi.org/10.1016/B978-0-12-820589-1.00026-9

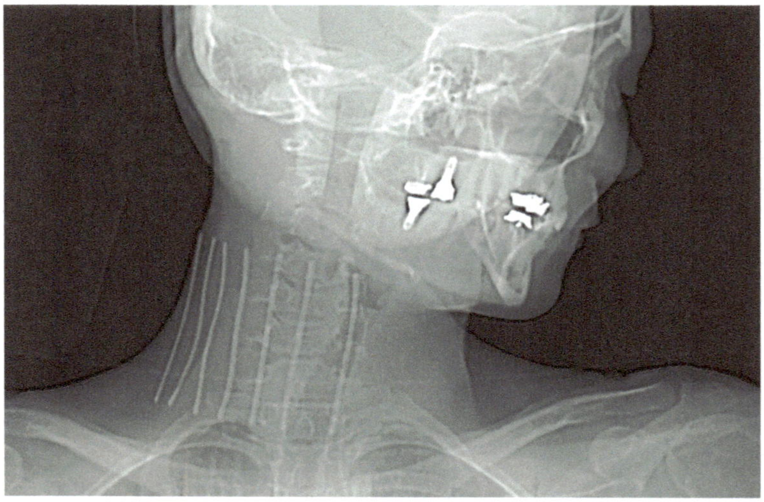

FIG. 1 A skin grid marker that is placed over the target area.

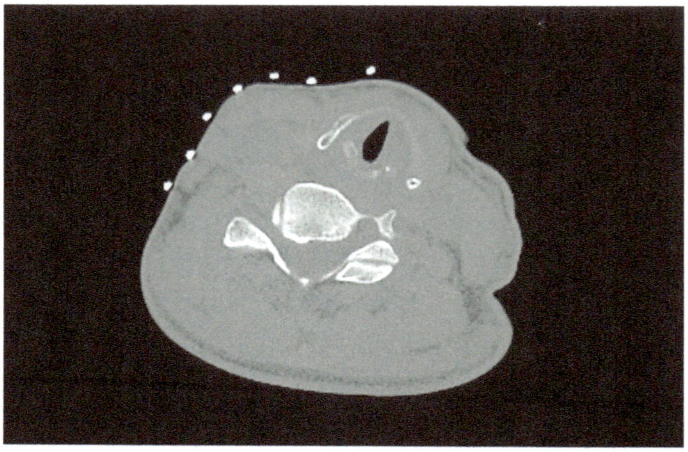

FIG. 2 Preliminary axial image with a skin grid marker to identify the optimum needle entry point at the target area.

(2) *Needle-guidance phase*: The interventionalist gradually advances the needle until target area is reached (Fig. 3). There are two types of CT scanners available to perform CT-guided interventions (White et al., 1980).
   (a) Conventional CT scan: After the needle is advanced, the interventionalist typically leaves the CT suite, the technician then obtains the predetermined axial images, typically a central image slice at the needle trajectory, and two additional images cranial and caudal to the central image slice. These additional slices are obtained to determine whether any cranial or caudal deviation of the needle is present from the predetermined needle trajectory. Ideally, the entire needle trajectory should remain in the central and single image slice. The interventionalist reviews the images for needle trajectory, plans the next step for needle guidance, and reenters the CT suite. The interventionalist does not necessarily need to wear a lead shielding during the procedure. And, this cycle is repeated every time a new CT image is required until the target point is reached and the procedure is completed (White et al., 1980).
   (b) CT Fluoroscopy (CTF): This is more advanced technology and less commonly available than the conventional CT scan. With this technology, the interventionalist may stay in the room to do a quick-check image with a foot pedal during the procedure for appropriate correction and stepwise guidance of the needle to the target. Therefore, the interventionalist is required to wear appropriate lead shielding during the procedure Although there is an option to

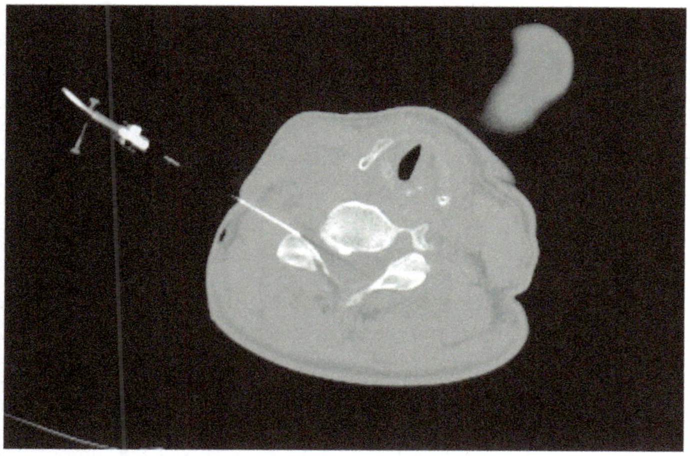

FIG. 3 The needle placement at the target area in the posterior neural foramen.

perform a continuous CT imaging with CTF similar to continuous fluoroscopy imaging, this mode is rarely used to reduce the radiation exposure. Instead, a quick-check CT image is preferred to avoid excessive radiation exposure (White et al., 1980).

(3) *Contrast injection phase*: Once the predetermined target area is reached, the needle tip must be confirmed under CT images. CT scan also allows for injection of small amount of air as a contrast, instead of a contrast agent, and the air will be visible under CT imaging. This is another advantage of CT guidance over fluoroscopy if in case the patient had history of allergic reaction to contrast agent. Most of the time, provided that there is no contraindication to contrast agent use, an injection of a small volume of contrast agent is preferred, as the liquid contrast agent will better mimic a subsequent medication injection. By obtaining new images at this time, appropriate distribution of contrast is verified (Fig. 4). If the desired distribution is not achieved, the needle position is further corrected. The same steps need to be repeated before the optimum needle position is achieved (White et al., 1980).

Epidural injections

The ultimate choice of approach or technique (interlaminar vs transforaminal) to use should be made by the treating physician by balancing potential risks and benefits with each technique for each patient (Bogduk, 2013; Rathmell et al., 2015).

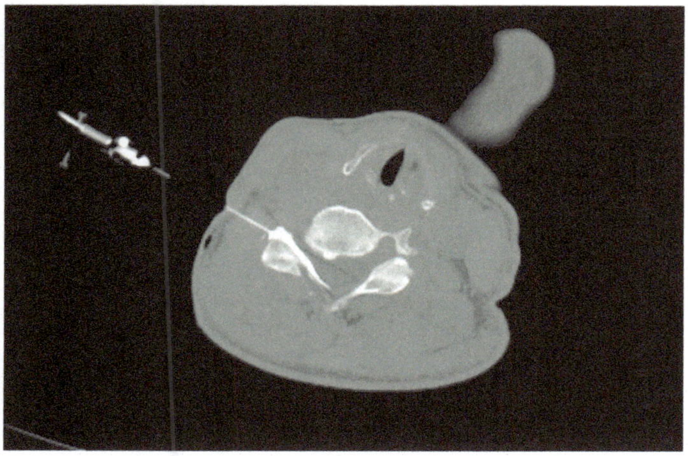

FIG. 4 Contrast injected through the needle and the spread of the contrast medially to the neural foramen and epidural space.

Interlaminar epidural injections

CT-guided interlaminar ESIs are performed with a similar technique for cervical, thoracic, and lumbar levels. The patient is placed in a prone position on the CT table. Mild spine flexion is recommended for better visualization of the interlaminar space (Gangi et al., 1998). After proper positioning of the patient, CT scanning is then utilized to visualize the anatomy of the target area (Gangi et al., 1998). Appropriate needle entry point and the other parameters such as needle length, trajectory, and angulation are determined as described earlier. Local anesthetic is injected to skin and subcutaneous tissues at the needle entry point (Gangi et al., 1998). For interlaminar approach, we recommend slight paramedian approach ipsilateral to the more symptomatic side. The following are the justifications for this recommendation:

(1) Ligamentum flavum has been shown in anatomical studies to have a developmental midline defect. This is due to failure of ligamentum flavum to fuse at the midline during the developmental stage in some patients. This midline defect in ligamentum flavum is very common in the cervical and upper thoracic levels (Lirk et al., 2003). Inability to feel the change in resistance with "the loss of resistance" technique commonly used in interlaminar epidural injections may lead to inadvertent dural puncture and associated complications.
(2) Ligamentum flavum is thicker in the paramedian area which can improve the tactile recognition of engagement of blunt tip epidural needle to the ligamentum flavum before "the loss of resistance" is felt.
(3) It has been shown on studies that the medications delivered via paramedian approach tend to stay in the ipsilateral epidural space without crossing to the contralateral side (Gungor & Aiyer, 2017), and one such clinical study showed that the medication delivered via paramedian approach may have similar outcomes to those of transforaminal approach after ESIs (Candido, Raghavendra, Chinthagada, Badiee, & Trepashko, 2008).
(4) If bilateral symptoms are present requiring epidural injection, bilateral approach to interlaminar ESI may be considered, as the delivery of medication is mainly confined to ipsilateral epidural space when a small volume of injectate is used (Gungor & Aiyer, 2017).

Depending on the clinician's preference, a 17- to 22-gauge blunt tip epidural needle is slowly advanced toward the target. Loss of resistance to air or normal saline is used during advancement of the epidural needle depending on the preference of the clinician. There is usually an increased resistance applied to the plunger when the needle tip engaged into the ligamentum flavum. At this stage, the needle is advanced with minimal movements and with intermittent checking of "loss of resistance" to plunger. Once the loss of resistance is felt, the needle tip should be in the epidural space. There should be no fluids aspirated from the epidural needle. If any clear fluid suggestive of cerebrospinal fluid or blood is observed, the procedure should be aborted, and the patient should be monitored closely. It is recommended that the procedure should be rescheduled at a later date (Bogduk, 2013; Rathmell et al., 2015). If there is a negative response to gentle aspiration of the needle, a small amount of contrast (1–2 mL) should be injected to verify the appropriate epidural spread, and to rule out any intravascular or intrathecal uptake (Gangi et al., 1998). More importantly, intrathecal contrast spread is easily identifiable with CT guidance, allowing the provider to be aware if their needle is in the cerebrospinal fluid (CSF) (Gangi et al., 1998). Based on literature, <1 mL of contrast medium is required to confirm the accurate needle placement. Hence, this is another advantage of CT guidance, as less volume of contrast agent is needed, when compared to fluoroscopy-guided interventions (Cho, 2010; Gangi et al., 1998).

CT guidance is effective for interlaminar epidural injections, as it can help to avoid bone structures and gain access to the epidural space even in a patient population with severe degenerative changes in the spinal anatomy and with technically difficult access to the interlaminar space (Gangi et al., 1998). In addition, CT guidance allows for better visualization of pathological structures such as synovial cysts, epidural scarring, and thecal sac deformity (Gangi et al., 1998).

Interlaminar ESIs can be performed without contrast in patients with a documented contraindication to use of contrast media such as significant history of contrast allergy. However, it is important to note that this should only be attempted by physicians who are experienced with "loss of resistance" technique (Bogduk, 2013; Rathmell et al., 2015). In these circumstances, particulate steroids are contraindicated and only nonparticulate steroids should be used (Bogduk, 2013; Rathmell et al., 2015). In addition to reviewing preliminary CT images, cervical interlaminar ESI should not be attempted without reviewing prior MRI images at the target level to confirm that there is adequate epidural space for needle entry (Bogduk, 2013; Rathmell et al., 2015). Cervical interlaminar ESIs are recommended to be performed at C7-T1, but preferably not higher than the C6-C7 level (Bogduk, 2013; Rathmell et al., 2015).

Cervical transforaminal epidural steroid injections

Epidural injections are commonly performed for cervical radiculopathy (Vorobeychik et al., 2016). Interlaminar approach is more commonly utilized for cervical epidural injections. Selective nerve root blocks and transforaminal approaches may sometimes be preferred over interlaminar approach for diagnostic and/or therapeutic purposes. Epidural adhesions and scar tissue tend to develop as part of the normal healing process following spinal surgery via posterior approach. These adhesions increase the risk of inadvertent dural puncture with use of interlaminar approach and prevent the distribution of medications to the area of pathology. Therefore, the main advantage of transforaminal approach is that it allows for the steroid medication to be injected near the nerve root and close to the area of pathology (Hoang et al., 2013). CT guidance offers the advantage of improved soft-tissue visualization of vascular structures abundant in the cervical spine (Hoang et al., 2013). Various studies describe different sites of needle placement that is thought to provide optimal pain relief (Hoang et al., 2013). Cho describes a method that utilizes an extraforaminal technique with CT guidance (Cho, 2010). Wolter et al. discuss positioning the patient prone versus supine allowing them to place the needle extraforaminally from a dorsal approach (Wolter, Mohadjer, Berlis, et al., 2009). Finally, Wagner and Cyteval et al. describe an approach where the needle is located at the outer edge of the posterior foramen (Cyteval, Thomas, Decoux, et al., 2004; Wagner, 2005). As a result, there is no concrete evidence as to which approach is superior for cervical transforaminal ESI.

Technique

The anterolateral approach in supine position is widely utilized. After proper positioning of the patient, preliminary CT images are obtained to visualize the targeted area (Gangi et al., 1998) (Figs. 1 and 2). Appropriate needle entry point and the other parameters such as needle length, trajectory, and angulation are determined as described earlier in this manuscript. Local anesthetic is injected to skin and subcutaneous tissues at the needle entry point. Usually 25-gauge needle is preferred. Low-volume extension tubing is usually attached to the needle and the extension tubing is connected to a syringe at the distal end. The needle, extension tubing, and syringe are prefilled with contrast agent prior to the procedure. Extension tubing is recommended for all transforaminal ESIs to minimize movement of the needle tip once it has reached its appropriate target (Bogduk, 2013; Rathmell et al., 2015). This is to avoid unnecessary needle manipulation once the needle tip is at the target area. Once the needle is positioned at the posterior aspect at the entrance of the foramen (Fig. 3), gentle aspiration is performed to rule out any presence of blood or cerebrospinal fluid. Then, 0.2–0.5 mL of contrast is injected. Real-time imaging is recommended during contrast injection to determine any vascular uptake for fluoroscopy-guided injections, and similar mode is also available in new-generation CTF scanners. The risks versus benefit should be determined by the interventionalist, as CTF increases the risk of radiation exposure. Following contrast injection, new axial CT scan images are obtained to verify the needle tip position and the contrast spread. Appropriate contrast spread should be along the nerve root and partially entering into the epidural space medially (Fig. 4). Complete extraforaminal spread, intrathecal spread, or intravascular spread should be ruled out (Hoang et al., 2013). The contrast spread allows for interventionalist to better assess distribution of steroid medication. After injection of contrast, the syringe is detached and syringe containing steroid is attached and negative aspiration is repeated at this stage before the steroid medication is injected. Cervical transforaminal ESIs should not be performed in patients with a documented contraindication to use of contrast medium such as significant history of contrast allergy (Bogduk, 2013; Rathmell et al., 2015). Only nonparticulate steroid is recommended for cervical transforaminal steroid injection due to the risk of vascular infarct secondary to intraarterial injection that can potentially result in catastrophic complications (Bogduk, 2013; Rathmell et al., 2015).

Interestingly, there are limited reports of vascular injection with cervical transforaminal steroid injections under CT guidance in the literature. Hoang et al. demonstrated that intravascular injections are uncommon (only 1 case) with CT imaging as the carotid artery, vertebral artery, and jugular vein can be visualized on cross-sectional imaging and thus avoided (Vorobeychik et al., 2016), although small segmental arteries may not be easily visualized. Comparatively, vascular structures cannot be directly visualized under fluoroscopy unless they are calcified, resulting in this modality having a potentially higher intravascular injection risk compared to CT-guided injections (Hoang et al., 2013).

Thoracic transforaminal epidural steroid injections

Interlaminar approach is more commonly utilized for thoracic epidural injections. Selective nerve root blocks and transforaminal approach may sometimes be preferred over interlaminar approach for diagnostic and/or therapeutic purposes. Epidural scar tissue and adhesions develop following the posterior thoracic spinal surgery. These adhesions increase the risk of inadvertent dural puncture during interlaminar approach and prevent the distribution of medications to the area of pathology. Therefore, the main advantage of this type of transforaminal approach is that it allows for the steroid medication to be injected near the nerve root and close to the area of pathology (Hoang et al., 2013). CT guidance offers the advantage of improved soft-tissue visualization of vascular structures, pleura, and lungs (Hoang et al., 2013). This is important to note as this injection technique involves risk of pneumothorax. Intraarterial injection risk is also higher and may lead to catastrophic spinal cord infarct. Thus, the clinician should be very familiar with the anatomy of this region

(Mandell, Czuczman, Gaviola, Ghazikhanian, & Cho, 2017). The artery of Adamkiewicz is a branch of the aorta and supply anterior spinal artery via anterior radiculomedullary arteries. The origin of artery of Adamkiewicz arises usually between T9 and L5, and most commonly at T9 and very rarely below L2 (23.5% occurrence) (Mandell et al., 2017).

Technique

The patient is placed in a prone position on the CT table. Appropriate needle entry point and the other parameters such as needle length, trajectory, and angulation are determined as described earlier in this manuscript. Local anesthetic is injected to skin and subcutaneous tissues at the needle entry point (Gangi et al., 1998). Usually, a 25-gauge needle is recommended for thoracic transforaminal epidural injections. However, some clinicians prefer 22-gauge needle due to its stiffness and better steerability of this size needle. Low-volume extension tubing is usually attached to needle and the extension tubing is connected to a syringe at the distal end. The needle, extension tubing, and syringe are prefilled with contrast agent prior to the procedure. Extension tubing is recommended for all transforaminal ESIs to minimize movement of the needle tip once it has reached its appropriate target (Bogduk, 2013; Rathmell et al., 2015). This is to avoid unnecessary needle movements once the needle tip is at the target area. The interventionalist should maintain the needle posterior to the rib to avoid possible puncture of pleura leading to pneumothorax (Rosas & Gilula, 2010). Once the needle is positioned at the posterior aspect at the entrance of the foramen, gentle aspiration is performed to rule out any presence of blood or cerebrospinal fluid. Then, 0.2–0.5 mL of contrast is injected. Real-time imaging is recommended during contrast injection to determine any vascular uptake for fluoroscopy-guided injections and this mode is also available in new-generation CTF scanners. The risks versus benefit should be determined by the interventionalist, as CTF increases the risk of radiation exposure. Following contrast injection, new axial CT scan images are obtained to verify the needle tip position and the contrast spread. Appropriate contrast spread should along the nerve root and partially entering into the epidural space medially. Complete extraforaminal spread, intrathecal spread, or intravascular spread should be ruled out (Hoang et al., 2013). The contrast spread allows for interventionalist to better assess distribution of steroid medication. After injection of contrast, the syringe is detached and the syringe containing steroid is attached and negative aspiration is repeated at this stage before the steroid medication is injected. Due to risk of vascular infarct secondary to intraarterial injection, nonparticulate steroid is recommended for thoracic transforaminal steroid injections (Bogduk, 2013; Rathmell et al., 2015).

Rosas and Gilula were able to review 198 thoracic foraminal nerve root blocks and ESIs under fluoroscopic guidance; however, many of the complications they observed were similar to what can be observed under CT guidance (Rosas & Gilula, 2010). These minor complications included vasovagal response, headache, lightheadedness, and transient pain. One of the major complications that should be monitored for is pneumothorax, which in this case was only observed in one patient; however, patient recovered with spontaneous resolution and did not require chest tube placement (Rosas & Gilula, 2010). CT-guided epidural injection may be advantageous as it is more sensitive in detecting small pneumothorax in suspected cases postprocedure as compared to fluoroscopy or chest-X ray imaging.

Lumbar transforaminal steroid injections

As mentioned earlier for thoracic epidural injections, the origin of artery of Adamkiewicz may extend to L5 level, but most commonly above the L2 level (Mandell et al., 2017). This is important to note as this injection technique involves risk of intraarterial injection specifically in the upper lumbar levels. Thus, the clinician should be very familiar with anatomy of this region and avoid particulate steroids above L3 level (Mandell et al., 2017).

Technique

The patient is placed in a prone position on the CT table. Technically, the most challenging lumbar spinal level is usually the lowermost lumbar level at L5-S1 where the L5 nerve roots exit the foramina. There are at least two reasons: First, the iliac crest may be in the needle trajectory most often requiring the angulation of needle. Second, most significant degenerative changes such as severe facet arthropathy, synovial cysts, and osteophytes are present in the lower lumbar spinal levels. These types of anatomical changes may pose technical challenges, with extra planning for needle trajectory during preliminary CT images. Appropriate needle entry point and the other parameters such as needle length, trajectory, and angulation are determined as described earlier in this manuscript. Local anesthetic is injected to skin and subcutaneous tissues at the needle entry point. A 22-gauge needle is recommended for lumbar transforaminal epidural injections. As mentioned earlier, using a 25-gauge needle is not recommended as the flexibility of this smaller gauge needle interferes with the steerability of the needle during needle guidance phase due to target point being deeper than cervical and thoracic levels. Proceeding with lumbar transforaminal steroid injection can be carried out with two techniques. The first is known as the "safe triangle" method, with CT imaging helping the clinician target the anterior epidural space within the anterior margin of the neural foramen. Essentially, the "triangle" anatomically is bordered by the inferior margin of the pedicle, the exiting nerve root, and an imaginary line drawn inferiorly from the anterior margin of the pedicle (Mandell et al., 2017). This approach is also known as the subpedicular or supraneural approach, and was initially described as "safe" due to the relatively low anatomic risk of nerve injury (Mandell et al., 2017). The other approach is known as the Kambin triangle, which is also a triangular area located over the dorsal part of the intervertebral disk. Kambin triangle is bordered by the superior endplate of the lower vertebral body inferiorly, the hypotenuse is the exiting nerve root and the posterior margin of the triangle is an imaginary line formed by the endplate inferiorly and the superior articulating facet superiorly (Mandell et al., 2017). With CT guidance, the target is also the anterior epidural space that is located dorsal to the intervertebral disk (Mandell et al., 2017). Similarly, to the safe triangle approach, contrast injection should be able to confirm epidural and perineural flow. The interventionalist must be cautious not to advance the needle too ventrally as it can lead to intradiscal injection (Mandell et al., 2017). Once the needle is positioned at the posterior aspect at the entrance of the foramen, gentle aspiration is performed to rule out any presence of blood or cerebrospinal fluid. Extension tubing is recommended for all transforaminal ESIs to minimize movement of the needle tip after it has reached its appropriate target (Bogduk, 2013; Rathmell et al., 2015). Then, 0.2–0.5 mL of contrast is injected. Subsequently, new axial CT scan images are obtained to verify the needle tip position and the contrast spread. Appropriate contrast spread should be visualized along the nerve root and partially entering into the epidural space medially. Complete extraforaminal spread, intrathecal spread, or intravascular spread should be ruled out (Hoang et al., 2013). The contrast spread allows for interventionalist to better assess distribution of steroid medication. After injection of contrast, the syringe is detached and the syringe containing steroid is attached and negative aspiration is repeated at this stage prior to injection of steroid medication. Due to risk of vascular infarct secondary to intraarterial injection, nonparticulate steroid, such as dexamethasone, is recommended for lumbar transforaminal ESIs as a first-line medication (Bogduk, 2013; Rathmell et al., 2015). If inadequate clinical response is obtained by using dexamethasone after the first injection, particulate steroids may also be used for transforaminal ESIs in the lower lumbar levels as a second-line therapy. Transforaminal ESIs can be performed without contrast in patients with a documented contraindication to contrast media such as history of contrast allergy, but in these circumstances, only preservative-free, particulate-free steroids should be used (Bogduk, 2013; Rathmell et al., 2015).

Park et al. reviewed 100 cases and found no difference in functional improvement between the Kambin triangle approach when compared to the safe triangle approach. Intraarterial and intradisc injection occurred in both techniques (intraarterial was reported in 4% Kambin triangle and 12% for safe triangle, intradisc injection was reported in 2% for Kambin triangle and 4% for safe triangle) (Park et al., 2011). Another study by Levi et al. retrospectively reviewed 257 transforaminal steroid injections that compared Kambin versus "safe" triangle approaches. The investigators found that with the Kambin triangle approach, complication rates were 4.7% intradiscal injection rate, 3.1% intrathecal injection rate, and 6.6% intravascular injection rate (Levi, Horn, & Corcoran, 2016). In comparison, the rate of complications with "safe" triangle technique was lower, with literature reporting intradiscal injection between 0.17% and 2.3%, intrathecal injections at 0.04%, and intravascular being 11.2% (Mandell et al., 2017).

In their systematic review, Bui and Bogduk critically analyzed four studies that investigated the use of CT guidance for lumbar transforaminal ESIs (Bui & Bogduk, 2013). However, the authors discuss in their systematic review a particular study by Lutze et al. (Lutze, Stendel, Vesper, & Brock, 1997). Lutze et al. studied 80 patients who underwent either CT-guided ($n = 40$) or fluoroscopy-guided ($n = 40$) ESI for lumbar radiculopathy (Bui & Bogduk, 2013). This head-to-head study involved patients that were followed up at 1 and 6 months, and the head-to-head comparison showed that at 6 months, CT-guided injections had a significantly better outcome ($P < .001$) (Lutze et al., 1997). It should be noted, however, that the authors of this study state that "the differences in the median values among the two groups were not large enough to exclude the possibility that the difference is due to random sampling variability" (Lutze et al., 1997). However, due to the lack of data by Lutze et al. to support this statement, Bui and Bogduk suggest that the "differences observed were neither statistically nor clinically significant" (Bui & Bogduk, 2013). In studies reviewed by Bui and Bogduk, the first study they reviewed was conducted by Lee et al. that treated 140 patients with lumbar radiculopathy that showed a decrease in mean pain score from 7.3 to 3.9 and 3.8 at 1 and 3 months, respectively (Lee, Lin, Hwang, Howing, & Wang, 2005). The second study reviewed treated 37 patients with lumbar radicular pain due to foraminal stenosis or disk herniation, and at 6 months, 23 patients had 50% relief with an average pain score of 1/10 (Wewalka, Abdelrahimsai, Wiseinger, & Uher, 2012). The other study reviewed 42 patients with lumbar radicular pain due to spinal stenosis, and the authors reported a statistically significant decrease of median pain scores from 9 to 2.5 at 6 months posttreatment (Karaeminogullari, Sahin, Boyvat, et al., 2005). However, Bui and Bogduk noted in their review that no data from this study were reported to allow for calculation of success rates of the injection (Berger, Dousset, Delmer, et al., 1999). The final study in this review analyzed a study by Berger et al., with 139 patients having lumbar transforaminal steroid injections, and the remaining patients having either thoracic or cervical transforaminal steroid injections (Berger et al., 1999). Berger et al. report that 68 patients had long-lasting relief, with a follow-up time period of 14.8 ± 9.1 months (Berger et al., 1999). Additionally, 34 patients were reported to have had temporary relief (Berger et al., 1999). However, Bui and Bogduk mentioned the shortcoming in this study was that the magnitude of relief was not reported, and therefore, it was difficult to assess the success of the CT-guided injections (Bui & Bogduk, 2013). Overall, Bui and Bogduk did not feel that the studies were robust, or had enough statistical analysis to provide enough evidence in support of transforaminal steroid injections under CT guidance (Bui & Bogduk, 2013).

Radiation safety

Radiation exposure should be taken into consideration with CT-guided lumbar epidural injections. For background information, there are two types of radiation effects on humans: stochastic (such as cancer induction or genetic impact) and deterministic (tissue necrosis, gastrointestinal damage, erythema, or hematopoietic damage) (Bui & Bogduk, 2013). Clinical symptoms from radiation exposure usually occur at doses greater than 0.5 Gy (Bui & Bogduk, 2013). It is thought that leukemia is associated with chronic radiation exposures as low as 50–100 cGY, and a linear correlation between dose and incidence in doses ranging between 100 and 500 cGY (Bui & Bogduk, 2013). Other cancer types are also a risk with ionizing radiation, which include thyroid, bone, and lung (Bui & Bogduk, 2013). Therefore, for both patients and clinicians, ionizing radiation should be a concern when utilizing CT guidance for epidural injections.

For interventional spine procedures, such as epidural injections, there have been a significant number of publications investigating the radiation exposure with these specific procedures. It should be noted that radiation exposure varies greatly, as there are different factors that contribute to exposure which include technique, proceduralist, and CT scanner settings (Artner, Lattig, Reichel, & Cakir, 2012). Artner et al. suggest in their publication that while the radiation exposure of a standard CT protocol was 1.5–3.53 mSv, this could be reduced between 0.22 and 0.43 mSv if a low-dose protocol was utilized (Artner et al., 2012).

Shepherd et al. reported a mean dose length product (DLP) for all procedures decreased from 1458 ± 1022 to 199 ± 101 mGy cm ($P < .001$) (Shepherd, Hess, Chin, Gould, & Dillon, 2011). Schmid et al. compared the radiation doses in CT- and fluoroscopy-guided epidural injections using an Alderson Rando phantom at the lumbar segment L4/5 (Schmid, Schmitz, Borchardt, et al., 2006). They described effective doses of 1.51–3.53 mSv (4–10 scans) for CT guidance for standard protocols and 0.22–0.43 mSv (4–10 scans) for low-dose protocols (Schmid et al., 2006). Hoang et al. calculated mean radiation doses of 0.85 mSv for lumbar ESIs using conventional fluoroscopy guidance and 3.35 mSv for CT-fluoroscopy guidance, when a full diagnostic lumbar scan was performed as a part of the procedure (Hoang, Yoshizumi, Toncheva, et al., 2011). Richards et al. calculated a relative risk ratio for inducing a cancer of 1:3200 for diagnostic scans of the whole lumbar spine in standard diagnostic protocol and 1:200,000 for low-dose protocol with an estimated effective dose of 0.1 mSv. Narrowing the area of interest to dorsal structures from L3 to L5 (effective dose of 3.5 mSv) in standard protocol also reduced the risk to 1:5200 (Richards, George, Metelko, & Brown, 2010). Therefore, it is evident that utilizing certain techniques and settings can help modify and alleviate radiation exposure. Other methods and protocols include: switching from helical to axial image acquisition, reducing tube current, narrowing the area of interest, and reducing scatter exposure (use of lead drape and radiation protection gloves) (Hoang et al., 2013). There have seen several large-scale studies that indicate and give evidence for the radiation safety dose of CT-guided epidural injections. One retrospective study by Artner et al. investigated 1286 lumbar epidural injections under CT guidance, and reviewed what would be the most effective radiation dose without decreasing significantly image quality (Artner et al., 2012). The results showed that 3–6 scans were needed for each procedure, with an average intervention time of 5–10 min. Consequently, for epidural lumbar injections, a mean effective radiation dose of 1.34 mSv (CI 95%, 1.30–1.38) was calculated (Artner et al., 2012).

Complications

While CT-guided epidural injections are generally quite safe, possible complications need to be discussed. Depriester et al. performed over 1529 CT-guided cervical and lumbar transforaminal injections, with 24.4% being cervical and the other 75.6% lumbar (Depriester, Setbon, Larde, et al., 2012). The complications that were observed include blood pressure surges (0.2%) in three cases (Depriester et al., 2012). Vasovagal syncope occurred in 2.3% (35 cases) of the study population; however, all resolved spontaneously after leg elevation (Depriester et al., 2012). Allergic reaction occurred in 0.3% (5 cases) of the population, with main presenting symptoms being dyspnea and facial erythema (Depriester et al., 2012). These were treated and reaction resolved quickly after given IV injection of 200 mg of hydrocortisone. Finally, reproduction of pain during the procedure was quite frequent (65%) in the lumbar region, which also at times included paresthesia production (Depriester et al., 2012). Finally, the authors noted that a certain number of patients had increased radicular pain after the injection, which was similar to the report of the case series by Huston et al. that reported an incidence of 37.4% (Huston, Slipman, & Garvin, 2005). Bui and Bogduk in their review stated that of the eight reported cases of spinal cord infarction following lumbar transforaminal injection, five involved the use of CT guidance (Bui & Bogduk, 2013).

This was also seen in a study by Atluri et al. that reported over 18 cases of paralysis after lumbar transforaminal ESIs at every level of the lumbar spine as well as T12-S1 utilizing both CT and fluoroscopic guidance (Atluri, Glaser, Shah, & Sudarshan, 2013). Therefore, it is critical to be aware that utilization of CT guidance does not eliminate the risk of potential intravascular injection. This may be attributed to either unavailability of real-time image guidance during contrast injection in conventional CT scanners, or less frequent use of real-time imaging during contrast injection prior to steroid medication injection in CTF due to concerns of excessive radiation exposure. Real-time contrast injection is recommended in guidelines, prior to steroid medication injection for fluoroscopy-guided interventions (Bogduk, 2013; Rathmell et al., 2015).

Conclusion

CT-guided epidural injections are an excellent alternative when considering the type of image guidance. Overall, this technique is accurate and safe when carrying out an epidural injection. It allows for precise visualization of the spine anatomy, as well as for proper placement of the needle in the desire anatomical location. Most importantly, CT-guided epidural injections can be most beneficial in patients with difficult anatomy, such as minimal epidural space access or severe spinal stenosis. However, as noted earlier, the major limitation for CT-guided epidural injections is increased radiation exposure. It should be noted that much of the data analyzed regarding radiation safety are almost 10 years old, and therefore, updated data on this section are warranted to provide clinicians with more accurate clinical analysis regarding safety (Table 1). Nevertheless, with proper technique and skilled expertise, this issue can be mitigated. It should also be noted that cost and availability are other major limiting factors for CT-guided injections, and, as a result, will unlikely to become the gold standard among interventionalists in the near future. Nevertheless, this is a safe and effective method to carry out epidural injections, and, if resources allow, use of CT guidance may be preferable in selected group of patients for ESIs.

Applications to other areas

None.

Other agents of interest

None.

TABLE 1 Summary of CT-guided interventions.

Advantages:
Better visualization of anatomical structures
Air or less dose of contrast can be used
Disadvantages:
Increased radiation exposure
Generally longer procedure time
Expensive and less readily available than fluoroscopy
Inability to perform real-time contrast injection

Mini-dictionary of terms

ESI: epidural steroid injection
CT: computed tomography
CTF: computed tomography fluoroscopy
CSF: cerebrospinal fluid
MRI: magnetic resonance imaging
DLP: dose length product

Key facts

Computed tomography is an alternative image guidance while performing ESIs.

Summary points

- CT guidance allows for precise visualization of the spine anatomy, as well as for proper placement of the needle in the desire anatomical location.
- CT guidance may be preferable in selected group of patients with difficult anatomy, such as minimal epidural space access or severe spinal stenosis.
- Cost, availability, and increased radiation exposure are the major limiting factors for CT-guided injections when compared to fluoroscopy-guided ESIs.

References

Artner, J., Lattig, F., Reichel, H., & Cakir, B. (2012). Effective dose of CT-guided epidural and periradicular injections of the lumbar spine: A retrospective study. *The Open Orthopaedics Journal*, *6*, 357–361.

Atluri, S., Glaser, S. E., Shah, R. V., & Sudarshan, G. (2013). Needle position analysis in cases of paralysis from transforaminal epidurals: Consider alternative approaches to traditional technique. *Pain Physician*, *16*(4), 321–334.

Berger, O., Dousset, V., Delmer, O., et al. (1999). Evaluation de l'efficacite des infiltrations foraminales de croticoides guides sous tomodensitometrie, dans le traitement des radiculagies par conflit foraminal. *Journal de Radiologie*, *80*, 917–925.

Bogduk, N. (Ed.). (2013). *Practice guidelines for spinal diagnostic and treatment procedures* International Spine Intervention Society.

Bui, J., & Bogduk, N. (2013). A systematic review of the effectiveness of CT-guided, lumbar transforaminal injection of steroids. *Pain Medicine*, *14*(12), 1860–1865.

Candido, K. D., Raghavendra, M. S., Chinthagada, M., Badiee, S., & Trepashko, D. W. (2008). A prospective evaluation of iodinated contrast flow patterns with fluoroscopically guided lumbar epidural steroid injections: The lateral parasagittal interlaminar epidural approach versus the transforaminal epidural approach. *Anesthesia & Analgesia*, *106*(2), 638–644.

Chandra, R. V. (2013). Computed tomography guidance for spinal intervention: Basics of technique, pearls, and avoiding pitfalls. *Pain Physician*, *16*, 369–377.

Cho, C. H. (2010). Cervical nerve injection: Computed tomography guidance with intravenous contrast and extraforaminal needle placement. Series of seven consecutive case reports. *The Spine Journal*, *10*, e1–e6.

Cyteval, C., Thomas, E., Decoux, E., et al. (2004). Cervical radiculopathy: Open study on percutaneous periradicular foraminal steroid infiltration performed under CT control in 30 patients. *AJNR. American Journal of Neuroradiology*, *25*, 441–445.

Depriester, C., Setbon, S., Larde, A., et al. (2012). CT-guided transforaminal cervical and lumbar epidural injections. *Diagnostic and Interventional Imaging*, *93*(9), 704–710.

Gangi, A., Dietemann, J. L., Mortazavi, et al. (1998). CT guided interventional procedures for pain management in the lumbosacral spine. *Radiographics*, *18*, 621–633.

Gungor, S., & Aiyer, R. (2017). A computed tomography-guided demonstration of the unilateral distribution of low-volume epidural Injectate. *Pain Practice*, *17*(7), 952–955.

Hoang, J. K., Massoglia, D. P., Apostol, C. D., Lascola, J. D., Eastoowd, & Kranz, P. G. (2013). *American Journal of Neuroradiology*, *34*(3), 688–692.

Hoang, J. K., Yoshizumi, T. T., Toncheva, G., et al. (2011). Radiation dose exposure for lumbar spine epidural steroid injections: A comparison of conventional fluoroscopy data and ct fluoroscopy techniques. *American Journal of Roentgenology*, *197*(4), 778–782.

Huston, C. W., Slipman, C. W., & Garvin, C. (2005). Complications and side effects of cervical and lumbosacral selective nerve root injections. *Archives of Physical Medicine and Rehabilitation*, *86*, 277–283.

Karaeminogullari, O., Sahin, O., Boyvat, F., et al. (2005). Transforaminal epidural steroid injection under computed tomography guidance in relieving lumbosacral radicular pain. *Acta Orthopaedica et Traumatologica Turcica*, *39*, 416–420.

Lee, K. S., Lin, C. L., Hwang, S. L., Howing, S. L., & Wang, C. K. (2005). Transforaminal periradicular infiltration guided by CT for unilateral sciatica—An outcome study. *Clinical Imaging*, *29*(3), 211–214.

Levi, D., Horn, S., & Corcoran, S. (2016). The incidence of intradiscal, intrathecal and intravascular flow during the performance of retrodiscal (infraneural) approach for lumbar transforaminal epidural steroid injections. *Pain Medicine*, *17*(8), 1416–1422.

Lirk, P., Kolbitsch, C., Putz, G., Colvin, J., Colvin, H. P., Lorenz, I., … Moriggl, B. (2003). Cervical and high thoracic ligamentum flavum frequently fails to fuse in the midline. *Anesthesiology*, *99*(6), 1387–1390.

Lutze, M., Stendel, R., Vesper, J., & Brock, M. (1997). Periradicular therapy in lumbar radicular syndromes: Methodology and results. *Acta Neurochirurgica (Wien)*, *139*, 719–724.

Mandell, J. C., Czuczman, G. J., Gaviola, G. C., Ghazikhanian, V., & Cho, C. (2017). The lumbar neural foramen and transforaminal epidural steroid injections: An anatomic review with key safety considerations in planning the percutaneous approach. *American Journal of Roentgenology*, *209*, W26–W35.

Park, J. W., Nam, H. S., Cho, S. K., Jung, H. J., Lee, B. J., & Park, Y. (2011). Kambin's triangle approach of lumbar transforaminal epidural injection with spinal stenosis. *Annals of Rehabilitation Medicine*, *35*(6), 833–843.

Rathmell, J. P., Benzon, H. T., Dreyfuss, P., Huntoon, M., Wallace, M., Baker, R., … Buvanendran, A. (2015). Safeguards to prevent neurologic complications after epidural steroid injections consensus opinions from a multidisciplinary working group and national organizations. *Anesthesiology*, *122*(5), 974–984.

Richards, P. J., George, J., Metelko, M., & Brown, M. (2010). Spine computed tomography doses and cancer induction. *Spine*, *35*(4), 430–433.

Rosas, H. G., & Gilula, L. A. (2010). Performing thoracic transforaminal injections: A new technique. *Radiology*, *254*(2), 595–600.

Schmid, G., Schmitz, A., Borchardt, D., et al. (2006). Effective dose of CT and fluoroscopy-guided perineural/epidural injections of the lumbar spine: A comparative study. *Cardiovascular and Interventional Radiology*, *29*(1), 84–91.

Shepherd, T. M., Hess, C. P., Chin, C. T., Gould, R., & Dillon, W. P. (2011). Reducing patient radiation dose during CT-guided procedures: Demonstration in spinal injections for pain. *American Journal of Neuroradiology*, *32*(10), 1776–1782.

Vorobeychik, Y., Sharma, A., Smith, C. C., Miller, D. C., Stojanovic, M. P., Lobel, S. M., … Standards Division of the Spine Intervention Society. (2016). The effectiveness and risks of non-image-guided lumbar interlaminar epidural steroid injections: A systematic review with comprehensive analysis of the published data. *Pain Medicine*, *17*(12), 2185–2202.

Wagner, A. L. (2005). CT fluoroscopic-guided cervical nerve root blocks. *AJNR. American Journal of Neuroradiology*, *26*, 43–44.

Wewalka, M., Abdelrahimsai, A., Wiseinger, G. F., & Uher, E. M. (2012). CT-guided transforaminal epidural injections with local anesthetic, steroid, and tramadol for the treatment of persistent lumbar radicular pain. *Pain Physician*, *15*, 153–159.

White, A. H., Derby, R., & Wynne, G. (1980). Epidural injections for the diagnosis and treatment of low back pain. *Spine*, *5*, 67–68.

Wolter, T., Mohadjer, M., Berlis, A., et al. (2009). Cervical CT-guided, selective nerve root blocks: Improved safety by dorsal approach. *AJNR. American Journal of Neuroradiology*, *30*, 336–337.

Chapter 27

Chronic pain: Linking deep brain stimulation and sensory functional MRI

Witold H. Polanski and Johann Klein
Department of Neurosurgery, University Hospital Carl Gustav Carus, of the Technical University of Dresden, Dresden, Germany

Abbreviations

CT	computer tomography
DBS	deep brain stimulation
DTI	diffusion tension imaging
fMRI	functional magnetic resonance imaging
PLIC	posterior limb of the internal capsule
PVG	periventricular gray
STC	spinothalamocortical tract
VPL	nucleus ventralis posterolateralis of the thalamus

Introduction

Chronic pain is a challenging healthcare issue, as about 1.5% of the population suffer from medically refractory pain (Bouhassira, Lanteri-Minet, Attal, Laurent, & Touboul, 2008). It is associated with impairment in cognition and attention (Moriarty, McGuire, & Finn, 2011) as well as decline of social functioning due to frequent dependence on opioids (Gureje, Von, Simon, & Gater, 1998). Particularly, severe medically refractory neuropathic pain occurs in 25% of patients after brachial plexus injury (Narakas, 1978). These patients often suffer from continuous burning or throbbing pain which seems to be associated with thalamic neuroplasticity and shooting pain attacks due to hyperactivity of the dorsal horn (Parry, 1980, 1984). After failure of pharmacotherapy and physical therapy, neuromodulatory treatment options like spinal cord stimulation, motor cortex stimulation, and deep brain stimulation (DBS) may have a favorable effect (Farrell, Green, & Aziz, 2018).

History of deep brain stimulation for chronic pain

DBS is an established therapy for movement disorders and has been performed in patients with chronic pain since the early 1960s (Gol, 1967). Thereby, leads are implanted into the brain to stimulate specific areas, and thus, to modulate brain circuit activity arising from pathological synaptic interactions between multiple nuclei (Johnson, Miocinovic, McIntyre, & Vitek, 2008; McIntyre & Anderson, 2016). The implanted DBS system allows for patient-individual programming of stimulation parameters (e.g., electrode polarity, frequency, stimulation amplitude, impulse width) (Fig. 1A and B). Several studies reported inconsistent pain reduction as well as unwanted stimulation-dependent effects with DBS (for review see Farrell et al., 2018). The first trial conducted in 1976 was limited by poor enrolment, a high attrition rate, a very heterogeneous patient population, and inconsistent deep brain targets (Coffey, 2001). Consequently, DBS for chronic pain was branded investigational as off-label use.

Target determination for DBS was originally based on standard stereotactic coordinates from atlases (Harary & Cosgrove, 2019; Schaltenbrand, 1959) using the ventricular system within the brain (ventriculography) as a point of reference. The introduction of computer tomography (CT) in 1972 and magnetic resonance imaging (MRI) in 1977 allowed for a more specific visualization of the brain anatomy. But as standard coordinates were still used for target determination, interindividual anatomic differences were not considered. This harbored the risk of incorrect lead placement, leading to limited effectiveness and unwanted stimulation effects.

The Neurobiology, Physiology, and Psychology of Pain. https://doi.org/10.1016/B978-0-12-820589-1.00027-0

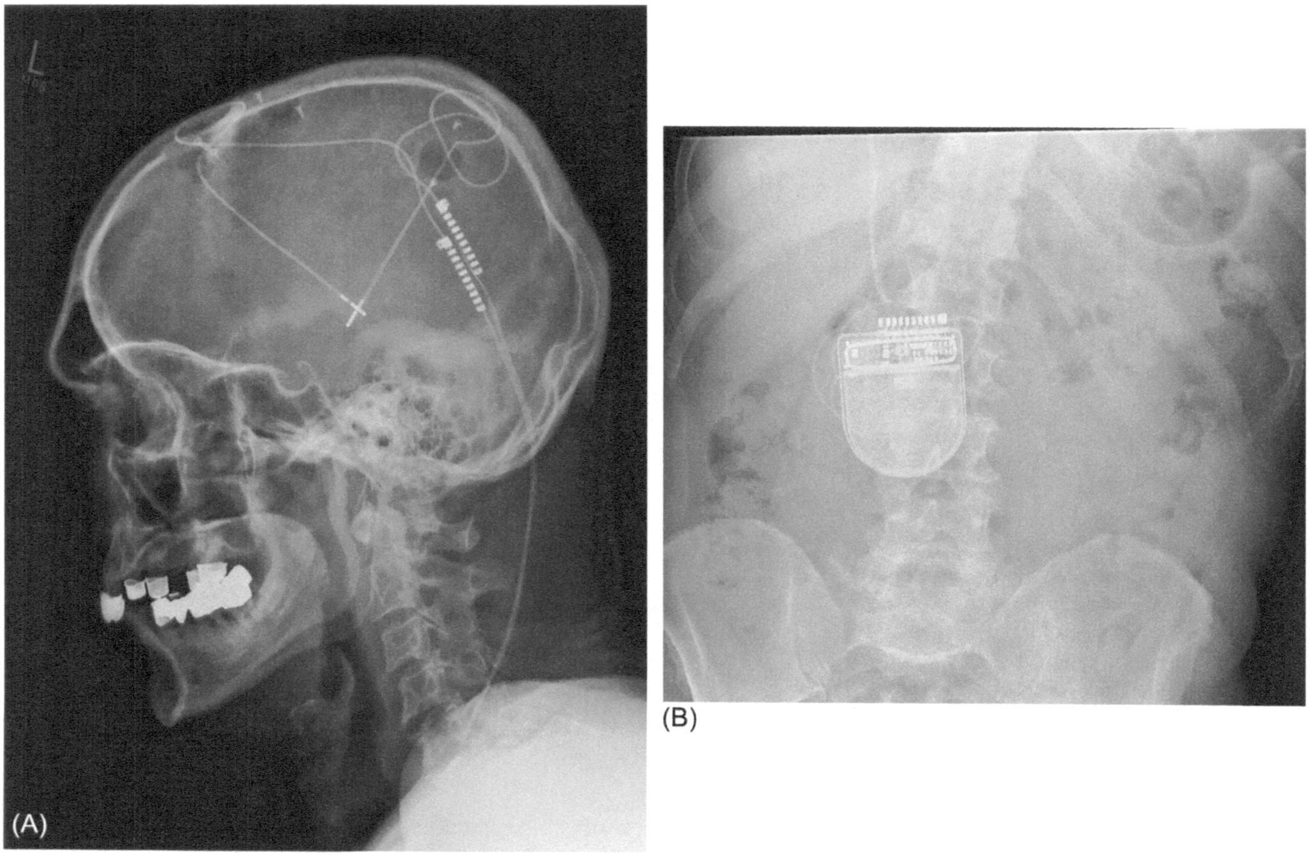

FIG. 1 X-rays of the DBS system. Implanted intracerebral electrodes in the posterior limb of the internal capsule (PLIC) and nucleus ventralis posterolateralis (VPL) are connected with subcutaneous extensions (A) and the stimulator for deep brain stimulation (B).

Central pain processing

The target regions for DBS in chronic pain vary, whereby the sensory nuclei of the thalamus (nucleus ventralis posterolateralis, VPL and nucleus ventralis posteromedialis, VPM) (Mark & Ervin, 1965; Turnbull, Shulman, & Woodhurst, 1980), the periventricular gray (PVG) (Hosobuchi, Adams, & Linchitz, 1977), and the posterior limb of the internal capsule (PLIC) (Adams, Hosobuchi, & Fields, 1974) seem to be promising regions. The ventral posterior nucleus complex of the thalamus seems to be the main relay for sensory-discriminative nociceptive processing of pain. It is connected to various cortical and subcortical structures (Montes et al., 2005) as well as the spinal cord. In chronic pain patients, a hypoperfusion of the thalamus was observed (Di Piero et al., 1991) and could be restored through motor cortex stimulation, leading to pain reduction (Peyron, Faillenot, Mertens, Laurent, & Garcia-Larrea, 2007). The VPL showed abnormal firing patterns in chronic pain patients due to an absence of normal sensory input (Gerhart, Yezierski, Fang, & Willis, 1983). The PVG is part of a descending pathway that modulates nociception by reducing spinal cord excitability (Mayer & Liebeskind, 1974; Porreca, Ossipov, & Gebhart, 2002). Tracey et al. showed that distraction reduced pain intensity and increased PVG activity (Tracey et al., 2002). Thereby, the stimulation of the PVG was generally recommended for treatment of nociceptive pain, while the stimulation of internal capsule or the thalamus was preferred for neuropathic pain (Bittar et al., 2005). Various other brain regions including the limbic system are involved in descending modulation and constitute the pain neuromatrix (Melzack, 1990, 2001). Within the neuromatrix theory, chronic pain is not considered a consequence of continuous noxious stimulation. It is rather caused by complex processing of information influenced by existing synaptic architecture, which is determined by sensory and also genetic factors. Limbic structures such as the anterior cingulate cortex, the prefrontal cortex, and the nucleus accumbens play a crucial role in placebo analgesic effects (Wager et al., 2004) since they alter the affective component of pain through modulation of inhibitory projections to the medial thalamus and the dorsal horn neurons (Lorenz, Minoshima, & Casey, 2003).

Modern visualization of DBS targets

Despite modern MRI, the mentioned targets cannot be visualized directly. Therefore, the targeting is performed indirectly by using the third ventricle or the border of the internal capsule as reference points. To address this limitation, newer developments like diffusion tensor imaging (DTI) and functional MRI (fMRI) were implemented to allow for patient-specific target delineation with regard to anatomical variability. In a pilot study, Polanski et al. used fMRI for the visualization of the cortical sensory region in patients suffering from chronic neuropathic pain after brachial plexus injury. This was performed while stimulating the painful region with a surgical brush during the fMRI acquisition. Afterward, the activated cortex region and the sensory thalamus were defined as primary regions of interest for fibertracking in the DTI and the visualized tract was targeted for DBS. The combination of DTI and sensory fMRI-based individual DBS targeting of the PLIC with stimulation of the PVG or the VPL led to a mean reduction in permanent pain of 59.1% and in pain attacks of 86.2% (Polanski et al., 2019) (Figs. 2 and 3). Comparable studies revealed a general pain reduction of 40% (Abreu et al., 2017) and 43% (Boccard et al., 2017); however, the differentiation between permanent pain and pain attacks so far was not described in the available literature. While the stimulation of the PVG and VPL reduced permanent pain but had no effect on pain attacks, the stimulation of the PLIC was highly efficient for the pain attacks. The rationale for the combination of differential stimulation sites were synergistic analgesic effects of more than one anatomical site of stimulation (Bittar et al., 2005; Nandi, Aziz, Carter, & Stein, 2003). fMRI- and DTI-based targeting of specific targets including the PLIC should be considered for a patient-individual approach to treating chronic pain syndromes by DBS in the future. As revealed by reconstruction of the lead coordinates and fusion with DTI data, stimulation within the PLIC seems to modulate fibers of the spinothalamocortical tract (STC) (Hong, Son, & Jang, 2010). It has been reported to also alleviate thalamic pain (Hunsche et al., 2013). During pain processing, the STC exerts an enhanced descending influence on ventral posterior thalamic nuclei (Wang et al., 2007) and is involved in the sensory discriminative aspect of pain (Marchand, 2008). The stimulation of patients with plexus injury requires a precisely selected stimulation site to allow for discrimination of somatotopic pain regions. This is in contrast to thalamic pain, which affects the entire contralateral half of the body. fMRI- and DTI-based DBS within the PLIC combined with PVG or VPL stimulation showed promising results in patients with chronic neuropathic pain after brachial plexus injury (Polanski et al., 2019).

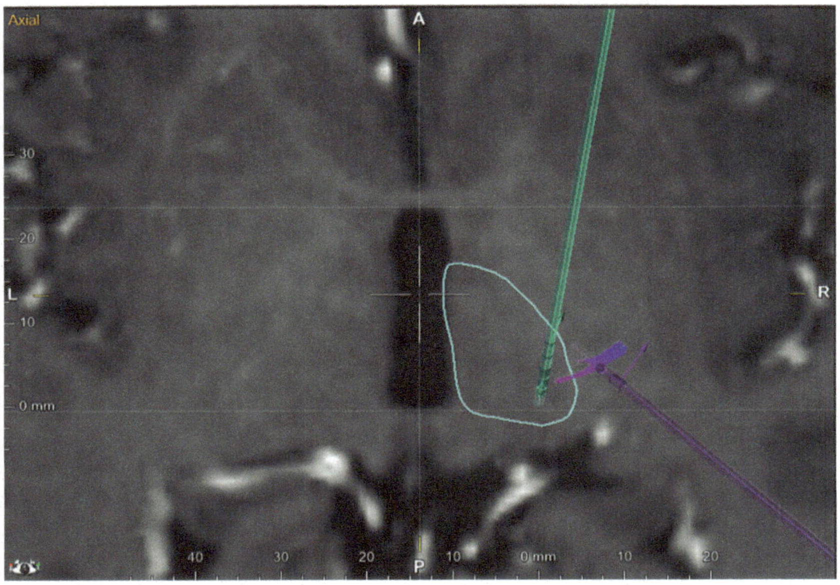

FIG. 2 Axial CT with planned leads for DBS. Axial CT at the level of the anterior and posterior commissure with marked thalamus *(blue)* and the spinothalamocortical tracts *(violet)* in the posterior limb of the internal capsule (PLIC). The implanted leads in the nucleus ventralis posterolateralis (VPL, *green*) and the PLIC were reconstructed in the preoperative MRI after fusion with a postoperative CT with BrainLab iPlan 3.0 (BrainLab AG, Feldkirchen, Germany).

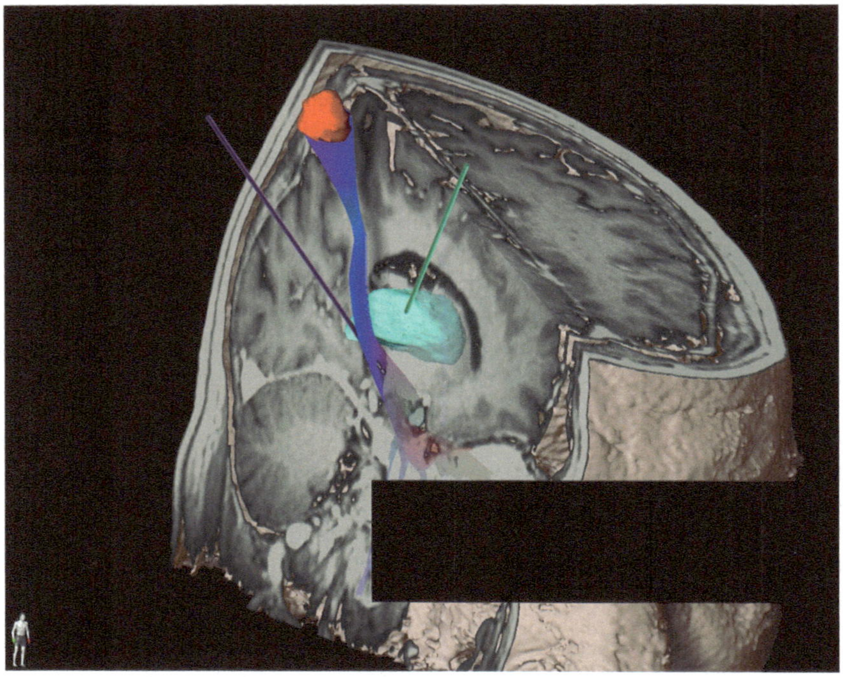

FIG. 3 3D reconstruction of the planned trajectories and the targeted structures. The planned electrode trajectory for the posterior limb of the internal capsule (PLIC, *violet*) and for the nucleus ventralis posterolateralis (VPL, *green*) of the thalamus *(blue)*. The postcentral sensory fMRI activation of the pain region (hand) is marked *red*, the spinothalamocortical tract was reconstructed from this area using diffusor tensor imaging. BrainLab iPlan 3.0 (BrainLab AG, Feldkirchen, Germany) was used for fMRI analysis, fibertracking, and trajectory planning.

Application to other areas

Several studies showed a superiority of DTI-based DBS compared to conventional targeting methods for the treatment of depression (Coenen et al., 2018), obsessive-compulsive disorder (Coenen et al., 2017), and pain (Coenen, Kieselbach, Mader, & Reinacher, 2015; Hunsche et al., 2013). The combination of standard MRI sequences with DTI also improved the therapeutic effects of DBS in movement disorders (Seddighi, Seddighi, & Nikouei, 2017) as the stimulated fibers could be targeted more precisely (Nowacki, Schlaier, Debove, & Pollo, 2018). The combination of DTI and fMRI is becoming increasingly important in cognitive and clinical neuroscience. The advantage of combining both methods is the opportunity to examine the relationship between anatomical structures and their functions and to visualize functionally related networks (Schonberg, Pianka, Hendler, Pasternak, & Assaf, 2006). Furthermore, combining DTI and fMRI can be used to facilitate structural connectivity studies under pathological conditions (Whitwell et al., 2011) and for guidance of surgical interventions (Zhu et al., 2014). Therefore, this combined method may lead to a more patient-individual approach with better clinical outcomes. Future connectivity studies based on DTI and fMRI data may further advance our understanding of functional and dysfunctional networks within the brain.

Other methods of interest

Since DBS is classified as an investigational procedure, other established neuromodulation techniques in different areas can also be used for pain treatment. Options like motor cortex stimulation and spinal cord stimulation show beneficial effects in chronic pain patients. The motor cortex serves as a convenient "entry point" for neuromodulation since stimulation of the motor cortex modulates distant areas as detected by increased regional blood flow in various structures like the thalamus, the insula, the frontal lobe, the cingulate gyrus, and the upper brainstem (Garcia-Larrea & Peyron, 2007). Furthermore, motor cortex stimulation triggers the corticothalamic output by changing the concentration of various neurotransmitters in the limbic system, brainstem, and spinal cord resulting in the modulation of emotional aspects of pain by inhibitory effects on these pathways (DosSantos, Ferreira, Toback, Carvalho, & DaSilva, 2016; Garcia-Larrea et al., 1999). Also, the anterior cingulate cortex seems to be involved in pain processing via neural plasticity, since it has projections deeply related to the motor cortex (Zhuo, 2008). The implantation of epidural or subdural electrodes led to a pain reduction in chronic pain patients with a response rate of 72%, while noninvasive techniques like transcranial magnetic stimulation or transcranial direct current stimulation showed a response rate of 45% with fewer risks for the patients (Klein et al., 2015; Lima & Fregni, 2008). The reason for a smaller effect of noninvasive stimulation may be the short duration of stimulation with less than five sessions. Both techniques are promising therapeutic options for medically refractory chronic pain.

Shealy, who described paresthesia due to electrical stimulation of the dorsal column, performed the first spinal cord stimulation in 1967 (Shealy, Taslitz, Mortimer, & Becker, 1967). Since then, this method has been used worldwide with rapid clinical implementation of novel methods including high-frequency stimulation and burst stimulation. The gate-control therapy of pain, proposed by Melzack and Wall (1965), spurred the development of spinal cord stimulation as a target for neuromodulation. Clinical studies demonstrated benefits in patients with neuropathic pain (Linderoth & Foreman, 2017) and there is evidence that also nociceptive pain can be accessed by spinal cord stimulation (Yang et al., 2020). The pain relief after spinal cord stimulation is thought to be mediated through modulation of maladaptive response to local injury in the segmental and supraspinal levels. There is evidence that spinal cord stimulation modulates the activity of the spinothalamic tract (Foreman, Beall, Coulter, & Willis, 1976) with connections to the PVG and the reticular formations (Jundi, Saade, Banna, & Jabbur, 1982; Saade, Salibi, Banna, Towe, & Jabbur, 1983). As different neuromodulation techniques target various areas within the pain matrix, the neurosurgeon is challenged with defining the best treatment modality for the individual patient.

Mini-dictionary of terms

Deep brain stimulation (DBS): Deep brain stimulation has been an established treatment option for movement disorders for 30 years. During the procedure, leads are implanted into the brain to stimulate specific areas and modulate brain circuit activity arising from pathological synaptic interactions.

Periventricular gray (PVG): The PVG is a small nucleus around the cerebral aqueduct within the tegmentum of the midbrain. It plays a crucial role in enkephalin-dependent descending pain.

Nucleus ventralis posterolateralis (VPL): The VPL is one of 13 nuclei within the thalamus. It is a relay for nociceptive, tactile, and proprioceptive information from the body.

Posterior limb of the internal capsule (PLIC): The PLIC is an area of white matter within the brain that contains ascending and descending axons of sensory fibers of the thalamus, acoustic fibers, fibers of the optic radiation, and fibers from the occipital and temporal lobes to the pontine nuclei.

Sensory functional MRI: During sensory functional MRI, the brain activity is measured by detecting changes of the blood flow. During fMRI, specific tactile stimulation of body regions is performed in order to activate and visualize the corresponding sensory cortex.

Diffusion tensor imaging: This is a technique to visualize neuronal tracts. It is based on anisotropic water diffusion along axons.

Key facts

Key facts of deep brain stimulation

- The first DBS was performed for parkinsonian tremor and was reported by Albe Fessard 1963.
- The procedure is an established therapy option for Parkinson's disease, essential tremor, dystonia, or obsessive-compulsive disorder and is experimentally used in psychiatric disorders like depression, anxiety, or dementia.
- Although the exact therapeutic mechanism is unknown, the effects may be explainable by rebalancing neural circuits.
- Over 160,000 patients worldwide have been treated with DBS for a variety of neurological and psychiatric conditions as of 2018.
- Although DBS for chronic pain has been performed since the early 1960s, the initial FDA approval for chronic pain was rejected 1989 because of inconsistent pain reduction in different studies.

Key facts of sensory functional MRI

- The group of Seji Ogawa discovered the fMRI technique in 1990 but the underlying hypothesis was not confirmed until 2001 by Nikos Logothetis and his group.
- The first fMRI in human was performed by John Belliveau after visual stimulation in 1991.
- In contrast to conventional MRI, functional MRI is based on oxyhemoglobin instead of water molecules.

- fMRI can be compared to other functional imaging methods with spatial and temporal resolution like positron emission tomography, near infrared spectroscopy, electroencephalography, and magnetoencephalography.
- Usually, the stimuli for fMRI are presented via visual, auditory, or locomotive tasks but can involve more complex tests with odors, tastes, emotions, memory, and thoughts.

Key facts of diffusion tensor imaging (DTI)

- DTI is an MRI imaging technique that measures changes of extracellular water diffusion in white matter and thus can reconstruct trajectories of white matter tracts.
- It shows not only the orientation of fibers but also quantifications of integrity and structural connectivity.
- The validity of this method is limited due to poor resolution, inaccuracies, possible misregistrations with anatomic images, and the inability to resolve complex fiber crossing.
- Various studies have shown improved outcomes of patients who underwent DBS using DTI for target determination.

Summary points

- Established therapy options like motor cortex stimulation, spinal cord stimulation, or transcranial magnetic stimulation are promising for medically refractory chronic pain patients and worth further research.
- DBS for chronic pain is branded investigational as off-label use.
- The inconsistent pain reduction with DBS in several studies can be explained with poor enrolment, high attrition rates, a very heterogeneous patient population, and inconsistent deep brain targets.
- The visualization and targeting of specific areas of the pain matrix for DBS can be improved by fMRI and DTI. This may help to improve the clinical results.
- The advantage of combining both methods is the opportunity to examine the relationship between anatomical structures and their functions and to visualize more functionally related networks under pathological conditions.
- Future connectivity studies based on DTI and fMRI data may result in an advanced understanding of functional and dysfunctional networks within the brain and lead to more efficient targeting.

References

Abreu, V., Vaz, R., Rebelo, V., Rosas, M. J., Chamadoira, C., Gillies, M. J., … Pereira, E. A. C. (2017). Thalamic deep brain stimulation for neuropathic pain: Efficacy at three years' follow-up. *Neuromodulation*, *20*(5), 504–513. https://doi.org/10.1111/ner.12620.

Adams, J. E., Hosobuchi, Y., & Fields, H. L. (1974). Stimulation of internal capsule for relief of chronic pain. *Journal of Neurosurgery*, *41*(6), 740–744. https://doi.org/10.3171/jns.1974.41.6.0740.

Bittar, R. G., Kar-Purkayastha, I., Owen, S. L., Bear, R. E., Green, A., Wang, S., & Aziz, T. Z. (2005). Deep brain stimulation for pain relief: A meta-analysis. *Journal of Clinical Neuroscience*, *12*(5), 515–519. https://doi.org/10.1016/j.jocn.2004.10.005.

Boccard, S. G. J., Prangnell, S. J., Pycroft, L., Cheeran, B., Moir, L., Pereira, E. A. C., … Aziz, T. Z. (2017). Long-term results of deep brain stimulation of the anterior cingulate cortex for neuropathic pain. *World Neurosurgery*, *106*, 625–637. pii: S1878-8750(17)31081-1 https://doi.org/10.1016/j.wneu.2017.06.173.

Bouhassira, D., Lanteri-Minet, M., Attal, N., Laurent, B., & Touboul, C. (2008). Prevalence of chronic pain with neuropathic characteristics in the general population. *Pain*, *136*(3), 380–387. pii: S0304-3959(07)00449-6 https://doi.org/10.1016/j.pain.2007.08.013.

Coenen, V. A., Kieselbach, K., Mader, I., & Reinacher, P. C. (2015). Diffusion tensor magnetic resonance imaging (DTI) tractography-guided deep brain stimulation in neuropathic pain. *Acta Neurochirurgica*, *157*(4), 739–741. https://doi.org/10.1007/s00701-015-2356-1.

Coenen, V. A., Sajonz, B., Reisert, M., Bostroem, J., Bewernick, B., Urbach, H., … Madler, B. (2018). Tractography-assisted deep brain stimulation of the superolateral branch of the medial forebrain bundle (slMFB DBS) in major depression. *NeuroImage: Clinical*, *20*, 580–593. https://doi.org/10.1016/j.nicl.2018.08.020.

Coenen, V. A., Schlaepfer, T. E., Goll, P., Reinacher, P. C., Voderholzer, U., Tebartz van Elst, L., … Freyer, T. (2017). The medial forebrain bundle as a target for deep brain stimulation for obsessive-compulsive disorder. *CNS Spectrums*, *22*(3), 282–289. https://doi.org/10.1017/S1092852916000286.

Coffey, R. J. (2001). Deep brain stimulation for chronic pain: Results of two multicenter trials and a structured review. *Pain Medicine*, *2*(3), 183–192. https://doi.org/10.1046/j.1526-4637.2001.01029.x.

Di Piero, V., Jones, A. K., Iannotti, F., Powell, M., Perani, D., Lenzi, G. L., & Frackowiak, R. S. (1991). Chronic pain: A PET study of the central effects of percutaneous high cervical cordotomy. *Pain*, *46*(1), 9–12. https://doi.org/10.1016/0304-3959(91)90026-t.

DosSantos, M. F., Ferreira, N., Toback, R. L., Carvalho, A. C., & DaSilva, A. F. (2016). Potential mechanisms supporting the value of motor cortex stimulation to treat chronic pain syndromes. *Frontiers in Neuroscience*, *10*, 18. https://doi.org/10.3389/fnins.2016.00018.

Farrell, S. M., Green, A., & Aziz, T. (2018). The current state of deep brain stimulation for chronic pain and its context in other forms of neuromodulation. *Brain Sciences*, *8*(8). https://doi.org/10.3390/brainsci8080158. pii: brainsci8080158.

Foreman, R. D., Beall, J. E., Coulter, J. D., & Willis, W. D. (1976). Effects of dorsal column stimulation on primate spinothalamic tract neurons. *Journal of Neurophysiology*, *39*(3), 534–546. https://doi.org/10.1152/jn.1976.39.3.534.

Garcia-Larrea, L., & Peyron, R. (2007). Motor cortex stimulation for neuropathic pain: From phenomenology to mechanisms. *NeuroImage*, *37*(Suppl. 1), S71–S79. https://doi.org/10.1016/j.neuroimage.2007.05.062.

Garcia-Larrea, L., Peyron, R., Mertens, P., Gregoire, M. C., Lavenne, F., Le Bars, D., … Laurent, B. (1999). Electrical stimulation of motor cortex for pain control: A combined PET-scan and electrophysiological study. *Pain*, *83*(2), 259–273. https://doi.org/10.1016/s0304-3959(99)00114-1.

Gerhart, K. D., Yezierski, R. P., Fang, Z. R., & Willis, W. D. (1983). Inhibition of primate spinothalamic tract neurons by stimulation in ventral posterior lateral (VPLc) thalamic nucleus: Possible mechanisms. *Journal of Neurophysiology*, *49*(2), 406–423. https://doi.org/10.1152/jn.1983.49.2.406.

Gol, A. (1967). Relief of pain by electrical stimulation of the septal area. *Journal of the Neurological Sciences*, *5*(1), 115–120. pii: 0022-510X(67)90012-3.

Gureje, O., Von, K. M., Simon, G. E., & Gater, R. (1998). Persistent pain and well-being: a World Health Organization Study in Primary Care. *JAMA*, *280* (2), 147–151. pii: joc72070.

Harary, M., & Cosgrove, G. R. (2019). Jean Talairach: A cerebral cartographer. *Neurosurgical Focus*, *47*(3), E12. https://doi.org/10.3171/2019.6.FOCUS19320.

Hong, J. H., Son, S. M., & Jang, S. H. (2010). Identification of spinothalamic tract and its related thalamocortical fibers in human brain. *Neuroscience Letters*, *468*(2), 102–105. pii:S0304-3940(09)01420-7 https://doi.org/10.1016/j.neulet.2009.10.075.

Hosobuchi, Y., Adams, J. E., & Linchitz, R. (1977). Pain relief by electrical stimulation of the central gray matter in humans and its reversal by naloxone. *Science*, *197*(4299), 183–186. Retrieved from http://www.ncbi.nlm.nih.gov/pubmed/301658.

Hunsche, S., Sauner, D., Runge, M. J., Lenartz, D., El, M. F., Treuer, H., … Maarouf, M. (2013). Tractography-guided stimulation of somatosensory fibers for thalamic pain relief. *Stereotactic and Functional Neurosurgery*, *91*(5), 328–334. pii: 000350024 https://doi.org/10.1159/000350024.

Johnson, M. D., Miocinovic, S., McIntyre, C. C., & Vitek, J. L. (2008). Mechanisms and targets of deep brain stimulation in movement disorders. *Neurotherapeutics*, *5*(2), 294–308. https://doi.org/10.1016/j.nurt.2008.01.010.

Jundi, A. S., Saade, N. E., Banna, N. R., & Jabbur, S. J. (1982). Modification of transmission in the cuneate nucleus by raphe and periaqueductal gray stimulation. *Brain Research*, *250*(2), 349–352. https://doi.org/10.1016/0006-8993(82)90429-2.

Klein, M. M., Treister, R., Raij, T., Pascual-Leone, A., Park, L., Nurmikko, T., … Oaklander, A. L. (2015). Transcranial magnetic stimulation of the brain: Guidelines for pain treatment research. *Pain*, *156*(9), 1601–1614. https://doi.org/10.1097/j.pain.0000000000000210.

Lima, M. C., & Fregni, F. (2008). Motor cortex stimulation for chronic pain: Systematic review and meta-analysis of the literature. *Neurology*, *70*(24), 2329–2337. https://doi.org/10.1212/01.wnl.0000314649.38527.93.

Linderoth, B., & Foreman, R. D. (2017). Conventional and novel spinal stimulation algorithms: Hypothetical mechanisms of action and comments on outcomes. *Neuromodulation*, *20*(6), 525–533. https://doi.org/10.1111/ner.12624.

Lorenz, J., Minoshima, S., & Casey, K. L. (2003). Keeping pain out of mind: The role of the dorsolateral prefrontal cortex in pain modulation. *Brain*, *126*(Pt 5), 1079–1091. https://doi.org/10.1093/brain/awg102.

Marchand, S. (2008). The physiology of pain mechanisms: From the periphery to the brain. *Rheumatic Diseases Clinics of North America*, *34*(2), 285–309. pii: S0889-857X(08)00028-8 https://doi.org/10.1016/j.rdc.2008.04.003.

Mark, V. R., & Ervin, F. R. (1965). Role of thalamotomy in treatment of chronic severe pain. *Postgraduate Medicine*, *37*, 563–571. Retrieved from http://www.ncbi.nlm.nih.gov/pubmed/14285799.

Mayer, D. J., & Liebeskind, J. C. (1974). Pain reduction by focal electrical stimulation of the brain: An anatomical and behavioral analysis. *Brain Research*, *68*(1), 73–93. https://doi.org/10.1016/0006-8993(74)90534-4.

McIntyre, C. C., & Anderson, R. W. (2016). Deep brain stimulation mechanisms: The control of network activity via neurochemistry modulation. *Journal of Neurochemistry*, *139*(Suppl. 1), 338–345. https://doi.org/10.1111/jnc.13649.

Melzack, R. (1990). Phantom limbs and the concept of a neuromatrix. *Trends in Neurosciences*, *13*(3), 88–92. https://doi.org/10.1016/0166-2236(90)90179-e.

Melzack, R. (2001). Pain and the neuromatrix in the brain. *Journal of Dental Education*, *65*(12), 1378–1382. Retrieved from https://www.ncbi.nlm.nih.gov/pubmed/11780656.

Melzack, R., & Wall, P. D. (1965). Pain mechanisms: A new theory. *Science*, *150*(3699), 971–979. https://doi.org/10.1126/science.150.3699.971.

Montes, C., Magnin, M., Maarrawi, J., Frot, M., Convers, P., Mauguiere, F., & Garcia-Larrea, L. (2005). Thalamic thermo-algesic transmission: Ventral posterior (VP) complex versus VMpo in the light of a thalamic infarct with central pain. *Pain*, *113*(1–2), 223–232. https://doi.org/10.1016/j.pain.2004.09.044.

Moriarty, O., McGuire, B. E., & Finn, D. P. (2011). The effect of pain on cognitive function: A review of clinical and preclinical research. *Progress in Neurobiology*, *93*(3), 385–404. pii: S0301-0082(11)00003-7 https://doi.org/10.1016/j.pneurobio.2011.01.002.

Nandi, D., Aziz, T., Carter, H., & Stein, J. (2003). Thalamic field potentials in chronic central pain treated by periventricular gray stimulation – A series of eight cases. *Pain*, *101*(1–2), 97–107. https://doi.org/10.1016/s0304-3959(02)00277-4.

Narakas, A. (1978). Surgical treatment of traction injuries of the brachial plexus. *Clinical Orthopaedics and Related Research*, (133), 71–90. Retrieved from http://www.ncbi.nlm.nih.gov/pubmed/688719.

Nowacki, A., Schlaier, J., Debove, I., & Pollo, C. (2018). Validation of diffusion tensor imaging tractography to visualize the dentatorubrothalamic tract for surgical planning. *Journal of Neurosurgery*, *130*(1), 99–108. https://doi.org/10.3171/2017.9.JNS171321.

Parry, C. B. (1980). Pain in avulsion lesions of the brachial plexus. *Pain*, *9*(1), 41–53. pii:0304-3959(80)90027-5.

Parry, C. B. (1984). Pain in avulsion of the brachial plexus. *Neurosurgery*, *15*(6), 960–965. Retrieved from http://www.ncbi.nlm.nih.gov/pubmed/6334817.

Peyron, R., Faillenot, I., Mertens, P., Laurent, B., & Garcia-Larrea, L. (2007). Motor cortex stimulation in neuropathic pain. Correlations between analgesic effect and hemodynamic changes in the brain. A PET study. *Neuroimage*, *34*(1), 310–321. https://doi.org/10.1016/j.neuroimage.2006.08.037.

Polanski, W. H., Zolal, A., Klein, J., Kitzler, H. H., Schackert, G., Eisner, W., & Sobottka, S. B. (2019). Somatosensory functional MRI tractography for individualized targeting of deep brain stimulation in patients with chronic pain after brachial plexus injury. *Acta Neurochirurgica*. https://doi.org/10.1007/s00701-019-04065-2.

Porreca, F., Ossipov, M. H., & Gebhart, G. F. (2002). Chronic pain and medullary descending facilitation. *Trends in Neurosciences*, *25*(6), 319–325. https://doi.org/10.1016/s0166-2236(02)02157-4.

Saade, N. E., Salibi, N. A., Banna, N. R., Towe, A. L., & Jabbur, S. J. (1983). Spinal input pathways affecting the medullary gigantocellular reticular nucleus. *Experimental Neurology*, *80*(3), 582–600. https://doi.org/10.1016/0014-4886(83)90309-6.

Schaltenbrand, G. (1959). *Einfuehrung in die stereotaktischen Operationen mit einem Atlas des menschlichen Gehirns. Introduction to stereotaxis with an atlas of the human brain*. Stuttgart: Thieme.

Schonberg, T., Pianka, P., Hendler, T., Pasternak, O., & Assaf, Y. (2006). Characterization of displaced white matter by brain tumors using combined DTI and fMRI. *Neuroimage*, *30*(4), 1100–1111. https://doi.org/10.1016/j.neuroimage.2005.11.015.

Seddighi, A. S., Seddighi, A., & Nikouei, A. (2017). Deep brain nucleus targeting in Parkinson's disease and essential tremor by image guided surgery using neuronavigation system with tractography and volume of tissue of activated assessment. *Hellenic Journal of Nuclear Medicine*, *20*(Suppl), 14–19. Retrieved from https://www.ncbi.nlm.nih.gov/pubmed/29324910.

Shealy, C. N., Taslitz, N., Mortimer, J. T., & Becker, D. P. (1967). Electrical inhibition of pain: Experimental evaluation. *Anesthesia & Analgesia*, *46*(3), 299–305. Retrieved from https://www.ncbi.nlm.nih.gov/pubmed/6067264.

Tracey, I., Ploghaus, A., Gati, J. S., Clare, S., Smith, S., Menon, R. S., & Matthews, P. M. (2002). Imaging attentional modulation of pain in the periaqueductal gray in humans. *Journal of Neuroscience*, *22*(7), 2748–2752. https://doi.org/10.1523/JNEUROSCI.22-07-02748.2002.

Turnbull, I. M., Shulman, R., & Woodhurst, W. B. (1980). Thalamic stimulation for neuropathic pain. *Journal of Neurosurgery*, *52*(4), 486–493. https://doi.org/10.3171/jns.1980.52.4.0486.

Wager, T. D., Rilling, J. K., Smith, E. E., Sokolik, A., Casey, K. L., Davidson, R. J., … Cohen, J. D. (2004). Placebo-induced changes in FMRI in the anticipation and experience of pain. *Science*, *303*(5661), 1162–1167. https://doi.org/10.1126/science.1093065.

Wang, J. Y., Chang, J. Y., Woodward, D. J., Baccala, L. A., Han, J. S., & Luo, F. (2007). Corticofugal influences on thalamic neurons during nociceptive transmission in awake rats. *Synapse*, *61*(5), 335–342. https://doi.org/10.1002/syn.20375.

Whitwell, J. L., Avula, R., Master, A., Vemuri, P., Senjem, M. L., Jones, D. T., … Josephs, K. A. (2011). Disrupted thalamocortical connectivity in PSP: A resting-state fMRI, DTI, and VBM study. *Parkinsonism & Related Disorders*, *17*(8), 599–605. https://doi.org/10.1016/j.parkreldis.2011.05.013.

Yang, F., Duan, W., Huang, Q., Chen, Z., Ford, N., Gao, X., … Guan, Y. (2020). Modulation of spinal nociceptive transmission by sub-sensory threshold spinal cord stimulation in rats after nerve injury. *Neuromodulation*, *23*(1), 36–45. https://doi.org/10.1111/ner.12975.

Zhu, D., Zhang, T., Jiang, X., Hu, X., Chen, H., Yang, N., … Liu, T. (2014). Fusing DTI and fMRI data: A survey of methods and applications. *Neuroimage*, *102*(Pt 1), 184–191. https://doi.org/10.1016/j.neuroimage.2013.09.071.

Zhuo, M. (2008). Cortical excitation and chronic pain. *Trends in Neurosciences*, *31*(4), 199–207. https://doi.org/10.1016/j.tins.2008.01.003.

Chapter 28

Neurocognition and placebo analgesia: Linking in functional magnetic resonance imaging

Sara Palermo
Neuroradiology Unit, Neurological Institute C. Besta IRCCS Foundation, Milan, Italy

Abbreviations

ACC	anterior cingulate cortex
ALE	activation likelihood estimation
fMRI	functional magnetic resonance imaging
MACM	meta-analytic connectivity model
MCC	midcingulate cortex
MPFC	medial prefrontal cortex
PA	placebo analgesia
PAG	periaqueductal gray
PET	positron emission tomography
PID	pain intensity difference
ROI	region of interest
SI	primary somatosensory cortex
SII	secondary somatosensory cortex
VBM	voxel-based morphometry

Introduction

The placebo effect is not only a powerful means of testing new pharmacological treatments and the therapeutic ritual, but also a way of investigating the relationship between our mind and body. In the history of human thought, the mind has long been regarded as a completely different structure from the body. It was even thought that the substance of which it is made could not be the same as that of physical bodies. Descartes, one of the greatest thinkers of the 17th century, was one of the supporters of the mind-body dualism and of the idea that the mind, not being a physical object, could not be studied with the laws and methods used by the sciences.

From the end of the 19th century, this belief has faded more and more thanks to the birth of disciplines such as neuropsychology and experimental neuroscience and the development of new and increasingly sophisticated investigation techniques that allow researchers to study the functioning of the brain and of the mind as we study the body and natural phenomena. In particular, a new era has opened thanks to neuroimaging techniques, and now neuroscientists can explore not only the structure and functioning of the brain, but also the role of neurodiversity: assessing how functioning changes in relation to factors such as age, sex, personality, and culture in which people live. The starting postulate consists in the equivalence between psychic life and brain life.

Among the numerous scientific evidences that falsify the mind-body dualism, there is one of the most intriguing phenomena. This is the placebo effect. Before the second half of the 20th century, when Stuart Wolf first defined the placebo effect, medical knowledge did not believe that placebo had any impact on clinical outcomes. Before the second half of the 20th century, medical knowledge did not believe that placebo had any impact on clinical outcomes. An article entitled "The Pharmacology of Placebo" (Wolf, 1959) first defined placebo and presented a "picturesque contradiction": by definition, pharmacology studies drugs and the ways in which the phenomena induced by these substances take place in the organism,

The Neurobiology, Physiology, and Psychology of Pain. https://doi.org/10.1016/B978-0-12-820589-1.00028-2

having as its purpose the use of those same substances for therapeutic purposes. At the time, however, there was a widespread belief that the global effect of a treatment was due both to a pharmacological action and to "other forces involved and the circumstances in which the administration takes place" (Wolf, 1950).

Over the past 20 years, studies on the placebo effect have multiplied, also thanks to the identification of the psycho-neuro-endocrine-immunological mechanisms that regulate the mind-body relationship (Gay & Bishop, 2014). This research activity has allowed to develop evidence-based treatments that exploit placebo responses in clinical practice. Indeed, the placebo effect is not only reflected in a subjective experience, but it is associated with measurable physiological changes. Nonetheless, the psychobiological mechanisms underlying this phenomenon have long remained unclear. Little has been known about the occurrence of placebo with respect to "time" and neurocognitive predictors. Which emotional and cognitive factors induce positive expectations? How do these factors interact and manage to modulate the physical experience of the symptom and the result of the therapy? What are the neural underpinnings of these neurocognitive aspects? These are intriguing and complex questions.

Understanding the placebo effect: From the biological approach to the advent of neuroimaging techniques

A new definition of placebo effect appeared at the beginning of the 21st century: a change in the body or mind that occurs as a result of the symbolic meaning attributed to an event or object in the health sector (Brody, 2000). When administering a placebo, which is a simulation, but full of symbolic meaning, positive expectations of well-being and healing are induced. These expectations can influence the body in such a way as to produce beneficial effects (Benedetti, 2009). To date, placebo can therefore be understood as a powerful mechanism of interaction between mind and body (Amanzio & Palermo, 2016; Palermo, Benedetti, Costa, & Amanzio, 2015).

The placebo effect can be elicited both by learning processes (classical conditioning) and by high-level cognitive processes (expectations and beliefs that the patient develops during treatment). These are two solid interpretative models, which although different are not self-excluding (Palermo, 2018). The placebo effect is the psychobiological effect due to the positive psychosocial context that accompanies a therapy (Benedetti, 2009, 2013). The fundamental element of any treatment context would seem to be represented by the "doctor-patient relationship" (Benedetti, 2013), a relationship that can be treated in evolutionary, physiological, and neuroscientific terms (Frisaldi, Giudetti, Pampallona, & Benedetti, 2014). From a physiological and neurobiological perspective, this relationship can be divided into at least four phases: (1) feeling sick, (2) seek relief, (3) meet the therapist, and (4) receive therapy (Benedetti, 2013; Frisaldi et al., 2014). The only ritual of the therapeutic act can elicit healing, due to the patient's expectations of benefit. From an evolutionary point of view, the function of this relationship is to promote the cessation of suffering at the simple meeting with a doctor. It is a "socio-neuronal system" that has evolved to ensure psychological and social support for the weakest, while guaranteeing the elimination of discomfort and suffering from simple social contact, that was, the meeting with the "healer" (Benedetti, 2013; Frisaldi et al., 2014).

The study of the mechanisms underlying the mind-brain dialogue (Etkin, Pittenger, Polan, & Kandel, 2005)—and changes in brain structure/functioning following an experience—has created great expectation on the possibility of understanding the neurobiological equivalents of the effects that are observed following the placebo treatment (the so-called placebo response). Specifically, the advent of modern and noninvasive neuroimaging techniques provided the unique opportunity to study the neural mechanisms of placebo responses (Amanzio & Palermo, 2016; Benedetti, 2014).

Electrophysiological and brain imaging techniques are considered the main experimental approaches: electroencephalography, positron emission tomography (PET), and functional magnetic resonance imaging (fMRI) are the most widely applied methods. Magnetic resonance-based measures (such as arterial spin labeling, diffusion tensor imaging, spectroscopy, and volumetric imaging) and resting state functional connectivity have increasingly been used to assess pain-related changes in the brain. Importantly, neuroimaging techniques allow scientists to observe the neurobiological mechanisms of the placebo effect and could 1 day improve clinical trials and treat different clinical conditions (Kuehn, 2005). Indeed, the placebo effect shows how some mental mechanisms can activate physiological responses that influence the internal state of the body. The belief of being treated, for example, can decrease chronic stress levels. If this happens, the immune system's response improves, improving the organic reaction to disease. In other cases, the belief of take a painkiller increases the production of opioid neuropeptides (endorphins), which really have the effect of decreasing pain. Not surprisingly most of the knowledge about the neurocognitive and neural underpinnings of the placebo effect comes from the field of pain and analgesia. To date, placebo analgesia (PA) is the most carefully examined and best neurobiologically characterized placebo response (Tracey, 2010). The interest in placebo analgesia is increasing: from 1957 to 2020 almost

5000 studies have been indexed on PubMed. How many of these have tried to clarify which functional changes accompany the onset of the placebo effect? A cursory PubMed search with the search terms "(placebo analgesia) AND fMRI NOT review" on May 29, 2020 resulted in 97 articles. These numbers suggest how complex it is to do neuroimaging research in this field. fMRI studies cannot be implemented without first considering some pain characteristic and parameters that profoundly affect the evaluation of placebo analgesia. The understanding of this phenomenon cannot be separated from the understanding of pain, in its physiological, evolutionary, and conservative meaning (Palermo, 2018). Key factors in the design, analysis, description, and evaluation of an fMRI experiment must be thoroughly considered and planned a priori to obtain a robust, reproducible result (Moayedi, Salomons, & Atlas, 2018).

Pain modulatory mechanisms relevant for the neuroimaging study of PA

Pain is characterized by a sensorial-discriminatory component to which an affective-motivational and cognitive-evaluative component is associated, relating to the activation of thought processes in which the causes of pain are sought to remedy them through behavioral planning (Ingvar, 1999). Neuroimaging techniques have made it possible to identify a brain network whose activation pattern is related to the perception of pain as previously supposed by Ingvar (Legrain, Iannetti, Plaghki, & Mouraux, 2011). This network is actually characterized by two primary functional nodes: a node more tied to the discriminatory-sensory aspects of pain which has the somatosensory cortex as its functional substrate; a node more tied to the experiential and emotional components which is associated with the cingulate cortex, the limbic system, and the insula. The cingulate cortex and the insula constitute an empathic-reminiscent system of painful experience, which makes it possible to understand and share the pain of others (Palermo, 2017). Pain also involves some areas assigned to executive functions, through a rich ascending and descending network (Palermo et al., 2015), controlled by neurotransmitters such as dopamine, serotonin, cholecystokinin, norepinephrine, neurokinin 1, the system of the oxytocin and opioids. Legrain et al. (2011) argue that this functional network is involved in the detection, orientation of attention and reaction to salient events. It could therefore represent a basic mechanism through which any event relevant to body integrity would be detected, regardless of the sensory channel through which it is transmitted (Legrain et al., 2011). This neurocognitive process would involve the construction of a multimodal cortical representation of the body and the surrounding space. Assuming that this network acts as a defensive system that signals threats that are potentially harmful to the body, the emphasis of the research is no longer placed on the quality of sensation aroused by harmful stimuli but on the neurocognitive functions required by the occurrence of these potential threats (Legrain et al., 2011).

Brain regions involved in different classes of mental functions (cognition, emotion, motivation, and sensation) are functionally associated in the context of nociception and generate the pain experience (Tracey, 2010; Tracey & Johns, 2010). The interactions among these neural hubs provide a means by which emotional and motivational signals can alter pain experience and perception through interactions with the descending pain modulation system (Ossipov, Morimura, & Porreca, 2014). The latter is the main way through which cognitive and contextual influences (both internal and external) can modify a painful experience (Tracey et al., 2002; Valet et al., 2004; Wiech, Ploner, & Tracey, 2008). Indeed, the descending pathways involved in modulating the pain signal are of direct relevance for understanding the placebo effect.

Pain and nociceptive stimuli in neuroimaging study of PA

The fMRI research protocol design must take into account the type of pain at the baseline (present/absent, nociceptive or neuropathic), duration (acute, chronic), location (superficial somatic, deep somatic, visceral), temporal characteristics (constant, floating, recurrent), and intensity. An overlooked aspect is that the subjective perception of pain intensity is not related to the type/extent of tissue damage, but it depends on the interaction of physical, emotional, and cultural factors.

Some researchers apply nociceptive stimuli to elicit pain in healthy subjects and/or in patients. Experimental pain models in humans offer the opportunity to explore the neural pain system—and the mechanisms underlying it—in controlled settings. They are also ideal for testing analgesic compounds (Reddy, Naidu, Rani, & Rao, 2012) and for designing neuroimaging research paradigms. Standardized stimuli of a different nature (primarily: chemical, electrical, mechanical, thermal)—and of frequency, intensity and duration established a priori—can be applied to the skin, muscles, and viscera for a differentiated assessment of various pain pathways and mechanisms (Reddy et al., 2012). Thermal pain is the most common and most well-established stimulus used in neuroimaging (Duerden & Albanese, 2013). Additional acute experimental pain models such as bladder filling, muscular hypertonic saline injections, ischemia, and visceral/rectal distention might be more appropriate for inquiries about the neural bases of musculoskeletal or visceral chronic pain disorders (Moayedi et al., 2018). Regardless of the choice made, the researcher must be able to establish a tactile threshold, a pain threshold (which is relatively constant across subjects for a given stimulus), and the maximum pain tolerance (which varies much more widely across subjects and depends on prescribed medications. Once these fundamental parameters have been acquired on the individual, it will therefore be possible to establish them for the whole experimental sample. There are two typical paradigms (Palermo et al., 2015): fixed-intensity (a pain experiment that used the same level of stimulation for all participants) or subject-dependent (an experiment that used dissimilar stimuli for each subject).

Pain evaluation and temporal phases in neuroimaging study of PA

The evaluation of the variation in perceived pain is the most subject to bias in neuroimaging study of PA. Pain perception is intrinsically subjective and pain estimate relies mainly on the verbal report (Bromm, 1984). Furthermore, pain evaluation tools are likely to be interpreted in different ways by different examiners, based on the criteria that have been decided to apply (Farrar, Portenoy, Berlin, Kinman, & Strom, 2000). The numeric rating and visual analogue scales are commonly used to assess the present intensity of acute pain during fMRI acquisition (Bendinger & Plunkett, 2016). They are reliable, sensitive to change, and easy to administer verbally or digitally. Indeed, are applicable for unidimensional assessment of pain and PA in most experimental settings (Hjermstad et al., 2011). Pain intensity difference (PID, using a numerical rating 0–10 or 0–100 scale) and percentage pain intensity difference (PID%, 0%–100% scale) allow to calculate the PA accordingly. The detection of the pain intensity variation following the administration of the placebo will be arranged at a specific time in the temporal continuum of nociception (Amanzio, Benedetti, Porro, Palermo, & Cauda, 2013; Amanzio & Palermo, 2016; Kong et al., 2008) (Fig. 1):

- *Phase 1: expectation.* Before the pain begins, the expectation (anticipation of relief) can modulate the perception of subsequent painful stimuli. In an fMRI paradigm, this corresponds to the "predicted analgesia" period (the period between the start of the scan and the start of the stimulus or the period prior to each trial in event-related fMRI paradigms).
- *Phase 2: administration of the nociceptive stimulus.* During the administration of painful stimuli, placebo treatment—being associated with previous knowledge about the analgesic effects of medical treatments—could inhibit the input signals of harmful stimuli. This component is linked to brain activity between the beginning of the painful stimulation and its cessation.
- *Phase 3: subjective assessment of the intensity of perceived pain (appraisal).* After the cessation of the painful stimulation, previous knowledge on the effect of the treatment can subconsciously alter the evaluation of experienced pain intensity.

Sample characteristic relevant for the neuroimaging study of PA

Researchers must control not only the experimentally induced pain and the instants in which to request the evaluation of the pain intensity, they must also guarantee the homogeneity of the experimental sample according to the experimental question to be answered. Neuroimaging requires unique restrictions on subjects enrollment, and many of these constraints can determine a burden on fMRI experiments with pathological populations. The exploration of the neurocognitive functions associated with PA is more rigorous if the model (healthy and/or pathological) is accompanied by a well-known neurobiological and neurophysiological phenotype. The possible influences of the pathology on nociception must be known at the outset. Some pathologies such as neuropathies and chronic pain could cause bias in the evaluation of the pain

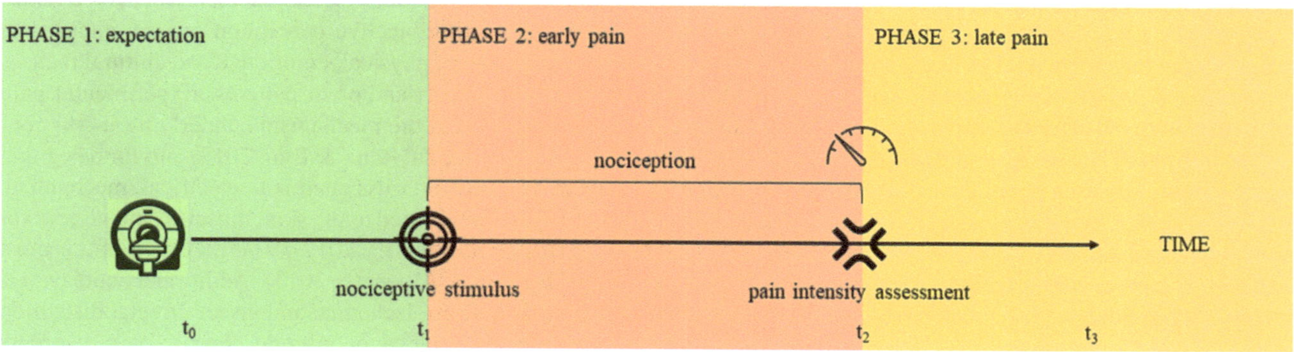

FIG. 1 Temporal continuum in the study of pain and PA in fMRI. There are three stages of placebo analgesia that can be considered: expectation, early pain, and late pain.

experienced in the fMRI acquisition phase due to the "painful background noise." Indeed, each case should be carefully evaluated. For example, irritable bowel syndrome or functional gastrointestinal disorder can cause discomfort; however, they do not harm the intestines or lead to any serious disease concomitant therapies (Amanzio et al., 2013). Indeed, several fMRI study on pain and PA compare patients with pain disorders with matched control patients to compare the fMRI results between groups. If the samples are not selected to match nonpain-related characteristics, the occurrence must take into consideration that distinct fMRI patterns of activations/deactivations between groups reflect processes other than the pain disorder of interest (Moayedi et al., 2018).

A fundamental question is whether the subjective pain intensity reduction perceived in PA is associated with real changes in the brain activity associated with the pain experience. And, if so, favored and supported by what cognitive factors?

Neurocognitive predictors in neuroimaging study of PA: Attention, expectation, and reappraisal

Nociception is particularly influenced by attention processes which allocate processing resources to salient internal/external events—magnifying subsequent physiological/behavioral responses—to attenuate reactions to not relevant stimuli (Corbetta & Shulman, 2002; Palermo et al., 2015). Pain retrieves attention *per se* and cannot be easily overlooked. fMRI research can apply paradigms in which subjects are actively engage in distracting tasks to circumvent this trend. "To be distracted from pain" lowers the recruitment of brain areas typically related to the sensorial, cognitive, and affective evaluation of the nociceptive stimulus, primary and secondary somatosensory cortices (SI and SII), such as the anterior cingulate cortex (ACC), insula and thalamus (Wiech et al., 2008). Interaction analyses highlighted neural hubs with activations higher than what would have been expected from a simple summation of distraction tasks and pain: for example, ACC, the prefrontal cortex and the periaqueductal gray (PAG) (Bantick et al., 2002; Valet et al., 2004). The functional network here described subserves an opioid-mediated descending modulatory pathway mainly acting at the level of the spinal cord dorsal-horn.

Expectations about upcoming salient events are compared with sensory information from the bottom-up and allow to adapt sensory, motor, and cognitive systems to adjust behavioral responses. A harmful stimulus can *(dis)*confirm these expectations in two directions: it may be more intense or less intense than expected. Expectation can therefore prejudice perception so that the stimulation intensity evaluation respects the expected intensity level, being classified as lower or higher than the actual nociceptive stimulus (Wiech et al., 2008) (Fig. 2).

A relief expectation can induce PA even without the physical administration of a placebo. As no placebo is provided, this kind of PA should be more appropriately described as a "placebo-like" effect (Amanzio et al., 2013). Simple verbal suggestions can be used as placebo/nocebo manipulations to bias subjects' expectations in fMRI studies on pain (Tracey, 2010). It was thus possible to demonstrate that the expectation of a decrease in pain can reduce both the subjective pain

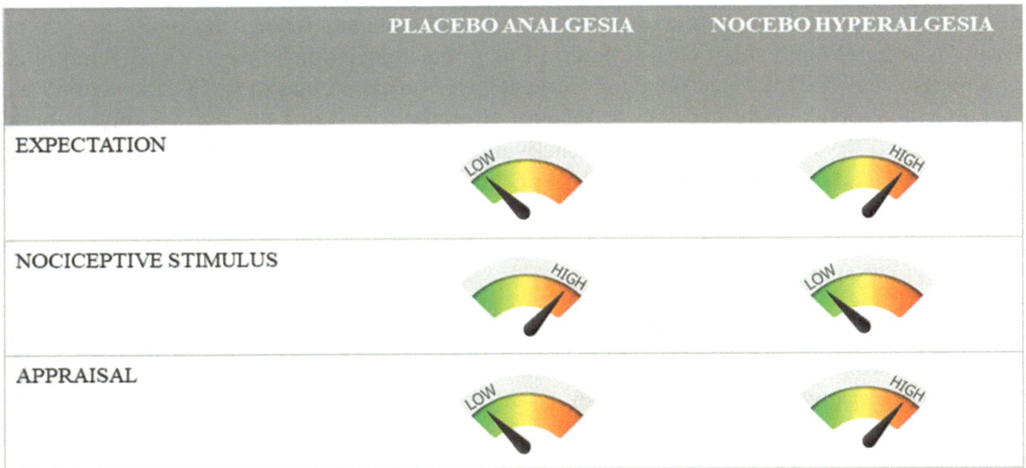

FIG. 2 Interaction among expectation, nociceptive stimulus and appraisal. Placebo analgesia response follows when an expectation of low-level pain is induced, and the subject received stimulation at a higher level than expected. Similarly, the expectation of high-level pain (nocebo effect) might lead to an increase in perceived pain intensity (hyperalgesia).

intensity evaluation and the activations in the pain matrix (Koyama, McHaffie, Laurienti, & Coghill, 2005). Considering fMRI findings, the PA expectation urges the recruitment of a bilateral cerebral network characterized by the left anterior cingulate, right precentral, and lateral prefrontal cortex and the left PAG (Amanzio et al., 2013). Expectation is the most important factor in PA (Benedetti, 2013) and the decreased effectiveness of hidden treatments represents one of the best evidences of the crucial role of expectation (Benedetti, 2013; Palermo et al., 2019).

Pain is usually perceived as threatening; however, the extent of the menace depends upon the belief in coping resources owned by each person. Interestingly, coping resources also benefit when they are not fielded or are illusory. The underlying mechanism would seem to lie in a change in the meaning or reevaluation of a noxious event so that it becomes less threatening because control is possible (Arntz & Schmidt, 1989). Perceived internal control, which triggers reappraisal processes able to modify the experience of pain, can be experimentally induced by allowing participants to stop a painful stimulation. This kind of reappraisal has shown to induce analgesic effects through the recruitment of a functional network involving the insula, ACC, SII, and the right anterior ventrolateral prefrontal cortex (Salomons, Johnstone, Backonja, & Davidson, 2004; Wiech et al., 2006).

These findings suggest that attention, expectation, and reappraisal can be considered neurocognitive features of PA.

The neuroimaging meta-analytic approach to the study of PA and related neurocognitive factors

Quantitative metaanalysis techniques, such as Activation Likelihood Estimation (ALE), offer information about the anatomical reliability of fMRI results reported in a set of neuroimaging studies within the existing literature on a topic (Laird et al., 2005). To deepen the data analysis, it is therefore possible to associate a meta-analytic connectivity model (MACM) (Eickhoff et al., 2011; Laird et al., 2013). The resulting output images of the MACM analysis can be used for an automated behavioral analysis of selected region of interest (ROI). Through this procedure, it is possible to consider interoception, emotion, perception, action, cognition, as well as their related subcategories (Lancaster et al., 2012; Palermo et al., 2015).

Thanks to ALE metaanalysis, an increased activity associated with the PA expectation has been corroborate in the ACC, the PAG, and the lateral prefrontal cortex (Amanzio et al., 2013). The latter resulted selectively activated in the PA expectation phase but not during early pain suggesting that anticipatory responses may reflect a neurocognitive process that is not confined to pain (Amanzio et al., 2013). Indeed, the lateral prefrontal cortex is a cognitive control hub, implicated in voluntary switching after a delay, prospective memory, and monitoring processes. This rostral processing system is cross-domain (Burgess, Scott, & Frith, 2003) and capable of guiding behavior in situations where the course of action is not obvious or predetermined a priori (Burgess, Veitch, de Lacy Costello, & Shallice, 2000; Goel & Grafman, 2000). Moreover, ACC is involved in action monitoring (Morese, Mario, & Palermo, 2018), which is particularly important in situations that may put may constitute a threat (Palermo, 2017). This functional network supports the mental representation of an impending noxious stimulus. Several studies have proposed that the anticipation of pain triggers a specific neuronal system, distinct from that involved in the perception of pain. Indeed, during PA in early pain, a complex pattern of activations—that is likely to reflect a structural-functional association between PA and emotional regulation—was detected (Amanzio et al., 2013). Functional recruitment covers areas such as the ACC, the medial/lateral prefrontal cortices, the left inferior parietal lobule and postcentral gyrus, the anterior insula (AI), the thalamus, hypothalamus, PAG, and pons (Amanzio et al., 2013). At the same time, deactivations were observed in the left mid- and posterior cingulate cortex, superior temporal and precentral gyri, in the left AI and right posterior insula, in the claustrum and putamen, and in the right thalamus and caudate body (Amanzio et al., 2013). These deactivations reflect an "authentic reduction" of pain intensity perception. Therefore, it seems possible that cerebral networks involved in nociception are downregulated in parallel with behavioral analgesia.

A suggestive hypothesis suggests that the brain network involved in pain anticipation—having ACC and AI as core neural hubs—can be involved in initiating executive processes for monitoring the processing of salient information, so as to avoid interference from irrelevant stimuli, in the experimental paradigms of anticipation of pain (Palermo et al., 2015). The ACC and AI coactivation networks largely overlap, involving virtually the same areas. MACM allowed to identify different behavioral domains associated with pain anticipation in the preselected ROIs. For both ROIs, brain functioning during the expectation phase was highly associated with the behavioral domains of action (imagination, inhibition, and execution), emotion (fear), and perception (pain and interception) (Palermo et al., 2015). ACC and AI seem to play a key role in representing conceptual information relevant for survival and in transducing perceptions into physiological and affective-behavioral responses (Roy, Shohamy, & Wager, 2012).

Neuroimaging findings in the study of PA in Alzheimer's disease

Although PA is well described in normal condition, research on analgesia on pathological population—especially in neurodegenerative disorders—is pivotal, causing a little understanding of the neurocognitive mechanisms and neural underpinnings that underlie it (Matthiesen, Lunde, Wohlert Kjær, Carlino, & Vase, 2019).

Considering what previously stated about the existence of a general system supporting PA in the expectation phase, an analysis of the internal context (executive-attention and metacognitive-executive functions, cognitive flexibility, prospective memory, and action-monitoring) appears to indicate different responses in the working memory counterpart (Damasio, 1996; Wager et al., 2004), which seems to be compromised in major neurocognitive disorders such as Alzheimer's disease (Amanzio et al., 2011), in which loss of expectancy abilities disrupts PA, making analgesic therapies less effective (Benedetti et al., 2006). PA neural correlates in this kind of patients are still not perfectly identified. An intriguing possibility is that executive (dys)functions of the medial prefrontal cortex (MPFC) may be implicated (Palermo et al., 2019), since a flanker test on healthy subjects previously revealed a placebo-induced transient change in prefrontal error monitoring and control functions (Koban, Brass, Lynn, & Pourtois, 2012). One of the most effective fMRI event-related paradigms to investigate some of these functions (action-monitoring and response-inhibition) in neurocognitive disorders (Palermo, Stanziano, & Morese, 2018) is the GO-NoGO task (Braver, Barch, Gray, Molfese, & Snyder, 2001). This fMRI paradigm allowed to demonstrate the existence of a relationship among executive dysfunctions, lower PA effect and hypoactivation of the MPFC, especially significant in the midcingulate cortex (MCC) (Palermo et al., 2019). These findings have been found not to be altered by disease-specific gray matter atrophy, suggesting a "special purely" functional role played by the MCC. This brain area acts as a core neural hub for self-monitoring, set-shifting, and response-inhibition abilities (Morese et al., 2018; Palermo, 2017; Palermo et al., 2018). Indeed, metacognitive-executive dysfunction might have a role in expectancy-based PA pathogenesis. This interpretation is supported by previous research on healthy subjects, demonstrating that expectancy-based PA can be affected by PFC dysfunctionality (Krummenacher, Candia, Folkers, Schedlowski, & Schönbächler, 2010).

Future directions

Neuroimaging is demonstrating its enormous potential as a "biomarker," an indicator of the possibility of being a PA responder/nonresponder. The growing interest in neurodiversity (understood as sociodemographic, psychological, cultural, socioeconomic characteristics) will provide further insights into the mechanisms underlying the PA. Future neuroimaging studies should use multimodal techniques by combining with the knowledge already available in other areas of cognitive neuroscience such as self-awareness and executive cognitive control, mentalizing and Theory of Mind, social decision-making, and interactions. These directions could provide a biological basis for a neurocognitive model on pain modulation, which can make predictions about interindividual differences and its manipulation through pharmacological and psychological interventions.

More research in the field of neurodegenerative disorders is a must. It will be desirable to understand which neurocognitive factors afflict PA and which neurodegenerative disorders can benefit from placebo effect—and from the optimized perception of the treatment—to adjust painkilling interventions accordingly.

Applications to other areas

The chapter focuses on considerations that are unique to neuroimaging studies of PA in humans, including study design and analysis, inferences that can be drawn from these studies, and future directions of the approach. Brain and mind dimensions must dialogue to be able to grasp the clinical history of a patient, which is also determined by the dynamic intersection of an internal context (cognition, emotion, and internal articulation of meaning) and an external context (physical and social environment). All of this can be grasped through the lens of the neurocognitive approach. While recognizing the value of the placebo effect in the approach to the patient, it could also be a promoter of confounding effects on the results of clinical trials. A better understanding of the neurobiological mechanisms underlying the placebo effect could therefore help scientists improve the design of clinical trials and clinicians optimize treatments with a view to personalized medicine. The study of PA could be particularly useful in the neuropsychiatric field (schizophrenic, obsessive-compulsive, and autistic spectrum), as well as in the context of chronic diseases, neurological and neuromuscular diseases, or oncological diseases.

Chronicity deserves a special mention, especially in the elderly. The prevalence of chronic pain grows with aging. A careful study of PA in successful physiological aging, in typical cognitive aging and frailty elderly, could open us to the enhancement of nonpharmacological treatment strategies and to exploit placebo as an add-on to pharmacological treatments existing today.

Other agents of interest

In this chapter, we have described detailed mechanisms of how placebo analgesia works, including neural underpinnings recruitment and neurocognitive associated factors. Other nonpharmacological pain treatments include:

Psychological interventions
Hypnosis
Relaxation and imagination techniques
Autogenic training
Distraction techniques
Physical activity
Massage
Chiropractic
Osteopathic
Physiokinesitherapy
Acupuncture
Thermotherapy and cryotherapy
Radiotherapy
Ozone
Neuro-damaging techniques

Nonpharmacological therapies promote significant decreases in the pharmacological load (already high in the case of patients in polypharmacology) and reduce the risk for the patient of overdose of drugs that can affect potential depressive and anxiety-related states related to the presence of pain. In this way, they improve the quality of life of patients in a concrete way.

Mini-dictionary of terms

Activation likelihood estimation: a quantitative voxel-based metaanalysis technique aimed at determining above-chance convergence of activation probabilities across different fMRI experiments.

Event-related paradigm: experimental design used to detect changes in the hemodynamic response to neural activity in response to presented stimuli individually and randomly (events).

fMRI: a neuroimaging technique that consists in the use of magnetic resonance imaging to evaluate brain functionality in a complementary way to morphological imaging. fMRI provides a probabilistic map of functionally eloquent brain areas.

Meta-analytic connectivity model: a technique for determining functional connectivity of specific brain structures by identifying patterns of coactivation across thousands of subjects.

Neuroimaging: set of technological tools and experimental procedures for the visualization of the brain in vivo, both in its structural/anatomical details (morphological neuroimaging) and during the execution of particular motor and cognitive tasks (functional neuroimaging).

Pain threshold: the lowest intensity at which a given stimulus is perceived as painful.

Pain tolerance: the greatest level of pain that a subject is prepared to endure.

Placebo effect: any improvement in a symptom or physiological condition following placebo treatment.

Placebo response: individual's neurobiological and psychophysiological response to an inert substance or a sham treatment.

Key facts of functional magnetic resonance imaging

- It is a noninvasive biomedical imaging technique, which provides a map of functionally eloquent brain areas.
- It measures brain activity by detecting changes associated with blood flow.

- The technique can localize activity to within millimeters but, using standard techniques, no better than within a window of a few seconds.
- It is used for both clinical and research purposes.
- Virtually free of side effects and with very few contraindications.
- Understanding the relation between psychological processes and brain function, is made particularly difficult by the fact that psychological and neurocognitive processes are not directly observable.
- Both block and event-related designs are based on the subtraction paradigm, which assumes that specific cognitive processes can be added selectively in different conditions.
- It is not possible to argue that individual brain areas are solely responsible for specific mental functions.
- A certain type of response in each area does not express the specificity of the activation stimulus of the same.
- Reverse inference is common in public presentations of imaging research.

Summary points

- The placebo effect is a way of investigating the mind-body relationship.
- Neuroimaging research can provide quantitative measures of the neurocognitive factors associated with placebo treatment.
- Placebo analgesia is a fruitful model to explore the neurocognitive and neural underpinnings of the placebo effect.
- The understanding of this phenomenon cannot be separated from the understanding of pain.
- Key factors in the design, analysis, description, and evaluation of an fMRI experiment must be thoroughly considered.
- Attention, expectation, reappraisal, metacognitive-executive functions, and emotional regulations seem to be the neurocognitive factors involved.
- A highly distributed supramodal system aimed at the processing of information and the self-regulation of adaptive responses to the environment exist.
- Future neuroimaging studies should use multimodal techniques and focus on research in the field of neurodiversity and neurodegenerative disorders.

References

Amanzio, M., Benedetti, F., Porro, C. A., Palermo, S., & Cauda, F. (2013). Activation likelihood estimation meta-analysis of brain correlates of placebo analgesia in human experimental pain. *Human Brain Mapping*, *34*(3), 738–752.

Amanzio, M., & Palermo, S. (2016). Conceptualizing placebo as active component and adjunct in psychological treatment. In S. Maltzman (Ed.), *Oxford handbook of treatment processes and outcomes in counseling psychology*. Oxford, UK: Oxford University Press.

Amanzio, M., Torta, D. M. E., Sacco, K., Cauda, F., D'Agata, F., Duca, S., et al. (2011). Unawareness of deficits in Alzheimer's disease: Role of the cingulate cortex. *Brain*, *134*, 1061–1076.

Arntz, A., & Schmidt, A. J. M. (1989). Perceived control and the experience of pain. In A. Steptoe, & A. Appels (Eds.), *Stress, personal control and health* (pp. 131–162). NJ, USA: John Wiley & Sons.

Bantick, S. J., Wise, R. G., Ploghaus, A., Clare, S., Smith, S. M., & Tracey, I. (2002). Imaging how attention modulates pain in humans using functional MRI. *Brain*, *125*(Pt 2), 310–319.

Bendinger, T., & Plunkett, N. (2016). Measurement in pain medicine. *BJA Education*, *16*(9), 310–315.

Benedetti, F. (2009). *Placebo effects*. Oxford: Oxford University Press (chapter 2).

Benedetti, F. (2013). Placebo and the new physiology of the doctor-patient relationship. *Physiological Reviews*, *93*(3), 1207–1246.

Benedetti, F. (2014). Placebo effects: From the neurobiological paradigm to translational implications. *Neuron*, *84*, 623–637.

Benedetti, F., Arduino, C., Costa, S., Vighetti, S., Tarenzi, L., Rainero, I., & Asteggiano, G. (2006). Loss of expectation-related mechanisms in Alzheimer's disease makes analgesic therapies less effective. *Pain*, *121*(1–2), 133–144. https://doi.org/10.1016/j.pain.2005.12.016.

Braver, T. S., Barch, D. M., Gray, J. R., Molfese, D. L., & Snyder, A. (2001). Anterior cingulate cortex and response conflict: Effects of frequency, inhibition and errors. *Cerebral Cortex*, *11*, 825–836.

Brody, H. (2000). The placebo response. Recent research and implications for family medicine. *The Journal of Family Practice*, *49*(7), 649–654.

Bromm, B. (1984). The measurement of pain in man. In B. Bromm (Ed.), *Pain measurement in man* (pp. 3–13). Amsterdam: Elsevier.

Burgess, P. W., Scott, S. K., & Frith, C. D. (2003). The role of the rostral frontal cortex (area 10) in prospective memory: A lateral versus medial dissociation. *Neuropsychologia*, *41*, 906–918.

Burgess, P. W., Veitch, E., de Lacy Costello, A., & Shallice, T. (2000). The cognitive and neuroanatomical correlates of multitasking. *Neuropsychologia*, *38*, 848–863.

Corbetta, M., & Shulman, G. L. (2002). Control of goal-directed and stimulus-driven attention in the brain. *Nature Reviews Neuroscience*, *3*, 201–215.

Damasio, A. R. (1996). The somatic marker hypothesis and the possible functions of the prefrontal cortex. *Philosophical Transactions of the Royal Society, B: Biological Sciences*, *351*, 1413–1420.

Duerden, E. G., & Albanese, M. C. (2013). Localization of pain-related brain activation: A meta-analysis of neuroimaging data. *Human Brain Mapping*, *34*, 109–149.

Eickhoff, S. B., Bzdok, D., Laird, A. R., Roski, C., Caspers, S., Zilles, K., & Fox, P. T. (2011). Co-activation patterns distinguish cortical modules, their connectivity and functional differentiation. *NeuroImage*, *57*(3), 938–949. https://doi.org/10.1016/j.neuroimage.2011.05.021.

Etkin, A., Pittenger, C., Polan, H. J., & Kandel, E. R. (2005). Toward a neurobiology of psychoterapy: Basic science and clinical application. *Journal of Neuropsychiatry and Clinical Neurosciences*, *17*(2), 145–158.

Farrar, J. T., Portenoy, R. K., Berlin, J. A., Kinman, J. L., & Strom, B. L. (2000). Defining the clinically important difference in pain outcome measures. *Pain*, *88*(3), 287–294.

Frisaldi, E., Giudetti, L., Pampallona, A., & Benedetti, F. (2014). Psychology, neurobiology, ethics of the placebo effects and of the doctor-patient relationship. *Giornale Italiano di Farmacoeconomia e Farmacoutilizzazione*, *6*(4), 17–22.

Gay, C. W., & Bishop, M. D. (2014). Research on placebo analgesia is relevant to clinical practice. *Chiropractic & Manual Therapies*, *22*(1), 6.

Goel, V., & Grafman, J. (2000). Role of the right prefrontal cortex in illstructured planning. *Cognitive Neuropsychology*, *17*, 415–436.

Hjermstad, M. J., Fayers, P. M., Haugen, D. F., Caraceni, A., Hanks, G. W., Loge, J. H., et al. (2011). Studies comparing numerical rating scales, verbal rating scales, and visual analogue scales for assessment of pain intensity in adults: A systematic literature review. *Journal of Pain and Symptom Management*, *41*(6), 1073–1093.

Ingvar, M. (1999). Pain and functional imaging. *Philosophical Transactions of the Royal Society of London. Series B, Biological Sciences*, *354*, 1347–1358.

Koban, L., Brass, M., Lynn, M. T., & Pourtois, G. (2012). Placebo analgesia affects brain correlates of error processing. *PLoS One*, *7(11)*, e49784.

Kong, J., Gollub, R. L., Polich, G., Kirsch, I., Laviolette, P., Vangel, M., ... Kaptchuk, T. J. (2008). A functional magnetic resonance imaging study on the neural mechanisms of hyperalgesic nocebo effect. *The Journal of Neuroscience*, *28*(49), 13354–13362. https://doi.org/10.1523/JNEUROSCI.2944-08.2008.

Koyama, T., McHaffie, J. G., Laurienti, P. J., & Coghill, R. C. (2005). The subjective experience of pain: Where expectations become reality. *Proceedings of the National Academy of Sciences*, *102*, 12950–12955.

Krummenacher, P., Candia, V., Folkers, G., Schedlowski, M., & Schönbächler, G. (2010). Prefrontal cortex modulates placebo analgesia. *Pain*, *148*(3), 368–374.

Kuehn, B. M. (2005). Pain studies illuminate the placebo effect. *JAMA*, *294*(14), 1750–1751.

Laird, A. R., Eickhoff, S. B., Rottschy, C., Bzdok, D., Ray, K. L., & Fox, P. T. (2013). Networks of task co-activations. *NeuroImage*, *80*, 505–514. https://doi.org/10.1016/j.neuroimage.2013.04.073.

Laird, A. R., Fox, P. M., Price, C. J., Glahn, D. C., Uecker, A. M., Lancaster, J. L., et al. (2005). ALE meta-analysis: Controlling the false discovery rate and performing statistical contrasts. *Human Brain Mapping*, *25*, 155–164.

Lancaster, J. L., Laird, A. R., Eickhoff, S. B., Martinez, M. J., Fox, P. M., & Fox, P. T. (2012). Automated regional behavioral analysis for human brain images. *Frontiers in Neuroinformatics*, *6*, 23. https://doi.org/10.3389/fninf.2012.00023.

Legrain, V., Iannetti, G. D., Plaghki, L., & Mouraux, A. (2011). The pain matrix reloaded: A salience detection system for the body. *Progress in Neurobiology*, *93*(1), 111–124.

Matthiesen, S. T., Lunde, S. J., Wohlert Kjær, S., Carlino, E., & Vase, L. (2019). Placebo analgesia effects across central nervous system diseases: What do we know and where do we need to go? *Pain Reports*, *4*(3), e717.

Moayedi, M., Salomons, T. V., & Atlas, L. Y. (2018). Pain neuroimaging in humans: A primer for beginners and non-imagers. *The Journal of Pain*, *19(9)*, 961.e1–961.e21.

Morese, R., Mario, S., & Palermo, S. (2018). Commentary: Metacognition and perspective-taking in Alzheimer's disease: A mini-review. *Frontiers in Psychology*, *9*, 2010.

Ossipov, M. H., Morimura, K., & Porreca, F. (2014). Descending pain modulation and chronification of pain. *Current Opinion in Supportive and Palliative Care*, *8*(2), 143–151.

Palermo, S. (2017). Reduced self-awareness across pathologies: Involvement of the anterior cingulate cortex. In A. Costa, & E. Villalba (Eds.), *Vol. 28. Horizons in Neuroscience Research* (pp. 137–176). Happauge, New York: Nova Science Publishers.

Palermo, S. (2018). Placebo analgesia, pain anticipation, and the possible neural correlates of nocebo effect. *Rivista Internazionale di Filosofia e Psicologia*, *9*(3), 259–279.

Palermo, S., Benedetti, F., Costa, T., & Amanzio, M. (2015). Pain anticipation: An activation likelihood estimation meta-analysis of brain imaging studies. *Human Brain Mapping*, *36*(5), 1648–1661.

Palermo, S., Rainero, I., Stanziano, M., Vase, L., D'Agata, F., Rubino, E., et al. (2019). A novel neurocognitive approach for placebo analgesia in neurocognitive disorders. *Experimental Gerontology*, *118*, 106–116.

Palermo, S., Stanziano, M., & Morese, R. (2018). Commentary: Anterior cingulate cortex and response conflict: Effects of frequency, inhibition and errors. *Frontiers in Behavioral Neuroscience*, *12*, 171.

Reddy, K. S., Naidu, M. U., Rani, P. U., & Rao, T. R. (2012). Human experimental pain models: A review of standardized methods in drug development. *Journal of Research in Medical Sciences*, *17*(6), 587–595.

Roy, M., Shohamy, D., & Wager, T. D. (2012). Ventromedial prefrontalsubcortical systems and the generation of affective meaning. *Trends in Cognitive Science*, *16*, 147–156.

Salomons, T. V., Johnstone, T., Backonja, M. M., & Davidson, R. J. (2004). Perceived controllability modulates the neural response to pain. *Journal of Neuroscience*, *24*(32), 7199–7203.

Tracey, I. (2010). Getting the pain you expect: Mechanisms of placebo, nocebo and reappraisal effects in humans. *Nature Medicine*, *16*, 1277–1283. https://doi.org/10.1038/nm.2229.

Tracey, I., & Johns, E. (2010). The pain matrix: Reloaded or reborn as we image tonic pain using arterial spin labelling. *Pain*, *148*, 359–360.

Tracey, I., Ploghaus, A., Gati, J. S., Clare, S., Smith, S., Menon, R. S., et al. (2002). Imaging attentional modulation of pain in the periaqueductal gray in humans. *Journal of Neuroscience*, *22*, 2748–2752.

Valet, M., Sprenger, T., Boecker, H., Willoch, F., Rummeny, E., Conrad, B., et al. (2004). Distraction modulates connectivity of the cingulo-frontal cortex and the midbrain during pain—An fMRI analysis. *Pain*, *109*, 399–408.

Wager, T. D., Rilling, J. K., Smith, E. E., Sokolik, A., Casey, K. L., Davidson, R. J., et al. (2004). Placebo-induced changes in FMRI in the anticipation and experience of pain. *Science*, *303*, 1162–1167.

Wiech, K., Kalisch, R., Weiskopf, N., Pleger, B., Stephan, K. E., & Dolan, R. J. (2006). Anterolateral prefrontal cortex mediates the analgesic effect of expected and perceived control over pain. *The Journal of Neuroscience*, *26*(44), 11501–11509.

Wiech, K., Ploner, M., & Tracey, I. (2008). Neurocognitive aspects of pain perception. *Trends in Cognitive Sciences*, *12*, 306–313.

Wolf, S. (1950). Effects of suggestion and conditioning on the action of chemical agents in human subjects; the pharmacology of placebos. *The Journal of Clinical Investigation*, *29*, 100–109.

Wolf, S. (1959). The pharmacology of placebos. *Pharmacological Reviews*, *11*, 689–704.

Chapter 29

Linking the cortex, functional spectroscopy, and pain: Features and applications

Wolnei Caumo[a], Janete S. Bandeira[a], and Jairo Alberto Dussan-Sarria[b,c]

[a]*Pain & Neuromodulation Laboratory, Hospital de Clinicas de Porto Alegre, Porto Alegre, RS, Brazil,* [b]*Universidade FEEVALE, Novo Hamburgo, RS, Brazil,* [c]*Clínica Hamburguesa de Anestesiologia, Novo Hamburgo, RS, Brazil*

Abbreviations

ADHD	attention-deficit hyperactivity disorder
aMC	anterior motor cortex
BOLD	blood oxygen level-dependent
CA	cortical activation
CAI	cortical area of interest
CW	continuous-wave
DCM	dynamic causal modeling
DLPFC	dorsal lateral prefrontal cortex
DSM-5	diagnostic and statistical manual of mental disorders
EEG	electroencephalography
fNIRS	functional near infrared spectroscopy
fMRI	functional magnetic resonance imaging
FD	frequency domain
FC	functional connectivity
O2Hb	oxygenated hemoglobin
HHb	deoxygenated hemoglobin
HRF	hemodynamic response function
IC	insular cortex
LTD	long-term depression
LTP	long-term potentiation
MEG	magnetoencephalography
M1	primary motor cortex
MC	motor cortex
MRI	magnetic resonance imaging
mPFC	medial prefrontal cortex
DOT	optical diffusion tomography
NIR	near infrared spectroscopy
NAc	nucleus accumbens
NMDA	*N*-methyl-D-aspartate acid
PAG	periaqueductal gray
PFC	prefrontal cortex
pMC	posterior motor cortex
PMC	premotor cortex
RVM	rostroventral medulla
rACC	rostral anterior cingulate cortex
rTMS	repeated transcranial magnetic stimulation
PET	positron emission tomography

The Neurobiology, Physiology, and Psychology of Pain. https://doi.org/10.1016/B978-0-12-820589-1.00029-4

S1/SI	primary somatosensory cortex
RAGT	robot-assisted gait training
S2/SII	secondary somatosensory cortex
SMA	supplementary motor area
SMC	sensorimotor cortex
SSC	somatosensory cortex
tDCS	transcranial direct current stimulation
TMS	transcranial magnetic stimulation
TD	time-domain

Basic concepts

According to the new definition of pain, *it is an unpleasant sensory and emotional experience associated with, or resembling that associated with, actual or potential tissue damage* (Raja et al., 2020). Acute pain appears in response to tissue damage and has a physiological and adaptive function. While chronic pain is the one that lasts beyond three months of the expected time to resolve the condition or disease, an results from poorly adaptive neuroplasticity processes. The maladaptive neuroplasticity encompasses the impaired functioning of neurons and circuits in nociceptive pathways, with the following changes: increased neuronal excitability, improved synaptic efficacy, and reduced inhibition. This cascade of processes is part of the central sensitization syndrome (CSS) (Latremoliere & Woolf, 2009).

According to the CSS concept, chronic pain is a multifaceted experience produced by characteristic "neuro-signature" patterns generated by sensory inputs. Still, it may be caused by the abnormal function of circuitry of a widespread network of neurons that produce reverberating patterns, self-generating and sustaining chronic pain and their correlated symptoms. The pain is a multidimensional experience, involving somatosensory, cognitive-evaluative, affective-motivational, and social aspects. The functional plasticity in neural circuits of pain has provided mechanistic insights, including changes in molecules, synapses, and neurobiological systems engaged in perception and behavior (Prescott, Ma, & De Koninck, 2014). However, to progress in pain research, we need to work together to understand the mechanisms of pain, develop novel approaches to management, and have a clear view of where are knowledge gaps to target our research endeavors most effectively. This approach can help the clinician to integrate the knowledge in neuroscience to improve the clinical research and care at bedside. This way, this chapter aims to present the functional near infrared spectroscopy as a tool to study pain processing, some of its features, and potential applications in the research and clinical setting.

Cortical functions related to pain processing

Chronic pain is a maladaptive neuroplasticity process involving cortical regions and classical areas of the pain neuromatrix circuit. The pain is the result of an orchestrated process with somatosensory, cognitive, affective and social components. In the last decades, there was progress in the understanding of the anatomy and pathways related to the nociceptive system mapping these neural connections by functional imaging, such as magnetic resonance imaging (fMRI), positron emission tomography (PET), and near-infrared-spectroscopy (fNIRS). These techniques assess the recruitment of neural networks by looking at the blood flow within specified regions of the brain. It is crucial to have well-defined tasks (paradigms) to determine which target areas would be probed and the best stimuli for each research question. A proper definition of stimulus, the oscillation of its intensity, and duration can improve the precision and reproducibility of the experiments. Ideally, these measures could be correlated with other parameters, such as biomarkers of neuroplasticity, clinical picture, genetics, electrophysiology, and pharmacology (Gatchel, Peng, Peters, Fuchs, & Turk, 2007). Other measures that can be used together with functional neuroimage are physiologic reactions (i.e., autonomic), such as limb withdrawal, and an increase in heart rate indicating infracortical responses at the spinal cord, brainstem, hypothalamus, and thalamus levels (Slater, Fitzgerald, & Meek, 2007). Other behaviors such as facial actions and vocalization (i.e., cry or verbal complaints) are valued as suitable proxy measures of the pain experienced in those who cannot communicate verbally. However, it is unclear whether these cues accurately reflect pain perception and awareness involving cortical processes.

Although it is recognized that the cerebral cortex is involved in various dimensions of pain perception, the debate now focuses on determining the specific pain-related functions within each cortical region and how they interact. The most constantly activated brain regions in acute pain studies are the insula, anterior cingulate, thalamus, and basal ganglion (Apkarian, Bushnell, Treede, & Zubieta, 2005). The primary and secondary somatosensory cortices provide discriminative sensory information when activated by noxious stimuli. The prefrontal cortex (PFC), together with the amygdala, hypothalamus, hippocampus, thalamus, periaqueductal gray matter (PAG), and basal ganglia, is involved in the evaluative and

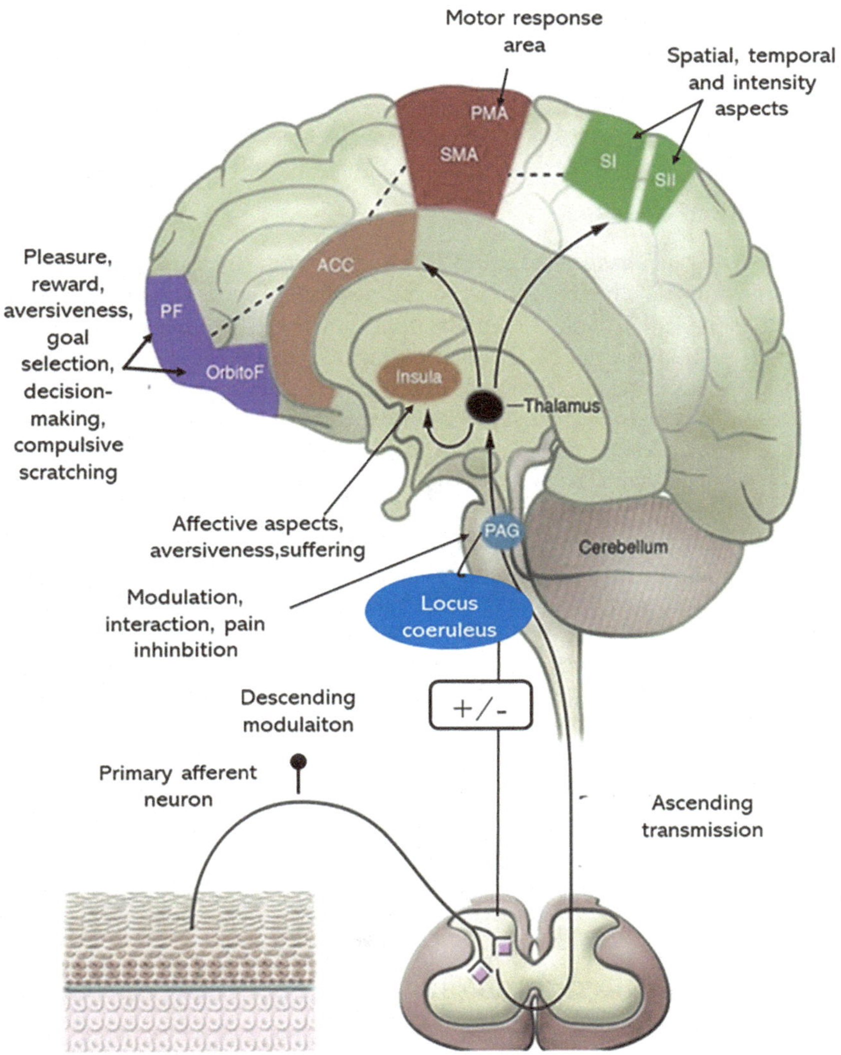

FIG. 1 Pain and emotion pathways processing at cortical and infracortical areas. Cortical areas involved in pain processing.

cognitive dimensions of pain (Cieslik et al., 2013; Davis et al., 2017). The primary motor cortex (M1) plays a significant role in pain modulation. The increase in cortical excitability in the M1 promotes a modulatory effect into intracortical, thalamic, and descending neurons. The cortical and infracortical areas involved in pain processing are presented in Fig. 1.

Role of the motor cortex in pain processing

The motor cortex (MC) processing is overlapped with those areas associated with pain neuromatrix. MC is divided into anterior (aMC) and posterior (pMC). The MC receives inputs from thalamic relay nuclei, somatosensory cortex, premotor cortex, and sensory association areas from the cortex, basal ganglia, and cerebellum. The pMC is involved in orienting the body in response to sensory stimuli and motor execution. While the aMC is associated with the motor function (Mutschler et al., 2009), pain processing, fear-avoidance (Vogt, 2005), cognitive control, and emotion processing. The aMC receives projections from the spinothalamic system transmitting nociceptive information to the cortex. A preclinical study showed that MC suppressed activity in the primary somatosensory cortex and prefrontal cortex. Also, the activation of MC can be achieved by global pain network modulation. These processes involve corticocortical and thalamocortical loops rather than merely activating the primary M1. Studies of functional neuroimaging have shown that MC activates supraspinal areas involved in the perception and emotional appraisal of pain, working together with several structures, including the lateral thalamus, ACC, anterior insula, and PAG (García-Larrea et al., 1999). The analgesic effect of MC is correlated with the release of endogenous opioids in the ACC, insula, and PAG (Zortea et al., 2019). In a preclinical experiment of neuropathic pain, the MC stimulation reversed the central and peripheral pain (Lucas, Ji, & Masri, 2011), activating the limbic system and PAG and inhibiting the thalamic nuclei and spinal nociceptive neurons (Pagano et al., 2011). It has been hypothesized that MC induces analgesia by activating the descending analgesic pathways.

Role of the PFC in pain processing

Acute and chronic pain induce relevant changes in the PFC gene expression, structure, activity, and connectivity. The PFC role in pain processing is integrated with other brain areas, including primary and secondary somatosensory cortices, insular cortex (IC), ACC, and thalamus (Apkarian et al., 2005). The discriminative pain intensity involves connections of the ventrolateral pathways, from the insular cortex to PFC. In contrast, the discrimination of the spatial aspects of pain consists of the activation of the posterior parietal cortex and the right PFC (Oshiro, Quevedo, McHaffie, Kraft, & Coghill, 2009). The posterior insula is connected to the primary somatosensory cortex (S1), the secondary somatosensory cortex (SII), ventrolateral PFC, and orbitofrontal cortex (Wiech, Jbabdi, Lin, Andersson, & Tracey, 2014). Activation of the ACC and medial prefrontal cortex (mPFC) is associated with increased activity on the PAG (Peyron & Fauchon, 2019). The mPFC is also related to modulation of pain catastrophizing (Seminowicz & Davis, 2006), reduction of pain-induced sympathetic activity (Perlaki et al., 2015), and decrease in facial expressions of pain (Karmann et al., 2016). The PAG is the main control center for descending pain modulation. The PAG originates mainly from PFC projections, and its input from the mPFC, the dorsomedial convexity of the medial wall extending into the ACC, and the anterior insular cortex. According to fMRI data, the lateral and dorsolateral subregions of the PAG are connected to PFC, which is implicated in executive functions. In contrast, the ventrolateral PAG is linked to brain regions associated with descending pain modulation, including the ACC and upper pons/medulla (Coulombe, Erpelding, Kucyi, & Davis, 2016). Under physiological conditions, an increase of the functional connectivity between the thalamus and mPFC influences working memory, and the PFC is involved in empathy for pain (Laneri et al., 2017).

The cortical dysfunction in chronic pain patients includes a molecular, circuit, and anatomical changes. In the prefrontal cortex at the molecular level, there is a decrease in the glutamate and reductions in the gray matter density. The increase of sympathetic activity reduces the cerebral flow and the reorganization of white matter connectivity, especially to the basal ganglia and insula. Also, the literature has observed fractional anisotropy and fiber track density to multiple pain regions, including the anterior ACC (Shiers & Price, 2020).

Nature of brain regions recruited during pain: Insights from human imaging studies

Neuroimaging techniques permit to assess the neuronal activity on different spatial and temporal scales. We will focus on imaging approaches that are of interest to address the contribution of brain circuits to pain perception, particularly findings related to the fNIRS. Functional neuroimage provides an excellent starting point to understand how the perception of pain is created in the brain across diverse brain regions in the context of an acute painful stimulus, resting activity changes in chronic pain states, as well as alterations in the brain produced by painkillers and neuromodulatory interventions (i.e., transcranial neuromodulation, acupuncture, exercise, etc.).

As the fMRI, the fNIRS is a technique of functional neuroimaging that assesses the cortical activation by the blood flow (BOLD) signals. fNIRS permits an assessment of spatiotemporal activity levels of the brain at rest or in response to certain stimuli or self-generated motor actions (Friston, 2009). This technique provides an indirect readout of neuronal activity based on the BOLD signal. The fNIRS measures relative changes in the oxyhemoglobin (O2Hb) and deoxyhemoglobin (HHb) regional concentration. These alterations reflect increased tissue perfusion due to higher metabolic demands by augmented neuronal activity (Scholkmann et al., 2014). This method is ideally suited to determine activity patterns across the entire brain at a spatiotemporal resolution that allows inferences on the brain region that is activated and to relate these activities to individual brain states (e.g., the subject consciously receiving a specific stimulus, or generates a movement).

The biological rationale behind near-infrared spectroscopy

Neuronal electrical activity demands adenosine triphosphate (ATP), which is provided by oxidative metabolism. It is so to the point that the cerebral metabolic rate of oxygen consumption (CMRO2) generally parallels cortical electrical activity. To attend such demands, the cerebral blood flow (CBF) varies with metabolic activity, increasing inflow in the cortical regions with higher neuronal requirements. Thus, neuronal firing triggers a regional response that includes an increase

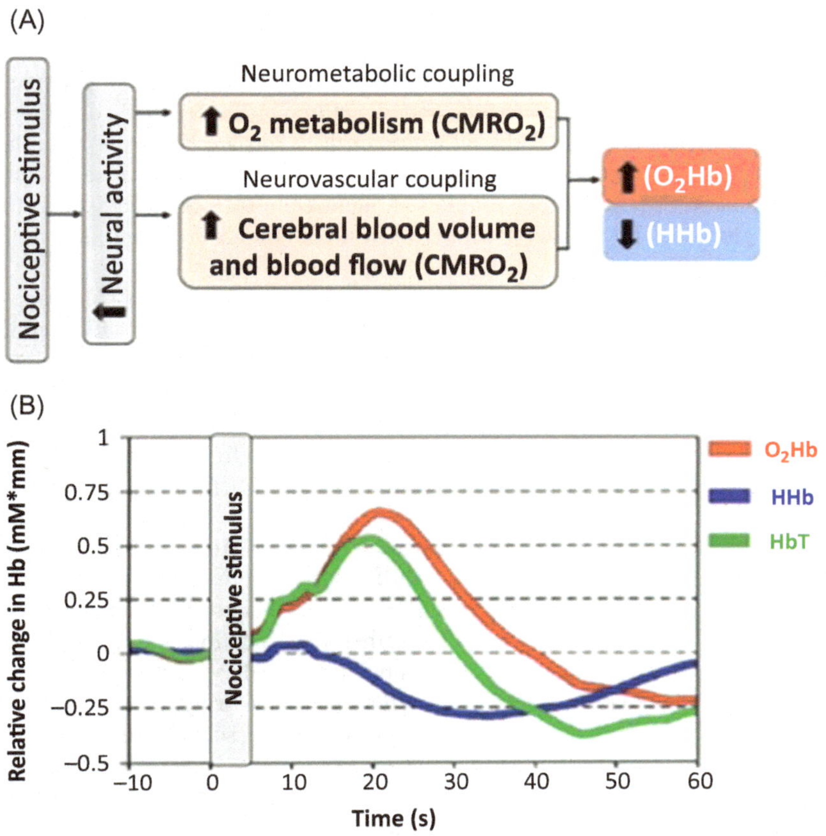

FIG. 2 An illustrative figure of the BOLD signal and neurovascular coupling evoked by a nociceptive stimulus. (A) Neural activation by a nociceptive stimulus with an increase of neurometabolic activity and O2Hb consumption and consequent increase on the rates of HHb. (B) BOLD signal evoked from an increase of neural activation after a nociceptive stimulus.

in CMRO2, in CBF, in blood volume, and in saturation of oxygen (higher concentration of O2Hb and a lower concentration of HHb). This interaction is called neurovascular coupling. The neurovascular coupling allows us to accept that understanding hemodynamical regional phenomena informs brain functioning (electrochemical processes). Fig. 2 presents the increased neurometabolic activity related to the neural activation by nociceptive stimulus and the BOLD signal in the curves of O2Hb and HHb.

Technically, spectroscopy is the analysis of the interaction between a portion of the electromagnetic field and matter. In the particular case of brain study (our matter of interest), the near-infrared spectrum of the light (approximately 650–950 nm) penetrates the human body (up to a few centimeters in the human head) and is either scattered or absorbed. Fortunately, knowing the characteristics of a light source, measuring its attenuation allows us to infer the concentration of relevant chromophores in our body (namely, O2Hb, and HHb) using the modified Beer-Lambert law. Although it is challenging measuring light attenuation in biological tissues, the combination of different light sources and receptors improves resolution. Although it is challenging to measure light attenuation in biological tissues, various light sources and receptors enhance resolution. The location of the light sources and detectors in the human scalp can be correlated to electroencephalography (EEG) areas and cortical gyri, providing references for anatomical correlations of the hemodynamic changes detected. Since the technique allows us to assess relative changes in cortical hemodynamics when neuronal demands occur, the term functional NIRS was adopted. Thus, to understand cortical neuronal firing regions, we illuminate a human head with multiple light sources and measure hemodynamic and metabolic (oxygen) changes.

fNIRS neuroimaging: Technical aspects, advantages, limitations, and applications

The use of fNIRS to examine the cortical tissue was initially proposed by Jöbsis (1977), who was the first researcher to demonstrate the efficiency of infrared light penetration into the skull in monitoring cortical activation. The early studies used to assess simple motor and sensory tasks and to show the applicability of the technique to study brain activity in a noninvasive manner, in each area of the cortex. Since then, a growing number of studies have been investigating more complex cognitive functions. Among these was as a method to assess the cortical response to algogenic stimuli.

fNIRS is a technique that uses infrared light, carried by an optical fiber, which radiates light to the scalp of the person under examination. The light emitted by the sources propagates through the skull and is captured by the detectors, drawing a banana-shaped curve on the cortex's surface structure. Then, changes in the reflection of the lights are measured, with the possibility of distinguishing between the levels of oxygenation and blood deoxygenation. This is possible because all activation of a specific area of the brain due to some function's performance requires an increase in metabolic activity and a consequent increase in the supply of oxygen and glucose. The hemoglobin present in the blood absorbs light, allowing fNIRS detectors to capture changes in the intensity of reflected light.

The positioning of light sources and detectors on the scalp reproduces electrodes placement for electroencephalography following the the 10 × 20 system (Jasper, 1958), according to the cortical area of interest. This system is widely used in studies with electroencephalography (EEG). Some surveys have associated the fNIRS system with magnetic resonance imaging (MRI), to more precisely allow the comparison of data obtained with this measure and functional magnetic resonance imaging (fMRI). However, the fusion of fNIRS and fMRI data has been useful to identify when and where hemodynamic signals are changing in human brains. The fNIRS optodes and their corresponding optical lightpath are placed on the similar locations of EEG electrodes. The arrangement of fNIRS sources and detectors projected on the brain is presented in Fig. 3.

The setup required for fNIRS is light in comparison to functional magnetic resonance imaging (MRI), magnetoencephalography (MEG), or even computerized tomography (CT). In practical terms, an fNIRS setup resembles EEG, in that the subject wears a cap with sensors. Moreover, fNIRS equipment could be as light and portable as a laptop that can be carried in a back-pack, as the subject performs a task with his cap on. These features strengthen the "functional" essence of the technique, as it could be used to study cortical activity in more naturalistic environments and behaviors.

Considering that the idea is to capture signals of hemodynamic changes that reflect neuronal firing, a target cortical region (or areas) and tasks that elicit activation of such circuits should be chosen. The corresponding scalp is then populated with an array of light sources and detectors (channels), and the subject is ready for the experimental session (Fig. 4A and B). Given that relative changes are detected, usual protocols propose to begin with some sort of "resting" state, to further compare them with "events," or tasks, that elicit the cortical activity of interest. Such tasks can be as diverse as human brain capabilities: visualization (ask the patient to imagine or visualize a situation), motor (hand movements or walking), cognitive (calculation or conversation), auditory (tones or music), visual (looking at images or scenes), olfactory (smelling the scent of coffee, flowers, or rotten food), gustatory (tasting chocolate), or somatosensory (pain).

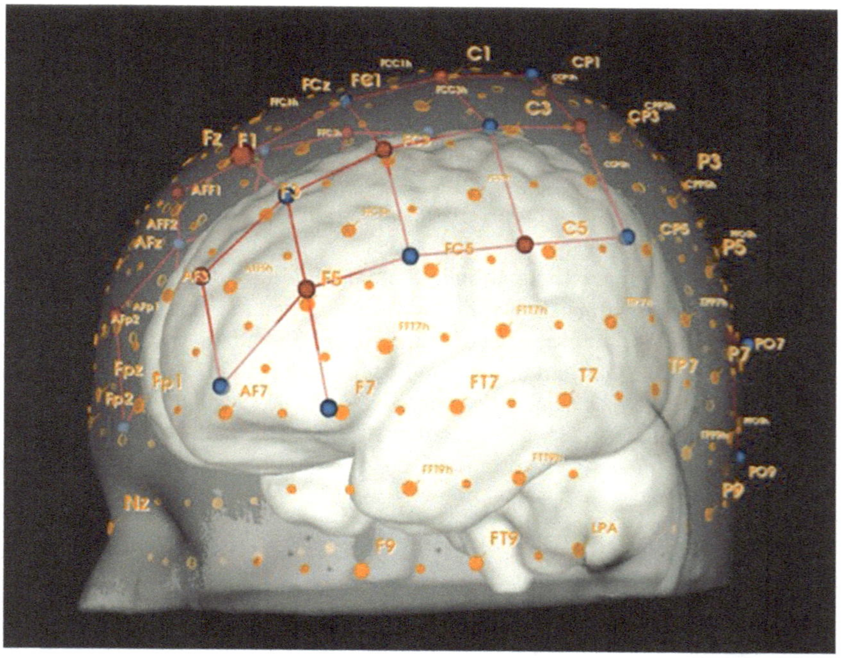

FIG. 3 Cerebral cortex activity is inferred by positioning fNIRS optodes on the scalp following the electro-encephalography (EEG) 10 × 20 system. The figure represents EEG 10 × 20 system (*orange*) and fRNIS optodes placement. Detectors depicted in *blue*, sources in *red*.

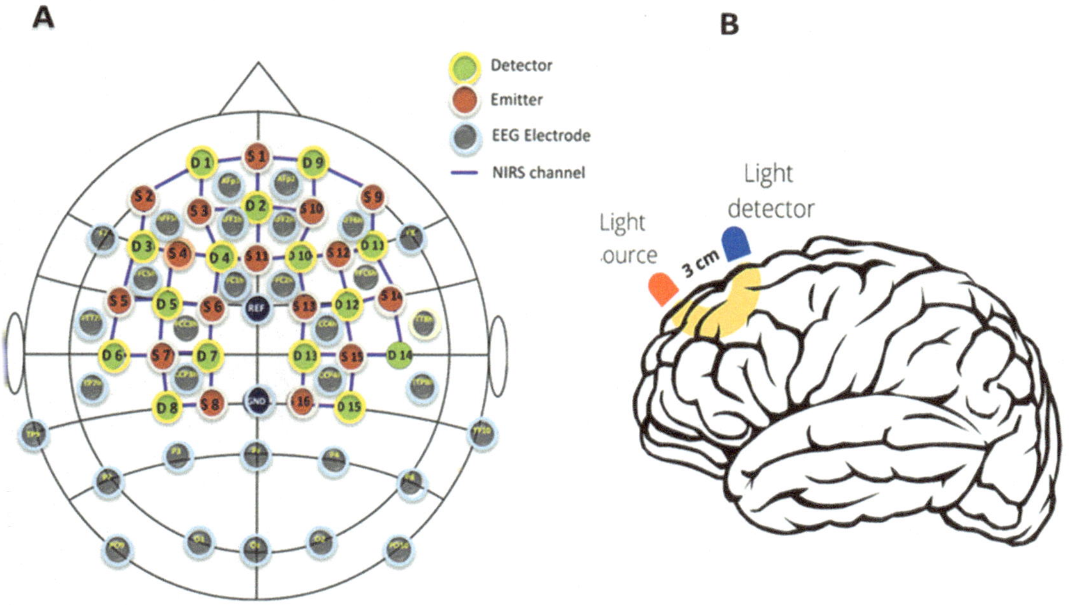

FIG. 4 Montage, emitter, and detectors, according to EEG. (A) Montage. Emitter *(red)*, detectors *(light green)*, in correspondence with 10/10 electroencephalography (EEG) system. Channels formed are in *gray*. The montage on the MC area and PFC area. (B) Near-infrared light does not significantly disperse on the dermis and bone tissues, so it can penetrate in an elliptic format known as "banana shape" to cortical tissue, between a source and detector optode, forming a measuring channel. If the distance between light source and detector would be around 3 cm, the penetration of light can get to 2 cm deep. *(Adapted from Naseer, N., & Hong, K. S. (2015). fNIRS-based brain-computer interfaces: A review.* Frontiers in Human Neuroscience. *https://doi.org/10.3389/fnhum.2015.00003.)*

The hemodynamic characteristics of the reference state in the cortical areas of interest are further compared with the one elicited during the events, and inferences of neuronal activities are achieved. Since the human brain is not a conglomerate of isolated functional units that work as on-off switches, cortical areas interact continuously, and their hemodynamic states vary accordingly. Their correlation in time and space can also be estimated and is used to infer connectivity. Such interactions can be tracked, measured, and studied, as far as the fNIRS channels can reach them. The anatomical, physiological, physical, and mathematical concepts involved in selecting an adequate experimental montage, execution, data processing, and analysis are beyond the scope of the present chapter.

The main advantages of fNIRS rely on light sources, usually diodes, which are safe. Light exposure in the scalp could increase temperature less than 0.5°C. To carry the light to the sources, and from the scalp (detectors), fiber optics are employed. They are adaptable to different head positions, facilitating its portability and use in more naturalistic environments (compared with fMRI, MEG, or CT-Scans). Also, fNIRS can be used simultaneously with other research tools, such as EEG, transcranial direct current stimulation (tDCS), transcranial and magnetic stimulation (TMS). Moreover, more than one subject can be studied simultaneously, setting up an fNIRS system for two participants as they (for example) have a conversation, therapy, or perform a task. Due to its "event-related" characteristics, it provides reliable temporal information (which facilitates inference of causality), but poor spatial resolution.

The spectroscopy device has emitters and detectors that are attached to the scalp. Several models use near-infrared, safe, and noninvasive wavelengths, usually between 650 and 950 nm. The near-infrared light does not suffer significant dispersion of the dermal and bone layers, penetrating in an elliptical shape (banana shape) up to about 2 cm in depth, between an optode emitter and an optode detector, forming reading channels (Naseer & Hong, 2015). These channels can be correlated to EEG areas, cortical gyri, Brodmann areas, or another cerebral spatial resolution. The most usual separation between the light source and the detector is 3 cm. Nonetheless, increment this to 4 cm can increase the penetration depth of the light sampled tissues. To provide physiological (noise) data within the superficial layers, the short separation channels used are located within 1 cm of the source. This approach can then regress the activity from the long separation channel, resulting in a signal corresponding to activity solely within deep brain tissues. As fNIRS depends on the refraction of the light, it only allows assessing cortical structures close to the scalp (up to 2 cm depth). Also, depending on the optodes montage (scalp location, number of channels, the distance between them), spatial resolution is compromised. Sensor placement on hair is limited because it causes interference with the NIR light. Optode placement precision and consistent interrogation of the same brain regions are a major concern with NIRS. Ideally, the accuracy of the interrogated cerebral region by the NIRS device should be verified by simultaneous MRI because the ability to ascertain the exactitude of the target area is limited. This drawback could be improved using multichannel NIRS devices that cover multiple regions of the scalp. Combining the strengths of NIRS with those of MRI would also be greatly beneficial. fNIRS signals reflect combined synaptic activity from inhibitory interneurons from the pyramidal cortex, unable to differentiate inhibitory or excitatory synapses. Regarding the development of scientific protocols, there is still a lot of space for the development of mathematical models for data analyses, as most of the statistical models are parametric, while studies using more natural environments could demand research protocols less rigid (i.e., not "carbox paradigms"), that might not fit current analyses methods.

The sequence of steps to perform neuroscience studies using fNIRS

To improve the quality of the neuroscience research using the fNIRS, it is recommended that the protocols' follow a sequence of steps presented in Fig. 5.

Step 1: Planning. According to the research question, an experimental paradigm is design. To increase statistical power, consider using repetitions of blocks of stimuli or event-related paradigms. fNIRS measures are then collected during the execution of the designed experiment. To better evaluate the presence of hemodynamic responses, the experimental task might be interspersed with contrast conditions or stimuli or rest periods.

Step 2: Data acquisition. fNIRS comprises the placement of light sources and detectors (i.e., "optodes") on the participants' scalp. For adults, due to the "banana-shape" light dispersion, consider to keep 3 cm of distance between sources and detector . The fNIRS signal measured by the detector, e.g., the cold light intensity signal, originates from the tissue volume located below the source. The frequency of the acquisition, the number of channels, and the sampling depend on the fNIRS instrument used.

Step 3: Preprocessing. Data are usually visually inspected to assess signals' quality (e.g., the presence of significant motion artifacts, and heartbeat oscillations indicating a good optical coupling between the optodes and the scalp). The time-series are converted into changes in attenuation (or optical density, OD) and then into concentration changes of O2Hb and HHb (1O2Hb and 1HHb), usually through the modified Beer-Lambert law (Duncan et al., 2014). To extract information without any source of variability in the O2Hb and HHb, signals not related to the task-evoked hemodynamic activity needs to be removed or minimized. For mitigating the noise, there is digital filtering, which is a mathematical procedure applied to discrete time-series to reduce or enhance specific properties of the input signals (e.g., frequency ranges). Filters are divided into three classes:

1. High-pass filters can remove high-frequency components above the cut-off frequency.
2. Low-pass filters remove low-frequency components below the cut-off frequency.
3. Band-pass filters preserve the frequency range between a more moderate and a higher cut-off rate.

Step 4: Statistical inferences (if applicable). The preprocessed O2Hb and HHb signals evoked by the task or stimulus can be examined using different statistical approaches. One of the common statistical frameworks employed is the General Linear Model (GLM). The GLM advantage is its more statistical power since it consideres the entire fNIRS time-series. However, the disadvantage of using the GLM is that it assumes a specific predefined hemodynamic response function. Such premise should be carefully revised in some populations, for example, in neonates, as hemodynamical response might differ across brain regions. It is essential to mention that this is only one example among several statistical approaches that can be applied to analyze the fNIRS data.

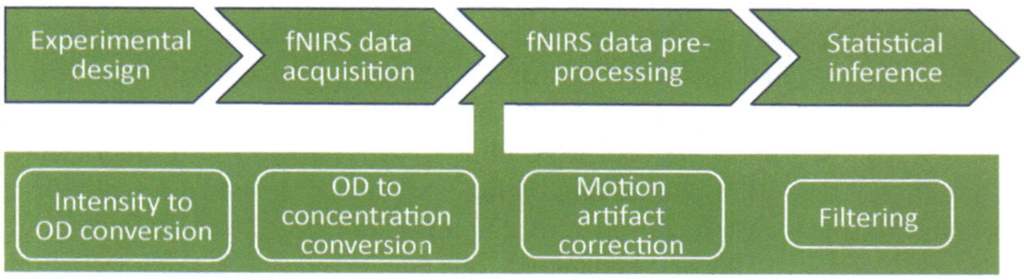

FIG. 5 Steps recommended for protocols' using fNIRS in neuroscience. Typical sequence of neuroscience study using fNIRS. Optical density (OD).

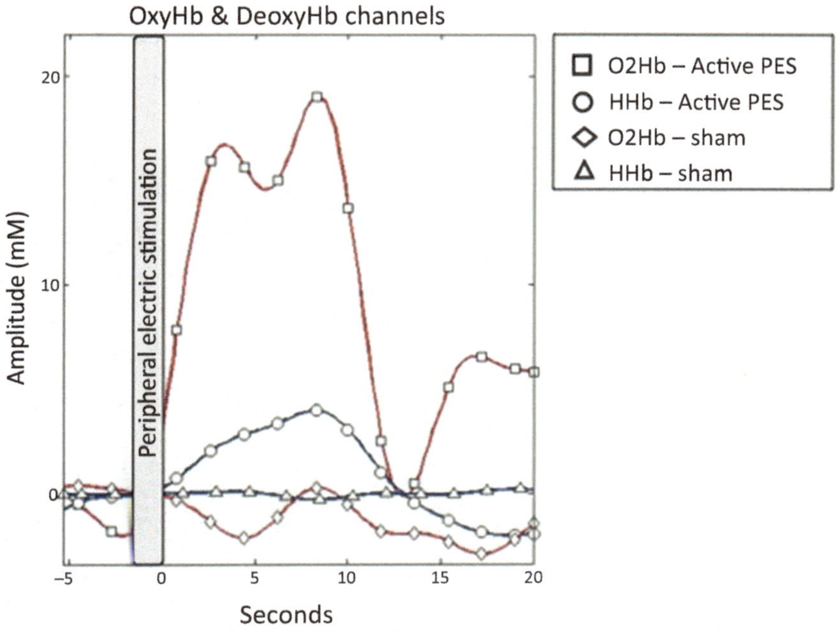

FIG. 6 Curve of O2Hb and HHb across the time after peripheral electric stimulation (PES). NIRS device was placed to record the curve of O2Hb and HHb changes for active and sham PES on the right prefrontal cortex. The active peripheral intervention consisted of 10 Hz electrical nonpainful stimulus in burst current, 10 s ON and 20 s OFF, for 12 min, generating 24 blocks of hemodynamic curves in response to electrical current on unilateral right accessory spinal nerve stimulation. *(From Bandeira, J.S., Antunes, L.d.C., Soldatelli, M. D., Sato, J.R., Fregni, F., & Caumo, W. (2019). Functional spectroscopy mapping of pain processing cortical areas during non-painful peripheral electrical stimulation of the accessory spinal nerve. Frontiers in Human Neuroscience, 13(200). https://doi.org/10.3389/fnhum.2019.00200.)*

Dynamic trace pattern evoked by electrical stimulation

Cortical hemodynamic response to pain stimulation seems to have a characteristic pattern. Fig. 6 is presented a fNIRS measurement of an experiment conducted in our Laboratory. Line A represents the O2Hb concentration, which presented a transient increase after unilateral electrical current stimulation of the right accessory spinal nerve followed by O2Hb reduction to baseline. The HHb, on the other hand, decreased slightly. The time to reach the lowest HHb (35 s) was longer than required to achieve the highest O2Hb (20 s).

Near-infrared spectroscopy is an evaluation instrument that measures concentrations of a compound in a sample, based on its absorption spectrum of the infrared (or near-infrared) light of the electromagnetic spectrum emitted by the device. For biological tissues formed by substances that reflect and scatter light in different ways, the calculations of the substances' concentrations are made based on the modified Beer-Lambert law. fNIRS is functional optical spectroscopy of the neuroimaging method used to monitor cortical function in real-time, by calculating the concentrations of O2Hb, HHb, and total hemoglobin tHb.

Cerebral activity produces increased oxygen consumption, accompanied by an increase in cerebral blood flow due to neurovascular coupling, followed by the return to baseline response activity, known as blood oxygen level-dependent activity (BOLD) (Phillips, Chan, Zheng, Krassioukov, & Ainslie, 2015). This makes it possible to measure blood oxygenation changes, since O2Hb is converted into HHb during neural activity, reflecting the cerebral hemodynamic response. This is reflected in changes in O2Hb and HHb measures in the observed region, which can be detected noninvasively by fNIRS. There are several types of near-infrared imaging techniques. Continuous wave (CW), frequency domain (FD), and time domain (TD) are mostly used. They have different light emission parameters and varying amounts of information to be dealt with the right data analyses. Fig. 7 presents the hemodynamic activation pattern in time sequence after peripheral electrical stimulation.

Applications of fNIRS in pain research

In the field of pain research, fNIRS could be used for several purposes, among them, as a tool to search for cortical pain "signatures." Also, it could be used to study cortical connectivity upon physical, psychological, or social pain. Furthermore, it could be used to track pain signals while under anesthesia or in social environments. On the other hand, fNIRS could also be used to address brain hemodynamics, irrespective of cortical correlation.

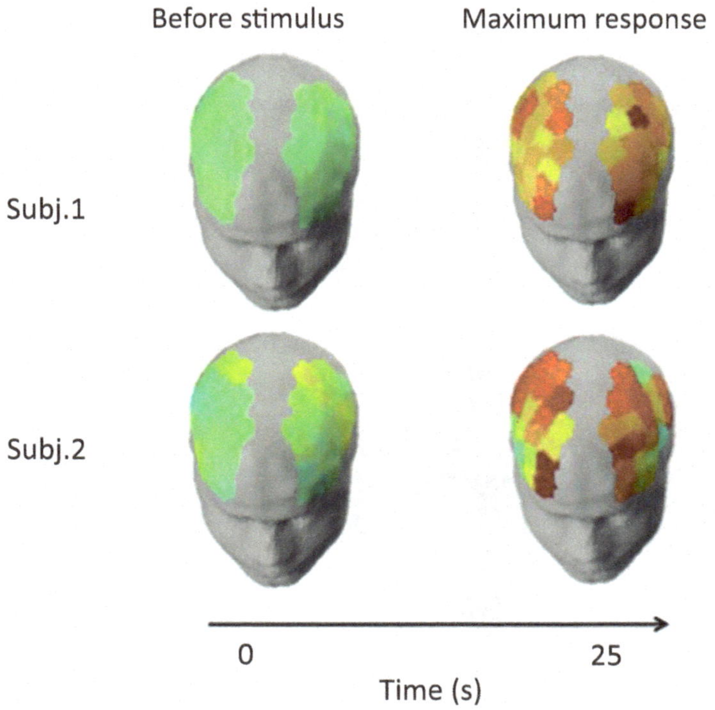

FIG. 7 fNIRS activation patterns in time sequence. Graphic representation frames on a 3D head surface model in two subjects, showing O2Hb changes until its maximum response after the electric stimulus. *Red pattern* represents activation, *blue* is for deactivation response. *(Data from Laboratory of Pain and Neuromodulation, Porto Alegre, Brazil (Caumo W).)*

Although they were not studying cortical networks specifically, in 2006, Akin and Bilensoy (Akin & Bilensoy, 2006) used fNIRS to study patients with migraines without aura. Using a breath-holding task (in a pain-free period), they observed that patients had decreased cerebrovascular reactivity to this metabolic perturbation as compared with healthy controls. Later, in 2010, Viola et al. (Viola, Viola, Litterio, Buongarzone, & Fiorelli, 2010) studied patients with prolonged aura using transcranial Doppler and NIRS. During prolonged migraine aura, they found cortical hypoperfusion areas corresponding to decreased metabolic demand in the topography of aura symptoms. These findings in the migraine research topic demonstrate the versatility of using fNIRS with a different scope, but still pain research.

In 2009, Becerra et al. (Becerra et al., 2009), with the Pain and Analgesia Imaging Neuroscience, demonstrated an S1 cortex response pattern upon innocuous and noxious heat stimuli as the early fNIRS steps toward the search for cortical representations of pain. Re et al. (2013) reported a pattern of cortical activation proportional to electromyography stimulation intensity in the contralateral sensorimotor cortex. Similarly, in the same year, Barati et al. (2013), from Drexel University, reported a proportional pattern of cortical activation in the forehead, with a cold pressor test paradigm as pain stimulation. Further, in 2013, Lee et al. from Kaohsiung Medical University (Lee et al., 2013) and the Advanced Research Institute for Bioscience observed different pain-related (pressure type, induced with a push-pull digital force gauge) and itch-related (histamine-induced) blood flow responses in the frontal cortex. In Japan, too, Sakuma et al. (2014) demonstrated that gingival pain caused O2Hb levels to decrease in the superior frontal gyrus, the middle frontal gyrus, and the orbital part of the superior, middle, and inferior frontal gyrus, on the brain side contralateral to the pain load. Üçeyler et al. from the University of Würzburg (Üçeyler et al., 2015) in 2015 used fNIRS to demonstrate that upon unilateral pressure pain stimulation, patients with fibromyalgia presented increased bilateral cortical activation when compared with healthy controls, and more potent contralateral activity over the dPFC when compared with patients with major depression. Using pressure, too, but this time applied to the lumbar spine, Vrana et al. from University Hospital of Balgrist in Zurich (Vrana, Meier, Hotz-Boendermaker, Humphreys, & Scholkmann, 2016a, 2016b), in 2016, reported characteristic changes in [O2Hb] in the SMA and S1, but not after brushing stimulus. Such changes were also present in patients with chronic low back pain but were not different from those observed in healthy volunteers. Whether various body sites could elicit different cortical responses, Yennu et al. from the University of Texas at Arlington in 2016 (Yennu, Tian, Gatchel, & Liu, 2016) found that thermal stimulation at three distinct body parts activated the PFC with consistent temporal and spatial patterns, without clear spatial resolution. Similarly, Hu et al. from the University of Michigan (Hu et al., 2018) in 2018 assessed S1 and PFC for clinical orofacial pain and observed that patients' clinical pain could be predicted by their baseline functional connectivity between S1 and PFC.

Aasted et al. from Harvard Medical School in 2016 (Aasted et al., 2016) confirmed deactivations on frontal cortices upon electrical pain stimulation in healthy volunteers. Further, in the same year and from Harvard Medical School too, Kussman et al. (2016) demonstrated the feasibility of using fNIRS during general anesthesia to probe signals of pain stimulation during actual surgical procedures. In their study, they observed the marked decrease in O2Hb 4–6 s after painful stimuli (cardiac ablation), which mirrored those seen with noxious stimulation in awake, healthy volunteers during sedation colonoscopy, and functional magnetic resonance imaging activations in response to pain. Further, they (Peng et al., 2018) tested prefrontal response to oral morphine in healthy subjects. Using fNIRS they detected an attenuation effect of morphine in the pain signal within this region, as well as in the contralateral S1, that did not appear upon innocuous stimulation.

Striving to use fNIRS as a tool to find cortical markers of pain, Pourshoghi et al. from Drexel University in 2016 (Pourshoghi, Zakeri, & Pourrezaei, 2016) published the application of a functional data analysis on fNIRS that classified signals from the frontal cortex of subject submitted to the cold pressor task, and correctly identified low-pain and high-pain signs with 94% accuracy. Sun et al. (2019), from the Fourth Military Medical University in China, in 2019, used fNIRS as part of the assessment of patients with spinal cord injury who received repeated transcranial magnetic stimulation (rTMS) to treat neuropathic pain. In their experiment, researchers probed cortical activation of bilateral M1, the primary somatosensory cortex (S1), premotor cortex (PMC), and prefrontal cortex (PFC) during the handgrip task. However, no relevant pain relief (nor cortical activation changes) was observed during the first two weeks of rTMS and in the subsequent four weeks after the activation of M1 and PMC was remarkably suppressed, while pain scores were reduced.

It is also essential to combine fNIRS with other physiological measures such as noninvasive blood pressure and psychophysical pain measures (i.e., thermal, electric stimulus, stress, etc.). An integrative approach can view how systemic effects and peripheral stimulus could affect cerebral hemodynamic. According to Scholkmann, Hafner, Metz, Wolf, and Wolf (2017), these effects should be investigated because it is improbable that cerebral hemodynamic changes are purely associated with brain activation. For example, a group from Tokyo, Yasui et al. (2010) reported concomitant use of fNIRS and electrocardiography to test cortical and autonomic responses to heat and steam. They probed the whole cortex available at the scalp. They found that during the steam-generating sheet, the O2Hb decreased in the anterior dorsal region of the medial prefrontal cortex (mPFC), while parasympathetic activity increased. Their findings pointed out that mPFC receives convergent inputs from the limbic system, and outputs to the thalamus and brainstem involved in emotional control and response to stress. So, they suggest that the hypoactivation observed could be related to the state of relaxation induced by the intervention studied. In Texas, Pollonini, Miao, and Ahn (2020) examined the effect of transcranial direct current stimulation (tDCS) in patients with knee osteoarthritis. They assessed the bilateral motor and somatosensory cortices with functional near-infrared spectroscopy (fNIRS) in response to thermal pain. They observed that tDCS increased activation in these regions (supporting the notion that this therapy increases cortical excitability) and decreased pain.

Some authors postulate that fNIRS is a preferable tool to assess cortical activation induced by electrical stimulation because it is less sensitive to electrical interference than other neuroimaging techniques. The fNIRS allows an individual and temporal assessment of the cortical hemodynamic response continuously since the changes are rapid within the stimulation period. The fNIRS signals reflect the combined synaptic activity of inhibitory interneurons in the pyramidal cortex, regardless of the excitatory or inhibitory character, also modulating pyramidal postsynaptic structures and other cortical targets (Tanosaki et al., 2001). Studies with fNIRS have helped differentiate responses to innocuous, painful, and even paresthetic stimuli. The last research managed to relate the change in the amplitude of the HbO and HHb curve with the pain scores on the Visual Analogue Scale (VAS) reported by the volunteers after the application of painful stimulus applied to the right thumb. To evaluate the prefrontal and motor cortex response to nonpainful peripheral electrical stimulation as a tool for pain treatment, Bandeira et al. (2019) found cortical activation of these areas with 10 Hz active but not sham stimulation, measured by a 40-channel fNIRS system. Therefore, a variety of study configurations can be made on pain research, using fNIRS technology.

A wide variety of tools employing physiological parameters linked to pain are currently undergoing investigation. The fNIRS can provide an objective and robust assessment of cortical activation evoked by pain. Although the fNIRS is not established as a pain measure, it is used to assess brain activity in neonates and children (Wolfberg & du Plessis, 2006). It is also currently used to study the temporal and spatial localization of cortical activity in response to other stimuli, such as vision, sound, language, and taste. Another advantage of fNIRS is a lack of exposure to ionizing radiation, which allows for repeated use over extended periods. This technique is promising as a tool for independently assessing pain in adults and children. It is particularly useful when self-reporting is not possible. The fNIRS employment as a continuous bedside monitor would be of particular use in critical care or intraoperative settings.

Recent advances in neuroimaging techniques led to increased knowledge in neuroscience, helping to understand the aspects of physiology and pathophysiology of pain. Neuroimaging and its relationship with the level of pain and behavioral symptoms indicate that these measures can be useful as markers of gravity and as parameters to measure the response to treatment. In the functional approach, the image can be acquired in a resting state (reference measure), in paradigms such as the Default Mode Network (DMN) (Boissoneault, Sevel, Letzen, Robinson, & Staud, 2017; Davis et al., 2017), during simple tasks such as tests with alternating movements of the fingers (finger tapping) or while the subject receives a stimulus and/or performs a given cognitive task.

Some instruments evaluate brain electrical activity directly, such as the diagnostic TMS using the evoked potential engine (MEP) and other parameters and the electroencephalogram (EEG), while others measure metabolic activity through the cerebral hemodynamic response, which indirectly reflects the increase in neuronal electrical activity, such as functional magnetic resonance imaging (fMRI) and functional near-infrared spectroscopy (fNIRS). Among the advantages of using fMRI are excellent spatial resolution and the ability to assess deep structures, but it is an expensive test, performed in a restricted environment, with magnetic radiation and noise. The advantages of using fNIRS for the study of brain areas and their respective function are its cost, portability, temporality, and possibility of use during daily activities, but it does not evaluate infracortical layers. Other evaluative instruments, such as positron emission tomography (PET), optical diffusion tomography (DOT), and magnetoencephalography (MEG), are also used in several studies with the same objective.

fNIRS and cortical connectivity in pain research

Brain connectivity is considered as an essential tool for understanding the human brain function. It provides unique information complementing the brain activation. The anatomical connectivity is related to how dendrites and axons brain are anatomically interconnected. However, anatomical connectivity does not give information about the activation dynamics at the time. In contrast, the functional connectivity (FC) refers to temporal correlations to how it distributed neural networks without considering the anatomical substratum. Upon this perspective, the FC is a statistical pattern that does not aim to determine direct or indirect interactions among those brain structures where these activities occur. The effective connectivity shows how regions interact. The FC evaluates how the brain coordinates, in space and time, the events of distinct brain areas, following the principle of functional integration. Effective connectivity might influence how a neural system exerts on another, either at the synaptic level or at the level of a large scale. Differently, the FC has a priori model defined most often as a neurobiological and causal model concerning interactions between functional structures of the brain. While effective connectivity gives some insight into the nature of the connections and their strengths, the FC analyses are useful for identifying patterns of how brain regions interact (which areas are coupled) and compare patterns, especially between groups. Also, it is important to consider that changes in connectivity depend on known or observable events specified during analysis, which require that researchers make assumptions regarding the relationship between experimental timing and potential functional reorganization in brain networks.

The experience of pain is a dynamic phenomenon that varies significantly between individuals, even among patients with chronic pain who have ascribed the same medical diagnosis. In contrast, pain perception fluctuates over time, either in healthy individuals and in chronic pain disorders. Also, pain can change cognitive processes, attention, and cognitive performance. The FC varies dynamically across time and is an approach to account for this variance. It can analyze how coordinated neural activity between different regions evolves and changes across time and helps to understand how networks of brain regions dynamically communicate, construct, and modulate acute and chronic pain.

In 2019, Gökçay et al. (Gökçay, Eken, & Baltaci, 2019) used fNIRS to study patients with a confirmed diagnosis of somatic symptom disorder according to the 5th version of the Diagnostic and Statistical Manual of Mental Disorders (DSM-5). They examined cortical activity in response to painful and control stimuli (pressure threshold with electronic Von Frey and brush stimulation) in 19 patients and compared them with 21 healthy volunteers. In their experiment, they observed changes in O2Hb in the postcentral gyrus and performed a sliding window correlation to obtain dynamic functional connectivity (dFC) changes. After appropriate processing of these data, according to the different conditions studied, they performed hyperparameter optimized classifications, and found that the most discriminative and accurate markers of the Somatic Symptom Disorder were the connections between right superior temporal gyrus and left angular gyrus (rSTG-lAG), right middle frontal gyrus and left supramarginal gyrus (rMFG-lSMG), and right middle temporal gyrus and left middle frontal gyrus (rMTG-lMFG).

In acute pain, a study showed the relationship between the rostral anterior cingulate cortex (rACC) and the antinociceptive descending pain modulatory network in healthy volunteers (Bingel, Lorenz, Schoell, Weiller, & Büchel, 2006). Interestingly, previous work found higher rACC activation in response to noxious stimulation during the placebo relative to control (Wager et al., 2004). A study with healthy subjects using painful laser stimuli found that the connectivity of rACC with rostral anterior cingulate cortex covaried more with activity in the periaqueductal gray (PAG) and amygdala under placebo than control. Given the role of the PAG and the amygdala in descending antinociceptive control and effective responses to pain (Simons et al., 2014), the rACC may play a role in experienced analgesia through its association with subcortical antinociceptive networks. A later study found that the rACC-PAG was abolished with opioid-antagonist naloxone, demonstrating that the functional coupling between these regions is opioid-dependent (Eippert et al., 2009). For instance, a study with healthy subjects using dynamic causal modeling (DCM) observed that descending effective connectivity from DLPFC to the PAG significantly diminished placebo analgesia (Sevel, Craggs, Price, Staud, & Robinson, 2015). Additionally, it found that modulatory connectivity between these regions during pain stimulation (relative to rest) predicted the placebo analgesia over two weeks later. In healthy subjects using dynamic causal modeling (DCM) method with data acquired during the cold pain (thermal stimuli of 0°C, cold pain), it was observed that effective connectivity increased in the right-(R)PFC to mid-(M)-PFC. In contrast, they found a significant decrease in both MPFC to LPFC and LPFC to mPFC connectivity. In a clinical sample of dental clinical pain, a series of consecutive descending cold stimuli (32°C to 0°C) increased FC of cognitive-emotional regions (PFC) with the effected S1 orofacial area contralateral. The baseline FC between S1 and PFC predicted patients' clinical pain experience.

Remarks and future directions

In sum, fNIRS is less expensive than other neuroimaging modalities as the fMRI. It can provide hemodynamic signals with high temporal sampling in a range of contexts and populations in neuroscience. fNIRS provides insights of the brain function in our everyday lives in health, disease, and rehabilitation. It allows to study aspects related to behavior, pain, cognition, and neurorehabilitation. For example, it is a suitable method to understand how performance in tightly controlled cognitive tasks relates to our everyday lives and can help diagnose and treat clinical conditions. It is an appropriate method for assessing cognition in the developing brain, such as face and language processing, and functional specialization within the visual, auditory, and sensorimotor systems (Pinti et al., 2018). It is also a feasible technique to evaluate executive and language dysfunctions and neurodevelopmental disorders (i.e., autism and attention-deficit hyperactivity disorder, etc.). In this way, fNIRS has shown clear potential to monitor the neurorehabilitation progress of motor and cognitive functions.

Several studies showed that in the field of pain, the fNIRS could help understand the pain processing in different brain areas. The connectivity analysis of fNIRS data permits an integrative view of neural networks of varying brain areas and how temporal correlations are distributed in neural networks and how regions interact. The fNIRS is a promising technique for exploring brain activations in descriptive and explanatory studies. The fNIRS is a practical tool to assess the cortical function related to pain upon a biopsychosocial perspective (nociceptive, pain, suffering, and pain behavior). Especially when used in combination with other methods available, for example, for multimodal imaging (e.g., fNIRS-fMRI, fNIRS-EEG), it can offer us unique perspectives on brain activation and connectivity. Multimodal approaches gain a more nuanced and comprehensive description of the pain processing to enhance each performance and compensate for each disadvantage. Although it has been possible to explore the cortical function in the processing of different pain components, most of the studies were conducted with healthy subjects. The literature does not give clear evidence that allows drawing convincing and generalized conclusions. Thus, further studies should investigate functional and structural brain abnormalities that may account for clinical pain state as well as changes in the neural networks involved in the pain processing by the neuromodulation and pharmacological therapies of pain. Also, the fNIRS opens an avenue to an integrative assessment of functional images with complementary methods including transcranial magnetic stimulation, EEG, etc.

Applications to other areas

fNIRS offers a broad range of possibilities to understand the neuroscience mechanisms in our everyday lives in health, disease, and over the lifespan. Considering that the fNIRS measures are based on the BOLD signal, it is possible to assess multiple cerebral cortex locations. The main advantage of fNIRS in behavior science is its flexibility to examine the neural response to effective touch or imitation behavior to record synchronously to the sequence of the alternated by resting period. Also, hyperscanning permits to record signals from two, three, or four participants simultaneously. Additionally, this technique does not have long lists of exclusionary conditions of many behaviors, situations, and mental processes that fMRI cannot currently examine. For the neurocognitive science field, it offers the possibility to record brain processes involving top-down control of freely moving participants, for example. The studies with these measures have been conducted on healthy adults and patients with neurological deficits (e.g., Parkinson's disease and mild cognitive impairments). Also, the coupled use of EEG-fNIRS provides detailed spatiotemporal information correlated with brain activity and connectivity during the robot-assisted gait training (RAGT) rehabilitation. This assessment during RAGT might help understand neuroplastic changes based on the underlying network activity during a voluntary drive for motor learning (Berger, Horst, Müller, Steinberg, & Doppelmayr, 2019).

There are wearable fNIRS devices that are lightweight, compact, and more robust to motion artifacts than the conventional fNIRS. For the outdoor assessment, in naturalistic environments, detector saturation should be used due to the stray sunlight. The fNIRS is also used to record the cortical processing of interpersonal interaction or naturalistic behavior, such as speaking versus signaling with a hand or in situations involving a quiet environment or in an open environment. Another applicability has been to examine the neural substrates of emotional dysregulation and executive function deficits in individuals with attention-deficit hyperactivity disorder (ADHD) during developmental ages. In patients with major depression (MDD), decreased O2Hb values were observed during a verbal fluency task (VFT) in the bilateral frontal and temporal cortices compared with healthy subjects. Also, depressive patients with higher O2Hb activation during VFT before treatment tend to show considerable improvement in depressive symptoms after treatment (Tomioka et al., 2015). Apart from the assessment on MDD, there is a growing evidence that is correlated with the altered NIRS in different psychiatric disorders. For example, a series of NIRS assessments in schizophrenia or autism spectrum disorders indicated that prefrontal cortex activity is closely related to the social cognitive function. In the anesthesia with sevoflurane, the fNIRS accuracy on the PFC was 94.7% to identify an emergency state automatically during the anesthesia (Hernandez-Meza et al., 2018). Furthermore, this technique might evaluate any behavior, mental process, or state of mind outside the laboratory with applicability in several fields within cognitive neuroscience (e.g., social interaction, creativity, self-initiated adaption to novel situations, etc.). The brain area most assessed using the fNIRS on behavior and cognition science is the PFC. The main reason to justify this choice is that the PFC is a large region involved with voluntary action, decision-making, and dealing with open-ended situations and, for one technical view, presents less hair of the skull.

Mini-dictionary of terms

Optode emitter produces the incident light from an emitter that will travel into brain tissue.

Optode detector is an optical sensor device that optically measures a specific substance, usually with the aid of a chemical transducer.

Optical density (OD): In spectroscopy, OD is the measure of absorbance and is defined as the ratio of the intensity of light falling upon a material and the intensity transmitted.

Blood oxygen level-dependent (BOLD) is a term used to describe imaging signals and methods that measure changes in hemoglobin's magnetic properties accompanying changes in blood oxygen levels.

Key factors related to fNIRS

- A technical approach to use in multimodal pain assessment
- A secure technique to evaluating the effect of pain and emotion on the brain in real time
- A method to assess the cortical function in the behavioral studies in naturalistic setting
- A suitable method to assess neuroplasticity state in the clinical neuroscience field
- A method to assess the cortical function in subject's unconscious, under anesthesia, and neonates

Summary points

The fNIRS is a suitable approach to map the brain function in the following conditions:

- To assess in dynamic real time the cortical processing of effective touch or imitation behavior, personal and interpersonal, of signals from two or more persons simultaneously during many behaviors, situations, and mental processes
- To examine the functional interaction among brain areas to study pain or cognition in a controlled experiment of laboratory or informal assessment in a naturalistic environment
- To understand neoplastic changes in the neural network function during everyday lives in health, disease, and the aspects of behavior, pain, cognition, and rehabilitation
- To comprehend performance in brain areas during motor rehabilitation, neurodevelopmental disorders, executive, and language dysfunctions
- To assess cognition in the developing brain, such as face and language processing, functional specialization within the visual, auditory, and sensorimotor systems

References

Aasted, C. M., Yücel, M. A., Steele, S. C., Peng, K., Boas, D. A., Becerra, L., et al. (2016). Frontal lobe hemodynamic responses to painful stimulation: A potential brain marker of nociception. *PLoS One*, *11*(11), e0165226. [Internet]. Available from http://dx.plos.org/10.1371/journal.pone.0165226.

Akin, A., & Bilensoy, D. (2006). Cerebrovascular reactivity to hypercapnia in migraine patients measured with near-infrared spectroscopy. *Brain Research*, *1107*(1), 206–214.

Apkarian, A. V., Bushnell, M. C., Treede, R. D., & Zubieta, J. K. (2005). Human brain mechanisms of pain perception and regulation in health and disease. *European Journal of Pain*, *9*(4), 463–484.

Bandeira, J. S., Antunes, L.d. C., Soldatelli, M. D., Sato, J. R., Fregni, F., & Caumo, W. (2019). Functional spectroscopy mapping of pain processing cortical areas during non-painful peripheral electrical stimulation of the accessory spinal nerve. *Frontiers in Human Neuroscience*, *13*(200). https://doi.org/10.3389/fnhum.2019.00200. https://www.frontiersin.org/article/10.3389/fnhum.2019.00200/full.

Barati, Z., Shewokis, P. A., Izzetoglu, M., Polikar, R., Mychaskiw, G., & Pourrezaei, K. (2013). Hemodynamic response to repeated noxious cold pressor tests measured by functional near infrared spectroscopy on forehead. *Annals of Biomedical Engineering*, *41*(2), 223–237.

Becerra, L., Harris, W., Grant, M., George, E., Boas, D., & Borsook, D. (2009). Diffuse optical tomography activation in the somatosensory cortex: Specific activation by painful vs. non-painful thermal stimuli. *PLoS One*, *4*(11), 1–5.

Berger, A., Horst, F., Müller, S., Steinberg, F., & Doppelmayr, M. (2019). Current state and future prospects of EEG and fNIRS in robot-assisted gait rehabilitation: A brief review. *Frontiers in Human Neuroscience*. https://doi.org/10.3389/fnhum.2019.00172.

Bingel, U., Lorenz, J., Schoell, E., Weiller, C., & Büchel, C. (2006). Mechanisms of placebo analgesia: rACC recruitment of a subcortical antinociceptive network. *Pain*, *120*, 8–15.

Boissoneault, J., Sevel, L., Letzen, J., Robinson, M., & Staud, R. (2017). Biomarkers for musculoskeletal pain conditions: Use of brain imaging and machine learning. *Current Rheumatology Reports*, *19*, 5.

Cieslik, E. C., Zilles, K., Caspers, S., Roski, C., Kellermann, T. S., Jakobs, O., et al. (2013). Is there one DLPFC in cognitive action control? Evidence for heterogeneity from co-activation-based parcellation. *Cerebral Cortex*, *23*, 2677–2689.

Coulombe, M. A., Erpelding, N., Kucyi, A., & Davis, K. D. (2016). Intrinsic functional connectivity of periaqueductal gray subregions in humans. *Human Brain Mapping*, *37*, 1514–1530.

Davis, K. D., Flor, H., Greely, H. T., Iannetti, G. D., MacKey, S., Ploner, M., et al. (2017). Brain imaging tests for chronic pain: Medical, legal and ethical issues and recommendations. *Nature Reviews. Neurology*.

Duncan, A., Meek, J. H., Clemence, M., Elwell, C. E., Fallon, P., Tyszczuk, L., et al. (2014). Measurement of cranial optical path length as a function of age using phase resolved near infrared spectroscopy. *Pediatric Research*, 1–11.

Eippert, F., Bingel, U., Schoell, E. D., Yacubian, J., Klinger, R., Lorenz, J., et al. (2009). Activation of the opioidergic descending pain control system underlies placebo analgesia. *Neuron*, *63*, 533–543.

Friston, K. J. (2009). Modalities, modes, and models in functional neuroimaging. *Science*, *326*, 399–403.

García-Larrea, L., Peyron, R., Mertens, P., Gregoire, M. C., Lavenne, F., Le Bars, D., et al. (1999). Electrical stimulation of motor cortex for pain control: A combined PET-scan and electrophysiological study. *Pain*, *83*, 259–273.

Gatchel, R. J., Peng, Y. B., Peters, M. L., Fuchs, P. N., & Turk, D. C. (2007). The biopsychosocial approach to chronic pain: Scientific advances and future directions. *Psychological Bulletin*, *133*, 581–624.

Gökçay, D., Eken, A., & Baltaci, S. (2019). Binary classification using neural and clinical features: An application in fibromyalgia with likelihood-based decision level fusion. *IEEE Journal of Biomedical and Health Informatics*, *23*(4), 1490–1498.

Hernandez-Meza, G., Izzetoglu, M., Osbakken, M., Green, M., Abubakar, H., & Izzetoglu, K. (2018). Investigation of optical neuro-monitoring technique for detection of maintenance and emergence states during general anesthesia. *Journal of Clinical Monitoring and Computing*, *32*, 147–163.

Hu, X., Racek, A. J., Bellile, E., Nascimento, T. D., Bender, M. C., Toback, R. L., et al. (2018). Brain functional changes before, during, and after clinical pain. *Journal of Dental Research*, *97*(5), 523–529.

Jasper, H. (1958). The ten twenty electrode system of the international federation. *Electroencephalography and Clinical Neurophysiology*, *52*, 3–6.

Jöbsis, F. F. (1977). Noninvasive, infrared monitoring of cerebral and myocardial oxygen sufficiency and circulatory parameters. *Science*, (80).

Karmann, A. J., Maihöfner, C., Lautenbacher, S., Sperling, W., Kornhuber, J., & Kunz, M. (2016). The role of prefrontal inhibition in regulating facial expressions of pain: A repetitive transcranial magnetic stimulation study. *The Journal of Pain*, *17*, 383–391.

Kussman, B. D., Aasted, C. M., Yücel, M. A., Steele, S. C., Alexander, M. E., Boas, D. A., et al. (2016). Capturing pain in the cortex during general anesthesia: Near infrared spectroscopy measures in patients undergoing catheter ablation of arrhythmias. *PLoS One*, *11*(7), 1–13.

Laneri, D., Krach, S., Paulus, F. M., Kanske, P., Schuster, V., Sommer, J., et al. (2017). Mindfulness meditation regulates anterior insula activity during empathy for social pain. *Human Brain Mapping*, *38*, 4034–4046.

Latremoliere, A., & Woolf, C. J. (2009). Central sensitization: A generator of pain hypersensitivity by central neural plasticity. *The Journal of Pain*, *10*(9), 895–926. [Internet]. Available from. http://www.ncbi.nlm.nih.gov/pubmed/19712899.

Lee, C.-H., Sugiyama, T., Kataoka, A., Kudo, A., Fujino, F., Chen, Y.-W., et al. (2013). Analysis for distinctive activation patterns of pain and itchy in the human brain cortex measured using near infrared spectroscopy (NIRS). *PLoS One*, *8*(10), e75360.

Lucas, J. M., Ji, Y., & Masri, R. (2011). Motor cortex stimulation reduces hyperalgesia in an animal model of central pain. *Pain*, *152*, 1398–1407.

Mutschler, I., Wieckhorst, B., Kowalevski, S., Derix, J., Wentlandt, J., Schulze-Bonhage, A., et al. (2009). Functional organization of the human anterior insular cortex. *Neuroscience Letters*, *457*, 66–70.

Naseer, N., & Hong, K. S. (2015). fNIRS-based brain-computer interfaces: A review. *Frontiers in Human Neuroscience*. https://doi.org/10.3389/fnhum.2015.00003.

Oshiro, Y., Quevedo, A. S., McHaffie, J. G., Kraft, R. A., & Coghill, R. C. (2009). Brain mechanisms supporting discrimination of sensory features of pain: A new model. *The Journal of Neuroscience*, 29, 14924–14931.

Pagano, R. L., Assis, D. V., Clara, J. A., Alves, A. S., Dale, C. S., Teixeira, M. J., et al. (2011). Transdural motor cortex stimulation reverses neuropathic pain in rats: A profile of neuronal activation. *European Journal of Pain*, *15*, 268. e1-14.

Peng, K., Yücel, M. A., Steele, S. C., Bittner, E. A., Aasted, C. M., Hoeft, M. A., et al. (2018). Morphine attenuates fNIRS signal associated with painful stimuli in the medial frontopolar cortex (medial BA 10). *Frontiers in Human Neuroscience*, *12*, 1–17.

Perlaki, G., Orsi, G., Schwarcz, A., Bodi, P., Plozer, E., Biczo, K., et al. (2015). Pain-related autonomic response is modulated by the medial prefrontal cortex: An ECG-fMRI study in men. *Journal of the Neurological Sciences*, *349*, 202–208.

Peyron, R., & Fauchon, C. (2019). Functional imaging of pain. *Revue Neurologique*, *175*, 38–45.

Phillips, A. A., Chan, F. H., Zheng, M. M. Z., Krassioukov, A. V., & Ainslie, P. N. (2015). Neurovascular coupling in humans: Physiology, methodological advances and clinical implications. *Journal of Cerebral Blood Flow and Metabolism*, *36*, 647–664.

Pinti, P., Tachtsidis, I., Hamilton, A., Hirsch, J., Aichelburg, C., Gilbert, S., et al. (2018). The present and future use of functional near-infrared spectroscopy (fNIRS) for cognitive neuroscience. *Annals of the New York Academy of Sciences*, *1464*, 5–29.

Pollonini, L., Miao, H., & Ahn, H. (2020). Longitudinal effect of transcranial direct current stimulation on knee osteoarthritis patients measured by functional infrared spectroscopy: A pilot study. *Neurophotonics*, *7*(02), 1.

Pourshoghi, A., Zakeri, I., & Pourrezaei, K. (2016). Application of functional data analysis in classification and clustering of functional near-infrared spectroscopy signal in response to noxious stimuli. *Journal of Biomedical Optics*, *21*(10), 101411. [Internet]. Available from http://biomedicaloptics.spiedigitallibrary.org/article.aspx?articleID=2522054.

Prescott, S. A., Ma, Q., & De Koninck, Y. (2014). Normal and abnormal coding of somatosensory stimuli causing pain. *Nature Neuroscience*, *17*, 183–191.

Raja, S. N., Carr, D. B., Cohen, M., Finnerup, N. B., Flor, H., Gibson, S., et al. (2020). The revised International Association for the Study of Pain definition of pain. *Pain*, *161*, 1976–1982. [Internet]. Publish Ah. Available from https://journals.lww.com/10.1097/j.pain.0000000000001939.

Re, R., Muthalib, M., Contini, D., Zucchelli, L., Torricelli, A., Spinelli, L., et al. (2013). Cerebral cortex activation mapping upon electrical muscle stimulation by 32-channel time-domain functional near-infrared spectroscopy. In *Advances in experimental medicine and biology* (pp. 441–447). [Internet]. Available from http://link.springer.com/10.1007/978-1-4614-7411-1_59.

Sakuma, S., Inamoto, K., Higuchi, N., Ariji, Y., Nakayama, M., & Izumi, M. (2014). Experimental pain in the gingiva and its impact on prefrontal cortical hemodynamics: A functional near-infrared spectroscopy study. *Neuroscience Letters*, *575*, 74–79. [Internet]. Available from https://doi.org/10.1016/j.neulet.2014.05.040.

Scholkmann, F., Hafner, T., Metz, A. J., Wolf, M., & Wolf, U. (2017). Effect of short-term colored-light exposure on cerebral hemodynamics and oxygenation, and systemic physiological activity. *Neurophotonics*, *4*, 045005.

Scholkmann, F., Kleiser, S., Metz, A. J., Zimmermann, R., Mata Pavia, J., Wolf, U., et al. (2014). A review on continuous wave functional near-infrared spectroscopy and imaging instrumentation and methodology. *NeuroImage*, *85*.

Seminowicz, D. A., & Davis, K. D. (2006). Cortical responses to pain in healthy individuals depends on pain catastrophizing. *Pain*, *120*, 297–306.

Sevel, L. S., Craggs, J. G., Price, D. D., Staud, R., & Robinson, M. E. (2015). Placebo analgesia enhances descending pain-related effective connectivity: A dynamic causal modeling study of endogenous pain modulation. *The Journal of Pain*, *16*, 760–768.

Shiers, S., & Price, T. J. (2020). Molecular, circuit, and anatomical changes in the prefrontal cortex in chronic pain. *Pain*, *161*, 1726–1729.

Simons, L. E., Moulton, E. A., Linnman, C., Carpino, E., Becerra, L., & Borsook, D. (2014). The human amygdala and pain: Evidence from neuroimaging. *Human Brain Mapping*, *35*, 527–538.

Slater, R., Fitzgerald, M., & Meek, J. (2007). Can cortical responses following noxious stimulation inform us about pain processing in neonates? *Seminars in Perinatology*, *31*, 298–302.

Sun, X., Long, H., Zhao, C., Duan, Q., Zhu, H., Chen, C., et al. (2019). Analgesia-enhancing effects of repetitive transcranial magnetic stimulation on neuropathic pain after spinal cord injury: An fNIRS study. *Restorative Neurology and Neuroscience*, *37*(5), 497–507.

Tanosaki, M., Hoshi, Y., Iguchi, Y., Oikawa, Y., Oda, I., & Oda, M. (2001). Variation of temporal characteristics in human cerebral hemodynamic responses to electric median nerve stimulation: A near-infrared spectroscopic study. *Neuroscience Letters*, *316*, 75–78.

Tomioka, H., Yamagata, B., Kawasaki, S., Pu, S., Iwanami, A., Hirano, J., et al. (2015). A longitudinal functional neuroimaging study in medication-naïve depression after antidepressant treatment. *PLoS One*. https://doi.org/10.1371/journal.pone.0120828.

Üçeyler, N., Zeller, J., Kewenig, S., Kittel-Schneider, S., Fallgatter, A. J., & Sommer, C. (2015). Increased cortical activation upon painful stimulation in fibromyalgia syndrome. *BMC Neurology*, *15*(1), 210. [Internet]. Available from http://www.biomedcentral.com/1471-2377/15/210.

Viola, S., Viola, P., Litterio, P., Buongarzone, M. P., & Fiorelli, L. (2010). Pathophysiology of migraine attack with prolonged aura revealed by transcranial Doppler and near infrared spectroscopy. *Neurological Sciences*, *31*(Suppl. 1), 165–166.

Vogt, B. A. (2005). Pain and emotion interactions in subregions of the cingulate gyrus. *Nature Reviews. Neuroscience*, *6*, 533–544.

Vrana, A., Meier, M. L., Hotz-Boendermaker, S., Humphreys, B. K., & Scholkmann, F. (2016a). Different mechanosensory stimulations of the lower back elicit specific changes in hemodynamics and oxygenation in cortical sensorimotor areas—A fNIRS study. *Brain and Behavior: A Cognitive Neuroscience Perspective*, *6*(12), 1–18.

Vrana, A., Meier, M. L., Hotz-Boendermaker, S., Humphreys, B. K., & Scholkmann, F. (2016b). Cortical sensorimotor processing of painful pressure in patients with chronic lower back pain—An optical neuroimaging study using fNIRS. *Frontiers in Human Neuroscience*, *10*, 1–18.

Wager, T. D., Rilling, J. K., Smith, E. E., Sokolik, A., Casey, K. L., Davidson, R. J., et al. (2004). Placebo-induced changes in fMRI in the anticipation and experience of pain. *Science*, (80).

Wiech, K., Jbabdi, S., Lin, C. S., Andersson, J., & Tracey, I. (2014). Differential structural and resting state connectivity between insular subdivisions and other pain-related brain regions. *Pain*, *155*, 2047–2055.

Wolfberg, A. J., & du Plessis, A. J. (2006). Near-infrared spectroscopy in the fetus and neonate. *Clinics in Perinatology*, *33*, 707–728.

Yasui, H., Takamoto, K., Hori, E., Urakawa, S., Nagashima, Y., Yada, Y., et al. (2010). Significant correlation between autonomic nervous activity and cerebral hemodynamics during thermotherapy on the neck. *Autonomic Neuroscience: Basic & Clinical*, *156*(1–2), 96–103. [Internet]. Available from https://doi.org/10.1016/j.autneu.2010.03.011.

Yennu, A., Tian, F., Gatchel, R. J., & Liu, H. (2016). Prefrontal hemodynamic mapping by functional near-infrared spectroscopy in response to thermal stimulations over three body sites. *Neurophotonics*, *3*(4), 045008.

Zortea, M., Ramalho, L., Alves, R. L., Alves, C. F.d. S., Braulio, G., Torres, I. L.d. S., et al. (2019). Transcranial direct current stimulation to improve the dysfunction of descending pain modulatory system related to opioids in chronic non-cancer pain: An integrative review of neurobiology and meta-analysis. *Frontiers in Neuroscience*. https://doi.org/10.3389/fnins.2019.01218.

Chapter 30

Muscle origins of myofascial pain syndrome

Eva María Martínez-Jiménez[a], Daniel López-López[b], Carlos Romero-Morales[c], Victoria Mazoteras-Pardo[d], Marta Elena Losa-Iglesias[a,e], David Rodríguez-Sanz[a], Marta San-Antolín-Gil[f], Ricardo Becerro-de-Bengoa-Vallejo[a], and César Calvo-Lobo[a]

[a]*Department of Nursing, School of Nursing, Physiotherapy and Podiatry, University Complutense of Madrid, Madrid, Spain,* [b]*Research, Health and Podiatry Group, Department of Health Sciences, Faculty of Nursing and Podiatry, University of A Coruña, Ferrol, Spain,* [c]*Faculty of Sport Sciences, Universidad Europea de Madrid, Madrid, Spain,* [d]*Research Group "ENDOCU", Department of Nursing, Physiotherapy and Occupational Therapy, Faculty of Physiotherapy and Nursing of Toledo, University of Castilla-La Mancha, Toledo, Spain,* [e]*Department of Nursing, Faculty of Health Sciences, Universidad Rey Juan Carlos, Alcorcón, Spain,* [f]*Department of Psychology, Universidad Europea de Madrid, Madrid, Spain*

Abbreviations

ATP	adenosine triphosphate
BK	bradykinin
CGRP	peptide related to the calcitonin gene
EMG	electromyography
IL	interleukins
MPS	myofascial pain syndrome
MTrPs	myofascial trigger points
TNF-α	tumor necrosis factor alpha
VC	vasoconstriction

Introduction

The myofascial pain syndrome (MPS) generates the most frequent cause of musculoskeletal pain (Hidalgo-Lozano et al., 2010) and it is caused by the presence of myofascial trigger points (MTrPs) in muscle tissue that could be classified into two different types (active or latent myofascial trigger point) depending on its clinical characteristics (Sanz et al., 2016). If they generate spontaneous pain without pressure compression, stretch, overload, or contraction of the muscle, they are called active myofascial trigger point and their nociceptive symptoms increase when they are compressed. Those who need pressure compression, stretch, overload, or contraction of the muscle are called latent myofascial trigger points (Bron, Franssen, Wensing, & Oostendorp, 2007; Dor & Kalichman, 2017).

Because the trigger point alone generates myofascial pain syndrome without the need for associated pathologies, this syndrome is considered a specific diagnosis and primary dysfunction (Mense & Gerwin, 2010). There are other pathologies that can associate a myofascial pain syndrome secondarily, such as visceral disorders as endometriosis (Weiss, 2001), cystitis (Doggweiler-Wiygul, 2004), dysmenorrhea, and prostatitis (Jarrell, 2004). Traumatic pathologies like lateral epicondylalgia (Shmushkevich & Kalichman, 2013), arthrosis (Bajaj, Bajaj, Graven-Nielsen, & Arendt-Nielsen, 2001; Dor & Kalichman, 2017), whiplash injury (Freeman, Nystrom, & Centeno, 2009), idiopathic neck pain (Fernández-de-las-Peñas, Alonso-Blanco, & Miangolarra, 2007), shoulder pain (Hidalgo-Lozano et al., 2010), arm muscles (Fernández-De-Las-Peñas et al., 2012), and temporomandibular pain (Fernández-De-Las-Peñas et al., 2010) have shown a relationship with the presence of MTrP.

Nociception in myofascial pain syndrome is local and referred (Travell & Simons, 2004). The complete pathophysiological mechanism that produces referred pain is not exactly known (Fernández-De-Las-Peñas et al., 2012). Similarly, there are different theories about the origin of MTrP. Currently, evidence supports the hypothesis that MTrPS are associated with dysfunctional motor end plates due to the presence of spontaneous electrical activity or endplate noise that the

The Neurobiology, Physiology, and Psychology of Pain. https://doi.org/10.1016/B978-0-12-820589-1.00030-0

treatment with dry needling of the trigger points eliminates (Ge, Fernández-de-las-Peñas, & Yue, 2011). The possible association of muscle spindles in etiology has also been studied (Ge, Serrao, Andersen, Graven-Nielsen, & Arendt-Nielsen, 2009). Both the dysfunctional motor ends plates and the muscle spindles could explain local pain and could be the origin of the pathophysiology of referred pain of MTrPs.

Nociception in muscle tissue

Studies on muscle physiology show that both mechanical and chemical stimuli produce muscle pain due to stimulation of muscle nociceptors (Mense, 2009). Muscle nociceptors are free nerve endings whose axon presents varicosities with the neuropeptide substance P and the peptide related to the calcitonin gene (CGRP) in vesicles. These neuropeptides are released when the nerve fibers are electrically active by a mechanic or quimical stimulus (Mense & Gerwin, 2010). Thus, injection of neuropeptides and inflammatory mediators has produced local and reported muscle pain in experimental studies (Babenko et al., 1999; Graven-Nielsen, 2006).

The stimulation of the nociceptors produces in the first place the liberation of the neuropeptides on the part of the vesicles with substance P and CGRP of the nerve ending. These neuropeptides cause vasodilation and increase the permeability of the microcirculation. If the amount of substance P and CGRP is large enough, mast cells release histamine, platelets secrete serotonin, endothelial cells secrete prostaglandins (PG), and increased bradykinin (BK) occur (Massaad et al., 2004).

These chemicals (PG, BK, histamine, and serotonin) are also substances that produce a muscle-damaging stimulus and stimulate the muscular nociceptors, producing a positive feedback system in the body of muscular nociceptive irritation. Because the increase in BK stimulates the release of tumor necrosis factor (TNF-α) and interleukins, which in turn increase the amount of BK, which enhance peripheral sensitization (Calixto, Cabrini, Ferreira, & Campos, 2000).

Higher concentrations of substance P, bradykinin, CGRP, TNF-α, interleukins, and norepinephrine have been found in active trigger points than in latent trigger points, and in turn higher in both groups than in pain-free muscle sites (Shah et al., 2008; Shah, Phillips, Danoff, & Gerber, 2005). The sustained contraction of the fibers of the taut bands and/or the chemical stimulus produces the release of these nociceptors (Gerwin, 2008).

There is an absence of blood flow within the trigger point further this vasodilation around the TP that generates the release of neuropeptides that entails a decrease in the oxygenation of the tissue inside the TP (Sikdar, Ortiz, Gebreab, Gerber, & Shah, 2010; Fig. 1).

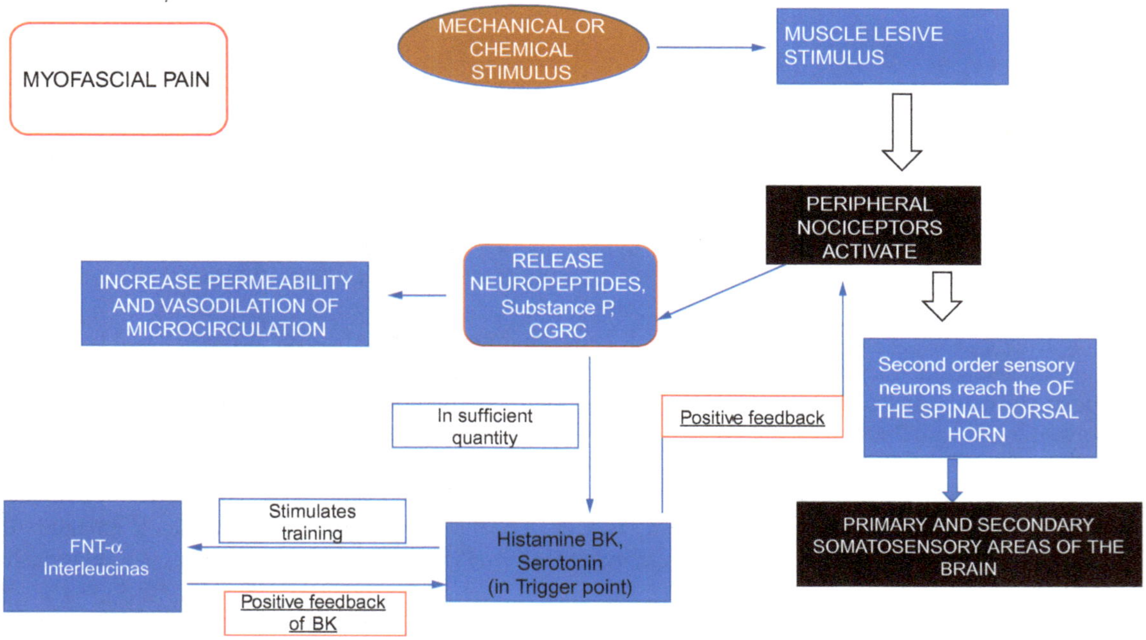

FIG. 1 Origin of myofascial pain. The chemical or mechanical stimulus generates an activation of the nociceptors of the muscular tissue. This activation of the nociceptors activates the sensitization of the spinal dorsal horn neuron and a secretory liberation of neuropeptides, substance P, and CGRP of nerve ending. In sufficient quantity, of Substance P and CGRP mast cells release histamine, platelets secrete serotonin, endothelial cells secrete prostaglandins (PG), and bradykinin (BK), which can also stimulate nociceptors (positive feedback).

Referred pain and peripheral sensitization process

Local pain is usually called primary pain and it has felt as the same area of injured tissues (Giamberardino, 2003). The pain that characterizes MTrPS is referred pain. It is deep pain of undefined territory and difficult to locate that occurs in areas far from injured tissues. It is the typical pain of ligaments, muscles, ligaments, tendons, joints, or internal organs (Hogan & Norby, 2010; Murray, 2009). Hypertonic saline injection into the muscle of healthy subjects has been a procedure used to localize muscle MTrP-specific referred pain areas. Thanks to this technique, it would be possible to determine the map of the anatomical localization of the referred pain elicited (Alonso-Blanco et al., 2011).

The referred pain generates a map of the own pain for each structure. That is why the typical pain in myocardial infarction, appendicitis, or MTrP of internal twin can be defined such as pain at the right shoulder in gallbladder inflammation (Aasvang et al., 2015). This is due to the sensory input of the neurons of the dorsal horn receive information from the similar areas in all subjects (Gerwin, 1994). Large maps of referred pain and intensity are correlated with higher sensitizations (Fernández-De-Las-Peñas & Dommerholt, 2014).

The referred pain is usually expressed with hyperalgesia and trophic changes, specifically hypotrophy (Giamberardino, 2003) due to the presence of nociceptors in the muscles. Neuropeptides and inflammatory mediators activate muscle nociceptors that generate a reduction of the pain threshold and an increase in the response of the peripheral nociceptors (Fernández-De-Las-Peñas et al., 2012). That is, the same stimulus generates a greater nociceptive response. This phenomenon is called peripheral sensitization (Fernández-De-Las-Peñas & Dommerholt, 2014). The study of Shah et al. (2008) confirm the peripheral sensitization process, since there have been higher concentrations of nociceptors (neuropeptides and inflammatory mediators) in muscle areas far from the active trigger point of the trapezius compared to people without a MTrP.

This sensitization affects sensitive nociceptive and nonnociceptive fibers. The nociceptive hypersensitivity of the MTrP or hyperalgesia and the nonnociceptive hypersensitivity called allodynia occur in the MTrP; therefore, it shows alteration of both fibers (Li, Ge, Yue, & Arendt-Nielsen, 2009). In addition, the sensory fibers of proprioception information are involved. Proprioceptive fiber compression of MTrP, not in other areas, has been shown to increase the pain threshold to pressure and referred pain, confirming its involvement in the pathophysiology of pain, and involvement of proprioceptive fibers in this peripheral sensitization (Wang et al., 2010).

Proprioceptive fibers' disturbance symptoms explain why patients with epicondylitis report difficulties in writing and fine motility activities (Travell & Simons, 2004).

Sympathetic facilitation of muscle pain

The studies on animal and human MTrP related to the sympathetic nervous system are consistent in stating that a sympathetic facilitation would explain its results (Chen, Chen, Kuan, Chung, & Hong, 1998; Chung, Ohrbach, & McCall, 2004; McNulty, Gevirtz, Hubbard, & Berkoff, 1994).

The sympathetic nervous system increases the efferent discharge, and an increase in frequency and amplitude is observed in the electromyogram of the muscular MTrP (Chung et al., 2004; McNulty et al., 1994). This explains that higher levels of norepinephrine have been found in active MTrP (Shah et al., 2005) and decreased cutaneous blood flow due to the vasoconstrictive effect of the sympathetic system (Zhang, Ge, Yue, Kimura, & Arendt-Nielsen, 2009).

Ge, Fernández-de-las-Peñas, Madeleine, and Arendt-Nielsen (2008) affirm that sympathetic facilitation lowers the pain threshold to the pressure of the MTrP and lowers the referred pain threshold as well as increase local and referred pain, producing a sensitization of the MTrP (Fig. 2).

The physiological mechanisms of this phenomenon are unknown, although the increase in ATP and norepinephrine seems to be related (Gerwin, Dommerholt, & Shah, 2004).

Sympathetic facilitation explains the symptoms of autonomic disturbances observed for many years. Cutaneous atrophy and alteration in sweating are typical signs in the affected muscle area of patients diagnosed with MPS, which have been described since the beginning of the description of the MTrP by the authors (Travell & Simons, 2004).

Central sensitization

Central sensitization is an increased excitability of central nervous system. Neurons are induced by the presence of MTrP. The symptoms are hyperalgesia, allodynia, increased EMG in others and in TP muscle (Fernández-De-Las-Peñas & Dommerholt, 2014).

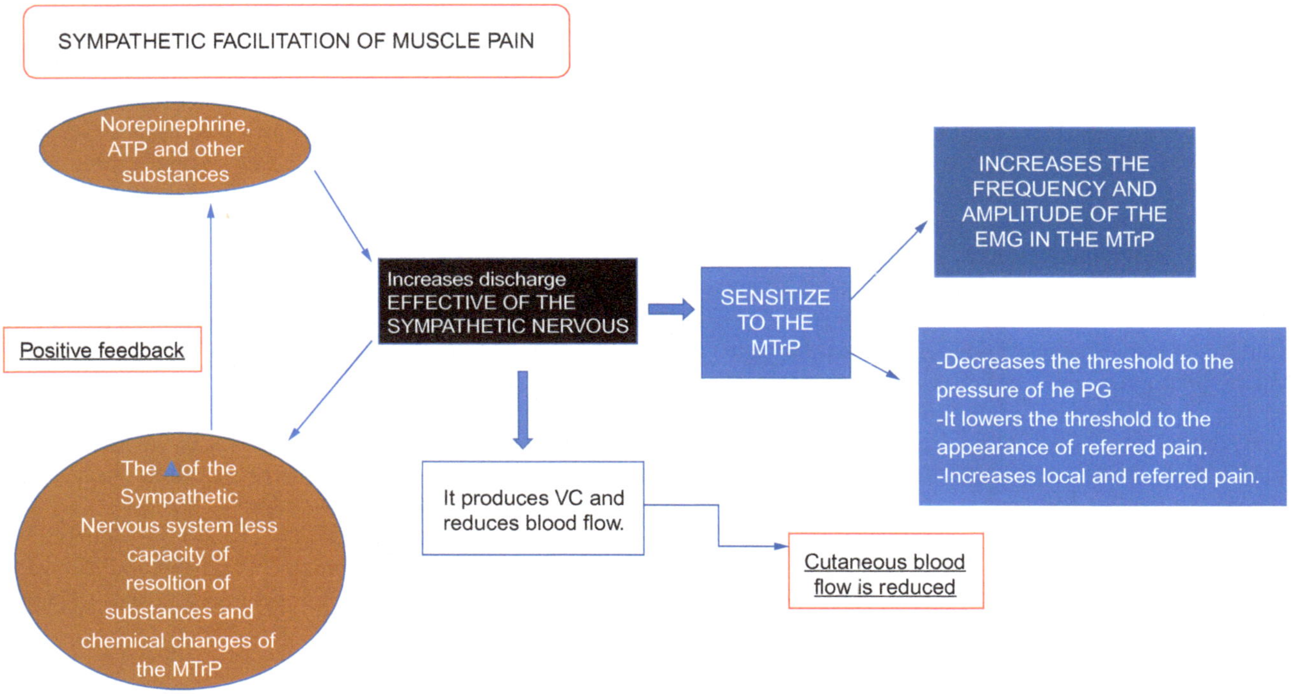

FIG. 2 Sympathetic facilitation of muscle pain. ATP and norepinephrine are involved in the activation of the sympathetic facilitation in MTrP by unknown, but positively feedback mechanisms. The increase discharge of sympathetic nervous system produces sensitization of MTrP generating a EMG change, a local and referred pain increased, and a decreased threshold.

Nociceptive-free nerve endings are the axons of the first sensory neuron that synapse in the dorsal horn of the medulla with the second-order neuron. This neuron forms part of the central nervous system and is responsible for transmitting information to the somatosensory cortex (Patton & Thibodeau, 2010).

Current evidence defines central sensitization as an awareness process produced by the MTrP in the related spinal dorsal horn (Meng, Ge, Wang, & Yue, 2015). Two important variables that can sensitize neurons in the spinal cord: first is the presence of multiple MTrP in one or multiple muscles and second is the duration of the nociceptive stimulus produced by MTrP. Thus, greater number of MTrPs and duration of referred pain are proportionally related to central sensitization. Greater durations and intensities of muscle pain are related to greater areas of referred pain (Graven-Nielsen, McArdle, et al., 1997; Fig. 3).

Spinal cord sensitization is due to the plasticity of neural connections, and MTrP pain produces changes in the dorsal horn of spinal cord more effectively than other tissues without MTrP since they have a greater number of sensory neurons (Kuan, Hong, Chen, Chen, & Chien, 2007). Changes in the spinal dorsal horn associated with MTrP increase the expression of glutamate receptor (mGluR1α/mGluR5/NMDAR1), and diameter of neurons associated with MTrP was smaller than normal. These results indicate that A fibers participated in the development and maintenance of the central sensitization (Meng et al., 2015; Fig. 3).

Subsequent to the spinal cord neurons, the information on the referred pain from the MTrP reaches cortical levels in the areas of primary and secondary somatosensory areas of the brain, such as the amygdala, the central cingulate gyrus (Dommerholt, 2011). Imaging studies on neural connections in patients with MTrP observed activity of the primary and secondary somatosensory cortex, lower parietal cortex, and middle part of the insula with increased activity, suggesting that MTrP nociception is processed in various brain areas, which may enhance its sensation. In addition, there is pathological hypoactivity of the hippocampus indicating that a dysfunctional stress response maintains and favors hyperalgesia (Niddam, 2009; Niddam, Chan, Lee, Yeh, & Hsieh, 2008). These cortical changes in the sensory pathways of pain and their processing induce the appearance of characteristic symptoms of central sensitization such as hyperalgesia (increased perception of pain in a stimulus that is painful under normal conditions) and allodynia (painful perception of a stimulus that is not healthy in health conditions) and EMG increase of other muscles (Dommerholt, 2011; Fig. 4).

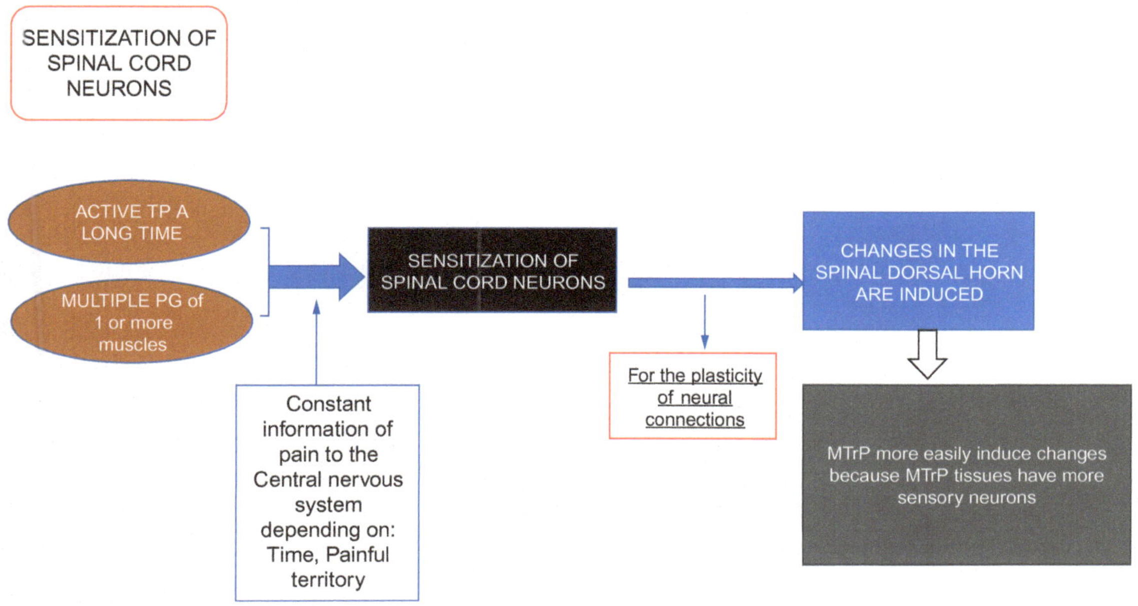

FIG. 3 Sensitization of spinal cord neurons. In the process of sensitization of spinal cord neurons, the number of MTrP, reference extension of pain, and time of onset of pain intervene proportionally, because muscle tissue have a lot of sensory neurons that change easily spinal dorsal horn (increase the expression of glutamate receptor (mGluR1α/mGluR5/NMDAR1) and diameter of neurons associated with MTrP was smaller than normal).

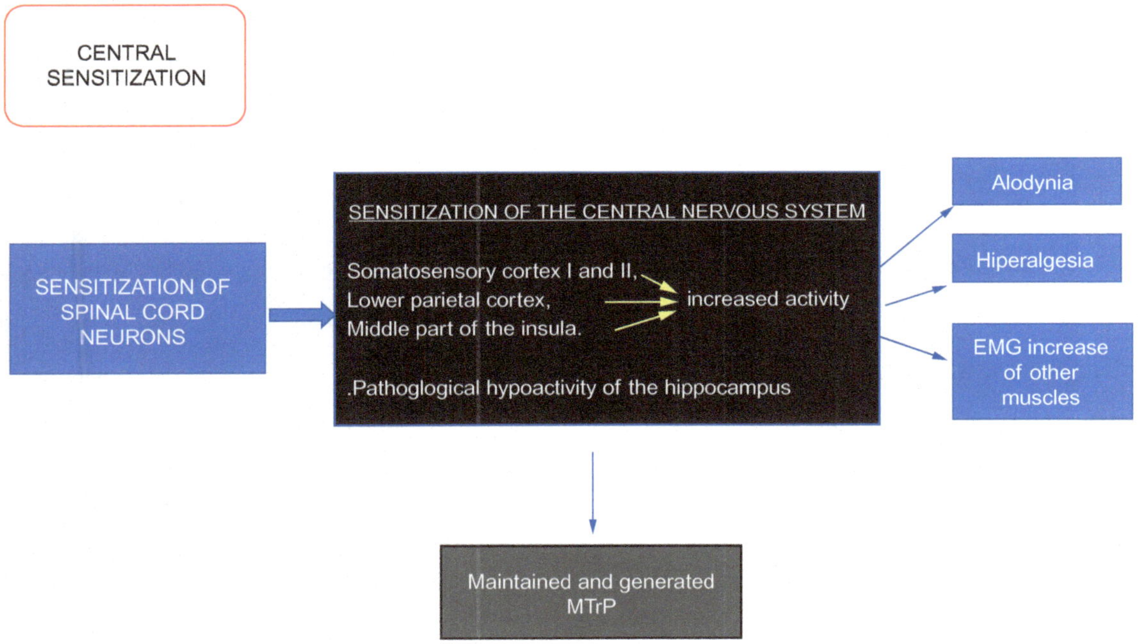

FIG. 4 Central sensitization process. The central sensitization process begins with sensitization of the spinal cord neurons. The information that reaches the sensitized neuron of the spinal cord produces an increase in the activity of certain areas of the brain (somatosensory cortex I, II, inferior parietal cortex, middle part of the insula) and produces hypoactivity of the hippocampus when processed. This generates as a consequence the symptoms of allodynia, hyperalgesia and increased EMG activity, in addition to perpetuating the MTrP.

Referred pain is maintained and fueled by a phenomenon of central sensitization because it has observed in clinical trials that anesthetization of the referred pain area reduces pain (Rubin, Gandevia, Henderson, & Macefield, 2009). Furthermore, central sensitization can produce the activation of MTrPs and their regulation, such as decreased area of referred pain after use of infusions of the *N*-methyl-D-aspartate (NMDA) antagonist ketamine in fibromyalgia (Graven-Nielsen et al., 2000; Fig. 4).

It is considered that the referred muscular pain can be facilitated by a dysfunction of the mechanisms of descending pain inhibition characteristic of peripheral sensitization (Arendt-Nielsen & Ge, 2010).

Different studies show that central sensitization is a process that can be reversible in MTrP with dry needling technique (Freeman et al., 2009; Giamberardino, 2003).

Clinical applications

Muscle pain must be understood as a complex pathophysiological mechanism that requires special attention. Understanding that its repercussion of pain will be greater if there is a greater territory of referred pain, and a greater number of MTrP are present and with a longer duration (Graven-Nielsen, McArdle, et al., 1997).

Probably, higher levels of referred pain require more sessions and intensity of treatment. So, in long-term MTrP or pain intensity, they benefit from the dry needling treatment for its effectiveness, due to different studies show that central sensitization is a process that can be reversible in MTrP with dry needling (Freeman et al., 2009; Giamberardino, 2003).

Conclusion

Currently, the nerve pathways of any type of referred pain have not been precisely defined. Although all the defined models of referred pain do agree that the nociceptive neurons of the dorsal horn of the medulla or brainstem receive convergent inputs from various tissues, which due to these different inputs the upper brain center cannot interpret their origin (Graven-Nielsen, Arendt-Nielsen, Svensson, & Jensen, 1997).

Muscle pain, in addition to being referred pain, has sensitization characteristics of peripheral nociceptors, which may be facilitated by an increase in sympathetic activity and by an alteration of the mechanisms of descending pain inhibition (Arendt-Nielsen & Ge, 2010).

Application to other areas

There are other causes of referred pain that can be benefited from the current knowledge about MPS. In patients with fibromyalgia, whose disease is centrally sensitized, some, but not all, dry needling treatment of MTrP reduces nociception, pain, and generalized sensitivity (Ge, Wang, et al., 2011).

Other agents of interest

Knowledge of the pathophysiology of muscle pain can be used to perform human experimental pain models for assessing the therapeutic efficacy of analgesic drugs (Olesen, Andresen, Staahl, & Drewes, 2012).

Mini-dictionary of terms

Myofascial pain syndrome: A group of symptoms taken by sensory alteration (dysesthesias with the presence of pain), motor alteration (loss of strength and stretching capacity), autonomic dysfunction (of the sympathetic nervous system) generated by the presence of trigger points in the muscle (Fernández-De-Las-Peñas et al., 2012; Travell & Simons, 2004).

Trigger point: Palpable and hyperirritable nodule area within a group of muscle fibers of greater tension (called taut band) present in the striated muscle, which generate referred pain (Sanz et al., 2016).

Referred pain: It is deep pain of undefined territory and difficult to locate that occurs in areas far from the origin of the pain. It is the typical pain of ligaments, muscles, ligaments, tendons, joints, or internal organs (Hogan & Norby, 2010; Murray, 2009).

Active trigger point: The referred and local pain produced by this type of trigger point is indicated by the patient as his pain (Travell & Simons, 2004).

Latent trigger point: The referred and local pain produced by this type of trigger point when pressed is a new pain by the patient with whom he is not familiar (Travell & Simons, 2004).

Peripheral sensitization: Peripheral sensitization is described as a reduction in the pain threshold and an increase in responsiveness of the peripheral nociceptors (Fernández-De-Las-Peñas & Dommerholt, 2014).

Central sensitization: Central sensitization is an increased excitability of central nervous system induced by the presence of TP whose symptoms are hyperalgesia, allodynia, increased EMG in others, and in MTrP muscle.

Key facts of myofascial pain

- MPS is characterized by local pain and referred pain.
- The referred pain generates a map of the own pain for each structure.
- Referred pain is due to the sensory input of the neurons of the dorsal horn receive information from the similar areas in all subjects.
- Sensitization of muscle nociceptors starts the entire pathophysiology of pain.
- Nociceptors sensitization generates peripheral sensitization as well as central sensitization that perpetuate the syndrome.
- Sympathetic facilitation favors the sensitization of MTrP and the lowering of the pain threshold.

Summary points

- This chapter is focused on explaining the current knowledge of the pathophysiology of the origin of muscle pain.
- Muscle pain generates a variety of local and distant symptoms, so it is necessary to call it syndrome, MPS.
- MPS is made up of local and referred pain.
- Referred pain is not exclusive to MPS and has characteristics common to other referred pain.
- All referred pain is due to sensitization of the central nervous system.

References

Aasvang, E. K., Werner, M. U., & Kehlet, H. (2015). Referred pain and cutaneous responses from deep tissue electrical pain stimulation in the groin. *British Journal of Anaesthesia*. https://doi.org/10.1093/bja/aev170.

Alonso-Blanco, C., Fernández-de-las-Peñas, C., Fernández-Mayoralas, D. M., Isabel de-la-Llave-Rincón, A., Pareja, J. A., Svensson, P., et al. (2011). Prevalence and anatomical localization of muscle referred pain from active trigger points in head and neck musculature in adults and children with chronic tension-type. *Pain Medicine*, *12*. Wiley Periodicals, Inc. Headachep me_1204 1453.1463 https://academic.oup.com/painmedicine/article-abstract/12/10/1453/1889633.

Arendt-Nielsen, L., & Ge, H. Y. (2010). *Pathophysiology of referred muscle pain* (pp. 51–59). Jones and Bartlett Publishers. https://vbn.aau.dk/en/publications/pathophysiology-of-referred-muscle-pain.

Babenko, V., Graven-Nielsen, T., Svensson, P., Drewes, A. M., Jensen, T. S., & Arendt-Nielsen, L. (1999). Experimental human muscle pain and muscular hyperalgesia induced by combinations of serotonin and bradykinin. *Pain*, *82*(1), 1–8. https://doi.org/10.1016/S0304-3959(99)00026-3.

Bajaj, P., Bajaj, P., Graven-Nielsen, T., & Arendt-Nielsen, L. (2001). Trigger points in patients with lower limb osteoarthritis. *Journal of Musculoskeletal Pain*, *9*(3), 17–33. https://doi.org/10.1300/J094v09n03_03.

Bron, C., Franssen, J., Wensing, M., & Oostendorp, R. A. B. (2007). Interrater reliability of palpation of myofascial trigger points in three shoulder muscles. *Journal of Manual and Manipulative Therapy*, *15*(4), 203–215. https://doi.org/10.1179/106698107790819477.

Calixto, J. B., Cabrini, D. A., Ferreira, J., & Campos, M. M. (2000). Kinins in pain and inflammation. *Pain*, *87*(1), 1–5. https://doi.org/10.1016/S0304-3959(00)00335-3.

Chen, J. T., Chen, S. M., Kuan, T. S., Chung, K. C., & Hong, C. Z. (1998). Phentolamine effect on the spontaneous electrical activity of active loci in a myofascial trigger spot of rabbit skeletal muscle. *Archives of Physical Medicine and Rehabilitation*, *79*(7), 790–794. https://doi.org/10.1016/S0003-9993(98)90357-4.

Chung, J. W., Ohrbach, R., & McCall, W. D. (2004). Effect of increased sympathetic activity on electrical activity from myofascial painful areas. *American Journal of Physical Medicine & Rehabilitation*, *83*(11), 842–850. https://doi.org/10.1097/01.PHM.0000143399.34798.A7.

Doggweiler-Wiygul, R. (2004). Urologic myofascial pain syndromes. *Current Pain and Headache Reports*, *8*(6), 445–451. Springer https://doi.org/10.1007/s11916-004-0065-1.

Dommerholt, J. (2011). Dry needling—Peripheral and central considerations. *Journal of Manual and Manipulative Therapy*. https://doi.org/10.1179/106698111X13129729552065.

Dor, A., & Kalichman, L. (2017). A myofascial component of pain in knee osteoarthritis. *Journal of Bodywork and Movement Therapies*, *21*(3), 642–647. Churchill Livingstone https://doi.org/10.1016/j.jbmt.2017.03.025.

Fernández-de-las-Peñas, C., Alonso-Blanco, C., & Miangolarra, J. C. (2007). Myofascial trigger points in subjects presenting with mechanical neck pain: A blinded, controlled study. *Manual Therapy*, *12*(1), 29–33. https://doi.org/10.1016/j.math.2006.02.002.

Fernández-De-Las-Peñas, C., & Dommerholt, J. (2014). Myofascial trigger points: Peripheral or central phenomenon? *Current Rheumatology Reports*. https://doi.org/10.1007/s11926-013-0395-2.

Fernández-De-Las-Peñas, C., Galán-Del-Río, F., Alonso-Blanco, C., Jiménez-García, R., Arendt-Nielsen, L., & Svensson, P. (2010). Referred pain from muscle trigger points in the masticatory and neck-shoulder musculature in women with temporomandibular disorders. *Journal of Pain*, *11*(12), 1295–1304. https://doi.org/10.1016/j.jpain.2010.03.005.

Fernández-De-Las-Peñas, C., Gröbli, C., Ortega-Santiago, R., Fischer, C. S., Boesch, D., Froidevaux, P., et al. (2012). Referred pain from myofascial trigger points in head, neck, shoulder, and arm muscles reproduces pain symptoms in blue-collar (Manual) and white-collar (Office) workers. *Clinical Journal of Pain*, *28*(6), 511–518. https://doi.org/10.1097/AJP.0b013e31823984e2.

Freeman, M. D., Nystrom, A., & Centeno, C. (2009). Chronic whiplash and central sensitization; an evaluation of the role of a myofascial trigger points in pain modulation. *Journal of Brachial Plexus and Peripheral Nerve Injury*, *4*(1), e13–e20. https://doi.org/10.1186/1749-7221-4-2.

Ge, H. Y., Fernández-de-las-Peñas, C., Madeleine, P., & Arendt-Nielsen, L. (2008). Topographical mapping and mechanical pain sensitivity of myofascial trigger points in the infraspinatus muscle. *European Journal of Pain*, *12*(7), 859–865. https://doi.org/10.1016/j.ejpain.2007.12.005.

Ge, H. Y., Fernández-de-las-Peñas, C., & Yue, S. W. (2011). Myofascial trigger points: Spontaneous electrical activity and its consequences for pain induction and propagation. *Chinese Medicine*, *6*, 13. BioMed Central https://doi.org/10.1186/1749-8546-6-13.

Ge, H. Y., Serrao, M., Andersen, O. K., Graven-Nielsen, T., & Arendt-Nielsen, L. (2009). Increased H-reflex response induced by intramuscular electrical stimulation of latent myofascial trigger points. *Acupuncture in Medicine*, *27*(4), 150–154. https://doi.org/10.1136/aim.2009.001099.

Ge, H. Y., Wang, Y., Fernández-de-las-Peñas, C., Graven-Nielsen, T., Danneskiold-Samsøe, B., & Arendt-Nielsen, L. (2011). Reproduction of overall spontaneous pain pattern by manual stimulation of active myofascial trigger points in fibromyalgia patients. *Arthritis Research and Therapy*, *13* (2). https://doi.org/10.1186/ar3289.

Gerwin, R. D. (1994). Neurobiology of the myofascial trigger point. *Bailliere's Clinical Rheumatology*, *8*(4), 747–762. https://doi.org/10.1016/S0950-3579(05)80046-9.

Gerwin, R. D. (2008). The taut band and other mysteries of the trigger point: An examination of the mechanisms relevant to the development and maintenance of the trigger point. *Journal of Musculoskeletal Pain*, *16*(1–2), 115–121. Taylor & Francis https://doi.org/10.1080/10582450801960081.

Gerwin, R. D., Dommerholt, J., & Shah, J. P. (2004). An expansion of Simons' integrated hypothesis of trigger point formation. *Current Pain and Headache Reports*, *8*(6), 468–475. https://doi.org/10.1007/s11916-004-0069-x.

Giamberardino, M. A. (2003). Referred muscle pain/hyperalgesia and central sensitisation. *Journal of Rehabilitation Medicine, Supplement*, *41*, 85–88. https://doi.org/10.1080/16501960310010205.

Graven-Nielsen, T. (2006). Fundamentals of muscle pain, referred pain, and deep tissue hyperalgesia. *Scandinavian Journal of Rheumatology*, *35*(Suppl. 122), 1–43. Taylor and Francis A.S https://doi.org/10.1080/03009740600865980.

Graven-Nielsen, T., Arendt-Nielsen, L., Svensson, P., & Jensen, T. S. (1997). Quantification of local and referred muscle pain in humans after sequential i. m. injections of hypertonic saline. *Pain*, *69*(1–2), 111–117. https://doi.org/10.1016/S0304-3959(96)03243-5.

Graven-Nielsen, T., Aspegren Kendall, S., Henriksson, K. G., Bengtsson, M., Sörensen, J., Johnson, A., et al. (2000). Ketamine reduces muscle pain, temporal summation, and referred pain in fibromyalgia patients. *Pain*, *85*(3), 483–491. https://doi.org/10.1016/S0304-3959(99)00308-5.

Graven-Nielsen, T., McArdle, A., Phoenix, J., Arendt-Nielsen, L., Jensen, T. S., Jackson, M. J., et al. (1997). In vivo model of muscle pain: Quantification of intramuscular chemical, electrical, and pressure changes associated with saline-induced muscle pain in humans. *Pain*, *69*(1–2), 137–143. https://doi.org/10.1016/S0304-3959(96)03270-8.

Hidalgo-Lozano, A., Fernández-De-Las-Peñas, C., Alonso-Blanco, C., Ge, H. Y., Arendt-Nielsen, L., & Arroyo-Morales, M. (2010). Muscle trigger points and pressure pain hyperalgesia in the shoulder muscles in patients with unilateral shoulder impingement: A blinded, controlled study. *Experimental Brain Research*, *202*(4), 915–925. https://doi.org/10.1007/s00221-010-2196-4.

Hogan, M. C., & Norby, S. M. (2010). Evaluation and management of pain in autosomal dominant polycystic kidney disease. *Advances in Chronic Kidney Disease*, *17*(3), e1. NIH Public Access https://doi.org/10.1053/j.ackd.2010.01.005.

Jarrell, J. (2004). Myofascial dysfunction in the pelvis. *Current Pain and Headache Reports*, *8*(6), 452–456. https://doi.org/10.1007/s11916-004-0066-0.

Kuan, T. S., Hong, C. Z., Chen, J. T., Chen, S. M., & Chien, C. H. (2007). The spinal cord connections of the myofascial trigger spots. *European Journal of Pain*, *11*(6), 624–634. https://doi.org/10.1016/j.ejpain.2006.10.001.

Li, L. T., Ge, H. Y., Yue, S. W., & Arendt-Nielsen, L. (2009). Nociceptive and non-nociceptive hypersensitivity at latent myofascial trigger points. *Clinical Journal of Pain*, *25*(2), 132–137. https://doi.org/10.1097/AJP.0b013e3181878f87.

Massaad, C. A., Safieh-Garabedian, B., Poole, S., Atweh, S. F., Jabbur, S. J., & Saadé, N. E. (2004). Involvement of substance P, CGRP and histamine in the hyperalgesia and cytokine upregulation induced by intraplantar injection of capsaicin in rats. *Journal of Neuroimmunology*, *153*(1–2), 171–182. https://doi.org/10.1016/j.jneuroim.2004.05.007.

McNulty, W. H., Gevirtz, R. N., Hubbard, D. R., & Berkoff, G. M. (1994). Needle electromyographic evaluation of trigger point response to a psychological stressor. *Psychophysiology*, *31*(3), 313–316. https://doi.org/10.1111/j.1469-8986.1994.tb02220.x.

Meng, F., Ge, H. Y., Wang, Y. H., & Yue, S. W. (2015). A afferent fibers are involved in the pathology of central changes in the spinal dorsal horn associated with myofascial trigger spots in rats. *Experimental Brain Research*, *233*(11), 3133–3143. https://doi.org/10.1007/s00221-015-4382-x.

Mense, S. (2009). Algesic agents exciting muscle nociceptors. *Experimental Brain Research*, *196*(1), 89–100. https://doi.org/10.1007/s00221-008-1674-4.

Mense, S., & Gerwin, R. D. (2010). *Muscle Pain: Understanding the Mechanisms*. Berlin, Heidelberg: Springer-Verlag. https://doi.org/10.1007/978-3-540-85,021-2_2.

Murray, G. M. (2009). Guest editorial: Referred pain. *Journal of Applied Oral Science*, *17*(6), 1. Bauru School of Dentistry, University of Sao Paulo https://doi.org/10.1590/S1678-77572009000600001.

Niddam, D. M. (2009). Brain manifestation and modulation of pain from myofascial trigger points. *Current Pain and Headache Reports*, *13*(5), 370–375. Springer https://doi.org/10.1007/s11916-009-0060-7.

Niddam, D. M., Chan, R. C., Lee, S. H., Yeh, T. C., & Hsieh, J. C. (2008). Central representation of hyperalgesia from myofascial trigger point. *NeuroImage*, *39*(3), 1299–1306. https://doi.org/10.1016/j.neuroimage.2007.09.051.

Olesen, A. E., Andresen, T., Staahl, C., & Drewes, A. M. (2012). Human experimental pain models for assessing the therapeutic efficacy of analgesic drugs. *Pharmacological Reviews*, *64*(3), 722–779. https://doi.org/10.1124/pr.111.005447.

Patton, K. T., & Thibodeau, G. A. (2010). *Anatomy & physiology*. Mosby/Elsevier.

Rubin, T. K., Gandevia, S. C., Henderson, L. A., & Macefield, V. G. (2009). Effects of intramuscular anesthesia on the expression of primary and referred pain induced by intramuscular injection of hypertonic saline. *Journal of Pain*, *10*(8), 829–835. https://doi.org/10.1016/j.jpain.2009.01.327.

Sanz, D. R., Lobo, C. C., López, D. L., Morales, C. R., Marín, C. S., & Corbalán, I. S. (2016). Interrater reliability in the clinical evaluation of myofascial trigger points in three ankle muscles. *Journal of Manipulative and Physiological Therapeutics*, *39*(9), 623–634. https://doi.org/10.1016/j.jmpt.2016.09.002.

Shah, J. P., Danoff, J.v., Desai, M. J., Parikh, S., Nakamura, L. Y., Phillips, T. M., et al. (2008). Biochemicals associated with pain and inflammation are elevated in sites near to and remote from active myofascial trigger points. *Archives of Physical Medicine and Rehabilitation*, *89*(1), 16–23. https://doi.org/10.1016/j.apmr.2007.10.018.

Shah, J. P., Phillips, T. M., Danoff, J.v., & Gerber, L. H. (2005). An in vivo microanalytical technique for measuring the local biochemical milieu of human skeletal muscle. *Journal of Applied Physiology*, *99*(5), 1977–1984. https://doi.org/10.1152/japplphysiol.00419.2005.

Shmushkevich, Y., & Kalichman, L. (2013). Myofascial pain in lateral epicondylalgia: A review. *Journal of Bodywork and Movement Therapies*, *17*(4), 434–439. https://doi.org/10.1016/j.jbmt.2013.02.003.

Sikdar, S., Ortiz, R., Gebreab, T., Gerber, L. H., & Shah, J. P. (2010). Understanding the vascular environment of myofascial trigger points using ultrasonic imaging and computational modeling. In *2010 Annual international conference of the IEEE Engineering in Medicine and Biology Society, EMBC'10, 2010* (pp. 5302–5305). https://doi.org/10.1109/IEMBS.2010.5626326.

Travell, J. G., & Simons, D. G. (2004). *Dolor y disfunción miofascial: El manual de los puntos gatillo*. Panamericana.

Wang, Y. H., Ding, X. L., Zhang, Y., Chen, J., Ge, H. Y., Arendt-Nielsen, L., et al. (2010). Ischemic compression block attenuates mechanical hyperalgesia evoked from latent myofascial trigger points. *Experimental Brain Research*, *202*(2), 265–270. https://doi.org/10.1007/s00221-009-2129-2.

Weiss, J. M. (2001). Pelvic floor myofascial trigger points: Manual therapy for interstitial cystitis and the urgency-frequency syndrome. *Journal of Urology*, *166*(6), 2226–2231. https://doi.org/10.1016/S0022-5347(05)65539-5.

Zhang, Y., Ge, H. Y., Yue, S. W., Kimura, Y., & Arendt-Nielsen, L. (2009). Attenuated skin blood flow response to nociceptive stimulation of latent myofascial trigger points. *Archives of Physical Medicine and Rehabilitation*, *90*(2), 325–332. https://doi.org/10.1016/j.apmr.2008.06.037.

Part III

Psychology and behavior

Chapter 31

Behavioral markers of pain: Understanding the cognitive, motor, and societal interactions in the pain experience

S.A. Holmes[a,b], A. Quinlan[a], and M.E. Pierce[c]

[a]*Department of Anesthesiology, Critical Care and Pain Medicine, Boston Children's Hospital, Boston, MA, United States,* [b]*Instructor in Anesthesia, Harvard Medical School, Boston, MA, United States,* [c]*Translational Research Center for TBI & Stress Disorders, VA Boston Health Care System, Boston University School of Medicine, Boston, MA, United States*

Abbreviations

BRIEF	behavior rating inventory of executive function
CPT-II	cognitive performance test
GABA	gamma-aminobutryric acid
MRI	magnetic resonance imaging

As of 2020, the International Association for the Study of Pain defines the presence of pain through self-reporting of symptoms (Raja et al., 2020). The aim of this chapter is to provide a holistic view and explore the basis through which we understand pain behaviors. We end the chapter with recommendations for how pain behavior should be addressed in the future in light of the technological advancement of objective tools.

Psychology and cognition

Fear avoidance

The fear avoidance model outlines how pain is experienced in a healthy state and how this interaction with pain may become abnormal and develop chronic characteristics. Acute pain is an aversive stimulus that protects an organism from noxious stimuli. The physical presence of a painful stimulus (e.g., mechanical, thermal, or chemical) activates neural threat responses. These responses are designed for the acute response in so much as they evolved to decrease overtime as the threat neutralizes and safety is reestablished (Heathcote et al., 2020; Simons, 2016). However, in the context of chronic pain, fear and pain are triggered from stimuli that should otherwise be nonthreatening, leading to incorrect and prolonged signaling which results in an ongoing painful experience (Simons, 2016). Following a painful inciting event, the fear avoidance model describes the self-perpetuating cycle by which pain related fear is maintained and leads to the avoidance of activity, higher functional disability, increased pain, and further catastrophizing and goes on to suggest that the chronic nature of pain is the result of maladaptive behaviors in relation to an inciting injury (Asmundson et al., 2012). Unlearning a fear response, referred to as fear extinction, involves learning an inhibitory response to the fear stimulus (Myers, 2006). Early work on fear extinction therapies, like exposure therapy that are targeted at threat-learning circuitry (e.g., limbic-executive networks) were found to be less effective in adolescents than in adult patients (Heathcote et al., 2020); however, this work is relatively young and is receiving more attention.

Pain catastrophizing

Although it is unknown why certain individuals react maladaptively to pain, pain catastrophizing relates to the way an individual experiences pain, and how these negative cognitive and emotional processes can influence behavior in the future

The Neurobiology, Physiology, and Psychology of Pain. https://doi.org/10.1016/B978-0-12-820589-1.00031-2

(Friedrichsdorf et al., 2016). Pain catastrophizing is a negative cognitive attributional style in which an individual hyper focuses on the experience of pain (Pielech et al., 2014). Fear of pain and pain catastrophizing are central behavioral markers of pediatric chronic pain. Adolescents with self-reported high levels of pain-related fear and distress demonstrated increased activation of threat-learning brain regions (e.g., amygdala), when compared to healthy controls and adolescents in the low pain group (Heathcote et al., 2020), demonstrating a neural basis for this pain phenotype. Pain catastrophizing is directly linked to poor physical functioning, experiencing nonpainful stimuli as painful, and increased depression and anxiety (Pielech et al., 2014) and may be influenced by parental behaviors in the context of their child's pain (Simons, 2016). As such, a dynamic relationship is shared between child and parent, and is further complicated by the extent to which underlying pain regions of the brain are active and remain active in this cohort.

Cognitive performance

It is well accepted that children with chronic pain frequently miss school due to pain; however, the extent to which this reflects changes in a child's cognitive abilities is unclear. It has been observed that approximately 50% of an adolescent chronic pain cohort showed abnormal executive functioning or attention span in at least one category. Similarly, a correlation between the intensity of self-reported pain and attention deficits was also observed, drawing attention to the extent of pain and cognitive deficits. High levels of pain catastrophizing were associated with increased difficulty on tasks of working memory, exhibited both by the BRIEF and CPT-II assessments (Weiss et al., 2018). Consistent with these findings, a separate study applied the Brief Test of Attention in hospitalized patients with Complex Regional Pain Syndrome-I and revealed attention and working memory as the cognitive functions most affected in the pediatric pain population (Cruz et al., 2011). That is, among parents whose child missed school due to pain, 44% reported their child's grades declining (Dick & Riddell, 2010).

Despite significant academic disruption in this population, in a study that looked at cognitive functioning and academic achievement in youth with chronic pain (Ho et al., 2008), children who presented with pain demonstrated average or above average academic skills and cognitive ability when compared to that of their peers (Ho et al., 2008). Despite reports that upwards of 51% of children with chronic pain have trouble attending school (Palermo et al., 2014), this high rate may be attributed mostly to the social implications of chronic pain, rather than changes in intellectual capability. These results signified that despite chronic pain, intellectual and learning capabilities may remain unimpaired in children and adolescents with a pain diagnosis (Ho et al., 2008). Furthermore, a separate study performed in an inpatient rehabilitation setting (Cruz et al., 2011) determined that children with chronic pain had intact neuropsychological performance and intellectual functioning (Cruz et al., 2011) indicating a change in cognitive functioning itself is not the cause of poor school performance in this population. A critical role for environment is underscored by these findings. Further research is needed to evaluate the impact of school absences as the primary cause of self-reported academic decline in pediatric chronic pain patients.

Comorbidities and chronic pain

Chronic pain is accompanied by diverse comorbidities that affect both adult and child populations. Generally, pain can share neuronal pathways with factors such as anxiety (Zhuo, 2016), and a person with chronic pain may exhibit changes in basic reward and aversive pathways within their brain that can make them more prone toward suicidality (Elman et al., 2013) as well as posttraumatic stress disorder (Brennstuhl et al., 2015; Kind & Otis, 2019). As such is imperative to understand such comorbidities as they develop and evolve in children and integrate with pain behaviors.

Comorbid behavioral and psychological conditions are prevalent in children with chronic pain. The most common psychiatric disorders seen in this population are internalizing conditions, namely anxiety and depression (Noel et al., 2016). One study aimed at demonstrating the prevalence of comorbid psychiatric disorders in children with chronic pain looked at pediatric patients with recurrent tension headaches, migraine headaches, and chronic abdominal pain. Each patient group presented with a higher rate of behavioral or emotional comorbidities (75.5%, 65.6%, and 52%, respectively) when compared with the control group (21%) (Machnes-Maayan et al., 2014). The high prevalence of emotional disorders cooccurring with chronic pain is likely due to overlapping neurobiological factors such as genes, shared brain networks, hormones, and inflammation (Vinall et al., 2016). Additionally, both types of disorders have mutual cognitive, behavioral, and biological mechanisms which lead to the maintenance of such conditions and perpetual psychosocial dysfunction (Vinall et al., 2016). Researchers have found that children who experience somatic complaints (e.g., headaches) over multiple developmental periods were at highest risk for developing comorbid psychological disorders (Shanahan et al., 2015), indicating the need for immediate, individualized treatment of pediatric pain to reduce physical and psychological severity.

While high rates of psychiatric disorders are consistently seen in children with chronic pain (Vinall et al., 2016), an important question that warrants further research is whether the development of pain leads to mental health conditions, or whether emotional disorders play a causal role in the occurrence of pain syndromes.

Not only are children who experience chronic pain or somatic symptoms during childhood more likely to have a psychiatric comorbidity in youth, this population is at elevated risk of developing mental health conditions in adulthood as well (Vinall et al., 2016). Even if a child experiences remission of their symptoms in childhood or adolescence, these individuals remain at elevated risk for psychiatric conditions in the future (Vinall et al., 2016). Researchers have found that 21.1% of adults who had chronic pain in childhood experienced anxiety disorders and similarly 24.5% experienced depression in adulthood (Noel et al., 2016), emphasizing the need to understand such diversity to pain behaviors.

Motor behavior

Motor response to pain

One of the most significantly recognized perspectives on pain and motor behavior is the Pain Adaptation Model (Lund et al., 1991). In this model, pain is considered an aversive stimulus, and such attempts are made by the body to minimize factors that would produce, or continue, the painful experience. On a physiological level, this amounts to a reduction in agonist musculature and excitation of antagonist muscle groups. For example, if you were to sustain injury to your biceps, the natural reaction would be to inhibit the use of this muscle and tone up the triceps. This simple model demonstrates a braking mechanism that acts to "freeze" the situation—the action itself does not improve the injury and attempts to prevent any further injury. At the brain level, inhibitory and excitatory changes are observed in terms of neurotransmitters, modulating expression of glutamate (excitatory) and inhibitory (GABA) factors in specific regions of the motor cortex associated with the agonist-antagonist pairings (Neigh et al., 2018). Interestingly, bilateral changes in the level of cortical excitability are observed during such injury events, showing how pain on one side of the body can implicate both sides of the brain. Moreover, in persons with unilateral wrist/hand pain, modulating levels of cortical excitability were precited to underlie decreases in cognitive function as well as bilateral sensorimotor differences. These were measured in tests such as pressure pain thresholds, grip strength, and the Purdue Peg Board Test (Pelletier et al., 2019). Notably, this aligns as well with the preservation of motor units in the presence of chronic pain which appears to remain intact (Tsay & Giummarra, 2016); alluding to a focus on changes in the discharge rate of motor neurons. To mitigate the behavioral impact of a muscle group being inhibited, additional recruitment of muscle groups charged with similar mechanical objects may be undertaken (Hodges & Tucker, 2011) at the expense of more energy and reduced quality of movement in an adaptive manner as movements are relearned. The extent of such relearning however is likely to correlate with significant disability in movement time or functional disability.

Changing our motor behaviors

At a behavioral level, the changes at the brain and spinal cord resulting from an acute or chronic pain setting amount to significant disability and behavioral disruptions. In the presence of pain, an acute painful stimulus has the capacity to influence both the preparation and upcoming movement execution stages (Neigh et al., 2018). In these early movement stages, information in terms of visual, tactile, and proprioceptive input is integrated to understand the environment and join that with intended movement goals. Experiencing pain, even acutely, modifies these motor programs in line with the Pain Adaptation Model (Lund et al., 1991). During the learning phases of these movements, an increase in attention may be driven toward the part of the body that is experiencing pain. This attention has both positive and negative potential. As outlined in a capsaicin model (a form of experimentally induced pain), an increase in attention toward an affected area has the potential to improve the motor learning process (Dancey et al., 2016)—learning in an antinociceptive manner by enhancing movement sense (Sole et al., 2015). This should be noted in light of highly cited research on the impact of high levels of attention/arousal. That is, although elevated levels of attention can have a positive effect on factors such as learning and performance, allowing persons to adapt and be engaged with task demands, if attention is too high, this offers diminishing returns (Yerks & Dodson, 1908). The length at which a person is exposed to a painful stimulus is directly related to the length of impairment (Huot-Lavoie et al., 2019). Pain can also reflect a reduction in the anticipatory strategy (i.e., movement preparation) and difficulty in transferring motor skills due to pain-related changes in motor strategy (Bouffard et al., 2016). Learning in the context of pain is thus a significant obstacle and is associated with maintaining this pain-adaptive strategy over time (Lamothe et al., 2014). Any attempts at rehabilitation or mitigating the negative long-term effects of pain on behavior require relearning motor sequences (Dancey et al., 2019).

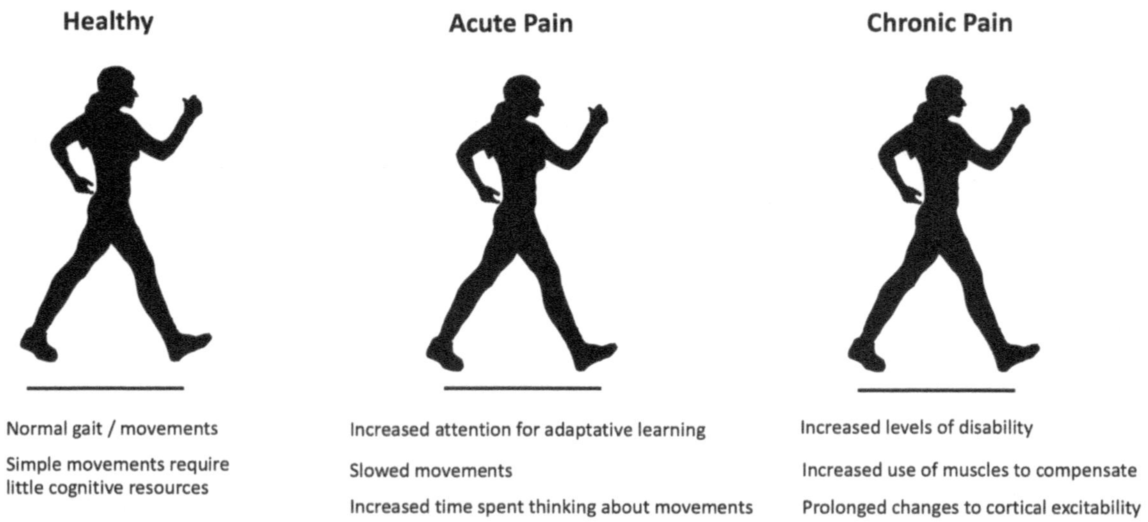

FIG. 1 Stages of motor behavior impacted by pain. Behavioral effects on motor behaviors can be observed even with exposure to an acute painful stimulus. The length of time experiencing pain has a direct effect on motor behaviors and the resulting disability.

The impact that pain has on motor behavior is clear with significant influence over how the brain not only performs but learns motor behaviors (see Fig. 1 for summary). Both acute and chronic pain can produce significant changes in motor behavior and lead to disability and even cross over into other realms such as cognitive difficulties and sensory abnormalities, and overlap between physical and social pain (Sturgeon & Zautra, 2016). Although there are some therapeutic options to mitigate or reverse some of the behavioral effects of pain, research has remained sparse in this section of the pain literature and requires significant attention.

Social interactions and pain behavior convention

Social interaction

Human interaction is a core component to our species. Interactions among family, friends, coworkers, and other members of society define our behaviors by either enabling or disabling the ability to express ourselves. The extent to which we engage in these interactions and discussions has a measurable impact on how we interact with pain-oriented events. Here, we focus on how our personal interaction with others shapes our pain behaviors.

Pain is contagious. Because of the empathetic nature of human behavior, pain has the capacity to be socially transmitted (Martin et al., 2015), meaning that how an individual may approach a painful stimulus may be influenced by who they are interacting with. Moreover, even if individuals do not receive an aversive pain signal, it is possible to be primed—making them more likely to experience pain in a situation that did not previously elicit a pain response (Ji et al., 2018). Sense of power in a given situation will actually enhance sensory processing of painful stimuli (Martin et al., 2015). The amount of empathy from a partner may influence pain reporting and will also depend on their attachment style (Hurter et al., 2014). Taking an example from healthcare, physicians are more likely to empathize with patients when it aligns with their "overt goals" but are less likely to engage in spontaneous empathy (Spring et al., 2019). In a pain sense, this is likely a low transference setting for pain behaviors; however, it likely has a significant negative impact on how we express our pain behaviors in a healthcare setting.

Our communities have a direct impact on pain perception. When modulated in the context of a laboratory setting, social exclusion and perceived social support may increase and reduce the severity how individuals perceive the severity of an acute experimental painful stimulus (Brown et al., 2003; Master et al., 2009). As found by Oliveira et al. (2015), in persons with low back pain, the level of social isolation was negatively impacted in approximately 39.2% of participants. At baseline, it was found that levels of social isolation predicted greater levels of pain catastrophizing, anxiety, and depression; however, they did not have the same predictive value at 6 months with social isolation only predicting the extent of disability—inferring a time-dependent relationship. In a separate study Johnson and Dunbar (2016) observed that higher pain tolerance was correlated with larger self-reported social group size. Chronic pain is linked with poorer relationships and self-imposed isolation (Snelling, 1994), decreasing the number of individuals in a person's network and regular pain is

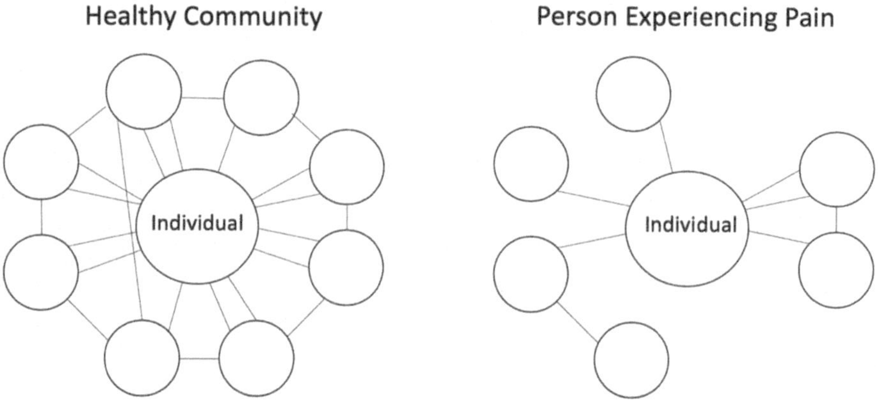

FIG. 2 The effect of pain on social structures. Each *circle* and *line* represents a person and their ability to interact with another, respectively. For persons experiencing chronic pain, the feeling of pain can lead to isolation and the erosion of social structures which have shown to have positive effects on quality of life.

associated with having lower numbers in friend groups, poor peer relationship quality, victimization, and social problems in the family unit (Forgeron et al., 2010, 2011). Together, findings underscore a relationship wherein pain has a negative impact on our social environment and the extent to which we interact with others (Fig. 2).

Community structure—Access to resources

Physical infrastructure is lacking. Globally, the prevalence of pain around the world is one of, if not the, most significant healthcare condition among children and adults. For example, the prevalence of chronic back pain ranges from 18.0% to 24.0% and has been reported to be as high as 38% (Henschke et al., 2015) and a 1-year incidence as high as 36.0% (Hoy et al., 2010). For older patients, the prevalence of chronic pain is 25%–76% and jumps to 93% in persons living in residential care (Abdulla et al., 2013). In the United States, the economic impact of pain is felt in terms of lost productivity and ranged from \$560 to \$635 billion dollars in 2010—depending on calculation method—and contributed to greater costs than cancer and diabetes combined (Gaskin & Richard, 2011). Pain has the greatest negative impact on quality of life when compared with other health-related conditions while contributing the most toward disability (Henschke et al., 2015). This amounts to significant challenges for lower income areas and communities as they face the conflicting demands of needing to treat pain patients and the rising costs in providing pain treatment services.

Social structures are flawed. When reporting pain levels and seeking treatment, race has been found to modulate the importance of trustworthiness on empathy (Mathur et al., 2010). Despite African American and Hispanic communities showing different levels of pain thresholds and unpleasantness than persons who are Caucasia, they are less likely to receive pain medication (and lower doses), wait longer for pain medicine, and be undertreated for pain conditions (Mossey, 2011). Moreover, in a cohort of patients with low back pain, African Americans reported greater pain and disability than Caucasians but were rated by their clinicians as having less (Anderson et al., 2009). As such, flaws in how persons reach out to healthcare and express their pain behaviors can be due to their socioeconomic status as well as the racial inequalities. Undertreating of pain conditions can have significant consequences and lead to comorbidities such as reduced mobility, loss of strength, sleep disturbances, immune system impairment, increased susceptibility to disease, and medication dependence (Wang et al., 2013). It is imperative for individualized medicine moving forward to address the knowledge gaps of prior generations and provide equal access to healthcare resources.

Biological sex and gender

From a societal standpoint, how we approach the aspect of biological sex and gender has changed dramatically in the past decade. In the context of the pain field, there has been a generous amount of research that has shown how biological aspects of sex can interact with how we perceive a painful stimulus as well as how we change our behaviors. For example, it was found by Fillingim et al. (2009) in a national representative sample of migraine headache sufferers that the prevalence of most common forms of pain is higher in woman and that women display enhanced sensitivity to most forms of experimentally induced pain. The concept of perception of masculinity and femininity has also been shown to modulate the interaction between pain and social context (Keogh, 2015). Males have been shown to increase their pain tolerance only when in the presence of other male and not female friends (Edwards et al., 2017; Walsh et al., 2017). This may derive from differences in biological sex in terms of empathy, and more specifically empathy for pain. As outlined by Singer and colleagues (Singer, 2004), empathy for pain can be regarded as observations of others in pain eliciting similar neuronal responses as if the person observing the pain were experiencing pain themselves. Females have been reported to have greater empathy than males (Han et al., 2008) and express pain symptoms more (Han et al., 2017) than their male counterparts. However, we must begin to ask how this affects individuals who identify as transgender. To date, little to no research has been done on how pain levels are experienced in this cohort, despite the growing number of individuals who identify as transgender. It will be a critical focus for future research to identify how factors such as hormonal therapy, loss of social support, and the physical tolls of surgery affect this population.

What is coming in the next 5–10 years?

The major challenge to understanding the role of pain behaviors in the coming decade is understanding their role in patient identification. That is, research is focusing increasingly on the use of objective pain criteria. However, simply applying such objective criteria is currently divorced from our modern subjective understanding of pain, and therefore we have no definitive method of saying that what we are observing, either on MRI, or through physiology, is in fact a pain event. Research into subconscious processing of painful stimuli has yielded extremely fruitful results, but most importantly demonstrates that integrating objective criteria into our understanding of pain behaviors can help address pain diagnostics. To this point, medical imaging techniques such as magnetic resonance imaging, positron emission tomography, and magnetic resonance spectroscopy all show promising results for being able to show changes in cortical thickness, brain metabolism, and inflammation. However, how these techniques are integrated with behavior is a secondary issue that will require significant address. Higher level computing techniques including machine learning represent formidable candidates for their ability to integrate high volumes of data from diverse modalities. Effective application of these techniques will enable the combined use of behavioral and physiological data for individualized pain medicine and integrate both the power of objective data and the personalized pain experience.

Conclusion

In conclusion, the subjective perception of pain is a gold standard for understanding pain conditions in both modern clinical and research avenues. As reviewed in this chapter and shown in Fig. 3, behavioral elements are diverse and need to be approached from different vantage points. Childhood and parental approaches to pain levels are frequently unique and need to be addressed in equal regard, recognizing the child's agency in terms of their own feelings, but also parental perspectives of their behaviors that they may not be aware of. Two of the more dominant themes that are likely to emerge in the coming years are a focus on the motor-related effects of pain behaviors and an integration of objective tools into clinical and research use to define a pain state. This field is continuing to evolve in new ways every day with sights set on personalized medicine and preventing the acute to chronic transition of pain behaviors.

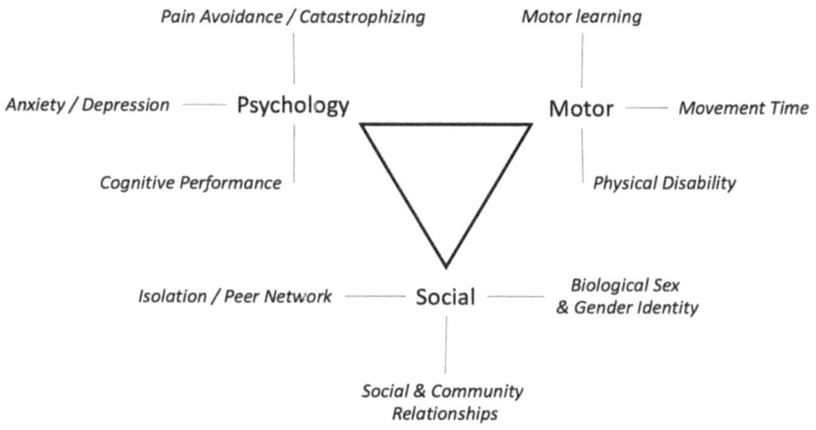

FIG. 3 Domains of pain behavior: three dominant themes in pain behaviors are cognitive performance, motor behaviors and disability, and social interactions. They each contribute independently toward a person's overall behavioral response and expression of pain; however, they may act synergistically to further cognitive, physical, and social challenges.

Application to other areas

Pain behaviors are observed in all conditions, both acute and chronic, that involve a component of pain. The behaviors outlined in this chapter are directly applicable toward understanding the neuroscience underlying the pain experience. Methods such as structural and functional brain imaging provide insight into healthy and unhealthy brain states; however, they require complementary behavioral data in order to understand the significance of any observed deviance from normal. Brain-based data provide unique insight into subliminal processing of painful stimuli, where behaviors may be too small to measure or observe or may be absent entirely. Indeed, behavioral data must be understood in a domain-oriented fashion, integrating the complex nature of pain and its capacity to interfere with lower order processes such as vision and higher order processes such as cognition. The material outlined in this chapter is meant as a broad understanding of what can be observed and attributed toward the pain experience in terms of behavior. As pain behaviors become more engrained from the acute to chronic pain experience, they display unique profiles. What initially can be understood in simple evolutionary terms in terms of a discrete pain stimulus may ultimately require the integration of knowledge about socioeconomics, peer support groups, methods of approaching gender identity, and access to medical resources to properly understand. As our understanding of the diversity of pain behaviors grows, we also need consider new and better measurement tools and not simply transition toward completely objective metrics that are divorced from the pain experience. Integration with other modalities will improve our understanding of pain medicine and allow us to transition toward a personalized medicine approach.

Other agents of interest

This chapter focused on three domains of understanding pain behaviors: cognition, motor, and social. Although these are arguably the most influential, we were not able to describe in depth the subdomains that are critical toward fully understanding how pain influences one's behaviors. We invite the reader to explore the elements of the pain learning experience including fear of pain and difficulties associated with the extinction process. Also, there is a plethora of information available regarding how pain contributes to conditions such as posttraumatic stress disorder and how this disorder is not restricted to veteran populations, impacting both female and male cohorts. To this point, we need to begin working with concepts of gender identity for persons in pain in a more constructive and evidence-based way. To date, little information is available regarding the transgender experience of pain and how this evolves as persons' transition through medical and societal challenges. We invite the reader to consider every domain they encounter in light of how a patient's personal perception and biases may lead them to experience that event. How we each experience a painful stimulus will be unique and will likely vary from one time point to another.

Mini-dictionary of terms

Pain adaptation: After a painful stimulus, the degree to which our behaviors are modified as a direct result of pain exposure. Adaptation can either be in a way that reduces the occurrence of pain (e.g., a pain reflex arc) or prolongs it (maladaptive; e.g., reducing social interactions).

Fear avoidance model: A psychiatric model developed using musculoskeletal pain that outlines how chronic pain develops as a result of attention and avoidance-based behaviors.

Fear of pain: A construct outlining the extent to which a single, or multiple, pain experiences shape human behavior and offers insight into the extent of disability resulting from the pain experience.

Pain catastrophizing: Describing or perceiving a painful event in exaggerated terms and to ruminate about the pain experience. Feelings of helplessness are also commonly experienced.

Motor behavior: How our brains control our muscles to interact with the world. The coordinated series of movements between large and small muscles groups will define our agility and disability and can be mediated by the extent of acute and chronic pain being managed.

Key facts

- Pain can have a negative effect on cognitive performance and frequently presents with other comorbidities.
- Scoring high on pain catastrophizing can have a negative impact on behavior and is associated with pain chronicity.
- Pain has an inhibitory influence over motor behaviors and can lead to significant physical disability.

- Social interactions among patients and their friends, families, and others in the community are diminished in response to pain.
- Community structures are critical toward improving and maintaining healthy pain expression which requires addressing issues such as transgender and racial inequalities.

Summary points

- Pain can have a very diffuse and dynamic impact on behaviors.
- Pain is a perceptual event and is intimately related with a person's own psychology.
- Use of neuroimaging can help understand the organic basis of pain behaviors and drivers of pain chronification.
- Pain can be present without any observed tissue pathology underscoring the need to understand how pain behaviors impact activity of daily living.

References

Abdulla, A., Adams, N., Bone, M., Elliott, A. M., Gaffin, J., Jones, D., et al. (2013). Guidance on the management of pain in older people. *Age and Ageing*, *42*(1), i1–i57.

Anderson, K. O., Green, C. R., & Payne, R. (2009). Racial and ethnic disparities in pain: Causes and consequences of unequal care. *Journal of Pain*, *10*(12), 1187–1204. https://doi.org/10.1016/j.jpain.2009.10.002.

Asmundson, G. J., Noel, M., Petter, M., & Parkerson, H. A. (2012). Pediatric fear-avoidance model of chronic pain: Foundation, application and future directions. *Pain Research and Management*, *17*(6), 397–405. https://doi.org/10.1155/2012/908061.

Bouffard, J., Bouyer, L. J., Roy, J.-S., & Mercier, C. (2016). Pain induced during both the acquisition and retention phases of locomotor adaptation does not interfere with improvements in motor performance. *Neural Plasticity*, *2016*, 1–9. https://doi.org/10.1155/2016/8539096.

Brennstuhl, M.-J., Tarquinio, C., & Montel, S. (2015). Chronic pain and PTSD: Evolving views on their comorbidity. *Perspectives in Psychiatric Care*, *51*(4), 295–304. https://doi.org/10.1111/ppc.12093.

Brown, J. L., Sheffield, D., Leary, M. R., & Robinson, M. E. (2003). Social support and experimental pain. *Psychosomatic Medicine*, *65*(2), 276–283. https://doi.org/10.1097/01.PSY.0000030388.62434.46.

Cruz, N., O'Reilly, J., Slomine, B. S., & Salorio, C. F. (2011). Emotional and neuropsychological profiles of children with complex regional pain syndrome type-I in an inpatient rehabilitation setting. *Clinical Journal of Pain*, *27*(1), 27–34. https://doi.org/10.1097/AJP.0b013e3181f15d95.

Dancey, E., Murphy, B., Andrew, D., & Yielder, P. (2016). Interactive effect of acute pain and motor learning acquisition on sensorimotor integration and motor learning outcomes. *Journal of Neurophysiology*, *116*(5), 2210–2220. https://doi.org/10.1152/jn.00337.2016.

Dancey, E., Yielder, P., & Murphy, B. (2019). The interactive effect of tonic pain and motor learning on corticospinal excitability. *Brain Sciences*, *9*(3), 63. https://doi.org/10.3390/brainsci9030063.

Dick, B. D., & Riddell, R. P. (2010). Cognitive and school functioning in children and adolescents with chronic pain: A critical review. *Pain Research and Management*, *15*(4), 238–244. https://doi.org/10.1155/2010/354812.

Edwards, R., Eccleston, C., & Keogh, E. (2017). Observer influences on pain: An experimental series examining same-sex and opposite-sex friends, strangers, and romantic partners. *Pain*, *158*(5), 846–855. https://doi.org/10.1097/j.pain.0000000000000840.

Elman, I., Borsook, D., & Volkow, N. D. (2013). Pain and suicidality: Insights from reward and addiction neuroscience. *Progress in Neurobiology*, *109*, 1–27. https://doi.org/10.1016/j.pneurobio.2013.06.003.

Fillingim, R. B., King, C. D., Ribeiro-Dasilva, M. C., Rahim-Williams, B., & Riley, J. L. (2009). Sex, gender, and pain: A review of recent clinical and experimental findings. *Journal of Pain*, *10*(5), 447–485. https://doi.org/10.1016/j.jpain.2008.12.001.

Forgeron, P. A., King, S., Stinson, J. N., McGrath, P. J., MacDonald, A. J., & Chambers, C. T. (2010). Social functioning and peer relationships in children and adolescents with chronic pain: A systematic review. *Pain Research and Management*, *15*(1), 27–41. https://doi.org/10.1155/2010/820407.

Forgeron, P. A., McGrath, P., Stevens, B., Evans, J., Dick, B., Finley, A. G., et al. (2011). Social information processing in adolescents with chronic pain: My friends don't really understand me. *Pain*, *152*(12), 2773–2780. https://doi.org/10.1016/j.pain.2011.09.001.

Friedrichsdorf, S., Giordano, J., Desai Dakoji, K., Warmuth, A., Daughtry, C., & Schulz, C. (2016). Chronic pain in children and adolescents: Diagnosis and treatment of primary pain disorders in head, abdomen, muscles and joints. *Children*, *3*(4), 42. https://doi.org/10.3390/children3040042.

Gaskin, D., & Richard, P. (2011). The economic costs of pain in the United States. In *Relieving pain in America: A blueprint for transforming prevention, care, education, and research*. National Academy Press.

Han, S., Fan, Y., & Mao, L. (2008). Gender difference in empathy for pain: An electrophysiological investigation. *Brain Research*, *1196*, 85–93. https://doi.org/10.1016/j.brainres.2007.12.062.

Han, X., He, K., Wu, B., Shi, Z., Liu, Y., Luo, S., et al. (2017). Empathy for pain motivates actions without altruistic effects: Evidence of motor dynamics and brain activity. *Social Cognitive and Affective Neuroscience*, *12*(6), 893–901. https://doi.org/10.1093/scan/nsx016.

Heathcote, L. C., Timmers, I., Kronman, C. A., Mahmud, F., Hernandez, J. M., Bentley, J., et al. (2020). Brain signatures of threat–safety discrimination in adolescent chronic pain. *Pain*, *161*(3), 630–640. https://doi.org/10.1097/j.pain.0000000000001753.

Henschke, N., Kamper, S. J., & Maher, C. G. (2015). The epidemiology and economic consequences of pain. *Mayo Clinic Proceedings*, *90*(1), 139–147. https://doi.org/10.1016/j.mayocp.2014.09.010.

Ho, G. H. Y., Bennett, S. M., Cox, D., & Poole, G. (2008). Brief report: Cognitive functioning and academic achievement in children and adolescents with chronic pain. *Journal of Pediatric Psychology*, *34*(3), 311–316. https://doi.org/10.1093/jpepsy/jsn077.

Hodges, P. W., & Tucker, K. (2011). Moving differently in pain: A new theory to explain the adaptation to pain. *Pain*, *152*(Suppl), S90–S98. https://doi.org/10.1016/j.pain.2010.10.020.

Hoy, D., Brooks, P., Blyth, F., & Buchbinder, R. (2010). The epidemiology of low back pain. *Best Practice & Research Clinical Rheumatology*, *24*(6), 769–781. https://doi.org/10.1016/j.berh.2010.10.002.

Huot-Lavoie, M., Ting, W. K.-C., Demers, M., Mercier, C., & Ethier, C. (2019). Impaired motor learning following a pain episode in intact rats. *Frontiers in Neurology*, *10*, 927. https://doi.org/10.3389/fneur.2019.00927.

Hurter, S., Paloyelis, Y., Williams, A. C. d. C., & Fotopoulou, A. (2014). Partners' empathy increases pain ratings: Effects of perceived empathy and attachment style on pain report and display. *Journal of Pain*, *15*(9), 934–944. https://doi.org/10.1016/j.jpain.2014.06.004.

Ji, R.-R., Nackley, A., Huh, Y., Terrando, N., & Maixner, W. (2018). Neuroinflammation and central sensitization in chronic and widespread pain. *Anesthesiology*, *129*(2), 343–366. https://doi.org/10.1097/ALN.0000000000002130.

Johnson, K. V.-A., & Dunbar, R. I. M. (2016). Pain tolerance predicts human social network size. *Scientific Reports*, *6*(1), 25267. https://doi.org/10.1038/srep25267.

Keogh, E. (2015). Men, masculinity, and pain. *Pain*, *156*(12), 2408–2412. https://doi.org/10.1097/j.pain.0000000000000328.

Kind, S., & Otis, J. D. (2019). The interaction between chronic pain and PTSD. *Current Pain and Headache Reports*, *23*(12), 91. https://doi.org/10.1007/s11916-019-0828-3.

Lamothe, M., Roy, J.-S., Bouffard, J., Gagné, M., Bouyer, L. J., & Mercier, C. (2014). Effect of tonic pain on motor acquisition and retention while learning to reach in a force field. *PLoS One*. *9*(6)e99159https://doi.org/10.1371/journal.pone.0099159.

Lund, J. P., Donga, R., Widmer, C. G., & Stohler, C. S. (1991). The pain-adaptation model: A discussion of the relationship between chronic musculoskeletal pain and motor activity. *Canadian Journal of Physiology and Pharmacology*, *69*(5), 683–694. https://doi.org/10.1139/y91-102.

Machnes-Maayan, D., Elazar, M., Apter, A., Zeharia, A., Krispin, O., & Eidlitz-Markus, T. (2014). Screening for psychiatric comorbidity in children with recurrent headache or recurrent abdominal pain. *Pediatric Neurology*, *50*(1), 49–56. https://doi.org/10.1016/j.pediatrneurol.2013.07.011.

Martin, L. J., Hathaway, G., Isbester, K., Mirali, S., Acland, E. L., Niederstrasser, N., et al. (2015). Reducing social stress elicits emotional contagion of pain in mouse and human strangers. *Current Biology*, *25*(3), 326–332. https://doi.org/10.1016/j.cub.2014.11.028.

Master, S. L., Eisenberger, N. I., Taylor, S. E., Naliboff, B. D., Shirinyan, D., & Lieberman, M. D. (2009). A picture's worth: Partner photographs reduce experimentally induced pain. *Psychological Science*, *20*(11), 1316–1318. https://doi.org/10.1111/j.1467-9280.2009.02444.x.

Mathur, V. A. Harada, T., Lipke, T., & Chiao, J. Y. (2010). Neural basis of extraordinary empathy and altruistic motivation. *NeuroImage*, *51*(4), 1468–1475. https://doi.org/10.1016/j.neuroimage.2010.03.025.

Mossey, J. M. (2011). Defining racial and ethnic disparities in pain management. *Clinical Orthopaedics and Related Research®*, *469*(7), 1859–1870. https://doi.org/10.1007/s11999-011-1770-9.

Myers, K. M. (2006). Different mechanisms of fear extinction dependent on length of time since fear acquisition. *Learning & Memory*, *13*(2), 216–223. https://doi.org/10.1101/lm.119806.

Neigh, G., Bekhbat, M., & Rowson, S. (2018). Neuroimmunology: Behavioral effects. In Oxford University Press.

Noel, M., Groenewald, C. B., Beals-Erickson, S. E., Gebert, J. T., & Palermo, T. M. (2016). Chronic pain in adolescence and internalizing mental health disorders: A nationally representative study. *Pain*, *157*(6), 1333–1338. https://doi.org/10.1097/j.pain.0000000000000522.

Oliveira, V. C., Ferreira, M. L., Morso, L., Albert, H. B., Refshauge, K. M., & Ferreira, P. H. (2015). Patients' perceived level of social isolation affects the prognosis of low back pain: Social isolation and low back pain. *European Journal of Pain*, *19*(4), 538–545. https://doi.org/10.1002/ejp.578.

Palermo, T. M., Valrie, C. R., & Karlson, C. W. (2014). Family and parent influences on pediatric chronic pain: A developmental perspective. *American Psychologist*, *69*(2), 142–152. https://doi.org/10.1037/a0035216.

Pelletier, R., Paquette, É., Bourbonnais, D., Higgins, J., Harris, P. G., & Danino, M. A. (2019). Bilateral sensory and motor as well as cognitive differences between persons with and without musculoskeletal disorders of the wrist and hand. *Musculoskeletal Science and Practice*, *44*, 102058. https://doi.org/10.1016/j.msksp.2019.102058.

Pielech, M., Ryan, M., Logan, D., Kaczynski, K., White, M. T., & Simons, L. E. (2014). Pain catastrophizing in children with chronic pain and their parents: Proposed clinical reference points and reexamination of the Pain Catastrophizing Scale measure. *Pain*, *155*(11), 2360–2367. https://doi.org/10.1016/j.pain.2014.08.035.

Raja, S. N., Carr, D. B., Cohen, M., Finnerup, N. B., Flor, H., Gibson, S., et al. (2020). The revised International Association for the Study of Pain definition of pain: Concepts, challenges, and compromises. *Pain*. https://doi.org/10.1097/j.pain.0000000000001939 Publish Ahead of Print.

Shanahan, L., Zucker, N., Copeland, W. E., Bondy, C. L., Egger, H. L., & Costello, E. J. (2015). Childhood somatic complaints predict generalized anxiety and depressive disorders during young adulthood in a community sample. *Psychological Medicine*, *45*(8), 1721–1730. https://doi.org/10.1017/S0033291714002840.

Simons, L. E. (2016). Fear of pain in children and adolescents with neuropathic pain and complex regional pain syndrome. *Pain*, *157*, S90–S97. https://doi.org/10.1097/j.pain.0000000000000377.

Singer, T. (2004). Empathy for pain involves the affective but not sensory components of pain. *Science*, *303*(5661), 1157–1162. https://doi.org/10.1126/science.1093535.

Snelling, J. (1994). The effect of chronic pain on the family unit. *Journal of Advanced Nursing*, *19*(3), 543–551.

Sole, G., Osborne, H., & Wassinger, C. (2015). The effect of experimentally-induced subacromial pain on proprioception. *Manual Therapy*, *20*(1), 166–170. https://doi.org/10.1016/j.math.2014.08.009.

Spring, V. L., Cameron, C. D., McKee, S., & Todd, A. R. (2019). Intentional and unintentional empathy for pain among physicians and nonphysicians. *Social Psychological and Personality Science*, *10*(4), 440–448. https://doi.org/10.1177/1948550618771890.

Sturgeon, J. A., & Zautra, A. J. (2016). Social pain and physical pain: Shared paths to resilience. *Pain Management*, *6*(1), 63–74. https://doi.org/10.2217/pmt.15.56.

Tsay, A. J., & Giummarra, M. J. (2016). Position sense in chronic pain: Separating peripheral and central mechanisms in proprioception in unilateral limb pain. *Journal of Pain*, *17*(7), 815–823. https://doi.org/10.1016/j.jpain.2016.03.005.

Vinall, J., Pavlova, M., Asmundson, G., Rasic, N., & Noel, M. (2016). Mental health comorbidities in pediatric chronic pain: A narrative review of epidemiology, models, neurobiological mechanisms and treatment. *Children*, *3*(4), 40. https://doi.org/10.3390/children3040040.

Walsh, J., Eccleston, C., & Keogh, E. (2017). Sex differences in the decoding of pain-related body postures. *European Journal of Pain*, *21*(10), 1668–1677. https://doi.org/10.1002/ejp.1072.

Wang, Y., Kennedy, J., Caggana, M., Zimmerman, R., Thomas, S., Berninger, J., et al. (2013). Sickle cell disease incidence among newborns in New York State by maternal race/ethnicity and nativity. *Genetics in Medicine*, *15*(3), 222–228. https://doi.org/10.1038/gim.2012.128.

Weiss, K. E., Harbeck-Weber, C., Zaccariello, M. J., Kimondo, J. N., Harrison, T. E., & Bruce, B. K. (2018). Executive functioning in pediatric chronic pain: Do deficits exist? *Pain Medicine*, *19*(1), 60–67. https://doi.org/10.1093/pm/pnx020.

Yerks, R. M., & Dodson, J. D. (1908). The relation of strength of stimulus to rapidity of habit-formation. *Journal of Comparative Neurology and Psychology*, *18*, 459–482. https://doi.org/10.1002/cne.920180503.

Zhuo, M. (2016). Neural mechanisms underlying anxiety–chronic pain interactions. *Trends in Neurosciences*, *39*(3), 136–145. https://doi.org/10.1016/j.tins.2016.01.006.

Chapter 32

Adverse life events, sensitization of spinal nociception, and chronic pain risk

Jamie L. Rhudy and Natalie Hellman
The University of Tulsa, Department of Psychology, Tulsa, OK, United States

Abbreviations

ACEs adverse childhood experiences
ALEs adverse life events
CPM conditioned pain modulation
CS central sensitization
ELA early life adversity
ELS early life stress
EMG electromyography
NFR nociceptive flexion reflex
PTSD posttraumatic stress disorder
QST quantitative sensory tests
TS temporal summation

Adverse life events: A definition

Adverse life events (ALEs) are major stressors occurring at any time throughout the lifespan and pose a significant physical or psychological threat to safety and/or security. This term includes experiences during childhood (typically <18 years; also known as adverse childhood events [ACEs], early life stress [ELS], or early life adversity [ELA]) and those in adulthood. ALEs are not daily hassles or minor stressors (e.g., stuck in traffic). ALEs can be a single brief event (e.g., car crash), repetitive (e.g., domestic violence), or occur continuously for years (e.g., emotional neglect). Many events can be characterized as ALEs; however, it is worth noting that the majority of the literature on ALEs and pain focuses on ALEs such as abuse, neglect, and events involving direct exposure to a serious threat (e.g., combat, assault, accidents). Relatively little research has explored pain and ALEs such as homelessness, unemployment, or poverty.

Adverse life events and health

In their landmark study of almost 14,000 adults, Felitti et al. (1998) examined the relationship between ACEs and leading causes of death. ACEs were defined as abuse (physical, sexual, emotional), neglect (physical, emotional), and household dysfunction (domestic violence, parental incarceration, mental illness/substance abuse by household member). They found a dose-dependent effect; as the number of ACEs increased, so did rates of self-reported ischemic heart disease, cancer, chronic bronchitis/emphysema, hepatitis/jaundice, skeletal fractures, and poor self-rated health (Felitti et al., 1998). Importantly, these effects are replicable (Hughes et al., 2017).

Adverse life events and chronic pain

ACEs are also associated with chronic pain (e.g., Sachs-Ericsson, Sheffler, Stanley, Piazza, & Preacher, 2017). Although many studies of ACEs and chronic pain are cross-sectional, results from a longitudinal study of over 7000 participants found that ACEs were associated with an increased risk for chronic widespread pain in adulthood (Jones, Power, & Macfarlane, 2009). Further, the number of ACEs, rather than the type, seems to promote pain risk (You, Albu, Lisenbardt, & Meagher, 2019).

The Neurobiology, Physiology, and Psychology of Pain. https://doi.org/10.1016/B978-0-12-820589-1.00032-4

Importantly, the impact of adversity on pain is not limited to childhood adversity. Adulthood adversity (i.e., $\geq$18 years of age) is associated with chronic back pain (Tesarz et al., 2015), headaches/migraines (Arcaya et al., 2017), and functional somatic syndromes such as fibromyalgia, chronic widespread pain, temporomandibular disorder, and irritable bowel syndrome, to name a few (Afari et al., 2014). Similar to ACEs, adulthood ALEs appear to have a dose-dependent effect on chronic pain risk (Sledjeski, Speisman, & Dierker, 2008).

The impact of adversity (stress) on nociception in animals

Animal studies are important because they provide experimental controls that would be unethical in humans. However, they are limited to studying the impact of stress/adversity on nociception (pain signaling) and nociceptive behaviors, because it is difficult to assess the psychological experience of pain in animals. A full review is beyond the scope of this chapter, but an interested reader is referred to reviews by Jennings, Okine, Roche, and Finn (2014) and Burke, Finn, McGuire, and Roche (2017). Much of this work has focused on the effects of an acute stressor on nociceptive responding immediately following the stressor (e.g., stress-induced hyperalgesia); thus, results may not provide insights into the long-term consequences of adversity. By contrast, early life stress (ELS) models expose the animal pup and/or mother to stressors (e.g., maternal separation) and then test the pup during adulthood. Most ELS studies are designed to examine the impact of ELS on stress systems (Baram et al., 2012); fewer examine nociceptive systems specifically (Burke et al., 2017).

In general, these studies show ELS sensitizes nociceptive systems to increase vulnerability for persistent hyperalgesia in adulthood, but findings depend on the model of ELS and the type of pain assay. The most consistent finding is that ELS produces hyperalgesia to colorectal distension, a model of irritable bowel syndrome (Burke et al., 2017). One noteworthy study by Alvarez, Green, and Levine (2013) demonstrated that ELS produces hyperalgesia on a measure of rodent muscle pain, but that exposure to a noise stressor as adults led to a delayed (after 10+ days) enhancement of this hyperalgesia. These effects were partially mediated by interleukin-6. Thus, ELS leads to nociceptive sensitization that is compounded by heightened stress-reactivity in adulthood. This is preclinical evidence for the dose-dependent effect of ALEs on pain vulnerability and suggests a role of immunity/inflammation in the effect.

Pain systems and their assessment in humans

Pain transmission involves the activation of peripheral nerves (nociceptors; i.e., Aδ- and C-fibers) that carry nociceptive information to the dorsal horn of the spinal cord. From there, several tracts ascend to the brain that allow distribution to multiple supraspinal regions that process the sensory, affective, and cognitive components of pain, as well as other regions that promote pain-related behaviors and physiological arousal (e.g., Apkarian, Bushnell, Treede, & Zubieta, 2005). Thus, pain and nociception can be amplified along these ascending pathways by sensitization of peripheral nociceptors, spinal neurons, or supraspinal regions associated with pain processing.

There are also brain-to-spinal cord circuits that allow nociception to be inhibited or facilitated at the spinal level. The most highly studied descending circuit includes the periaqueductal gray and its outputs to the rostroventromedial medulla and dorsolateral pons tegmentum; these regions have descending fibers within the dorsolateral funiculus that modulate spinal nociception (Fields, Basbaum, & Heinricher, 2006). Supraspinal connections between this descending circuit and areas such as the amygdala, hypothalamus, and anterior cingulate cortex are believed responsible for modulation of pain by cognitive-affective mechanisms (Bushnell, Ceko, & Low, 2013). Although this brain-to-spinal cord circuit has received the most attention, nociception modulation can occur via multiple paths, including completely supraspinal mechanisms (e.g., cortico-cortical circuits) (Piché, Arsenault, & Rainville, 2009). Thus, pain and nociception can be amplified by impairments in inhibitory processes, increases in facilitatory processes, or both.

There are several implications of pain mechanisms: (1) sensitization of peripheral, spinal, or supraspinal nociceptive circuitry can promote chronic pain, (2) ascending nociception is under dynamic control from inhibitory and facilitatory processes that can promote chronic pain (i.e., augmented facilitation, impaired inhibition), (3) modulatory effects can occur at multiple levels of the neuraxis, and (4) nociception may or may not result in pain; therefore, nociception and pain do not necessarily correlate. Given these, experimental human pain assessments (i.e., quantitative sensory tests, QST) have been designed to measure different aspects of these nociceptive processes.

The main distinction is between *static* and *dynamic* QST. Static QST provide a snapshot of pain/nociception at a given point in time. Examples include the application of noxious stimuli to assess the intensity/amount that first evokes pain (i.e., pain threshold) or that is maximally tolerable (i.e., pain tolerance). Another method is to apply a suprathreshold stimulus and then measure the evoked pain. Static measures reflect the final perceptual output of the ascending pain system at a specific time in the individual's life. But, different stimulus modalities and assessment methods can drive different aspects of the pain system (different nociceptors, tissues, etc.); therefore, responses to static measures do not correlate highly (e.g., Bhalang, Sigurdsson, Slade, & Maixner, 2005).

By contrast, dynamic QST assess how pain/nociception is modulated over time, and some evidence suggests they are better predictors of chronic pain risk than static measures (Yarnitsky, Granot, & Granovsky, 2014). Two of the most common are temporal summation of pain (TS-pain) and conditioned pain modulation (CPM). TS-pain involves the administration of a series of suprathreshold stimuli that are of the same intensity. When the interval between stimuli in the series is short (i.e., ≤ 3 s) this results in an increase (summation) in pain across the stimulus series, even though the stimulus intensity is kept constant. This summation is believed to be mediated by temporary hyperexcitability of nociceptive spinal neurons (i.e., windup in animal models) (Eide, 2000). Enhanced TS-pain (larger increases in pain across the stimulus series) is associated with central sensitization (hyperresponsive central nociceptive neurons, a risk factor for chronic pain) and chronic pain (Yarnitsky et al., 2014).

CPM involves the presentation of a phasic painful test stimulus prior to and during a tonic painful conditioning stimulus. The application of the conditioning stimulus generally evokes a reduction/inhibition of pain evoked by the test stimulus. This pain-inhibits-pain effect is believed to reflect descending, brain-to-spinal cord, inhibitory mechanisms, but may also involve cortico-cortical mechanisms that work in parallel (Piché et al., 2009). Impairments in CPM-related inhibition are associated with chronic pain (Yarnitsky et al., 2014).

Adversity and pain processing in humans

Studies examining the influence of adversity on human pain fall into three general categories: (1) studies that examine the relationship between ALEs and chronic/clinical pain, (2) studies that examine the relationship between posttraumatic stress disorder (PTSD; a psychological disorder stemming from exposure to certain types of ALEs) and pain, and (3) studies that examine the relationship between ALE exposure and pain processing in healthy, pain-free participants.

There are limitations to what can be learned from studying ALEs in chronic pain patients, because the effect of ALEs on pain processing may be obscured by disease processes. For example, central sensitization and hyperalgesia are commonly observed in chronic pain patients regardless of the etiology (e.g., ALEs, physical injury); thus, comparisons between pain patients with or without ALEs may not produce significant group differences if both groups are hyperalgesic.

Additionally, studies of the association between PTSD and pain processing focus on the psychological consequence of ALE exposure rather than on ALE exposure itself. Although psychological reactions to ALEs are important, they may not fully account for the ALEs-pain risk relationship. For example, numerous studies have shown that ALEs promote psychopathology, such as depression, anxiety, and PTSD; nonetheless, psychopathology only partially explains the impact of ALEs on pain (e.g., Sachs-Ericsson et al., 2017). Consistent with this notion, studies linking ALEs to pain risk show that the relationships remain even when psychological reactions are controlled (e.g., Sturycz et al., 2019; You et al., 2019). For these reasons, this review focuses on ALE exposure in humans who are currently pain-free so that any observed ALE-pain relationships are not confounded by disease status, health care inequities, or other consequences of having chronic pain. These studies are summarized in Table 1.

Fillingim and Edwards (2005) assessed heat and ischemia pain threshold and tolerance, as well as TS-pain to heat in college students with and without a history of childhood abuse. They found no group differences, except that those with a history of abuse reported *less* pain unpleasantness (hypoalgesia) during the TS-pain procedure, but no differences in summation. Granot et al. (2011) examined responses to static heat pain measures in women with and without a history of sexual abuse. Sexual abuse was associated with higher pain thresholds (hypoalgesia), higher ratings of painful stimuli (hyperalgesia), and a greater tendency to terminate tonic painful stimuli (hyperalgesia).

A series of studies conducted by Hellman and colleagues examined the impact of sexual assault history on pain processing in Native Americans and non-Hispanic White participants from the *Oklahoma Study of Native American Pain Risk (OK-SNAP)* (Note: ethnic differences were not found in any OK-SNAP results reviewed here). One study assessed individuals with a history of sexual assault, individuals with ALE exposure without sexual assault, and individuals without ALE exposure. Sexual assault history was associated with lower cold pain thresholds (hyperalgesia), increased sensitivity on cool detection threshold, and higher ischemia pain tolerances (hypoalgesia) (Hellman et al., 2019). ALE exposure (with and without sexual assault) was associated with lower cold pain tolerances (hyperalgesia). A second study examined TS-pain and used similar groups of participants (Hellman et al., 2019). Notably, the series of painful electric stimuli also assessed temporal summation of the nociceptive flexion reflex (TS-NFR). The NFR is a spinal reflex assessed from electromyography of the hamstring muscle in response to Aδ-fiber activation and is used as a physiological correlate of spinal nociception. The size of the reflex summates in response to the painful stimulus series, so TS-NFR is an objective measure

TABLE 1 Studies examining ALEs and pain mechanisms in healthy, pain-free participants.

Authors	Year	Participants	Sex	Adversity dose assessed?	Adversity assessment	Nonpain QST	Nonpain QST results	Static pain measures	Static results	Static implication	Dynamic pain measures	Dynamic results	Dynamic implication	Comments
Fillingim and Edwards	2005	College students: $N = 34$ (21 women and 13 men) with childhood abuse history, vs 76 without abuse history (35 women and 41 men)	M/F	N	Sexual-physical abuse history questionnaire assessed for childhood abuse	N/A	N/A	Heat pain threshold or tolerance, ischemia pain threshold or tolerance	No group diff on heat pain threshold, heat pain tolerance, ischemia pain threshold, ischemia pain tolerance	No effect of ALEs	TS-pain	Participants without abuse history showed increased pain unpleasantness ratings to TS	Hypoalgesia	Did not include a measure of ALE severity or dose
Granot, Somer, Zisman-Ilani, Beny, Sadger, Mirkin, … Yovell,	2011	All Caucasian: $N = 42$, 21 women with sexual abuse history undergoing treatment vs 21 healthy women matched by age, education, ethnicity	F	N	Not described for patients or controls	N/A	N/A	Heat pain threshold; ratings of phasic and tonic suprathreshold heat	SA group had higher pain thresholds; SA reported higher pain to phasic stimuli and were more likely to terminate tonic stimuli	Mixed: hypoalgesia on threshold; hyperalgesia on ratings	N/A	N/A	N/A	Not clear if controls were exposed to ALEs that were not sexual abuse, and all abuse survivors undergoing treatment
Hellman, Sturycz, Lannon, Ross, Toledo, Kuhn, … Rhudy	2019	Community participants from OK-SNAP: 47 with a history of sexual assault, 47 with a history of nonsexual assault ALE exposure (matched on the number of nonsexual assault ALEs and age), and 62 individuals without ALE exposure	M/F	N	Life events checklist—4; range 0–17	Cool detection threshold	Lower thresholds = more sensitive	Cold pain threshold or tolerance; ischemia pain threshold or tolerance	ALE exposure (SA and non-SA) associated with lower cold tolerances; SA associated with lower cold pain threshold and higher ischemia pain tolerance; no diffs on ischemia pain threshold	Hyperalgesia on cold measures; hypoalgesia on ischemia	N/A	N/A	N/A	No differences between Native Americans and non-Hispanic Whites were noted

Hellman, Sturycz, Huber, Lannon, Toledo, Kuhn, … Rhudy	2019	Community participants from OK-SNAP; 46 individuals with a history of sexual assault, 46 individuals with ALE exposure, but without a history of sexual assault (matched on the number of nonsexual assault ALEs and age), and 37 individuals without ALE exposure	M/F	N	Life events checklist—4; range 0–17	N/A	N/A	N/A	N/A	N/A	TS-electric pain; TS-nociceptive flexion reflex (TS-NFR)	No differences between groups	No effect of sexual assault or nonsexual assault ALE exposure	No differences between Native Americans and non-Hispanic Whites were noted
Hellman, Sturycz, Lannon, Kuhn, Güereca, Toledo, … Rhudy	2019	Community participants from OK-SNAP; 32 individuals with a sexual assault history, 32 individuals with ALE exposure but no sexual assault history (matched on age, race, sex, and mean number of nonsexual assault ALEs), and 40 individuals without ALE exposure	M/F	N	Life events checklist—4; range 0–17	N/A	N/A	Nociceptive flexion reflex (NFR) threshold	No differences between groups	No effect of ALES on static spinal nociception	CPM of pain; CPM of NFR	No group differences on CPM of pain; ALE exposure associated with impaired CPM of NFR	ALE exposure disrupts inhibition of spinal nociception	No differences between Native Americans and non-Hispanic Whites were noted

Continued

TABLE 1 Studies examining ALEs and pain mechanisms in healthy, pain-free participants—cont'd

Authors	Year	Participants	Sex	Adversity dose assessed?	Adversity assessment	Nonpain QST	Nonpain QST results	Static pain measures	Static results	Static implication	Dynamic pain measures	Dynamic results	Dynamic implication	Comments
Hellman, Kuhn, Lannon, Payne, Sturycz, Palit, … Rhudy	2018	Community participants from OK-SNAP; 33 individuals with a history of sexual assault and 33 matched individuals with no ALE exposure (matched on age, race, sex, and mean number of nonsexual assault ALEs)	M/F	N	Life events checklist—4; range 0–17	N/A	N/A	N/A	N/A	N/A	Emotional modulation of pain and NFR	No group diffs in pain modulation; SA associated with no emotional modulation of NFR	SA disrupts modulation of spinal nociception	No differences between Native Americans and non-Hispanic Whites were noted
You and Meagher	2016	College students: low group <5 ALEs ($n=75$) and high group 10+ ALEs ($n=51$) adversity	M/F	Y	Early trauma inventory self report-short form (11 items; 5 for emotional abuse, 5 physical abuse, 6 for sexual abuse; scores range from 0 to 27)	N/A	N/A	Heat pain threshold	No differences between groups	No effect	Temporal summation of pain	Greater TS-pain and slower decay in high adversity group	ALEs are associated with spinal sensitization	Type of adversity not related to outcomes or depression
You, Creech, and Meagher	2016	College women; $N=32$ healthy young women; individuals reporting no trauma or 1 SD above the mean on the trauma screen were sampled	F	Y	Childhood trauma events scale and recent traumatic events scale	N/A	N/A	N/A	N/A	N/A	Area of secondary hyperalgesia and pain intensity following topical capsaicin administration	Stressful life events predicted larger areas of secondary hyperalgesia ($b=0.40$, $SEb=1.7$, $P=.023$, $R2=0.16$), but did not predict the intensity of secondary hyperalgesia ($P=.771$)	ALEs are associated with spinal sensitization	

You and Meagher	2018	College students; male and females with low (<5 ALEs; $n = 31$) and high (>9 ALEs; $n = 31$) childhood adversity	M/F	Y	Early trauma inventory self report short form (range 0–27)	N/A	N/A	N/A	N/A	N/A	Area of secondary hyperalgesia and pain intensity following topical capsaicin administration; TS-heat pain	ALEs associated with greater secondary allodynia/ hyperalgesia; no effect of ALEs on TS-pain	ALEs associated with spinal sensitization	Secondary allodynia effect partially explained by PTSD symptoms
You, Creech, Vichaya, Young, Smith, and Meagher	2014	College women; 41 controls (21 no disclosure, 20 written emotional disclosure of ALE) vs 46 ALE history (18 no disclosure, 28 written emotional disclosure of ALE)	F	Y	Childhood trauma questionnaire and recent traumatic events scale	N/A	N/A	N/A	N/A	N/A	Area of secondary hyperalgesia following topical capsaicin administration	Did not report baseline differences, but disclosure reduced secondary hyperalgesia at 1-month postdisclosure	ALEs associated with spinal sensitization	Tested the effect of disclosure on secondary hyperalgesia in persons with ALEs
Sturycz, Hellman, Payne, Kuhn, Hahn, Lannon, … Rhudy	2019	Community participants from OK-SNAP; $N = 246$ men ($n = 111$) and women ($n = 135$)	M/F	Y	Life events checklist—4; range 0–17	N/A	N/A	Nociceptive flexion reflex (NFR) threshold	No association between ALEs and NFR threshold	No effect of ALES on static spinal nociception	TS-pain; TS-NFR	No association between ALEs and TS-pain; ALES associated with higher TS-NFR; ALEs associated with higher overall pain ratings	ALEs associated with spinal sensitization and hyperalgesia	Analyses controlled for psychological stress, psychological symptoms, and negative and positive affect; No differences between Native Americans and non-Hispanic Whites were noted

Continued

TABLE 1 Studies examining ALEs and pain mechanisms in healthy, pain-free participants—cont'd

Authors	Year	Participants	Sex	Adversity dose assessed?	Adversity assessment	Nonpain QST	Nonpain QST results	Static pain measures	Static results	Static implication	Dynamic pain measures	Dynamic results	Dynamic implication	Comments
Kell, Hellman, Huber, Lannon, Kuhn, Sturycz, … Rhudy	2021	Community participants from OK-SNAP; $N=246$ men ($n=120$) and women ($n=126$)	M/F	Y	Life events checklist—4; range 0–17	N/A	N/A	N/A	N/A	N/A	CPM of pain; CPM of NFR	No association between ALEs and CPM-pain; ALES associated with impaired CPM-NFR; ALEs not associated with overall pain ratings	ALEs associated with spinal sensitization	Analyses controlled for general health, body mass index, sex, age, blood pressure, sleep quality, psychological stress, psychological symptoms; No differences between Native Americans and non-Hispanic Whites were noted

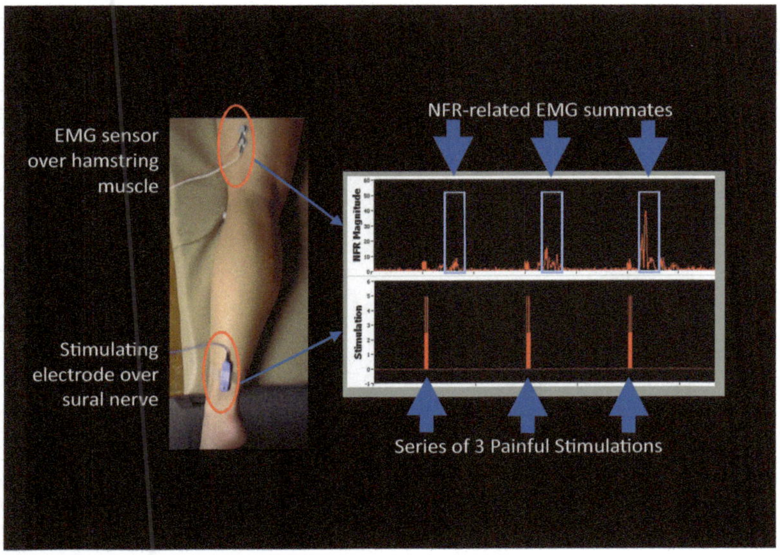

FIG. 1 Temporal summation of the nociceptive flexion reflex (TS-NFR). A series of painful stimulations of the same intensity are delivered over the sural nerve and the NFR size is measured from EMG of hamstring muscle. The stimulus series leads to an increase in size (summation) of the NFR. Greater summation = greater spinal sensitization.

of spinal sensitization (see Fig. 1). No group differences on either TS-pain or TS-NFR were found, suggesting that sexual assault and nonsexual assault ALEs are not associated with spinal sensitization. A third study examined the impact of CPM in similar groups (Hellman et al., 2019). Both pain and NFR were assessed to examine inhibition of pain and spinal nociception, respectively. NFR inhibition was impaired in both ALE exposed groups (with and without sexual assault), but pain inhibition did not differ between groups. Moreover, a static measure of NFR (i.e., NFR threshold) was also not different between groups. This study indicates that descending, brain-to-spinal cord inhibitory circuity was impaired by ALE exposure, but that supraspinal (e.g., cortico-cortical) pain inhibitory mechanisms were still intact to inhibit the perception of pain. Finally, a fourth study assessed emotional modulation of pain and NFR (a unique dynamic measure) in individuals with a history of sexual assault and matched individuals with no ALE exposure (Hellman et al., 2018). Pleasant, neutral, and unpleasant pictures were presented during which painful electric stimulation were delivered to evoke pain and the NFR. This paradigm generally results in inhibited pain/NFR during pleasant pictures and enhanced pain/NFR during unpleasant pictures. They found that emotions modulated pain in both groups, but that emotional modulation of NFR was absent in the group with a sexual assault history. This was interpreted to mean that sexual assault impairs the brain-to-spinal cord modulation circuitry.

A series of studies by You and colleagues examined the dose-dependent effect of ALEs. Instead of categorizing people into groups based on a type of ALE (e.g., sexual assault), or the presence or absence of ALES, the authors assessed the impact of the number of childhood ALEs (similar to ACEs research). You and Meagher (2016) studied college students with low and high exposure to ALEs and found no group differences in heat pain thresholds, but greater TS-pain (spinal sensitization). This effect remained after controlling for depression. In a follow-up study, You, Creech, and Meagher (2016) used a novel dynamic measure to assess spinal sensitization in women. Capsaicin is a chemical irritant found in chili peppers that, when applied to the skin, evokes burning pain and inflammation. Sensitization in the skin where capsaicin is applied is due to sensitization of peripheral nociceptors, but sensitization in the surrounding skin reflects expansion of receptive fields in spinal neurons (referred to as secondary hyperalgesia/allodynia; see Fig. 2). Thus, assessing the *area* of secondary hyperalgesia assesses an inflammatory marker of spinal sensitization. The authors found that higher exposure to ALEs was associated with a larger area of secondary hyperalgesia, but not with greater pain intensity in the area of secondary hyperalgesia. This suggests a dose-dependent effect of adversity on spinal sensitization without a concomitant increase in pain. This result was replicated in a sample of men and women, but they did not replicate their previous finding of enhanced TS-pain (You & Meagher, 2018). Interestingly, this same group demonstrated that the expanded area of secondary hyperalgesia can be reversed by a brief (20 min), 3-session emotional disclosure intervention in which participants wrote about their most stressful experience (You et al., 2014).

Considering these dose-dependent effects of ALEs, we reanalyzed OK-SNAP data to examine the influence of ALE exposure on dynamic pain measures. In the first study, Sturycz et al. (2019) found that the number of ALEs was significantly related to greater TS-NFR (physiological correlate of spinal sensitization), but not TS-pain (Fig. 3). These findings

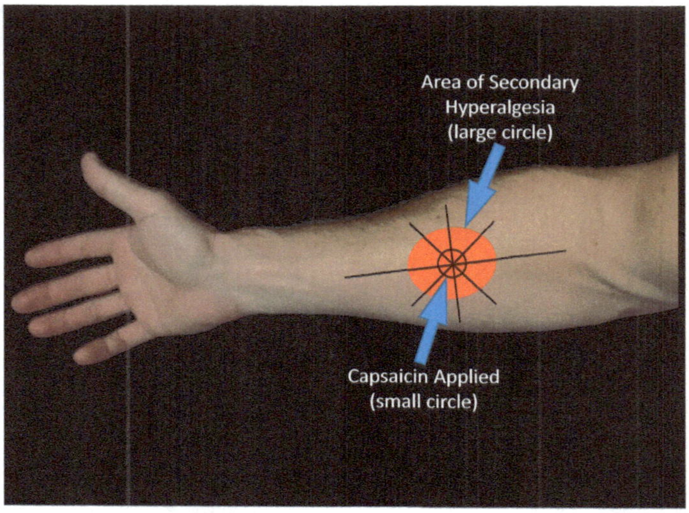

FIG. 2 Capsaicin-related secondary hyperalgesia. Capsaicin is applied to a small area of the forearm *(small circle)*. After several minutes, capsaicin is removed and then the area of secondary hyperalgesia is assessed by mechanical stimulation of the skin using a monofilament *(large circle)* along lines radiating from the center of capsaicin application. Larger area = greater spinal sensitization.

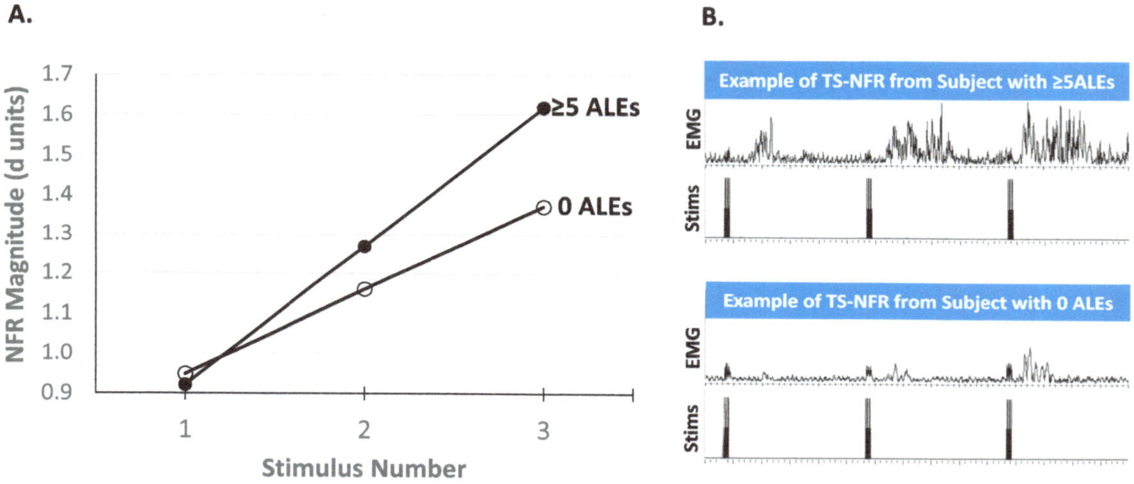

FIG. 3 ALEs and TS-NFR. Greater ALE exposure is associated with greater summation of NFR suggesting greater spinal sensitization (Panel A). Examples of TS-NFR EMG from a subject with 0 ALEs and a subject with ≥5 ALEs (Panel B). *(Adapted from Sturycz, C. A., Hellman, N., Payne, M. F., Kuhn, B. L., Hahn, B., Lannon, E. W., … Rhudy, J. L. (2019). Race/ethnicity does not moderate the relationship between adverse life experiences and temporal summation of the nociceptive flexion reflex and pain: Results from the Oklahoma Study of Native American Pain Risk.* Journal of Pain, 20, *941–955, with permission.)*

were not due to psychological stress, psychological symptoms, or affective state. The second study by Kell et al. (2020) examined CPM of pain and NFR and found that ALEs were related to impairment of NFR inhibition. Specifically, intact NFR inhibition was observed in those with 0–1 ALE, no NFR inhibition was observed in those with 2–4 ALEs, and NFR was facilitated in those with ≥5 ALEs. By contrast, there was no relationship between ALEs and pain inhibition. These findings will be discussed in the next section.

Adversity, latent sensitization, and pain risk

Fig. 4 summarizes the effects of ALEs on pain processing in pain-free individuals. Results from studies on pain-free individuals suggest exposure to ALEs is associated with spinal sensitization as assessed by several dynamic measures (TS-NFR, secondary hyperalgesia). Results are even more consistent when the dose of ALEs is considered. Further, evidence from Kell et al. implies that descending inhibition of spinal nociception is impaired, even transitioning from spinal

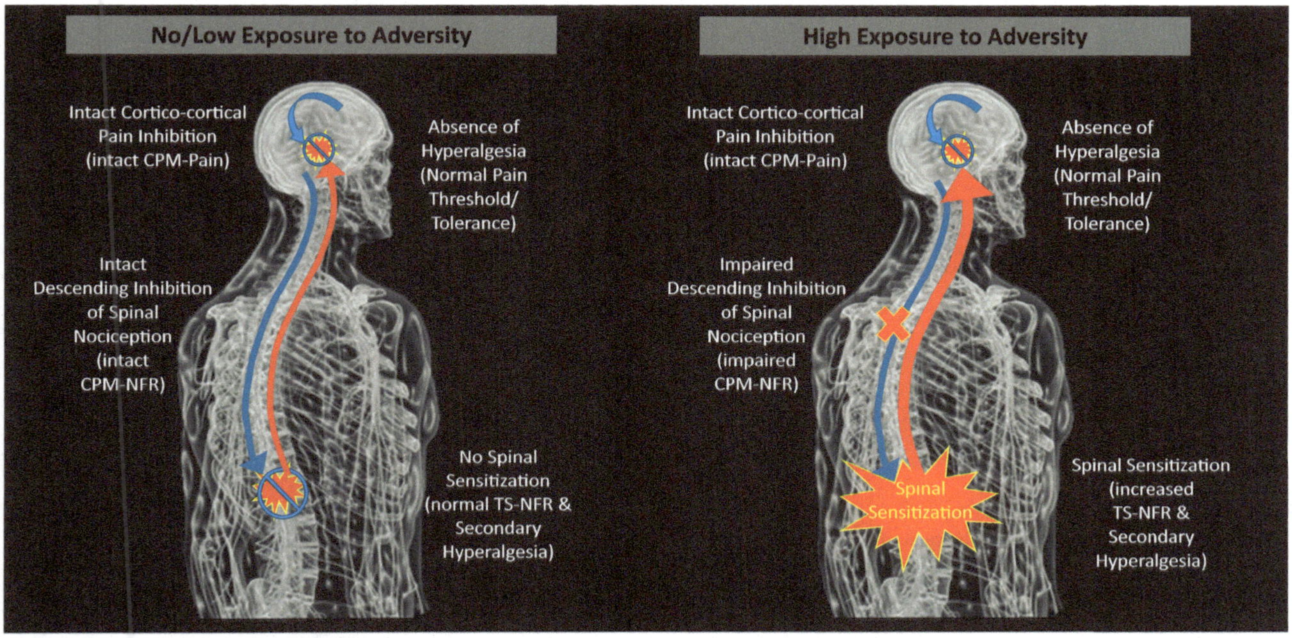

FIG. 4 Model linking ALEs and latent sensitization. Under conditions of low ALE exposure, there is no spinal sensitization and inhibitory mechanisms are intact, so no hyperalgesia is present. However, cumulative exposure to ALEs promotes spinal sensitization (increased TS-NFR/secondary hyperalgesia, impaired inhibition of spinal nociception), but spinal sensitization does not result in in hyperalgesia because pain inhibition keeps amplified spinal nociception suppressed (i.e., latent sensitization).

inhibition to spinal facilitation following high levels of ALE exposure. At the same time, mechanisms of pain inhibition (perhaps cortico-cortical circuits) are not impaired by ALEs. This could be a compensatory mechanism that dampens amplified spinal nociception thus preventing hyperalgesia from being reliably detected on static measures of pain perception (e.g., threshold, tolerance). The lack of effect on pain inhibition (CPM of pain) or static measures suggest that spinal sensitization may be an early sign of chronic pain risk.

This ALE-related pain phenotype is similar to the phenomenon of "latent sensitization" observed in animals (Taylor & Corder, 2014). Latent sensitization occurs following an injury/stressor and represents amplification of spinal nociception that outlasts the injury/stressor without concurrent evidence of hyperalgesia. The absence of hyperalgesia is due to compensatory mechanisms that inhibit pain (Corder et al., 2013; Taylor & Corder, 2014). But, when the inhibitory mechanisms are antagonized (e.g., opioid blockade), this unmasks the latent spinal sensitization and produces hyperalgesia (Corder et al., 2013; Le Roy et al., 2011). For example, Corder et al. (2013) injected the paws of mice with a chemical irritant that produces an inflammatory injury and hyperalgesia that remits after 10 days. However, administration of naltrexone (opioid antagonist) after day 10 blocked endogenous opioid inhibition and reinstated hyperalgesia, thus unveiling the latent sensitization. Sham-injured mice did not show hyperalgesia following naltrexone. Importantly, latent sensitization can be observed for months following the initial injury. Moreover, descending brain-to-spinal cord facilitation may promote the latent sensitization (Taylor & Corder, 2014), although peripheral mechanisms may also contribute (Alvarez et al., 2013). Thus, exposure to ALEs may increase risk for chronic pain by producing a latent spinal sensitization that is masked by compensatory inhibitory mechanisms.

Implications and future directions

Fig. 5 depicts a hypothetical model for the relationships between ALEs, latent sensitization, and chronic pain risk. If correct, there are several issues that require investigation. First, it is not clear that ALEs lead to spinal sensitization. One possibility is that ALEs elicit an inflammatory response that triggers latent sensitization that is worsened by subsequent ALEs. Indeed, ALE exposure is associated with increased inflammation (Baumeister, Akhtar, Ciufolini, Pariante, & Mondelli, 2016) in a dose-dependent manner (Hostinar, Lachman, Mroczek, Seeman, & Miller, 2015). Additionally, as noted previously, inflammatory responses partially mediate some forms of ELS-induced hyperalgesia in rodents (Alvarez et al., 2013).

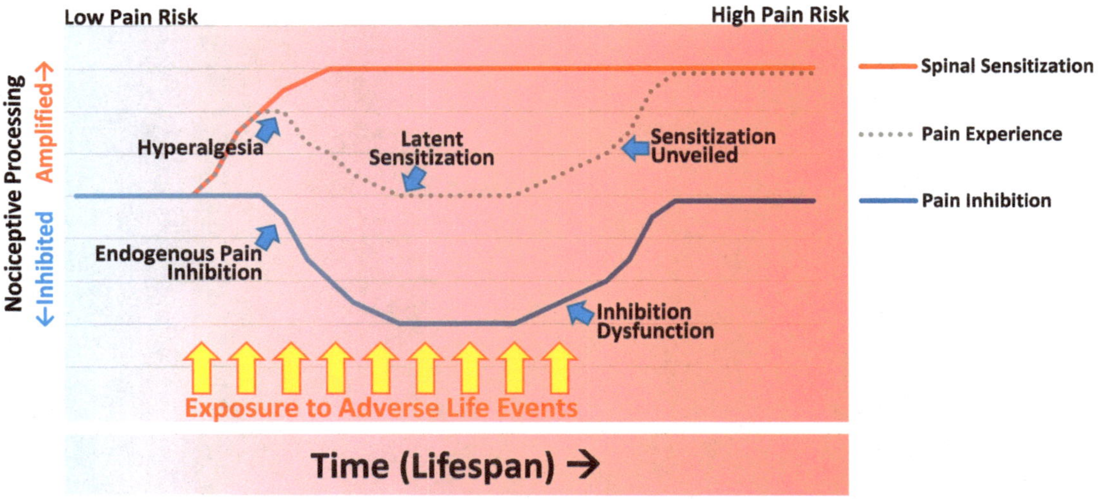

FIG. 5 Hypothetical time course for the effect of ALEs on latent sensitization and chronic pain risk. ALE exposure leads to spinal sensitization and a temporary hyperalgesia that is offset by compensatory endogenous pain inhibition that suppresses hyperalgesia and makes spinal sensitization latent. However, a dysfunction of inhibitory processes, perhaps by additional ALE exposure, fatigue of inhibitory systems, or advanced age, may lead to hyperalgesia and chronic pain.

Moreover, a study of >14,000 participants found that C-reactive protein (indicator of systemic inflammation) was the stress-related process most strongly associated with pain risk (Slade, Sanders, & By, 2012).

It is also important to identify what causes the transition from latent sensitization to pain chronification. One possibility is that subsequent exposure to ALEs eventually impairs pain inhibition, unveils the latent sensitization, and promotes hyperalgesia. Consistent with this, chronic pain episodes can be triggered by stress (e.g., Larauche, Mulak, & Taché, 2012), hyperalgesia following latent sensitization can be reinstated by stressors (Le Roy et al., 2011), and ALE exposure can diminish endogenous opioid activity (Lovallo, Acheson, Vincent, Sorocco, & Cohoon, 2018). Moreover, a recent study found that corticotropin-releasing hormone (stress hormone released by the hypothalamus) unveiled latent sensitization and produced hyperalgesia in rodents (Chen, Taché, & Marvizón, 2018).

Alternatively, since compensatory inhibitory mechanisms normally keep spinal sensitization latent by suppressing hyperalgesia, eventually these inhibitory mechanisms may fatigue or become dysfunctional. This could explain why CPM pain inhibition declines with age (Hackett, Naugle, & Naugle, 2019), perhaps a result of wear-and-tear across the lifespan. Moreover, if pain inhibition is the last bastion against pain chronification, this could explain why impaired CPM pain inhibition assessed prior to surgery is one of the best prospective predictors of postsurgical chronic pain (Yarnitsky et al., 2014).

It is also interesting to consider the potential implications for drug misuse. If compensatory inhibitory mechanisms start to fail, this could motivate the use of exogenous substances (e.g., opiates) to counteract the gradual unveiling of latent sensitization. This means latent sensitization could identify persons at high risk for chronic pain and substance misuse.

And finally, if ALEs promote latent sensitization, research is needed to identify ways of reversing it to avoid pain chronification. As noted earlier, You et al. (2014) found that writing about ALEs for 20 min a day over three sessions was enough to reduce spinal sensitization. Similarly, another study found that a 5 min, 8-session cognitive-behavioral intervention could reduce spinal sensitization (Salomons, Moayedi, Erpelding, & Davis, 2014). Further, our lab and others have shown that relaxation interventions can be used to increase inhibition of spinal nociception (Emery et al., 2006; Rhudy, Hellman, Sturycz, Toledo, & Palit, 2020). Thus, these interventions could be important starting points, either together or in isolation, to stop the transition to chronic pain.

Summary

Evidence supports an accumulated stress hypothesis (i.e., dose-dependent effect) that promotes latent spinal sensitization that is kept in check by compensatory pain inhibitory mechanisms that suppress hyperalgesia. With subsequent exposure to ALEs, these compensatory mechanisms may fail, ultimately leading to hyperalgesia and pain chronicity. However, this hypothesis requires testing by prospective, longitudinal studies of pain-free individuals that assess the relationship between ALEs and dynamic measures of pain and spinal nociception over time.

Applications to other areas

Evidence suggests that adverse life events (ALEs) lead to spinal sensitization that is not reliably accompanied by hyperalgesia as measured by perceptual measures of pain. The lack of hyperalgesia may stem from compensatory pain inhibitory mechanisms (e.g., endogenous opioids) that keep amplified spinal nociception from being perceived within supraspinal regions. As a result, persons with high ALE exposure may be at risk for developing chronic pain if those compensatory mechanisms fail. This has implications for populations at higher risk of exposure to adversity (e.g., those residing in lower socioeconomic neighborhoods). It could also have implications for analgesic effectiveness in ALE exposes populations. For example, if the endogenous opioid system is actively engaged to suppress spinal sensitization, these individuals may not respond as well to exogenously administered opioids.

Other agents of interest

Not applicable.

Mini-dictionary of terms

- Adverse life events (ALEs)—major stressors that can occur throughout the lifespan and pose a significant physical/psychological threat to safety and/or security. In studies of early life, these are called adverse childhood experience (ACEs) or early life stress (ELS).
- Central sensitization—a persistent state of hyperresponsivity of central nociceptive neurons that initiate and maintain chronic pain.
- Dynamic QST Measures—measures of pain/nociception that assess the ability to regulate nociception/pain (e.g., conditioned pain modulation, temporal summation).
- Latent Sensitization—sensitization of nociceptive neurons that does not result in hyperalgesia due to compensatory inhibition mechanisms that block pain amplification
- Nociceptive Flexion Reflex (NFR)—a pain-related spinal reflex (e.g., reflexive foot withdrawal from a sharp tack) that assesses spinal nociception.
- Secondary Hyperalgesia—pain in tissue surrounding an injury that is due to expansion of spinal neuron receptive fields. Area of the secondary hyperalgesia is used to assess the spinal nociception.
- Static QST Measures—measures of pain/nociception that assess pain at a single moment in time (e.g., pain threshold/tolerance).

Key facts about adversity

- ALEs include many types of threatening events, such as violence, bereavement, relationship threats, social/environmental stress, natural/manmade disasters, and illness/injury to oneself or loved ones (Seery, Holman, & Silver, 2010).
- On average, people in the United States experience 7.69 ALEs in their lifetime (Seery et al., 2010).
- ALEs are associated with risky health behaviors (e.g., smoking, drug use) but this does not fully explain the relationship between ALEs and health (Hughes et al., 2017).
- Persons who experience ALEs have an increased risk of developing chronic pain (You et al., 2019).
- Psychological interventions can effectively treat ALE-related pain and distress (Bosco, Gallinati, & Clark, 2013).

Summary points

- Adverse life events (ALEs) are major stressors that pose a psychological/physical threat to safety and/or security.
- As the number of ALEs increases, so does the risk for negative health outcomes, including chronic pain.
- Persons exposed to more ALEs (dose-dependent) show greater spinal sensitization.
- ALE-related spinal sensitization may not produce hyperalgesia due to compensatory pain inhibitory processes that suppress it; therefore, sensitization is latent.
- Latent sensitization may ultimately lead to chronic pain if endogenous inhibitory processes fail; thus, latent sensitization may be an early sign of chronic pain risk.

Conflict of interest

The authors report no conflicts of interest.

References

Afari, N., Ahumada, S. M., Wright, L. J., Mostoufi, S., Golnari, G., Reis, V., & Cuneo, J. G. (2014). Psychological trauma and functional somatic syndromes: A systematic review and meta-analysis. *Psychosomatic Medicine*, *76*(1), 2–11. https://doi.org/10.1097/PSY.0000000000000010.

Alvarez, P., Green, P. G., & Levine, J. D. (2013). Stress in the adult rat exacerbates muscle pain induced by early-life stress. *Biological Psychiatry*, *74*(9), 688–695.

Apkarian, A. V., Bushnell, M. C., Treede, R.-D., & Zubieta, J.-K. (2005). Human brain mechanisms of pain perception and regulation in health and disease. *European Journal of Pain*, *9*(4), 463–484.

Arcaya, M. C., Lowe, S. R., Asad, A. L., Subramanian, S., Waters, M. C., & Rhodes, J. (2017). Association of posttraumatic stress disorder symptoms with migraine and headache after a natural disaster. *Health Psychology*, *36*(5), 411.

Baram, T. Z., Davis, E. P., Obenaus, A., Sandman, C. A., Small, S. L., Solodkin, A., & Stern, H. (2012). Fragmentation and unpredictability of early-life experience in mental disorders. *American Journal of Psychiatry*, *169*(9), 907–915.

Baumeister, D., Akhtar, R., Ciufolini, S., Pariante, C. M., & Mondelli, V. (2016). Childhood trauma and adulthood inflammation: A meta-analysis of peripheral C-reactive protein, interleukin-6 and tumour necrosis factor-α. *Molecular Psychiatry*, *21*(5), 642–649.

Bhalang, K., Sigurdsson, A., Slade, G. D., & Maixner, W. (2005). Associations among four modalities of experimental pain in women. *The Journal of Pain*, *6*(9), 604.

Bosco, M. A., Gallinati, J. L., & Clark, M. E. (2013). Conceptualizing and treating comorbid chronic pain and PTSD. *Pain Research and Treatment*, *2013*, 1–10.

Burke, N. N., Finn, D. P., McGuire, B. E., & Roche, M. (2017). Psychological stress in early life as a predisposing factor for the development of chronic pain: Clinical and preclinical evidence and neurobiological mechanisms. *Journal of Neuroscience Research*, *95*(6), 1257–1270.

Bushnell, M. C., Ceko, M., & Low, L. A. (2013). Cognitive and emotional control of pain and its disruption in chronic pain. *Nature Reviews. Neuroscience*, *14*(7), 502–511. https://doi.org/10.1038/nrn3516.

Chen, W., Taché, Y., & Marvizón, J. C. (2018). Corticotropin-releasing factor in the brain and blocking spinal descending signals induce hyperalgesia in the latent sensitization model of chronic pain. *Neuroscience*, *381*, 149–158.

Corder, G., Doolen, S., Donahue, R. R., Winter, M. K., Jutras, B. L., He, Y., … Storm, D. (2013). Constitutive μ-opioid receptor activity leads to long-term endogenous analgesia and dependence. *Science*, *341*(6152), 1394–1399.

Eide, P. K. (2000). Wind-up and the NMDA receptor complex from a clinical perspective. *European Journal of Pain*, *4*(1), 5–15. https://doi.org/10.1053/eujp.1999.0154.

Emery, C. F., Keefe, F. J., France, C. R., Affleck, G., Waters, S., Fondow, M. D., … Stainbrook, D. (2006). Effects of a brief coping skills training intervention on nociceptive flexion reflex threshold in patients having osteoarthritic knee pain: A preliminary laboratory study of sex differences. *Journal of Pain and Symptom Management*, *31*(3), 262–269.

Felitti, V. J., Anda, R. F., Nordenberg, D., Williamson, D. F., Spitz, A. M., Edwards, V., … Marks, J. S. (1998). Relationship of childhood abuse and household dysfunction to many of the leading causes of death in adults. The adverse childhood experiences (ACE) study. *American Journal of Preventive Medicine*, *14*(4), 245–258. https://doi.org/10.1016/s0749-3797(98)00017-8.

Fields, H. L., Basbaum, A. I., & Heinricher, M. M. (2006). Central nervous system mechanisms of pain modulation. In S. B. McMahon, & M. Koltzenburg (Eds.), *Textbook of pain* (5th ed., pp. 125–143). Philadelphia: Elsevier/Churchill Livingstone.

Fillingim, R. B., & Edwards, R. R. (2005). Is self-reported childhood abuse history associated with pain perception among healthy Young women and men? *Clinical Journal of Pain*, *21*(5), 387–397.

Granot, M., Somer, E., Zisman-Ilani, Y., Beny, A., Sadger, R., Mirkin, R., … Yovell, Y. (2011). Characteristics of response to experimental pain in sexually abused women. *The Clinical Journal of Pain*, *27*(7), 616–622.

Hackett, J., Naugle, K. E., & Naugle, K. M. (2019). The decline of endogenous pain modulation with aging: A meta-analysis of temporal summation and conditioned pain modulation. *The Journal of Pain*. https://doi.org/10.1016/j.jpain.2019.09.005.

Hellman, N., Kuhn, B. L., Lannon, E. W., Payne, M. F., Sturycz, C. A., Palit, S., … Rhudy, J. L. (2018). Emotional modulation of pain and spinal nociception in sexual assault survivors. *Psychosomatic Medicine*, *80*(9), 861–868.

Hellman, N., Sturycz, C. A., Huber, F. A., Lannon, E. W., Toledo, T. A., Kuhn, B. L., … Rhudy, J. L. (2019). Do sexual assault survivors exhibit spinal cord (central) sensitization of incoming pain signals? In *Paper presented at the International Society for Traumatic Stress Studies, Boston, MA*.

Hellman, N., Sturycz, C. A., Lannon, E. W., Kuhn, B. L., Güereca, Y. M., Toledo, T. A., … Rhudy, J. L. (2019). Conditioned pain modulation in sexual assault survivors. *The Journal of Pain*, *20*(9), 1027–1039.

Hellman, N., Sturycz, C. A., Lannon, E. W., Ross, E. N., Toledo, T. A., Kuhn, B. L., … Rhudy, J. L. (2019). Pain processing in sexual assault survivors. In *Paper presented at the European Society for Traumatic Stress Studies, Rotterdam, The Netherlands*.

Hostinar, C. E., Lachman, M. E., Mroczek, D. K., Seeman, T. E., & Miller, G. E. (2015). Additive contributions of childhood adversity and recent stressors to inflammation at midlife: Findings from the MIDUS study. *Developmental Psychology*, *51*(11), 1630.

Hughes, K., Bellis, M. A., Hardcastle, K. A., Sethi, D., Butchart, A., Mikton, C., … Dunne, M. P. (2017). The effect of multiple adverse childhood experiences on health: A systematic review and meta-analysis. *The Lancet Public Health*, *2*(8), e356–e366.

Jennings, E. M., Okine, B. N., Roche, M., & Finn, D. P. (2014). Stress-induced hyperalgesia. *Progress in Neurobiology*, *121*, 1–18.

Jones, G. T., Power, C., & Macfarlane, G. J. (2009). Adverse events in childhood and chronic widespread pain in adult life: Results from the 1958 British birth cohort study. *Pain*, *143*(1–2), 92–96. https://doi.org/10.1016/j.pain.2009.02.003.

Kell, P. A., Hellman, N., Huber, F. A., Lannon, E. W., Kuhn, B. L., Sturycz, C. A., … Rhudy, J.L. (2021). The relationship between adverse life events and endogenous inhibition of pain and spinal nociception: Findings from the Oklahoma Study of Native American Pain Risk (OK-SNAP). *Journal of Pain*, *S1526-5900*(21), 00187. https://doi.org/10.1016/j.jpain.2021.03.146. 00187-5. (Online ahead of print).

Larauche, M., Mulak, A., & Taché, Y. (2012). Stress and visceral pain: From animal models to clinical therapies. *Experimental Neurology*, *233*(1), 49–67.

Le Roy, C., Laboureyras, E., Gavello-Baudy, S., Chateauraynaud, J., Laulin, J.-P., & Simonnet, G. (2011). Endogenous opioids released during non-nociceptive environmental stress induce latent pain sensitization via a NMDA-dependent process. *The Journal of Pain*, *12*(10), 1069–1079.

Lovallo, W. R., Acheson, A., Vincent, A. S., Sorocco, K. H., & Cohoon, A. J. (2018). Early life adversity diminishes the cortisol response to opioid blockade in women: Studies from the family health patterns project. *PLoS One*, *13*(10), e0205723.

Piché, M., Arsenault, M., & Rainville, P. (2009). Cerebral and cerebrospinal processes underlying counter irritation analgesia. *Journal of Neuroscience*, *29* (45), 14236–14246.

Rhudy, J. L., Hellman, N., Sturycz, C. A., Toledo, T. A., & Palit, S. (2020). Modified biofeedback (Conditioned Biofeedback) promotes anti-nociception by increasing the nociceptive flexion reflex threshold and reducing temporal summation of pain: A controlled trial. *The Journal of Pain*, *21*, 663–676. PMID: 31683023.

Sachs-Ericsson, N. J., Sheffler, J. L., Stanley, I. H., Piazza, J. R., & Preacher, K. J. (2017). When emotional pain becomes physical: Adverse childhood experiences, pain, and the role of mood and anxiety disorders. *Journal of Clinical Psychology*, *73*(10), 1403–1428. https://doi.org/10.1002/jclp.22444.

Salomons, T. V., Moayedi, M., Erpelding, N., & Davis, K. D. (2014). A brief cognitive-behavioural intervention for pain reduces secondary hyperalgesia. *Pain*, *155*(8), 1446–1452.

Seery, M. D., Holman, E. A., & Silver, R. C. (2010). Whatever does not kill us: Cumulative lifetime adversity, vulnerability, and resilience. *Journal of Personality and Social Psychology*, *99*(6), 1025.

Slade, G. D., Sanders, A. E., & By, K. (2012). Role of allostatic load in sociodemographic patterns of pain prevalence in the US population. *The Journal of Pain*, *13*(7), 666–675.

Sledjeski, E. M., Speisman, B., & Dierker, L. C. (2008). Does number of lifetime traumas explain the relationship between PTSD and chronic medical conditions? Answers from the National Comorbidity Survey-Replication (NCS-R). *Journal of Behavioral Medicine*, *31*(4), 341–349.

Sturycz, C. A., Hellman, N., Payne, M. F., Kuhn, B. L., Hahn, B., Lannon, E. W., … Rhudy, J.L. (2019). Race/ethnicity does not moderate the relationship between adverse life experiences and temporal summation of the nociceptive flexion reflex and pain: Results from the Oklahoma Study of Native American Pain Risk. *Journal of Pain*, *20*, 941–955.

Taylor, B. K., & Corder, G. (2014). Endogenous analgesia, dependence, and latent pain sensitization. In *Behavioral neurobiology of chronic pain* (pp. 283–325). Springer.

Tesarz, J., Gerhardt, A., Leisner, S., Janke, S., Treede, R.-D. D., & Eich, W. (2015). Distinct quantitative sensory testing profiles in nonspecific chronic back pain subjects with and without psychological trauma. *Pain*, *156*(4), 577–586. https://doi.org/10.1097/01.j.pain.0000460350.30707.8d.

Yarnitsky, D., Granot, M., & Granovsky, Y. (2014). Pain modulation profile and pain therapy: Between pro-and antinociception. *Pain*, *155*(4), 663–665.

You, D. S., Albu, S., Lisenbardt, H., & Meagher, M. W. (2019). Cumulative childhood adversity as a risk factor for common chronic pain conditions in young adults. *Pain Medicine*, *20*(3), 486–494.

You, D. S., Creech, S. K., & Meagher, M. W. (2016). Enhanced area of secondary hyperalgesia in women with multiple stressful life events: A pilot study. *Pain Medicine*, *17*(10), 1859–1864.

You, D. S., Creech, S. K., Vichaya, E. G., Young, E. E., Smith, J. S., & Meagher, M. W. (2014). Effect of written emotional disclosure on secondary hyperalgesia in women with trauma history. *Psychosomatic Medicine*, *76*(5), 337–346.

You, D. S., & Meagher, M. W. (2016). Childhood adversity and pain sensitization. *Psychosomatic Medicine*, *78*(9), 1084–1093.

You, D. S., & Meagher, M. W. (2018). Childhood adversity and pain facilitation. *Psychosomatic Medicine*, *80*(9), 869–879.

Chapter 33

Cognitive-affective modulation of pain: The placebo and nocebo phenomena and their impact on pain treatment☆

Sergiu Albu[a], Hans Linsenbardt[b,c], and Mary W. Meagher[d]

[a]*Guttmann Brain Health Institute, Barcelona, Spain,* [b]*Battlefield Pain Management, U.S. Army Institute of Surgical Research, San Antonio, TX, United States,* [c]*Oak Ridge Institute for Science and Education, Oak Ridge, TN, United States,* [d]*Department of Psychology, Texas A&M University, College Station, TX, United States*

Abbreviations

ACC	anterior cingulate cortex
ACT	acceptance and commitment therapy
AMYG	amygdala
CBT	cognitive behavioral therapy
CCK	cholecystokinin
CMA	cingulate motor area
DA	dopamine
DLPFC	dorsolateral prefrontal cortex
INS	insula
NAc	nucleus accumbens
OFC	orbitofrontal cortex
PAG	periaqueductal gray matter
PFC	prefrontal cortex
S1	primary somatosensory cortex
S2	secondary somatosensory cortex
THAL-VL	ventrolateral thalamus

Introduction

Affective and cognitive processes play a significant role in pain modulation, which, along with sensory processes, determine the global experience of pain (Fig. 1). Therefore, the wide variability in pain experience observed between individuals is partly explained by cognitive and affective modulation of incoming nociceptive information.

An extensive body of research has elucidated distinct neural pathways that encode the multidimensional experience of pain. Neuroimaging studies of the "neurologic signature" of pain have identified an extensive brain network that is activated in response to noxious stimulation. This includes brain areas "specific" to pain [the ventrolateral thalamus (THAL-VL), the secondary somatosensory cortex (S2) and the dorsal posterior insula (dpINS)], as well as "nonspecific" areas [the anterior insula (aINS) and the anterior cingulate cortex (ACC)], which are activated by many psychological processes (Wager et al., 2013). For example, expectations can modulate subjective and neural responses to pain in nociceptive-specific portions of INS, ACC, and THAL. In addition, pain expectancy effects involve responses in "nonspecific" areas that are less reliably activated by noxious stimulation, including the dorsolateral prefrontal cortex (DLPFC) and the orbitofrontal cortex (OFC) (Atlas & Wager, 2012). Furthermore, research in primates has revealed that most spinothalamic fibers project into the INS (41%), the S2 cortex (29%), and the cingulate motor area (CMA) (24%), and only 4% of the

☆ The views expressed in this book chapter are those of the author(s) and do not reflect the official policy or position of the U.S. Army Medical Department, Department of the Army, DoD, or the U.S. Government.

The Neurobiology, Physiology, and Psychology of Pain. https://doi.org/10.1016/B978-0-12-820589-1.00033-6

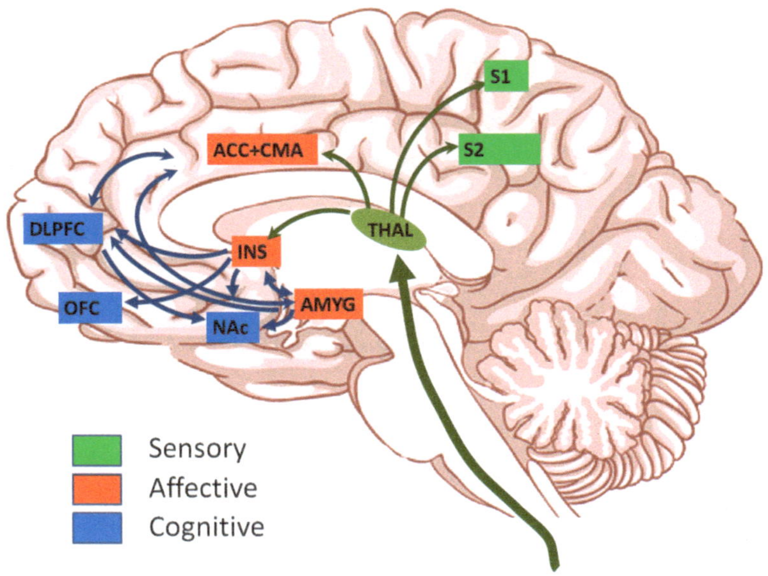

FIG. 1 Key cortical targets of the spinothalamic tract and brain structures involved in supraspinal pain processing. *ACC*, anterior cingulate cortex; *AMYG*, amygdala; *CMA*, cingulate motor area; *DLPFC*, dorsolateral prefrontal cortex; *NAc*, nucleus accumbens; *OFC*, orbitofrontal cortex; *THAL*, thalamus; *S1*, primary somatosensory cortex; *S2*, secondary somatosensory cortex.

fibers project into the primary somatosensory cortex (S1) (Dum, Levinthal, & Strick, 2009). This is not surprising as the main role of acute pain is to detect noxious stimuli and trigger reflexive motor withdrawal movements from the noxious stimulus whereas pain-induced negative affect promotes avoidance learning and defensive behaviors. Although the aINS and the ACC have been commonly considered nonnociceptive brain structures, their contribution to the construction of the affective dimension of pain and pain-related behavior is fundamental. Their role in avoidance learning and defensive behaviors is closely related to the activity in the prefrontal cortex (PFC), the hippocampus, and the nucleus accumbens (NAc), which perform pain-related attention, valuation, and emotion regulation.

Descending pain modulation systems also provide important pathways through which emotional and cognitive processes alter pain by regulating incoming nociceptive signals. The DLPFC and the ACC represent the main sources of cortical afferents to the periaqueductal grey matter and the rostral ventromedial medulla leading to descending inhibition and facilitation of incoming nociceptive signals at the level of the spinal cord; therefore, this could partly explain how pain can be modified by attention, expectations, and emotions (Tracey & Mantyh, 2007). In addition, pain relief, as a reward and motivational process, implies effective endogenous opioid release in the ACC and a downstream dopaminergic signaling in the NAc. Increased negative motivational value of nociception and decreased reward value of pain relief may lead to chronic pain states (Porreca & Navratilova, 2017). In general, increasing positive affect and expectations, directing attention away from pain, or reducing anxiety can decrease pain, whereas opposite processes can enhance pain perception. These opposite effects of cognitive-affective modulation of noxious stimuli refer to placebo and nocebo phenomena as positive (hypoalgesia) and negative (hyperalgesia) results of pain processing, respectively.

Chronic pain is related to anatomical and functional alterations within brain areas implicated in cognitive-affective modulation of pain such as the DLPFC and medial PFC, the ACC, and the INS, and defective endogenous pain modulation, in addition to potential peripheral and central sensitization processes (Bushnell, Ceko, & Low, 2013). Clinicians and researchers have been taking advantage of cognitive-affective pain modulation using multiple behavioral methods [e.g., cognitive behavioral therapy, acceptance and commitment therapy, emotion awareness and expression therapy, and mindfulness-based stress reduction] (Kerns, Sellinger, & Goodin, 2011; Lumley et al., 2017; McCracken & Vowles, 2014; Turner et al., 2016) to improve chronic pain management. However, pain states should be evaluated as multidimensional experiences, which both allow for identification of psychological determinants of persistent pain and facilitate design of personalized therapeutic interventions based on biopsychosocial approach in patients with chronic pain.

Psychological mechanisms of placebo analgesia and nocebo hyperalgesia

"Nonspecific" effects of medication and nonpharmacological pain therapies are important determinants of variability in response to treatment. Neurobiological mechanisms of placebo and nocebo effects have been widely studied in pain medicine. Currently, a growing body of evidence has documented their impact on a variety of health outcomes, including orthopedic surgery (Moseley et al., 2002), brain stimulation in Parkinson's patients (Benedetti et al., 2016), and response to antidepressant (Pecina et al., 2015).

In the context of administration of an inert substance or sham intervention, placebo hypoalgesia refers to a decrease in pain perception, whereas nocebo hyperalgesia refers to an increase in pain. These are expressions of two opposite, "nonspecific" effects on pain modulation. Early research attributed these phenomena to positive and negative expectations built upon verbal instructions about the intervention and contextual cues (e.g., healthcare environment, doctor/nurse-patient relationship, or the use of medical devises). Expectation of negative future outcomes tend to increase anxiety whereas expectation of future positive outcomes tend to reduce anxiety and trigger psychological behaviors related with reward mechanisms (Benedetti, Carlino, & Pollo, 2011). Although expectations are considered more complex and dynamic cognitions about future events, they are often determined by previous conditioning and therapeutic experiences (Colloca & Miller, 2011).

Verbal instructions about the anticipated effects of therapeutic intervention can induce placebo and nocebo effects to both inert and pharmacologically active substances ("active placebo"). Before administration of a potent opioid (remifentanil), verbal suggestions about either an expected enhanced analgesia or exacerbation of pain produced significant, opposite effects on perceived pain (Bingel et al., 2011). Although in this case expectation was manipulated through verbal instructions, other factors such as the awareness of the treatment being administered, previous knowledge of the effects of potent pain-killers (through personal experiences or learning from others), environmental cues, or social interactions may also have contributed to the expected outcome. Prior positive experience with an active drug can increase expectations and induce placebo analgesia through Pavlovian/classical conditioning mechanisms. *The conditioned placebo/nocebo effect* (conditioned response) involves learning that a previously neutral cue (conditioned stimulus) predicts the occurrence of an active drug (unconditioned stimulus) by repeatedly pairing the conditioned with the unconditioned stimulus. Although placebo and nocebo responses can be learned through unconscious processes (Babel, 2019; Brascher, Kleinbohl, Holzl, & Becker, 2017; Egorova et al., 2015; Jensen et al., 2012), manipulation of cognitive expectation during conditioning can alter the magnitude of the response (Benedetti, Carlino, & Pollo, 2011). Furthermore, prior positive and negative experiences with treatments can enhance or diminish subsequent learning of conditioned placebo responses (Colloca & Benedetti, 2006). Other forms of implicit learning, such as *observational learning* (observing benefit or harm of interventions in another person), require no reinforcement and can act independently from conditioning and expectations. The magnitude of a placebo effect on an individual can be influenced by observation of analgesic benefits of a placebo manipulation on others (Colloca & Benedetti, 2009). Nevertheless, the magnitude of response depends on context cues, social interactions between patients and healthcare providers, and is prone to interferences from emotional and cognitive states. Patients' perception of doctors' empathy and competence, expectations of treatment efficacy, and high desire for pain relief can reduce anxiety and increase positive expectancy of analgesia. On the other hand, patients' perception of lack of empathy, along with personal negative beliefs and biases about treatment or uncertainty about the expected outcomes, can increase anxiety and reduced analgesia. Of particular relevance for nocebo hyperalgesia, social interactions, emotions, and cognitive states can influence important dynamic variations in the context-dependent expectancies. Pain catastrophizing, empathy traits, and the gender of the observed person undergoing pain-related intervention in observational learning paradigms all predict enhanced pain perception, highlighting the role of psychological traits (Swider & Babel, 2013; Vogtle, Barke, & Kroner-Herwig, 2013). Fig. 2 depicts an integrated model of the cognitive-affective mechanisms of placebo analgesia. Psychological interventions minimizing nocebo and optimizing placebo should be considered in clinical practice to improve pain relief (Darnall & Colloca, 2018).

Neurobiological and neurochemical mechanisms of placebo analgesia and nocebo hyperalgesia

Considering the relevance of expectation and conditioning for placebo effects, early research suggested that cognitive networks play a mechanistic role in triggering opioid- and nonopioid-mediated pain suppression (Amanzio & Benedetti, 1999). Indeed, neuroimaging studies revealed dynamic changes in brain activity related to anticipation of pain experiences. Anticipation of analgesia was associated with early activation of brain areas involved in the cognitive-affective evaluation of pain (increased activity in the OFC and DLPFC, ACC and decreased activity in the pINS and S2) triggering descending inhibition [increased activity in the periaqueductal gray matter (PAG)] (Wager et al., 2004). Expectancy-induced placebo analgesia was also related to significant activation of the NAc, which correlated with suppression of pain intensity and pain affect (Zubieta et al., 2005). The NAc, which represents a central component of the brain's reward pathways, integrates dopamine (DA) and opioid-mediated pathways involved in pain relief. Variation in activity of the reward-aversion circuit to pain relief may contribute to individual differences in pain modulation, as high expectation of pain relief induced by placebo analgesia, which was associated with greater DA and μ-opioid activation in the NAc (Scott et al., 2008). In contrast,

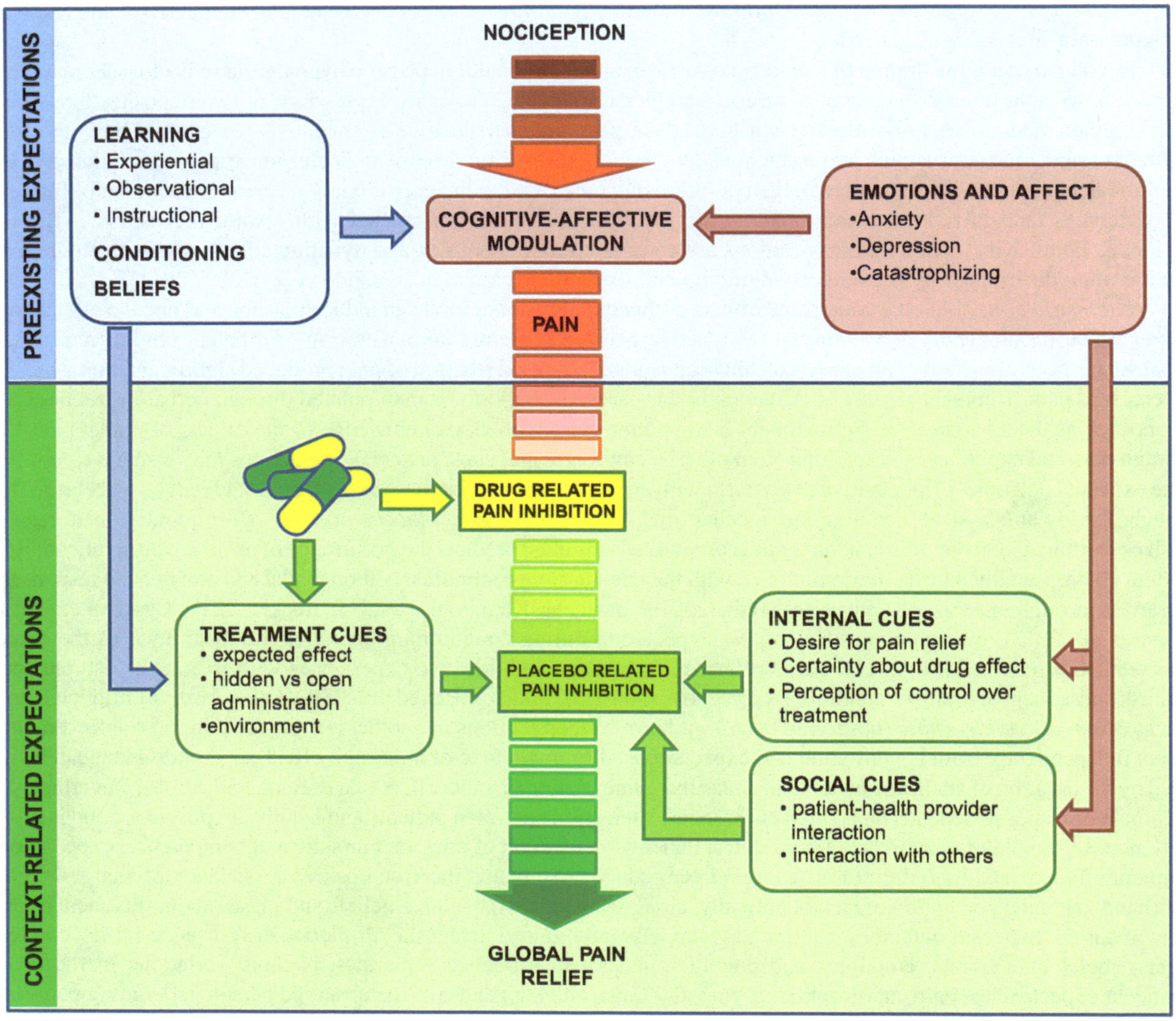

FIG. 2 An integrated model of placebo analgesia. Contribution of preexisting and context-related expectations and emotions to global pain relief.

nocebo hyperalgesia was related to opposite effects (Scott et al., 2008). Importantly, DA has no direct analgesic effects, as administration of DA-antagonists (Becker et al., 2013; Wrobel, Wiech, Forkmann, Ritter, & Bingel, 2014) or DA precursor levodopa (Zunhammer et al., 2017) has no significant effects on placebo analgesia. Rather, DA mediates positive and negative reinforcement learning and motivation to seek pain relief, depending upon activation of the PFC and the endogenous analgesic systems (Jarcho et al., 2016). Together, these findings suggest that placebo analgesia is determined both by early cognitive, emotional, and motivational appraisal of anticipated pain and by engagement of endogenous opioid and non-opioid systems. A robust placebo effects seems to require appropriate conceptual beliefs about the future occurrence of an event, maintained in PFC networks, that are supported by experience-dependent learning in striatal and brainstem circuits (Wager & Atlas, 2015). In contrast, expectancy-induced nocebo hyperalgesia seems to result predominantly from enhanced activity in the medial pain system and the hippocampus, which are responsible for emotional and cognitive aspects of pain perception (Kong et al., 2008), whereas the opioid system seems to be less relevant to nocebo hyperalgesia (Benedetti & Amanzio, 1997). Furthermore, differences in placebo and nocebo phenomena are not related to differential brain activity changes in pain-related brain regions. Rather, positive and negative expectancies seem to engage different brain networks (Figs. 3 and 4) that are each determined by separate cognitive constructs and behavioral patterns (Freeman et al., 2015).

Other nonopioid systems, such as the endocannabinoid system, the cholecystokinin (CCK) system, and the hormones vasopressin and oxytocin, have no direct analgesic effect, but rather are involved in emotional and behavioral mechanisms

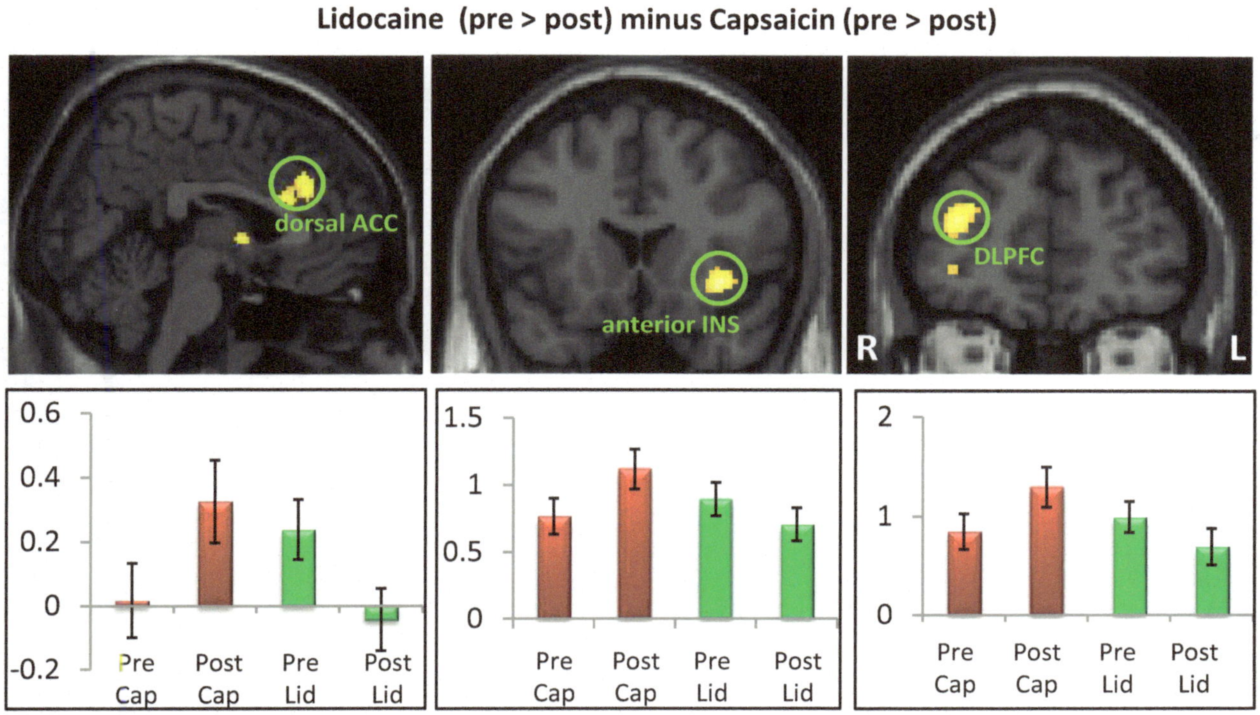

FIG. 3 Brain regions activation during pain administration in positive expectancy relative to negative expectancy. The bar indicates the peak beta of each brain regions pre- and posttreatment. *Cap*, capsaicin; *ACC*, anterior cingulate cortex; *DLPFC*, dorsolateral prefrontal cortex; *INS*, insula; *L*, left side; *Lid*, lidocaine; *R*, right side. *(From Freeman, S., Yu, R., Egorova, N., Chen, X., Kirsch, I., Claggett, B., et al. (2015). Distinct neural representations of placebo and nocebo effects.* NeuroImage, 112, *197–207. Fig. 4. Page 202, Copyright (2015), with permission from Elsevier.)*

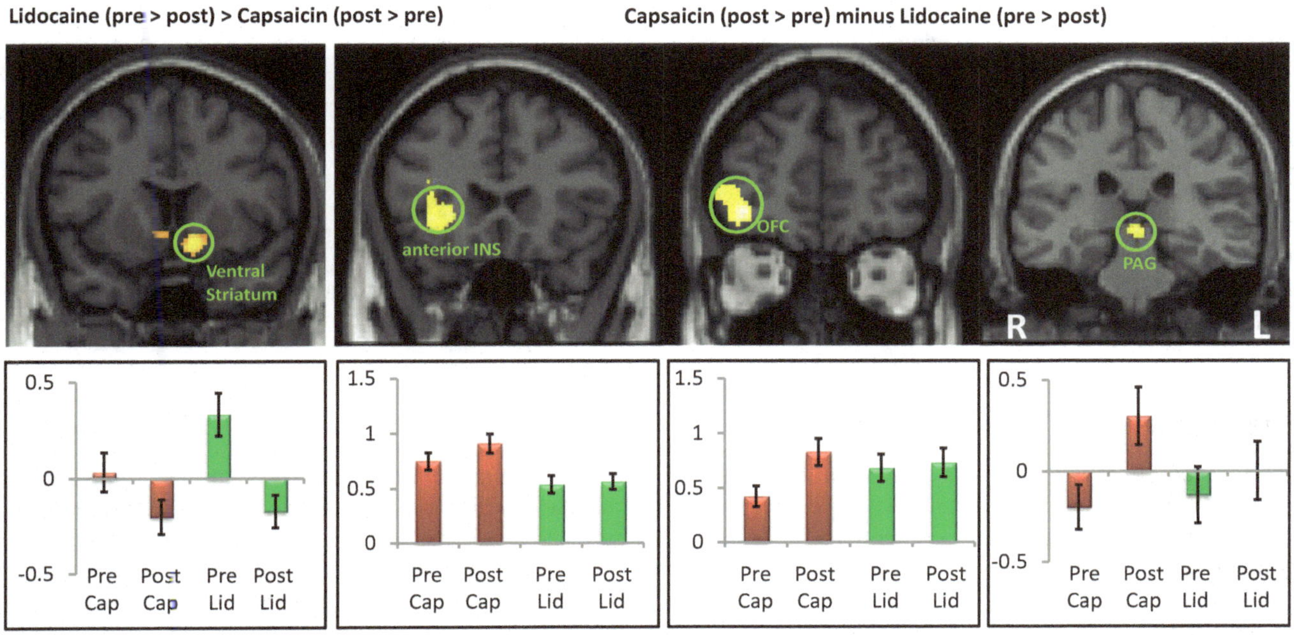

FIG. 4 Brain regions activation in the opposite direction during pain administration. The bar indicates the peak beta of each brain regions pre- and posttreatment. *Cap*, capsaicin; *INS*, insula; *OFC*, orbitofrontal cortex; *PAG*, periaqueductal gray matter; *L*, left side; *Lid*, lidocaine; *R*, right side. *(From Freeman, S., Yu, R., Egorova, N., Chen, X., Kirsch, I., Claggett, B., et al. (2015). Distinct neural representations of placebo and nocebo effects.* NeuroImage, 112, *197–207. Fig. 5. Page 203, Copyright (2015), with permission from Elsevier.)*

TABLE 1 Neurochemical mechanisms of placebo and nocebo.

Mediator	Sensory component	Cognitive-affective component	Placebo	Nocebo
Opioids	Major role in endogenous analgesia	Role in learning and affect regulation.	+	
Endocannabinoids	Modulatory role on the opioidergic and dopaminergic system	Inhibition of the affective dimension of pain. Regulate stress response and hedonic reward states.	+	
Dopamine	No direct pain relief	Role in reward processing and learning mechanisms. Enhances pain relief expectancy.	+	
Cholecystokinin	Suppresses opioid analgesia	Antagonistic effects on the dopaminergic system. Increases anxiety and suppresses positively motivated behaviors.		+
Oxytocin-Vasopressin	No direct pain relief	Role in stress-related analgesia. Regulate social behavior and interaction. Inhibition of the affective component of pain.	+	
Cortisol	No direct pain relief	Role in stress regulation and stress-related pain modulation. Role in associative learning and memory.	+	+

of pain modulation. The endocannabinoid system plays a major role in mood regulation, learning, and reward processing, and mediates some aspects of nonopioid mechanisms of placebo analgesia (Benedetti, Amanzio, Rosato, & Blanchard, 2011). On the other hand, CCK is a neuropeptide involved in the modulation of anxiety and stress-related behaviors, with antagonistic effects on opioid-mediated analgesia (Benedetti, Amanzio, & Maggi, 1995). Thus, the CCK system seems to play a central role in expectancy-induced nocebo hyperalgesia by increasing anticipatory anxiety (Benedetti, Amanzio, Vighetti, & Asteggiano, 2006). Nevertheless, the magnitude of anticipatory anxiety may increase activity in the hypothalamic-pituitary-adrenal axis and stress-related analgesia. Vasopressin can boost placebo analgesia, and human studies have shown that the effect is significantly larger compared to oxytocin; however, effects were only observed in women with lower cortisol levels and dispositional anxiety (Colloca, Pine, Ernst, Miller, & Grillon, 2016). This may reflect the relationship between social behavior and anxiety on placebo analgesia. In clinical settings, however, multiple systems, including previous expectations, treatment cues, social interactions, and emotions, may trigger multiple overlapping pain-modulation mechanisms, potentially accounting for high interindividual response to pain treatment. Table 1 summarizes neurochemical mechanisms of placebo and nocebo phenomena.

The relevance of placebo and nocebo effects for pain treatment in clinical practice

The placebo effect is recognized as a major contributor to pain treatment success and placebo-controlled clinical trials represent the "gold standard" of evidence-based medicine for testing the efficacy and safety of new drugs. Despite the focus on placebo, nocebo is known to potentially limit therapeutic effects of analgesic drugs or worsen and prolong pain symptoms, which both increase treatment costs and reduce patients' quality of life. Therefore, a better understanding of the clinical implications and impacts of cognitive-affective mechanisms for pain management could allow for development of strategies to optimize placebo effects and reduce nocebo effects in pain treatments (Bartels et al., 2017; Darnall & Colloca, 2018; Manai, van Middendorp, Veldhuijzen, Huizinga, & Evers, 2019). As we previously mentioned, positive and negative expectations play a major role in pain modulation. For example, pain catastrophizing contributes to pain amplification, with research suggesting that it may do so by reducing descending inhibitory pain modulation and altering brain function (Blasini, Corsi, Klinger, & Colloca, 2017; Seminowicz et al., 2013). Importantly, psychological interventions are effective in managing pain catastrophizing and other psychological factors known to amplify pain and reduce treatment effectiveness (Salomons, Moayedi, Erpelding, & Davis, 2014; Seminowicz et al., 2013). A "Clinical

Toolkit" has been developed to guide clinicians regarding evidence-based strategies for minimizing nocebo and optimizing placebo effects (Darnall & Colloca, 2018). Although some of these skillsets can be easily applied by practitioners from diverse backgrounds, others require specialized training in pain psychology (e.g., cognitive behavioral therapy for pain).

Preexpectations are built upon instructions about treatment and interventions, previous personal experience with treatments, observation of the effects of intervention in others, and previous beliefs and biases about treatments and health professionals. For example, positive treatment expectancies doubled the analgesic benefit of the potent opioid remifentanil, whereas negative treatment expectancies abolished its analgesic effects (Bingel et al., 2011). On the other hand, observing the effect of treatment in others (observational social learning) produced further pain relief to first-hand experience with treatment, which was similar in magnitude to the observed placebo hypoalgesia (Colloca & Benedetti, 2009). Furthermore, positive or negative personal experiences with pain treatments may increase or decrease expectations of efficacy of subsequent treatments (either for the same or similar drugs). This is particularly relevant for patients with chronic pain who take one or several drugs with different therapeutic efficacy over large periods of time. Negative treatment experiences determine carry-over effects that generalize across therapeutic approaches and the route of administration, independently of conscious expectations (Zunhammer et al., 2017). Trait anxiety and depression may increase susceptibility for negative carry-over of treatment experience to subsequent treatments (Kessner et al., 2014), which may have implications for patients with chronic pain and associated mood disorders.

Expectations generated in clinical context are more dynamic due to interference from internal and external (environmental and social) cues. Internal cues include feelings of certainty about the drug's effect, perception of control over treatment and desire for pain relief. The "open-hidden paradigm" of drug administration demonstrates the interaction between external and internal cues. Individuals may respond differently to drugs administered openly (e.g., instructing patients about the powerful pain-killer being administered) versus in a hidden manner (e.g., by an automatic infusion machine). Amanzio and colleagues (Amanzio, Pollo, Maggi, & Benedetti, 2001) showed that open drug administration can trigger placebo and nocebo responses, which may alter the effectiveness of analgesia. In contrast, hidden injections of the nonopioid ketorolac were less effective and required a higher dose of medication to reduce pain by 50% compared with open injections. Similarly, hidden interruption of morphine administration prolonged analgesia compared to when patients were told that administration was being stopped (Colloca, Lopiano, Lanotte, & Benedetti, 2004).

Social interactions with healthcare providers and others in clinical environments can modulate patient's expectations and individual's emotional and cognitive state. Perceived physician's competence and empathy can enhance the effectiveness of treatment being administered (Czerniak et al., 2016), whereas the presence of a friend observer during pain procedures can increase pain threshold and tolerance (Edwards, Eccleston, & Keogh, 2017).

The correct assessment and optimization of these factors that enhance placebo analgesia and minimize nocebo hyperalgesia (Table 2) may critically increase the efficacy of pain treatments.

TABLE 2 Strategies to optimize placebo and reduce nocebo.

Mechanisms	Enhance placebo	Reduce nocebo
Instructional learning	Inform about the reasons a treatment is being done. Explain the underlying mechanisms.	Provide information of nocebo. Explain possible misattribution of side effects.
	Inform about the expected outcomes.	Avoid setting unrealistic treatment goals.
	Focus on expected positive outcomes.	Discuss potential side effects in a positive framing.
Experiential and observational learning	Identify previous first-hand experience and knowledge about effective treatments with a similar mechanism of action.	Identify previous failure of medical treatments and explain differences with the current treatment.
Drug-related cues	Use open drug administration.	Avoid unlabeled pill box.
	Inform about treatment changes and reasonable positive expectations.	Avoid changing from branded to generic drugs and if needed offer the patient to choose.
Internal cues	Identify and reinforce positive treatment effects.	Inform if treatment response is not immediate, instead is build-up after repeated or prolonged administration.
	Use pain diaries emphasizing reduction in pain intensity and duration of pain relief.	Avoid pain diaries overemphasizing pain worsening periods and side effects.

Continued

TABLE 2 Strategies to optimize placebo and reduce nocebo—cont'd

Mechanisms	Enhance placebo	Reduce nocebo
Social cues (patient-health provider interaction)	Express competence, empathy and trust. Express realistic expectations about treatment efficacy.	Avoid expressing nervousness, being noncollaborative or judgmental. Avoid spending too much time on patient's related documentation.
	Check patient's understanding of treatment goals. Include the patient in the decision-making to strengthen compliance.	Answer questions to avoid doubts. Avoid using technical language.
	Provide in person or over the phone follow up. Highlight successes.	Provide other sources of evidence based information on the treatment. Have a plan for noneffective treatment.
Social cues (interaction with others)	Encourage being accompanied by supportive friend/relative during procedures.	Avoid limiting supportive interactions in health-care settings.
Emotions and affect	Identify optimism and effective self-control of pain-related distress. Consider mindfulness to promote copping strategies.	Identify symptoms of anxiety, depression and catastrophizing. Consider cognitive behavioral therapy to enhance positive distraction and cognitive restructuring.

Applications to other areas

Placebo and nocebo phenomena related to nonopioid drugs and implications for pain treatment

"Nonspecific" effects of active drugs (placebo and nocebo effects) are also observed with administration of nonopioid medication (nonsteroidal antiinflammatory drugs, antiepileptic, and antidepressive drugs) used for treatment of chronic pain. However, the magnitude of the placebo effect can differ in patients compared to healthy individuals (Charron, Rainville, & Marchand, 2006). Several factors may reduce placebo and increase nocebo effects in people with chronic pain, including peripheral and central sensitization of nociceptive afferents, failure to activate the PFC, and the descending endogenous pain inhibitory mechanisms or enhanced descending facilitatory mechanisms (Tracey & Mantyh, 2007). On the other hand, chronic pain is frequently associated with negative mood, difficulties with emotion regulation, and catastrophic thinking (Darnall & Colloca, 2018). Patients with anxiety and depression are more likely to develop chronic pain, and conversely persistent pain increases the risk of comorbid mood disorders (Fernandez et al., 2017; Tunks, Crook, & Weir, 2008). A previous negative experience with pain treatment interventions, anticipatory anxiety, sustained attention to negative cues, and somatization (a heightened awareness of bodily sensation) may determine a shift from placebo analgesia to nocebo hyperalgesia, which may greatly impact treatment efficacy, adherence to treatment and likelihood of experiencing nonspecific side effects.

Other agents of interest

Role of cognitive behavioral therapy in placebo/nocebo and pain treatment

Psychological interventions such as cognitive behavioral therapy (CBT) and mindfulness may contribute to pain relief. Placebo analgesia related to CBT could be especially relevant in the management of chronic pain as results in most RCTs of pharmacological treatments report limited positive effects compared to the placebo control groups.

CBT focuses on correcting negative thoughts using behavioral and verbal techniques that may improve pain treatment outcomes, including reduction in opioid consumption, misuse, and abuse. CBT is especially beneficial for patients with chronic pain who report high catastrophizing, anxiety, and depression symptoms. The therapeutic effect is related to a significant reduction of catastrophizing immediately posttreatment and the effect can be maintained for several months. Neuroimaging studies revealed that an 11-week CBT intervention for coping with chronic pain reduced pain catastrophizing and resulted in increased grey matter volume in the bilateral DLPFC, posterior parietal cortex, and subgenual ACC, which support the role of CBT in cognitive and affective reappraisal of pain and enhanced descending modulation of pain (Seminowicz et al., 2013). On the other hand, mindfulness practices, which promote nonjudgmental, present-focused awareness of one's experiences, reduce pain and enhance coping by altering its meaning, interpretation, and appraisal (Zeidan & Vago, 2016). This helps the individual shift from a passive (escape and avoidance) to an active (acceptance, cognitive, and affective control) experience of pain by engaging multiple brain mechanisms that change with increasing meditative practice (Garland et al., 2019; Zeidan & Vago, 2016). With increased practice, a decoupling occurs between the sensory experience and the meaning of the pain to the individual. Acceptance and commitment therapy (ACT) is a form of CBT that combines acceptance and mindfulness approaches along with committed action and behavior change strategies to increase the patient's psychological flexibility and active engagement in life (McCracken & Vowles, 2014). Importantly, it is a process-based therapy with a broad focus on cognitive, emotional, social, and motivational factors. ACT is especially helpful for chronic pain patients who develop fear, anxiety, and avoidance behaviors. These patients become focused on their pain, which leads to reduced engagement in rewarding activities, which in turn leads to increased emotional distress and more pain. Rather than putting their lives on hold to avoid pain and emotional distress, the goal of ACT is to help the individual to begin to live their lives more fully by aligning their behavior with their deepest values. Outcome research indicates that ACT increases physical and social activity and decreases anxiety, depression, and disability, while also reducing pain-related healthcare utilization (Feliu-Soler et al., 2018; McCracken & Vowles, 2014; Serbic & Pincus, 2017). By decreasing psychological distress and the focus on avoiding pain sensations, ACT may help reduce somatization and susceptibility to the nocebo effect (Manai et al., 2019). Together, mindfulness and ACT may improve self-control over pain and emotional regulation, and they may have the potential to increase positive expectancy by shifting the focus away from avoiding pain to increasing their activity and function, empowering patients to see how they can improve their lives.

Mini-dictionary of terms

- Placebo hypoalgesia refers to decreased pain perception that appears in the context of pain treatment but is unrelated to the pharmacological effects of the drug.
- Nocebo hyperalgesia represents an increase in pain sensitivity, a nonspecific effect observed in the context of pain treatment/interventions.
- Expectation (or expectancy) represents specific conscious reasoning about the likelihood of future events (or the effects of interventions), which is based on instructions, previous experiences, and social observation.
- Anticipatory anxiety is a physiological and affective response to an uncertain threatening context or an emotionally aversive future event.
- Instructional learning is a form of learning the effects and responses to interventions through explicit verbal or written suggestions.
- Social learning is a form of knowledge acquisition that occurs through observation of the effects of intervention in others.
- Conditioned analgesia/hyperalgesia represents learning-based response of pain modulation induced by repeated pairing of an initially neutral stimulus (inert substance/intervention) with an active drug/intervention, which then leads to analgesic/hyperalgesic response when the inert substance/intervention is administered.
- Endogenous pain modulatory systems encompass brain and spinal cord structures that provide facilitation or inhibition of nociceptive inputs.
- Pain catastrophizing is characterized by negative thoughts and feelings (rumination, magnification, and helplessness) that may enhance pain experiences and increase the risk of developing chronic pain.
- Open-hidden paradigm represents drug administration when the patient is given information about the medication being administered compared to a drug administration by an automated system when the patient is unaware of a drug being given.
- Cognitive behavioral therapy is a form of psychotherapy that modifies maladaptive cognitions (thoughts and beliefs) and behaviors, improving emotion regulation and pain.
- Mindfulness practices and acceptance and commitment therapy promote awareness and acceptance of one's experiences and coping strategies, with a shift from passive to active coping with pain and related distress.

Key facts of placebo analgesia

- Placebo analgesia is determined by an early cognitive, emotional, and motivational appraisal of anticipated painful stimuli rather than an early suppression of nociceptive processing.
- Placebo analgesia engages endogenous opioid systems that have direct pain inhibitory effects and nonopioid systems that are involved in the emotional and behavioral mechanisms of pain modulation.

- Dispositional anxiety and stress-related responses increase activity in the hypothalamic-pituitary-adrenal axis, which may boost vasopressin-induced placebo analgesia in women.
- Previous positive experiences with pain treatments may enhance placebo analgesia.
- Classical conditioning procedures can produce placebo analgesia effects that are stronger than verbal suggestions of pain relief alone.
- Observing pain relief effects of an intervention in others can induce comparable placebo analgesia in a first-hand experience pain intervention.
- Having positive expectancies of pain relief via verbal instructions may double the analgesic benefit of potent opioids as compared to no expectation condition.
- Open injections of nonopioid drugs were more effective and required a smaller dose of medication to reduce pain by 50% compared with hidden injections.
- Patients' perception of doctors' empathy, competence along with positive expectations of treatment efficacy can enhance placebo analgesia.
- Failure to activate the prefrontal cortex and the descending endogenous pain inhibitory mechanisms may reduce placebo analgesia and increase the risk for chronic pain.

Key facts of nocebo hyperalgesia

- Nocebo hyperalgesia is related to preferential activation of the medial pain system and the hippocampus which mediate the affective-cognitive components of pain.
- Anxiety, depression, and catastrophizing may increase susceptibility for negative expectations and nocebo hyperalgesia.
- The endogenous opioid system is less relevant for mechanisms of nocebo hyperalgesia.
- The cholecystokinin system plays a central role in anticipatory anxiety and pain facilitation following nocebo suggestions.
- Previous negative experiences with pain treatments may determine carry-over effects that generalize across therapeutic approaches and administration route, independent of conscious expectations.
- Hidden injections of the nonopioid ketorolac were less effective and required a higher dose of medication to reduce pain by 50% compared with open injections.
- Negative expectations resulting from discontinuance of drug administration can determine fast disappearance of pain relief effects.
- Cognitive behavioral therapy reduces pain catastrophizing in patients with chronic, noncancer pain.
- Mindfulness improves coping strategies, self-control over pain, and emotional regulation to reduce negative expectations of pain treatment.

Summary points

- Cognitive and emotional factors play a major role in the psychosocial dimension of pain and explain inter- and intraindividual variability in pain perception and response to treatment.
- Distinct cognitive and behavioral processes underlying the placebo analgesia and nocebo hyperalgesia are related to a differential engagement of cognitive-affective brain networks, rather than to changes in pain-related brain regions (pain matrix).
- Nonopioid systems, such as the dopamine, cholecystokinin, and endocannabinoid systems together with the hormones vasopressin and oxytocin have no direct analgesic effect, but rather are involved in learning of behaviors related to pain relief, stress-related reactions to pain anticipation, and mediation of anticipatory anxiety.
- The prefrontal cortex and the anterior cingulate cortex play a central role in opioid-mediated endogenous pain inhibition triggered by cognitive and emotions processes.
- Expectations, as major determinants of placebo and nocebo-related pain modulation, are generated through verbal instructions, previous first-hand or observed effects of treatment, and social interactions.
- Maladaptive cognitive processes (catastrophizing) and negative affect (anxiety and depression) may increase susceptibility for development of chronic pain and to limit response to treatment.

References

Amanzio, M., & Benedetti, F. (1999). Neuropharmacological dissection of placebo analgesia: Expectation-activated opioid systems versus conditioning-activated specific subsystems. *The Journal of Neuroscience*, *19*(1), 484–494.

Amanzio, M. Pollo, A., Maggi, G., & Benedetti, F. (2001). Response variability to analgesics: A role for non-specific activation of endogenous opioids. *Pain*, *90*(3), 205–215.

Atlas, L. Y., & Wager, T. D. (2012). How expectations shape pain. *Neuroscience Letters*, *520*(2), 140–148.

Babel, P. (2019). Classical conditioning as a distinct mechanism of placebo effects. *Frontiers in Psychiatry*, *10*, 449.

Bartels, D. J. P., van Laarhoven, A. I. M., Stroo, M., Hijne, K., Peerdeman, K. J., Donders, A. R. T., et al. (2017). Minimizing nocebo effects by conditioning with verbal suggestion: A randomized clinical trial in healthy humans. *PLoS One*, *12*(9), e0182959.

Becker, S., Ceko, M., Louis-Foster, M., Elfassy, N. M., Leyton, M., Shir, Y., et al. (2013). Dopamine and pain sensitivity: Neither sulpiride nor acute phenylalanine and tyrosine depletion have effects on thermal pain sensations in healthy volunteers. *PLoS One*, *8*(11), e80766.

Benedetti, F., & Amanzio, M. (1997). The neurobiology of placebo analgesia: From endogenous opioids to cholecystokinin. *Progress in Neurobiology*, *52*(2), 109–125.

Benedetti, F., Amanzio, M., & Maggi, G. (1995). Potentiation of placebo analgesia by proglumide. *Lancet*, *346*(8984), 1231.

Benedetti, F., Amanzio, M., Rosato, R., & Blanchard, C. (2011). Nonopioid placebo analgesia is mediated by CB1 cannabinoid receptors. *Nature Medicine*, *17*(10), 1228–1230.

Benedetti, F., Amanzio, M., Vighetti, S., & Asteggiano, G. (2006). The biochemical and neuroendocrine bases of the hyperalgesic nocebo effect. *The Journal of Neuroscience*, *26*(46), 12014–12022.

Benedetti, F., Carlino, E., & Pollo, A. (2011). How placebos change the patient's brain. *Neuropsychopharmacology*, *36*(1), 339–354.

Benedetti, F., Frisaldi, E., Carlino, E., Giudetti, L., Pampallona, A., Zibetti, M., et al. (2016). Teaching neurons to respond to placebos. *The Journal of Physiology*, *594*(19), 5647–5660.

Bingel, U., Wanigasekera, V., Wiech, K., Ni Mhuircheartaigh, R., Lee, M. C., Ploner, M., et al. (2011). The effect of treatment expectation on drug efficacy: imaging the analgesic benefit of the opioid remifentanil. *Science Translational Medicine*, *3*(70). 70ra14.

Blasini, M., Corsi, N., Klinger, R., & Colloca, L. (2017). Nocebo and pain: An overview of the psychoneurobiological mechanisms. *Pain Reports*, *2*(2), e585.

Brascher, A. K., Kleinbohl, D., Holzl, R., & Becker, S. (2017). Differential classical conditioning of the nocebo effect: Increasing heat-pain perception without verbal suggestions. *Frontiers in Psychology*, *8*, 2163.

Bushnell, M. C., Ceko, M., & Low, L. A. (2013). Cognitive and emotional control of pain and its disruption in chronic pain. *Nature Reviews. Neuroscience*, *14*(7), 502–511.

Charron, J., Rainville, P., & Marchand, S. (2006). Direct comparison of placebo effects on clinical and experimental pain. *The Clinical Journal of Pain*, *22*(2), 204–211.

Colloca, L., & Benedetti, F. (2006). How prior experience shapes placebo analgesia. *Pain*, *124*(1–2), 126–133.

Colloca, L., & Benedetti, F. (2009). Placebo analgesia induced by social observational learning. *Pain*, *144*(1–2), 28–34.

Colloca, L., Lopiano, L., Lanotte, M., & Benedetti, F. (2004). Overt versus covert treatment for pain, anxiety, and Parkinson's disease. *Lancet Neurology*, *3*(11), 679–684.

Colloca, L., & Miller, F. G. (2011). The nocebo effect and its relevance for clinical practice. *Psychosomatic Medicine*, *73*(7), 598–603.

Colloca, L., Pine, D. S., Ernst, M., Miller, F. G., & Grillon, C. (2016). Vasopressin boosts placebo analgesic effects in women: A randomized trial. *Biological Psychiatry*, *79*(10), 794–802.

Czerniak, E., Biegon, A., Ziv, A., Karnieli-Miller, O., Weiser, M., Alon, U., et al. (2016). Manipulating the placebo response in experimental pain by altering doctor's performance style. *Frontiers in Psychology*, *7*, 874.

Darnall, B. D., & Colloca, L. (2018). Optimizing placebo and minimizing nocebo to reduce pain, catastrophizing, and opioid use: A review of the science and an evidence-informed clinical toolkit. *International Review of Neurobiology*, *139*, 129–157.

Dum, R. P., Levinthal, D. J., & Strick, P. L. (2009). The spinothalamic system targets motor and sensory areas in the cerebral cortex of monkeys. *The Journal of Neuroscience*, *29*(45), 14223–14235.

Edwards, R., Eccleston, C., & Keogh, E. (2017). Observer influences on pain: An experimental series examining same-sex and opposite-sex friends, strangers, and romantic partners. *Pain*, *158*(5), 846–855.

Egorova, N., Park, J., Orr, S. P., Kirsch, I., Gollub, R. L., & Kong, J. (2015). Not seeing or feeling is still believing: Conscious and non-conscious pain modulation after direct and observational learning. *Scientific Reports*, *5*, 16809.

Feliu-Soler, A., Montesinos, F., Gutierrez-Martinez, O., Scott, W., McCracken, L. M., & Luciano, J. V. (2018). Current status of acceptance and commitment therapy for chronic pain: A narrative review. *Journal of Pain Research*, *11*, 2145–2159.

Fernandez, M., Colodro-Conde, L., Hartvigsen, J., Ferreira, M. L., Refshauge, K. M., Pinheiro, M. B., et al. (2017). Chronic low back pain and the risk of depression or anxiety symptoms: Insights from a longitudinal twin study. *The Spine Journal*, *17*(7), 905–912.

Freeman, S., Yu, R., Egorova, N., Chen, X., Kirsch, I., Claggett, B., et al. (2015). Distinct neural representations of placebo and nocebo effects. *NeuroImage*, *112*, 197–207.

Garland, E. L., Brintz, C. E., Hanley, A. W., Roseen, E. J., Atchley, R. M., Gaylord, S. A., et al. (2019). Mind-body therapies for opioid-treated pain: A systematic review and meta-analysis. *JAMA Internal Medicine*, *180*(1), 91–105.

Jarcho, J. M., Feier, N. A., Labus, J. S., Naliboff, B., Smith, S. R., Hong, J. Y., et al. (2016). Placebo analgesia: Self-report measures and preliminary evidence of cortical dopamine release associated with placebo response. *NeuroImage: Clinical*, *10*, 107–114.

Jensen, K. B., Kaptchuk, T. J., Kirsch, I., Raicek, J., Lindstrom, K. M., Berna, C., et al. (2012). Nonconscious activation of placebo and nocebo pain responses. *Proceedings of the National Academy of Sciences of the United States of America*, *109*(39), 15959–15964.

Kerns, R. D., Sellinger, J., & Goodin, B. R. (2011). Psychological treatment of chronic pain. *Annual Review of Clinical Psychology*, *7*, 411–434.

Kessner, S., Forkmann, K., Ritter, C., Wiech, K., Ploner, M., & Bingel, U. (2014). The effect of treatment history on therapeutic outcome: Psychological and neurobiological underpinnings. *PLoS One*, *9*(9), e109014.

Kong, J., Gollub, R. L., Polich, G., Kirsch, I., Laviolette, P., Vangel, M., et al. (2008). A functional magnetic resonance imaging study on the neural mechanisms of hyperalgesic nocebo effect. *The Journal of Neuroscience*, *28*(49), 13354–13362.

Lumley, M. A., Schubiner, H., Lockhart, N. A., Kidwell, K. M., Harte, S. E., Clauw, D. J., et al. (2017). Emotional awareness and expression therapy, cognitive behavioral therapy, and education for fibromyalgia: A cluster-randomized controlled trial. *Pain*, *158*(12), 2354–2363.

Manai, M., van Middendorp, H., Veldhuijzen, D. S., Huizinga, T. W. J., & Evers, A. W. M. (2019). How to prevent, minimize, or extinguish nocebo effects in pain: A narrative review on mechanisms, predictors, and interventions. *Pain Reports*, *4*(3), e699.

McCracken, L. M., & Vowles, K. E. (2014). Acceptance and commitment therapy and mindfulness for chronic pain: Model, process, and progress. *The American Psychologist*, *69*(2), 178–187.

Moseley, J. B., O'Malley, K., Petersen, N. J., Menke, T. J., Brody, B. A., Kuykendall, D. H., et al. (2002). A controlled trial of arthroscopic surgery for osteoarthritis of the knee. *The New England Journal of Medicine*, *347*(2), 81–88.

Pecina, M., Bohnert, A. S., Sikora, M., Avery, E. T., Langenecker, S. A., Mickey, B. J., et al. (2015). Association between placebo-activated neural systems and antidepressant responses: Neurochemistry of placebo effects in major depression. *JAMA Psychiatry*, *72*(11), 1087–1094.

Porreca, F., & Navratilova, E. (2017). Reward, motivation, and emotion of pain and its relief. *Pain*, *158*(Suppl 1), S43–S49.

Salomons, T. V., Moayedi, M., Erpelding, N., & Davis, K. D. (2014). A brief cognitive-behavioural intervention for pain reduces secondary hyperalgesia. *Pain*, *155*(8), 1446–1452.

Scott, D. J., Stohler, C. S., Egnatuk, C. M., Wang, H., Koeppe, R. A., & Zubieta, J. K. (2008). Placebo and nocebo effects are defined by opposite opioid and dopaminergic responses. *Archives of General Psychiatry*, *65*(2), 220–231.

Seminowicz, D. A., Shpaner, M., Keaser, M. L., Krauthamer, G. M., Mantegna, J., Dumas, J. A., et al. (2013). Cognitive-behavioral therapy increases prefrontal cortex gray matter in patients with chronic pain. *The Journal of Pain*, *14*(12), 1573–1584.

Serbic, D., & Pincus, T. (2017). The relationship between pain, disability, guilt and acceptance in low back pain: A mediation analysis. *Journal of Behavioral Medicine*, *40*(4), 651–658.

Swider, K., & Babel, P. (2013). The effect of the sex of a model on nocebo hyperalgesia induced by social observational learning. *Pain*, *154*(8), 1312–1317.

Tracey, I., & Mantyh, P. W. (2007). The cerebral signature for pain perception and its modulation. *Neuron*, *55*(3), 377–391.

Tunks, E. R., Crook, J., & Weir, R. (2008). Epidemiology of chronic pain with psychological comorbidity: Prevalence, risk, course, and prognosis. *Canadian Journal of Psychiatry*, *53*(4), 224–234.

Turner, J. A., Anderson, M. L., Balderson, B. H., Cook, A. J., Sherman, K. J., & Cherkin, D. C. (2016). Mindfulness-based stress reduction and cognitive behavioral therapy for chronic low back pain: Similar effects on mindfulness, catastrophizing, self-efficacy, and acceptance in a randomized controlled trial. *Pain*, *157*(11), 2434–2444.

Vogtle, E., Barke, A., & Kroner-Herwig, B. (2013). Nocebo hyperalgesia induced by social observational learning. *Pain*, *154*(8), 1427–1433.

Wager, T. D., & Atlas, L. Y. (2015). The neuroscience of placebo effects: Connecting context, learning and health. *Nature Reviews. Neuroscience*, *16*(7), 403–418.

Wager, T. D., Atlas, L. Y., Lindquist, M. A., Roy, M., Woo, C. W., & Kross, E. (2013). An fMRI-based neurologic signature of physical pain. *The New England Journal of Medicine*, *368*(15), 1388–1397.

Wager, T. D., Rilling, J. K., Smith, E. E., Sokolik, A., Casey, K. L., Davidson, R. J., et al. (2004). Placebo-induced changes in FMRI in the anticipation and experience of pain. *Science*, *303*(5661), 1162–1167.

Wrobel, N., Wiech, K., Forkmann, K., Ritter, C., & Bingel, U. (2014). Haloperidol blocks dorsal striatum activity but not analgesia in a placebo paradigm. *Cortex*, *57*, 60–73.

Zeidan, F., & Vago, D. R. (2016). Mindfulness meditation-based pain relief: A mechanistic account. *Annals of the New York Academy of Sciences*, *1373*(1), 114–127.

Zubieta, J. K., Bueller, J. A., Jackson, L. R., Scott, D. J., Xu, Y., Koeppe, R. A., et al. (2005). Placebo effects mediated by endogenous opioid activity on mu-opioid receptors. *The Journal of Neuroscience*, *25*(34), 7754–7762.

Zunhammer, M., Ploner, M., Engelbrecht, C., Bock, J., Kessner, S. S., & Bingel, U. (2017). The effects of treatment failure generalize across different routes of drug administration. *Science Translational Medicine*, *9*(393), eaal2999.

Chapter 34

Nociception-related behavioral phenotypes in adult zebrafish

Fabiano V. Costa[a], Luiz V. Rosa[a], Allan V. Kalueff[b,c], and Denis B. Rosemberg[a]

[a]*Department of Biochemistry and Molecular Biology School of Life Sciences, Federal University of Santa Maria, Santa Maria, Rio Grande do Sul, Brazil,* [b]*School of Pharmacy, Southwest University, Chongqing, China,* [c]*Ural Federal University, Yekaterinburg, Russia*

Abbreviations

TRP	transient receptor potential ion channels
TRPA1a	transient receptor potential ankyrin1a
TRPA1b	transient receptor potential ankyrin1b
TRPV1	transient receptor potential vaniloid-1
zDOPa/b	zebrafish δ opioid a/b receptor
zKOP	zebrafish κ opioid receptor
zMOP	zebrafish μ opioid receptor
zNOP	zebrafish nociceptin opioid receptor

Introduction

Pain evokes various responses in vivo that serve to communicate unpleasant experiences at various levels (Keefe & Pryor, 2007). Nociception is the neural processes of encoding and processing noxious stimuli, whereas pain is the conscious perception of nociception, and suffering is the negative emotional responses to nociception and pain (Fig. 1) (Fordyce et al., 1984). However, accurate analyses and measurement of pain-related behaviors in animal models can be challenging, since pain is a complex experience that involves not only transduction of nociceptive stimuli but also cognitive and emotional processing (Blackburn-Munro, 2004; Piel, Kroin, van Wijnen, Ranjan, & Im, 2014; Tracey & Mantyh, 2007). Together, this raises the question of how do we measure pain-related behaviors in subjects that cannot express their feelings? Moreover, can we verify if our observations match with the subjective experience of these individuals? To address this knowledge gap, here we discuss various behaviors related to pain, focusing on a novel experimental model organism, such as the zebrafish (*Danio rerio*), in translational pain research.

Nociception and pain in animal models

Unlike humans, laboratory animal models have less complex neural processing networks, making it difficult to correlate nociception with the individual subjective experience (Festing & Patel, 2005; Thomas, 2005). Bateson (1991) has suggested several clear criteria for the ability to feel pain in various species, noting that only animals that met all these criteria should be considered sentient organisms. Briefly, they should possess (i) nociceptors; (ii) converging nociceptor pathways to the brain; (iii) brain structures analogous to the human cerebral cortex that processes pain; as well as (iv) express opioid receptors and endogenous opioid substances in a nociceptive neural system; (v) display lower adverse behavioral and physiological effects after administration of analgesics or anesthetics; (vi) avoid potentially painful stimuli. Sneddon and colleagues have suggested an additional criterion for animal sentience, namely that animals should suspend normal behavior for a prolonged period, instead of showing a reflexive response, with changes in behavior that reflect signs of "discomfort" (Sneddon, Elwood, Adamo, & Leach, 2014).

The Neurobiology, Physiology, and Psychology of Pain. https://doi.org/10.1016/B978-0-12-820589-1.00034-8

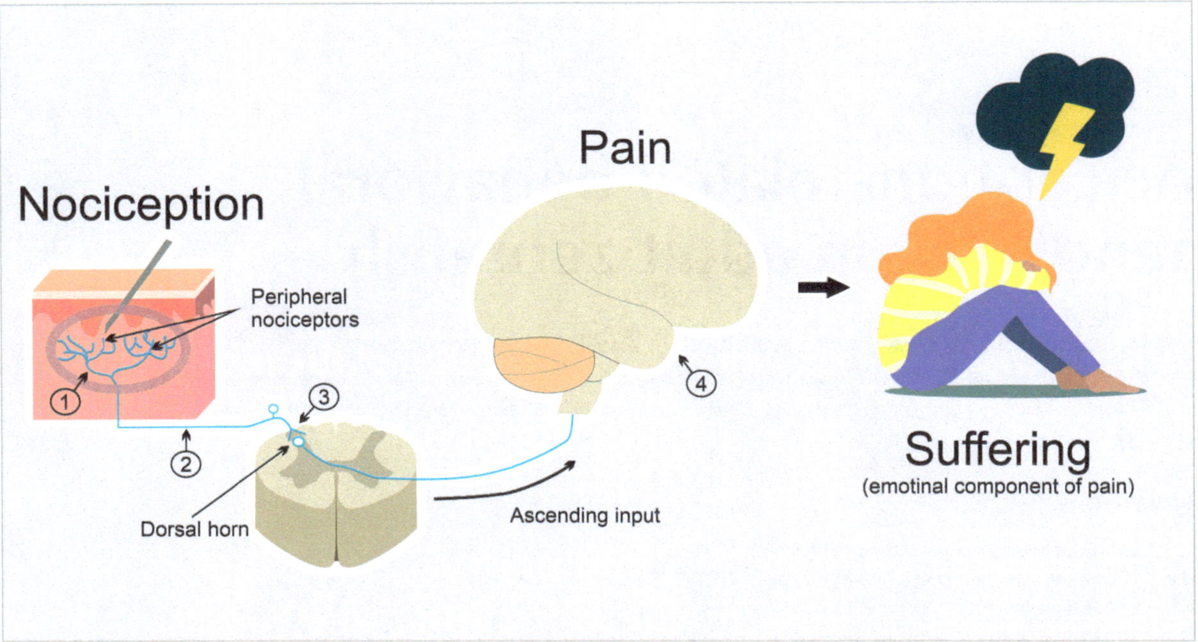

FIG. 1 Simplified diagram of the ascending pain pathway. Nociceptive stimulus is recognized by nociceptors. This signal travels to the dorsal horn of the spinal cord where it is sent to the brain for conscious perception of nociceptive stimuli. Finally, subjects exhibit negative emotional responses to nociception and pain, thereby causing suffering. 1—Transduction; 2—Transmission; 3—Modulation; 4—Perception (Basbaum, Bautista, Scherrer, & Julius, 2009).

Zebrafish as animal model for translational pain research

Although rodents are widely used in pain research (Meotti, Coelho, & Santos, 2010; Stevenson, Bilsky, & Negus, 2006), novel alternative experimental models, such as zebrafish, are also becoming important. For example, they can increase our understanding of evolutionarily conserved, "core" mechanisms underlying nociception and related behaviors (Du et al., 2017; Sneddon, Braithwaite, & Gentle, 2003b). Indeed, zebrafish are becoming critical in probing pain-related mechanisms and behaviors (Gonzalez-Nunez & Rodriguez, 2009), based on their evolutionarily conserved mechanisms involved in nociception, and robust, well-characterized behaviors (Fig. 2) that are bidirectionally sensitive to various algogens and analgesic drugs (Bao et al., 2019; Costa et al., 2019; Kalueff et al., 2013; Sneddon et al., 2003b; Taylor et al., 2017).

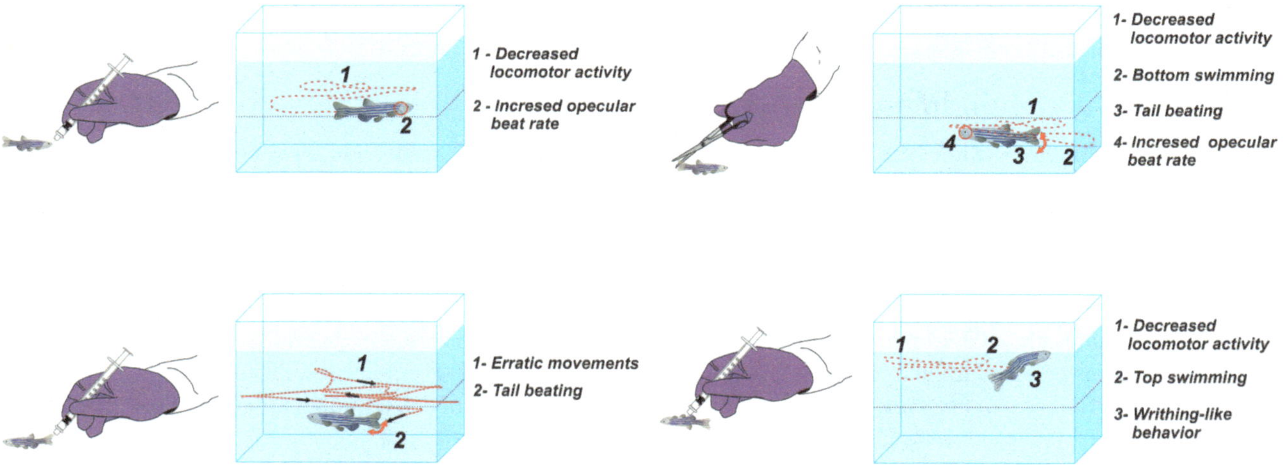

FIG. 2 Behavioral neurophenotyping of pain in zebrafish models. A brief summary of the main experimental protocols to assess pain in adult zebrafish and their respective behavioral phenotypes.

Orthologs to human receptors involved in nociception have already been characterized in zebrafish, including the μ opioid receptor (zMOP; encoded by *opriml*) (Barrallo, Gonzalez-Sarmiento, Alvar, & Rodriguez, 2000), the κ opioid receptor (zKOP; encoded by *oprkl*) (Alvarez et al., 2006), the two functional copies of zebrafish δ opioid receptor (zDOP; encoded by *oprdla* and *oprdlb*) (Barrallo, Gonzalez-Sarmiento, Porteros, Garcia-Isidoro, & Rodriguez, 1998; Pinal-Seoane et al., 2006), the nociceptin opioid receptors (zNOP, encoded by *oprll*) (Rivas-Boyero et al., 2011), as well as the transient receptor potential ion channels (TRPA1a, TRPA1b, and TRPV1) (Prober et al., 2008; Saito & Shingai, 2006) and acid-sensing ion channels (Paukert et al., 2004). Overt electrical activity and transcriptional changes in the midbrain and forebrain occur after harmful stimuli in teleost fishes, including zebrafish (Braithwaite & Boulcott, 2007; Nordgreen, Tahamtani, Janczak, & Horsberg, 2014; Reilly, Quinn, Cossins, & Sneddon, 2008). Moreover, while structural anatomy of nociceptive fibers in zebrafish has not been examined in detail, other teleost species possess typical nociceptors (e.g., A-delta and C-fibers) (Sneddon, 2002), collectively suggesting that this taxon fully adheres to the Bateson-Sneddon's criteria of sentience listed above.

Analyses of specific behavioral phenotypes expressed in the presence of algogens is a cornerstone for an adequate assessment of nociception-related responses in vivo. Unlike rodents, which have multiple well-characterized pain-like behaviors (Bobinski, Teixeira, Sluka, & Santos, 2018; Bonin, Bories, & De Koninck, 2014; De Rantere, Schuster, Reimer, & Pang, 2016; Gonzalez-Cano et al., 2017; Martins et al., 2018), such phenotypes in zebrafish remain poorly understood (Table 1), necessitating further translational research in this organism.

Zebrafish-based pain models

Following the characterization of polymodal nociceptors in teleost fish (Sneddon, Braithwaite, & Gentle, 2003a), the search for new approaches to assess pain responses in zebrafish has begun. The first protocol reported in zebrafish involved the

TABLE 1 Selected examples of zebrafish-based pain models in the AB strain (see Fig. 2 for a graphic summary).

Noxious agents	Treatment	Behavioral endpoints	References
Acetic acid (5%)	Injected (lips)	Decreased locomotor activity, increased opercular beat rate, and rocking behavior	Reilly et al. (2008)
Acetic acid (1%)[a]	Injected (tail)	Increased both erratic movements and tail beating	Maximino (2011)
Acetic acid (5 and 10%)	Injected (lips)	Decreased locomotor activity	Correia, Cunha, Scholze, and Stevens (2011)
Fin clipping (30 s)	Removal of caudal posterior tissue	Increased opercular beat rate, increased tail beating and decreased locomotor activity	Schroeder and Sneddon (2017)
Formalin (0.1%)	Injected (lips and tail)	Decreased locomotor activity	Magalhaes et al. (2017)
Histamine (0.1, 0.5, 1 and 2 mg/kg)	Injected (lips)	Decreased locomotor activity	Taylor et al. (2017)
Complete Freund's adjuvant	Injected (lips)	Decreased locomotor activity	Taylor et al. (2017)
Cinnamaldehyde (10, 20 and 40 mM)	Injected (lips)	Decreased locomotor activity	Taylor et al. (2017)
Alila isothiocyanate (1.02 M)	Injected (lips)	Decreased locomotor activity	Taylor et al. (2017)
Acetic acid (2.5, 5, 10 and 15%)	Injected (lips)	Decreased locomotor activity	Taylor et al. (2017)
Acetic acid (0.5, 1, 2.5 and 5%)	Injected (peritoneum)	Decreased locomotor activity, increased time spent on top area and altered body curvature	Costa, Rosa, et al. (2019)
Fin clipping	Removal of caudal posterior tissue	Increased time at the bottom, decreased locomotor activity and more clustered shoals	Thomson, Al-Temeemy, Isted, Spencer, and Sneddon (2019)

[a]*In the SF strain.*

application of 5 μL of 5% acetic acid onto the fish lips, causing hypolocomotion, increased opercular beat rate, and an aberrant "rocking" behavior (Reilly et al., 2008), as briefly summarized in Fig. 2.

Altered locomotion is an important behavioral endpoint which indicates poor welfare (Kristiansen et al., 2004), and is similar to the immobility observed in other animals (e.g., chickens) after a noxious stimulation (Gentle, 1992). Albeit useful for evaluating pain-like behavior in adult zebrafish (Correia et al., 2011; Curtright et al., 2015; Taylor et al., 2017), this endpoint does not necessarily represent specific pain-like behavior, since it may indicate states of stress, anxiety, nonspecific sedation, toxicity and/or a pro-depressant behavior (Kalueff et al., 2013).

Thus, different approaches to induce local nociception in fish species, such as administration of algogens close to the fins (Maximino, 2011), or the caudal fin stimulation procedure (Schroeder & Sneddon, 2017), have been proposed. Both protocols induce erratic movements (swimming in a zig-zag unpredictable pattern) and increased tail beating (tail-beat movements that do not lead to propulsion in the water, Fig. 2), suggesting a local nociceptive effect. Additionally, the writhing-like phenotype has been recently reported as a novel behavioral endpoint to measure local pain (Costa, Canzian, Stefanello, Kalueff, & Rosemberg, 2019; Costa, Rosa, et al., 2019). Such behavior can be elicited by a single intraperitoneal injection of 5 μL 5% acetic acid, producing marked changes in body posture (resembling the writhing phenotype in rodents), as well as hypolocomotion. Interestingly, the writhing-like behavior of zebrafish (Fig. 2) reflects a direct activation of nociceptors following acetic acid administration, and is also sensitive to classical anti-pain medications (e.g., morphine and diclofenac).

Furthermore, nociceptive effects following the administration of different algogenic agents have also been reported in zebrafish. For example, administration of histamine, complete Freund's adjuvant, cinnamaldehyde, alila isothiocyanate, or formalin, lowers zebrafish locomotor activity (strikingly similar to effects in rodents) (Magalhaes et al., 2017; Taylor et al., 2017). However, unlike pain caused by these substances, there are no current protocols that measure mechanical allodynia (pain to nonnociceptive stimuli) or neuropathic pain in adult zebrafish, thus meriting further scrutiny. Finally, although measuring a single pain-related behavioral endpoint may yield conflicting or nonspecific results, the parallel assessment of several different behaviors and using alternative animal models and/or species can improve face and predictive validity of such models of pain.

Applications to other areas

Moreover, because zebrafish have sophisticated behaviors, automated video-tracking tools can be used to perform 3D reconstructions of the swimming traces (Cachat et al., 2011; Rosa et al., 2018; Thomson et al., 2019) to assess pain-related behaviors. The expression of all major opioid receptors, as well as the presence of well-characterized neurotransmitter systems, makes zebrafish a valuable tool to investigate central mechanisms underlying their pain pathobiology. Given high fertility rate of zebrafish and the transparency of embryos (Antinucci & Hindges, 2016; Sakai, Ijaz, & Hoffman, 2018), the analysis of molecular brain markers using whole-mount in situ hybridization and fluorescent probes can help determine how genes and proteins exert genetic and epigenetic regulation of pain. Finally, high genetic homology (>70%) with humans, evolutionarily conserved molecular targets, as well as small size and easy/inexpensive maintenance (Fontana, Mezzomo, Kalueff, & Rosemberg, 2018; Howe et al., 2013), make zebrafish models a powerful tool in translational pain research.

Other agents of interest

Furthermore, assessing specific behaviors can be important for elucidating neural circuitry and specific brain areas activated when zebrafish are exposed to noxious stimuli. Measuring the relative expression of *C-fos* (Chatterjee, Tran, Shams, & Gerlai, 2015), an immediate early gene that reflects neuronal activation (e.g., using whole-mount in situ hybridization and/or immunoistochemistry), can be promising to investigate how specific brain regions contribute to pain-like responses. Moreover, the spatiotemporal measurements of behavior not only in adult fish, but also using larval fish, may enhance probing how locomotion can be affected during and after the exposure to algogens.

Examining how genes influence nociception is still in its early stages in various organisms, including zebrafish (Bao et al., 2019; Barrallo et al., 2000). However, because zebrafish have a high degree of genetic conservation, this species offers a promising forward-genetics system to explore evolutionarily conserved candidate genes in nociception (Howe et al., 2013). Analyzing physiological parameters is a key strategy in elucidating mechanisms underlying pain responses. Likewise, because zebrafish express well-characterized aversive behaviors (e.g., bottom dwelling, freezing, erratic movements) (Kalueff et al., 2013), the correlation of fear/anxiety-like responses with specific pain-like behaviors can elucidate how stress influence pain in zebrafish and vice-versa.

The use of pharmacological tools also facilitates the discovery of novel mechanisms involved in zebrafish nociception. For example, both agonists (morphine) and antagonists (naloxone) of opioid receptors play a key role in opioidergic modulation of analgesia in zebrafish, regulating local pain (Bao et al., 2019; Demin et al., 2018; Taylor et al., 2017). Moreover, the suitability of zebrafish for medium-to-high throughput pharmacological screening (Stewart, Gerlai, & Kalueff, 2015) may accelerate preclinical discovery of novel drugs with analgesic properties.

Finally, using systems biology approaches to investigate transcriptome, proteome, and metabolome can be strategically incorporated into zebrafish-based pain models. Such omics-based approaches can also facilitate the translatability of zebrafish studies across species aiming to identify evolutionarily conserved pathways related to pain and nociception.

Mini-dictionary of terms

Nociceptors. Sensory receptors transducing and encoding noxious stimuli.

Teleosts The largest infraclass of *Actinopterygii*, the ray-finned fish, comprising around 96% of the existing fish species.

Formalin. Injection of formalin induces a concentration-dependent long-lasting (at least 14 days) mechanical allodynia (first stage) and hyperalgesia (second stage). At the former, formalin activates C-fibers, releasing substance P and bradykinin. At the latter, various pro-inflammatory mediators are released (e.g., histamine, prostaglandins, and serotonin).

Histamine. An important mediator of the inflammatory response.

Cinnamaldehyde. Organic compound found in cinnamon, an agonist of the TRPA1 receptor.

Alila isothiocyanate. An organic compound found in mustard oil, and agonist of the TRPA1 receptor.

Erratic movements. Fast left-right swimming with abrupt changes in velocity and direction of movement. This phenotype indicates increased fear/anxiety-like states.

Bottom dwelling. Movement to/preference toward the bottom of the tank, often in response to threat. This phenotype indicates increased fear/anxiety-like states.

Freezing. A complete cessation of movement (except for gills and eyes) by the fish while at the bottom of the tank. This phenotype indicates increased fear/anxiety-like states.

Key facts

Key facts of zebrafish

- The complete genome sequence of the zebrafish was published in 2013.
- The zebrafish is a tropical fish native to the Southeast Asia.
- The use of zebrafish as a model organism in biomedicine has begun in the 1960s.
- Zebrafish have a high (>70%) genetic homology to humans.
- Nearly 85% of genes known to be associated with human diseases have a zebrafish counterpart.

Summary points

- Unlike humans, laboratory animal models have less complex neural processing networks.
- The zebrafish has evolutionarily conserved mechanisms involved in nociception, and its behaviors are bidirectionally sensitive to various algogens and analgesic drugs.
- Overt electrical activity and transcriptional changes in the midbrain and forebrain occur after harmful stimuli in teleosts.
- Analyses of specific behavioral phenotypes expressed in the presence of algogens is key to adequately assess nociceptive responses in vivo.
- Because zebrafish have complex behaviors, automated video-tracking tools can be used, including 3D reconstructions of swimming traces, to assess pain-like behaviors.
- Zebrafish represent an important, translationally relevant model system to investigate the evolutionarily conserved neural bases of pain responses in vertebrates.

References

Alvarez, F. A., Rodriguez-Martin, I., Gonzalez-Nunez, V., Marron Fernandez de Velasco, E., Gonzalez Sarmiento, R., & Rodriguez, R. E. (2006). New kappa opioid receptor from zebrafish (*Danio rerio*). *Neuroscience Letters*, *405*(1–2), 94–99. https://doi.org/10.1016/j.neulet.2006.06.028.

Antinucci, P., & Hindges, R. (2016). A crystal-clear zebrafish for in vivo imaging. *Scientific Reports*, *6*, 29490. https://doi.org/10.1038/srep29490.

Bao, W., Volgin, A. D., Alpyshov, E. T., Friend, A. J., Strekalova, T. V., de Abreu, M. S., … Kalueff, A. V. (2019). Opioid neurobiology, neurogenetics and neuropharmacology in zebrafish. *Neuroscience*, *404*, 218–232. https://doi.org/10.1016/j.neuroscience.2019.01.045.

Barrallo, A., Gonzalez-Sarmiento, R., Alvar, F., & Rodriguez, R. E. (2000). ZFOR2, a new opioid receptor-like gene from the teleost zebrafish (*Danio rerio*). *Brain Research. Molecular Brain Research*, *84*(1–2), 1–6. https://doi.org/10.1016/S0169-328X(00)00152-2.

Barrallo, A., Gonzalez-Sarmiento, R., Porteros, A., Garcia-Isidoro, M., & Rodriguez, R. E. (1998). Cloning, molecular characterization, and distribution of a gene homologous to delta opioid receptor from zebrafish (*Danio rerio*). *Biochemical and Biophysical Research Communications*, *245*(2), 544–548. https://doi.org/10.1006/bbrc.1998.8496.

Basbaum, A. I., Bautista, D. M., Scherrer, G., & Julius, D. (2009). Cellular and molecular mechanisms of pain. *Cell*, *139*(2), 267–284. https://doi.org/10.1016/j.cell.2009.09.028.

Bateson, P. (1991). Assessment of pain in animals. *Animal Behaviour*, *42*, 827–839. https://doi.org/10.1016/S0003-3472(05)80127-7.

Blackburn-Munro, G. (2004). Pain-like behaviours in animals – How human are they? *Trends in Pharmacological Sciences*, *25*(6), 299–305. https://doi.org/10.1016/j.tips.2004.04.008.

Bobinski, F., Teixeira, J. M., Sluka, K. A., & Santos, A. R. S. (2018). Interleukin-4 mediates the analgesia produced by low-intensity exercise in mice with neuropathic pain. *Pain*, *159*(3), 437–450. https://doi.org/10.1097/j.pain.0000000000001109.

Bonin, R. P., Bories, C., & De Koninck, Y. (2014). A simplified up-down method (SUDO) for measuring mechanical nociception in rodents using von Frey filaments. *Molecular Pain*, *10*, 26. https://doi.org/10.1186/1744-8069-10-26.

Braithwaite, V. A., & Boulcott, P. (2007). Pain perception, aversion and fear in fish. *Diseases of Aquatic Organisms*, *75*(2), 131–138. https://doi.org/10.3354/dao075131.

Cachat, J., Stewart, A., Utterback, E., Hart, P., Gaikwad, S., Wong, K., … Kalueff, A. V. (2011). Three-dimensional neurophenotyping of adult zebrafish behavior. *PLoS One*, *6*(3). https://doi.org/10.1371/journal.pone.0017597, e17597.

Chatterjee, D., Tran, S., Shams, S., & Gerlai, R. (2015). A simple method for immunohistochemical staining of zebrafish brain sections for c-fos protein expression. *Zebrafish*, *12*(6), 414–420. https://doi.org/10.1089/zeb.2015.1147.

Correia, A. D., Cunha, S. R., Scholze, M., & Stevens, E. D. (2011). A novel behavioral fish model of nociception for testing analgesics. *Pharmaceuticals (Basel)*, *4*(4), 665–680. https://doi.org/10.3390/ph4040665.

Costa, F. V., Canzian, J., Stefanello, F. V., Kalueff, A. V., & Rosemberg, D. B. (2019). Naloxone prolongs abdominal constriction writhing-like behavior in a zebrafish-based pain model. *Neuroscience Letters*, *708*, 134336. https://doi.org/10.1016/j.neulet.2019.134336.

Costa, F. V., Rosa, L. V., Quadros, V. A., Santos, A. R. S., Kalueff, A. V., & Rosemberg, D. B. (2019). Understanding nociception-related phenotypes in adult zebrafish: Behavioral and pharmacological characterization using a new acetic acid model. *Behavioural Brain Research*, *359*, 570–578. https://doi.org/10.1016/j.bbr.2018.10.009.

Curtright, A., Rosser, M., Goh, S., Keown, B., Wagner, E., Sharifi, J., … Dhaka, A. (2015). Modeling nociception in zebrafish: A way forward for unbiased analgesic discovery. *PLoS One*, *10*(1). https://doi.org/10.1371/journal.pone.0116766, e0116766.

De Rantere, D., Schuster, C. J., Reimer, J. N., & Pang, D. S. (2016). The relationship between the rat grimace scale and mechanical hypersensitivity testing in three experimental pain models. *European Journal of Pain*, *20*(3), 417–426. https://doi.org/10.1002/ejp.742.

Demin, K. A., Meshalkina, D. A., Kysil, E. V., Antonova, K. A., Volgin, A. D., Yakovlev, O. A., … Kalueff, A. V. (2018). Zebrafish models relevant to studying central opioid and endocannabinoid systems. *Progress in Neuro-Psychopharmacology & Biological Psychiatry*, *86*, 301–312. https://doi.org/10.1016/j.pnpbp.2018.03.024.

Du, X., Yuan, B., Wang, J., Zhang, X., Tian, L., Zhang, T., … Zhang, F. (2017). Effect of heat-reinforcing needling on serum metabolite profiles in rheumatoid arthritis rabbits with cold syndrome. *Zhongguo Zhen Jiu*, *37*(9), 977–983. https://doi.org/10.13703/j.0255-2930.2017.09.018.

Festing, S., & Patel, T. (2005). The ethics of research involving animals: A review of the Nuffield Council on bioethics report from a research perspective. *Alternatives to Laboratory Animals*, *33*(6), 654–658. https://doi.org/10.1177/026119290503300603.

Fontana, B. D., Mezzomo, N. J., Kalueff, A. V., & Rosemberg, D. B. (2018). The developing utility of zebrafish models of neurological and neuropsychiatric disorders: A critical review. *Experimental Neurology*, *299*(Pt A), 157–171. https://doi.org/10.1016/j.expneurol.2017.10.004.

Fordyce, W. E., Lansky, D., Calsyn, D. A., Shelton, J. L., Stolov, W. C., & Rock, D. L. (1984). Pain measurement and pain behavior. *Pain*, *18*(1), 53–69. https://doi.org/10.1016/0304-3959(84)90126-x.

Gentle, M. J. (1992). Pain in birds. *Animal Welfare*, *1*, 235–247.

Gonzalez-Cano, R., Tejada, M. A., Artacho-Cordon, A., Nieto, F. R., Entrena, J. M., Wood, J. N., & Cendan, C. M. (2017). Effects of tetrodotoxin in mouse models of visceral pain. *Marine Drugs*, *15*(6). https://doi.org/10.3390/md15060188.

Gonzalez-Nunez, V., & Rodriguez, R. E. (2009). The zebrafish: A model to study the endogenous mechanisms of pain. *ILAR Journal*, *50*(4), 373–386. https://doi.org/10.1093/ilar.50.4.373.

Howe, K., Clark, M. D., Torroja, C. F., Torrance, J., Berthelot, C., Muffato, M., … Stemple, D. L. (2013). The zebrafish reference genome sequence and its relationship to the human genome. *Nature*, *496*(7446), 498–503. https://doi.org/10.1038/nature12111.

Kalueff, A. V., Gebhardt, M., Stewart, A. M., Cachat, J. M., Brimmer, M., Chawla, J. S., … Zebrafish Neuroscience Research, C. (2013). Towards a comprehensive catalog of zebrafish behavior 1.0 and beyond. *Zebrafish*, *10*(1), 70–86. https://doi.org/10.1089/zeb.2012.0861.

Keefe, F. J., & Pryor, R. W. (2007). Assessment of pain behaviors. In R. F. Schmidt, & W. D. Willis (Eds.), *Encyclopedia of pain* (pp. 136–138). Berlin, Heidelberg: Springer Berlin Heidelberg.

Kristiansen, T. S., Ferno, A., Holm, J. C., Privitera, L., Bakke, S., & Fosseidengen, J. E. (2004). Swimming behaviour as an indicator of low growth rate and impaired welfare in Atlantic halibut (*Hippoglossus hippoglossus L.*) reared at three stocking densities. *Aquaculture*, *230*, 137–151. https://doi.org/10.1016/S0044-8486(03)00436-8.

Magalhaes F. E. A., de Sousa, C., Santos, S., Menezes, R. B., Batista, F. L. A., Abreu, A. O., … Campos, A. R. (2017). Adult Zebrafish (*Danio rerio*): An alternative behavioral model of formalin-induced nociception. *Zebrafish*, *14*(5), 422–429. https://doi.org/10.1089/zeb.2017.1436.

Martins, D. F., Martins, T. C., Batisti, A. P., Dos Santos Leonel, L., Bobinski, F., Belmonte, L. A. O., … Santos, A. R. S. (2018). Long-term regular eccentric exercise decreases neuropathic pain-like behavior and improves motor functional recovery in an axonotmesis mouse model: The role of insulin-like growth Factor-1. *Molecular Neurobiology*, *55*(7), 6155–6168. https://doi.org/10.1007/s12035-017-0829-3.

Maximino, C. (2011). Modulation of nociceptive-like behavior in zebrafish (*Danio rerio*) by environmental stressors. *Psychology & Neuroscience*, *4*(1), 149–155. https://doi.org/10.3922/j.psns.2011.1.017.

Meotti, F. C., Coelho, I. D. S., & Santos, A. R. (2010). The nociception induced by glutamate in mice is potentiated by protons released into the solution. *The Journal of Pain*, *11*(6), 570–578. https://doi.org/10.1016/j.jpain.2009.09.012.

Nordgreen, J., Tahamtani, F. M., Janczak, A. M., & Horsberg, T. E. (2014). Behavioural effects of the commonly used fish anaesthetic tricaine methanesulfonate (MS-222) on zebrafish (*Danio rerio*) and its relevance for the acetic acid pain test. *PLoS One*, *9*(3). https://doi.org/10.1371/journal.pone.0092116, e92116.

Paukert, M., Sidi, S., Russell, C., Siba, M., Wilson, S. W., Nicolson, T., & Grunder, S. (2004). A family of acid-sensing ion channels from the zebrafish: Widespread expression in the central nervous system suggests a conserved role in neuronal communication. *The Journal of Biological Chemistry*, *279* (18), 18783–18791. https://doi.org/10.1074/jbc.M401477200.

Piel, M. J., Kroin, J. S., van Wijnen, A. J., Ranjan, K. C., & Im, H. J. (2014). Pain assessment in animal models of osteoarthritis. *Gene*, *537*(2), 184–188. https://doi.org/10.1016/j.gene.2013.11.091.

Pinal-Seoane, N., Martin, I. R., Gonzalez-Nunez, V., Marron Fernandez de Velasco, E., Alvarez, F. A., Sarmiento, R. G., & Rodriguez, R. E. (2006). Characterization of a new duplicate delta-opioid receptor from zebrafish. *Journal of Molecular Endocrinology*, *37*(3), 391–403. https://doi.org/10.1677/jme.1.02136.

Prober, D. A., Zimmerman, S., Myers, B. R., McDermott, B. M., Jr., Kim, S. H., Caron, S., … Schier, A. F. (2008). Zebrafish TRPA1 channels are required for chemosensation but not for thermosensation or mechanosensory hair cell function. *The Journal of Neuroscience*, *28*(40), 10102–10110. https://doi.org/10.1523/JNEUROSCI.2740-08.2008.

Reilly, S. C., Quinn, J. P., Cossins, A. R., & Sneddon, L. U. (2008). Behavioural analysis of a nociceptive event in fish: Comparisons between three species demonstrate specific responses. *Applied Animal Behaviour Science*, *114*, 248–259. https://doi.org/10.1016/j.applanim.2008.01.016.

Rivas-Boyero, A. A., Herrero-Turrion, M. J., Gonzalez-Nunez, V., Sanchez-Simon, F. M., Barreto-Valer, K., & Rodriguez, R. E. (2011). Pharmacological characterization of a nociceptin receptor from zebrafish (*Danio rerio*). *Journal of Molecular Endocrinology*, *46*(2), 111–123. https://doi.org/10.1530/JME-10-0130.

Rosa, L. V., Ardais, A. P., Costa, F. V., Fontana, B. D., Quadros, V. A., Porciuncula, L. O., & Rosemberg, D. B. (2018). Different effects of caffeine on behavioral neurophenotypes of two zebrafish populations. *Pharmacology, Biochemistry, and Behavior*, *165*, 1–8. https://doi.org/10.1016/j.pbb.2017.12.002.

Saito, S., & Shingai, R. (2006). Evolution of thermoTRP ion channel homologs in vertebrates. *Physiological Genomics*, *27*(3), 219–230. https://doi.org/10.1152/physiolgenomics.00322.2005.

Sakai, C., Ijaz, S., & Hoffman, E. J. (2018). Zebrafish models of neurodevelopmental disorders: Past, present, and future. *Frontiers in Molecular Neuroscience*, *11*, 294. https://doi.org/10.3389/fnmol.2018.00294.

Schroeder, P. G., & Sneddon, L. U. (2017). Exploring the efficacy of immersion analgesics in zebrafish using an integrative approach. *Applied Animal Behaviour Science*, *187*, 93–102. https://doi.org/10.1016/j.applanim.2016.12.003.

Sneddon, L. U. (2002). Anatomical and electrophysiological analysis of the trigeminal nerve in a teleost fish, *Oncorhynchus mykiss*. *Neuroscience Letters*, *319*(3), 167–171. https://doi.org/10.1016/s0304-3940(01)02584-8.

Sneddon, L. U., Braithwaite, V. A., & Gentle, M. J. (2003a). Do fishes have nociceptors? Evidence for the evolution of a vertebrate sensory system. *Proceedings of the Biological Sciences*, *270*(1520), 1115–1121. https://doi.org/10.1098/rspb.2003.2349.

Sneddon, L. U., Braithwaite, V. A., & Gentle, M. J. (2003b). Novel object test: Examining nociception and fear in the rainbow trout. *The Journal of Pain*, *4* (8), 431–440. https://doi.org/10.1067/S1526-5900(03)00717-X.

Sneddon, L. U., Elwood, R. W., Adamo, S. A., & Leach, M. C. (2014). Defining and assessing animal pain. *Animal Behaviour*, *97*, 201–212. https://doi.org/10.1016/j.anbehav.2014.09.007.

Stevenson, G. W., Bilsky, E. J., & Negus, S. S. (2006). Targeting pain-suppressed behaviors in preclinical assays of pain and analgesia: Effects of morphine on acetic acid-suppressed feeding in C57BL/6J mice. *The Journal of Pain*, *7*(6), 408–416. https://doi.org/10.1016/j.jpain.2006.01.447.

Stewart, A. M., Gerlai, R., & Kalueff, A. V. (2015). Developing highER-throughput zebrafish screens for in-vivo CNS drug discovery. *Frontiers in Behavioral Neuroscience*, *9*, 14. https://doi.org/10.3389/fnbeh.2015.00014.

Taylor, J. C., Dewberry, L. S., Totsch, S. K., Yessick, L. R., DeBerry, J. J., Watts, S. A., & Sorge, R. E. (2017). A novel zebrafish-based model of nociception. *Physiology & Behavior*, *174*, 83–88. https://doi.org/10.1016/j.physbeh.2017.03.009.

Thomas, D. (2005). Laboratory animals and the art of empathy. *Journal of Medical Ethics*, *31*(4), 197–202. https://doi.org/10.1136/jme.2003.006387.

Thomson, J. S., Al-Temeemy, A. A., Isted, H., Spencer, J. W., & Sneddon, L. U. (2019). Assessment of behaviour in groups of zebrafish (*Danio rerio*) using an intelligent software monitoring tool, the chromatic fish analyser. *Journal of Neuroscience Methods*, *328*, 108433. https://doi.org/10.1016/j.jneumeth.2019.108433.

Tracey, I., & Mantyh, P. W. (2007). The cerebral signature for pain perception and its modulation. *Neuron*, *55*(3), 377–391. https://doi.org/10.1016/j.neuron.2007.07.012.

Chapter 35

Pain, implantable pain devices, and psychosocial aspects of pain

Vishal Varshney[a], Thomas Rutledge[b], Maya D'Eon[c], and Krishnan Chakravarthy[d,e]
[a]*St. Paul's Hospital, Providence Healthcare, Vancouver, BC, Canada,* [b]*VA San Diego Healthcare System/UC San Diego Department of Psychiatry, Psychology Service, San Diego, CA, United States,* [c]*UCSD Department of Psychiatry, La Jolla, CA, United States,* [d]*Division of Pain Medicine, Department of Anesthesiology, University of California San Diego, La Jolla, CA, United States,* [e]*VA San Diego Health Care, San Diego, CA, United States*

Anatomy of pain pathways

Pain is defined as "an unpleasant sensory and emotional experience associated with actual or potential tissue damage, or described in terms of such damage" (Steeds, 2016). Inherent in this definition is an acceptance that pain can contribute to negative psychosocial experiences. The psychosocial impact of chronic pain has been well documented, as patients with chronic pain are two to five times more likely to develop depression, anxiety, and sleep disorders than in a generally healthy population (Haythornthwaite & Benrud-Larson, 2000). By understanding the anatomy of pain pathways, we can understand how pain contributes to psychosocial behaviors. This section aims to briefly summarize pain pathway anatomy.

Nociceptors are receptors in peripheral tissues that are activated specifically by painful stimuli. Painful stimuli are transduced by nociceptors into electrical and chemical signals, and transmitted from the periphery to the central nervous system, in a pathway called the ascending pain pathway or spinothalamic tract. This pathway involves four steps (transduction, transmission, modulation, and perception) and three orders of neurons (Fig. 1). First-order neurons traverse from peripheral nociceptors to the dorsal horn of the spinal cord, and their cell bodies are located in either the dorsal root ganglia or trigeminal ganglion. These neurons are responsible for transduction and transmission of pain signals. Second-order neurons traverse from the dorsal horn, which is divided into sections called laminae, to the medial and lateral thalamus. The thalamus is a key area for modulation of pain signals. Third-order neurons traverse from the thalamus to the cerebral cortex, where pain perception occurs. The laminae in the dorsal horn, labeled from I to V, contain several types of second-order neurons that respond to various types of noxious stimuli—laminae II and V are important for modulation and localization of pain. Neurons in lamina V receive convergent nonnoxious and noxious input via direct and indirect neuronal fiber inputs. These neurons are called wide dynamic range (WDR) neurons, in that they respond to a broad range of stimulus intensities. There is also commonly a visceral input to these WDR neurons, such that the resultant convergence of somatic and visceral input likely contributes to the phenomenon of referred pain. WDR neurons fire action potentials in a graded fashion, depending on stimulus intensity, and also exhibit "wind-up" (a short-lasting form of synaptic plasticity). During wind-up, repetitive stimulation of WDR neurons induces an increase of their evoked response and postdischarge with each stimulus. Because of this, pain therapies like Spinal Cord Stimulation (SCS) target WDR neurons and attempt to minimize their wind-up.

The dorsal horn and laminae of the spinal cord are also where another pain pathway, called the descending inhibitory tract, exerts its effect. Two neurotransmitters, norepinephrine and serotonin, are key to descending inhibition, where brainstem structures (like the periaqueductal gray and the nucleus raphe magnus) try to block pain transmission from the ascending pain pathway by either direct action on dorsal horn cells, inhibiting excitatory dorsal horn neurons or causing excitation of inhibitory neurons. The role of these specific neurotransmitters in modulating mood and anxiety cannot be overlooked when determining the role of pain in psychosocial behavior.

Our continuing understanding of how pain is processed now delineates the ascending and descending pathways even further, into three pathways—two ascending pain-evoking pathways, and at least one descending pain-inhibitory pathway. The two ascending pathways are labeled based on their anatomical location where they synapse in the thalamus, as lateral and medial pathways. The lateral pathway encodes the discriminatory and sensory component of pain, while the medial pathway encodes the motivational and affective component of pain—commonly described as "pain unpleasantness," or in

The Neurobiology, Physiology, and Psychology of Pain. https://doi.org/10.1016/B978-0-12-820589-1.00035-X

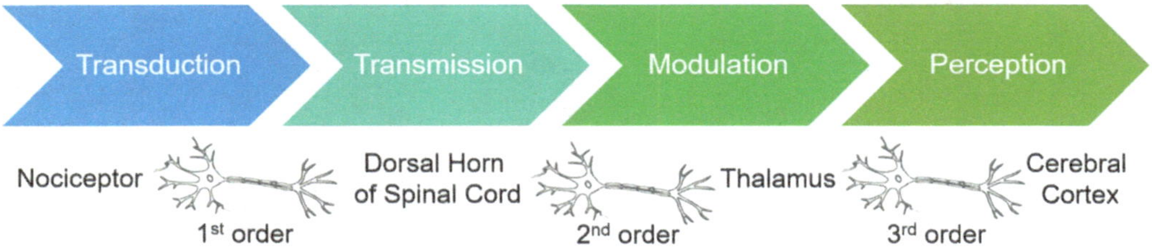

FIG. 1 Steps and orders of neurons involved in pain processing.

other words, "suffering." Given that pain is defined as an unpleasant sensory and emotional experience associated with actual or potential tissue damage, modulating both the sensory and affective components of pain through these pathways is important in chronic pain management. One such form of implantable electrical stimulation that has been demonstrated to modulate both the lateral and medial pain pathways is called nonlinear burst spinal cord stimulation, which will be discussed in detail later.

Psychosocial aspects of neuropathic pain

Patients considered for implantable pain device therapies present with high rates of comorbid psychiatric conditions. Many pain providers appreciate that conditions, such as clinical depression, are common among patients with chronic pain; however, they may still underestimate the degree to which this is true particularly for patients seen in their specialty pain clinics. For example, in one of the most widely cited reviews of depression rates among patients with chronic pain across different clinical settings (Bair, Robinson, Katon, & Kroenke, 2003), the authors found that comorbid depression was twice as prevalent (average of 52%) among patients in pain clinics compared to primary care clinics (average of 27%). Although the data are less precise concerning differences in related psychiatric conditions such as anxiety, research suggests similar patterns of disproportionately higher rates of psychiatric comorbidity among the relatively more chronic and disabled patients seen in pain clinics. For pain physicians, these statistics can be a sobering reminder of the importance of considering psychiatric characteristics among their patient to optimize treatment benefits.

Neuropathic pain and psychiatric impairment have bidirectional relationships (Fig. 2). Prospective studies, for example, indicate that patients with histories of poor psychiatric health are more likely to develop chronic pain (Ericsson et al., 2002). Alternatively, observational data indicate significant increases in mental illness among patients whose pain becomes chronic and that effective pain treatment often results in improved mental health (Hilderink, Burger, Deeg, Beekman, & Oude Voshaar, 2012). In some cases of neuropathic pain, both the pain symptoms and mental health conditions may even arise from the same causal event. A traumatic injury resulting in chronic neuropathic pain symptoms, for instance,

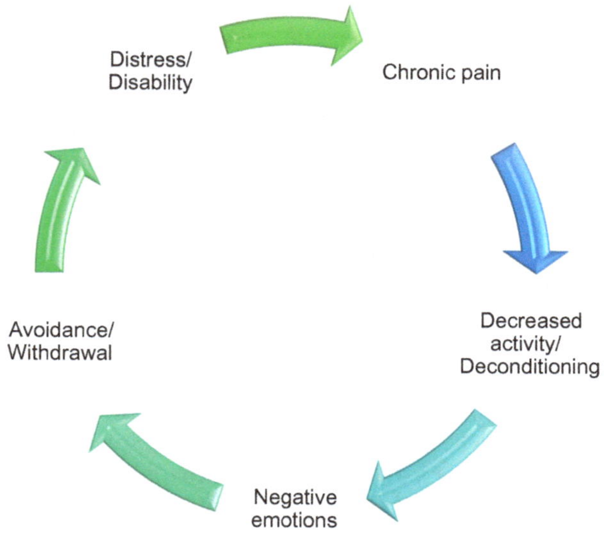

FIG. 2 The cyclical relationship between chronic pain and psychosocial stress.

may also produce posttraumatic stress symptoms. Regardless of the causal direction of the pain-psychiatric relationship, unattended psychiatric difficulties worsen responses to pain treatments, undermine treatment adherence, and increase functional impairment (Bair et al., 2003). An important role of the pain psychologist is to determine how pain and psychiatric factors are affecting the patient and help factor this information into their treatment plans.

Beyond even the psychiatric comorbidities most frequently associated with chronic pain conditions, the most common pathway through which psychosocial factors affect neuropathic pain treatment outcomes is through the patient's pain coping behaviors. During acute neuropathic injury, coping behaviors such as rest, taking time off work, and increased treatment-seeking behaviors are usually beneficial. In many cases, however, the invisible shift from acute pain to chronic pain results in the patient continuing to practice these acute pain coping behaviors long after the period in which they were adaptive (Bair et al., 2003). Patients may, in some cases, learn more effective chronic pain coping behaviors (e.g., pacing, making accommodations to work and leisure activities, practicing safe forms of exercise, decreasing reliance on medicines, etc.) on their own; when these changes have not occurred, however, the pain physician may need to include education, pain psychology referrals, or other forms of assistance to help patients develop improved coping behaviors as part of their implantable device treatments.

Increasingly, our understanding of pain-psychiatric relationships is informed by advances in neuroscience (Flor, 2014). With a growing body of imaging research demonstrating that pain and psychiatric functions are both regulated by brain regions such as the hippocampus, amygdala, and anterior cingulate cortex, and by overlapping neurotransmitters such as norepinephrine, GABA, and glutamate, there is strong evidence for a common biological basis in the comorbidity between chronic pain and mental health. Practically, pain providers can use these advancing neuroscience findings to assist their patients. For example, among patients with chronic neuropathic pain and comorbid depression and anxiety conditions, the pain provider may consider novel SCS waveforms, such as nonlinear burst stimulation (Bocci et al., 2018) that may have positive effects on pain perception and affective pathways. Alternatively, even with traditional SCS waveforms, the pain provider may supplement the SCS device with recommendations for mental health treatment support (research-supported treatment modalities include cognitive behavioral therapy for pain, acceptance and commitment therapy for pain, mindfulness-based stress reduction), regular physical exercise (producing mood benefits that rival gold standard psychotherapies and antidepressants), or SNRIs, such as duloxetine or venlafaxine (that favorably influence neurotransmitters involved in both depression and pain) (Qaseem, Wilt, McLean, Forciea, & for the Clinical Guidelines Committee of the American College of Physicians, 2017; Schuch et al., 2016).

Rationale for psychosocial evaluations

Psychosocial evaluations of candidates for implantable pain devices have been a recommended and integral part of clinical practice since the 1990s (Beltrutti, Lamberto, & Barolat, 2004; Gybels, Erdine, & Maeyaert, 1998). This longstanding practice is largely driven by research demonstrating associations between psychosocial factors, pain, and functional impairment (Block, Ohnmeiss, Guyer, Rashbaum, & Hochschuler, 2001; Verrils, Sinclair, & Barnard, 2016) and corroborated theoretically by the gate control theory, which suggests the impactful role of psychosocial factors in neuromodulation of pain (Melzack & Wall, 1965). Preprocedural psychosocial evaluations are typically performed by pain psychologists or mental health providers with training in behavioral medicine or health psychology. The format of a psychosocial evaluation includes components of (a) medical chart review; (b) clinical interview; and (c) questionnaire data.

Although there is unequivocal support for the interrelationship between psychosocial factors and the pain experience, the evidence is mixed regarding what specific psychosocial factors are reliably predictive of implantable pain device outcomes and should be included in psychosocial evaluations (Grinberg et al., 2019). As a result, there is inconsistency in the contents and implementation of the psychosocial evaluation for implantable devices. More broadly, however, several psychosocial factors have emerged as recommended to incorporate in psychosocial evaluations, including somatoform disorders and malingering, mood and anxiety conditions, substance abuse, treatment expectations and social isolation, suicidality and severe mental illness, cognitive impairment, personality disorders, and secondary gain (Grinberg et al., 2019).

Psychosocial evaluations are partly designed to improve candidate selection and partly to predict postimplantable device outcomes. Traditionally, "success" following an implantable pain device was defined by a reduction in self-reported pain intensity. However, with growing movement away from a narrow biomedical model of pain toward a broader biopsychosocial understanding of pain, there is a greater appreciation for the need to redefine outcomes of pain interventions to reflect the multifaceted nature of the pain experience and to include psychosocial changes and patient-defined goals (Turk et al., n.d.; Department of Veteran's Affairs, 2017; Deyo, Dworkin, & Amtmann, 2014; Dworkin et al., 2005). To integrate these pain management recommendations and address the above limitations that exist, a new model for conducting

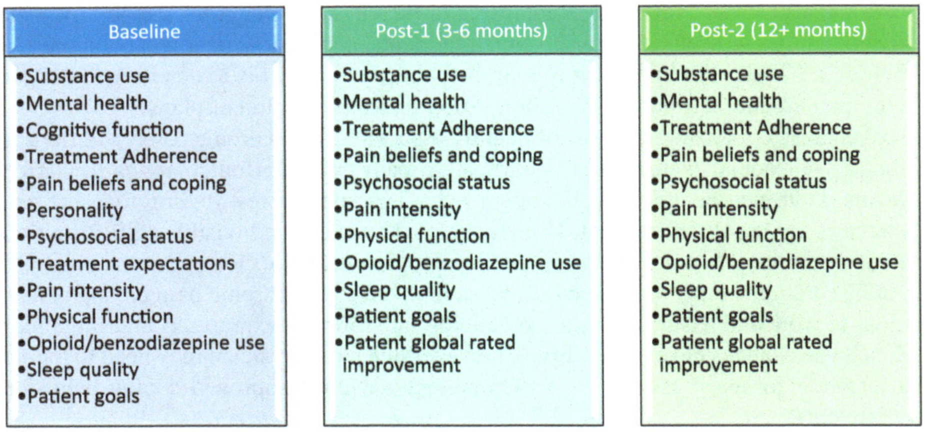

FIG. 3 Proposed model for psychosocial evaluations pre- and postdevice implantation.

psychosocial evaluations has been proposed (Fig. 3). In addition to assessing pain intensity, this model recommends viewing success as a function also of improvements in pain beliefs and ways of coping, physical functioning, medication use, patient-identified functional goals, mood symptoms, sleep, substance and alcohol use, personality, cognitive functioning, and treatment adherence. Optimally, patients should be assessed for progress on these pain and functional outcomes after 3–6 months and again after 12+ months to determine degree of sustained improvements.

Effect of spinal cord stimulation on pain processing

Spinal cord stimulation (SCS) is defined as the alteration of nerve activity through targeted delivery of a stimulus, such as electrical stimulation or chemical agents, to specific neurological sites in the body. Spinal cord stimulators consist of multiple electrodes, or leads, placed within the epidural space to electrically stimulate the dorsal column, the most posterior aspect of the spinal cord where pain pathways are located. These leads are tunneled subcutaneously to an implantable pulse generator (IPG), which contains a battery and microprocessor that delivers electricity at preset amplitudes, frequencies, and/or pulse widths to modulate pain signals.

The underlying mechanism of how SCS helps with pain has been largely attributed to the gate control theory, developed by Drs. Wall and Melzack in 1965 (Melzack & Wall, 1965). This theory suggests that activation of larger, nonpain conducting fibers (i.e., A-beta fibers) would override pain signals transmitted through the dorsal horn of the spinal cord. SCS technology has developed significantly over the last decade, and increasing research into this modality has also revealed newly proposed mechanisms of how SCS works to modulate pain, in addition to the gate control theory. These mechanisms involve anatomic, cellular, physiological, and chemical targets of SCS that alter how pain signals are transmitted, modulated, or perceived by the central nervous system. Tonic (or conventional), burst, and high-frequency stimulation differ based on frequencies, waveform patterns, and how charge transfer is balanced, and produce different patterns of activation of axons and adjacent neural tissues (Table 1) (Miller et al., 2016). This section will briefly review how newer modes of stimulation, such as burst and high frequency stimulation, are thought to utilize these newly identified mechanisms.

TABLE 1 Parameters used and suggested mechanisms of Tonic, Burst, and 10 kHz stimulation (Miller et al., 2016).

	Tonic (conventional)	Burst (nonlinear burst)	10 kHz
Amplitude (mA)	3.6–8.5	0.6	1.6–3.8
Frequency (Hz)	39–77	200	10,000
Pulse width (us)	347–591	1000	30
Suggested mechanism of pain relief	Gate control theory; inhibition of nociceptive sensorimotor reflex	Modulation of medial and lateral pain pathways; activation of anterior cingulate cortex	?Glial-neuronal interactions; attenuation of wind-up in WDR neurons

Tonic or conventional, stimulation has the been the primary form of SCS used for years to manage chronic pain conditions, like failed back surgery syndrome or complex regional pain syndrome. Tonic SCS, when active, induces paresthesias in the patient's pain region. While initially thought to work solely via the gate control theory, it is now thought that tonic stimulation utilizes the neurophysiologic interaction between the sensory and motor systems to modulate pain signals. The nociceptive sensorimotor reflex (NSR) is a polysynaptic spinal reflex considered to be an objective physiologic measure of nociception, and has been shown to correlate positively with perceived pain. Tonic SCS has demonstrated inhibition of this reflex, and correlates with efficacy of stimulation (Meuwissen et al., 2020). Tonic SCS has also been shown to decrease cortical excitability as measured via somatosensory evoked potentials (SSEP) and fMRI investigations. Many neuroanatomical reasons have been proposed for how this is accomplished, including inhibition of peripheral nociceptors, increased regional cerebral blood flow to pain regulatory centers, and increased activation of descending pain pathways. Neurochemical explanations for the mechanism of tonic SCS have also been established, as tonic SCS has been shown to increase levels of acetylcholine, serotonin, and endogenous opioids in animal studies (Jensen & Brownstone, 2019; Sdrulla, Guan, & Raja, 2018).

Despite being a newer modality, nonlinear burst SCS has been rapidly adopted by the pain medicine community as a preferable option to tonic SCS. This mode is paresthesia-free, meaning that patients do not feel paresthesias when the system is activated, which they prefer (De Ridder, Vanneste, Plazier, Van Der Loo, & Menovsky, 2010). Nonlinear burst SCS has also been shown to be superior to tonic SCS for neuropathic pain, with superiority specifically demonstrated in failed back surgery syndrome and radicular pain syndromes. As mentioned in "Anatomy of pain pathways" section, nonlinear burst SCS modulates both the lateral and medial pain pathways, modulating both the sensory components of pain in the lateral pathway, but also the motivational and affective component of pain in the medial pathway. Nonlinear burst SCS has also been shown to decrease WDR neuron activity in the dorsal horn (De Ridder et al., 2010; Meuwissen et al., 2020). These reductions in WDR firing are thought to be driven by some as-yet to be determined neurochemical signaling mechanism. Further research is required to elucidate the neurochemical mechanisms of action of burst stimulation. From a neurochemical perspective, nonlinear burst SCS has been demonstrated to activate the anterior cingulate cortex (ACC), responsible for descending pain inhibition, which suggests nonlinear burst SCS has both central and spinal effects through indirect activation of multiple neuronal pathways (Ahmed, Yearwood, De Ridder, & Vanneste, 2018; Meuwissen et al., 2020). The reason why nonlinear burst SCS activates the ACC is thought to be because mechanistically, following painful stimuli, medial thalamic cells fire in a burst pattern, thus potentiating the ACC (Kinfe et al., 2016; Sdrulla et al., 2018). Notably, the ACC also plays a substantial role in facilitating pain-induced attention, memory, fear, and anxiety behavior, suggesting that modulation of this pathway is crucial to attenuating the psychosocial aspects of pain behavior (Verbruggen, 2014).

High-frequency stimulation is another paresthesia-free mode of stimulation, and is a broad term meant to imply stimulation frequencies greater than that commonly used in tonic SCS (greater than 200 Hz). The most-studied frequency rate has been 10,000 Hz (10 kHz). This mode of stimulation was well studied in a population with failed back surgery syndrome presenting with both back and leg pain (Kapural et al., 2015). Many mechanisms have been proposed for how high-frequency stimulation provides pain relief, including axonal conduction block, desynchronization of axonal activity, and glial-neuronal interactions (Bocci et al., 2018). Preliminary studies have also demonstrated that high-frequency stimulation reduces wind-up in WDR neurons; however, less is known about its direct effects on mood, anxiety, and affective pain components.

Effects of modes of stimulation on psychosocial aspects of pain

While the benefits of SCS for pain have been well established, its objective effect on the psychosocial aspects of pain has only been recently validated. As previously mentioned, the psychosocial impact of pain is well documented, with psychological factors (like anxiety or depression) associated with increased pain and disability. In addition to providing pain relief, SCS can also significantly improve patients' psychological and functional outcomes. The mechanism of this, however, has not been clearly elucidated. Recently, the newer modes of stimulation, specifically burst SCS, have been shown to improve psychosocial aspects, as measured through psychometric studies and correlated with functional imaging, providing us with a clear mechanism of action of how psychological improvement occurs.

The EMP^{3}OWER study (Rosenberg et al., 2016) was a prospective, multicentered, observational study to assess psychological and functional outcomes in relation to pain relief with tonic SCS. Two hundred fifty-two patients (mean age 56.17 years) with chronic neuropathic pain diagnoses were followed over 12 months, and provided with various questionnaires pre- and post-SCS implantation to assess the impact of pain on psychological functioning. These included the short form-McGill Pain Questionnaire (short form-MPQ), Pain Disability Index (PDI), Short-Form 36 item health questionnaire

(SF-36), and the Pain Catastrophizing Scale (PCS). The short form-MPQ contains 22 descriptors of pain (i.e., throbbing, aching, itching), with higher scores indicating higher pain intensity. The PDI asks subjects to rate the severity of disruption caused by chronic pain, with higher scores indicating increased pain disability. The SF-36 is a validated measure of mental and physical health, with higher scores indicating better health. The PCS is a validated scale assessing how a patient thinks and feels about pain while experiencing pain, and assesses dimensions of pain catastrophizing, including magnification, helplessness, and rumination. This study confirmed effective pain control using tonic SCS, with a majority of subjects experiencing persistent pain relief over 50% for the 1-year duration of the study. The MPQ demonstrated reduction in perceived pain intensity throughout the study, with high levels of patient satisfaction with SCS. The PDI was significantly reduced, indicating a diminished disruption of daily activities, social activities, and work caused by chronic pain, which remained consistent at 6 and 12 months. The SF-36 documented improved scores for both mental and physical health from baseline. The PCS demonstrated a reduction of pain catastrophizing among individuals with chronic pain. The composite of significant improvement in these various psychometric studies suggests that the benefits of SCS therapy extend past just pain reduction. The mechanism of this, however, was not elucidated by this study, with the authors suggesting "multidirectional influences on neurobiology, neuroendocrinology, and psychological perceptions" being the main etiology of how tonic SCS improves psychosocial functioning.

In the previous section, mechanisms of action for pain relief pertaining to the different modes of SCS were discussed. Recent electrophysiological studies have shed light on how these modes of stimulation improve psychosocial aspects of pain, as this area continues to be established. Bocci et al. (Bocci et al., 2018) measured laser-evoked potentials (LEPs) and motor-evoked potentials (MEPs) to analyze how tonic, nonlinear burst, and high-frequency SCS affect central nervous system parameters. They confirmed that nonlinear burst SCS leads to a larger pain suppression than tonic stimulation, with nonlinear burst SCS having a novel mechanism of action at the cortical level. Specifically, they confirmed that nonlinear burst SCS has a greater effect on the ACC than both tonic and high-frequency SCS, with nonlinear burst SCS modulating a wider network comprising both the lateral and medial pain pathways. They also demonstrated nonlinear burst SCS modulates both the intracortical GABA(*b*)ergic and glutamatergic networks, which are both pathways recently implicated in mood disorders and as pharmacologic targets for mood-stabilizing agents.

The effect of nonlinear burst SCS on the affective part of pain processing is a prominent reason for its demonstrated improvement on psychosocial aspects of pain. This has been demonstrated with psychometric studies and multiple objective imaging modalities. Chakravarthy et al. (Chakravarthy, Malayil, Kirketeig, & Deer, 2019) demonstrated this in their recent systematic review and pooled analysis of real-world evidence and outcomes data of over 15 studies with more than 400 patients overall. They showed that patient-reported outcome measures, including depression, pain interference, and catastrophizing scales, were consistently improved with nonlinear burst SCS compared to tonic SCS. This effect was very pronounced, especially with the PCS, which improved to better than population norms.

EEG data have also demonstrated significantly higher activity in the dorsal thalamus with nonlinear burst SCS compared to other stimulation modes, further linking this mode of stimulation to the medial supraspinal pathway and the emotional, affective components to pain processing. When looking at blood oxygenation level-dependent (BOLD)-functional MRI (fMRI) imaging in failed back surgery syndrome (FBSS) patients during tonic SCS. Rasche et al. (Rasche et al., 2005) showed that tonic SCS alters mainly the lateral somatosensory pathway, and to a lesser extent, the structures of the medial affective pathway. A recent fMRI study by Meuwissen et al. (Meuwissen et al., 2020) showed that nonlinear burst SCS significantly increased BOLD signal levels in the raphe nuclei, nucleus accumbens, and caudate putamen in chronic neuropathic rats. These are all neural structures critically important in the regulation of mood and anxiety behaviors. Tonic SCS did not induce a significant increase in BOLD signal levels in these areas. These studies confirm that nonlinear burst SCS has potentially an equivalent, if not greater, impact on psychosocial aspects of mood and behavior than pain modulation.

Conclusions

Given the importance of recognizing pain in a biopsychosocial model, interventions that demonstrate more than just a reduction in pain intensity are crucial to a patient's overall management of their chronic pain. Modes of spinal cord stimulation have evolved substantially over the last decade, and they have demonstrated their ability to not only reduce pain intensity, but also improve affective aspects of pain in domains of pain disability, interference, catastrophizing, and physical and mental health. Of the modes of SCS available, nonlinear burst SCS has demonstrated the most promise in its ability to modulate both discriminatory and psychosocial aspects of pain. For future studies, affective measures such as the PDI, PCS, and depression scales should be recorded in conjunction with visual analog scales and functional improvement measures. This would be in accordance with recommendations of the Initiative on Methods, Measurement, and Pain Assessment in Clinical Trials group (IMMPACT) (Dworkin et al., 2005, 2008). Knowing that nonlinear burst SCS improves comorbid psychiatric symptoms associated with pain, this may also change how pain medicine physicians evaluate patients for SCS candidacy. In the future, comorbid psychological conditions may no longer be treated as possible exclusion criteria for SCS, for example, and instead be considered as primary outcomes along with changes in pain intensity and functional improvement.

Application to other areas

The link between pain and psychosocial disorders is well established. By being able to modulate both pain pathways and affective components of pain, implantable devices and modes of spinal cord stimulation also demonstrate an ability to regulate mood, anxiety, and other psychosocial comorbidities. This raises the hypothesis that implantable devices may also demonstrate efficacy in refractory psychiatric conditions. The evidence supporting neuromodulation therapies for psychiatric conditions is growing, especially with the use of transcranial magnetic stimulation and deep brain stimulation. They have largely been applied to refractory conditions, but are growing in indication to include personality disorders, posttraumatic stress disorder, and psychosis. Now, with the aforementioned evidence supporting nonlinear burst spinal cord stimulation for pain and psychosocial disorders, we can apply this to perhaps improve pain coping and catastrophizing, in addition to nonpain-related psychosocial disorders.

Other agents of interest

While nonlinear burst spinal cord stimulation has demonstrated an ability to modulate both pain and psychosocial aspects of pain, other modes of spinal cord stimulation may also be effective in achieving this. Glial cell modulation may demonstrate an ability to regulate mood and emotions, which may also improve the experience of pain perceived by patients. Determining ways to selectively stimulate psychological pathways in the periphery, such as through peripheral nerve stimulation, may also achieve success in modulating both pain and psychosocial aspects of pain. Last, differing modes of stimulation through deep brain stimulation may also achieve the balance of modulating pain signals and psychological pathways, recognizing that this may be effective in severe refractory cases.

Mini-dictionary of terms

Spinal cord stimulation: a therapy where electrical signals are sent to the dorsal columns of the spinal cord via electrodes placed in the epidural space for the purpose of modulating pain and pain signals.
Nonlinear burst: a form of stimulation that has been demonstrated to modulate both the lateral and medial pain pathways.
Medial pathway: an ascending pain pathway responsible for encoding the motivational and affective component of pain, commonly described as "pain unpleasantness," or in other words, "suffering."
Psychosocial: an approach by which individuals are assessed in the context of the combined influence that psychological factors and the surrounding social environment have on their physical and mental wellness and their ability to function.
Catastrophizing: an exaggerated negative mental set brought to bear during an actual or anticipated painful experience.
Psychiatric condition: a condition defined by criteria in the Diagnostic and Statistical Manual of Mental Disorders (DSM-5).
Coping behavior: a behavior practiced to mitigate an unpleasant sensation such as pain or negative emotions.

Key facts of pain, implantable devices, and psychosocial aspects of pain

- In ascending pain pathways, the lateral pathway encodes the discriminatory and sensory component of pain, while the medial pathway encodes the motivational and affective component of pain—commonly described as "pain unpleasantness," or in other words, "suffering."
- Nonlinear burst spinal cord stimulation is a mode of stimulation that modulates both the lateral and medial pain pathways, modulating both the sensory components of pain in the lateral pathway, but also the motivational and affective component of pain in the medial pathway.
- Rates of psychiatric disorders may be twice as high among patients with chronic pain seen in pain clinics compared to primary care settings.
- Pain and psychiatric symptoms share common neurotransmitters that may account for their frequent overlap in patients.
- Nonlinear burst SCS has potentially an equivalent, if not greater, impact on psychosocial aspects of mood and behavior than pain modulation.

Summary points

- Pain is associated with a substantial psychosocial impact.
- Implantable pain devices like spinal cord stimulation are being widely utilized for pain management.
- Psychosocial evaluations are one way to assess this extent when choosing patients for implantable pain devices.
- Nonlinear burst spinal cord stimulation is one mode of stimulation that has an anatomical and physiological basis for improving psychosocial aspects of chronic pain.
- For future studies, affective measures such as the PDI, PCS, and depression scales should be recorded in conjunction with visual analog scales and functional improvement measures.

References

Ahmed, S., Yearwood, T., De Ridder, D., & Vanneste, S. (2018). Burst and high frequency stimulation: Underlying mechanism of action. *Expert Review of Medical Devices*, *15*(1), 61–70. https://doi.org/10.1080/17434440.2018.1418662.

Bair, M., Robinson, R., Katon, W., & Kroenke, K. (2003). Depression and pain comorbidity: A literature review. *Archives of Internal Medicine*, *163*(20), 2433–2445.

Beltrutti, D., Lamberto, A., & Barolat, G. (2004). The psychological assessment of candidates for spinal cord stimulation for chronic pain management. *Pain Practice*, *4*(3), 204–221.

Block, A., Ohnmeiss, D., Guyer, R., Rashbaum, R., & Hochschuler, S. (2001). The use of presurgical psychological screening to predict the outcome of spine surgery. *The Spine Journal*, *1*(4), 274–282.

Bocci, T., De Carolis, G., Paroli, M., Barloscio, D., Parenti, L., Tollapi, L., … Sartucci, F. (2018). Neurophysiological comparison among tonic, high frequency, and burst spinal cord stimulation: Novel insights into spinal and brain mechanisms of action. *Neuromodulation*, *21*(5), 480–488. https://doi.org/10.1111/ner.12747.

Chakravarthy, K., Malayil, R., Kirketeig, T., & Deer, T. (2019). Burst spinal cord stimulation: A systematic review and pooled analysis of real-world evidence and outcomes data. *Pain Medicine*, *20*(1), S47–S57. https://doi.org/10.1093/pm/pnz046.

De Ridder, D., Vanneste, S., Plazier, M., Van Der Loo, E., & Menovsky, T. (2010). Burst spinal cord stimulation: Toward paresthesia-free pain suppression. *Neurosurgery*, *66*(5), 986–990. https://doi.org/10.1227/01.NEU.0000368153.44883.B3.

Department of Veteran's Affairs. (2017). *VA/DoD clinical practice guidelines for opioid therapy for chronic pain*.

Deyo, R., Dworkin, S., & Amtmann, D. (2014). Report of the National Institutes of Health task force on research standards for chronic low back pain. *Journal of Manipulative and Physiological Therapeutics*, *37*(7), 449–467.

Dworkin, R. H., Turk, D. C., Farrar, J. T., Haythornthwaite, J. A., Jensen, M. P., Katz, N. P., … Witter, J. (2005). Core outcome measures for chronic pain clinical trials: IMMPACT recommendations. *Pain*, *113*(1), 9–19. https://doi.org/10.1016/j.pain.2004.09.012.

Dworkin, R. H., Turk, D. C., Wyrwich, K. W., Beaton, D., Cleeland, C. S., Farrar, J. T., … Zavisic, S. (2008). Interpreting the clinical importance of treatment outcomes in chronic pain clinical trials: IMMPACT recommendations. *The Journal of Pain*, *9*(2), 105–121. https://doi.org/10.1016/j.jpain.2007.09.005.

Ericsson, M., Poston, W., II, Linder, J., Taylor, J. E., Haddock, C. K., & Foreyt, J. P. (2002). Depression predicts disability in long-term chronic pain patients. *Disability and Rehabilitation*, *24*(6), 334–340.

Flor, H. (2014). Psychological pain interventions and neurophysiology: Implications for a mechanism-based approach. *The American Psychologist*, *69*(2), 188–196.

Grinberg, A., D'Eon, M., Ellison, J., Rutledge, T., Castellanos, J., & Chakravarthy, K. (2019). A revised psychosocial assessment model for implantable pain devices to improve their evidence basis and consensus with updated pain management guidelines. *Pain Management*, *9*(2), 139–149.

Gybels, J., Erdine, S., & Maeyaert, J. (1998). Neuromodulation of pain: A consensus statement prepared in Brussels 16–18 January 1998 by the following task force of the European federation of IASP chapters (EFIC). *European Journal of Pain*, *2*(3), 203–209.

Haythornthwaite, J. A., & Benrud-Larson, L. M. (2000). *Psychological aspects of neuropathic pain*. Retrieved from http://journals.lww.com/clinicalpain/Fulltext/2000/06001/Psychological_Aspects_of_Neuropathic_Pain.17.aspx%5Cnhttp://journals.lww.com/clinicalpain/Abstract/2000/06001/Psychological_Aspects_of_Neuropathic_Pain.17.aspx.

Hilderink, P., Burger, H., Deeg, D., Beekman, A., & Oude Voshaar, R. (2012). The temporal relation between pain and depression: Results from the longitudinal aging study Amsterdam. *Psychosomatic Medicine*, *74*(9), 945–951.

Jensen, M. P., & Brownstone, R. M. (2019). Mechanisms of spinal cord stimulation for the treatment of pain: Still in the dark after 50 years. *European Journal of Pain*, *23*(4), 652–659. https://doi.org/10.1002/ejp.1336.

Kapural, L., Yu, C., Doust, M. W., Gliner, B. E., Morgan, D. M., Brown, L. L., … Burgher, A. H. (2015). Novel 10-kHz high-frequency therapy (HF10 therapy) is superior to traditional low-frequency spinal cord stimulation for the treatment of chronic back and leg pain. *Anesthesiology*, *123*(4), 851–860.

Kinfe, T. M., Pintea, B., Link, C., Roeske, S., Güresir, E., Güresir, Á., & Vatter, H. (2016). High frequency (10 kHz) or burst spinal cord stimulation in failed back surgery syndrome patients with predominant back pain: Preliminary data from a prospective observational study. *Neuromodulation*, *19*(3), 268–275. https://doi.org/10.1111/ner.12379.

Melzack, R., & Wall, P. (1965). Pain mechanisms: A new theory. *Science*, *150*(3699), 971–979.

Meuwissen, K. P. V., van der Toorn, A., Gu, J. W., Zhang, T. C., Dijkhuizen, R. M., & Joosten, E. A. J. (2020). Active recharge burst and tonic spinal cord stimulation engage different Supraspinal mechanisms: A functional magnetic resonance imaging study in peripherally injured chronic neuropathic rats. *Pain Practice*. https://doi.org/10.1111/papr.12879, papr.12879.

Miller, J. P., Eldabe, S., Buchser, E., Johanek, L. M., Guan, Y., & Linderoth, B. (2016). Parameters of spinal cord stimulation and their role in electrical charge delivery: A review. *Neuromodulation*, *19*(4), 373–384. https://doi.org/10.1111/ner.12438.

Qaseem, A., Wilt, T. J., McLean, R. M., Forciea, M. A., & for the Clinical Guidelines Committee of the American College of Physicians. (2017). Non-invasive treatments for acute, subacute, and chronic low back pain: A clinical practice guideline from the American College of Physicians. *Annals of Internal Medicine*, *166*(7), 514–530. https://doi.org/10.7326/M16-2367.

Rasche, D., Siebert, S., Stippich, C., Kress, B., Nennig, E., Sartor, K., & Tronnier, V. M. (2005). Spinal cord stimulation in failed-back-surgery-syndrome. Preliminary study for the evaluation of therapy by functional magnetic resonance imaging (fMRI). *Schmerz*, *19*(6), 497–500.

Rosenberg, J., Fabi, A., Candido, K., Knezevic, N., Creamer, M., Carayannopoulos, A., … Bennett, M. (2016). Spinal cord stimulation provides pain relief with improved psychosocial function: Results from EMP3OWER. *Pain Medicine*, *17*(12), 2311–2325. https://doi.org/10.1093/pm/pnw152.

Schuch, F., Vancampfort, D., Richards, J., Rosenbaum, S., Ward, P., & Stubbs, B. (2016). Exercise as a treatment for depression: A meta-analysis adjusting for publication bias. *Journal of Psychiatric Research*, *77*, 42–51.

Sdrulla, A. D., Guan, Y., & Raja, S. N. (2018). Spinal cord stimulation: Clinical efficacy and potential mechanisms. *Pain Practice*, *18*(8), 1048–1067. https://doi.org/10.1111/papr.12692.Spinal.

Steeds, C. E. (2016). The anatomy and physiology of pain. *Surgery (Oxford)*, *34*(2), 55–59. https://doi.org/10.1016/j.mpsur.2015.11.005.

Turk, D. C., Dworkin, R. H., Revicki, D., Harding, G., Burke, L. B., Cella, D., … Rappaport, B. A. (n.d.). Identifying important outcome domains for chronic pain clinical trials: An IMMPACT survey of people with pain. doi:https://doi.org/10.1016/j.pain.2007.09.002.

Verbruggen, H. (2014). Morphological complexity, plasticity, and species diagnosability in the application of old species names in DNA-based taxonomies. *Journal of Phycology*, *50*(1), 26–31. https://doi.org/10.1111/jpy.12155.

Verrils, P., Sinclair, C., & Barnard, A. (2016). A review of spinal cord stimulation systems for chronic pain. *Journal of Pain Research*, *9*, 481.

Chapter 36

Influence of psychological factors on myofascial pain

Marta San-Antolín-Gil[a], Daniel López-López[b], Ricardo Becerro-de-Bengoa-Vallejo[c], Marta Elena Losa-Iglesias[c,d], Carlos Romero-Morales[e], David Rodríguez-Sanz[c], Victoria Mazoteras-Pardo[f], Eva María Martínez-Jiménez[c], and César Calvo-Lobo[c]

[a]*Department of Psychology, Universidad Europea de Madrid, Madrid, Spain,* [b]*Research, Health and Podiatry Group, Department of Health Sciences, Faculty of Nursing and Podiatry, University of A Coruña, Ferrol, Spain,* [c]*Department of Nursing, School of Nursing, Physiotherapy and Podiatry, University Complutense of Madrid, Madrid, Spain,* [d]*Department of Nursing, Faculty of Health Sciences, Universidad Rey Juan Carlos, Alcorcón, Spain,* [e]*Faculty of Sport Sciences, Universidad Europea de Madrid, Madrid, Spain,* [f]*Research Group "ENDOCU", Department of Nursing, Physiotherapy and Occupational Therapy, Faculty of Physiotherapy and Nursing of Toledo, University of Castilla-La Mancha, Toledo, Spain*

Abbreviations

LTRs	local twitch responses
MPS	myofascial pain syndrome
MTrPs	myofascial trigger points

Introduction

Myofascial pain syndrome (MPS) may be defined as a set of sensory, motor, and autonomic signs and symptoms originated by myofascial trigger points (MTrPs) (Bron & Dommerholt, 2012). MPS includes a primary muscular dysfunction, associated with central sensitization and the extension of a certain segment within the medulla, giving rise to the concept of referred pain (Feelders et al., 1999). This phenomenon may occur in an acute or chronic and generalized or regional way (Ge & Arendt-Nielsen, 2011; Wang et al., 2012) and usually manifests itself in the same way during an effort or muscular activity as a familiar pattern for the subject (Travell, 2002). This pain is caused by the main source of the MPS, called as MTrPs (Travell, 2002), which may be defined as hypersensitive spots located in muscular-skeletal palpable taut bands, located within the muscles (Hong, 1996; Hong & Simons, 1998). Their compression may frequently cause sensitivity at the local level (Travell, 2002) and can arise due to a muscular metabolic alteration that produces, according to studies, an "energy crisis" (Sikdar, Ortiz, Gebreab, Gerber, & Shah, 2010) due to the existence of a prolonged contraction over time of the sarcomere, causing excessive pressure on the capillaries and producing an alteration in the production of required metabolites, as well as local hypoxia and tissue damage (Sikdar et al., 2010).

Myofascial trigger point types

PGMs are classified into two types, active and latent, both types with similar characteristics such as local twitch responses (LTRs), tenderness in a taut band, referred pain or spontaneous electrical activity, if clinically addressed by needle electromyography when the muscle is at rest (Travell, 2004). Both MTrPs present the capacity to produce central sensitization (Niddam, Chan, Lee, Yeh, & Hsieh, 2008; Xu, Ge, & Arendt-Nielsen, 2010). Regarding active MTrPs, these points may develop chronic musculoskeletal pain and spread of pain to different regions, increasing nociceptive stimuli (Ge, Fernández-de-Las-Peñas, & Yue, 2011). Central pain mechanisms may directly influence central sensitization and pain extension in a chronic way (Woolf, 2011). Last, it is also worth noting the influence of peripheral pain mechanisms (Graven-Nielsen & Arendt-Nielsen, 2010).

The Neurobiology, Physiology, and Psychology of Pain. https://doi.org/10.1016/B978-0-12-820589-1.00036-1

Relationship between psychological factors and myofascial pain

The relationship of physical pain and psychology presents evidence due to the image samples which show the overlaps of the anatomical regions associated with the emotional, affective, and cognitive characteristics and the sensory characteristics of the pain, as well as the areas affected during the course of a depressive disorder (Vadivelu et al., 2017).

Currently, there is an increase of studies that support the association of psychological, biological, and social factors with aspects related to symptoms, disability, and quality of life of people with chronic musculoskeletal pain (Ferrer-Peña, Moreno-López, Calvo-Lobo, López-de-Uralde-Villanueva, & Fernández-Carnero, 2019; Reis et al., 2019; Shahidi, Curran-Everett, & Maluf, 2015). In addition, psychological factors have been shown to be more associated with myofascial pain than joint disorders (Nifosì et al., 2007).

Studies show associations between musculoskeletal pain and fear-avoidance aspects, catastrophism, depression, anxiety, or personality among others (Deakon, 2012; Fritz & George, 2002; Reis et al., 2019; Vlaeyen, Kole-Snijders, Boeren, & van Eek, 1995). The maladaptive psychological characteristics can lead to worsening pain, to its maintenance developing chronic pain, and to a high disability in the daily life of the person suffering from pain. In contrast, psychological factors, more adaptive to the culture and environment, such as high adaptation to change, high resilience, easy acceptance of pain, high hope, or an optimistic vision, strongly promote adaptation to chronic or persistent pain like MPS (Vadivelu et al., 2017).

Indeed, our research group has investigated about the influence of psychological factors such as central sensitization and catastrophism (San-Antolín et al., 2019), fear-avoidance beliefs and kinesiophobia (San-Antolín et al., 2020), depression (San-Antolín et al., 2020), and personality traits (San-Antolín, Rodríguez-Sanz, & Calvo-Lobo, 2019) on patients who suffer from MPS (Fig. 1).

Personality traits and myofascial pain

The relationship between pain and personality traits has been studied over the years, although conclusive results are lacking due to the complexity and extension of this field. One of the diseases that reached clarifying conclusions about this topic is fibromyalgia. This disease may be associated with a specific personality profile, including dependent, demanding, and passive people. There have been many studies that address fibromyalgia and personality profile as controllers or perfectionists. Currently, it is assumed that people with chronic pain present a normal personality structure. The changes that may occur in this condition are a consequence of the impact of the disease itself on the life of the individual (Sternbach, 1976; Truyols, Pérez, Medinas, Palmer, & Sesé, 2008). Even so, it has not yet been established if certain types of personalities may predispose to myofascial pain (Blumer & Disease, 1982) or only modulate the perception and experience of pain (Sternbach, 1976).

Indeed, Hewitt et al. (Hewitt, Flett, & Ediger, 1996) developed a model showing that personality is the prelude to coping strategies to face different problems observed in a given individual, allowing to the subject to develop a greater or lesser degree of adaptation to the environment. This model assumes that, before episodes of pain, personality traits may determine which strategy to decrease or accentuate the characteristics of pain such as duration or intensity, as well as the development of chronic pain. As coping strategies, we can observe that catastrophism is one of those that help the development of chronic pain (Maestre, Esteve, Eva, & Martínez, 2001).

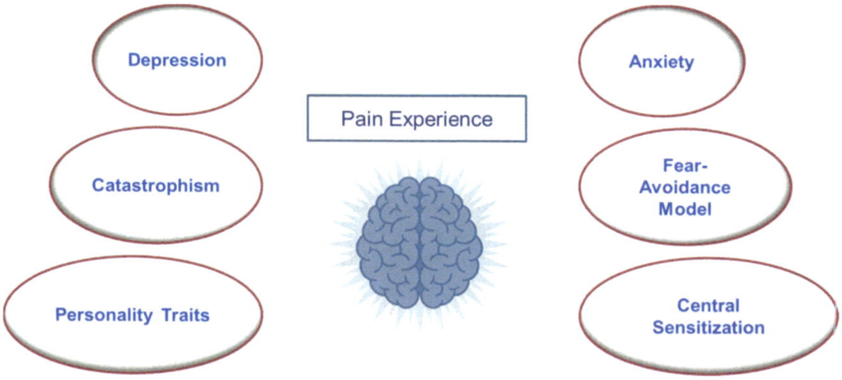

FIG. 1 Psychological factors and myofascial pain experience.

On the other hand, personality traits, such as extraversion or neuroticism, although they could present a relationship with the presence of MPS, did not seem to affect nociceptive sensory mechanisms at the biological level, although they may produce a notable influence on cognitive processes of interpretation and implications of painful signs (Harkins, Price, & Braith, 1989).

A growing number of authors suggested that psychological variables may not be only important in the experience of pain, but also may represent an enormous influence on the positive response of patients to treatment. Personality traits are usually exposed according to the model factors proposed by Eysenck in 1975 (Fig. 2), due to its good and complete representation of the individual's personality, being one of the most widely used criteria in research for the analysis of personality (Carrillo, Collado, & Rojo, 2005; Sandín, Valiente, Chorot, Olmedo Montes, & Santed Germán, 2002).

Thus, a pilot study carried out by our research group showed that patients who suffered from MPS may show higher scores on mainly neuroticism (Fig. 3) and possibly extraversion (Fig. 4) than healthy athletes. Despite being a pilot study and requiring studies with a larger sample size, these novel findings may suggest that both physical and psychological therapy could be key interventions in patients with MPS in order to improve functionality, prevent injuries, and optimize rehabilitation strategies (San-Antolín, Rodríguez-Sanz, & Calvo-Lobo, 2019).

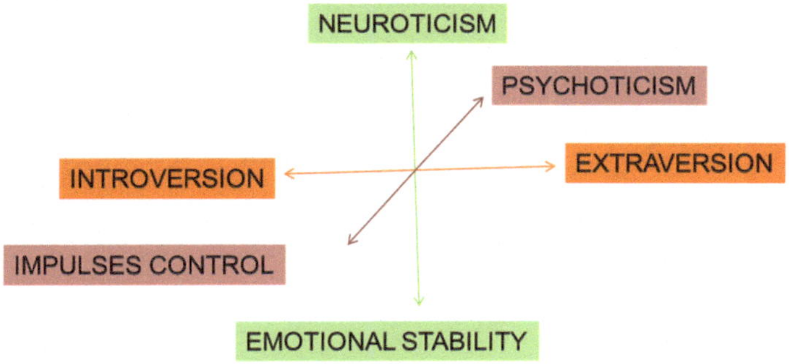

FIG. 2 Personality traits model proposed by Eysenck in 1975.

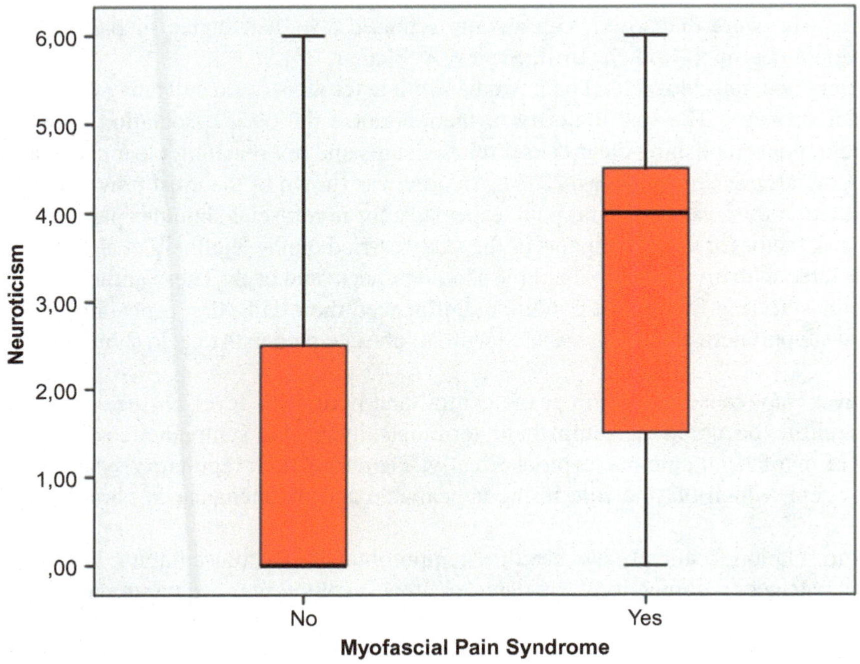

FIG. 3 Neuroticism differences between myofascial pain patients and healthy subjects of a pilot study.

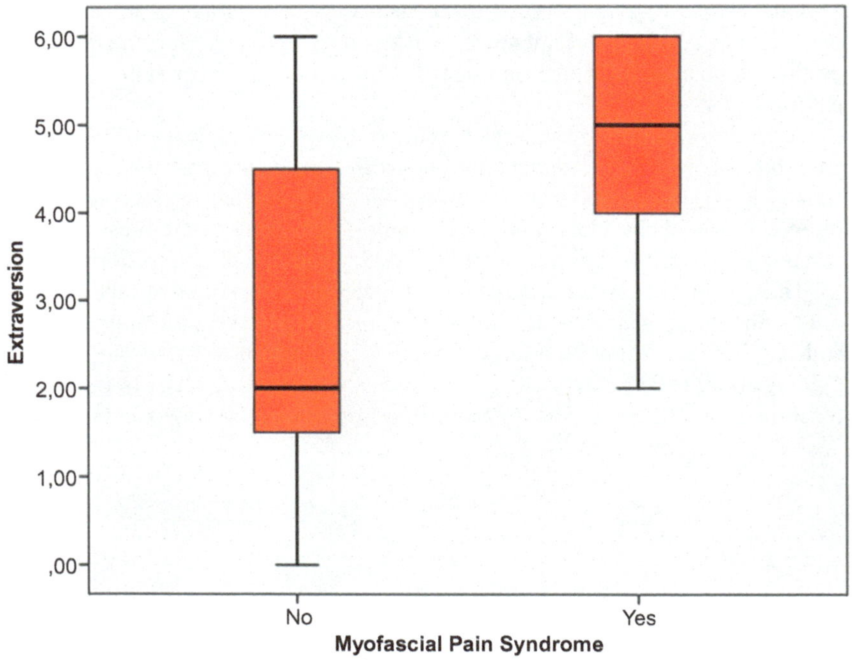

FIG. 4 Extraversion differences between myofascial pain patients and healthy subjects of a pilot study.

Anxiety and myofascial pain

There are many studies that suggest the influence between psychological factors such as anxiety and chronic pain. Pain experience may trigger anxious symptoms and emotional distress and a two-way interaction between both symptoms, changing throughout the life cycle and affected by factors related to pain such as its duration or intensity, and others such as situational and psychological characteristics of each individual. These findings show that there is a reciprocal relationship between anxiety and pain. People who were diagnosed with anxiety reported a higher degree of pain before the same stimulus compared to healthy people (Lerman, Rudich, Brill, Shalev, & Shahar, 2015).

Regarding the association between anxiety and musculoskeletal pain, studies with heterogeneous results may be found, depending on body areas and psychosocial stressors. The vast majority of them endorse the clear association between anxious symptoms and musculoskeletal pain, not establishing clear causal relationships and not reaching clear consensus regarding assessment or treatment (León González & Fornés Vives, 2015). Anxiety was shown as the most prevalent and psychologically most significant trait in relation to musculoskeletal pain, especially for myofascial shoulder pain. Stress may be considered as the most significant risk factor for the participants of the study carried out by Kjellberg et al. in 2007, showing that anxious symptoms were associated with myofascial pain of the shoulder, neck, and back. These authors determined that the anxiety presented by patients suffering from these conditions influenced their daily life, especially at the working environment. Finally, a poor social support network was associated with myofascial pain in the neck, shoulder, and back (Kjellberg & Wadman, 2007).

Recently, the idea that anxiety and stress may cause activation at the central and peripheral level has been proven. During these episodes, the release of epinephrine occurs in the sympathetic terminals, increasing sympathetic activation, and causing sensitization and excitability of myofascial pain nociceptors directly. Therefore, pain regulatory systems are altered, such as the ascending reticular system, which plays a role in the appearance and maintenance of chronic pain (Ohrbach & Michelotti, 2018).

One of the most-studied muscle areas in relation to anxiety has been the temporomandibular musculature. Events or mental states related to stress and anxiety may trigger parafunctions, bruxisms, or other involuntary movements at the mandibular level leading to myofascial pain in the area and understanding them as risk factors for the development of MPS (Nifosì et al., 2007). In this study, authors addressed some aspects that may contribute to the relationship of anxiety with myofascial pain, understanding as important factors to determine that anxiety contributes to pain development, pain may cause ineffective coping strategies to release pain and trigger behaviors of hypervigilance and somatosensory excitability.

Depression and myofascial pain

Depressive symptoms and pain are strongly associated. Both depression and sleep problems are very prevalent in people with chronic pain generating impairment in their quality of life, showing a high disability in daily activities. Sleep problems may also predict pain and depression in an individual (Lerman et al., 2015).

Chronic pain is currently considered to be a trigger for anxious or depressive symptoms (Tunks, Crook, & Weir, 2008). Sullivan concludes that depression is strongly associated with highly intense painful experiences and emotional distress in response to painful stimulation (Sullivan, Rodgers, & Kirsch, 2001). In relation to musculoskeletal pain, it was shown, in the same line that the interaction between chronic musculoskeletal pain and the depressive level is bidirectional, that is, individuals who present chronic musculoskeletal pain have a greater possibility of presenting depressive symptoms, while simultaneously observed in depressed people, greater probability of episodes of pain (Magni, Moreschi, & Rigatti-Luchini, 1994).

On the contrary, Okifuji et al. tried to determine if patients diagnosed with chronic pain did not suffer from depression. Indeed, fibromyalgia patients who presented trigger points were recruited and interviewed about their strategies for coping with pain. Authors concluded that there was not clear positive correlation between the pain suffered and the depressive symptoms, alluding to other psychological variables as causes of the onset of depression, such as cognitive distortions, little social support, or low standard of living (Okifuji, Turk, & Sherman, 2000).

The association of pain and depression has been previously reported including similar neurobiological alterations at a structural and functional level and the same pathophysiological mechanisms. Both depression and chronic pain presented a decrease in the neurotransmitter 5-hydroxytryptamine or serotonin, which sensitized sensory neurons and activates pain pathways. Neurotrophic factors were also decreased in both disorders, which caused an increase in the excitability of the second neuron and triggered central desensitization and, finally, alteration in the mechanisms of the hypothalamic-pituitary-adrenal axis. In this sense, depression was also associated, in addition to chronic pain, with anxiety, aversion, and fear, all of them mediated by monoamine neurotransmitters (Boakye et al., 2016).

Depression, together with anxiety, may be one of the reasons to explain why myofascial pain is generated. One of the theories argued that the musculature may be altered by certain thoughts and cause MTrPs. As we have previously exposed, the worsening of the MTrP may be influenced by the hypothalamic-pituitary-adrenal axis, which may also be altered in patients suffering from depression (Ruvalcaba Palacios & Domínguez Trejo, 2011). These same authors established a treatment in order to determine whether by modulating the autonomic activity of the participants who suffered from SDM at the lumbar level, they could modify the physical symptoms with respect to the PGM and the psychological symptoms that accompany the SDM. They were able to conclude that training sessions on improving autonomic functioning, improves mood, as well as other psychological variables such as anxiety or catastrophism (Ruvalcaba Palacios & Domínguez Trejo, 2011). MPS has also been associated with sleep problems, recently knowing that these problems in sleep hygiene are more determined by the depression associated with the MPS than by the myofascial pain itself (Dubrovsky et al., 2017).

Our research group showed that higher depression scores (Fig. 5) and levels were suffered from patients with gastrocnemius myofascial pain compared to healthy athletes. Furthermore, depression symptoms were predicted by the presence of myofascial pain (San-Antolín, Rodríguez-Sanz, López-López, et al., 2020).

Catastrophism and myofascial pain

As a consequence in the presence of catastrophism, pain, and MPS, experiences with greater intensity of pain, greater sensitivity to pain, lower mood, higher fear-avoidance patterns of certain daily activities, increased behaviors of pain, less effective coping strategies, greater disability, greater consumption of analgesic drugs, and greater number of sick leave have been observed (Edwards, Fillingim, Maixner, Sigurdsson, & Haythornthwaite, 2004; Olmedilla, Ortega, Boladeras, Abenza, & Esparza, 2008; Queraltó, 2005; Sullivan et al., 2001; Turner, Dworkin, Mancl, Huggins, & Truelove, 2001).

In relation to chronic pain, alterations in pain processes and their respective brain areas, such as memory, motor control, and attention in people with fibromyalgia, or less adherence to drug treatment have been found in patients with chronic rheumatic problems (Giesecke et al., 2004). These individuals present a worse prognosis for recovery, requiring more rehabilitation sessions (Stephens, Druley, & Zautra, 2002).

Regarding patients with chronic musculoskeletal pain, a clear alteration in attentional capacity was observed. In 2013, Dehaghani et al. carried out a study on this topic. Participants had to complete a task based on visual word detection. These authors demonstrated that people with musculoskeletal pain selectively paid more attention to words with painful overtones

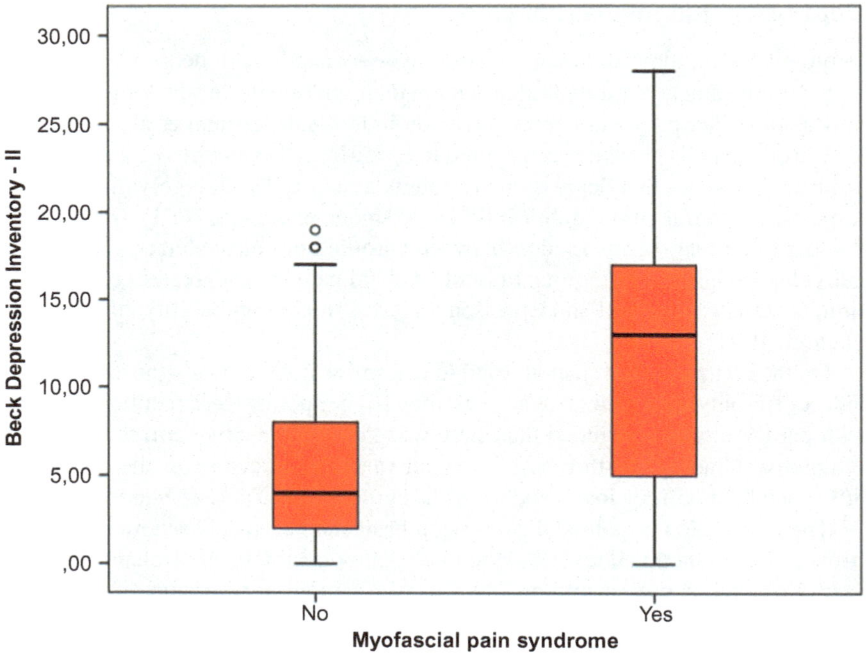

FIG. 5 Depression scores differences between myofascial pain patients and healthy subjects.

than to neutral words, modulated by the nature of the painful stimulus. Among these individuals, those who were more afraid of pain responded more slowly to those questions than people with less fear of pain with greater difficulty in disconnecting from them (Dehghani, Sharpe, & Nicholas, 2003).

From a psychological perspective, subjects susceptible to developing chronic pain may present responses of fear, hypervigilance, and anticipation, among other factors, on the appearance of possible painful symptoms, before episodes of acute pain. These characteristics may lead to the maintenance of pain beyond recovery from injury (Vlaeyen et al., 1995) and generate MPS in different muscle groups, demanding special attention to the catastrophic perception regarding myofascial pain (Velly et al., 2011).

Based on this relationship between catastrophic pain and MPS, we found studies that address this association with the aim to understand the neuropsychology of MPS. Patients who presented temporomandibular myofascial pain corroborated fearful attitudes toward pain, interference in patients' daily life, a low mood correlated with depression, and a strong association between dysfunctional pain and catastrophism (Turner et al., 2001). Indeed, a correlation was observed between high intracortical disinhibition and cortical excitability as a marker frequently observed in patients with MPS and psychological factors such as the presence of anxiety and catastrophism. This association was also reported in fibromyalgia and neuropathic pain. Furthermore, glutamatergic neurons in the cortex are known to have a presence in catastrophic pain mechanisms in patients with SDM (Thibaut, Zeng, Caumo, Liu, & Fregni, 2017; Volz et al., 2013). Finally, our research group determined that higher symptoms of central catastrophism, rumination, magnification, and helplessness were presented in patients with gastrocnemius myofascial pain with respect to healthy athletes (San-Antolín, Rodríguez-Sanz, Becerro-de-Bengoa-Vallejo, et al., 2019).

Pain fear-avoidance and kinesiophobia and myofascial pain

Regarding myofascial pain, the fear-avoidance model of musculoskeletal pain explains why some patients who have experienced acute pain could develop chronic pain. This model (Fig. 6) was developed by Vlaeyen and Linton in 2000, based on the Waddell theoretical model, among others (Leeuw et al., 2007; Waddell, Newton, & Henderson, 1993).

A concept related to fear-avoidance behavior in patients who suffer from pain is kinesiophobia, which may be defined as an emotion of intense fear, maladjusted to reality and, therefore, irrational and high inability to face physical activity or movements, preceded by a recurrence of the condition or injury or high cognitive vulnerability to pain (Larsson, Bodegård,

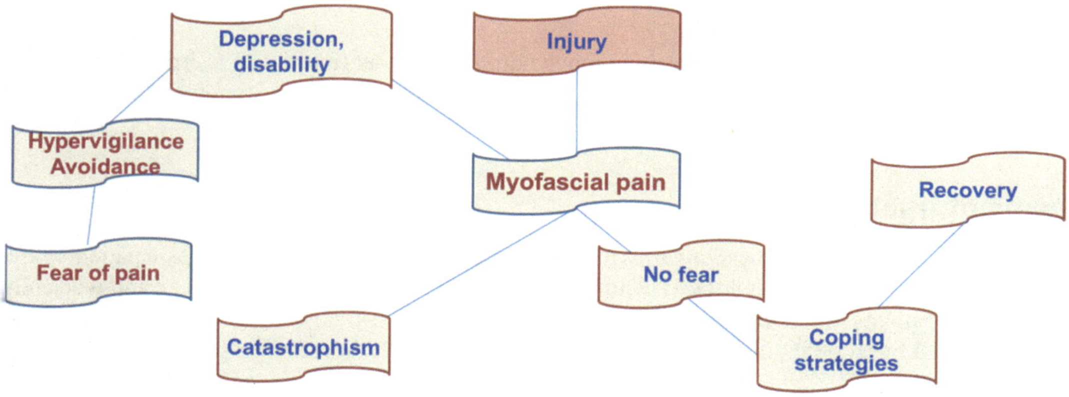

FIG. 6 Pain fear-avoidance model under myofascial pain condition.

Henriksson, & Oberg, 1990; Larsson, Ekvall Hansson, Sundquist, & Jakobsson, 2016; Mintken, Cleland, Whitman, & George, 2010).

People who suffered from shoulder MPS were afraid of their obligations or activities, both in their private and working life, according to their fear of motion and reinjury. Furthermore, they presented fear about the pain experience itself (Mintken et al., 2010). Finally, our research group reported that higher kinesiophobia levels, higher total and activity avoidance domain symptoms, although not for harm domain, and higher fear-avoidance beliefs total and domain symptoms for work and physical activity were presented in patients with gastrocnemius MPS with respect to healthy subjects. Greater kinesiophobia scores were predicted by higher fear-avoidance beliefs scores. Higher fear-avoidance beliefs were also predicted by the MPS presence and greater kinesiophobia levels (San-Antolín, Rodríguez-Sanz, Vicente-Campos, et al., 2020).

Central sensitization and myofascial pain

Regarding chronic pain, the complex mechanisms underlying central sensitization are present in chronic pain, which implies a treatment approach focused on long-term results and from a multimodal perspective, avoiding exclusively use pharmacological treatments (Nijs et al., 2019).

In people with musculoskeletal disorders, the presence of central sensitization may trigger a prolonged accentuation of synaptic excitability in the central nervous system nociceptive pathways, in many cases irreversibly (Graven-Nielsen & Arendt-Nielsen, 2010).

Currently, the term central sensitization syndrome has been used to refer to multiple conditions in which the presence of central sensitization is a common feature. These may include irritable bowel, fibromyalgia, temporomandibular disorder, restless leg syndrome or MPS, among others. This term highlights the necessity of a biopsychosocial perspective intervention (Yunus, 2008).

Patellar tendinopathy, lateral hip pain, or knee osteoarthritis have been related to central sensitization (Ferrer-Peña, Muñoz-Garcıa, Calvo-Lobo, & Fernandez-Carnero, 2019; Fingleton, Smart, Moloney, Fullen, & Doody, 2015; van Wilgen, Konopka, Keizer, Zwerver, & Dekker, 2013). Last, our research group determined that higher central sensitization symptoms were presented in patients with gastrocnemius myofascial pain with respect to healthy subjects. Nevertheless, a statistically significant presence of central sensitization syndrome was not presented after comparing both groups. Higher central sensitization scores were predicted by higher symptoms of catastrophism, female sex, and gastrocnemius myofascial pain presence (San-Antolín, Rodríguez-Sanz, Becerro-de-Bengoa-Vallejo, et al., 2019).

Clinical applications

Psychological interventions such as pain education or coping strategies should be considered in patients who suffer from MPS, especially under active MTrPs presence, in conjunction with physical therapy and other rehabilitation treatments in order to improve anxiety, depression, central sensitization, pain fear-avoidance, fear of motion, and catastrophism, among other key psychological factors (San-Antolín, Rodríguez-Sanz, Becerro-de-Bengoa-Vallejo, et al., 2019; San-Antolín, Rodríguez-Sanz, López-López, et al., 2020; San-Antolín, Rodríguez-Sanz, Vicente-Campos, et al., 2020).

Application to other areas

In addition to myofascial pain, other central sensitization syndromes such as fibromyalgia, irritable bowel syndrome, or temporomandibular disorders could be beneficed by the inclusion of psychological strategies into a multimodal intervention (Yunus, 2008).

Other agents of interest

The use of psychological self-reported questionnaires may be widely recommended in chronic pain patients exploration and clinical reasoning processes due to psychological factors could play a key role in the maintenance and worsening of central sensitization in these patients (San-Antolín, Rodríguez-Sanz, Becerro-de-Bengoa-Vallejo, et al., 2019; San-Antolín, Rodríguez-Sanz, López-López, et al., 2020; San-Antolín, Rodríguez-Sanz, Vicente-Campos, et al., 2020).

Mini-dictionary of terms

Catastrophism: This term may be defined as a set of cognitive and emotional factors which may facilitate to perceive a painful stimulus in a negative and exaggerated way.
Central sensitization: This term may be defined as an excitability of the central nervous system pathways.
Kinesiophobia: This term may be defined as fear of motion which may be measured by scores and levels.
Personality trait: This term may be defined as personality factors such as neuroticism, psychoticism, extraversion, and sincerity proposed by Eysenck in 1975.

Key facts of myofascial pain

- MPS may be considered as a set of motor, sensitive, and autonomic signs and symptoms secondary to MTrPs.
- The presence of MPS may be associated with psychological alterations.
- Psychological disorders may play a key role in the maintenance and worsening of the MPS prognosis.
- Psychological interventions should be considered under MPS conditions.
- Psychological interventions should be included into multimodal treatments of patients who suffer from MPS.

Summary points

- This chapter is focused on the influence of psychological factors on MPS.
- Personality traits may be associated with the presence of MPS, especially neuroticism.
- Higher anxiety and depression symptoms and levels may be presented in patients who suffer from MPS.
- Higher kinesiophobia and pain fear-avoidance symptoms may be shown in patients suffering from MPS.
- Higher catastrophism and central sensitization symptoms may be associated in patients with active MTrPs presence.
- Psychological interventions should be combined with multimodal treatments in order to improve the course, prognosis, and sequelae of MPS patients.

References

Blumer, D., & Disease, M. H. (1982). Chronic pain as a variant of depressive disease: The pain-prone disorder. *The Journal of Nervous and Mental Disease*, *170*(7), 381–406.

Boakye, P. A., Olechowski, C., Rashiq, S., Verrier, M. J., Kerr, B., Witmans, M., et al. (2016). A critical review of neurobiological factors involved in the interactions between chronic pain, depression, and sleep disruption. *Clinical Journal of Pain*, *32*(4), 327–336. https://doi.org/10.1097/AJP.0000000000000260.

Bron, C., & Dommerholt, J. D. (2012). Etiology of myofascial trigger points. *Current Pain and Headache Reports*, *16*(5), 439–444. https://doi.org/10.1007/s11916-012-0289-4.

Carrillo, J. M., Collado, S., & Rojo, N. (2005). Personalidad y tolerancia al dolor. *Clínica y Salud*, *16*(3), 223–236.

Deakon, R. T. (2012). Chronic musculoskeletal conditions associated with the cycling segment of the triathlon; prevention and treatment with an emphasis on proper bicycle fitting. *Sports Medicine and Arthroscopy Review*, *20*(4), 200–205. https://doi.org/10.1097/JSA.0b013e3182688fa0.

Dehghani, M., Sharpe, L., & Nicholas, M. K. (2003). Selective attention to pain-related information in chronic musculoskeletal pain patients. *Pain*, *105*(1–2), 37–46. https://doi.org/10.1016/S0304-3959(03)00224-0.

Dubrovsky, B., Janal, M. N., Lavigne, G. J., Sirois, D. A., Wigren, P. E., Nemelivsky, L., et al. (2017). Depressive symptoms account for differences between self-reported versus polysomnographic assessment of sleep quality in women with myofascial TMD. *Journal of Oral Rehabilitation*, *44* (12), 925–933. https://doi.org/10.1111/joor.12552.

Edwards, R., Fillingim, R. B., Maixner, W., Sigurdsson, A., & Haythornthwaite, J. (2004). Catastrophizing predicts changes in thermal pain responses after resolution of acute dental pain. *The Journal of Pain*, *5*(3), 164–170. https://doi.org/10.1016/j.jpain.2004.02.226.

Feelders, R. A., Swaak, A. J., Romijn, J. A., Eggermont, A. M., Tielens, E. T., Vreugdenhil, G., et al. (1999). Characteristics of recovery from the euthyroid sick syndrome induced by tumor necrosis factor alpha in cancer patients. *Metabolism, Clinical and Experimental*, *48*(3), 324–329.

Ferrer-Peña, R., Moreno-López, M., Calvo-Lobo, C., López-de-Uralde-Villanueva, I., & Fernández-Carnero, J. (2019). Relationship of dynamic balance impairment with pain-related and psychosocial measures in primary care patients with chronic greater trochanteric pain syndrome. *Pain Medicine*, *20* (4), 810–817. https://doi.org/10.1093/pm/pny160.

Ferrer-Peña, R., Muñoz-Garcıa, D., Calvo-Lobo, C., & Fernandez-Carnero, J. (2019). Pain expansion and severity reflect central sensitization in primary care patients with greater trochanteric pain syndrome. *Pain Medicine*, *20*(5), 961–970. https://doi.org/10.1093/pm/pny199.

Fingleton, C., Smart, K., Moloney, N., Fullen, B. M., & Doody, C. (2015). Pain sensitization in people with knee osteoarthritis: A systematic review and meta-analysis. *Osteoarthritis and Cartilage*, *23*(7), 1043–1056. https://doi.org/10.1016/j.joca.2015.02.163.

Fritz, J. M., & George, S. Z. (2002). Identifying psychosocial variables in patients with acute work-related low back pain: The importance of fear-avoidance beliefs. *Physical Therapy*, *82*(10), 973–983.

Ge, H.-Y., & Arendt-Nielsen, L. (2011). Latent myofascial trigger points. *Current Pain and Headache Reports*, *15*(5), 386–392. https://doi.org/10.1007/s11916-011-0210-6.

Ge, H.-Y., Fernández-de-Las-Peñas, C., & Yue, S.-W. (2011). Myofascial trigger points: Spontaneous electrical activity and its consequences for pain induction and propagation. *Chinese Medicine*, *6*(1), 13. https://doi.org/10.1186/1749-8546-6-13.

Giesecke, T., Gracely, R. H., Grant, M. A. B., Nachemson, A., Petzke, F., Williams, D. A., et al. (2004). Evidence of augmented central pain processing in idiopathic chronic low back pain. *Arthritis and Rheumatism*, *50*(2), 613–623. https://doi.org/10.1002/art.20063.

Graven-Nielsen, T., & Arendt-Nielsen, L. (2010). Assessment of mechanisms in localized and widespread musculoskeletal pain. *Nature Reviews Rheumatology*, *6*(10), 599–606. https://doi.org/10.1038/nrrheum.2010.107.

Harkins, S. W., Price, D. D., & Braith, J. (1989). Effects of extraversion and neuroticism on experimental pain, clinical pain, and illness behavior. *Pain*, *36* (2), 209–218. https://doi.org/10.1016/0304-3959(89)90025-0.

Hewitt, P. L., Flett, G. L., & Ediger, E. (1996). Perfectionism and depression: Longitudinal assessment of a specific vulnerability hypothesis. *Journal of Abnormal Psychology*, *105*(2), 276–280. https://doi.org/10.1037/0021-843X.105.2.276.

Hong, C. Z. (1996). Pathophysiology of myofascial trigger point. *Journal of the Formosan Medical Association*, *95*(2), 93–104.

Hong, C. Z., & Simons, D. G. (1998). Pathophysiologic and electrophysiologic mechanisms of myofascial trigger points. *Archives of Physical Medicine and Rehabilitation*, *79*(7), 863–872. http://www.ncbi.nlm.nih.gov/pubmed/9685106.

Kjellberg, A., & Wadman, C. (2007). The role of the affective stress response as a mediator of the effect of psychosocial risk factors on musculoskeletal complaints—Part 1: Assembly workers. *International Journal of Industrial Ergonomics*, *37*(4), 367–374. https://doi.org/10.1016/J.ERGON.2006.12.002.

Larsson, S. E., Bodegård, L., Henriksson, K. G., & Oberg, P. A. (1990). Chronic trapezius myalgia. Morphology and blood flow studied in 17 patients. *Acta Orthopaedica Scandinavica*, *61*(5), 394–398.

Larsson, C., Ekvall Hansson, E., Sundquist, K., & Jakobsson, U. (2016). Kinesiophobia and its relation to pain characteristics and cognitive affective variables in older adults with chronic pain. *BMC Geriatrics*, *16*(1), 128. https://doi.org/10.1186/s12877-016-0302-6.

Leeuw, M., Goossens, M. E. J. B., Linton, S. J., Crombez, G., Boersma, K., & Vlaeyen, J. W. S. (2007). The fear-avoidance model of musculoskeletal pain: Current state of scientific evidence. *Journal of Behavioral Medicine*, *30*(1), 77–94. https://doi.org/10.1007/s10865-006-9085-0.

León González, M., & Fornés Vives, J. (2015). Estrés psicológico y problemática musculoesquelética. Revisión sistemática. *Enfermeria Global*, *14*(2), 276–300.

Lerman, S. F., Rudich, Z., Brill, S., Shalev, H., & Shahar, G. (2015). Longitudinal associations between depression, anxiety, pain, and pain-related disability in chronic pain patients. *Psychosomatic Medicine*, *77*(3), 333–341. https://doi.org/10.1097/PSY.0000000000000158.

Maestre, C. R., Esteve, R., Eva, A., & Martínez, L. (2001). Neuroticismo, afrontamiento y dolor crónico. *Annals of Psychology*, *17*(1), 129–137.

Magni, G., Moreschi, C., & Rigatti-Luchini, S. (1994). Prospective study on the relationship between depressive symptoms and chronic musculoskeletal pain. *Pain*, *56*(3), 289–297.

Mintken, P. E., Cleland, J. A., Whitman, J. M., & George, S. Z. (2010). Psychometric properties of the fear-avoidance beliefs questionnaire and Tampa scale of Kinesiophobia in patients with shoulder pain. *Archives of Physical Medicine and Rehabilitation*, *91*(7), 1128–1136. https://doi.org/10.1016/j.apmr.2010.04.009.

Niddam, D. M., Chan, R.-C., Lee, S.-H., Yeh, T.-C., & Hsieh, J.-C. (2008). Central representation of hyperalgesia from myofascial trigger point. *NeuroImage*, *39*(3), 1299–1306. https://doi.org/10.1016/j.neuroimage.2007.09.051.

Nifosì, F., Violato, E., Pavan, C., Sifari, L., Novello, G., Guarda Nardini, L., et al. (2007). Psychopathology and clinical features in an Italian sample of patients with myofascial and temporomandibular joint pain: Preliminary data. *International Journal of Psychiatry in Medicine*, *37*(3), 283–300. https://doi.org/10.2190/PM.37.3.f.

Nijs, J., Leysen, L., Vanlauwe, J., Logghe, T., Ickmans, K., Polli, A., et al. (2019). Treatment of central sensitization in patients with chronic pain: Time for change? *Expert Opinion on Pharmacotherapy*, *20*(16), 1–10. https://doi.org/10.1080/14656566.2019.1647166.

Ohrbach, R., & Michelotti, A. (2018). The role of stress in the etiology of oral parafunction and myofascial pain. *Oral and Maxillofacial Surgery Clinics of North America*, *30*(3), 369–379. https://doi.org/10.1016/j.coms.2018.04.011.

Okifuji, A., Turk, D. C., & Sherman, J. J. (2000). Evaluation of the relationship between depression and fibromyalgia syndrome: Why aren't all patients depressed? *The Journal of Rheumatology*, *27*(1), 212–219.

Olmedilla, A., Ortega, E., Boladeras, A., Abenza, L., & Esparza, F. (2008). Relaciones entre catastrofismo ante el dolor y ansiedad competitiva en deportistas. *Annuary of Clinical And Health Psychology*, *4*, 45–51.

Queraltó, J. (2005). Análisis de los factores psicológicos moduladores del dolor crónico benigno. *The UB Journal of Psychology*, *36*(1), 37–60.

Reis, F., Guimarães, F., Nogueira, L. C., Meziat-Filho, N., Sanchez, T. A., & Wideman, T. (2019). Association between pain drawing and psychological factors in musculoskeletal chronic pain: A systematic review. *Physiotherapy Theory and Practice*, *35*(6), 533–542. https://doi.org/10.1080/09593985.2018.1455122.

Ruvalcaba Palacios, G., & Domínguez Trejo, B. (2011). Efectos psicológicos y físicos de la modulación autonómica en el dolor miofascial: un estudio aleatorizado. *Revista Mexicana de Análisis de La Conducta*, *37*(2), 99–115. https://doi.org/10.5514/rmac.v37.i2.26142.

San-Antolín, M., Rodríguez-Sanz, D., Becerro-de-Bengoa-Vallejo, R., Losa-Iglesias, M. E., Casado-Hernández, I., López-López, D., et al. (2019). Central sensitization and catastrophism symptoms are associated with chronic myofascial pain in the gastrocnemius of athletes. *Pain Medicine*. https://doi.org/10.1093/pm/pnz296.

San-Antolín, M., Rodríguez-Sanz, D., & Calvo-Lobo, C. (2019). Personality traits in athletes with myofascial pain syndrome in the gastrocnemius muscles: A pilot case-control study. *European Journal of Podiatry*, *5*(2), 39–46.

San-Antolín, M., Rodríguez-Sanz, D., López-López, D., Romero-Morales, C., Carbajales-Lopez, J., Becerro-de-Bengoa-Vallejo, R., et al. (2020). Depression levels and symptoms in athletes with chronic gastrocnemius myofascial pain: A case-control study. *Physical Therapy in Sport*, *43*, 166–172. https://doi.org/10.1016/j.ptsp.2020.03.002.

San-Antolín, M., Rodríguez-Sanz, D., Vicente-Campos, D., Palomo-López, P., Romero-Morales, C., Benito-de-Pedro, M., et al. (2020). Fear avoidance beliefs and Kinesiophobia are presented in athletes who suffer from gastrocnemius chronic myofascial pain. *Pain Medicine*. https://doi.org/10.1093/pm/pnz362.

Sandín, B., Valiente, R. M., Chorot, P., Olmedo Montes, M., & Santed Germán, M. A. (2002). Versión española del cuestionario EPQR-ABREVIADO (EPQR-A) (I): análisis exploratorio de la estructura factorial. *Revista de Psicopatología y Psicología Clínica*, *7*(3), 195–205. https://doi.org/10.5944/rppc.vol.7.num.3.2002.3933.

Shahidi, B., Curran-Everett, D., & Maluf, K. S. (2015). Psychosocial, physical, and neurophysiological risk factors for chronic neck pain: A prospective inception cohort study. *The Journal of Pain*, *16*(12), 1288–1299. https://doi.org/10.1016/j.jpain.2015.09.002.

Sikdar, S., Ortiz, R., Gebreab, T., Gerber, L. H., & Shah, J. P. (2010). Understanding the vascular environment of myofascial trigger points using ultrasonic imaging and computational modeling. In *2010 annual international conference of the IEEE engineering in medicine and biology, 2010* (pp. 5302–5305). https://doi.org/10.1109/IEMBS.2010.5626326.

Stephens, M. A. P., Druley, J. A., & Zautra, A. J. (2002). Older adults' recovery from surgery for osteoarthritis of the knee: Psychosocial resources and constraints as predictors of outcomes. *Health Psychology : Official Journal of the Division of Health Psychology, American Psychological Association*, *21*(4), 377–383.

Sternbach, R. A. (1976). *Psychophysiology of Pain*, *6*, 63–73. https://doi.org/10.2190/RFUR-TA4R-KXPB-18FL.

Sullivan, M. J. L., Rodgers, W. M., & Kirsch, I. (2001). Catastrophizing, depression and expectancies for pain and emotional distress. *Pain*, *91*(1–2), 147–154. https://doi.org/10.1016/S0304-3959(00)00430-9.

Thibaut, A., Zeng, D., Caumo, W., Liu, J., & Fregni, F. (2017). Corticospinal excitability as a biomarker of myofascial pain syndrome. *PAIN Reports*, *2*(3). https://doi.org/10.1097/PR9.0000000000000594, e594.

Travell, J. G. (2002). In J. G. Travell (Ed.), *Vol. 1. Myofascial pain and dysfunction*. Baltimore, MD: Williams & Wilkins (apropos of a muscles).

Travell, J. G. (2004). *Dolor y disfunción miofascial. El manual de los puntos gatillo. Mitad inferior del cuerpo*.

Truyols, M., Pérez, J., Medinas, M., Palmer, A., & Sesé, A. (2008). Psychological aspects relevant to chronic pain research and treatment. *Clínica y Salud*, *19*(3), 295–320.

Tunks, E. R., Crook, J., & Weir, R. (2008). Epidemiology of chronic pain with psychological comorbidity: Prevalence, risk, course, and prognosis. *Canadian Journal of Psychiatry*, *53*(4), 224–234. https://doi.org/10.1177/070674370805300403.

Turner, J. A., Dworkin, S. F., Mancl, L., Huggins, K. H., & Truelove, E. L. (2001). The roles of beliefs, catastrophizing, and coping in the functioning of patients with temporomandibular disorders. *Pain*, *92*(1–2), 41–51. https://doi.org/10.1016/S0304-3959(00)00469-3.

Vadivelu, N., Kai, A. M., Kodumudi, G., Babayan, K., Fontes, M., & Burg, M. M. (2017). Pain and psychology—A reciprocal relationship. *The Ochsner Journal*, *17*(2), 173–180.

van Wilgen, C. P., Konopka, K. H., Keizer, D., Zwerver, J., & Dekker, R. (2013). Do patients with chronic patellar tendinopathy have an altered somatosensory profile? – A quantitative sensory testing (QST) study. *Scandinavian Journal of Medicine and Science in Sports*, *23*(2), 149–155. https://doi.org/10.1111/j.1600-0838.2011.01375.x.

Velly, A. M., Look, J. O., Carlson, C., Lenton, P. A., Kang, W., Holcroft, C. A., et al. (2011). The effect of catastrophizing and depression on chronic pain—A prospective cohort study of temporomandibular muscle and joint pain disorders. *Pain*, *152*(10), 2377–2383. https://doi.org/10.1016/j.pain.2011.07.004.

Vlaeyen, J. W., Kole-Snijders, A. M., Boeren, R. G., & van Eek, H. (1995). Fear of movement/(re)injury in chronic low back pain and its relation to behavioral performance. *Pain*, *62*(3), 363–372. https://doi.org/10.1016/0304-3959(94)00279-n.

Volz, M. S., Medeiros, L. F., Tarragô, M. D. G., Vidor, L. P., Dall'Agnol, L., Deitos, A., et al. (2013). The relationship between cortical excitability and pain catastrophizing in myofascial pain. *The Journal of Pain*, *14*(10), 1140–1147. https://doi.org/10.1016/j.jpain.2013.04.013.

Waddell, G., Newton, M., & Henderson, I. (1993). A fear-avoidance beliefs questionnaire (FABQ) and the role of fear-avoidance beliefs in chronic low back pain and disability. *Pain*, *52*(2), 157–168.

Wang, C., Ge, H.-Y., Ibarra, J. M., Yue, S.-W., Madeleine, P., & Arendt-Nielsen, L. (2012). Spatial pain propagation over time following painful glutamate activation of latent myofascial trigger points in humans. *The Journal of Pain*, *13*(6), 537–545. https://doi.org/10.1016/j.jpain.2012.03.001.

Woolf, C. J. (2011). Central sensitization: Implications for the diagnosis and treatment of pain. *Pain*, *152*(3), S2–S15. https://doi.org/10.1016/j.pain.2010.09.030.

Xu, Y.-M., Ge, H.-Y., & Arendt-Nielsen, L. (2010). Sustained nociceptive mechanical stimulation of latent myofascial trigger point induces central sensitization in healthy subjects. *The Journal of Pain*, *11*(12), 1348–1355. https://doi.org/10.1016/j.jpain.2010.03.010.

Yunus, M. B. (2008). Central sensitivity syndromes: A new paradigm and group nosology for fibromyalgia and overlapping conditions, and the related issue of disease versus illness. *Seminars in Arthritis and Rheumatism*, *37*(6), 339–352. https://doi.org/10.1016/j.semarthrit.2007.09.003.

Part IV

Resources

Chapter 37

Recommended resources, sites, and research groups for the neuroscience of anesthetics and analgesics

Rajkumar Rajendram[a,b], Vinood B. Patel[c], and Victor R. Preedy[d]

[a]*College of Medicine, King Saud bin Abdulaziz University for Health Sciences, Riyadh, Saudi Arabia;* [b]*Department of Medicine, King Abdulaziz Medical City, King Abdullah International Medical Research Center, Ministry of National Guard Health Affairs, Riyadh, Saudi Arabia;* [c]*School of Life Sciences, University of Westminster, London, United Kingdom;* [d]*Department of Nutrition and Dietetics, Life Course Sciences, Faculty of Life Science and Medicine, King's College London, London, United Kingdom*

Abbreviations

3D	three dimensional
BMC	BioMed Central
DVCIPM	Defense and Veterans Center for Integrative Pain Management
IBDM	Institut de Biologie du Developpement de Marseille
NUSGRA	Newcastle Ultrasound Guided Regional Anaesthesia
NYSORA	New York School of Regional Anesthesia
SSRAUSA	Stamford School of Regional Anesthesia, United States of America
USA	United States of America
WIP	World Institute of Pain

Introduction

Every year millions of complex surgical procedures are performed worldwide. Today, patients undergoing these procedures expect to be pain free and have no recollection of the experience. This requires the administration of anesthetic agents to induce a reversible coma (i.e., absence of awareness) and analgesics to reduce pain sensation.

However, before the 19th century, surgery was only performed as a last resort. In that era, surgical procedures were performed without analgesia on awake patients. They caused severe pain and invoked an unspeakable terror that was associated with considerable risk. It is therefore not surprising that William Morton's use of inhaled diethyl-ether for intraoperative anesthesia in 1846 transformed surgical practice (Schlich, 2017).

In the 170 years since "Ether Day," anesthesia has progressed significantly. A large pantheon of anesthetic agents and analgesics are now available. Indeed, to ensure sufficient depth of anesthesia for surgery to be performed, the brain may be subjected a "cocktail" of anesthetics and analgesics. Yet, precisely how anesthetic agents and analgesics achieve their effects remains unclear.

Nonetheless, over the past 20 years, our understanding of the mechanisms of anesthesia has advanced in leaps and bounds. For many, but certainly not all, anesthetics, the characteristic that is most significant associated with potency is lipophilicity (the Meyer-Overton correlation; Kopp Lugli, Yost, & Kindler, 2009; Pavel, Petersen, Wang, Lerner, & Hansen, 2020). This association initially focused anesthetic research on membrane perturbations as the primary mechanism of action of inhaled anesthetic agents (Kopp Lugli et al., 2009; Pavel et al., 2020).

Over the last two decades, a chiral target has been suggested by the enantiomer selectivity of analgesics and anesthetics (Pavel et al., 2020). As a result, the concept that anesthetic agents may act by directly binding to ion channels has emerged (Franks, 2008; Hemmings Jr., Riegelhaupt, Kelz, et al., 2019; Pavel et al., 2020).

Photolabeling and mutagenic techniques have located anesthetic binding sites and increased understanding of the interactions between various anesthetic agents and their binding sites (Hemmings Jr. et al., 2019; Woll, Dailey, & Eckenhoff,

The Neurobiology, Physiology, and Psychology of Pain. https://doi.org/10.1016/B978-0-12-820589-1.00037-3

2018). How anesthetic agents bind these targets has been shown by crystallography (Loll, 2018). This knowledge is being exploited by physiologists studying the effects of anesthesia on neural circuits. Understanding the impact of anesthetic agents on various networks of neurons will pave the way to unraveling their effects on consciousness (Hemmings Jr. et al., 2019).

Progress has, undoubtedly, been made toward defining molecular targets of anesthetics and analgesics within the central nervous system. Light can be seen at the end of the laryngoscope... Yet, rather than homing in on a single target, many approaches have consistently revealed that anesthetic agents have several targets, each of which has many intrinsic binding sites (Hemmings Jr. et al., 2019; Woll et al., 2018). For example, photolabeling has revealed anesthetic binding targets on 10%–15% of detectable proteins (Hemmings Jr. et al., 2019; Woll et al., 2018). This equates to over 300 proteins (Hemmings Jr. et al., 2019). When undetected proteins are considered the true number of binding, targets could be 10-fold more (Hemmings Jr. et al., 2019). This is unsurprising. The plethora of effects of anesthetic agents on intraoperative physiology clearly illustrates the mechanistic multiplicity of these agents in vivo.

In recent years, an overwhelming number of novel methods and tools for investigation of the neuroscience of pain, anesthesia, and analgesia have been developed. So, there has been an explosion of knowledge and understanding in the neuroscience of pain, anesthesia, and analgesia. Despite this, extensive further research is required to fully elucidate how the complex actions of anesthetic agents on the molecular level translate into their behavioral effects in anesthetized patients (Hemmings Jr. et al., 2019). Regardless, it is now difficult even for experienced scientists to remain up-to-date. To assist colleagues who are interested in understanding more about the neuroscience of anesthesia and analgesia, we have therefore produced tables containing up-to-date resources in this chapter. The experts who assisted with the compilation of these tables of resources are acknowledged below.

Resources

Tables 1–8 list the most up-to-date information on the regulatory bodies (Table 1), journals (Table 2), books (Table 3), professional societies (Table 4), emerging technologies and platforms (Table 5), research groups and centers (Table 6), other resources (Table 7), and videos and other recordings (Table 8) that are relevant to an evidence-based approach to the neuroscience of pain, anesthetics, and analgesics.

Applications to other areas

This chapter lists the most up-to-date information on the regulatory bodies (Table 1), journals (Table 2), books (Table 3), professional societies (Table 4), emerging technologies and platforms (Table 5), research groups and centers (Table 6), other resources (Table 7), and videos and other recordings (Table 8) that are relevant to those interested in the neuroscience of pain, anesthetics, and analgesics.

TABLE 1 Regulatory bodies and relevant organizations.

Regulatory body or organization	Web address
American Academy of Orthopaedic Surgeons	https://orthoinfo.aaos.org/en/treatment/anesthesia-for-hip-and-knee-surgery/
American Association of Hip and Knee Surgeons	http://www.aahks.org/opioids
Centers for Disease Control and Prevention	https://www.cdc.gov/drugoverdose/prescribing/guideline.html
World Institute of Pain (WIP)	www.worldinstituteofpain.org
Pain Research Forum	https://www.painresearchforum.org/
Relief	https://relief.news/
Veterans Health Administration	https://www.va.gov/painmanagement/
Veterans Health Administration and Department of Defense	https://www.healthquality.va.gov/guidelines/Pain/cot/VADoDOTCPG022717.pdf

This table lists the regulatory bodies and organizations involved with the neuroscience of pain, anesthetics, and analgesics and associated specialties or interests. The links were accurate at the time of going to press but may move or alter. In these cases the use of the "Search" tabs should be explored at the parent address or site.

TABLE 2 Relevant journals publishing original research and review articles related to the neuroscience of pain, anesthetics, and analgesics.

Journal	Web address
Anesthesia	https://associationofanaesthetists-publications.onlinelibrary.wiley.com/journal/13652044
Anesthesia and Analgesia	https://journals.lww.com/anesthesia-analgesia/pages/default.aspx
Anesthesiology	https://pubs.asahq.org/anesthesiology
Anesthesiology Clinics	https://www.anesthesiology.theclinics.com/
BMC Anesthesiology	https://bmcanesthesiol.biomedcentral.com/
British Journal of Anesthesia	https://academic.oup.com/bja
British Journal of Pain	https://journals.sagepub.com/home/bjp
Current Opinion in Anaesthesiology	https://journals.lww.com/co-anesthesiology/pages/default.aspx
European Journal of Anaesthesiology	https://journals.lww.com/ejanaesthesiology/pages/default.aspx
European Journal of Pain	https://onlinelibrary.wiley.com/journal/15322149
Journal of Clinical Anesthesia	https://www.journals.elsevier.com/journal-of-clinical-anesthesia
Journal of Pain	https://www.sciencedirect.com/journal/the-journal-of-pain
Journal of Pain and Symptom Management	https://www.journals.elsevier.com/journal-of-pain-and-symptom-management
Journal of Pain Research	https://www.dovepress.com/journal-of-pain-research-journal
Journal of Palliative Medicine	https://home.liebertpub.com/publications/journal-of-palliative-medicine/41
Neurobiology of Pain	https://www.journals.elsevier.com/neurobiology-of-pain
Neuromodulation	https://onlinelibrary.wiley.com/journal/15251403
Pain	https://journals.lww.com/pain/pages/default.aspx
Pain Medicine	https://academic.oup.com/painmedicine
Regional Anesthesia and Pain Medicine	https://rapm.bmj.com/

Journals publishing original research and review articles related to the neuroscience of pain, anesthetics, and analgesics. Included in this list are the top 30 journals which have published the most number of articles on the neuroscience of pain, anesthetics, and analgesics over the past 5 years. Data derived from Scopus.

TABLE 3 Relevant books and selected articles.

Book title	Authors or editors	Publisher	Date	City and country
A Nation in Pain: Healing our Biggest Health Problem	Judy Foreman	Oxford University Press	2014	Oxford, United Kingdom
Acupuncture in Modern Medicine	Editor: Lucy L. Chen and Tsung O. Cheng	Intech	2013	Croatia
Atlas of Sonoanatomy for Regional Anesthesia and Pain Medicine	Manoj K. Karmakar, Edmund Soh, Victor Chee, Kenneth Sheah	McGraw-Hill Education	2018	New York, United States
Essentials of Regional Anesthesia	Alan David Kaye Richard D. Urman Nalini Vadivelu	Springer International Publishing AG	2018	Cham, Switzerland

Continued

TABLE 3 Relevant books and selected articles—cont'd

Book title	Authors or editors	Publisher	Date	City and country
Hadzic's Textbook of Regional Anesthesia and Acute Pain Management	Admir Hadzic	McGraw-Hill Education	2017	New York, United States
Opioid Receptors: Methods and Protocols	Santi M. Spampinato	Springer US	2020	New York, United States
Opioids in Medicine: A Comprehensive Review on the Mode of Action and the Use of Analgesics in Different Pain States	Enno Freye	Springer	2008	The Netherlands, Dordrecht
Pain: a review of three commonly used pain rating scales	Amelia Williamson and Barbara Hoggart	Blackwell Publishing Ltd., Journal of Clinical Nursing, 14, 798–804	2005	Hoboken, United States
Practical Management of Pain 5th Ed	Benzon, Rathmell, Wu, Turk, Argoff, Hurley	Elsevier	2013	Philadelphia, United States
Sigma Proteins: Evolution of the Concept of Sigma Receptors	Felix J. Kim, Gavril W. Pasternak	Springer	2017	
The Story of Pain: From Prayer To Painkillers	Joanna Bourke	Oxford University Press	2017	Oxford, United Kingdom
Translational Pain Research: From Mouse to Man.	Editors: Lawrence Kruger and Alan R. Light	CRC Press/Taylor & Francis	2010	Boca Raton, United States
Ultrasound Guided Regional Anesthesia	Stuart A. Grant David B. Auyong	Oxford University Press	2017	New York, United States
World Health Organization (WHO) Analgesic Ladder	Aabha A. Anekar; Marco Cascella	StatPearls Publishing LLC.	2020	United States of America

This table lists books and papers on the neuroscience of pain, anesthetics, and analgesics.

TABLE 4 Professional societies and other organizations covering pain, anesthetics, analgesics, or neuroscience.

Society name	Web address
American Academy of Pain Medicine	https://painmed.org/
American Society of Anesthesiologists	https://www.asahq.org/
American Society of Pain and Neuroscience	https://aspnpain.com/
American Society of Regional Anaesthesia and Pain Medicine	www.asra.com
Belgian Pain Society	https://www.belgianpainsociety.org/
European Society of Anesthesiology	https://www.esahq.org/
European Society of Regional Anaesthesia & Pain Therapy	https://esraeurope.org/
International Association for the Study of Pain	https://www.iasp-pain.org/
International Brain Research Organization	https://ibro.org
Malaysian Society of neuroscience	www.neuro.org.my
Peripheral Nerve Society	https://www.pnsociety.com/

TABLE 4 Professional societies and other organizations covering pain, anesthetics, analgesics, or neuroscience—cont'd

Society name	Web address
Society for Neuroscience	https://www.sfn.org
Spanish Society of Pain	https://www.sedolor.es/
Spine Intervention Society	https://www.spineintervention.org/
The American Academy of Pain Medicine	https://painmed.org/
The British Pain Society	https://www.britishpainsociety.org/
The European Pain Federation	https://europeanpainfederation.eu/
World Federation of Societies of Anesthesiologists	https://www.wfsahq.org/

This table lists some societies and organizations devoted to understanding the neuroscience of pain, anesthetics, and analgesics. Please note, occasionally the location of the websites or web address changes. In these cases, the use of the "Search" tabs should be explored at the parent address or site.

TABLE 5 Emerging techniques and platforms relevant to pain, anesthetics, or analgesics.

Organization or company or society name	Web address
Deep lab cut (machine learning for pain behavior)	http://www.mousemotorlab.org/deeplabcut
Pain Management Network	https://www.aci.health.nsw.gov.au/chronic-pain/for-everyone

This table lists some emerging technologies and platforms relevant to the neuroscience of pain, anesthetics, and analgesics. Please note, occasionally the location of the websites or web address changes. In these cases, the use of the "Search" tabs should be explored at the parent address or site.

TABLE 6 Research centers and groups investigating either pain, anesthetics, or analgesics.

Group or university name	Web address
College of Medicine, Tucson	https://pharmacology.arizona.edu/research-area/neuroscience-and-pain
Defense and Veterans Center for Integrative Pain Management (DVCIPM)	https://www.dvcipm.org/
Department of Neuropeptides, Mossakowski Medical Research Centre Polish Academy of Sciences	http://www.imdik.pan.pl/en/research-groups/departments/125-department-of-neuropeptide
Department of Pharmacology of Pain, Maj Institute of Pharmacology Polish Academy of Sciences	http://if-pan.krakow.pl/en/departments/Department-of-Pharmacology-of-Pain/
Institut de Biologie du Developpement de Marseille (IBDM)	http://www.ibdm.univ-mrs.fr/fr/equipe/specification-des-neurones-sensoriels/
King's College London, Wolfson Centre for Age-Related Diseases McMahon and Denk Labs	https://www.franziskadenk.com/
Stanford University Pain Management	Med.stanford.edu/pain.html
Thomas Nevian—Neuronal plasticity group (University of Bern)	https://physio.unibe.ch/~nevian/group/
University of Cambridge, Dept Pharmacology, Sensory neurophysiology and pain	https://www.phar.cam.ac.uk/research/Smith

This table lists some research centers and groups investigating the neuroscience of pain, anesthetics, and analgesics. Please note, occasionally the location of the websites or web address changes. In these cases, the use of the "Search" tabs should be explored at the parent address or site.

TABLE 7 Other resources of interest or relevance to examining either pain, anesthetics, or analgesics.

Name of resource or organization	Web address
Pain Exhibit—Art	http://painexhibit.org/en/
Physiopedia	https://www.physio-pedia.com/home/
SpineUniverse	https://www.spineuniverse.com/

This table lists some other resources relevant to the neuroscience of pain, anesthetics, and analgesics. Please note, occasionally the location of the websites or web address changes. In these cases, the use of the "Search" tabs should be explored at the parent address or site.

TABLE 8 Videos and other recordings of interest or relevance to either pain, anesthetics, or analgesics.

Name of video or subject	URL
"Confronting the Opioid Crisis by Developing New Measurements and Biomarkers of Pain"	https://videocast.nih.gov/watch=24904
"The Science of Pain"	https://www.youtube.com/watch?v=-TN1r25wAol
BLOCKJOCKS	https://www.youtube.com/user/BLOCKJOCKS/videos
London Society of Regional Anaesthesia (LSORA) Videos	https://www.youtube.com/channel/UCV8d6B_W6KmPoL_bWXeYiqQ/feed
Newcastle Ultrasound Guided Regional Anaesthesia (NUSGRA)	https://www.youtube.com/channel/UCxFY19blFhglJvZzDElaRqQ/featured
New York School of Regional Anesthesia (NYSORA) Continuing Medical Education	https://www.youtube.com/channel/UCaJ9YC4Zp5Rs7JL61J1BlBQ
Pain Doctor	https://www.youtube.com/c/Paindoctor/videos
Stamford School of Regional Anesthesia, United States of America (SSRAUSA)	https://www.youtube.com/user/SSRAUSA/featured
Videos by the New York School of Regional Anesthesia (NYSORA)	https://www.nysora.com/educational-tools/videos/

This table lists some sites where videos or other recordings of interest relevant to the neuroscience of pain, anesthetics, and analgesics can be found. Please note, occasionally the location of the websites or web address changes. In these cases, the use of the "Search" tabs should be explored at the parent address or site.

Hitherto, neuroscience was classified within the domain of biology. In the 21st century, the neuroscience is a multidisciplinary field of science that branches into a broad range of other disciplines. These related disciplines include, but are not limited to, mathematics, computer science, engineering, linguistics, philosophy, and psychology. Within the field of medicine; pain, anesthetics, and analgesics are of most interest to those specializing in anesthesia. However, these topics will also be of interest to several other healthcare professionals including nurses, physiotherapists, internists, hospitalists, emergency physicians, surgeons, and psychiatrists.

As a result, the nature of the resources presented in this chapter is such that they will be relevant to several other scientific and medical disciplines.

Mini-dictionary of terms

Anesthetic agents: medications that induce a reversible coma (i.e., loss of consciousness). These agents are usually administered to facilitate surgical procedures.

Analgesics: medications that reduce pain.

Crystallography: A technique which can determine the structure of a crystal (atomic and molecular). Crystalline atoms diffract a beam of X-rays into many directions. Measuring these beams' angles and intensities allows production of a 3D image of the crystal's electrons density. The positions of the crystal's atoms, their chemical bonds, etc. can be determined from this electron density map.

Meyer-Overton correlation: The potency of volatile anesthetics correlates with their lipid solubility. As a result, both Meyer and Overton proposed that anesthetics act by causing a nonspecific malfunction of the neuronal lipid bilayer. This theory has been superseded by more recent data suggesting that anesthetic agents may act on many different proteins.

Photolabeling: "labels" are attached to the active site of large molecules. This technique has been used to identify the binding sites of anesthetic agents.

Key points

- In 1846, William Morton's use of inhaled diethyl-ether for intraoperative anesthesia transformed surgical practice.
- While the clinical practice of anesthesia has advanced greatly in the past 170 years, the mechanisms of actions of these agents remain unknown.
- In recent years, an overwhelming number of novel methods and tools for investigation of the neuroscience of pain, anesthesia and analgesia have been developed.
- Progress has, undoubtedly, been made toward defining molecular targets of anesthetics and analgesics within the central nervous system.
- Despite great advances, vast amounts must still be learned before the mechanisms of the effects of anesthetics and analgesics on consciousness and pain are fully understood.

Summary points

- Every year millions of people worldwide undergo complex surgical procedures.
- The performance of these procedures requires administration of anesthetic agents to induce a reversible coma (i.e., absence of awareness) and analgesics to reduce pain sensation.
- Many anesthetic agents and analgesics are available, yet, precisely how these medications achieve their effects remains unclear.
- Rather than acting in on a single target, many approaches have revealed that anesthetic agents have several targets, each of which has many intrinsic binding sites.
- There has been an explosion of the knowledge and understanding in the physiology, neurobiology, and behavioral psychology related to pain. It is difficult even for experienced scientists to remain up-to-date.

Acknowledgments

We would like to thank the following authors for contributing to the development of this resource. We apologize if some suggested material was not included in this chapter or has been moved to different sections.

Aggarwal, Anuj
Baria, Ariel
Bellefroid, Eric
Chen, Antonia
Failo, Alessandro
Fattori, Victor
Ferraz, Camila R.
Hirose, Munetaka
Hong, Joe
Jaya, A. A. Gde Putra Semara
Kasanetz, Fernando
Laksono, Ristiawan
Lipiński, Piotr F.J.
Maier, Camelia
Manchope, Marilia F.
Nishijo, Hisao
Nuseir, Khawla
Phillips Commmander, Ryan C.
Porebski, Grzegorz
Trevisan, Gabriela
Tsuchiya, Masahiko
Verri, Jr., Waldiceu A.

References

Franks, N. P. (2008). General anaesthesia: From molecular targets to neuronal pathways of sleep and arousal. *Nature Reviews. Neuroscience*, *9*, 370–386.

Hemmings, H. C., Jr., Riegelhaupt, P. M., Kelz, M. B., et al. (2019). Towards a comprehensive understanding of anesthetic mechanisms of action: A decade of discovery. *Trends in Pharmacological Sciences*, *40*, 464–481.

Kopp Lugli, A., Yost, C. S., & Kindler, C. H. (2009). Anaesthetic mechanisms: Update on the challenge of unravelling the mystery of anaesthesia. *European Journal of Anaesthesiology*, *26*, 807–820.

Loll, P. J. (2018). Structural analysis of anesthetics in complex with soluble proteins. *Methods in Enzymology*, *603*, 3–20.

Pavel, M. A., Petersen, E. N., Wang, H., Lerner, R. A., & Hansen, S. B. (2020). Studies on the mechanism of general anesthesia. *Proceedings of the National Academy of Sciences of the United States of America*, *117*, 13757–13766.

Schlich, T. (2017). The history of anaesthesia and the patient-reduced to a body? *Lancet*, *390*(10099), 1020–1021.

Woll, K. A., Dailey, W. P., & Eckenhoff, R. G. (2018). Identification of general anesthetic target protein-binding sites by photoaffinity labeling and mass spectrometry. *Methods in Enzymology*, *602*, 231–246.

Index

Note: Page numbers followed by *f* indicate figures and *t* indicate tables.

N

R

S